THE M.G. WORKSHOP MANUAL

from the 'M'

to the 'TF'

THE

WORKSHOP MANUAL

FROM "M" TYPE to "T.F. 1500"

by

W. E. BLOWER

BentleyPublishers®
.com

BENTLEY PUBLISHERS® | Automotive Reference™

Bentley Publishers, a division of Robert Bentley, Inc.
1734 Massachusetts Avenue
Cambridge, MA 02138 USA
800-423-4595 / 617-547-4170

Information that makes the difference®

BentleyPublishers®.com

Updates and corrections

We will evaluate submissions and post appropriate editorial changes online as updates or tech discussion. Appropriate updates and corrections will be added to the book in future printings. Check for updates and corrections for this book before beginning work on your vehicle. See the following web address for additional information:

`http://www.BentleyPublishers.com/updates`

WARNING—Important Safety Notice

Do not use this manual for repairs unless you are familiar with basic automotive repair procedures and safe workshop practices. This manual illustrates the workshop procedures for some maintenance and service work. It is not a substitute for full and up-to-date information from the vehicle manufacturer or for proper training as an automotive technician. Note that it is not possible for us to anticipate all of the ways or conditions under which vehicles may be serviced or to provide cautions as to all of the possible hazards that may result.

We have endeavored to ensure the accuracy of the information in this manual. Please note, however, that considering the vast quantity and the complexity of the service information involved, we cannot warrant the accuracy or completeness of the information contained in this manual.

FOR THESE REASONS, NEITHER THE PUBLISHER NOR THE AUTHOR MAKES ANY WARRANTIES, EXPRESS OR IMPLIED, THAT THE INFORMATION IN THIS MANUAL IS FREE OF ERRORS OR OMISSIONS, AND WE EXPRESSLY DISCLAIM THE IMPLIED WARRANTIES OF MERCHANTABILITY AND OF FITNESS FOR A PARTICULAR PURPOSE, EVEN IF THE PUBLISHER OR AUTHOR HAVE BEEN ADVISED OF A PARTICULAR PURPOSE, AND EVEN IF A PARTICULAR PURPOSE IS INDICATED IN THE MANUAL. THE PUBLISHER AND AUTHOR ALSO DISCLAIM ALL LIABILITY FOR DIRECT, INDIRECT, INCIDENTAL OR CONSEQUENTIAL DAMAGES THAT RESULT FROM ANY USE OF THE EXAMPLES, INSTRUCTIONS OR OTHER INFORMATION IN THIS MANUAL. IN NO EVENT SHALL OUR LIABILITY WHETHER IN TORT, CONTRACT OR OTHERWISE EXCEED THE COST OF THIS MANUAL.

Your common sense and good judgment are crucial to safe and successful service work. Read procedures through before starting them. Think about whether the condition of your car, your level of mechanical skill, or your level of reading comprehension might result in or contribute in some way to an occurrence which might cause you injury, damage your car, or result in an unsafe repair. If you have doubts for these or other reasons about your ability to perform safe repair work on your car, have the work done at an authorized dealers or other qualified shop.

Library of Congress Catalog Card No. 75-33494
ISBN 978-0-8376-1746-6   Job code X017-03

First published 1952
First published in the United States 1970

ACKNOWLEDGEMENTS

The Author would like to express his appreciation of the assistance given by Percy Lay in providing photographs and drawings; Bill Lane for details on some of the very early models; Nevil Lloyd, Motor Racing Publications, Ltd., for his constant encouragement; Carburettors Ltd.; Shorrock Superchargers Ltd.; North Downs Engineering Co., Ltd.; and last, but by no means least, to Nina, who so quietly and constantly encouraged the compilation of this book.

The paper used in this publication is acid free and meets the requirements of American National Standard for Information Sciences—Permanence of Paper for Printed Library Material. (∞)

Manufactured in the United States of America

INTRODUCTION

The purpose of this book is to help all M.G. enthusiasts, whether they own the first " M " type Midget of the 1929 era or are fortunate enough to own the latest " TF " model.

It is not intended to describe in detail the comprehensive work of a complete overhaul, but to give information on the more normal maintenance and tuning to be undertaken by the amateur enthusiast.

It is the author's aim to explain as concisely as possible the functioning of each unit and yet provide at the same time adequate information to keep the car in tip-top tune. It is felt that all M.G. owners desire to " Maintain the Breed " and although they probably possess excellent mechanical knowledge, they lack necessary data to enable them to proceed with servicing and repair.

With this thought in mind, data sheets, lubrication charts and wiring diagrams have been included in order to assist in keeping the model in first-class running order.

Please read these warnings and cautions before proceeding with maintenance and repair work.

WARNING—

- Never work under a lifted car unless it is solidly supported on stands intended for the purpose. Do not support a car on cinder blocks, hollow tiles, or other props that may crumble under continuous load. Do not work under a car that is supported solely by a jack.

- If you are going to work under a car on the ground, make sure that the ground is level. Block the wheels to keep the car from rolling. Disconnect the battery ground strap to prevent others from starting the car while you are under it.

- Never run the engine unless the work area is well ventilated. Carbon monoxide kills.

- Friction materials such as brake and clutch linings or brake pads may contain asbestos fibers. Do not create dust by grinding, sanding or by cleaning with compressed air. Avoid breathing asbestos fibers and asbestos dust. Breathing asbestos may result in serious diseases, such as asbestosis or cancer, and cause severe injury and death.

- Tie long hair behind your head. Do not wear a necktie, scarf, loose clothing, or necklace when you work near machine tools or running engines. If your hair, clothing, or jewelry were to get caught in the machinery, severe injury could result.

- Disconnect the battery ground strap whenever you work on the fuel system or the electrical system. When you work around fuel, do not smoke or work near heaters or other fire hazards. Keep an approved fire extinguisher handy.

- Illuminate your work area adequately but safely. Use a portable safety light for working inside or under the car. Make sure its bulb is enclosed by a wire cage. The hot filament of an accidentally broken bulb can ignite spilled fuel or oil.

- Catch draining fuel, oil, or brake fluid in suitable containers. Do not use food or beverage containers that might mislead someone into drinking from them. Store flammable fluids away from fire hazards. Wipe up spills at once, but do not store the oily rags, which can ignite and burn spontaneously.

- Finger rings should be removed so that they cannot cause electrical shorts, get caught in running machinery, or be crushed by heavy parts.

- Keep sparks, lighted matches, and open flame away from the top of the battery. If hydrogen gas escaping from the cap vents is ignited, it will ignite gas trapped in the cells and cause the battery to explode.

- Always observe good workshop practices. Wear goggles when you operate machine tools or work with battery acid. Gloves or other protective clothing should be worn whenever the job requires it.

CAUTION—

- If you lack the skills, tools, and equipment, or a suitable workshop for any procedure described in this manual, we suggest you leave such repairs to an Authorized Dealer or other qualified shop. We especially urge you to consult your Authorized Dealer before attempting any repairs on a car still covered by the new-car warranty.

- British Leyland Motors is constantly improving its cars and sometimes these changes, both in parts and specifications, are made applicable to earlier models. Therefore part numbers listed in this manual are for reference only. Always check with your Authorized Parts Department for the latest information.

- Before starting a job, make certain that you have all necessary tools and parts on hand. Read all instructions thoroughly; do not attempt shortcuts. Use tools appropriate to the work and use only replacement parts meeting specifications. Makeshift tools, parts, and procedures will not make good repairs.

- Use pneumatic and electrical tools only to loosen threaded parts and fasteners. Never use such tools to tighten fasteners, especially on light alloy parts.

- Be mindful of the environment and ecology. Before you drain the crankcase, find out the proper way to dispose of the oil. Do not pour oil onto the ground, down a drain, or into a stream, pond, or lake. Consult local ordinances that govern the disposal of wastes.

CONTENTS

CHAPTER I

THE CHASSIS

CHASSIS

GENERAL DESCRIPTION.

All M.G. chassis are constructed from rigid box section side members supported at intervals by tubular cross members; on the 1¼ litre and TD models the front cross member incorporates the independent front suspension.

On all models suspension is by means of semi-elliptic springs except for the 1¼ litre and TD models where independent front suspension is introduced incorporating the use of coil springs.

On the M, D, F, J, L, P, K, N, KN, TA, and TB models the front ends of the front and rear springs are pivoted on fixed points, the rear ends being mounted in sliding trunnions.

On the TC, 1½, 2, and 2.6 litre models the front ends of the springs are mounted on fixed points, the rear ends on swinging shackles, as are the rear springs of the 1¼ litre and TD models.

The M, D, F, J, and L models were fitted with Hartford shock absorbers both front and rear.

The P, N, and KN models had Hartford shock absorbers fitted to the front axle, and Luvax hydraulic vane type to the rear.

Hydraulic vane type shock absorbers are fitted to both front and rear axles of the TA, TB, and 2 litre models.

The 1½ litre model, up to Chassis No. VA 1914 was fitted with the Luvax finger tip controlled type of shock absorber; the vane type shock-absorber being introduced at Chassis No. VA 1915.

The 2.6 litre, TC, 1¼ and TD models, are fitted with Girling Piston type Hydraulic Dampers.

REMOVAL OF FRONT AXLE. (M, D, F, J, P, L, N, KN, K, TA, TB, TC, 1½, 2 and 2.6 litre models.)

Block up the car under the chassis until the tyres are just touching or just clear of the ground. Remove the wheel securings and withdraw wheel from front hub. Disconnect the lower arms of the shock absorbers from their positions on the spring brackets; disconnect the brake cables from their anchorages on the brake back plates; on models fitted with hydraulic brakes disconnect the upper hoses where they join the copper pipes on the chassis.

Disconnect the drag link at the axle end; remove any clips securing electrical cables to the front axle. Disconnect the torque cables (TA, 1½, 2, and 2.6 litre models only). Undo the spring shackles at the front and rear end when axle complete with springs can be withdrawn from the car. On those models where the rear end of the spring is mounted in sliding trunnions it will be necessary to remove the lubricating pipe at the oil feed union on the securing nut, remove the securing hexagon nut when the spring at the rear end can be drawn out sideways complete with the rubber cover and outer half of the sliding trunnion.

DISMANTLING THE FRONT AXLE. (M, D, F, J, P, L, K, KN, N, TA, TB, TC, 1½, 2 and 2.6 litre models.)

Remove track rod from steering arm, and remove steering arms. Remove brake drums, front hubs (details as page 12) and brake back plates. Remove king pins (details as page 13) when stub axle complete can be withdrawn from the axle beam.

REASSEMBLY AND REFITTING AXLE. (M, D, F, J, P, L, K, KN, N, TA, TB, TC, 1½, 2, 2.6 litre models.)

Reassembly and replacement is carried out in the reverse manner to that set out above for removal.

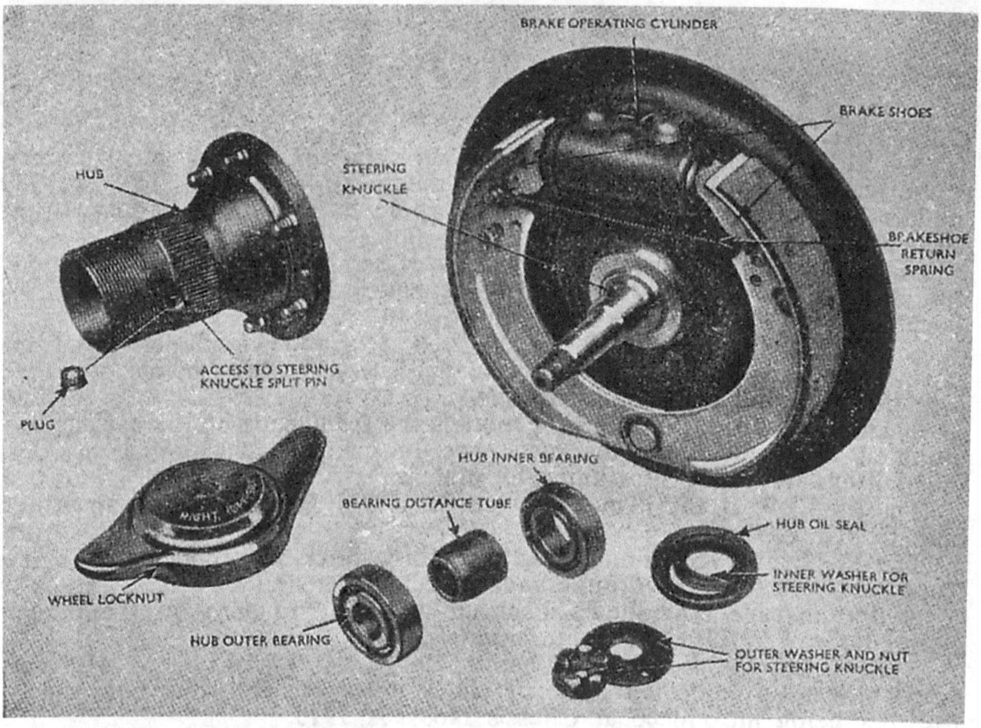

Fig. 1. The Front Hub Dismantled.

FRONT WHEEL BEARING. (M, D, F, J, P, L, N, K, KN, TA, TB, TC, 1½, 2, and 2.6 litre models.)

Removal and replacement: Jack up car at front end, remove road wheel and brake drum and take out greaser from inside hub. In the splined portion of the hub there is a small hole covered with a steel plug. Removal of this plug will give access to the split pin locking the castellated nut. Withdraw the split pin, remove the hub nut with a suitable box spanner and withdraw the outer washer. Then with a suitable extractor withdraw the hub complete with bearings. The outer ball race is smaller than the inner one and can be withdrawn from the housing; withdraw the oil seal and inner washer, and remove the inner bearing.

Replacement is carried out in the reverse manner to that detailed for removal. The bearings and distance piece will assemble without any difficulty, but it is advisable on tightening the castellated nut to spin the hub to make sure it rotates freely. Do not forget to replace the steel plug, fill the hub with grease and smear the hub splines with a little grease and graphite before replacing the road wheel.

REMOVAL AND REPLACEMENT OF KING PIN. (M, D, F, J, P, L, N, K, KN, TA, TB, TC, 1½, 2, and 2.6 litre models.)

Jack up car at front end and remove road wheel, brake drum and front hub. Disconnect brake cable from anchorage on brake back plate; on models fitted with hydraulic brakes disconnect the Lockheed hose at the point where it is connected to the pipe on the chassis.

Remove the bolts securing the brake back plate to the stub axle and remove back plate complete with brake gear.

Remove the metal cover and circlip from the top of the king pin (1½ litre only up to Chassis No. VA 1254).

Remove the bolt and washer from the top of the king pin (2.6 litre model only).

Remove the nut and washer from the cotter pin securing the king pin in position, then withdraw the cotter pin and press out the king pin.

Replacement is carried out in the reverse manner to that set out for removal.

NOTE:—Up to Chassis No. VA 1254 (1½ litre model) a thrust race was fitted to the top of the king pin. Commencing at Chassis No. VA 1255 this was superseded by a bronze thrust washer between the bottom of the axle beam eye and steering knuckle.

Earlier models can be converted by removing the thrust race, replacing the metal cover and circlip and fitting the thrust washer. If examination shows that replacement king pins are required, the later type should be fitted.

REMOVAL AND REPLACEMENT OF KING PIN BUSHES. (M, D, F, J, P, L, K, KN, N, TA, TB, TC, 1½, 2, and 2.6 litre models.)

Jack up front end of car, remove road wheel, brake drum, front hub, and brake back plate as detailed.

Remove king pin, withdraw stub axle and press out top and bottom king pin bushes.

Replacement is carried out in the reverse manner to that detailed for removal, but the following points should be noted:—

When new bushes have been fitted they should be reamered using a reamer of sufficient length to do the two at once thereby ensuring they are in line with each other, oil groove towards beam.

The king pin should not be fitted too tightly; it should be possible to press it in its bushes without the aid of tools. The clearance between the steel thrust washer and flanged bush should not exceed .004 in.

TORQUE CABLE ADJUSTMENT, FRONT AXLE. (TA, 1½, 2, and 2.6 litre models.)

When the front brakes are applied the torque cables absorb the torque reaction of the front axle. Adjust as follows: With the steering in the "straight ahead" position adjust the cables so that with pressure applied by the fingers midway between the fulcrum pins there is free up and down movement of ¾-1 in. or ⅜-½ in. either side of the centre line. Tighten the lock nuts on the cable yoke ends to maintain the adjustment.

FRONT AND REAR SPRINGS. (M, D, F, J, P, L, K, KN, N, TA, and TB models.)

Brass bushes are fitted to the front ends of the front springs, and Silentbloc bushes to the front ends of the rear springs. The front ends of the springs are pivoted upon fixed points. The rear ends slide in phosphor-bronze blocks fitted inside the ends of the rear spring cross tube; as the springs compress or expand the rear ends slide to and fro, the bronze bushes also being capable of rotating inside the chassis cross tube. A greaser connection is fed from the dash wall nipple and screwed into a distance piece, the whole assembly being completed by a large nut and rubber cover.

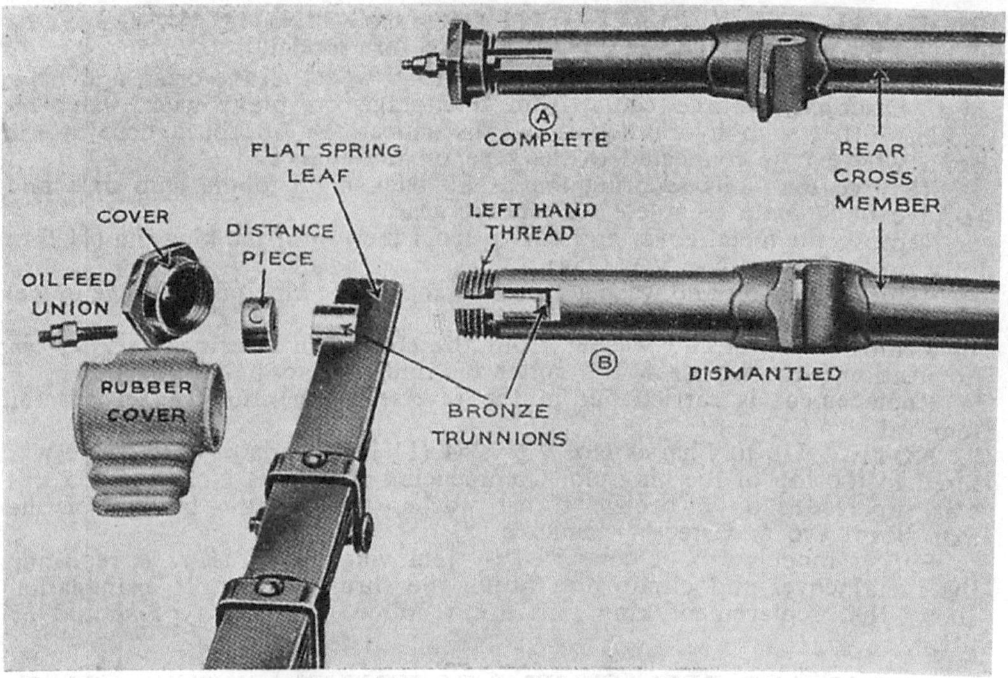

Fig. 2. Rear Spring Trunnion Bearing.

REMOVAL AND REPLACEMENT OF FRONT SPRINGS. (M, D, F, J, P, L, K, KN, N, TA, and TB models.)

Jack up the front end of the car and remove the road wheel. Disconnect the oil pipe from the oil nipple at the rear end of the spring. Disconnect the shock absorber arm from the bracket on the spring. Remove the nuts and withdraw the bolts securing the front spring to the axle beam. Remove the hexagon nut at the rear of the spring, and remove the front shackle pin. The spring, complete with rubber dust cover and one trunnion roller, can then be removed from the car.

Replacement is carried out in the reverse manner to that detailed for removal.

REMOVAL AND REPLACEMENT OF REAR SPRINGS (M, D, F, J, P, L, K, KN, N, TA, and TB models.)

Jack up the rear end of the car and remove the road wheel. Remove nuts from the " U " bolts securing the spring to the axle, supporting the centre of the spring with a jack, lower supporting jack and remove the " U " bolts. Disconnect the oil pipe from the oil nipple at the rear end of the spring. Remove the hexagon nut securing the rear end of the spring and the nut on the spring shackle pin. The spring, complete with rubber dust cover and one trunnion roller, can then be removed from the car.

Replacement is carried out in the reverse manner to that detailed for removal.

REPLACEMENT OF SHACKLE PIN BUSHES. (M, D, F, J, P, L, K, KN, N, TA, and TB models.)

Brass bushes are fitted to the front ends of the front springs only. When replacing them, it will be necessary to remove the road spring from the car, press out the old bush and press in the replacement. The replacement bushes will have to be reamered to suit the shackle pin.

Fig 3. *Special Tool T.*81 *for Compressing Rear Springs* (2 *and* 2.6 *Litre Models*).

ADJUSTMENT AND REPLACEMENT OF SPRING TRUNNION ROLLERS. (M, D, F, J, P, L, K, KN, N, TA, and TB Models.)

Remove the road spring as already detailed. Remaining trunnion roller can then be withdrawn easily from the rear spring cross member. When wear, which allows the road spring to move sideways, has taken place, the trunnion rollers will need adjustment. Normally it should be sufficient to remove the outer portion of the trunnion only. Then, with the aid of a lathe (or file, if extreme care is taken) remove from the split or inner end of the trunnion, slightly less metal than the amount the spring could actually move. If the sideways movement of the spring appears to be excessive, then it would be advisable to remove and modify both halves of the trunnion roller. If replacement rollers are fitted, ensure that they allow the spring to slide freely, without play, on the working portion of the main leaf. It may be necessary to ease the rollers slightly, and a very fine file or fine emery cloth should be employed to enable a good bearing surface to be obtained.

REMOVAL AND REPLACEMENT OF FRONT SPRINGS. (1½, 2, and 2.6 Litre Models.)

Jack up the front of the car and remove the road wheels. Disconnect the oil feed pipe to the front spring pin. Disconnect the shock absorber arm from the bracket on the spring. Remove the nuts and withdraw the bolts securing the front axle beam to the spring. Remove the front spring shackle at the front end of the spring, and the front spring pin at the rear end of the spring. The spring complete can then be removed from the car.

Replacement is carried out in the reverse manner to that detailed for removal.

REMOVAL AND REPLACEMENT OF REAR SPRINGS. (1½, 2, and 2.6 Litre Models.)

Jack up the rear end of the car and remove the road wheels. Disconnect the oil feed pipe to the rear spring shackle. Remove nuts from the " U " bolts securing the springs to the axle, supporting centre of the spring with a jack, lower supporting jack and remove " U " bolts. Remove the front spring pins and lower spring shackle pins on the 1½ and 2 litre model; remove the front spring pin and shackle pin plates on the 2.6 litre model. The spring can then be withdrawn from the car.

On reassembly a special tool, T81, is required for compressing the springs on the 2 and 2.6 litre models. Owing to its great free camber it cannot be reassembled in its free state. If suitable means of flattening the spring for reassembly are not available, it is suggested that the spring be stripped and fitted to the car one leaf at a time, using two jacks to squeeze the lower leaves upwards so that there is sufficient thread exposed for the nut to be started on the centre pinch bolt.

REPLACEMENT OF SHACKLE PINS AND BUSHES. (1½ litre model.)

Brass bushes are fitted to all spring eyes, with the exception of the front end of the rear springs, which are Silentbloc bushes. As the bushes are longer than the width of the spring eye, it is essential that when replacement bushes are fitted, an equal amount protrudes on either side of the eye. Brass washers are fitted over the protruding ends to provide a face surface for shackle plates. The bushes will require reamering to suit the shackle pin.

REPLACEMENT OF SHACKLE PINS AND BUSHES. (2 litre model.)

Brass bushes are fitted to all spring eyes, with the exception of the front end of the rear springs, which are Silentbloc bushes. There are two flanged bushes to each eye, pressed in from either side, with an oil retaining distance piece in the centre. The distance pieces should be removed with the bushes, the bushes also being reamered to suit the shackle pin.

SHACKLE PINS. (2.6 Litre Model.)

With the exception of the rear end of the near-side front spring, and both front ends of the rear spring, the shackle pins and bushes are of the screwed type.

Nuts at each side secure the shackle plates on the tapers of the screwed pins. Shake-proof spring washers are used, and it is essential that they are of the correct type. A special tool T 97, has been designed to ensure that the shackle pins are assembled in the correct position, that is, with an equal amount protruding on either side of the dumb iron. As the top and bottom pins must be assembled together, two tools will be required. The felt washers are specially prepared, and should the screwed pins be assembled out of position, these washers will be slack on one side and so tight on the other that they will be ineffective as an oil seal and dirt excluder.

KICK SHACKLE. (2.6 Litre Model.)

The rear end of the off-side front spring is attached to the chassis with what is known as a Kick Shackle. Its purpose is to prevent road shock from being transmitted to the steering wheel, and it is, therefore, essential that it be correctly set.

ADJUSTMENT OF THE KICK SHACKLE. (2.6 litre model.)

Fig. 5 shows the components in their respective positions. The coil springs are compressed to a position when the link can only move .020 in., this being obtained by fitting thrust washers between the coil springs and the chassis bracket. Under no circumstances should the adjustment differ.

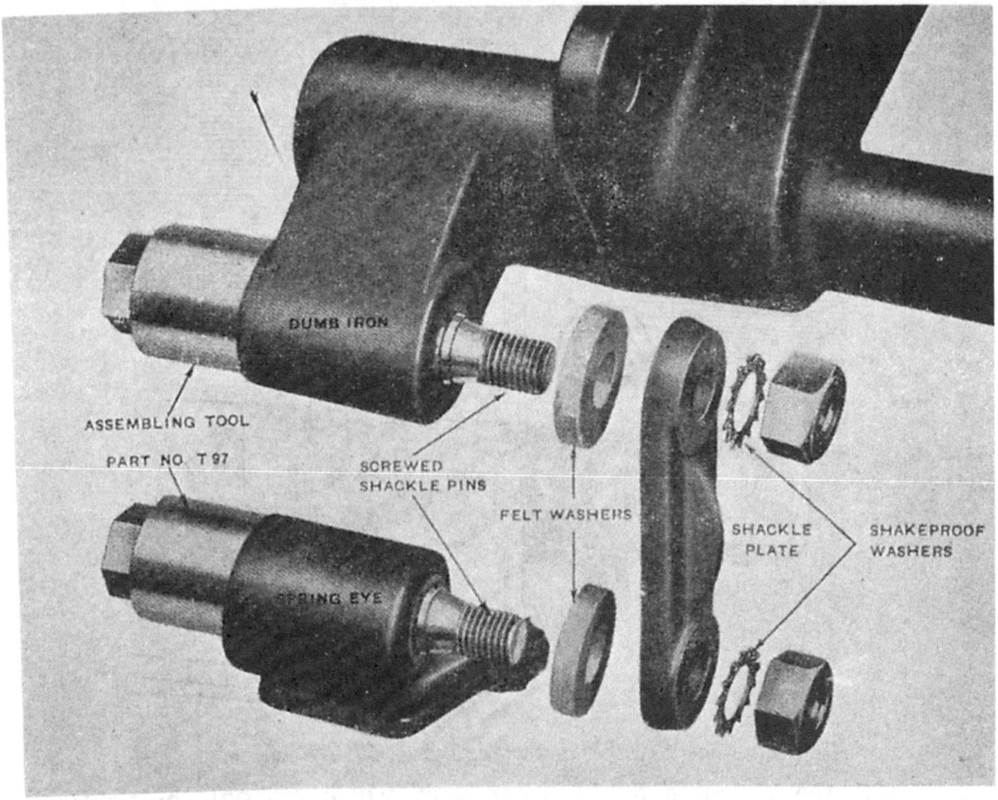

Fig. 4. Method of Assembling Shackle Pins (2.6 *Litre Model*).

FRONT ROAD SPRINGS AND SHACKLES. (TC Midget Model.)

The front ends of the springs are attached to the front end of the chassis by hardened steel suspension pins, which run directly in the spring eyes, being lubricated through grease gun nipples fitted in the outer ends.

Replacement pins will usually take up wear, unless it is excessive, when new top spring leaves will be required.

A swinging shackle is fitted at the rear end of each spring, fitted with four special rubber bushes placed into the spring eye and frame tube from either side, and clamped up by the shackle plates. The bushes are quite a loose fit, expanding when clamped into their housings. The bushes do not rotate on their surfaces, the angular movement being taken up by the rubber deflecting torsionally in itself.

When assembling the bushes they must maintain a central location so that the expansion of each half of the bush is equal. Insert each bush so that it protrudes equally on each side of the housing, and tighten up the shackle plates evenly. When central, the outer flanges of the bushes should all be of equal proportions.

REAR ROAD SPRINGS AND SHACKLES. (TC Midget.)

" Silentbloc " bushes are fitted to the front ends of the rear springs. Being enclosed in a metal housing, they can be pressed out of the spring eye and replaced as a unit. The working principle is the same as the front spring rubber bushes.

A rubber bushed shackle is fitted to the rear end of the spring, which is identical in operation to the front spring shackle. The same conditions

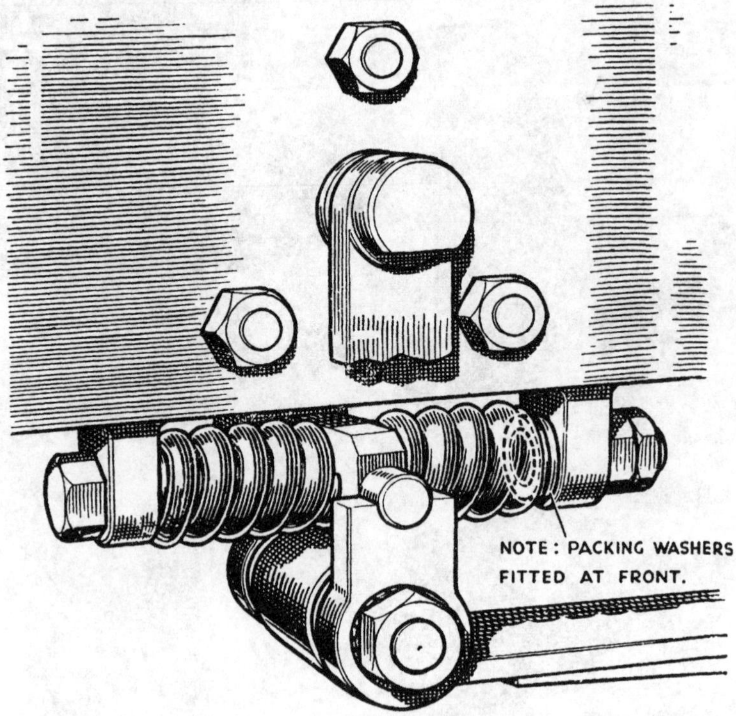

Fig. 5. Kick Shackle Arrangement (2.6 Litre Model).

apply, with the exception of the lower bush, which can only be tightened from one side. The bush will therefore have to be located a little to one side, to produce an equal pinch, which can only be achieved by trial and error.

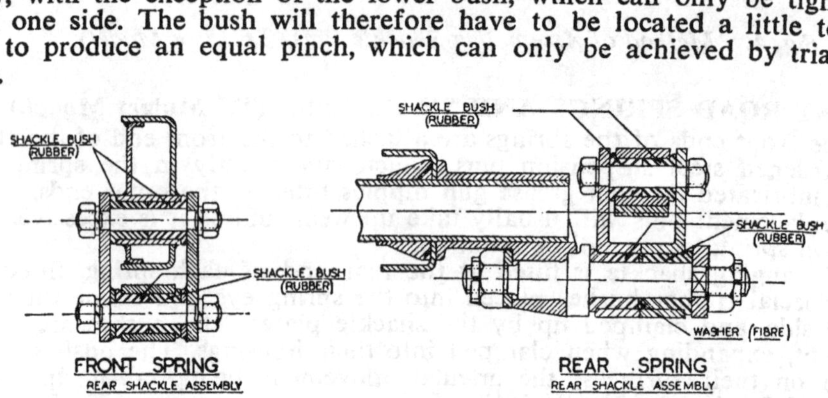

Fig. 6. Rear Shackle Assembly (Front and Rear Springs).

REAR SUSPENSION. (1¼ Litre and TD Midget Models.)

Semi-elliptic springs interleaved with rubber form the rear suspension. On no account should these springs be lubricated. The front ends of the springs are anchored in flexing rubber bushes, while the rear ends are mounted in similar bushes in swinging shackles. These four special rubber bushes are placed in the spring eye and frame tube from either side, and are clamped by the shackle plates. As the bushes are quite a loose fit, they will expand in their housings when clamped.

In the 1¼ Litre Saloon Model, an additional refinement is incorporated in the rear suspension, by the provision of a rubber-mounted lateral control link, fitted slightly above and to the rear of the rear axle. This has the effect of stabilising rear axle movement.

REMOVAL AND REPLACEMENT OF REAR SPRINGS. (TC and TD Midget and 1¼ Litre Models.)

Jack up the rear end of the car and remove the road wheel. Remove the nuts from the " U " bolts securing the spring to the axle, placing a small jack in the centre of the spring to take the load. Lower this latter jack and remove the " U " bolts. Remove the rear shackle pin nuts and spring washers of the top and bottom shackles, remove the shackle plates, pins and rubbers, and lower the rear end of the spring.

Remove the front shackle bolt, nut and spring washer, and remove the shackle bolt and fibre washers. The spring can then be lowered to the ground.

Replacement is carried out in the reverse manner to that detailed for removal.

NOTE:—When assembling the rubber bushes, it is essential for the swinging shackle to maintain a central location, so that the expansion of each half of the bush is equal. For Details, see Rear Road Springs and Shackles (TC Midget), page 17.

Finally, tighten all securing nuts when the car is in the laden position.

REMOVAL AND REPLACEMENT OF FRONT SPRINGS. (TC Midget.)

Jack up the front of the car and remove the road wheel. Support the centre of the spring with a small jack, and remove the nuts on the four bolts securing the front axle beam to the spring. Lower the jack and remove the four bolts. Remove the shackle pin nuts and spring washers of the top and bottom shackles, and shackle plates, pins and rubbers. The spring can then be lowered to the ground. Remove the front shackle bolt nut and spring washers, withdraw the shackle bolt, and the spring can then be removed from the car.

Replacement is carried out in the reverse manner to that detailed for removal.

LATERAL CONTROL LINK. (1¼ Litre Model.)

This device is mounted in rubber bushes; it is supported between two brackets located on the side of the chassis frame and rear axle casing. It consists of the control link tube " E," locknuts "A" and " C," adjuster nuts " B " and " F," fixing bolts " D " and " G," rubber bushes " H " and control cups " J."

ADJUSTMENT OF LATERAL CONTROL LINK. (1¼ Litre Model.)

This is accurately set before the car leaves the factory, and no adjustment should be necessary. Should adjustment or attention become necessary, proceed as follows:—

Weight the car at the rear end sufficiently to bring the control link parallel with the surface on which the car is standing. Grip the adjuster nut " B " and slacken off the locknuts "A" and " C ". Grip the fixing bolt " D " and slacken the adjuster " B ". Slacken the control link tube " E " from the fixing bolt " G ". Now adjust the adjuster nut " F " until the dimension of 1⅞ in. between the outer faces of the control link cups " J " is obtained, being equally disposed 15/16 in. each side of the bracket. Screw up the control link tube " E " tightly to the adjuster nut " F " and its spring washer. Tighten fixing bolt " D " until the dimension of 1 in. is obtained between the outer faces of the control link cups " J " and the centre of the bracket (see Fig. 7). Grip control link tube " E " and tighten the locknut " C ". The required dimension of 2 in. is obtained between the faces of the control link cups " J " by adjusting the adjuster nuts " B " and securely tightening the locknut " A ".

REMOVAL AND REPLACEMENT OF LATERAL CONTROL LINK. (1¼ Litre Model.)

Grip the fixing bolt " D ", slacken off the locknuts " C " and " A ", the adjuster nut " B ", and remove the fixing bolt from the control link

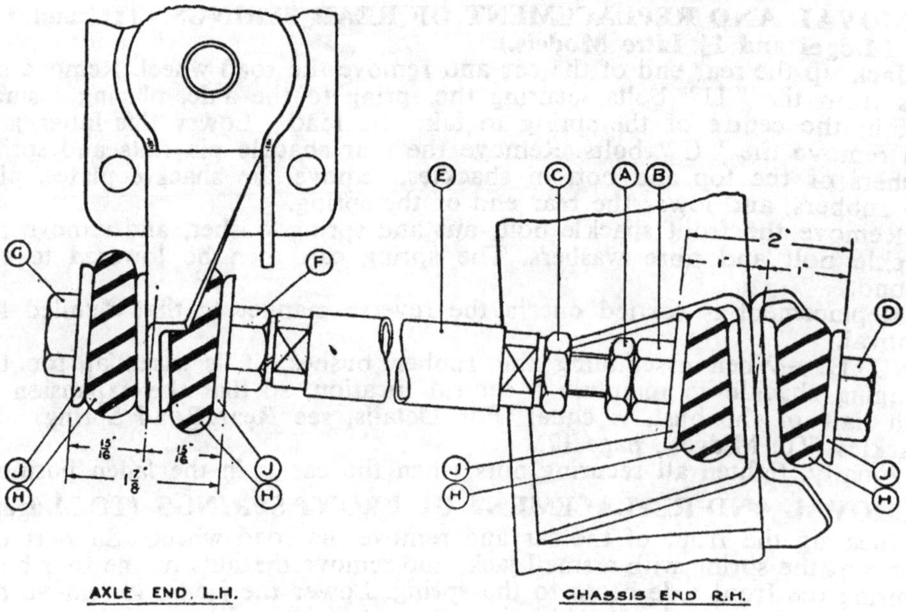

Fig. 7. Lateral Control Link.

tube " E ". Remove the control link tube from the fixing bolt " G " and it can then be withdrawn from the car. Grip bolt " G " and remove the nut " F ". The four bushes complete with the outer cups will now come away from the brackets.

Replacement is carried out in the reverse manner to that detailed for removal; after replacement it will be necessary to carry out the adjustment as detailed on Page 19.

FRONT SUSPENSION. (1¼ Litre and TD Models.)

The independent front suspension is of the wishbone type, incorporating coil springing.

Rubber bushes, requiring no lubrication, are provided for the inner mountings of the lower wishbones.

The top and bottom bearings of the king-pins are threaded, providing large bearing areas, and absorbing both thrust and journal loads.

REMOVING FRONT SUSPENSION. (1¼ Litre and TD Models.)

Jack up the front of the car until the road wheels are just clear of the ground. Remove the wheels. Place two jacks under the spring pans and jack up, taking some of the weight, until the hydraulic damper levers are clear of the rebound rubbers. Disconnect the hydraulic brake hoses. Slacken the tie-rod nuts, and screw the tie-rods out of the steering ball joints. Remove the cotters and nuts from the outer fulcrum bolts, drawing out the bolts and removing the front hub units complete; carefully preserve the thrust washers, rubber seals and retainers, and distance tubes. Remove the jacks from under the spring pans, press down the lower wishbone assemblies and remove the coil springs.

Remove the four bolts securing the spring pan to the levers, and remove the cotters, nuts and washers from the ends of the lower fulcrum pin. Slide off the levers and rubber bushes. Remove the bolts securing the lower fulcrum pins to the chassis cross member.

Remove the bolts securing the hydraulic dampers to the top of the chassis cross member. (The inner bolts are threaded into the cross member, the outer bolt heads being located inside the cross member).

The coil spring locating spigot plates are located inside the outer ends of the front cross member, being secured by one small bolt, and through which the outer damper bolts pass.

Dismantle the front hub units and king-pins as detailed on Pages 22 and 24 respectively.

REPLACING THE FRONT SUSPENSION. (1¼ Litre and TD Models.)

Bolt up the coil spring top locating plates inside the front cross member. Bolt on the hydraulic dampers, with the heads of the outer bolts inside the spring locating plate; these bolts have a long flat on the side of the head, which locates against the flange of the spring locating spigot plate, preventing the bolts from turning.

Bolt up the lower fulcrum pivot pins, with the nuts of the two rear inner bolts uppermost, and the six remaining bolts with the nuts below.

Fit the rubber bushes into the lower levers. These are quite a loose fit in the lever, but expand in their housings when clamped. Insert each bush so that it protrudes equally on each side of the housing, and then clamp up with the nut and washer, and fit the cotter pins (see Fig. 8).

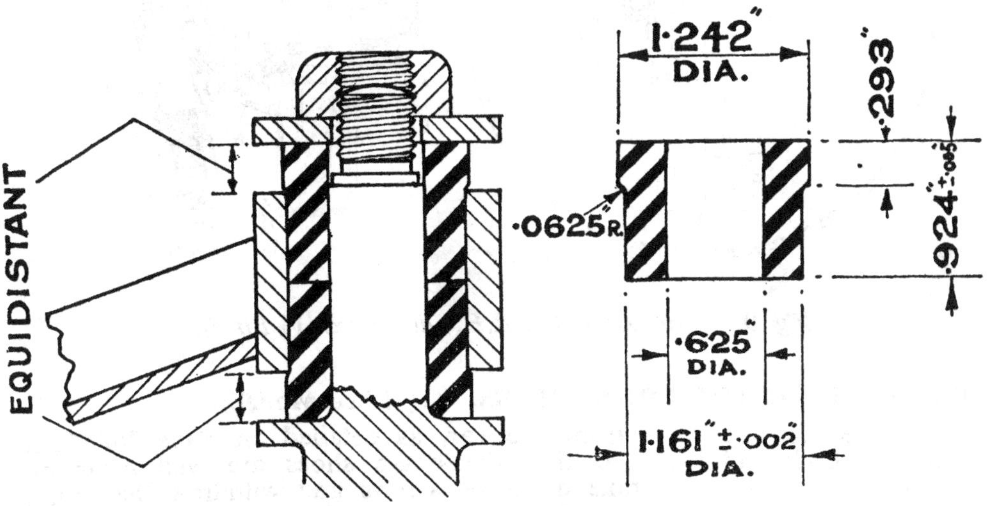

Fig. 8. *Correct Method of Clamping Rubber Bushes.*

Fit the spring pans between the levers, with the heads of the bolts inside the spring pan, tighten up the bolts, but leave them about half a turn slack. Smear each end of the coil springs with grease; press down the lower wishbone assemblies, and push the springs into the cross member and over the locating plates. Jack up the lower wishbone assemblies until they are approximately parallel to the ground.

Now assemble the hub units and king-pins as detailed on page 24. (Note—the stub-axles and nuts have right-handed threads for the right hand side and left-handed threads for the left hand side; the king-pin bearing threads follow the same rule).

Fit the front hub units to the suspension levers, ensuring that the thrust washers, rubber seals and retainers are in the correct order; tighten up the top and bottom slotted nuts, but leave them half a turn slack.

Connect up the hydraulic brake hoses. Screw the steering tie-rods into the outer steering ball joints, screwing the rods right in and then slackening off five complete turns; this will give a rough wheel alignment, rendering subsequent accurate alignment easier.

Bleed and adjust the brakes, and then replace the road wheels.

Lower all jacks; then bounce the car up and down at the front end a few times, to allow the suspension fulcrums to assume a free position. Tighten up the spring pan bolts, and tighten up and cotter the fulcrum bolts.

Check and adjust the front wheel alignment.

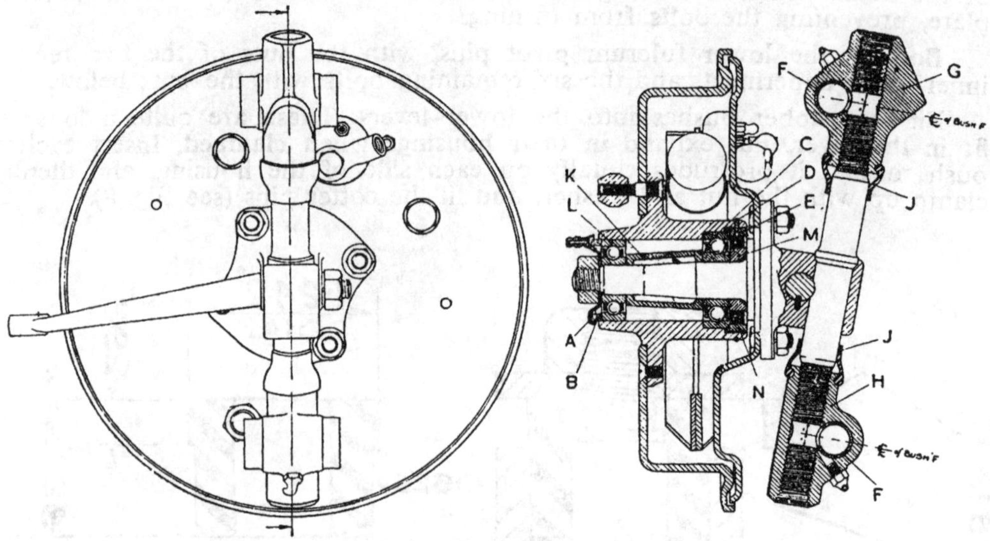

Fig. 9. Front Hub Unit R.H. Illustrated (1¼ Litre).

DISMANTLING THE FRONT HUBS. (1¼ Litre Model.)

Remove the hub units from the car, as detailed on Page 20. Turn back the brake adjusters so that the brake shoes are well clear of the drums. Remove the brake drum set screws and withdraw the drum. Remove the nuts and cotters from the stub axles. Remove the grease retainer plate and felt washer. Pull off the front hubs using extractor T.119. Remove the circlip and unscrew the locking ring complete with oil seal which is pressed inside. These rings have left hand threads on the left side and right hand threads on the right side. Their removal is facilitated if tool No. T.107 is used.

Using a soft drift, tap out the small outer bearing and distance piece, driving from the inside through the large bearing; then press out the large bearing.

Remove the three bolts and nuts and one set screw securing the brake back plates to the axle flange, and withdraw the backplates.

Dismantle the brake assembly.

Screw off king pin top and bottom swivel links. Remove the steering arm nuts and press out the arms from the stub axles. These are a taper fit with a Woodruff key (1¼ Litre and TD Models). Remove the steering ball joints from the steering arms and press out the king pins.

Fig. 10. *Removing Locking Nut from Front Hub* (1¼ *Litre*).

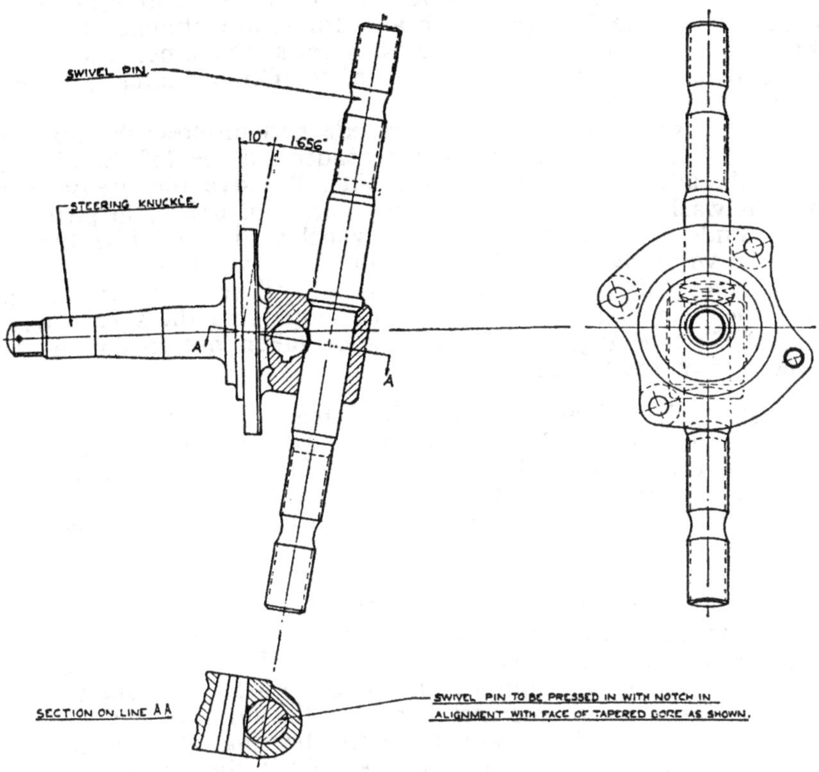

Fig. 11. *Assembly of Stub Axle and King Pin* (1¼ *Litre and TD Models*).

ASSEMBLY: Press the king pins into the stub axle; these should be a light press fit (1¼ Litre and TD Midget Models). Fit the steering arms and ball joints (both models). Oil the king pin threaded bearings, and screw on the swivel links, complete with the rubber seals, so that the bores of the bushes are exactly central with the grooves in the king pin; the lower one should be located correctly in position, as it cannot be turned once the brake back plate is fitted (both models).

Fit the brake assemblies to the back plates, and fit the back plates to the stub axles (both models).

Assemble the bearings in the hubs; press in the large bearing and screw in the locking ring, using tool No. T.107, already illustrated. Fit the locking circlip, locating in its original place; if by chance this is not possible, or new parts are fitted, drill through the hole in the hub, a new hole $\frac{7}{64}$ in. diameter and ⅛ in. deep, into the locking ring. It is preferable that this be done with the oil seal out, in which case the hole may be drilled right through the locking ring. Pack the hubs with the recommended grease, fit the distance pieces, and press in the outer bearings. Fit the distance washers to the stub axle, the chamfered side to the stub axle radius.

Fit the hubs to the stub axles, fit the felt washers and grease retainer plates, and tighten up the stub axle nuts firmly and cotter up.

Fit the brake drums to the hubs. Assemble the fulcrum pin distance tube, thrust washers and seals and containers; the hub units are then assembled complete.

REMOVAL OF FRONT BRAKE DRUM AND HUB. (TD Model.)

Remove the wheel disc and slacken off the road wheel nuts. Jack up the car at the front end, and remove the wheel nuts and road wheel.

On the earlier TD Models, the brake drums are attached to the wheel hubs by means of counter-sunk headed screws, the inner ends of which are peened over; these must not be disturbed. On the later models, composite drums and hubs are employed.

Remove the split pin from the stub axle and unscrew the nut, bearing in mind that these nuts are left hand threaded on the left hand side and right hand threaded on the right hand side. Remove the grease retaining disc and felt washer. Place the extractor, Part No. 68822, in position over the wheel studs and secure it by the wheel stud nuts. Use the central extractor and withdraw the brake drum and hub assembly.

As the inner ball race bearing spacer and oil seal will remain on the stub axle, they too must be withdrawn, by employing the special extractor, Part No. 68895 (see Fig. 12). Care must be taken not to damage the oil seal at the rear of the bearing.

IMPORTANT: When the front hub has been removed, the inner bearing, oil seal and hub distance washer must be removed from the stub axle, and replaced in the hub before it is refitted to the stub axle. If the hub is pressed on the shaft without first fitting the bearing and oil seal to it, the inner bearing will re-enter its housing, but the oil seal will only be pushed further from its correct position.

REPLACEMENT OF THE FRONT HUB. (TD Midget Model.)

Repack the bearings with grease before assembling the hub. Replace the bearing spacer with the chamfered side towards the small outer bearing. and then press the large outer bearing into position. Replace the oil seal and distance washer, with the metal face of the oil seal, and the recessed side of the distance washer, away from the bearing. Replace the hub on the stub axle, and fit a new felt washer. Refit the grease retainer and replace and tighten the hub nut.

Refit the road wheel. and lower the car to the ground.

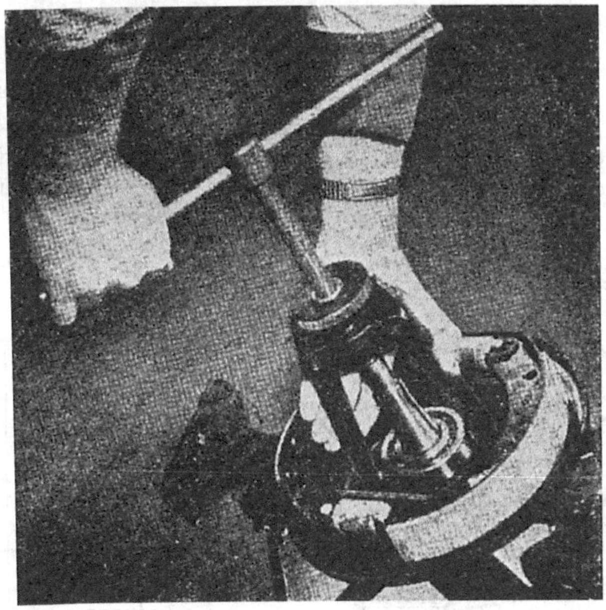

Fig. 12. *Withdrawing Inner Bearing Spacer and Oil Seal using extractor part No. 68895.*

REMOVAL AND REPLACEMENT OF FRONT COIL SPRINGS. (1¼ Litre and TD Midget Models.)

Jack up the front of the car until the wheels are clear of the ground, and remove the road wheel. Place an additional jack under the lower spring pan, and jack up until the hydraulic damper levers are clear of the rebound rubbers. Remove the lower fulcrum bolt; swing up the hub unit, and rest it on a suitable block. Release the jack from under the spring pan, press down the lower wishbone assembly and remove the coil spring.

Replacement is carried out in the reverse manner to that detailed for removal.

NOTE: Take care that the thrust washers, rubber seals and retainers are assembled in their correct order. Lubricate these parts and the fulcrum pins during and after assembly. Smear each end of the coil spring with grease.

GREASE LEAKAGE FROM FRONT HUBS. (1¼ Litre and TD Models.)

The front hubs on earlier models have been subject to grease leakage from the outside felt; later cars have a grease cap fitted to the end of the hub, thereby eliminating this trouble.

For cars without the grease cap, a special cap is available, retained in position by a spring and provided with a rubber sealing gasket, as indicated in Fig. 13 (TD Midget).

On the 1¼ Litre Model, the original oil retaining disc, with its felt washer, is removed from behind the hub retaining nut, and replaced by the flat washer, Part No. 500328. The new oil seal (Part No. 500336) is then fitted over the end of the hub, and the new grease retaining cap pushed over this. The new cap is retained in position by the flanged inner edge of the wheel, necessitating the removal of the wheel before the cap can be fitted. (See Fig. 14).

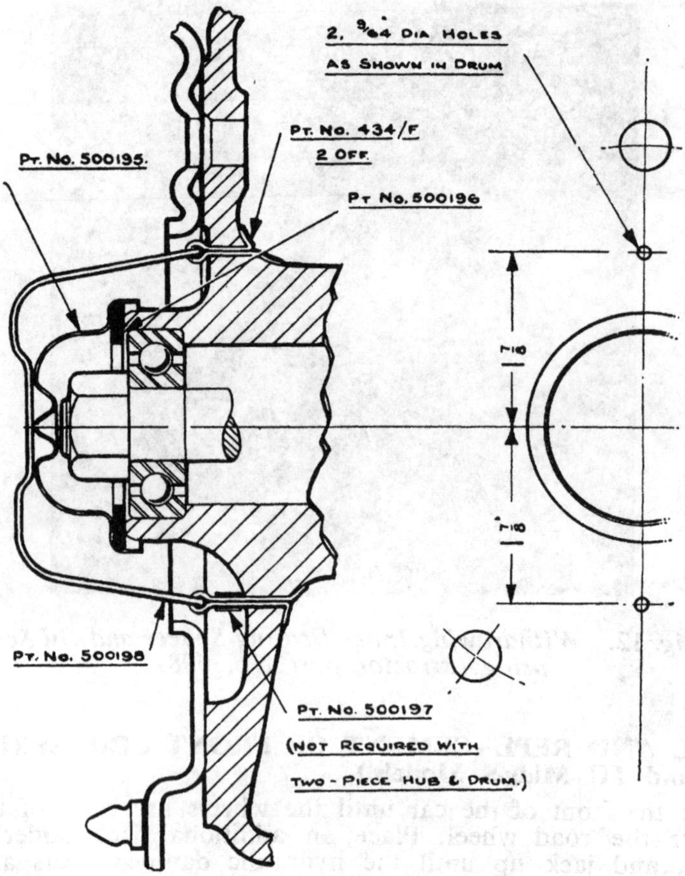

Fig. 13. Fitting of Oil Seal on Front Hub (TD Midget).

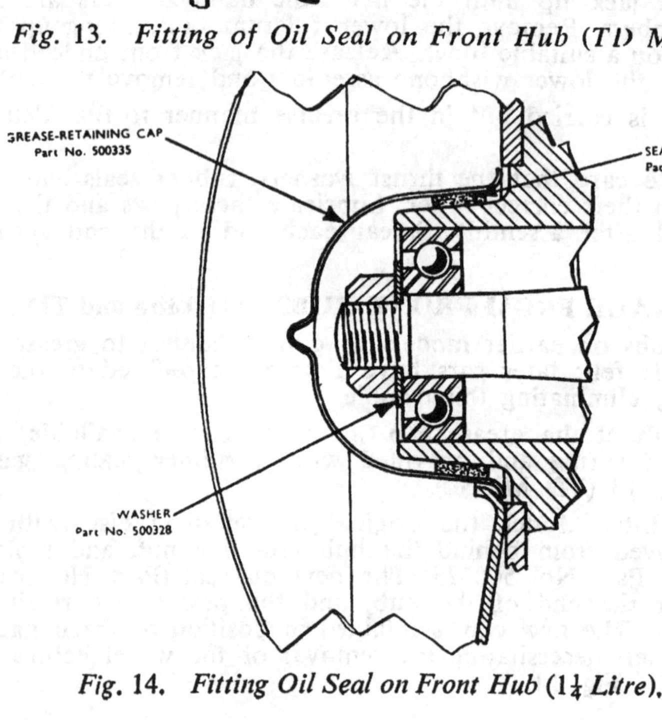

Fig. 14. Fitting Oil Seal on Front Hub (1¼ Litre).

SHOCK ABSORBERS. (M, D, F, J, K and L Models.)

Hartford type friction shock absorbers are fitted to both front and rear axles. It is recommended that no attempt be made to carry out repairs of any nature on these; should the shock absorbers become defective, it is considered advisable to obtain and fit service replacement units.

When leaving the Works, the shock absorbers are correctly set for average loads, but in time a little looseness will become apparent, rendering adjustment necessary.

Adjustment is effected by means of the large hexagon nut with the pointer attached, and is turned clockwise (ordinary right hand thread) to tighten the shock absorber. The dial is graduated 0, 2, 4, 8; the pointer should be turned not more than one degree at a time, testing repeatedly, preferably at speed on the road, until the best setting is found. It is important that the two shock absorbers on each axle are adjusted equally, and on no account should lubricant be applied.

FRONT SHOCK ABSORBERS. (P and N Models.)

Hartford friction type shock absorbers are fitted to the front axle, and the details given under Shock Absorbers (M.D.F., etc., Models) apply.

REAR SHOCK ABSORBERS. (P and N Models.)

Hydraulic shock absorbers are fitted to the rear axle. These should be topped up, if necessary, every 8,000 to 10,000 miles, with Girling Official Shock Absorber Fluid, to bring the level to within ¼ in. of the top.

The recuperator chamber should not be allowed to become empty, otherwise air will enter the chamber and impair the return action of the shock absorber. Do not fill up the chamber quite full, as the expansion due to temperature increase may set up a pressure which is harmful and unnecessary.

To increase the amount of damping, and to tighten the shock absorber, the regulator screw beneath the filler plug must be turned to the right, thus closing the valve. To decrease the damping, the screw is turned to the left, thus opening the valve and permitting a freer flow of fluid. A very fine degree of adjustment can be obtained, and it is sufficient to turn the regulator screw a quarter of a turn (i.e., one flat) at a time; ascertain, by road tests, if sufficient adjustment has been made. Care must be taken to make adjustments so that the damping action of both shock absorbers is equal.

If the original setting has been entirely lost, then by disconnecting the bottom end of the connecting link and moving the lever by hand, the shock absorbers can be balanced by individual adjustment. The best amount of damping can be determined by road test with the shock absorbers reconnected.

FRONT AND REAR SHOCK ABSORBERS. (TA and SA Models.)

The Girling Vane type shock absorbers fitted to the front and rear axles of the TA Midget are of the hydraulic type, and are designed to control the road springs both on compression and recoil.

To ensure the correct functioning of the shock absorbers, it is absolutely essential that the working chamber should always be completely full of the recommended fluid. By this means the whole of the mechanism is enclosed in another cylindrical chamber in which is maintained a reserve supply of fluid. Any shortage of fluid in the working chamber is automatically made good through valves in the lower wall of the cylinder; similarly, any air which may be present is expelled through a special air duct in the top of the reaction block. The oil valves fall automatically on to their seats when the shock absorber is doing no work, thus preventing loss of fluid or entry of air into the working cylinder.

ADJUSTMENT OF SHOCK ABSORBERS. (TA Midget Model.)

The method of increasing or decreasing the damping effect is identical to that for the rear shock absorbers fitted to the rear axle of the P and N models.

Ensure that identical adjustment is made to both shock absorbers, front and rear.

THE CONNECTING LINKS. (TA and SA Model.)

Tilting bearings, consisting of round blocks of rubber which grip the bearing portion of the connecting link, and which obviate the use of ball joints, are used in the connecting links between the lever arm and spring. Attention such as lubrication is therefore entirely eliminated.

SHOCK ABSORBER REPLENISHMENT. (TA and SA Model.)

Every 10,000 miles the recuperator chamber should be inspected by removing the filler plug on top of the case, and the fluid replenished if necessary. When the recuperator is three-quarters full there is an ample supply of fluid.

On no account should the chamber be allowed to become empty, for this allows air to enter, and so destroy the action of the shock absorber. On the other hand, the chamber should not be over-filled, as this will prevent the expansion due to temperature increase.

SHOCK ABSORBERS, FRONT AND REAR. (1¼ Litre Model, up to and including Chassis No. VA.1914.)

The shock absorbers fitted are of the hydraulic finger-tip controlled type, designed to control road springs on both compression and recoil. This system utilises the axle movement as the controlling factor for the amount of damping imposed on the spring by the shock absorber, with an over-riding control, so that the change in load and other variables can be provided for.

The four special shock absorbers are fitted with controllable valves operated by fluid pressure, the source of the pressure being the pump operated by the axle movement. The pressure is generated in proportion to the axle movement; the more the axle bounces in relation to the frame, the lighter the generated pressure in the system, with a consequent increase in damping capacity.

As the relative pressure between axle and frame decreases, so the pressure decreases, the damping capacity being reduced.

The over-riding control is a spill valve on the rear face of the engine shield, controlled by a lever on the steering column, which enables the generated pressure to be discharged into a reservoir at varying rates, depending on the position of the spill valve.

The best setting for the control lever is the "Mid" position, as rough roads tend to increase the damping automatically, smoother roads decreasing it. If the load or the speed is increased, a heavier damping effect is required, and the control lever should then be set to the "Max" position.

If the load is small and maximum flexibility is required, set the pointer to the "Min" position.

It will usually be found that the control requires to be set more towards the "Min" position in the winter than in the summer, other conditions being equal.

MAINTENANCE OF SHOCK ABSORBERS. (1½ Litre Series VA up to and including Chassis No. VA.1914.)

Every 8,000 to 10,000 miles unscrew the filler plug on the top of the damper casings and examine the fluid levels; the dampers should be topped up, if necessary, with Girling Official Hydraulic Shock Absorber Fluid, to the level of the filler plug.

The recuperator chamber should never be allowed to become empty, otherwise air enters the working chambers, impairing the action of the shock absorber.

MAINTENANCE OF FINGER-TIP CONTROL. (1½ Litre Series VA up to and including Chassis No. VA.1914.)

The essential parts ot the control are the spill valve and pressure pump, and the connecting pipe lines. The pressure pump is driven from the near side of the rear axle.

A filter in the bottom of the spill valve body can be cleaned by washing in petrol, after withdrawing the body, which is a press fit, from the valve chamber. A filter placed between the pipe and valve body can also be unscrewed and cleaned by washing in petrol.

If the pipe line is disconnected for any reason, oil will drain from the pipe line and will be replaced by air. It will therefore be necessary, on reassembly, to fill the pipe line "solidly" full of oil. To do this, first disconnect the spill valve outlet pipe from the bottom of the spill valve reservoir; then attach an oil gun (filled with the correct oil) to the end of the pipe line. Disconnect the unions on top of all four shock absorbers, then force oil into the pipe line until the ends of all four pipes are flowing with AIR-FREE oil. Reconnect the ends of all four pipes to the shock absorbers and, with the spill valve closed, raise the pressure in the system with the oilgun until the relief valve opens, discharging oil into the spill valve reservoir.

Once the filler cap is removed, oil can easily be seen surging past the cone valve.

With the pressure maintained in the system, the line should be carefully examined for leaks or weeping of oil at the joints. Every joint in the pipe line must be oiltight, and a new length of pipe should be fitted if there is the slightest suspicion of a crack. The pressure test is very important, for under working conditions the pipe line acts as a suction line, and the merest tendency to suck air will prove fatal to the correct functioning of the system.

A final check can be made by running the car on the road, to ensure that the desired control is obtained.

As an additional check, connect a pressure gauge to the spare plug and observe that the pressure rises as the car passes over an increasingly rough road. With the control in the "Mid" setting the pressure gauge should register 25 to 45 lbs. per sq. in. on a rough road, or from 50 to 70 lbs. per sq. in. on a very rough road.

It is imperative that the valve position is in correct relationship to that of the finger lever; this should be checked as follows:—With the finger lever in the "Max" position, a groove on the valve spindle should line up with a similar groove on the flexible drive support bracket on top of the reservoir; the valve will then be shut.

It is important to note that the fluid in the control system is entirely different to that as used in the shock absorbers. Once a month the reservoir of the spill valve should be inspected, and topped up if necessary, to within 1 in. of the top with Duckham's Shock Absorber Ride Control Oil.

SHOCK ABSORBERS. (1½ Litre Series VA, commencing at Chassis No. VA 1915.)

Both front and rear axles are controlled by piston-type hydraulic shock absorbers; as these dampers are set to give the correct amount of damping required, no further adjustment is required or provided for.

Every 8,000 to 10,000 miles, remove the filler plug positioned on the side of the unit, and, if necessary, top up to the level of the filler plug hole with Girling Official Piston Type Shock Absorber Fluid.

SHOCK ABSORBERS. (2.6 Litre Series WA.)

Piston type shock absorbers are fitted to this model, and no means of adjustment are provided. The only maintenance necessary is to replenish the fluid periodically to the correct level, and to check the securing bolts for tightness.

MAINTENANCE OF SHOCK ABSORBERS. (2.6 Litre Model.)

Every 8,000 to 10,000 miles, remove the filler plug positioned on the side of the unit about ¾ in. from the top, and replenish if necessary, with **Girling Official Piston Type Shock Absorber Thin Fluid. The correct fluid** level is obtained by filling right up to the plug hole. Prior to removing the filler plug, very carefully clean around the filling orifice with a clean brush dipped in petrol. Do not remove the cover plate on top of the unit.

This simple operation of "topping-up" must not be neglected, for if the low pressure chamber is allowed to become empty, air enters the cylinder, thereby impairing the action of the shock damper. If it is possible to move the lever for a short distance either side of the centre position before any resistance is offered, it is an indication that there is air in the pressure cylinder.

To overcome this fault, fill up to the correct level of the filler plug, disconnect the operating link between the axle and lever, and move the lever as far as it will go in one direction, hold it in this position for approximately five seconds, and then repeat the operation in the opposite direction. These operations should be repeated several times, to make sure that all air in the pressure cylinder is expelled.

A final check is to hold the lever in the central position, when there should be no idle movement before resistance is offered by the pistons.

HYDRAULIC DAMPERS. (TC and TD Midgets, and 1¼ Litre Saloon.)

The hydraulic dampers are double acting, resistance being offered to the compression and recoil of the road springs, and are accurately set before the car leaves the Works to give the amount of damping most suitable for the car. No adjustments are therefore necessary or provided for.

MAINTENANCE OF HYDRAULIC DAMPERS. (TC and TD Midgets. and 1¼ Litre Saloon.)

When in position on the car, maintenance is confined to periodic examination of the fluid level, and of the secure anchorage to the chassis, the fixing bolts being tightened as required.

LUBRICATION OF HYDRAULIC DAMPERS. (TC and TD Midgets, and 1¼ Litre Saloon.)

Every 12,000 miles the dampers should be topped up with Girling Official Piston Type Hydraulic Damper Thin Fluid, by removing the filled plug and filling up to the bottom of the filler plug hole.

The filler plug is positioned on top of the cover plate, and before removing the plug, carefully wipe the exterior of the damper with a clean rag, to prevent dirt entering through the filler hole.

This operation of topping up must not be neglected, for if the low pressure chamber is allowed to become empty, air enters the pressure cylinders, seriously impairing the action of the damper.

TESTING THE DAMPERS. (TC and TD Midgets, and 1¼ Litre Models.)

The resistance of the dampers can be checked by bouncing the car up and down. If the movement is uniform, no attention is required; should the resistance be erratic and free movement is felt, the damper should be removed for checking and topping up.

Indication of the resistance is carried out by bolting the damper to a plate held in a vice; this method avoids the possibility of distortion of the cylinder body. Hold the damper in the upright position and move the lever arm up and down through its full stroke when a moderate resistance throughout the stroke should be felt. Erratic resistance with free movement in the lever arm may well indicate lack of fluid.

To expel any air that may be present in the operating chambers, the fluid should be added while the lever arm is being worked up and down throughout its full stroke.

If there is no improvement after the addition of fluid, a replacement damper should be fitted.

If it is not possible to move the lever arm by hand, owing to there being too much resistance, a broken internal part or seized piston is normally indicated, and again in this case a replacement damper should be fitted.

With the exception of the TD and 1¼ Litre Models, it is felt unnecessary to detail instructions for the replacement of the front dampers, and consequently only the above models will be dealt with.

REMOVAL AND REPLACEMENT OF FRONT DAMPERS. (TD and 1¼ Litre Models.)

Place the car over a pit, place a jack under the lower spring pan, and jack up until the road wheel is clear of the ground.

Remove the road wheel, and remove the cotter and nut from the top outer fulcrum bolt, and withdraw the bolt. Swing out the hub unit from the upper wishbone or damper levers, and support it on a suitable stand, to prevent straining the brake hose.

Slacken the bolts securing the wing valance to the chassis at a point adjacent to the hydraulic dampers, which will give easier access to the inner fixing bolts (1¼ Litre Model only).

Spring back the wing valance and extract the two inner securing bolts, which are threaded into the front cross member (1¼ litre model only).

Remove the nuts and spring washers from the outer bolts, the heads of the bolts being located inside the cross member. Press the bolts through the base of the damper and remove the damper from the front cross member (1¼ Litre Model only).

Remove the four set screws securing the front damper and lift the damper from the front cross member (TD Midget Model only).

Replacement is carried out in the reverse manner to that detailed for removal.

NOTE:—1¼ Litre Model only: When replacing the damper, note that the two long outer fixing bolts have a long flat on one side of the head, which locates against the spring locating plate flange inside the front cross member, which prevents the bolt from turning.

To insert the bolts through the spring pan hole and spring coils, use a socket wrench with a 12 in. extension, one side of the socket wrench ground away, level with the flat on the bolt head.

PROPELLER SHAFT. (M, D, J, F, K, L, N, KN, and P Models.)

Hardy Spicer Universal joints are fitted to either end of the tubular propeller shaft. The front end is castellated, fitting into the castellated end of the universal joint, allowing for the end movement of the shaft brought about by the flexion of the springs and the rise and fall of the axle.

A nipple is provided for lubrication of the sliding coupling joint, access to the nipple being through a hole in the propeller shaft tunnel.

The universal joint consists of a plate carrying two jaws, a cross head with hardened steel roller bushes, and the opposite set of jaws attached to the propeller shaft (see Fig. 15).

A pressed steel cover and a spring-loaded dust excluding cover enclose the entire universal joint.

Semi-circular wire rings, locating in grooves in the hardened steel rollers, prevent the rollers from floating outwards. It is essential that this joint should be packed with Hardy Spicer grease.

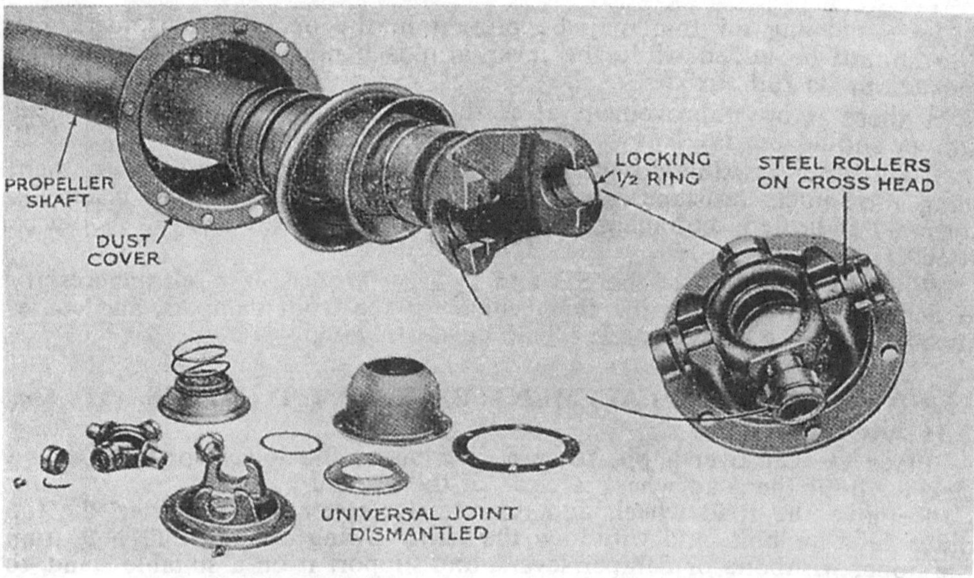

Fig. 15. *Various Compounds of the Hardy Spicer Universal Joint. (M, D, J, F, K, L, N, KN, and P Models.)*

PROPELLER SHAFT. (TA, TB, TC, TD Midget Models, and 1¼, 1½, 2, and 2.6 Litre Models, commencing at Chassis No. WA 0496 for 2.6 Litre Model.)

The needle bearing type of Hardy Spicer propeller shaft is fitted, consisting of a tubular shaft with a universal joint at each end. To accommodate axle movement, a splined sliding joint is packed with grease and fitted to the front and provision is made for additional lubrication.

Before disconnection of the propeller shaft, a very simple operation should be carefully carried out. This is to mark the two flanges, to assist in refitting to the original position, to retain balance.

To test for wear, that is wear on the thrust faces, test the lift in the joint; any circumferential movement of the shaft relative to the flange yokes indicates wear in the roller bearings or the splined shaft. A total of .004 in. circumferential movement, measured on the outside diameter of the spline, should not be exceeded.

DISMANTLING THE UNIVERSAL JOINTS.

Unscrew the dust cap and pull the sliding joint from the splined shaft. Pinch together the ears of the retaining circlips, and prise them from the grooves with a screwdriver. Hold the joint in the left hand, with the splined sleeve yoke on the top, and tap the radius of the yoke with a copper hammer, when the top bearing should begin to emerge. Turn the joint over and remove with the fingers.

If necessary, tap the bearing race from the inside with a small diameter bar, taking care not to damage the bearing race. Keep the joint in this position, to avoid dropping the needle rollers, repeating the operation for the opposite bearing.

To remove the splined sleeve yoke, rest the two exposed trunnions on wood blocks to protect their ground surfaces; tap the ears of the flange yoke to remove the remaining bearing. Treat the opposite end of the shaft in the same manner, and thoroughly wash all parts.

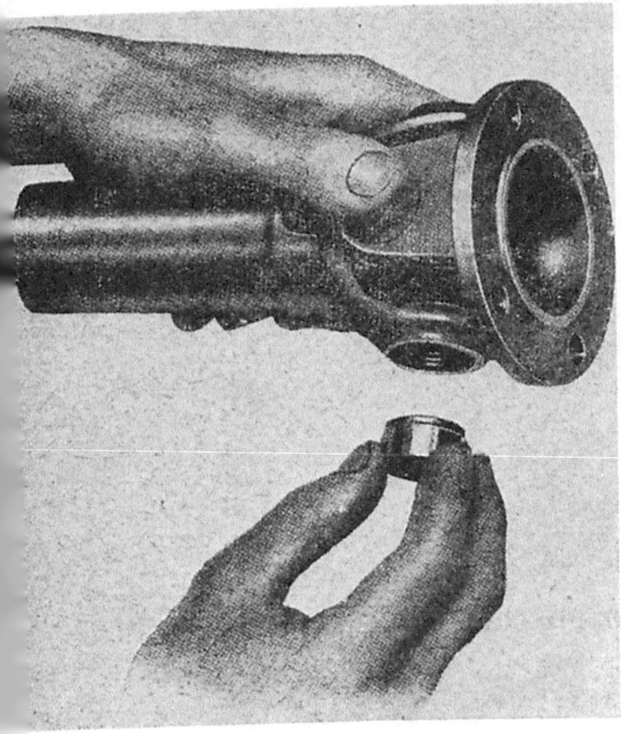

Fig. 16. *Withdrawing Needle Bearings.*

Fig. 17. *First Stage of Removing Universal Joint.*

REASSEMBLY OF PROPELLER SHAFT.

The drilled holes in the journal should be cleaned and filled with grease. Assemble the needle rollers in the bearing races and fill with oil.

Insert the spider in the flange yoke, and tap the bearing into position using a soft-nosed drift $\frac{1}{32}$ in. smaller in diameter than the hole in the yoke. It is absolutely essential that the bearing races are a tight fit on the yoke trunnions.

Firmly locate the circlips in their grooves. If the joint appears to bend, tap lightly with a wooden mallet, which will relieve any pressure of the bearings on the end of the journals.

Thread the dust cover and cork washer on to the splined shaft and replace the sliding joint, making sure that the trunnions on the sliding and fixed joints are in line. This latter is best checked by observing that the arrows marked on the splined yoke and the splined stub shaft are in line.

PROPELLER SHAFT. (2.6 Litre up to and including Chassis No. WA 0495.)

The needle bearing type of Propeller Shaft is fitted, and the details of the universal joints given on page 32 apply.

The twin propeller shaft is supported amidships by a roller bearing, rubber mounted on the chassis centre brace. The bearing itself is lubricated through the near-side semi-grouped nipples under the bonnet: refer to Fig. 19 for bearing lubrication.

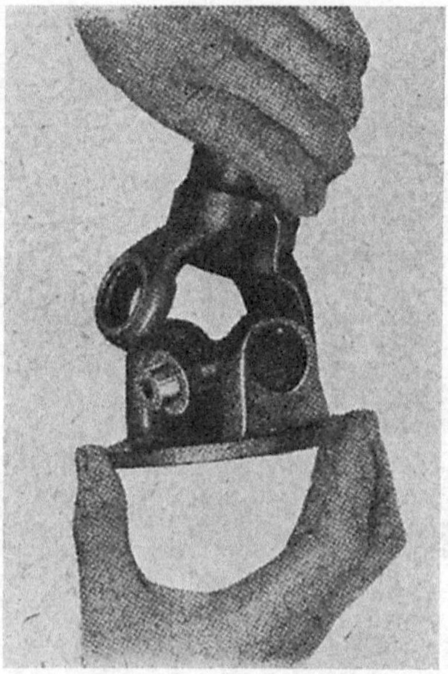

Fig. 18. *Separation of Spider Joint After Removal of Needle Bearing.*

Fig. 19. *Showing Amidship Bearing of Twin Propeller Shafts.* (2.6 *Litre, Ceased Chassis No. WA.* 0495.)

It is vitally important that under no consideration is the connection between the two propeller shafts broken. These shafts are balanced as a complete unit, and separation of the two flanges would necessitate rebalancing by the manufacturers.

Should it be necessary to remove the shaft, the centre bearing will have to be detached from the chassis.

When refitting this bearing, it is essential that it is secured in a position where it is possible to insert a .010 in. feeler between the shaft and its housing through the circumference and on either side; in this position it is just possible to move the bearing housing in any direction, made possible by its rubber mounting.

Finally, the bearing must be in the centre of the bearing surface on the shaft, which can be ascertained by moving the bearing on the shaft, fore and aft.

Fig. 20. Andrex Friction type Adjustable, Fully Enclosed Damper; Type T.E. I/N (TD Midget Mark II.)

SUPPLEMENTARY DAMPERS. (TD Midget, Mark II.)

The Andrex type T.E.I/N Suspension damper which is fitted as supplementary equipment, to both the front and rear of the Mark II TD Midget, is in principle a development of the well-known Hartford friction shock absorber and is totally enclosed. The construction consists of laminations of friction plates, moving and fixed, with discs of specially impregnated wood between them. The pressure on these plates, which are free to slide on the spindle, is applied by a plate spring.

This pressure is adjusted by means of a small set screw and lock nut located on the cover plate of the unit, approximately 2½ turns of which will increase the loading from minimum to the recommended maximum which it would be desirable to apply: The casing is filled with a heavy oil (Esso Cantona LK. 190) or similar, and once the desired settling is obtained, no further attention or servicing should be necessary.

The T.E. I/N Dampers are pre-set to a nominal loading of 16 lbs. before they leave the Works, the loading being measured by a slow pull with a spring balance on the end of the operating arm.

When fitted to the TD Midget Mark II the front dampers are set to 24 lbs. and the rear dampers to 22 lbs.

If it is desired to alter the setting a screw is provided on the damper near the upper oil seal. Always use two spanners when making the adjustment. To increase the damping or loading, hold the screw head with one spanner and slacken the lock nut with the other. Turn the screw clockwise about a ¼ of a turn; hold in this position and retighten locknut. This will increase the loading approximately 3 lbs. Repeat if necessary but do not over tighten, as this will produce hard suspension. Both off side and near side dampers must be adjusted the same amount to preserve the balance.

To reduce the loading, proceed as above but turn the screw in an anti-clockwise direction.

CHAPTER II

THE ENGINE

from the 'M'

to the 'TF'

CHAPTER II

THE ENGINE

ENGINE LUBRICATION. (M, D, and J Midget Models.)

The engine sump holds approximately one gallon of oil and is replenished through the filler on the off-side of the engine. Oil level readings are taken by means of the dipstick just beside the oil filler. There are two marks on the dipstick, the top one is the full mark and the lower one being the danger mark.

The oil level should be inspected daily and the level should be kept well up, never being allowed to fall lower than half-way between the " full " and " danger " mark. Overfilling should be avoided as this may lead to oiling up of the sparking plugs.

A gear type pump is carried at the front of the engine and driven from an inclined shaft. The pump sucks oil from the sump through the gauze strainer into the pump body and is delivered to the main bearing by an internal duct, the overhead valve gear being fed by an external pipe. A bye-pass relief valve is incorporated in the oil pump body.

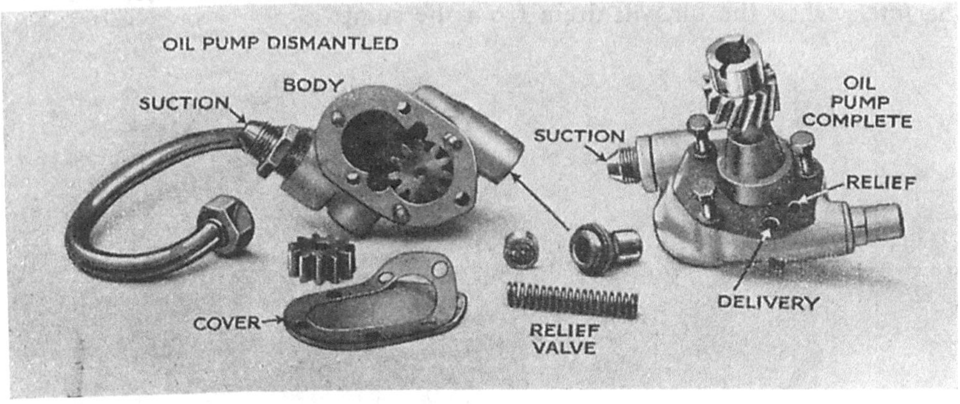

Fig. 1. *Oil Pump complete and dismantled.*

The relief valve consists of a cover enclosing a spring and maintaining the small piston up against the seating in the pump body until either the force of the oil through the pressure or cold non-fluidity forces the piston off its seating. Whenever this occurs there will be either the corresponding drop in oil pressure, or the release of excess oil will maintain the pressure at a point predetermined, this pressure being dependent upon the resistance of the spring. Pressure can be increased by substituting a stronger spring or by the introduction of washers in the cap, thereby increasing the tension of the spring. The pump draws its supply of oil from the bottom of the sump via an internal filter and a large diameter suction pipe, delivering oil at a high pressure into an internal duct, running across the off-side of the engine, whence it feeds into the forward end of another internal duct running to the rear crankshaft bearing. Oil grooves in the bearing and drilled passages in the crankshaft conduct oil to each big-end, and are fed consequently under high pressure.

From the pump delivery duct an external pipe extends vertically where it encounters a restrictor pin.

Oil passing the pin enters a drilled passage in the cylinder head which registers with a similar passage drilled in the front camshaft bearing bracket, communicating with the camshaft bearing itself and each of the valve rocker shafts. The shafts are drilled at intervals to correspond with the valve rockers, whose bushes are grooved, feeding drilled passages in the rocker arms which feed the faces.

The rocker shafts also feed the rear camshaft bearing and a duct in the forward camshaft bearing feeds the camshaft drive gears. Oil is drained back to the sump by a drain passage and external pipe at the rear and front end of the cylinder head, the front pipe lubricating the dynamo and inclined gears, finally passing through the crankshaft ball race on to the oil filter in the sump.

DRAINING THE SUMP. (M, D, and J Midget Models.)

It is preferable for the sump to be drained every 1,500 miles and it is recommended that this be done after returning from a long run when the oil is warm and more fluid. As the filter in the sump must eventually be removed for cleaning, it is recommended that this be removed at the same time for by doing so it will speed the operation.

Proceed as follows:—

Disconnect the main suction pipe from the filter at the front of the sump.

Unscrew the large hexagon plug attached to the internal filter, withdraw the filter, when the oil will drain from the sump.

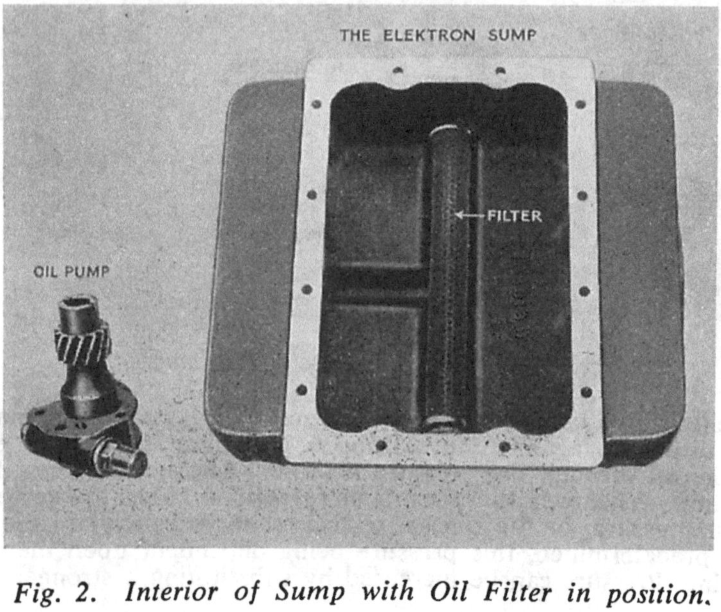

Fig. 2. Interior of Sump with Oil Filter in position.

After the oil has been drained the filter should be cleaned thoroughly and replaced and the main suction pipe reconnected.

When reconnecting the suction pipe to the threaded end of the filter great care must be taken to see that the pipe is fitted correctly, for incorrect fitting will cause an oil leak in the oil feed, which can lead to irreparable damage to the engine.

Refill the sump with the recommended grade of oil.

THE OIL RESTRICTOR PIN. (M, D, J, P, F, L, K, KN, and N Models.)

The purpose of the restrictor pin or oil metering pin is to regulate the oil feed to the overhead valve gear and on no account should the pin be filed, altered in shape or interfered with.

Every 500 miles the pin should be removed by passing a piece of wire hooked at its end, through the hole in the end of the pin.

The pin and its housing should be carefully cleaned before replacement.

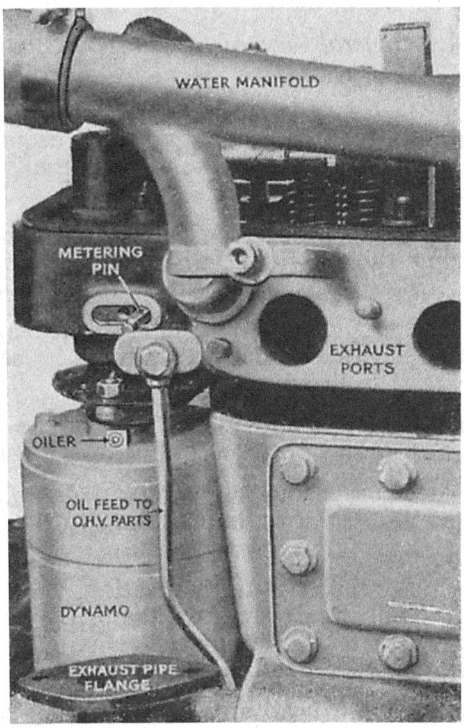

Fig. 3. Showing Metering Pin withdrawn from its housing in the cylinder head.

REMOVAL OF THE CYLINDER HEAD. (M, D, and J models.)

Open the radiator drain tap and while the water is draining proceed to:—

Remove the bonnet complete.

Detach the high-tension cables to the sparking plugs.

Remove the valve gear cover and do not mistake the "M.G." breathers at each end for nuts. They cannot be unscrewed.

Uncouple the petrol pipe, throttle and mixture controls and remove the carburetters. It is entirely unnecessary to dismantle the Induction Manifold.

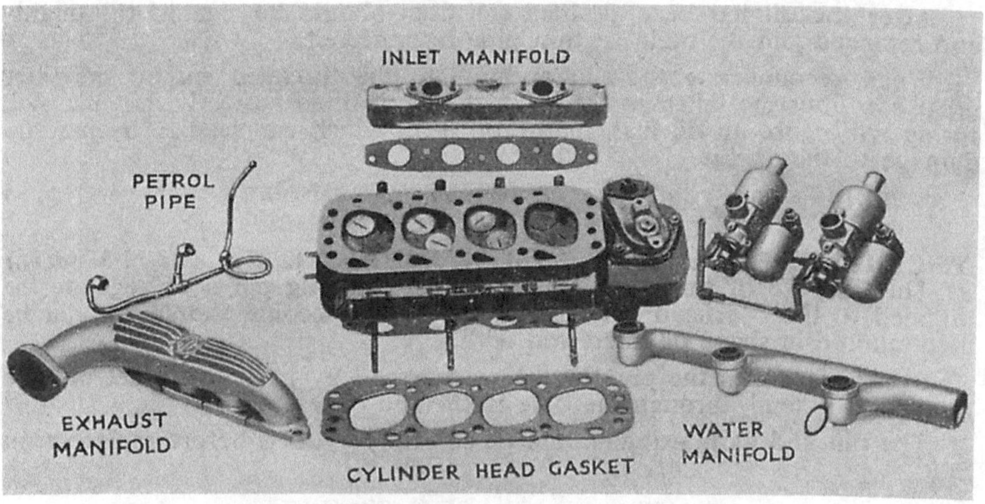

Fig. 4. General View of Head after Removal.

Release the water manifold by undoing the three attachment clips and detach the front exhaust pipe before removing the water and exhaust manifolds, in turn.

Uncouple the oil delivery pipe from the cylinder head (left-hand side forward) by unscrewing the single retaining bolt, and withdraw the oil restrictor to avoid the risk of loss when the head is removed.

Uncouple the oil return pipe at the front of the engine by unscrewing the two nuts only. It need not be withdrawn from its studs as it will draw straight off when the head is lifted.

Remove bodily the oil return pipe at the rear of the engine by unscrewing the retaining nuts at top and bottom, taking care not to lose the washers.

Remove the two bolts securing the circular flexible coupling to the dynamo drive yoke which will allow the coupling to be withdrawn with the cylinder head.

Slacken off the cylinder head nuts in rotation, half a turn at a time, until they are quite loose and finally remove them. Do not unscrew any one of these nuts completely before slackening the remainder as this will impose uneven strain upon the cylinder head, leading to distortion.

Break the joint between the head and the block by smartly tapping the sides of the head with a wooden mallet or with a hammer with a piece of wood interposed to take the blow.

If the joint does not break freely, insert a screwdriver or similar blunt instrument between the joint at the two places, one each side of the engine, where the gasket has been cut away. Do not insert the screwdriver too far and on no account should it be forced against the gasket which would be damaged as a result.

When the joint is broken, lift the head from the studs, withdrawing squarely.

Carefully lift the gasket straight off the cylinder head studs, keeping it parallel with the face of the cylinder block.

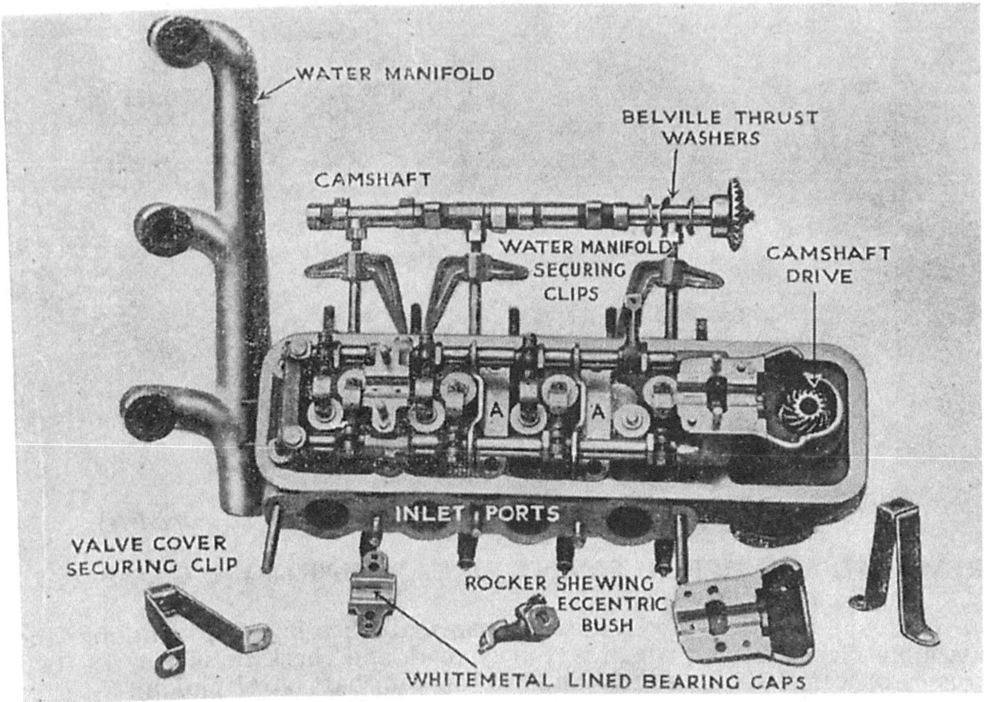

Fig. 5. Showing Cylinder Head after Camshaft Removal.

REPLACING THE CYLINDER HEAD. (M, D, J, P, F, L, K, KN, and N Models.)

It is recommended that new gaskets be fitted at all joints.

Coat both sides of the cylinder head gasket with an even film of Castile soap. Locate over the studs and gently push into position on the upper face of the cylinder block, using a short length of tubing to push the gasket in position, taking care to keep the gasket parallel with the face of the cylinder block.

Turn the camshaft until the timing marks on the spiral bevel gears coincide, and rotate the engine by means of the starting handle until the timing marks on the face of the flywheel (1/4 for the M, D, J, and P models, and 1/6 for the F, L, K, KN, and N models) show exactly in the centre of the clutch inspection aperture with the metal electrode on the distributor rotating arm pointing towards No. 1 contact stud.

Lower the head into position on the cylinder block when the two holes in the flexible coupling should almost coincide with those on the drive yoke of the dynamo.

Replace the cylinder head stud nuts, tightening up in the rotation given in Fig. 6 (M, D, J Models).

Replace flexible coupling bolts and remaining engine equipment, finally refilling the radiator with water.

Start up the engine allowing to run until thoroughly warm, switch off and tighten up cylinder head nuts.

After about 100 miles running remove the valve cover and again tighten up the cylinder head stud nuts.

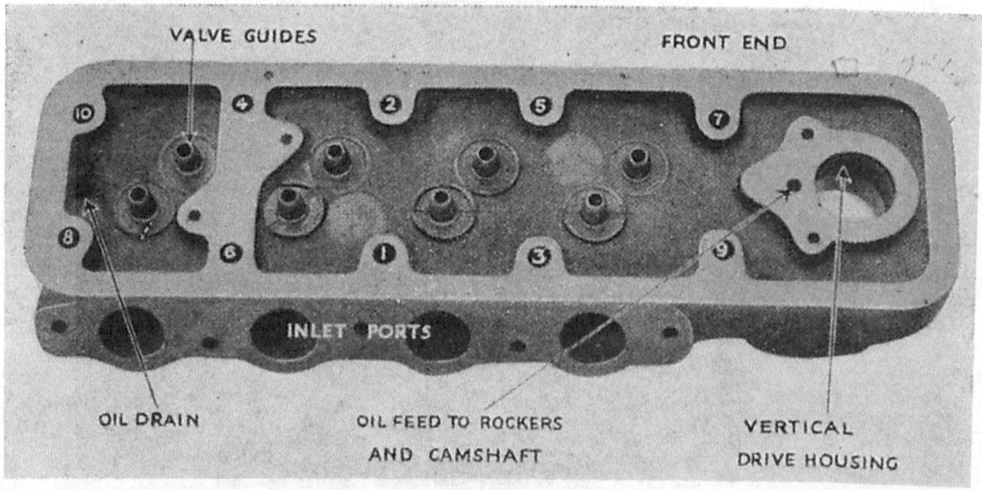

Fig. 6. Sequence for tightening cylinder head nuts (M, D, J models).

REMOVAL AND REPLACEMENT OF CAMSHAFT. (M, D, J, P, F, L, K, KN, and N Models.)

Remove camshaft by removing camshaft bearing caps, noting the assembly markings, and which way they read, and check timing marks (two crosses on camshaft bevel gear and one on camshaft bevel pinion).

These markings must marry together to make three crosses in a row when reassembled. The camshaft belleville and plain thrust washers go immediately behind the camshaft bevel gear.

Replacement is carried out in the reverse manner to that detailed for removal.

REMOVAL OF THE VALVES. (All Models.)

Remove head and camshaft as previously detailed and swing all the rocker arms clear of the valves. (M, D, J, P, F, L, K, KN, and N Models.)

The valve springs are secured by means of conical split cotters and to remove each valve place the cylinder head on a bench, combustion space downwards, with a block of wood or a suitable packing piece fitted into the combustion space below the valve being dealt with.

Depress the valve spring from above, remove the split cotter and release the pressure when the valve spring and cap can be withdrawn.

Do not forget to remove the synthetic rubber oil seals for the valve stems on the TA, TB, TC, TD, and 1¼, 1½, 2 and 2.6 Litre Models.

Remove the head from the bench, remove the wood block and withdraw the valves from their guides. Repeat this operation on all the remaining valves.

REPLACING THE VALVES. (All Models.)

When reassembling the valves ensure that they are inserted into their correct ports. First oil the valve stem with a little oil, insert the valve in its guide with the head resting on a wood packing block and replace the valve spring, with the valve cap resting on it.

Engage a tool on the cap and depress the spring to expose almost the whole of the cotter grooves in the upper end of the valve. Fit the synthetic rubber oil seal to the bottom of the groove. (TA, TB, TC, TD, and 1¼, 1½, 2 and 2.6 Litre Models.)

Insert the two conical cotters, small end downwards, and gradually release the spring. Make sure the cotters are seating correctly in their grooves.

Swing back the valve rocker arms on to the valves and replace the camshaft. (M, D, J, P, F, L, K, KN, and N Models.)

REMOVAL OF THE ROCKER SHAFT AND ROCKERS. (M, D, J, P. F. L, K, KN, and N Models.)

Remove cylinder head, remove camshaft and valves as previously detailed.

Remove the dowelled set screws securing the rear end of the rocker shafts in the rocker shaft supports.

Insert a ¼ in. B.S.F. bolt into the end of each rocker shaft and use as an extractor for withdrawing the rocker shafts.

As the shaft is withdrawn through each rocker, distance tube, etc., they should be removed and laid out in the correct assembly order, to ensure correct reassembly. Note also the position of the oil feed holes in the rocker shafts to ensure they also are replaced on their correct side. The tapped holes in the ends of the shafts must be placed to the rear of the cylinder head.

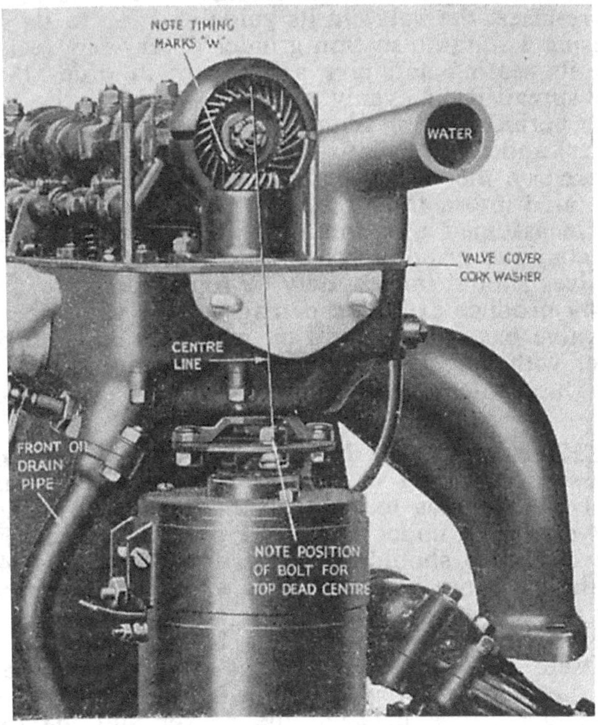

Fig. 7. Showing position of Dynamo Coupling and Camshaft Gears for T.D.C. (M, D, J, P, F, L, K, KN and N Models).

Replacement of the rocker shafts is carried out in the reverse manner to that detailed for removal.

REMOVING CARBON FROM THE HEAD. (All Models.)

Remove the carbon from the head using a screwdriver or other suitable blunt instrument, taking care not to damage the valve seatings. Wash the head and dry off if possible, by blowing thoroughly with compressed air rather than drying with a rag.

CLEANING THE PISTON HEAD. (All Models.)

It is important to remember that the pistons are aluminium alloy. Very great care therefore must be taken not to damage when cleaning. A very blunt instrument and no hard pressure is the best method and under no circumstances must an abrasive be used.

Do not remove the carbon around the top of the cylinder and around the piston edge as this forms an effective seal. While the piston is being cleaned insert an old piston ring in the top of the cylinder bore on top of the carbon.

Very carefully remove all particles of carbon ensuring that none gets into the water-ways or cylinder bores.

GRINDING IN THE VALVES. (All Models.)

Carefully clean each valve and examine for pitting. If they are in a pitted condition they should be refaced with a suitable grinder or replaced with new valves.

Each valve when fitted at the factory is numbered on its head and should be replaced in the corresponding valve port, No. 1 valve being fitted to the port nearest the front of the engine.

Lightly coat the bevelled face of the valve with fine or medium valve-grinding paste, re-insert the valve in its guide and lap to its seat, by a semi-rotary action, using a suitable suction grinder. Every few reciprocations raise the valve from its seating and give a half-turn in order that the grinding compound may spread itself evenly over the whole of the surface.

To assist this periodic lifting obtain a light coil spring (similar to a valve spring but lighter) and insert it into the port beneath the valve head. When pressure is released on the valve grinding tool the valve will pop up, when it can easily be rotated into a fresh position. Once the faces of both the valve and seating have assumed a clean, even, matt-surfaced appearance, it is unnecessary to continue grinding. A polished surface must not be expected.

Distorted valves should immediately be replaced for to attempt to grind them in will only produce extensive damage to the valve seating.

After each valve has been ground in, withdraw and thoroughly wash in paraffin, together with the valve seating and surrounding port, not forgetting the valve guides as some of the compound may have found its way into these.

ADJUSTING THE ROCKERS. (M, D, J, P, F, L, K, KN, and N Models.)

In the process of grinding in the valves a certain amount of metal is always removed which tends to reduce the clearance existing between the head of the valve and the under surface of the rocker arm. For the correct functioning the clearance should be set at not less than .004 in. inlet and exhaust valve, the clearance being checked when the engine has attained its normal running temperature.

On one side of the rocker is a hexagon bronze nut and a locking screw in the rocker itself. A steel locking nut is provided on the M type.

Slacken off the locking screw or nut with a $\frac{1}{8}$ in. Whitworth spanner and insert the feeler gauge between the rocker and the cam. The peak of the cam should be pointing upwards, when the adjusting spanner can be engaged on the bronze nut and rotated until the feeler gauge can just be withdrawn easily.

Hold the spanner exactly in this position and tighten up the steel locking screw or nut.

Again check the clearance to ensure that no movement of the setting took place while the lock screw or nut was being tightened.

When adjusting the rockers it is essential to ascertain that the thickened portion of the bronze nut is downwards and not upwards. If this is correctly assembled the clearance decreases with a downward movement of the adjusting spanner, but if incorrectly, a similar movement increases the clearance and the eccentric requires turning around, this being done by

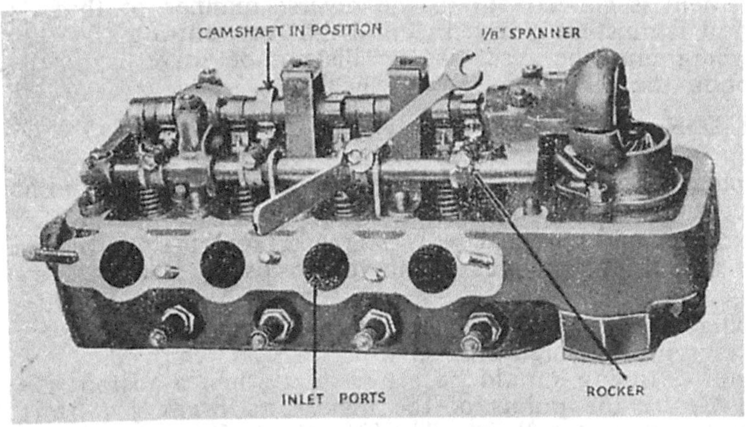

Fig. 8. Showing Spanner in position for purpose of Rocker Adjustment (Illustration purposes only).

continuing to rotate it by pressing the spanner handle downwards until the clearance does close.

The correct valve clearances are:—

M, D, and F Models.	Inlet and Exhaust. .004 in.
J Models.	Inlet and Exhaust. .006 in.
P, K, L, KN, and N Models.	Inlet. .006 in. Exhaust. .008 in.

REMOVAL AND REPLACEMENT OF OIL PUMP. (M, D, and J Models.)

On the J type Midget, drain the sump as detailed on page 42.

On the M and D Models remove the suction pipe from the sump to the pump and on all three models, remove the oil pump delivery pipe.

Remove the three set bolts (not the three nuts as these secure the oil pump cover plate) securing the pump to the front engine nose piece, and withdraw the pump complete.

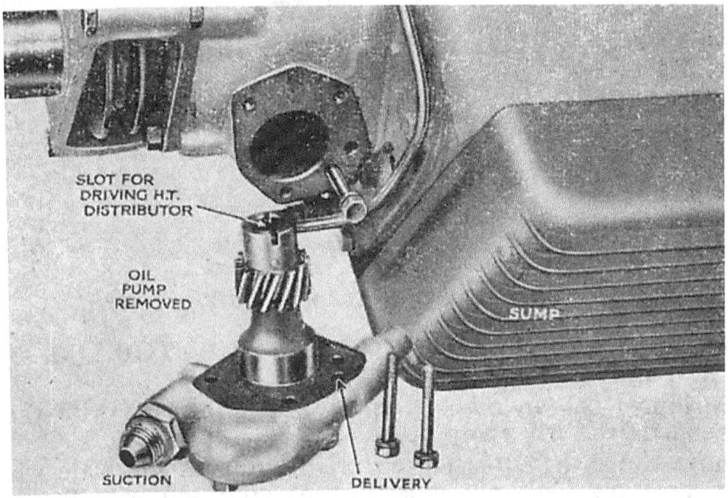

Fig. 9. View of Oil Pump Removed from Engine.

Replacement is carried out in the reverse manner to that detailed for removal, and it must be borne in mind that on refitting the oil pump the ignition timing must be reset as the distributor drive is taken from the female slot in the oil pump driven wheel.

DISTRIBUTOR. (M, D, J, P, F, L, K, KN, and N Models.)

Actually very little instruction need be given on this subject and apart from periodic examination of the points, contact breaker point gap and greasing, no other attention is required.

The distributor is provided with a greaser to lubricate the spindle and only high melting point grease should be used, one turn every 1,000 miles being sufficient.

The contacts can be inspected by unclipping the two springs when the cover carrying the high tension leads can be removed.

The ignition timing should be set so that when a piston is on top of Top Dead Centre the points of the make and break just start to open with the distributor at full retard. (KN, N. P, J Set Ignition Timing at 20° B.T.D.C.) The points may require adjustment from time to time, the clearance between the points should be .015 in./.020 in.

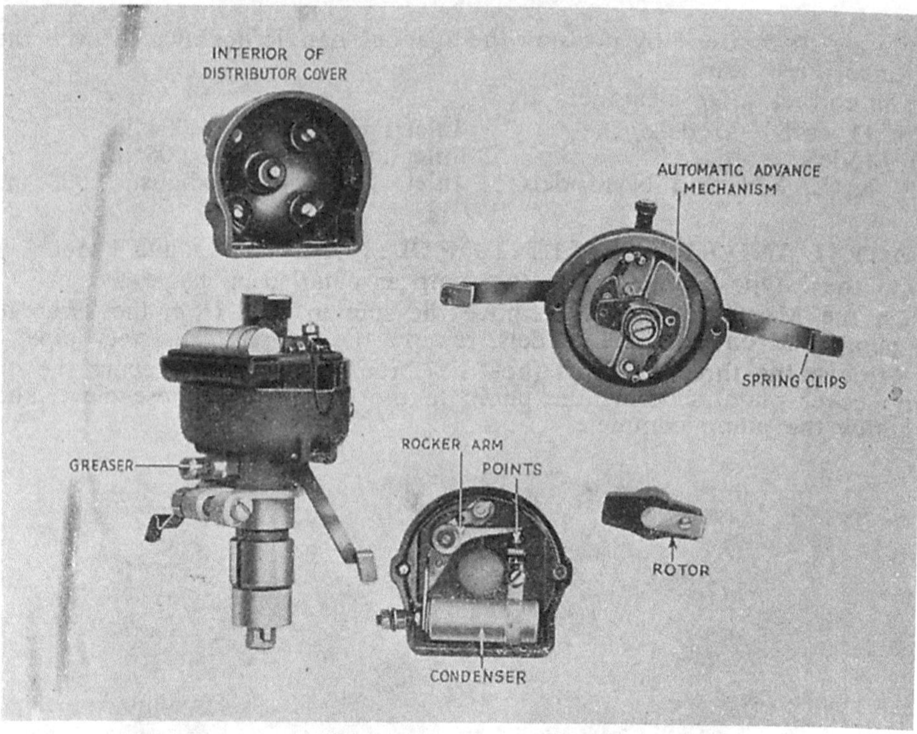

Fig. 10. View of Distributor with Cover Removed.

REMOVAL AND REPLACEMENT OF DISTRIBUTOR. (M, D, J, P, F, L, K, KN, and N Models.)

The distributor has an off-set tongue which registers with the slots in the upper end of the oil pump spindle.

To remove the distributor slacken off the nut on the plate which locks the unit in position. This nut must not be confused with the nut which locks the distributor to the advance plate.

Provided that after the cap, low-tension wires and controls (not J Model) have been removed the complete distributor may be replaced without retiming the engine.

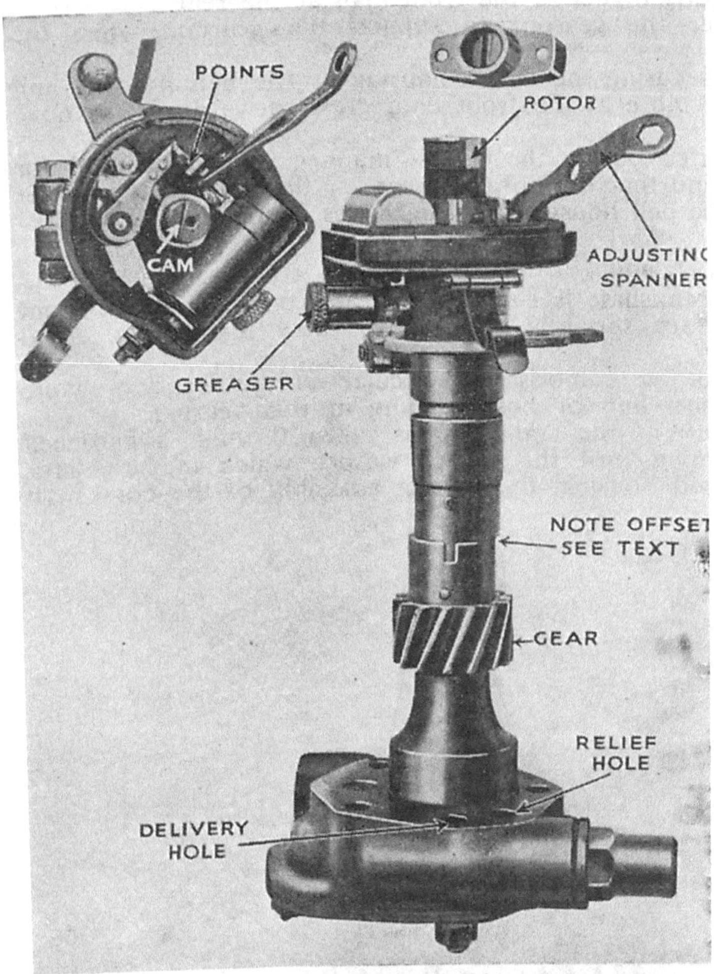

Fig. 11. *Assembly of Oil Pump and Distributor Removed from Engine.*

REMOVAL AND REPLACEMENT OF ENGINE. (M, D, and J Models.)

Disconnect the positive and negative cables to the dynamo.

Open the radiator drain tap and drain the water system.

Drain the sump as detailed on page 42.

Remove the bonnet complete, slacken off the jubilee clips securing the water hoses and remove all water hoses.

Remove the bolts securing the forked ends of the radiator tie rods on the bulkhead and the radiator and remove the tie rods.

Remove the bonnet side boards. (This does not apply to the J type Midget.)

Remove the two bolts securing the radiator to the radiator support bracket and lift radiator complete from car. Disconnect the petrol pipe to the carburetters and disconnect the slow-running and mixture controls and the accelerator controls at the carburetter end.

Disconnect the starter control, remove the securing bolts and remove the starter.

Remove two bolts securing the front exhaust pipe to the exhaust manifold flange, slacken off the front end of the silencer to the front exhaust pipe clip and remove the front exhaust pipe.

Remove the two rear engine mounting bolts, distance pieces, rubber mountings and fibre washers.

Place a sling around the engine so positioned that when the engine is lifted it will be slightly higher at the front than at the rear.

Place a jack under the gearbox to support the gearbox when the engine is removed.

Remove the bolts securing the clutch housing to the bell housing and lift the engine slightly more at the front end, draw forward, and remove from chassis frame.

Replacement is carried out in the reverse manner to that detailed for removal, bearing in mind that the fibre washers for the rear engine bearer are inserted between the bell housing and the chassis frame.

CRANKSHAFT. (M, D, and J Models.)

The two bearing crankshaft is supported at its front end in the front main bearing, a ball race specially designed to carry loads both radial and thrust directions.

The rear main bearing comprises a circular bronze bearing, white metal lined, the flywheel hub or boss running in this bearing.

The front end thrust of the crankshaft is taken through a hardened steel washer butting up against the bronze washer which in turn butts up against the front end housing, the correct assembly of the crankshaft

Fig. 12. *Showing Front End of Engine Dismantled. (M, D, J, P, F,L, K KN, and N Models.)*

in the housing being obtained by interposing a number of shims between the housing cover and the cast iron case.

The front end extension of the crankshaft is fitted with two gears, a helical bevel and a worm. The helical bevel gear drives the dynamo and the worm gear the oil pump and distributor.

It is strongly recommended that the shaft should not be reground owing to the fact that regrinding the journals to a smaller diameter weakens the structure and relative balance of the crankshaft.

This, coupled with the fact that a great number of these crankshafts are called upon to transmit exceedingly high engine r.p.m. and increased horse-power, makes it impossible to calculate the fatigue life.

Fig. 13. *Front End of Engine, Dynamo Removed.* (*M, D, J P, F, L, K, KN and N Models. Illustration Purposes only.*)

CRANKSHAFT REMOVAL. (M, D, and J Models.) ENGINE OUT OF FRAME.

Remove the sump and sparking plugs, withdraw the split pins from the big-end bearing bolts, remove the nuts and withdraw the bottom caps.

Withdraw the connecting rods and pistons, replacing the caps in their original positions and marking the rods, if necessary, to their corresponding bores.

Remove the clutch cover plate, which will give easy access to the clutch centre plate, forward pressure plate and clutch springs, facilitating their removal.

Remove the circlip securing the clutch spigot race and withdraw the spigot race.

To remove the crankshaft nut keep pin, a blow on one end with a blunt punch will tip it sufficiently to allow pliers to finally pull it out.

Now remove the crankshaft securing nut, but do not remove the four bolts securing the flywheel flange to the flywheel. These components must be withdrawn together.

Remove the four nuts for the flywheel securing bolts, place a suitable extractor in position, replace the four nuts and withdraw flywheel and flange complete.

Disconnect the oil feed plug to the rear main bearing.

Remove the six bolts securing the bell housing to the crankcase and withdraw bell housing.

Remove the bolts securing the vertical drive coupling and the bolts securing the dynamo base, disconnect the two leads to the terminals and remove dynamo.

Remove distributor and oil pump as detailed on pages 49 and 50.

Fig. 14. Showing Extractor Tool for Flywheel and Crankshaft Flange in Position (M, D, J, P, F, L, K, KN and N Models. Illustration Purposes only).

Fig. 15. Rear of Flywheel Bell Housing showi Rear Main Bearing Sleeve and Flywheel Flange (M, D, J, P, F, L, K, KN and N Models. Illustrati Purposes only).

Unscrew the starting handle dog nut, and front engine mounting plate.

Withdraw the worm wheel from the crankshaft by tapping it with a suitable soft tool (such as a piece of brass) through the dynamo housing.

It is a parallel fit on the crankshaft and is held in position by a key.

Using a suitable puller, remove the helical bevel pinion from the crankshaft. The holes are tapped $\frac{1}{4}$ in. B.S.F.

Remove the six bolts securing the front housing to the cylinder block when the housing complete with the front ball race can be removed.

The removal of the housing can be facilitated by inserting two $\frac{5}{16}$ in. B.S.F. bolts in the tapped holes and using with an ordinary puller.

The front main bearing can be removed from the housing after taking out the four fixing screws.

Fig. 16. View of Crankshaft Front End Components.

Principal parts to Fig. 16 are as follows:—

1. Helical bevel wheel.
2. Worm wheel for driving oil pump.
3. Distance column.
4. Thrust washers.
5. Shims.
6. Cover with thrust race.

REPLACEMENT OF THE CRANKSHAFT. (M, D, and J Models.)

Replacement is carried out in the reverse manner to that detailed for removal with attention being paid to the following points:—

1. Refit the clutch cover plate to the corresponding marks on the flywheel.
2. That the steel pads on the clutch forward pressure plate coincide with the pressure pins in the clutch cover plate.
3. That the tips on the flywheel nut keep pin face to the rear of the engine.
4. That the clutch is adjusted according to the details given in Chapter III.
5. Should the front main ball race be removed from the front housing, it will be found beneficial for future dismantling if the race is fitted with the smaller diameter half centre towards the front of the engine.

CRANKSHAFT OIL DUCT PLUGS. (M, D, J, P, F, L, K, KN, and N.)

It is of the utmost importance that crankshafts are balanced to a fine degree, both statically and dynamically, to eliminate bad vibration, harshness and periods of the engine.

As great care has been taken with the crankshafts fitted it is as well to draw attention to the plugs that seal the oil ducts.

From time to time, during the course of overhaul, the crankshaft oil ducts plugs have to be removed to enable the oil duct to be thoroughly cleaned and all sludge removed. If these plugs are damaged on removal they will have to be replaced with new ones, and great care should be exercised, as plugs of steel and aluminium are used, depending entirely upon which is necessary for the balancing of the crankshaft. If steel plugs are replaced where aluminium plugs are removed, the balance of the crankshaft will be seriously affected.

The length of the new plugs must coincide with the length of the old one.

It is always advisable to have a supply of both plugs readily available, to prevent any possibility of fitting the wrong type.

REMOVAL AND REPLACEMENT OF CONNECTING RODS AND PISTONS. ENGINE IN SITU. (M, D and J Models.)

Place the car over a pit or on a high-lift and drain the sump.

Remove the sump bolts and lower the sump.

Remove the split pins from the big-end bolts, remove the securing nuts and withdraw the bottom caps.

Withdraw the connecting rods complete with pistons, replacing the caps in their original positions and marking the rods, if necessary, to their corresponding bores.

Replacement is carried out in the reverse manner to that detailed for removal.

REMOVAL AND REPLACEMENT OF PISTONS. (M, D and J. Models.)

Remove connecting rods and pistons as detailed on page 55.

Place end caps into the ends of the hollow gudgeon pin, secure in a vice and remove the gudgeon pin clamp bolt and washer, and press out gudgeon pin. Piston can then be removed from connecting rod.

Replacement is carried out in the reverse manner to that detailed for removal.

NOTE.—Extreme care must be taken when refitting the piston to the bore to prevent the breaking of the piston rings.

REMOVAL AND REPLACEMENT OF PISTON RINGS. (All Models.)

If a special ring expander is not available, three pieces of thin steel, approximately .020 in. thick and ½ in. wide may be used.

Raise one end of the ring and slip the steel strip between the ring and the piston.

Move the strip round the piston and apply slight upwards pressure to the ring until it rests on the land above the ring grooves, when it can be eased off the piston. Do not remove the rings down over the piston skirt, always remove and replace from the top of the piston.

Before fitting new rings the piston grooves must be cleaned of carbon, but do not remove any metal during this process otherwise excessive side clearance will result.

NOTE.—New rings must be tested in the cylinder bore to ensure that the two ends do not butt together, but leave a gap of .006 in. to .010 in.

Where pistons are lapped the gap must be adjusted after lapping.

When checking the gap ensure that the ring is square in the bore by holding firmly on top of a piston temporarily inserted in the bore.

The rings must move freely in the grooves but there must be no vertical movement.

The gaps must be equally spaced around the piston with no two gaps opposite each other, for this would defeat their object in preventing oil from travelling straight up into the combustion space.

REMOVAL AND REPLACEMENT OF VERTICAL DRIVE ASSEMBLY. (M, D, J, P, F, L, K, KN, and N.)

Remove the valve cover, disconnect the positive and negative leads from their terminals on the dynamo.

Remove the vertical drive oil drain pipe.

Remove the dynamo as detailed in Chapter VII.

Remove the two nuts securing the vertical drive assembly to the front of the cylinder head and withdraw assembly complete with oil drain housing.

Replacement is carried out in the reverse manner to that detailed for removal.

NOTE.—It will be necessary to reset the timing on replacement.

DISMANTLING AND REASSEMBLING THE VERTICAL DRIVE Assembly. (M, D, J, P, F, L, K, KN and N Models.)

Remove the nut securing the vertical drive shaft at the bottom end.

Withdraw the vertical drive fork which is keyed to the shaft, taking care not to lose the Woodruff Key.

Withdraw oil thrower, and roller bearing and felt retainer plate and pull pinion shaft complete with Hyatt roller bearings and bearing sleeve through the top of the oil drain housing.

Reassembly is carried out in the reverse manner to that detailed for dismantling, noting that the oil thrower is fitted cup shape downwards.

MODIFICATION TO CRANKSHAFT. (Midget Series J.)

With a view to eliminating a slight roughness at certain engine speeds, a secondary ball race has been designed for fitting at the front end of the crankshaft. This additional ball race also forms a more efficient oil seal.

The modification in itself is not difficult to introduce, but the following points may be of assistance.

1. Slacken the engine fixing to enable the front end to be dropped sufficiently to allow the engine nosepiece and crankshaft thrust washers to be removed, taking care to support the front of the engine.

2. The crankshaft must be accurately tested for longitudinal movement and this must not exceed .008 in. If the movement exceeds .008 in. a new main ball race must be fitted, as the secondary race must not be allowed to take any thrust load.
3. Fig. 17 shows the assembly in detail but the method of shimming is very important and must be carried out very carefully.

Assemble all parts as shown in the illustration, less shims, and tighten the set screws holding the cover plate (MG464/102) into position, thus leaving a gap between the front housing and the rear portion of the nosepiece. Carefully measure the gap by feeler gauge, to which add a reading of .012 in., the total being the amount of shimming required between the faces. This will ensure that the slight end movement of the crankshaft will not allow any thrust load to be transmitted to the secondary bearing (x464/1).

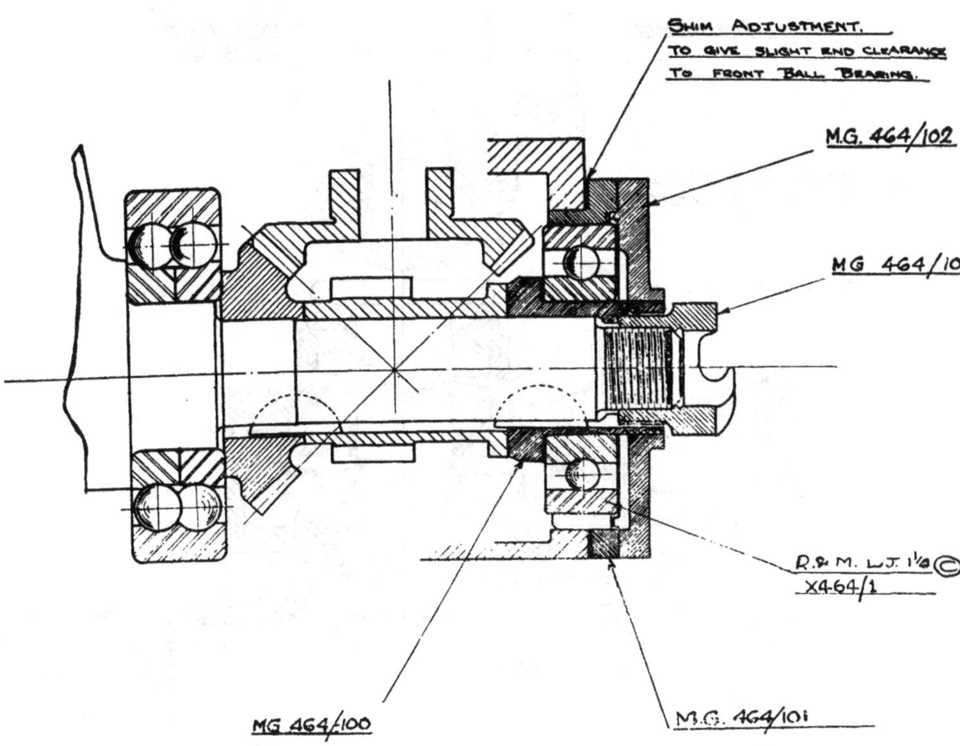

Fig. 17. *Secondary Ball Race for Crankshaft.*

OIL LEAK AT VERTICAL DRIVE ASSEMBLY. (P, PB, N, KN Models.)

Sources of Leakage.

1. Between the oil drain housing and vertical drive fork, marked "A" on the illustration.
2. Between the cylinder head face and oil drain housing face. marked "B" on the illustration.

Causes.

1. (*a*) The oil return thread of the vertical drive fork below the top of the oil drain housing gallery wall. Refer to illustration 1 (*a*).

(*b*) The oil drain housing gallery wall flat at the top instead of having a sharp edge (chamfer angle 60°). Refer to illustration 1 (*b*).

(*c*) Front oil drain pipe choked.

2. (*a*) Shims badly torn, as a result of being passed oved the stud threads, or buckled.

(*b*) The cylinder head and oil drain housing faces either damaged or bearing slight machining marks.

(*c*, The roller bearing and felt retainer plate standing proud of the oil drain housing, preventing the housing from seating on the shims. Refer to illustration 2 (*c*).

3. Too great a quantity of oil allowed to drain from the cylinder head into the vertical drive housing.

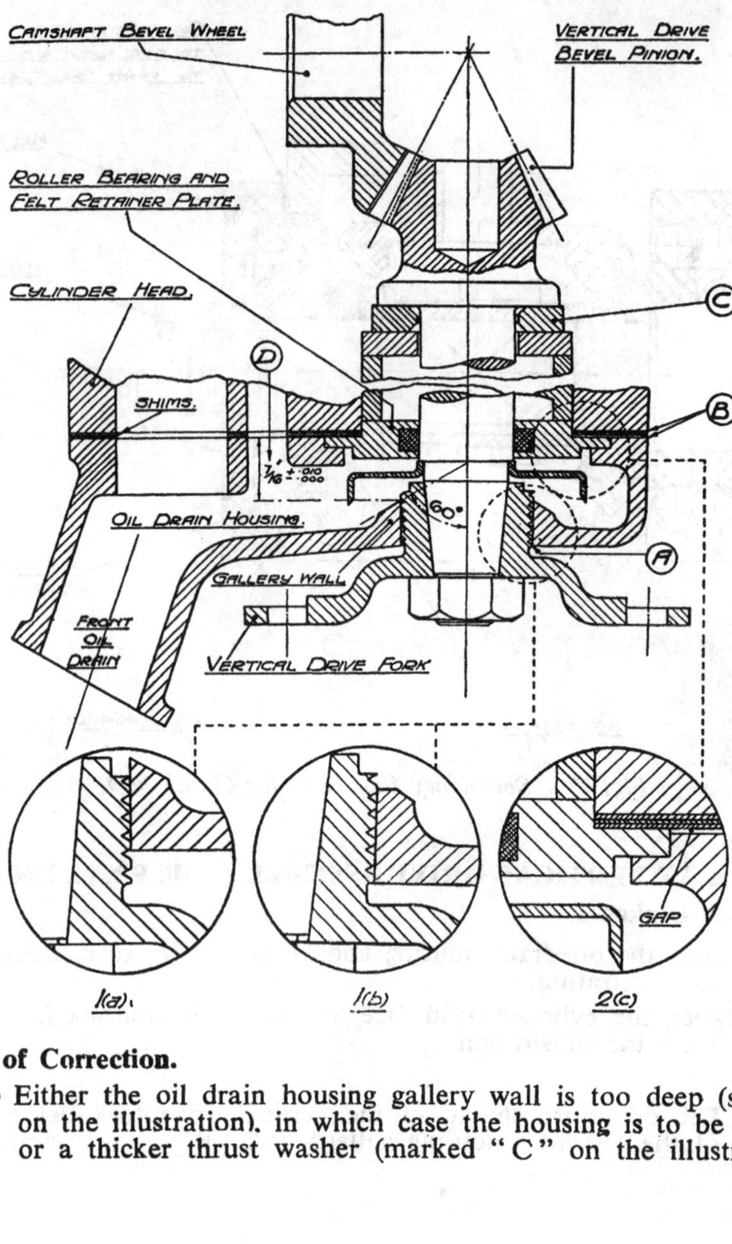

Method of Correction.

1. (*a*) Either the oil drain housing gallery wall is too deep (see " D " on the illustration), in which case the housing is to be renewed, or a thicker thrust washer (marked " C " on the illustration) is

to be fitted. The correct depth of the thrust washer is from .171 in. to .179 in. (Part Numbers P110/210 or P110/211). If it is found that a very thin washer has to be changed for one of the correct depth, then the difference between their depths is the required amount of extra shimming necessary between the oil drain housing and cylinder head faces, bearing in mind that the gears must mesh on their correct pitch line.

(*b*) If the oil drain housing gallery wall has a flat at the top instead of a sharp edge, or if this wall is too deep for the vertical drive fork, the housing should be replaced by a new one.

(*c*) Remove and clean thoroughly.

2. (*a*) Fit new shims.

(*b*) Lap the two faces together, using a fine abrasive.

(*c*) Renew housing.

3. Fit one shim (Part Number M.G.489/234) which has a $\frac{3}{8}$ in. diameter hole instead of a $\frac{27}{32}$ in. diameter hole, thereby controlling the quantity of oil which can pass into the vertical drive housing. This shim is .010 in. thick and it will be necessary to remove a shim or shims from the vertical drive housing equal to this thickness.

ENGINE LUBRICATION. (M.G. Midget Series P.)

The actual lubrication of the engine is very simple, but a number of points however do need periodical attention.

The gear type of oil pump fitted to the front of the power unit, draws its supply of oil from the sump through an internal gauze filter, and via an external pipe from the connection on the oil filter to the lower part of the oil pump itself.

The pump then delivers oil by means of an external pipe to the oil filter. Reference to Fig. 17a, shows the oil filter in its assembled and dismantled state, and the hole that can be seen on the removed oil filter marked "oil feed" couples up with a hole drilled in the crankcase, so that oil that is forced by the pump into the oil filter is cleansed and is delivered internally to the centre and rear main bearing. The oil feed passes through No. 2 union to deliver oil by means of an external pipe to the front main bearing and to the overhead valve gear. A by-pass relief valve is incorporated in the pump body which consists of a cover enclosing a spring and maintains the small piston up against the seating of the pump body until such time as either the force of the oil through the pressure, or cold non-fluidity, forces the piston off its seating.

Whenever this occurs the release of excess oil will maintain the oil pressure at its predetermined point, dependent upon the resistance of the spring. It is possible to increase the tension of the spring by introducing thin shims into the cap, but no such alteration should be effected without first obtaining advice from an M.G. recognised Distributor or Dealer.

Oil that is forced to the main bearings finds its way under pressure to the big-end bearings through holes drilled in the crankshaft, whence in the usual course it is thrown up by centrifugal force, finding its way upwards to lubricate the pistons and gudgeon pins.

The crankshaft itself is drilled so that oil can pass through the holes in the shaft from the main to the big end bearing. At the rear end of the crankshaft there is an oval slot with a hole in it, which serves as a duct for the oil at No. 4 big-end bearing. Oil is fed to Nos. 1, 2 and 3 big-end bearings through ducts in the front and centre main bearing.

As previously stated, oil is fed to the overhead valve gear, by means of an external pipe, where it encounters a restrictor pin fitted into the cylinder head itself.

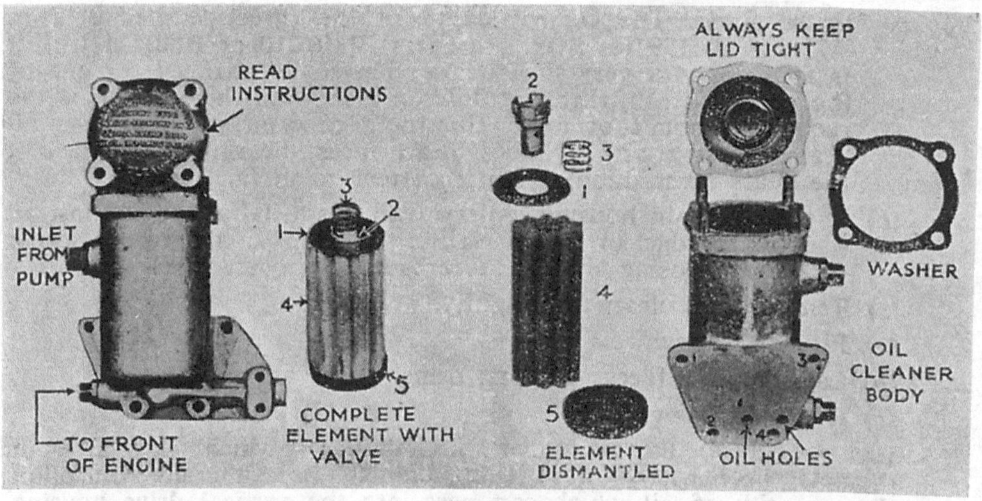

Fig. 17A. *The Tecalemit Oil Filter, shown in complete form and Dismantled. (P, L, K, KN and N Models.)*

The restrictor pin or metering pin is for the purpose of regulating the oil through to the overhead valve gear and on no account should it be damaged or interfered with.

Oil passing the pin enters a drilled passage in the cylinder head which registers with a similar passage drilled in the front camshaft bearing bracket communicating with a camshaft bearing itself and each of the valve rocker shafts. The valve rocker shaft is drilled at intervals to correspond with the valve rockers, and the bushes are grooved, feeding drilled passages in the rocker arms which lubricate the faces.

The rocker shaft in addition feeds the rear camshaft bearing with a generous supply of oil, and the ducts in the forward camshaft bearing impinge a constant stream of oil on the cam shaft drive gears. Surplus oil in the valve gear finds its way back to the engine sump down a drain passage and an external pipe at the rear of the cylinder block, and by a similar drain passage and pipe in the auxiliary drive cover at the forward end of the crankshaft, where it lubricates the dynamo drive and inclined shaft gears, finally passing through the crankshaft ball race on to the oil filter in the crankcase.

It is recommended that every 500 miles the restrictor pin should be withdrawn by passing a piece of stiff wire, hooked at its end, through the hole which can be seen at the end of the pin. The restrictor and its housing should be carefully cleaned before replacement, and we would stress again that on no account should the pin be filed, or its shape be altered or otherwise interfered with.

ENGINE LUBRICATION. (M.G. Magna Series F.)

The sump when full contains 6 qrts. of oil and the oil pump sucks the oil from the sump, through a gauze strainer into the pump body and is delivered to the main bearings by external piping and by a similar method to the overhead valve gear.

A gear type pump is used, fitted with a cover plate having a by-pass relief valve incorporated in the body of the pump. The relief valve consists of a cover which encloses a spring and maintains the small piston up against the seating in the pump body until such times as either the force of the oil through pressure or cold non-fluidity forces the piston off the seating. Whenever this occurs there will be either the corresponding drop in oil pressure or the release of excess oil will maintain the oil pressure at its recommended pressure, dependent upon the resistance of the spring. Obviously it is possible to increase the tension of the spring by introducing washers in the cap, or fitting a stronger spring, but no alteration should be effected without first obtaining the necessary advice from recognised M.G. Distributors and Dealers.

The oil is then delivered through the front end housing into the external pipe on the off side of the engine and by suitably drilling the crankcase, oil is forced to the main bearings and as the crankshaft is hollow drilled, the oil finds its way under pressure to the big-end bearings whence in the usual course it is thrown out by centrifugal force finding its way up to lubricate the pistons and gudgeon pins. The aluminium housing supporting the centre main bearing has two small holes drilled in it which register with feed holes cross-drilled in the main chamber. The webs of the crankshaft are drilled, and the holes are blanked up by plugs after the shaft is cross drilled to the journals. Oil then forces its way into the hollowed crankshaft and eventually finds its way to the back main bearing through a hole in the flywheel boss extension. There is a large groove inside the bearing which forms an oil pocket and this groove registers with a hole in the crankshaft. In order to prevent any leakage of lubrication to the rear main bearing housing there is an oil slinger with a form of return thread cut on the flywheel boss extension.

ENGINE LUBRICATION. (L Type Magna, K, KN, and N Type Magnette.)

The gear type oil pump as fitted to the engine, draws its supply of oil from the crankcase, through the medium of an internal gauze filter and external pipe connecting the bottom of the oil pump. Oil is then delivered from the pump to the oil filter and the two oil feed holes in the filter couple up with the holes drilled across the crankcase, so that the oil that is forced by the pump into the oil filter is cleansed and delivered internally to the centre and the rear main bearing. The oil passing through the filter delivers oil by the means of an external pipe to the front main bearing and the overhead valve gear. A cover plate is fitted on to the oil pump, and a by-pass relief valve is incorporated in the body of the pump. The relief valve consists of a cover which encloses a spring and maintains the small piston up against the seating in the pump body until such times as the force of the oil through pressure forces the piston off the seating. Whenever this occurs the release of excess oil will maintain the oil pressure at its predetermined pressure, dependent upon the resistance of the spring.

Obviously it is again possible to increase the tension of the spring by introducing washers into the cap or by obtaining and fitting a stronger spring but no alteration should be effected without first obtaining the advice of the M.G. recognised Distributors and Dealers.

Oil that is forced to the main bearings finds its way under pressure to the big-end bearings through holes drilled in the crankshaft, whence in the usual course it is thrown out by centrifugal force, finding its way upwards to lubricate the pistons and gudgeon pins.

The front main bearing and overhead valve gear are fed with oil by an external pipe from the near side feed of the oil pump. A restrictor pin limits the pressure at which oil is supplied to the cylinder head, and the surplus oil drains back to the sump through the oil return pipe, at the rear, centre and the forward end of the engine. The restrictor pin should not be filed or altered in any way and when the head is removed it should be taken out so that it is not lost and when reassembling it should be replaced.

The rear main bearing and the centre main bearing are fed from an oil pipe connecting up with a duct in the side of the crankcase fed from the oil filter.

The crankshaft itself is drilled so that oil passes through the holes in the shaft from the main to the big-end bearing. At the rear end of the crankshaft there is an oval slot with a hole in it, which serves as a duct for the oil to No. 6 big-end bearing.

DRAINING THE SUMP. (P Midget, L Magna, K, KN and N Magnette.)

It is preferable for the sump to be drained every 1,500 miles and it is recommended that this is done after returning from a journey when the oil is warm and more fluid.

Remove the drain plug in the centre, left-hand side for the P Midget and in the centre right-hand side for the L, K, KN and N Models, allowing the oil to drain from the sump.

DRAINING THE SUMP. (F Type Magna.)

The drain plug on this model forms also the main suction pipe connection for the oil pump and is attached to the internal suction filter.

As the filter in the sump must also be removed when the oil is drained, it is recommended that it be thoroughly cleaned before replacing.

Remove the main oil suction pipe from the connection on the filter plug, right-hand rear corner of the sump, unscrew the large hexagon plug nut attached to the filter, withdraw the filter, allowing the oil to drain from the sump.

When reconnecting the suction pipe to the threaded end of the filter, great care must be taken to see that the pipe is fitted correctly, for incorrect fitting will cause an oil leak in the oil feed, which can lead to irreparable damage to the engine.

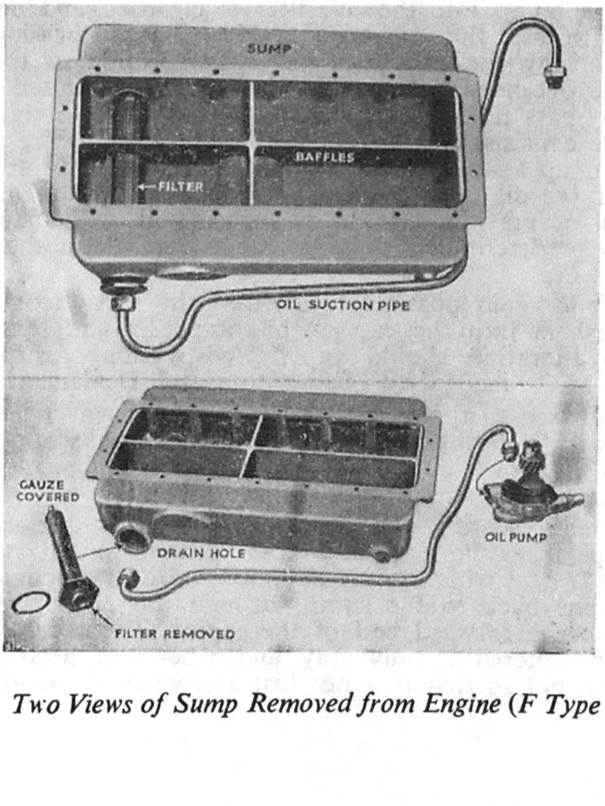

Fig .18. Two Views of Sump Removed from Engine (F Type Magna).

REMOVAL OF THE OIL FILTER. (P, K, L, KN and N Models.)

Disconnect, at the filter end, the feed pipe from the pump to the filter.

Disconnect, at the filter end, the oil feed pipe to the front main bearing and overhead valve gear.

Remove the four bolts securing the oil filter to the crankcase and withdraw filter.

DISMANTLING, REASSEMBLING AND REPLACING THE OIL FILTER. (P, L, K, KN and N Models.)

Remove the four nuts securing the oil filter cover to the oil filter body.

Remove the spring, relief valve and flat washer from the top of the filter element and withdraw the element.

Thoroughly clean the element before replacing.

Reassembly and replacement is carried out in the reverse manner to that detailed for dismantling and removal.

REMOVAL AND REPLACEMENT OF OIL PUMP. (P, F, L, K, KN and N Models.)

Disconnect the oil suction pipe at the pump end, and disconnect the oil pipe to the oil filter, also at the pump end.

Remove the set bolts securing the oil pump to the front engine nose piece, and withdraw the oil pump.

Replacement is carried out in the reverse manner to that detailed for removal.

NOTE:—When replacing the oil pump it may be necessary to re-time the engine as the distributor drive is taken from the female slot in the oil pump driven wheel.

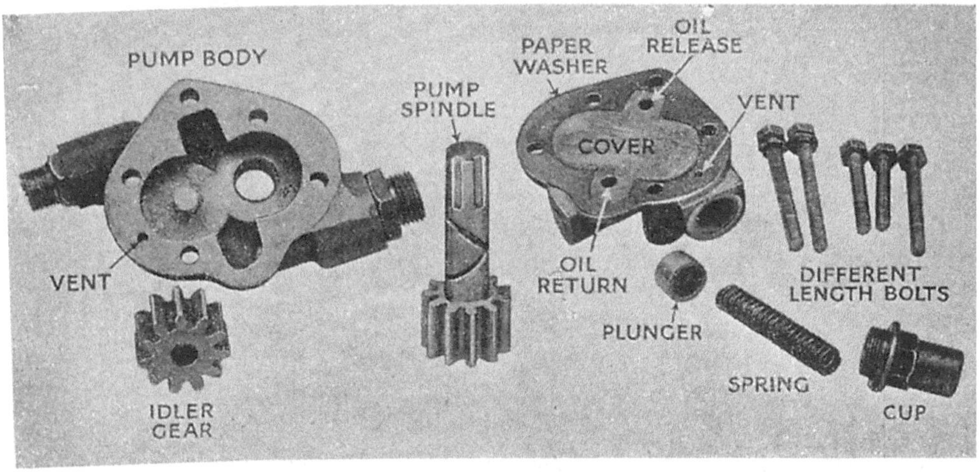

Fig. 19. The Oil Pump Components (P, F, L, K, KN and N Models).

METHOD OF DRIVING AND DISMANTLING OIL PUMP AND MAGNETO DRIVE. (K Models, where fitted.)

Reference to Fig. 20 shows the pump actually removed, the upper end of the pump spindle is castellated and registers inside the lower extension of the Magneto drive, which is shown in Fig. 21, coupled to the pump, this forming a single unit and driven by the gear which engages with the gear on the crankshaft.

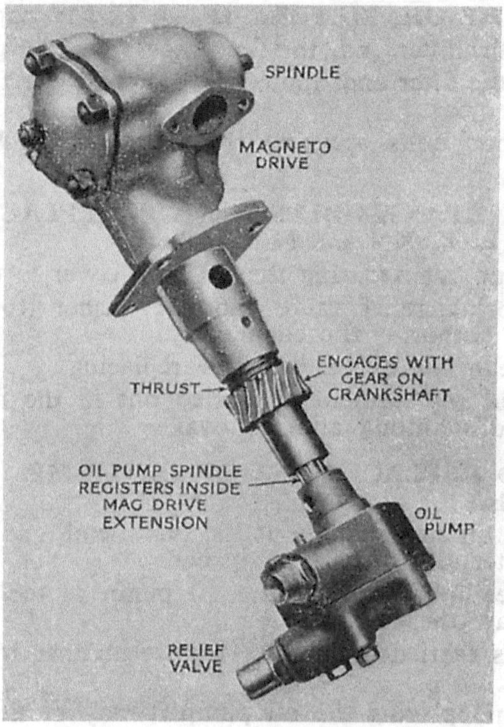

Fig. 20. Magneto Drive and Oil Pump Removed (K Model where Fitted).

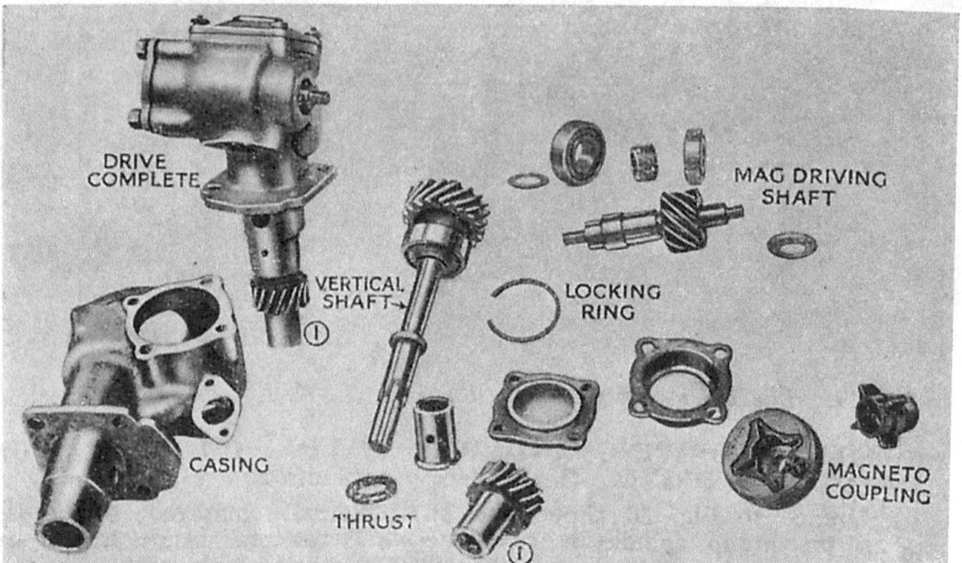

Fig. 21. Details of Magneto Drive (K Model where Fitted).

In order to remove the Magneto drive, undo the bolts, which secure it to the timing cover, disconnect the Magneto and give the housing a half turn in order to disengage the gears which are a true worm. If the Magneto drive is removed, it is necessary to remove the oil pump first, and when replacing the oil pump, care should be exercised to see that the splines register without force.

The Magneto drive runs on ball bearings and is well lubricated and it is therefore, highly improbable that it will ever need dismantling. The drive consists of the housing in which there are a pair of spiral gears attached to the vertical and horizontal shafts.

Fitted to the lower end of the vertical shaft is a worm wheel, and there is a thrust washer placed between the vertical gear and the shaft housing.

The Magneto coupling consists of a star coupling which is bolted to the Magneto shaft and registers inside a flexible body of the other half of the coupling, which is attached to the Magneto drive spindle.

The Magneto can be removed by slipping the two couplings out of engagement, after undoing the strap and set screws. It is as well, however, to mark how they came out so that they may be correctly re-engaged in relation to the correct firing order.

By slackening the nut on the vernier coupling, and rotating the latter slightly and then tightening the nuts, the amount of advance is increased.

WATER PUMP. (P Type Midget where fitted, K, L, KN and N Models.)

Reference to Fig. 22 shows the complete water pump after removal from the engine and also the pump dismantled. On the series P Midget the drive is taken from the worm gear on the front of the crankshaft which drives the oil pump and distributor.

On the K, L, KN and N Models, the drive is taken from a separate gear fitted to the front end of the crankshaft in front of the bevel which drives the dynamo.

The principal point on which care should be taken is the packing gland adjustment which consists of a brass bush screwed at one end to

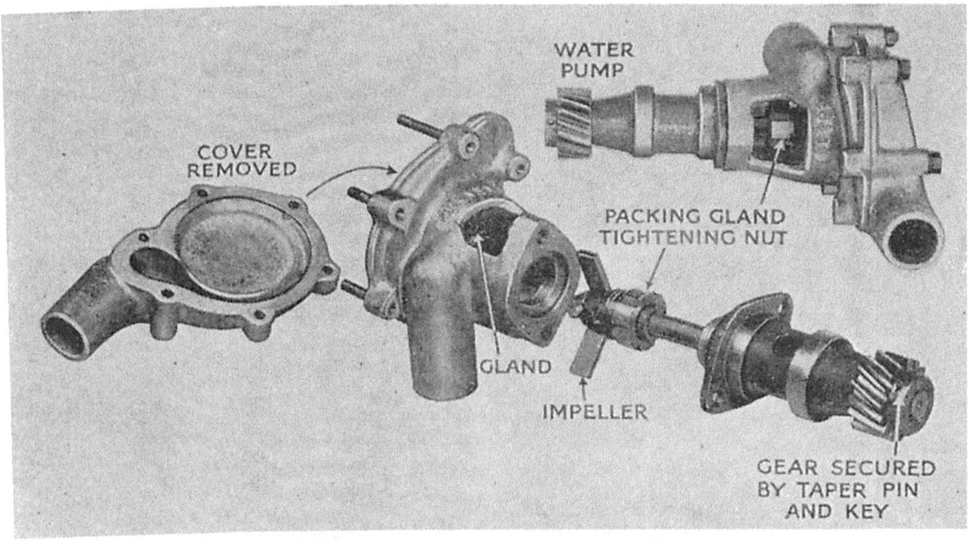

Fig. 22. Views of Water Pump (P Model where Fitted, L, K, KN and N Models).

register inside the pump body and hold the gland (consisting of a special water pump end packing), in position that will prevent water creeping along the pump spindle and leaking.

The packing gland nut can be tightened by inserting a piece of rod into the holes drilled in the nut and using the side of the pump body as a fulcrum.

The gland requires no lubrication and should only be tightened sufficiently to prevent leaks.

By removing the two nuts holding the pump in position and removing the rubber hose connections the pump can be withdrawn by giving a gentle pull.

When replacing care must be taken to see that the gears mesh before any pressure is applied.

REMOVAL OF CYLINDER HEAD. (P Model.)

Open the radiator drain tap and the small tap in the block and drain the cooling system.

Disconnect the high tension leads to the battery.

Remove the bonnet complete after removing the two rear bonnet hinge screws.

Detach the high tension cables to the sparking plugs and remove the plugs (to prevent breakage).

Disconnect the rev. counter drive from the rear of the camshaft and remove the valve gear cover. Do not mistake the two M.G. breathers at each end for nuts. They cannot be unscrewed.

Uncouple the petrol pipes from the float chambers, and disconnect the throttle and mixture controls.

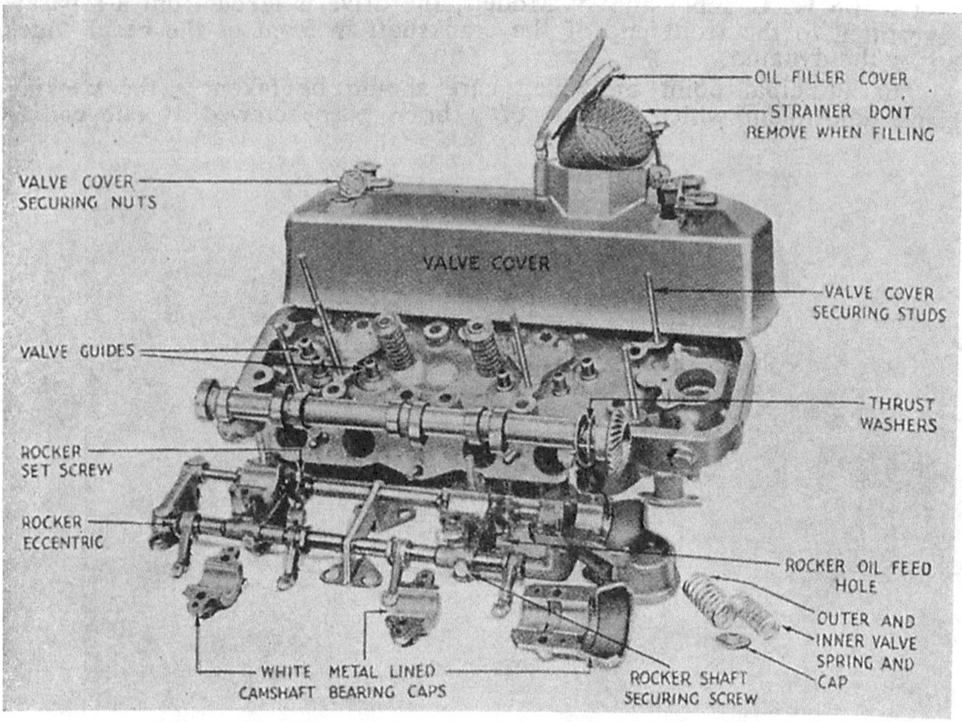

Fig. 23. Overhead Valve Gear in Dismantled Form (P Model).

Remove the bolts securing the carburetters to the induction manifold, and withdraw from manifold. There is no necessity to separate the carburetters.

Remove the nuts securing the induction manifold to the head, remove manifold. Release the water inlet and manifold by undoing the attachment clips.

Disconnect the exhaust pipe from the exhaust manifold by removing the three securing nuts.

Remove the four nuts securing the water manifold to the cylinder head and withdraw manifold.

Remove exhaust manifold from the cylinder head, after removing the five nuts securing the manifold to head.

Disconnect the oil delivery pipe from the cylinder head (left hand side forward) by unscrewing the single bolt and withdraw oil restrictor, to avoid risk of loss when head is removed.

Uncouple the oil return pipe at the front of the engine by unscrewing the two nuts only. It need not be withdrawn from its studs as it will draw off when the head is lifted.

Remove completely the intermediate and rear oil return pipes by unscrewing the retaining nuts at the top and bottom.

Remove the two bolts securing the circular flexible coupling to the dynamo drive yoke, allowing the coupling to be withdrawn with the cylinder head.

Slacken off the cylinder head nuts, in rotation, half a turn at a time until they can be removed by the fingers. Distortion of the head may take place if one nut is removed before slackening the others. (See Fig. 24.)

To break the joints lightly tap the sides of the head with a wooden mallet or with a hammer with a piece of wood interposed to take the blow.

The head can now be lifted from the block, withdrawing squarely over the studs.

Carefully lift the gasket over the studs, keeping parallel with the face of the block.

Replacement is carried out as per the instructions given on page 45.

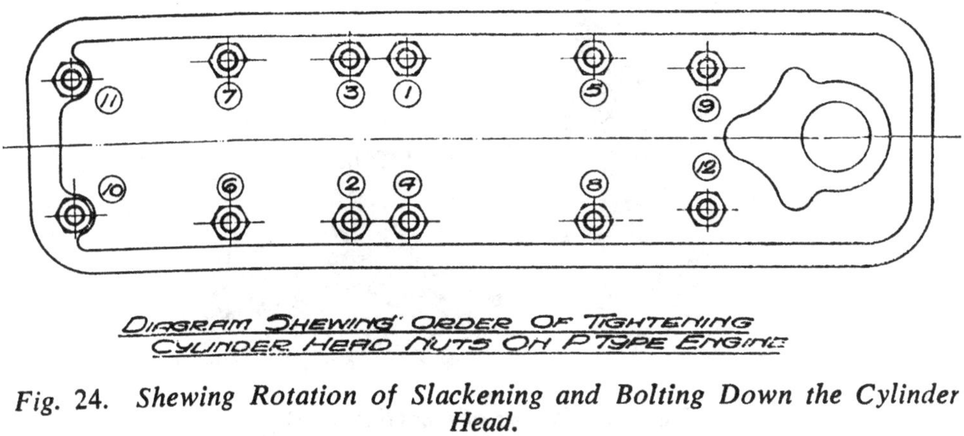

Fig. 24. Shewing Rotation of Slackening and Bolting Down the Cylinder Head.

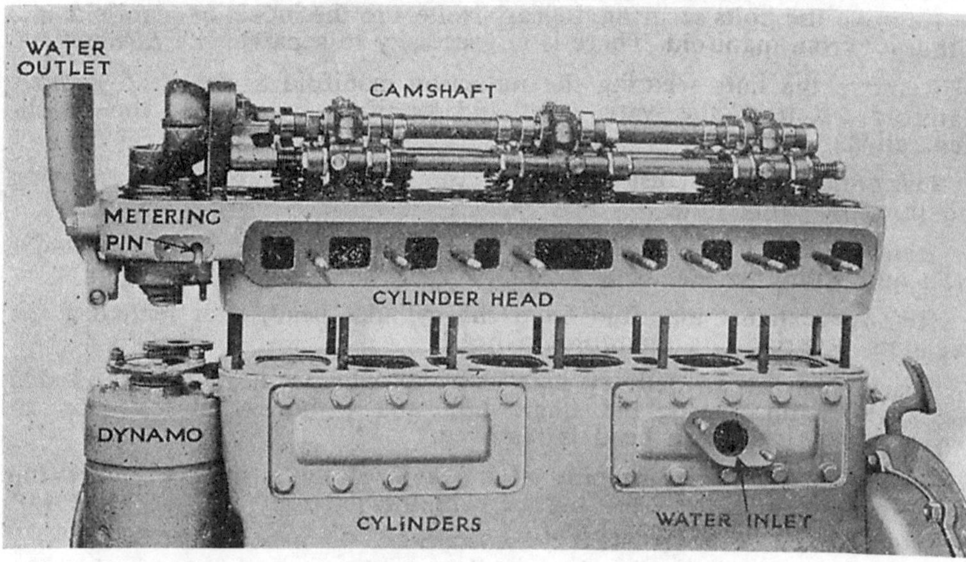

Fig. 25. Showing Head partially lifted from Block (F Model).

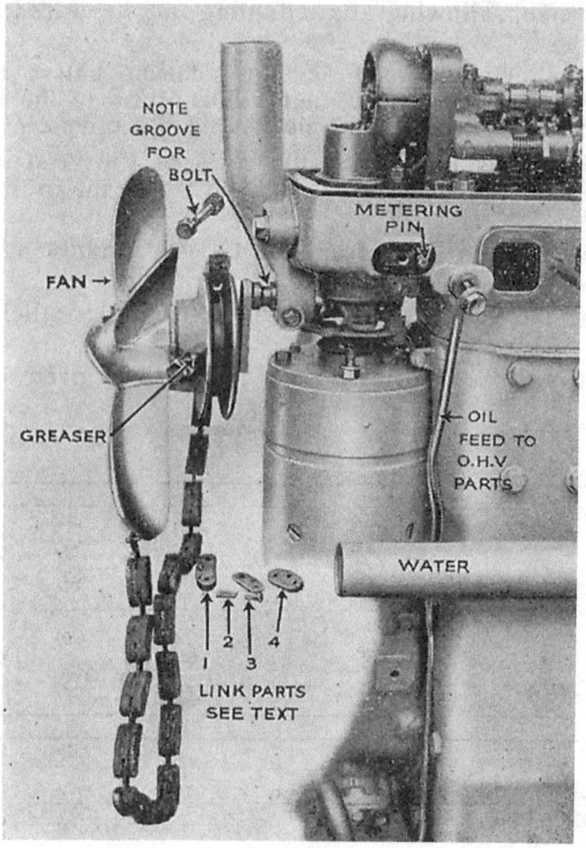

Fig. 26. Details of Fan (F Model).

REMOVAL OF CYLINDER HEAD. (F, L, K, KN and N Models.)

Drain the water system by opening the tap in the radiator and cylinder block.

Disconnect the high tension leads from the battery.

Remove the bonnet complete after taking out the two set screws securing the rear bonnet hinge screws.

Detach the leads to the sparking plugs and remove the sparking plugs, to prevent damage.

Disconnect the revolution counter drive from the rear end of the camshaft and remove the valve cover. Do not mistake the two "M.G." breathers for nuts. They cannot be unscrewed.

Uncouple the petrol pipes from the float chambers and disconnect the throttle and the mixture control.

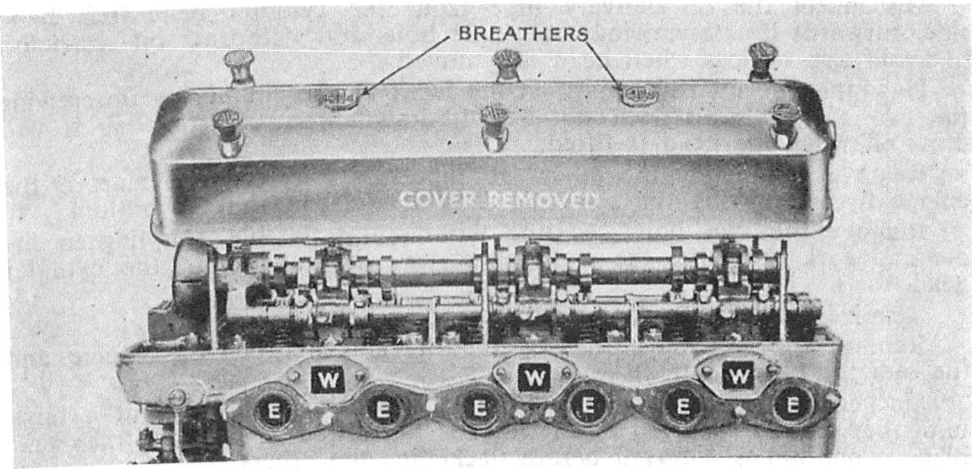

Fig. 27. Cylinder Head with Cover Removed.

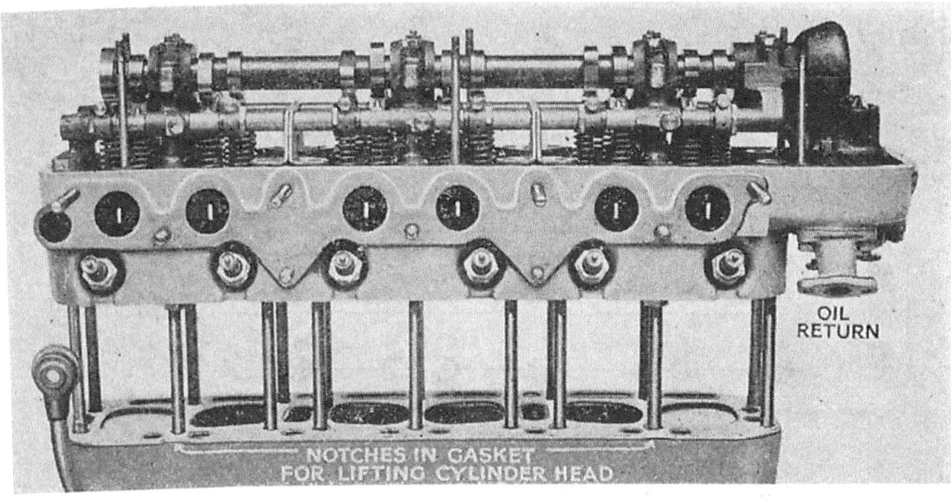

Fig. 28. Cylinder Head Partially Drawn (K and L Models).

Remove the carburetters, induction manifolds, and remove exhaust manifold after releasing exhaust pipe. (F type Magna only.)

Remove the nine nuts securing the induction manifold to the cylinder head and withdraw complete with carburetters. (L, K, KN and N Models.)

Release the water inlet and manifold by undoing the attachment clips (K, L, N and KN.)

Disconnect the exhaust pipe from the exhaust manifold by removing the six securing nuts. (K, L, N and KN.)

Remove the nuts securing the water manifold and withdraw manifold. (K, L, N and KN.)

Remove nuts securing exhaust manifold and withdraw manifold. (K, L, N and KN.)

Slacken off top and bottom hoses to radiator, remove fan belt and remove fan and water pipe complete. (F Type Magna only.)

Disconnect the oil delivery pipe from the cylinder head (left hand side, forward) by unscrewing the single bolts and withdraw oil restrictor, to avoid risk of loss when head is removed.

Uncouple the oil return pipe at the front of the engine, by unscrewing the two nuts only. It need not be withdrawn from its studs as it will draw off when the head is lifted.

Remove completely the oil return pipe in the centre and rear of the engine by unscrewing the retaining nuts at the top and the bottom.

Remove the two bolts securing the circular flexible coupling to the dynamo fork, allowing the coupling to be withdrawn with the cylinder head.

Remove the pipe to the oil pressure gauge.

Remove the two nuts securing the brackets between the engine and the radiator (where fitted).

Slacken the head nuts in rotation, a partial or half turn at a time, until they can be removed by the fingers. Distortion of the head may take place if one nut is removed before slackening the others.

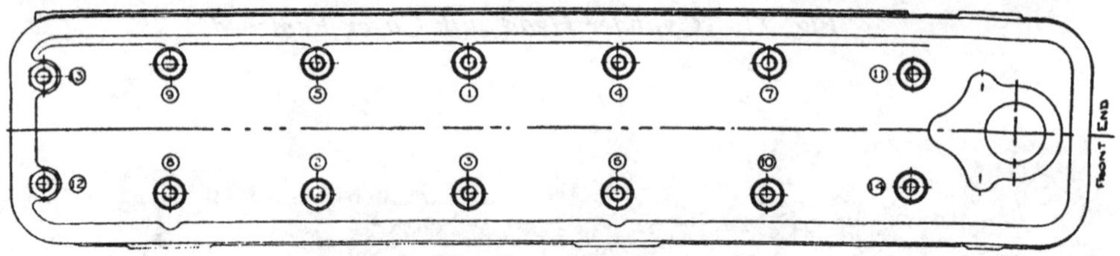

Fig. 29. Sequence for Tightening Cylinder Head Nuts (F Model).

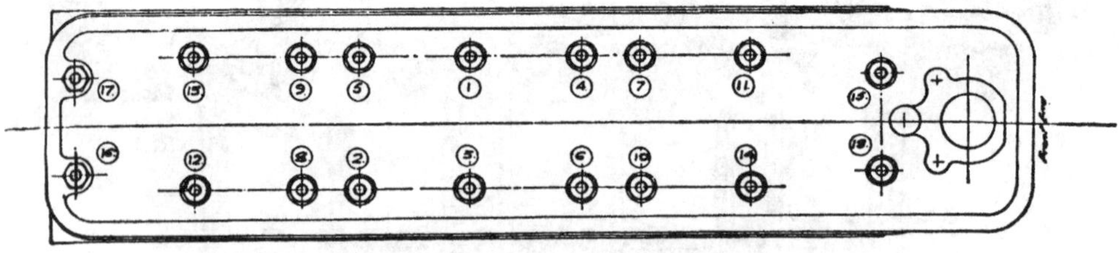

Diagram showing the order in which the cylinder head nuts should be screwed down on M.G. Magnette engines.

Fig. 30. Sequence for Slackening and Tightening Cylinder Head Nut (L, K, KN and N Models).

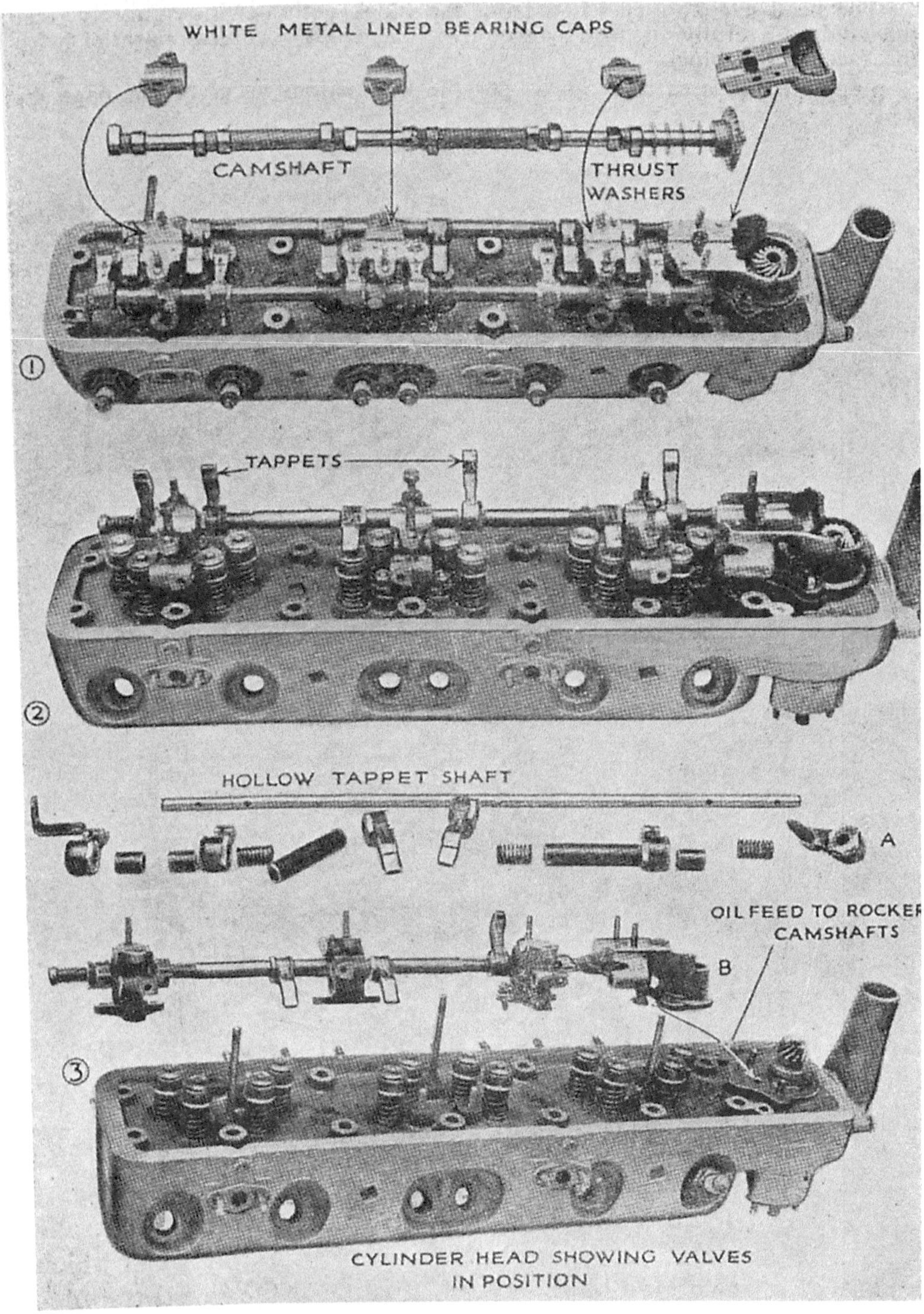

Fig. 31.—Cylinder Head in various stages of removal (F model).

To break the joint smartly tap the sides of the cylinder head with a wooden mallet or a hammer with a piece of wood interposed to take the blow.

The head can now be lifted from the block withdrawing squarely over the studs. Carefully lift the gasket over the studs, keeping parallel with the face of the block.

Replacement is carried out as per the instructions as given on page 45.

Fig. 32. Cylinder Head showing camshaft and rockers in position (K and L models).

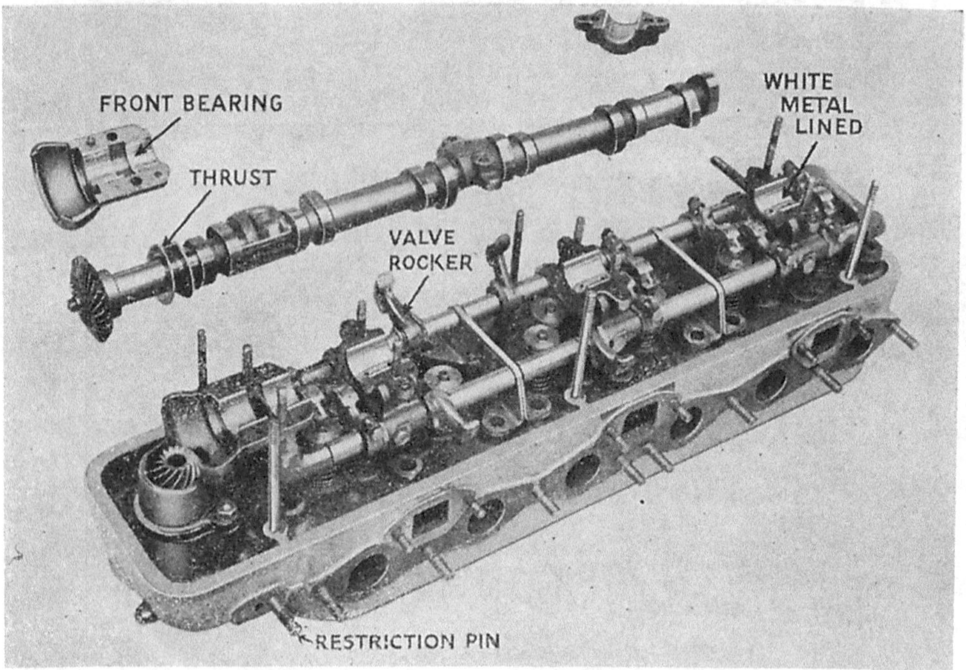

Fig. 33. Cylinder Head with camshaft removed (L and K models).

FAN. (F Type Magna.) Fig. 26.

The Fan is driven by a sectional leather belt with 28 links, and the boss of the fan which carries the bearing is provided with an oil nipple needing only occasional lubrication.

Over greasing may cause the grease to exude causing damage to the fan belt.

The fan spindle is attached to a bracket held in position by a bolt and nut in the extension of the water outlet pipe.

On slackening the nut the bracket is free to rotate in its housing in order to increase tension on the belt.

To remove a link, first slacken off the adjustment and by using a fine pointed screw driver, separate the links by removing the screw that holds the two together.

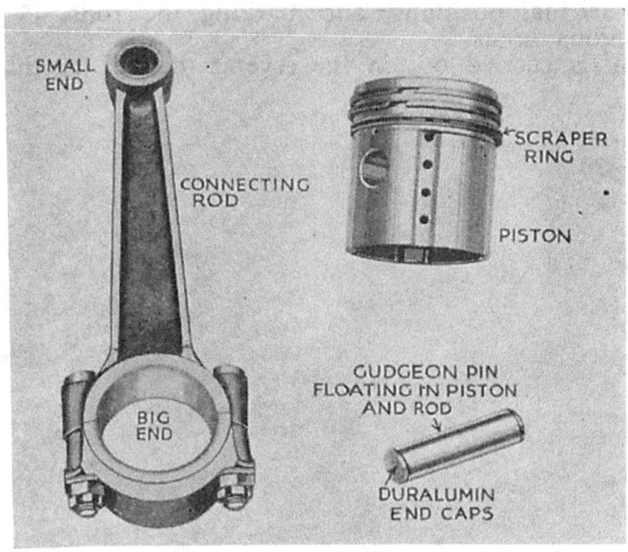

Fig. 34. *Piston and Connecting Rod (P, L, K, KN and N models).*

REMOVAL AND REPLACEMENT OF CONNECTING RODS AND PISTONS. (P, L, K, KN and N Models.)

Place car over a pit, on a high lift, or raise on trestles at front end.

Drain sump as detailed on page 62.

Disconnect oil suction pipes from the suction filter. (P, F, L, K, KN and N Models only.)

Remove the sump bolts and lower the sump.

Remove the cylinder head as detailed on pages 66 to 72.

Remove the split pins from the big-end bearing bolts, remove the nuts and withdraw the bottom cap.

Push the connecting rod gently up the bore until the piston and gudgeon pin are just above the level of the face of the block, and remove the gudgeon pin end cap. These caps are fitted to prevent the possibility of the gudgeon pin from scoring the bore, the gudgeon pin being free to float in piston and rod.

Gently press gudgeon pin through piston and remove piston from connecting rod. If pin is tight, use a suitable extractor.

Draw connecting rod down through the bore taking care not to damage the bore, and re-fit end cap.

It is advisable to mark the pistons and connecting rods with the corresponding bores.

Replacement is carried out in the reverse manner to that detailed for removal.

REMOVAL AND REPLACEMENT OF CONNECTING ROD AND PISTON. (F Type Magna.)

Place the car over a pit, on a hoist, or raise at front end on trestles. Drain the sump, remove the sump bolts and lower the sump.

Remove the split pins from the big end bearings, remove the securing nuts and withdraw the bottom cap.

Withdraw the connecting rods complete with pistons, replacing the caps in their original positions, and marking the rods, if necessary, to their corresponding bores.

Replacement is carried out in the reverse manner to that as detailed for removal.

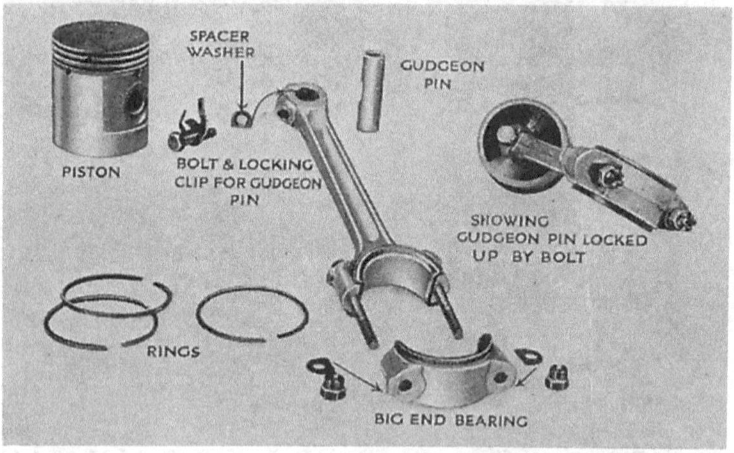

Fig. 35. Piston and Connecting Rod (F model).

REMOVAL AND REPLACEMENT OF ENGINE. (P, F, L, K, KN and N.)

Disconnect the positive and negative cables from the battery.

Drain the water system by opening the drain tap at the bottom of the radiator and in the right hand side of the cylinder block.

Drain the sump as detailed on page 62.

Remove the two screws securing the rear bonnet hinge and remove bonnet complete.

Slacken off the clips securing all water hoses and remove water hoses.

Remove the bolts securing the forked ends of the radiator tie rods to the bulk head and radiator and remove the rods.

Disconnect wiring to headlamps and remove headlamps from mounting brackets.

Remove the two nuts securing the bracket on the cylinder head to the radiator tie bar (where fitted).

Remove the bolts securing the headlamp mounting brackets to the radiator. Disconnect water thermometer from connection in radiator header tank.

Remove the nuts from the studs securing the radiator to the radiator bracket and lift radiator clear of chassis.

Disconnect the carburetter pipes, slow running and mixture controls and accelerator controls from the carburetters.

Disconnect the starter control at the starter, remove starter securing bolts and withdraw starter.

Disconnect oil thermometer connection from sump (where fitted).

Disconnect revolution counter drive (where fitted) from the rear of the cylinder head, and disconnect oil pressure gauge pipe.

Remove bolts securing front exhaust pipes to exhaust manifold(s) and withdraw pipe(s) from manifold(s).

Disconnect lead to high tension coil.

Place a sling around the engine and take the weight.

Remove the two front engine bearer bolts, distance pieces and rubber mountings.

Remove the two rear engine mounting bolts, distance pieces and rubber mounting, and place jack under gearbox to take gearbox load when engine is removed (F Type Magna Only).

Place jack under rear of engine to help support engine before removing bolts securing bell housing to clutch housing. (P, L, K, KN and N Models.) Remove bolts securing clutch housing to bell housing, lift engine, pull forward and lift from chassis frame.

Replacement is carried out in the reverse manner to that detailed for removal.

NOTE: On the N type Magnette, the engine removal is facilitated if the gearbox is first removed: Details see Chapter IV.

REMOVAL AND REPLACEMENT OF CRANKSHAFT. (P, F, L, K, KN and N Models.)

ENGINE OUT OF CHASSIS.

Drain and remove the sump as detailed on page 62.

Remove the cylinder head complete and withdraw the connecting rods and pistons, details as on pages 66 to 74.

Remove the clutch cover plate which will enable the remaining clutch components to be easily withdrawn.

Remove the circlip securing the clutch spigot race and remove the spigot race.

To remove the crankshaft nut keep pin, a blow on one end with a blunt punch, will tip it sufficiently to allow pliers to finally pull it out.

Now remove the crankshaft securing nut but do not remove the four bolts securing the flywheel flange to the flywheel. These components must be withdrawn together.

Remove the four nuts for the flywheel securing bolts, place a suitable extractor in position, replace the four nuts and withdraw flywheel complete with flange.

Disconnect the oil feed plug to the rear main bearing.

Remove the six bolts securing the flywheel housing to the crankcase and remove housing.

Remove the bolts securing the dynamo to the dynamo base, disconnect the two leads to the dynamo terminals and remove dynamo.

Remove distributor and oil pump as detailed on pages 50 and 63.

Unscrew the starting handle dog nut and front engine mounting plate.

Withdraw the water pump and gently tap the worm drive from the front end of the crankshaft, using a suitable drift such as a piece of brass, through the dynamo housing. It is a parallel fit on the crankshaft and is held in position by a key.

Using a suitable extractor remove the helical bevel pinion from the crankshaft. The holes in the gear are tapped ¼ in. B.S.F.

Remove the six bolts securing the front housing to the cylinder block, when the housing complete with the front main bearing, a ball race on the F type Magna and a white metal lined bearing on the P, L, K, KN and N Models, can be withdrawn from the cylinder block.

Removal of the front housing can be facilitated by inserting two 5/16 in. B.S.F. bolts in the tapped holes and use as with an ordinary puller.

Remove the four centre main bearings bolts and withdraw crankshaft through front end of the cylinder block, complete with the aluminium centre main bearing housing.

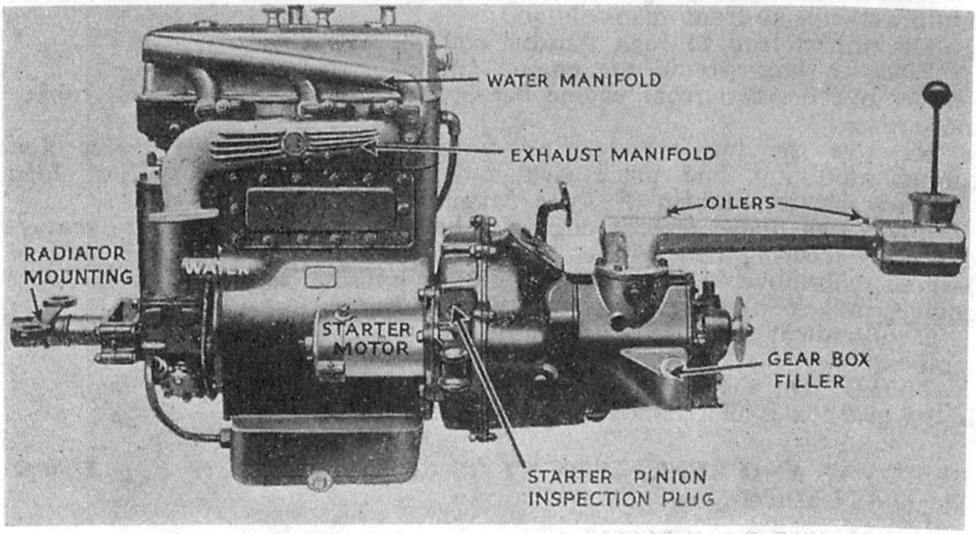

Fig. 36. L.H. view J type engine.

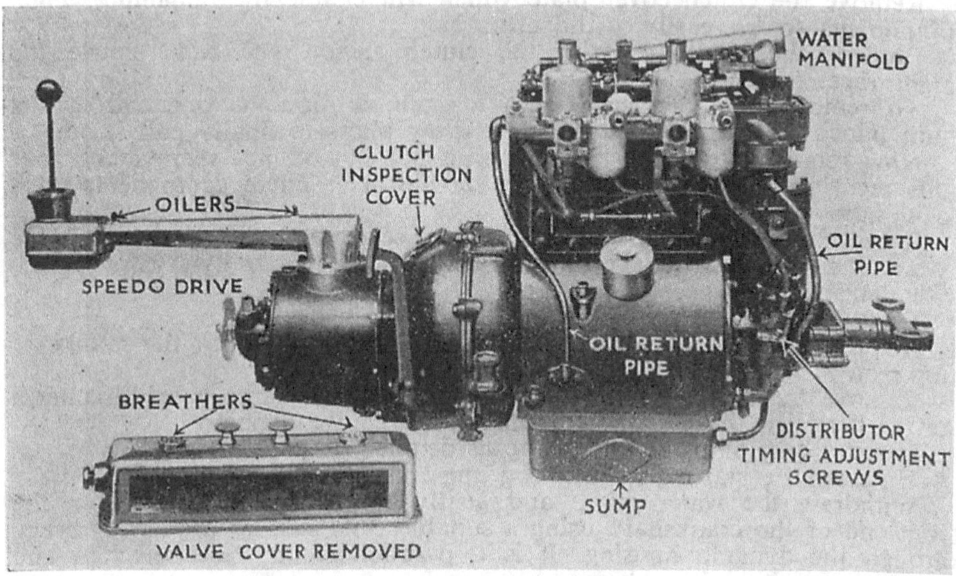

Fig. 37. R.H. view J type engine.

The centre main bearing can be removed from the crankshaft by removing the two small bolts holding the split bearing together at each side.

Replacement is carried out in the reverse manner to that detailed for removal and attention should be given to the following points.

1. Refit clutch cover plate to corresponding marks on flywheel.
2. The three steel bands on the clutch forward pressure plate coincide with the pressure pins in the clutch cover plate.
3. The tips of the flywheel nut keep pin face to the rear of the engine.
4. Adjust the clutch as detailed in Chapter III.

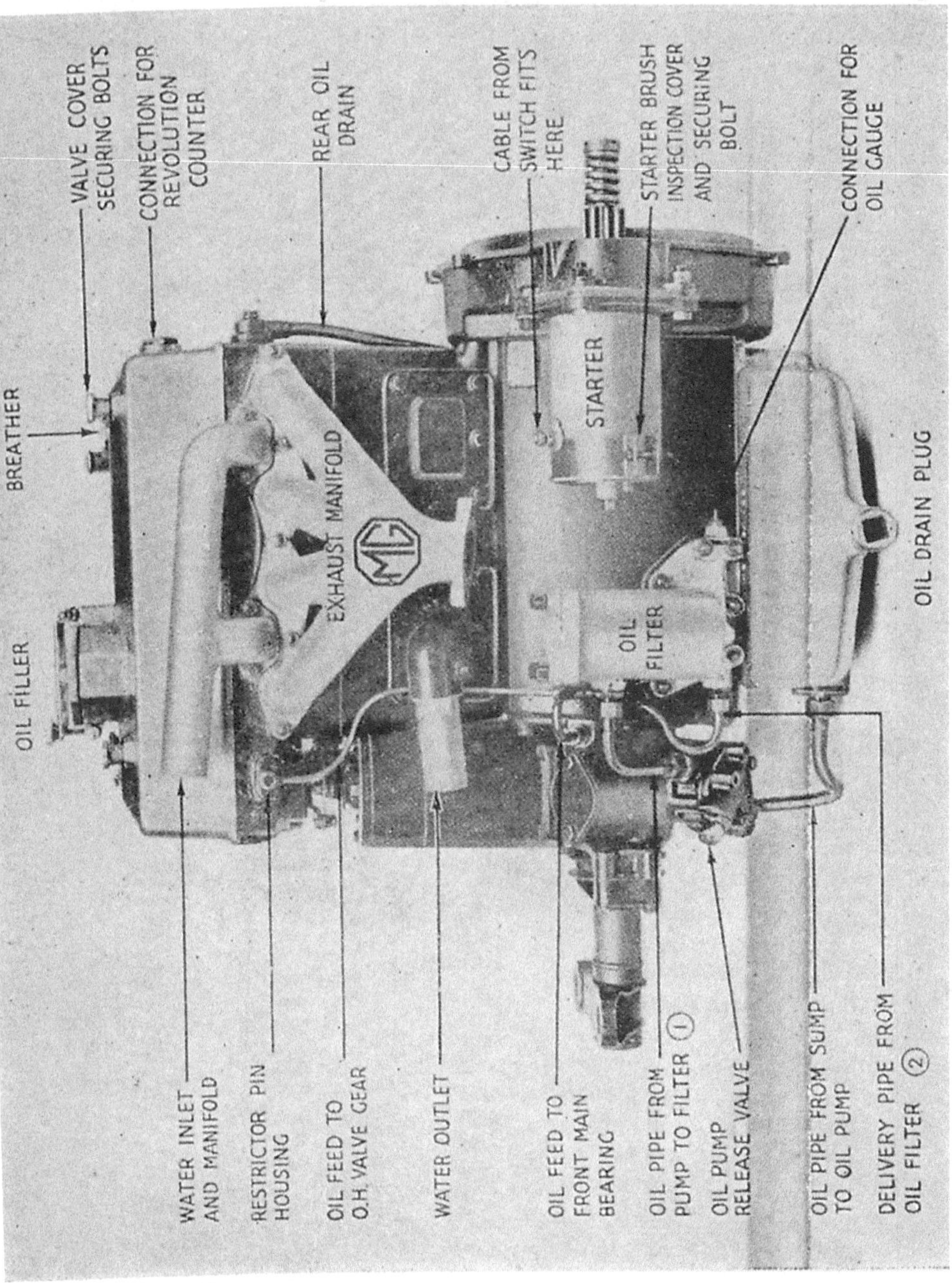

Fig. 38. L.H. view P type engine.

CRANKSHAFT END FLOAT. (P, N, KN and PB Models.)

Crankshaft end float is controlled by the white metal thrust washer.

The thrust ring has a fitting allowance of $\frac{1}{32}$ in. on the grooved face. An end float on the crankshaft of .003 in. is obtained by removing the surplus amount of metal from the grooved face of the thrust ring.

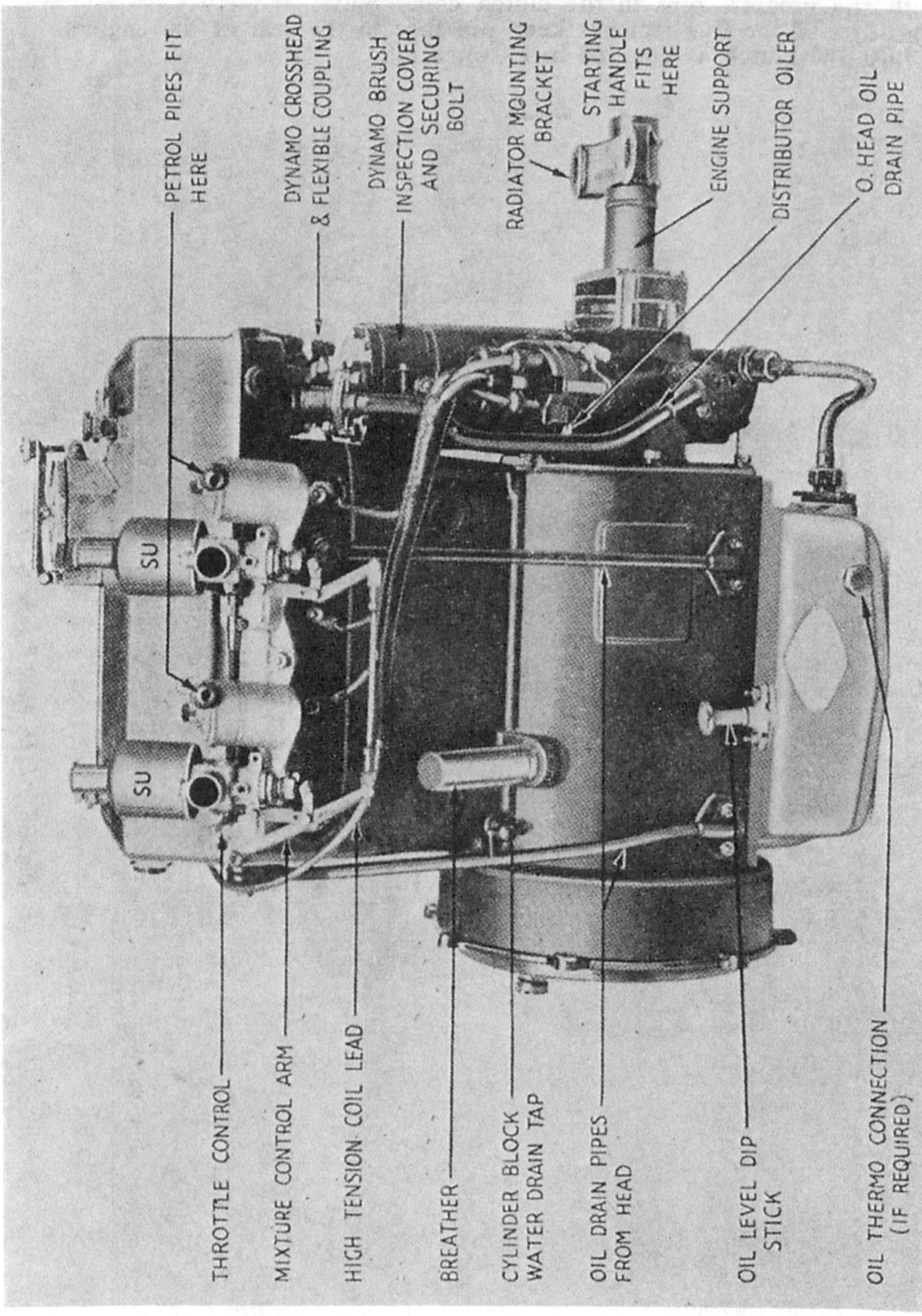

Fig. 39. R.H. view P type engine.

As an example, assuming the crankshaft has an end float of .025 in. then, as the correct end float is .003 in., there is an excessive amount .022 in. and the new thrust ring is required .022 in. thicker than the one removed.

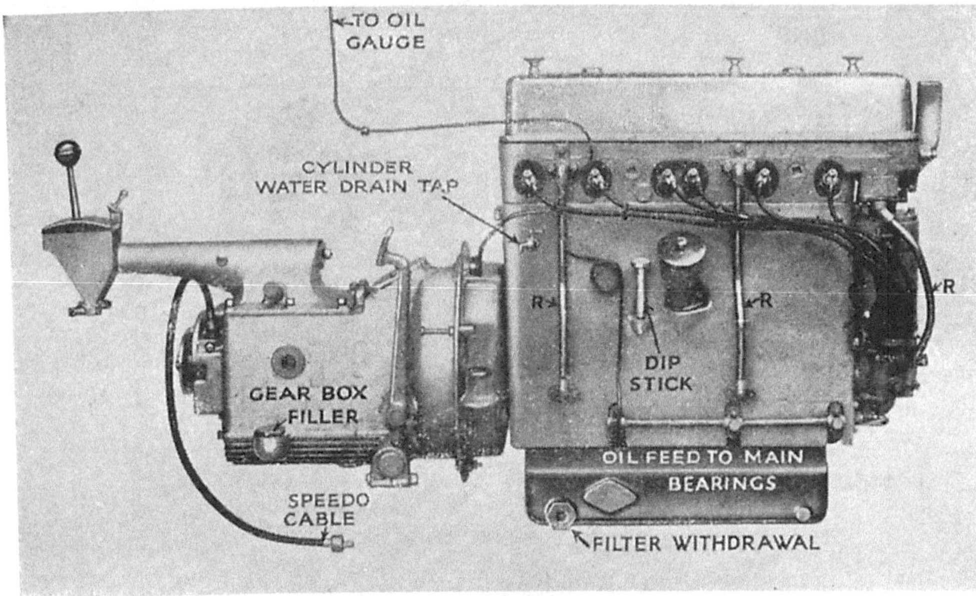

Fig. 40. *R.H. view F type engine.*

Fig. 41. *L.H. view F type engine.*

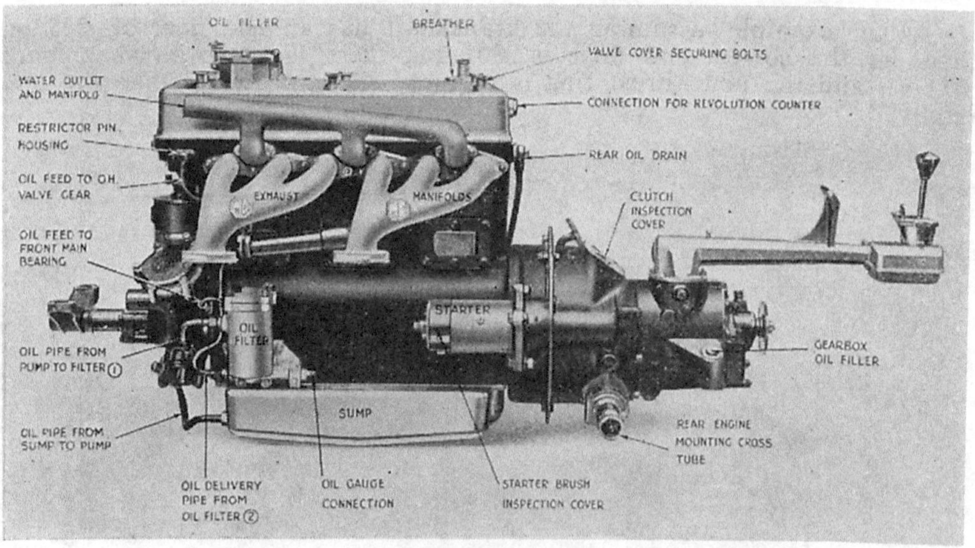

Fig. 42. L.H. view N type engine.

Fig. 43. R.H. view L type engine.

DESCRIPTION. (TA, TB, TC, 1¼, 1½, 2 and 2.6 Litre Models.)

The overhead valve push rod engine is built in unit construction with an oil immersed clutch on the TA, 2 Litre and early 1½ Litre Models, and a single dry plate clutch on the TB, TC, TD, 1¼, later 1½, and 2.6 Litre Models.

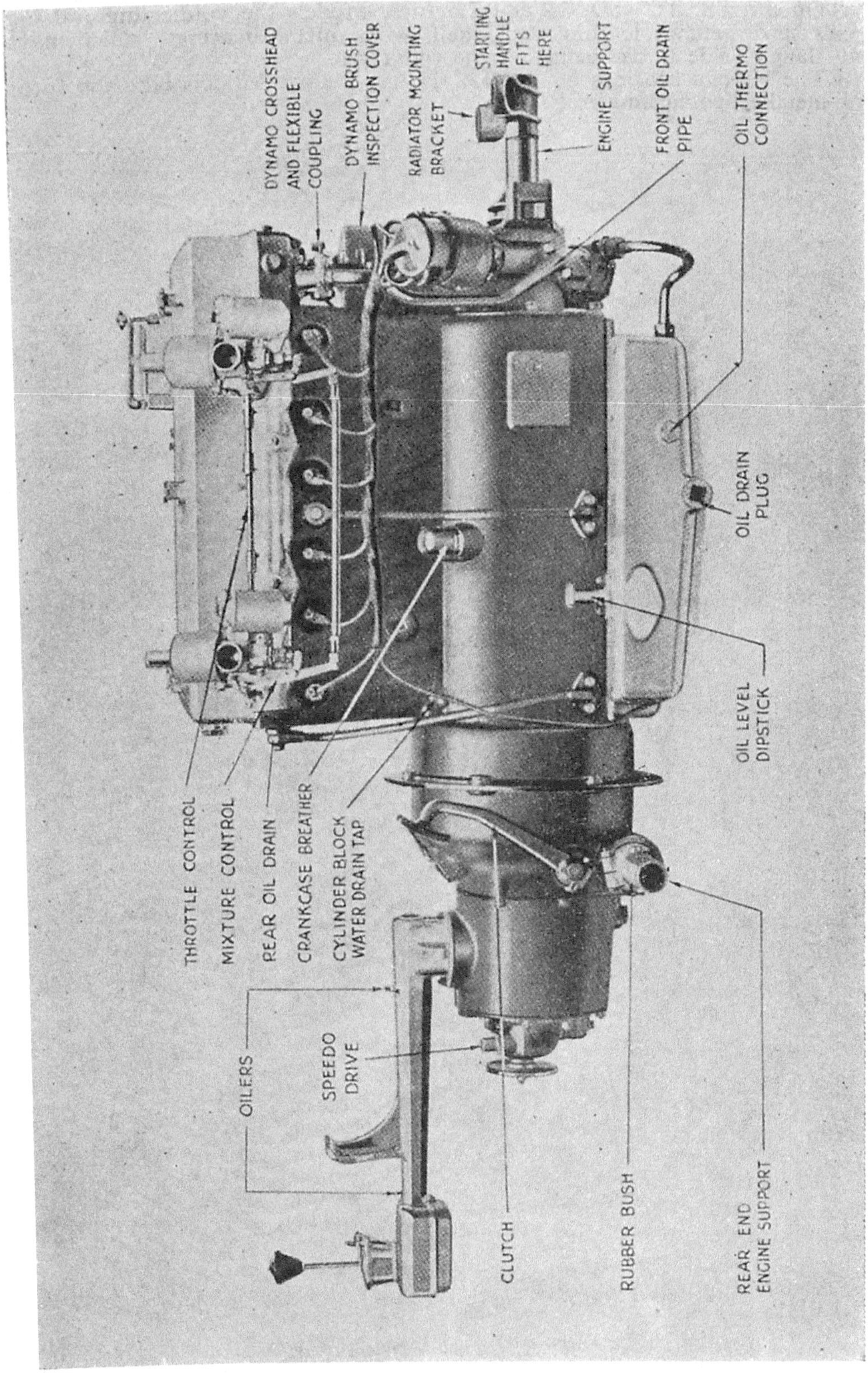

Fig. 44. R.H. view N type engine.

On the TB, TC, TD, $1\frac{1}{4}$ and 2.6 Litre Models the connecting rod big ends have renewable thin steel shell, white metal bearings, which have no flanges, side thrust being on the rod faces.

The big-end bearings on the TA, $1\frac{1}{2}$ and 2 Litre Models take the form of metalled connecting rods.

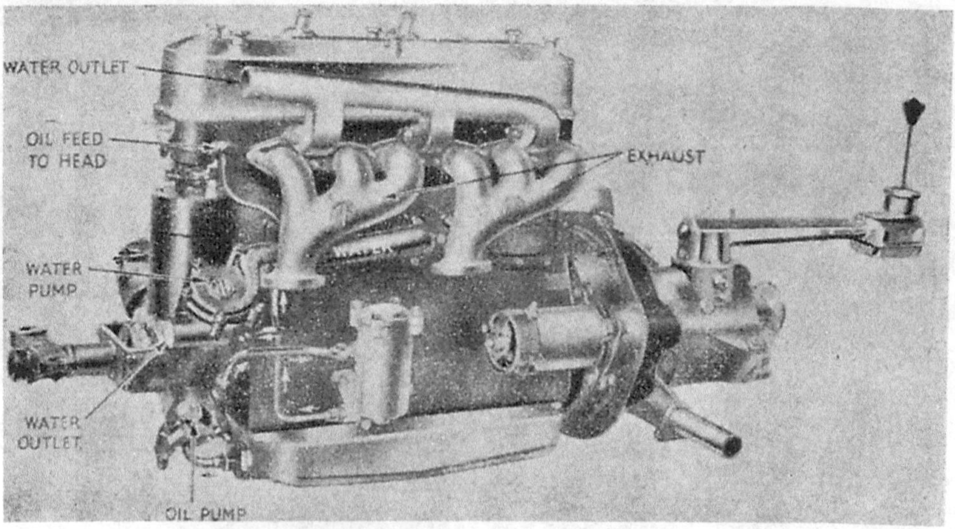

Fig. 45. L.H. view I type engine.

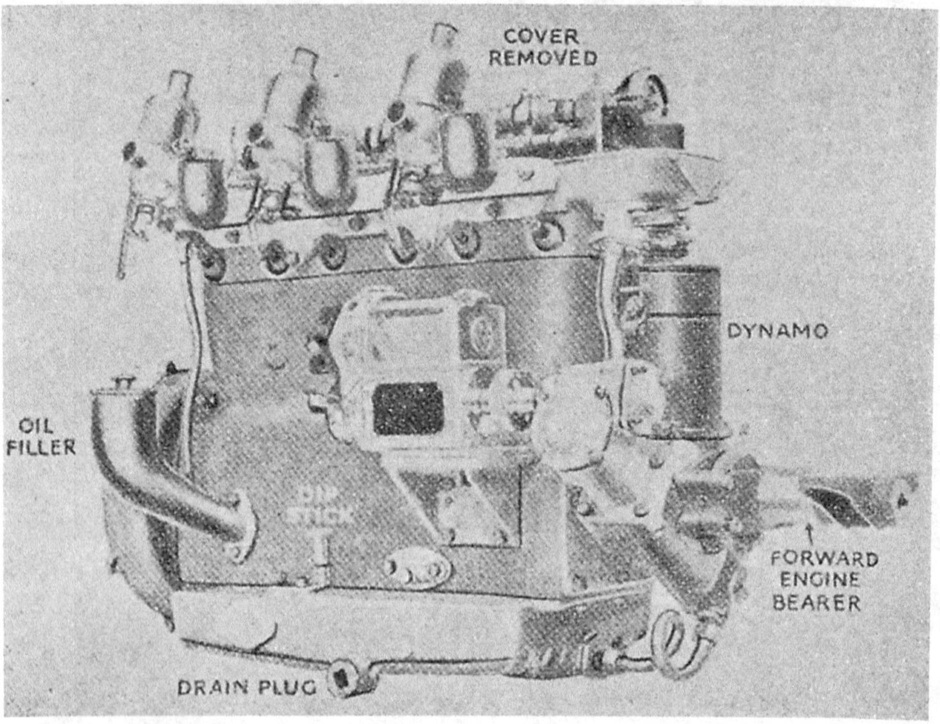

Fig. 46. R.H. view K type engine.

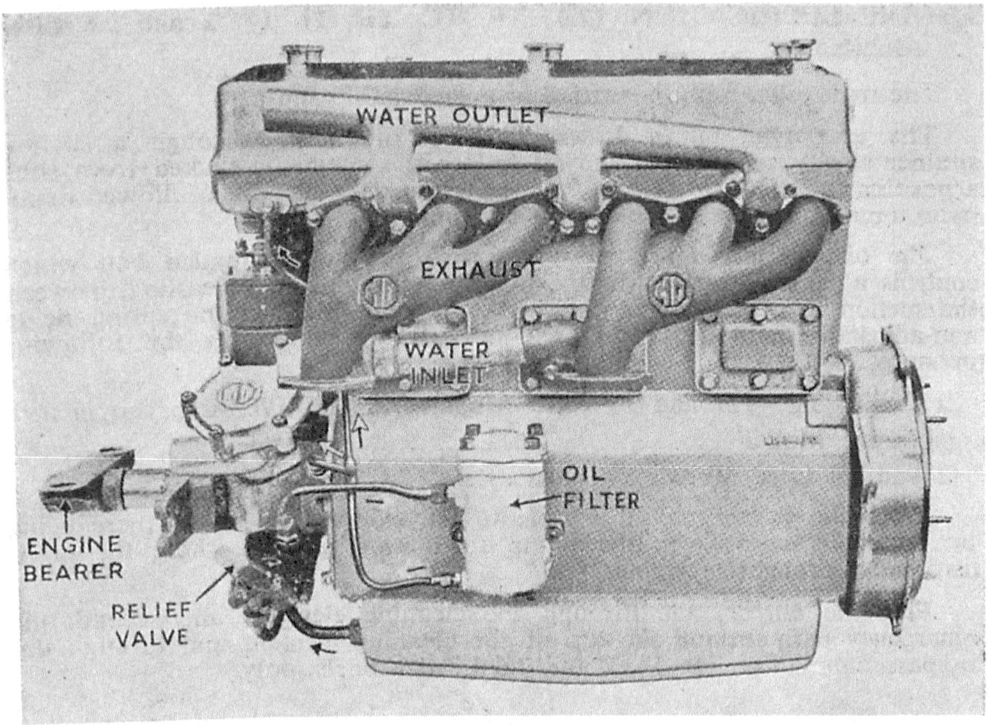

Fig. 47. L.H. view K type engine.

The little-end embracing the gudgeon pin, is slotted and fitted with a clamp screw and spring washer, locking it to the gudgeon pin.

The pistons are aluminium alloy, tin-coated, controlled expansion type, fitted with three rings (two compression and one slotted oil scraper).

On the TA Midget and the 2 Litre Models, up to and including engine numbers MPJG 696 and QPHG 1165, the pistons were fitted with four rings, of three different types, an additional oil scraper ring being fitted below the gudgeon pin.

The camshaft is driven from the crankshaft by a duplex roller chain. The three bearings for the camshaft of the TA, TB, TC, TD, 1¼, 1½ Litre and the four camshaft bearings for the 2 and the 2.6 Litre Models are fitted into the cylinder block and are fed with oil direct from the main gallery pipe. The chain on the TC, TD, 1¼ and 2.6 Litre is maintained in tension by a slipper block type of tensioner.

The camshaft operates the valves through the medium of the plunger type tappets which run direct in the cylinder block.

The main oil supply is carried in the detachable sump, the oil being drawn through a submerged filter in the sump by means of a gear type of oil pump driven from a helical gear on the camshaft, and delivered to the engine via an external oil filter.

The cylinder head is detachable.

The cooling water is circulated by means of an impellor mounted on the front wall of the cylinder block, and is driven in conjunction with the dynamo by a V belt from the pulley on the forward end of the crankshaft.

Carburation is by semi downdraught (downdraught on the 2 litre) S.U. carburetters fitted with a combination air cleaner, silencer and special manifold system. The fuel supply is by S.U. electric pump.

ENGINE LUBRICATION. (TA, TB, TC, TD, $1\frac{1}{4}$, $1\frac{1}{2}$, 2 and 2.6 Litre Models.)

The main oil supply is carried in a detachable sump.

The gear type pump draws oil from the sump through a suction strainer which ensures that the oil fed to the engine is sucked from such a position in the sump that any sludge formed in the oil is allowed maximum opportunity to settle to the bottom.

The oil pressure relief valve consists of a spring loaded ball which controls a passage formed in the oil pump bottom cover casting, between the suction and the delivery side of the pump pinions, the spring being non-adjustable and set to allow the ball to by-pass at the following pressures:—

TA, TB, TC, TD and $1\frac{1}{4}$ Litre Models	50/70 lbs. per sq. inch.
$1\frac{1}{2}$ Litre Model	40/60 ,, ,, ,, ,,
2 and 2.6 Litre Models	80 ,, ,, ,, ,,

From the pump the oil travels to the external oil filter, there being two possible ways from the pump to the oil gallery where the oil is distributed through the engine.

First the correct one through the cleaning element, and second, the emergency path around the top of the cleaning element and through the by-pass filter. (TA, TB, $1\frac{1}{2}$, 2 and 2.6 Litre Models, only.)

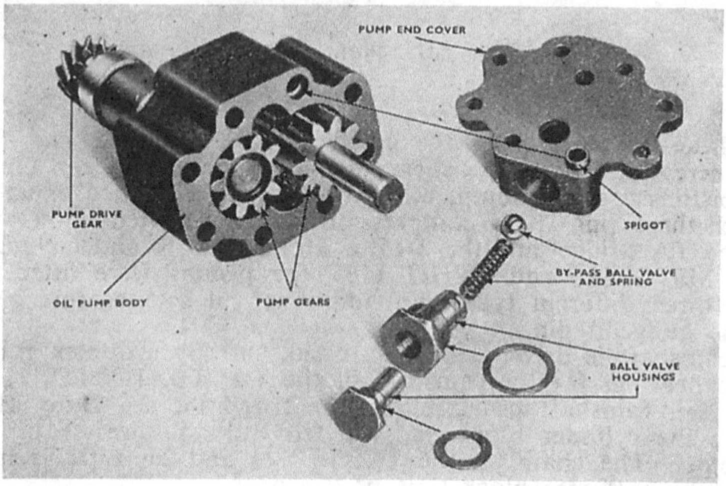

Fig. 48. Oil Pump dismantled (TA, TB, $1\frac{1}{2}$ and 2 Litre models).

The emergency path for the TC, TD and $1\frac{1}{4}$ Litre Models takes the form of a spring loaded relief valve housed in the cylinder block behind the pump body and oil can be distributed through this relief valve spring into the oil gallery.

The spring of the by-pass is such that, provided the filter is attended to periodically the valve remains permanently closed. Should the filter become clogged the by-pass will open, allowing unfiltered oil to enter the engine.

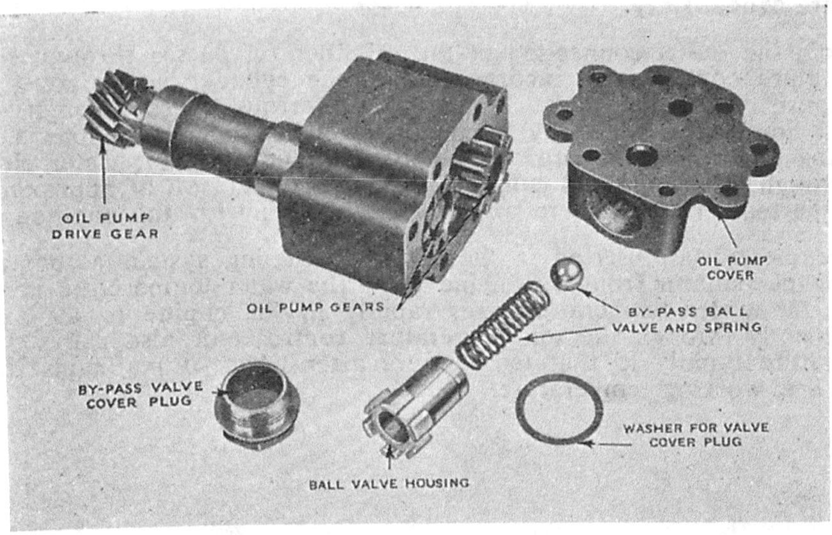

Fig. 49. *Oil Pump dismantled* (2.6 *Litre model*).

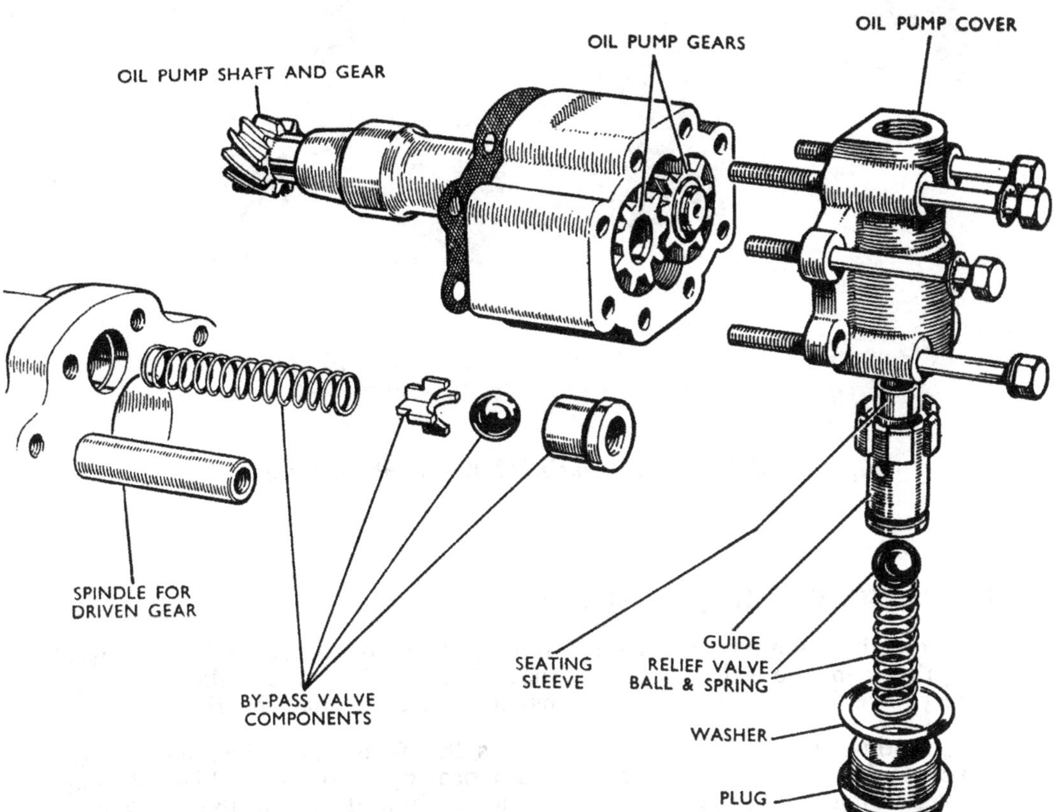

Fig. 50. *Oil Pump dismantled* (*TC, TD and* 1¼ *Litre models*).

2.6 Litre Model Only.

From the outlet connection of the oil filter oil passes through the oil temperature control unit, incorporated in the cylinder jacket cover plate and consisting of two large diameter non-ferrous pipes through which the oil flows, surrounded by a corrugated tube which functions as heat radiating fins. These fins are completely surrounded by the water circulating through the engine and consequently maintain the oil at approximately the same temperature. There are two advantages with this method.

The thermostat inserted in the water circulating system automatically isolates the radiator from the engine when the water temperature is below 70°C., the water temperature rises rapidly in the engine to 70°C., and consequently through the oil temperature control unit also raises the oil temperature rapidly to that temperature after which it maintains the oil at the best working temperature.

Fig. 51. *Oil Filter dismantled* (*TA, TB and* 1½ *Litre models*).

TA, TB, TC, TD, 1¼, 1½, 2 and 2.6 Litre Models.

From the " Outlet " connection the oil is led to the rear of the internal oil gallery in the side of the cylinder block. Passages drilled into this gallery pipe lead the oil to the camshaft and crankshaft bearings.

Counting from the front no. 1 feeds the front main bearing and the camshaft front bearing. The front main bearing feeds no. 1 big-end bearing via a ring cut in the white metal and the hole drilled in the crank web, which in turn feeds no. 1 cylinder wall through the spray hole drilled in the side of the big-end.

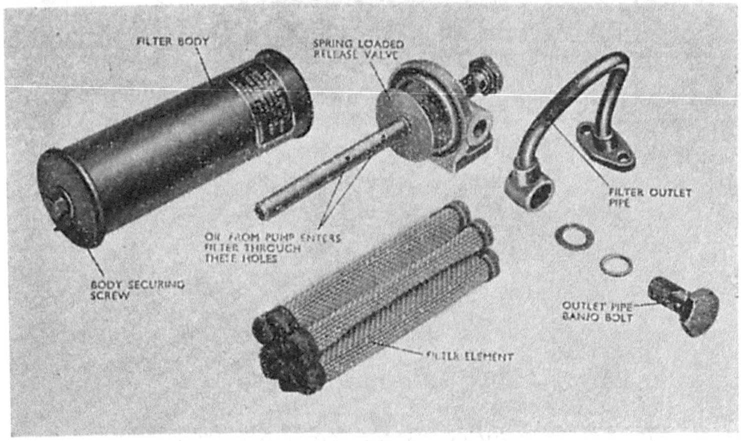

Fig. 52. *Oil Filter dismantled (2 Litre model).*

The camshaft front bearing has a forward leak to the camshaft chain wheel thrust face, and from there by three diagonal holes through the gear wheel boss to the inside of the wheel rim, whence radial holes take oil to the chain links.

Passage no. 2 feeds the camshaft second bearing and the centre main bearing (no. 2 centre bearing on the 2 and 2.6 Litre Models) this main bearing feeding nos. 2 and 3 big-ends as already detailed.

On the 2 and 2.6 Litre Models. Passage no. 3 feeds the third main bearing which in turn feeds nos. 4 and 5 big-ends and cylinder walls. The oil pump driving gears, which are in the middle of the third camshaft bearing, and the third camshaft bearing are also fed from here.

Passage no. 3 (no. 4 on 2 and 2.6 Litre) feeds the rear main bearing which in turn feeds no. 4 (no. 6 on 2 and 2.6 Litre Models) big-end and cylinder walls.

The vertical pipe which takes oil to the overhead valve gears is fed from the rear end of the gallery. The top union of the pipe communicates with passages drilled in the cylinder head to register with a hole drilled in the rear rocker shaft support, and so to the hollow rocker shaft. The rocker shaft is drilled at each rocker and oil passing the rocker bushes finds its way down the push rod drain passages to the sump.

On the models fitted with the oil immersed clutch, that is the TA Midget, 2 Litre, and 1½ Litre up to engine no. TPBG 1509, the rear main bearing feeds the clutch by a hole drilled up the centre of the crankshaft. Oil which passes the clutch plates escapes through gauze windows in the clutch cover plate, into the clutch housing where it is picked up by the teeth of the starter motor ring on the flywheel and thrown into the gallery on the nearside of the flywheel housing whence it drains back to the sump.

Fig. 53. Oil Filter dismantled (2.6 Litre model).

REMOVAL AND REPLACEMENT OF SUMP. (TA Midget.)

Drain the sump by removing the drain plug situated just below the body of the oil filter and remove the dipstick.

Raise the car at the front on trestles. Remove the nuts from the studs securing the front exhaust pipe flange to the exhaust manifold.

Remove the starter motor mounting bolts and withdraw the starter motor.

Remove the sump bolts and gently lower the sump to the ground.

Replacement is carried out in the reverse manner to that detailed for removal, ensuring that a new gasket is used and is a sound oil-tight fit.

REMOVAL AND REPLACEMENT OF SUMP. (1½ Litre Series VA Model.)

Drain the sump by removing the drain plug situated at the rear near-side of the sump, just below the oil filter.

Raise the car on trestles at the front end and remove the dipstick.

Slacken off the bolts securing the starter motor and swing over to give access to the sump bolts behind.

Remove the bolts securing the sump to the crankcase and gently lower the sump to the ground.

Replacement is carried out in the reverse manner to that detailed for removal, using a new sump gasket.

REMOVAL AND REPLACEMENT OF SUMP. (2 and 2.6 Litre Models.)

Drain the sump by removing the drain plug situated at the rear near-side of the sump just below the oil filter and remove the dipstick.

Raise the car at the front on trestles and disconnect the lead connected to the oil gauge rheostat fitted in the sump.

Remove the sump bolts and gently lower the sump to the ground.

Replacement is carried out in the reverse manner to that detailed for removal, using a replacement sump gasket.

REMOVAL AND REPLACEMENT OF SUMP. (TB, TC Midget.)

Drain the sump by removing the drain plug from the near-side of the sump just below the oil filter itself and remove the dipstick.

Remove the draglink to allow the sump to drop.

Remove the bolts securing the sump and gently lower sump to ground.

Replacement is carried out in the reverse manner to that detailed for removal, using a new gasket on replacement.

REMOVAL AND REPLACEMENT OF SUMP. (TD Midget and 1¼ Litre Saloon.)

Raise the car on a hydraulic jack or place the car over a pit.

Remove the exhaust system, remove the drain plug, and drain the oil from the sump. Remove the dipstick.

Release the clutch pedal pull-off spring from the return spring bracket.

Remove the split pin and clutch pin securing the intermediate clutch operating lever to the clutch operating rod.

Remove the two set bolts and spring washers securing the clutch cable abutment bracket to the sump, remove split pin and washer and slide off the intermediate clutch operating lever.

Remove the sump securing bolts and lower sump to the ground.

The anchorage for the engine fume pipe is attached to the first set screw below the crankcase and sump joint line on the near side of the flywheel housing.

Replacement is carried out in the reverse manner to that detailed for removal.

NOTE: Should it be necessary to fit a new sump gasket it should be cut as shown.

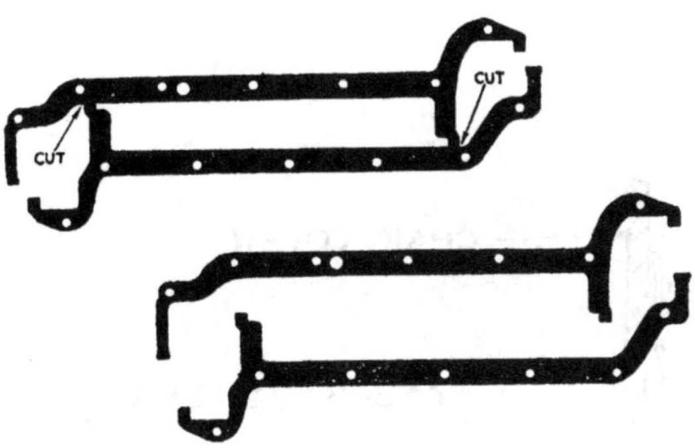

Fig. 54. Showing correct method of cutting Sump Gasket.

Ensure that the portion is left on which goes between the rear main bearing cap cork seal and crankcase. (See Fig. 54.)

If the cork composition packing ring in the groove of the rear main bearing cap is damaged fit a replacement.

Refer to Fig. 55 for correct fitting of this seal in conjunction with the sump gasket, to prevent the possibility of oil leaks.

If a replacement asbestos seal is fitted into the recess at the front of the sump, care should be taken that the ends of the new one are flush or a little above the face of the sump, and the sump gasket must go between the ends of the seal (see Fig. 56).

To ensure an oil-tight fit, the flanged edges of the sump and the joint faces of the cylinder block should be thoroughly cleaned before the sump is replaced. The sump flange can be coated with an even film of suitable jointing compound, but do not place jointing compound on the cylinder block flange.

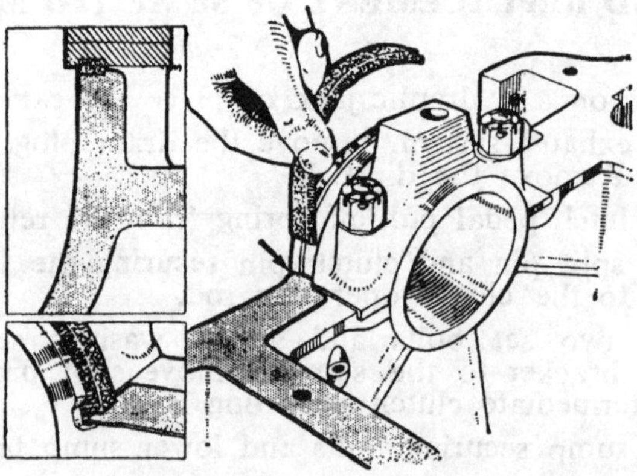

Fig. 55. Correct Fitting of the Rear Main Bearing Seal and Gasket.

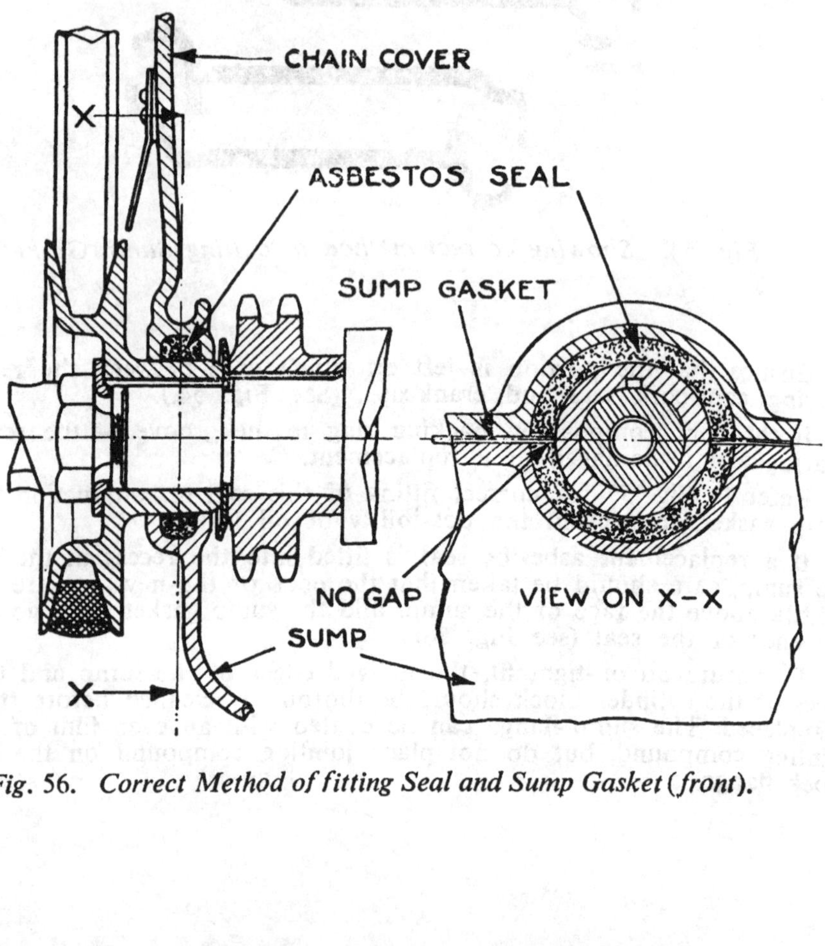

Fig. 56. Correct Method of fitting Seal and Sump Gasket (front).

REMOVAL AND REPLACEMENT OF OIL PUMP. (TA, TB, TC, TD, 1¼, 1½, 2 and 2.6 Litre Models.)

Drain the radiator and slacken off the top and bottom water hoses.

Remove the front engine mounting bolts and slightly jack up the front of the engine, which allows the pump to clear the frame member. (TA, TB, TC, TD, 1¼ and 1½ Litres only.)

Remove the oil pipe from the pump to the filter.

Remove the engine breather pipe (2 and 2.6 Litres only).

Remove the bolts securing the pump to the cylinder block, gently tap the side of the pump body and withdraw downwards.

Screw a suitable extractor into the end of the driven gear shaft and withdraw from the cylinder block.

The oil pump is replaced in the reverse manner to that detailed for removal care being taken to see that the abutting faces of the cylinder block and pump are clean before replacing.

Fit a new paper gasket between these joints. No gasket is fitted between the pump body and pump cover.

DISTRIBUTOR. (TA, TB, TC, TD, 1¼, 1½, 2 and 2.6 Litre Models.)

The distributor is provided with an automatic mechanism which varies the point of firing according to the requirements of the engine.

This is achieved by weights fitted in the distributor below the mounting for the contact breaker. As the speed of the engine varies these weights move under the action of centrifugal force and spring pressure and so advance or retard the cam which operates the contact breaker.

On no account should the tension of the springs be altered.

REMOVAL AND REPLACEMENT OF DISTRIBUTOR. (TA, TB, TC, TD, 1¼, 1½, 2 and 2.6 Litre Models.)

Remove the distributor cover and cables.

Remove the locking wire from the retaining screw and remove the actuating screw.

Withdraw the distributor from its bush in the cylinder block.

Replacement is carried out in the reverse manner to that detailed for removal.

When removing the distributor it is advisable to set the pistons on top dead centre and note the position of the rotor arm in relation to the distributor body.

TOP DEAD CENTRE MARK. (TA, TB, TC, TD, 1¼, 1½, 2 and 2.6 Litre.)

An indication arrow is fitted on the timing chain case, and a hole is drilled or a groove machined on the outer face of the crankshaft fan pulley. Turn engine by means of the starting handle until the hole or groove in the pulley lines up with the arrow on the timing chain cover for top dead centre, nos. 1 and 4 cylinders for TA, TB, TC, TD, 1¼ and 1½ Litre and nos. 1 and 6 cylinders for 2 and 2.6 Litre.

IGNITION TIMING ADJUSTMENT. (TA, TB, TC and 1¼ Litre Models.)

On replacing the distributor it must be retimed. Set the engine with nos. 1 and 4 on top dead centre. (See Fig. 57.)

Examine the valves to see which of the previously mentioned cylinders is starting its firing stroke, turn the distributor until the rotor is facing the appropriate segment and insert the distributor in is housing, "feeling" it in so that the nearest tooth is engaged. Turn the body about until the locking screw will enter, and lock.

Verify that the gap is properly set at .010 in./.012 in. and set the micrometer adjustment in the central position, slack off the pinch bolt and turn the head until the points are just breaking.

The correct starting point is "points" just breaking T.D.C. but may be varied not more than 1° either way. One division on the micrometer scale equals 4° (crankshaft) and each division is again divided into a definite number of "clicks" which are easily felt.

Fig. 57. Method of locating T.D.C. (illustration purposes only).

IGNITION TIMING ADJUSTMENT. (2 and 2.6 Litre Models.)

Proceed as per the Instructions given for the TA, TB, TC, 1½ Litre Models, bearing in mind that the correct starting point for these two models varies.

On the 2 Litre the correct starting point is 7° early and may be varied but not more than 1° either way.

On the 2.6 Litre the correct starting point is " points" just breaking 4° before T.D.C. This can of course be varied but not more than 1° either way. One division on the micrometer scale equals 4° (crankshaft), each division again being divided into a definite number of " clicks " which are easily felt. Therefore from the T.D.C. position the micrometer adjustment must be advanced one division.

On the 2.6 Litre Model nos. 1 and 6 cylinders can be brought to top dead centre by lining up the scribed line on the flywheel (marked 1 and 6) with the pointer fixed to the clutch housing after removing the clutch inspection cover. (See Fig. 58.)

IGNITION TIMING ADJUSTMENT. (TD Midget and 1¼ Litre Models.)

Proceed as per the instructions for the TA, TB, TC, etc., for replacing the distributor into its housing, set the contact breaker points to .010 in./.012 in. and check that the hole in the pulley still coincides with the arrow in the timing cover.

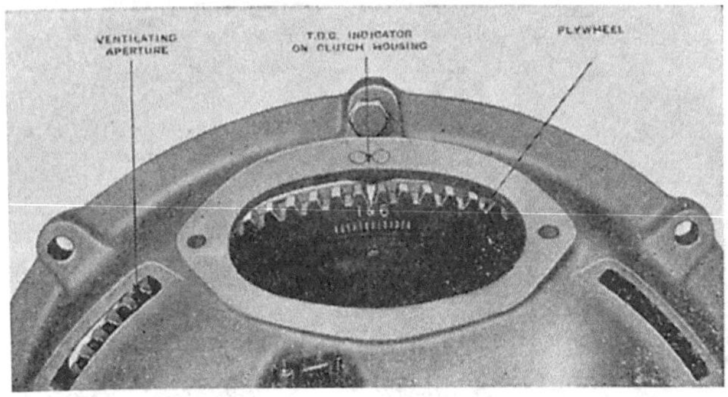

Fig. 58. *Method of obtaining T.D.C. from Flywheel.*

The contact breaker points should now just be commencing to open. If this should not be, release the clamping bolt at the base of the distributor and turn in an anti-clockwise direction until the points are fully closed, then turn carefully clockwise until the points just commence to open.

Securely tighten the clamp bolt, and re-check the timing to ensure that the setting has not been altered during the clamp bolt tightening operation.

NOTE: Before setting the timing on any model ensure that the automatic advance and retard mechanism is functioning correctly and is in the fully retarded position while the timing is set.

REMOVAL OF CYLINDER HEAD. (TA Midget.)

Disconnect battery.

Drain the water system by opening the tap at the bottom of the radiator and the tap in the off side of the cylinder block.

Take out the two screws securing the rear bonnet hinge and lift off bonnet complete.

Remove the bolts securing the headlamp brackets to the radiator and radiator stays to the radiator header tank.

Slacken off the water hoses by loosening one of the clips securing the rubber hose to the cast elbow bolted to the rear of the cylinder head and the uppermost clips securing the rubber hose above the thermostat.

Remove the two nuts securing the radiator to the cross member and lift radiator clear of the car.

Detach high tension cables from the sparking plugs, unclip the distributor head and remove high tension wiring complete.

Remove sparking plugs, petrol pipe, throttle and mixture control and uncouple the exhaust pipe from the exhaust manifold.

Disconnect the petrol pipes from the petrol pump and remove petrol pump from tool-box.

Remove the carburetter overflow pipes and remove the air cleaner and branch pipe.

Unbolt the carburetter from the induction manifold and remove induction and exhaust manifolds.

Remove the oil feed pipe to the overhead rocker gear.

Remove the valve cover.

Remove the push rods by the use of a special tool shown in Fig. 59.

Release the 10 cylinder head nuts a partial turn at a time until they are free for complete removal by hand.

Remove the cylinder head complete with rocker gear, fan and water pump.

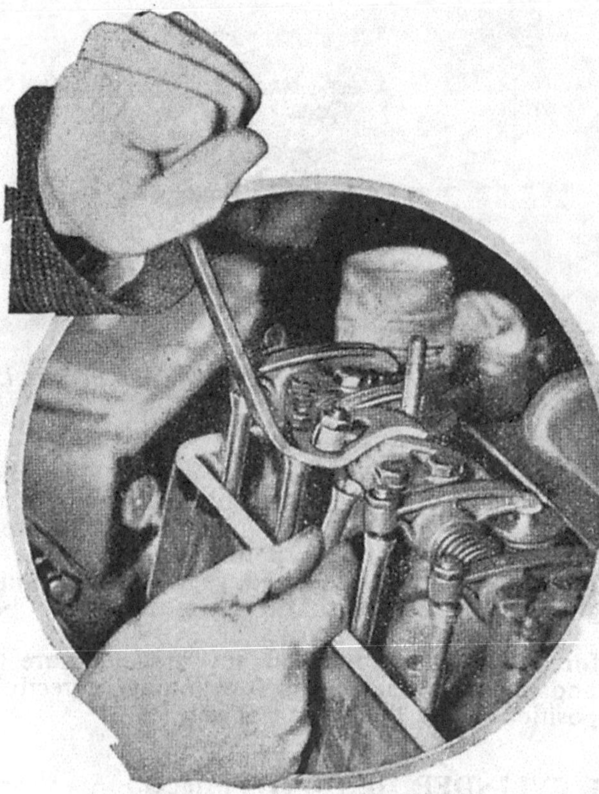

Fig. 59. Special Tool in use when removing push rod with Cylinder head intact

To facilitate breaking cylinder head joint, tap each side of the head with a hammer, with a piece of wood interposed to take the blow. When lifting the head a direct pull should be given, so that the head is pulled evenly up the studs.

REMOVING THE CYLINDER HEAD. (TB, TC, TD, 1¼, 1½, 2 and 2.6 Litre Models.)

Disconnect battery cables.

Open the water tap at the base of the radiator and the tap in the off side on the cylinder block to drain water system.

Remove the two screws at the rear end of the bonnet hinge and remove bonnet complete.

Detach the leads from the sparking plugs.

On the 2 Litre and 2.6 Litre they are carried in a fibre tube clipped to the cylinder head. It will be necessary to remove the screw securing the fibre tube and unclip the distributor head, and remove cables complete.

Remove sparking plugs.

Disconnect the throttle and mixture control and petrol pipes.

Uncouple the exhaust pipe from the manifold.

On the 2 and 2.6 Litre, disconnect the support holding the exhaust pipe to the cylinder block and let the pipe hang.

Disconnect the distributor suction control pipe from the rear carburetter and remove the pipe from its securing clips on the cylinder head. (2.6 Litre Series WA only.)

Remove the carburetter air cleaner(s) and branch pipe and carburetter(s) float chamber overflow pipe(s).

Unbolt the carburetter(s) from the induction manifold. (TB, TC, TD, 1¼ and 1½ Litre Models only.)

Remove the bolt clipping the exhaust pipe to the gearbox. (TD and 1¼ Litre only.)

On the 2 and 2.6 Litre Models the carburetter assembly and the exhaust manifolds are then held only by the five securing bridge pieces.

Remove the bridge pieces and lift away carburetter assembly complete with all attachments and the exhaust manifold. (2 and 2.6 Litre only.)

Remove the nuts securing the induction and exhaust manifold to the cylinder head, and withdraw manifold. (TB, TC, TD, 1¼ and 1½ Litre Models only.)

Loosen the top clips on the thermostat by-pass pipe.

Remove the top radiator hose and thermostat body.

Remove at the cylinder head the oil feed pipe to the rocker gear.

Slacken the fume pipe and side inspection cover, using a new gasket if necessary when replacing.

Uncouple the rubber hose water connection to the back of the cylinder head. (1½ and 2.6 Litre Models only.)

Remove the valve cover and rocker gear from the cylinder head and withdraw push rods. (TB, TC, TD, 1¼ Litre Models.)

Remove valve cover and using special tool part No. T.59 remove push rods. Fig. 59. (1½, 2 and 2.6 Litre Models.)

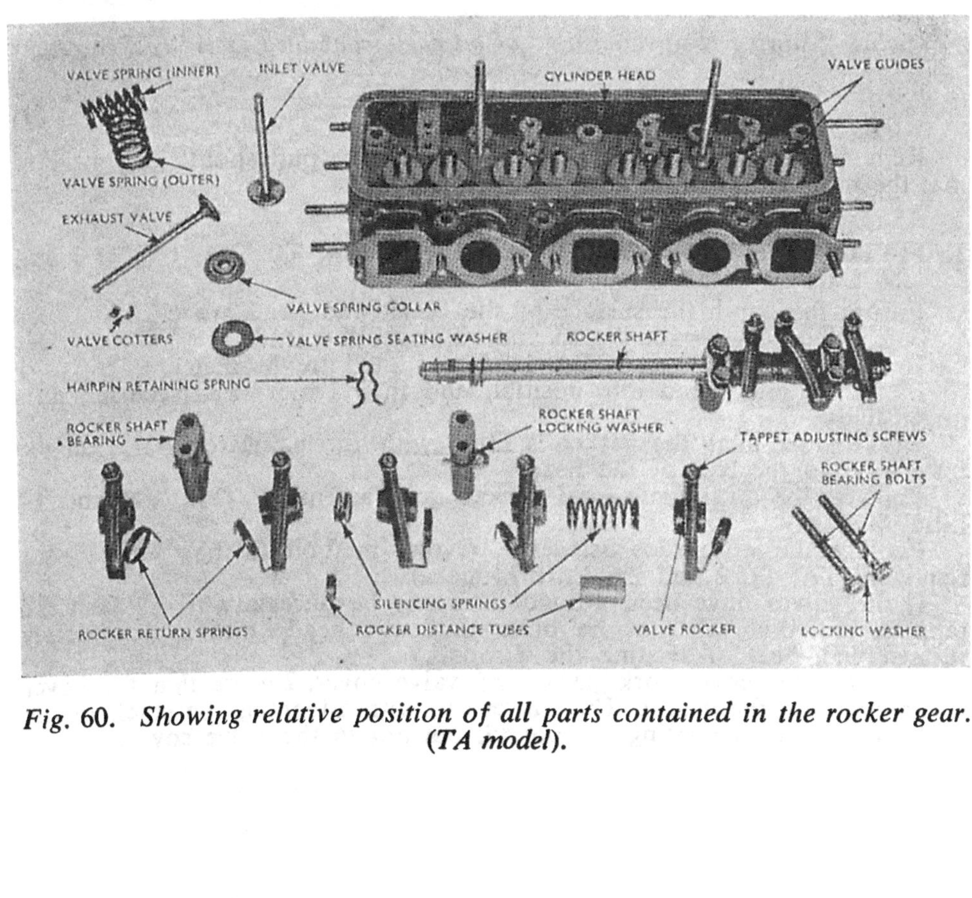

Fig. 60. Showing relative position of all parts contained in the rocker gear. (TA model).

Release the cylinder head stud nuts a partial turn at a time until they are free for complete removal by hand.

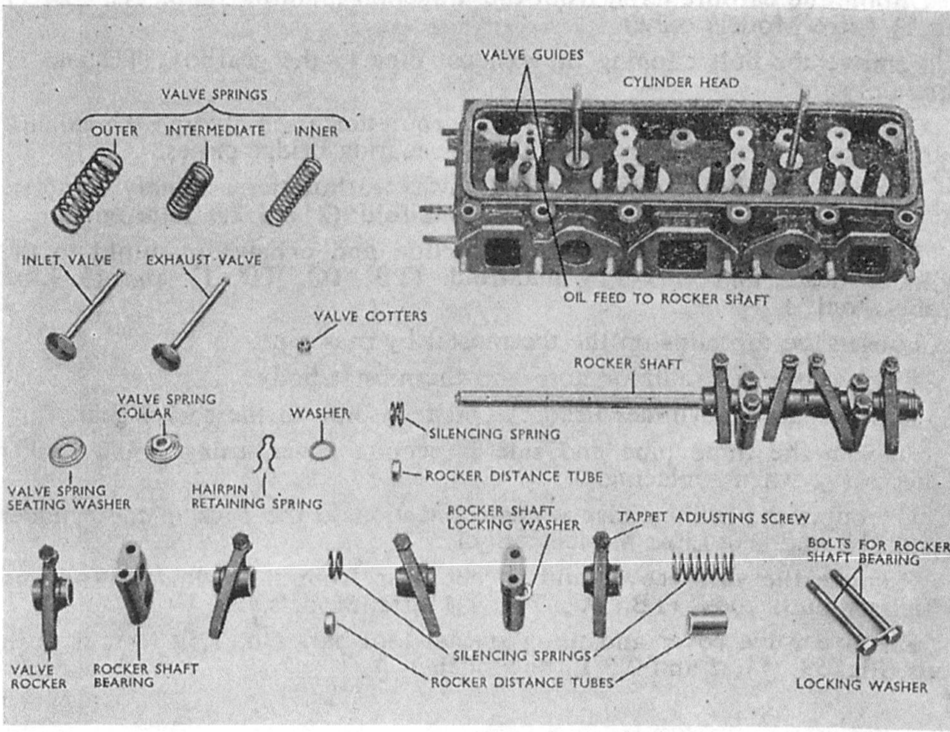

Fig. 61. Showing relative position of all parts contained in the rocker gear (1½ Litre).

Remove cylinder head. When lifting a direct pull should be given so that the head is pulled evenly up the studs.

REFITTING THE CYLINDER HEAD. (TA, TB, TC, TD, 1¼, 1½, 2 and 2.6 Litre Models.)

Ensure that both the surface of the block and head are clean.

Smear the gasket with grease and slip over the studs.

Replace the rocker gear. (TA, 1½, 2 and 2.6 Litre Models.)

Lower cylinder head into position and fit cylinder head securing nuts finger tight.

NOTE: See that the gasket is fitted with the elongated hole for the waterways to the rear of the head.

Replace the push rods and rocker gear assembly. (TB, TC and 1¼ Litre Models.)

Replace the push rods using the reverse method to that detailed for removal. (TA, 1½, 2 and 2.6 Litre Models.)

If the valves have been ground in it will be necessary to readjust the tappets, but this will not be done until the head, etc., is completely reassembled. See "Adjusting the Tappets."

Fit the valve cover cork gasket and valve cover. Ensure that the cover fits squarely on the gasket. The gasket can with advantage be stuck to the cylinder head with jointing compound, but not to the valve cover.

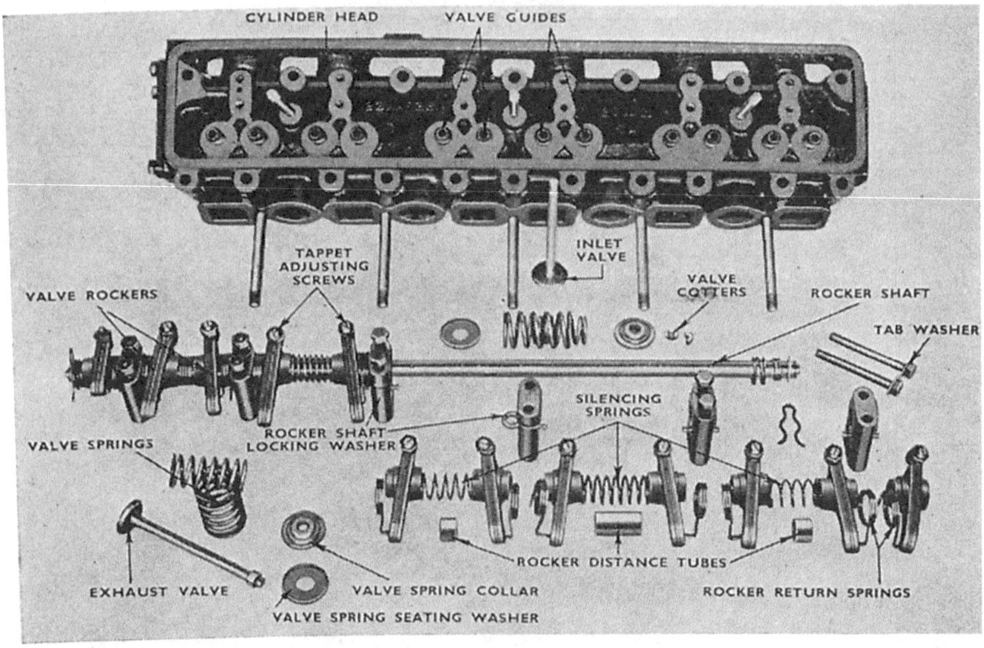

Fig. 62. Showing relative position of all parts contained in rocker gears, rocker return springs where not fitted on later models (2 Litre model).

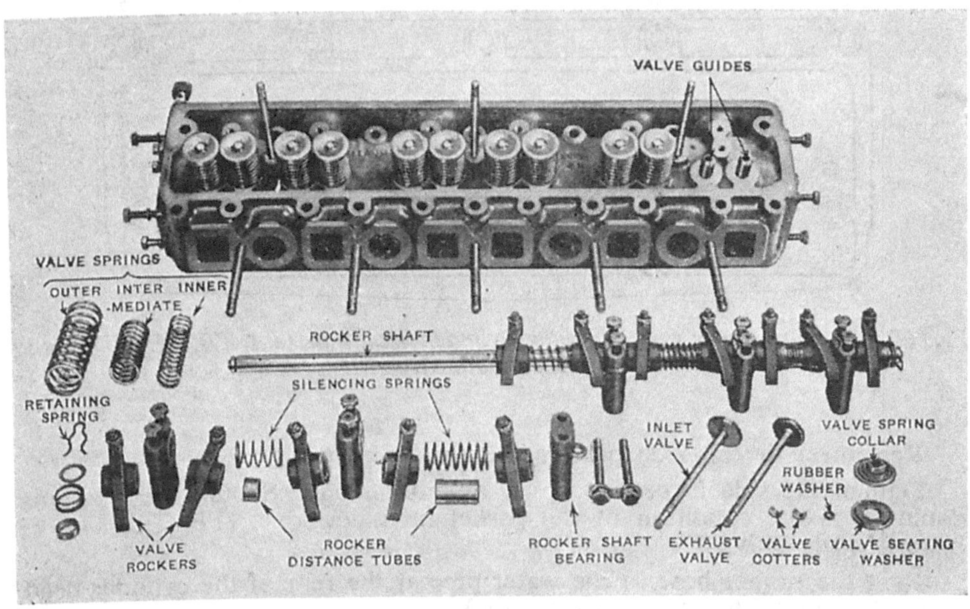

Fig. 63. Showing relative position of the parts contained in the rocker gear (2.6 Litre model).

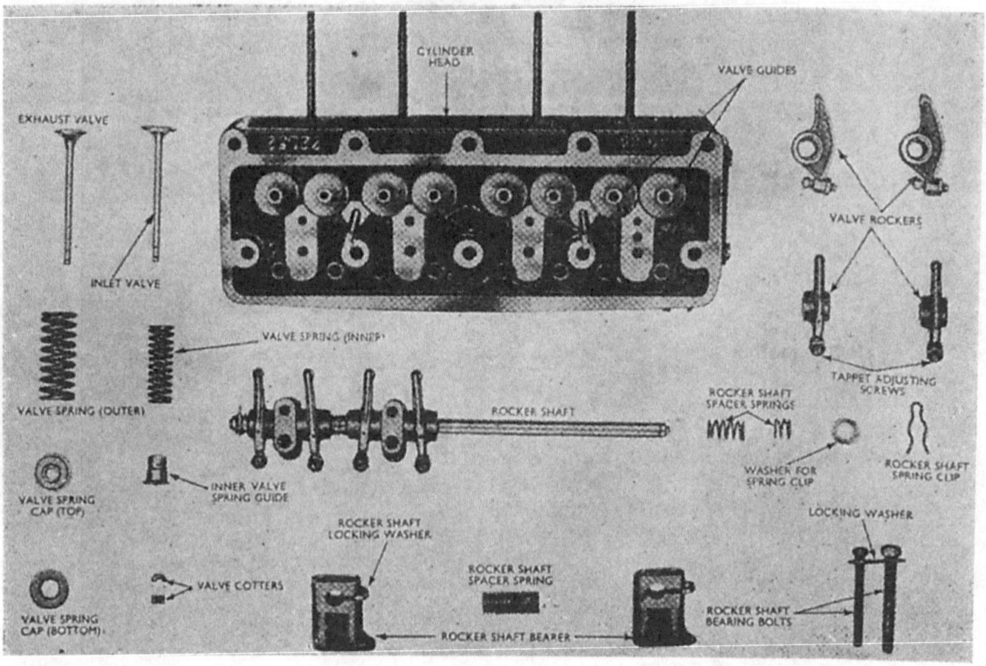

Fig. 64. The relative positions of the parts contained in the rocker gear (TB, TC, TD, 1¼ Litre models).

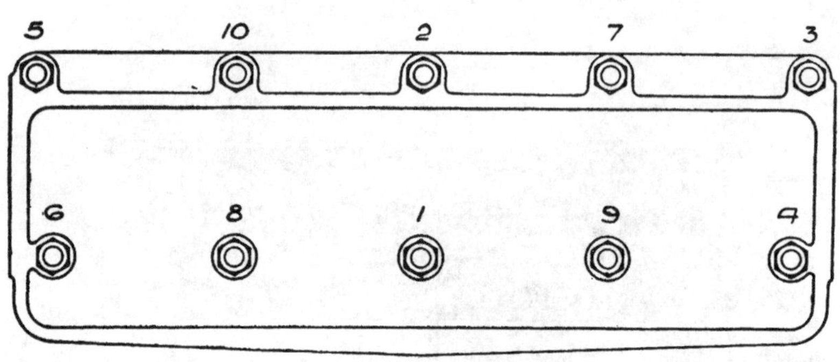

Fig. 65. Showing order for tightening cylinder head nuts (TA, TB, TC, TD and 1¼ Litre models).

Reconnect the oil feed pipe to the rocker gear.

Tighten the side inspection cover and fume pipe. Should there be any doubt as to the condition of the gasket fit a new one. (TB, TC, TD, 1¼ Litre Models only.)

Refit the rubber hose to the water pipe at the rear of the cylinder head. (TA, 1½, 2 and 2.6 Litre Models only.)

Refit the exhaust and induction manifolds complete with carburetter assembly, and evenly tighten down the securing nuts.

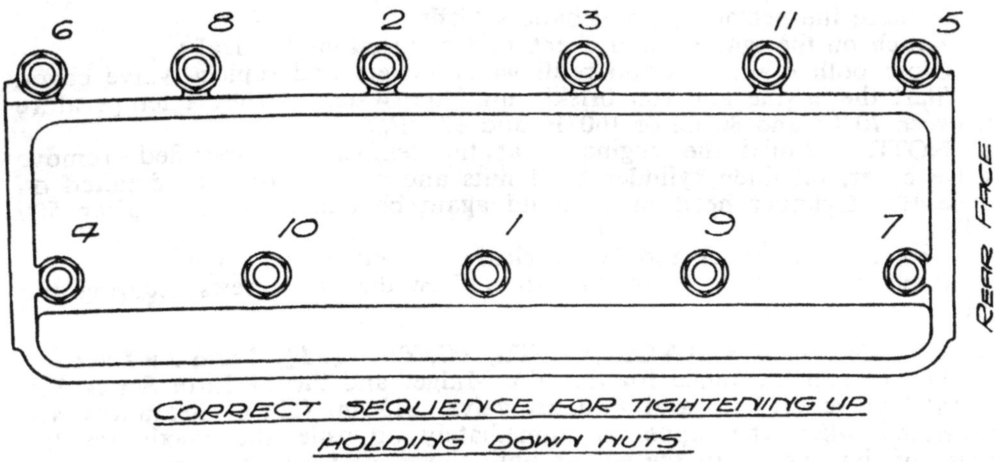

Fig. 66. *Correct sequence for tightening cylinder head nuts* (1½ *Litre model*).

If there is any doubt as to the condition of the manifold gaskets they should be replaced.

Fit the mixture controls and throttle controls.

Fit the exhaust pipe to the exhaust manifold, using a new gasket if necessary.

Clip up the exhaust pipe to the gearbox. (1¼ Litre and TD Midget only.)

NOTE: Pull up the securing nuts evenly a turn at a time.

Check and adjust the sparking plugs, replace and connect up the high tension leads to the plugs.

Refit the radiator. (TA Midget.)

Replace the thermostat body and radiator connecting hose.

Replace the distributor suction control pipe to the cylinder head and rear carburetter.

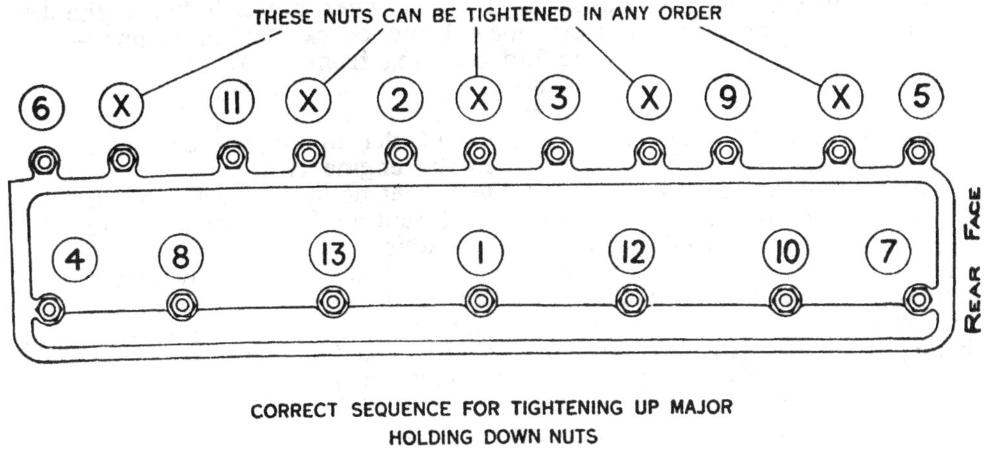

Fig. 67. *Correct sequence for tightening cylinder head nuts* (2 *and* 2.6 *Litre models*).

Replace the petrol pipe and battery leads.

Switch on the ignition and check fuel connections for leaks.

Close both drain taps and refill water system and replace valve cover.

Start the engine and run briskly until the water reaches a temperature between 70°C. and 80°C. or 160°F. and 175°F.

NOTE:—Whilst the engine is at the temperature specified, remove valve cover, retighten cylinder head nuts and adjust valves as detailed on page 100. Cylinder head nuts should again be tightened down after 500 miles.

Replace valve cover and fit air cleaner(s) and branch pipe.

Replace bonnet not forgetting to replace the two screws securing the rear bonnet hinge.

ADJUSTING THE VALVES. (TA, TB, TC, TD, 1¼, 1½, 2 and 2.6 Litre.)

The correct clearance for the TA Midget and the 1½ Litre Series VA is Inlet valve .010 in. and Exhaust valve .015 in. (Hot). Clearances are measured when the tappet is immediately opposite the maximum lift point of the cam. Turn the engine until the valve is at its maximum open position, then rotate the engine one complete turn. The later 1½ Litre Models had the tappet clearances set at .019 in. Inlet and Exhaust.

For the 2 and 2.6 Litre Models, the recommended tappet clearances are inlet and exhaust .015 in. set when hot.

As the back of the cams are truly circular it is not necessary that the cam should be exactly 180° to the tappet when checking clearance. The clearance can be checked, with the minimum turning of the engine, by adopting the following procedure. Turn the engine until No. 1 exhaust valve (the first counting from the front), is fully opened.

Then check the clearances on:—

No. 2 Exhaust The 4th counting from the front
„ 3 Inlet „ 5th „ „ „ „
„ 3 Exhaust „ 6th „ „ „ „
„ 5 Inlet „ 10th „ „ „ „
„ 6 Exhaust „ 12th „ „ „ „

Now give the engine two-thirds of a turn when No. 3 exhaust (the 6th counting from the front) is fully opened and check the clearances on:—

No. 1 Exhaust The 1st counting from the front
„ 2 Inlet „ 3rd „ „ „ „
„ 4 Exhaust „ 7th „ „ „ „
„ 6 Inlet „ 11th „ „ „ „

Again give the engine two-thirds of a turn when No. 2 Exhaust (the 4th counting from the front), is fully opened and check clearances on:—

No. 1 Inlet The 2nd counting from the front
„ 4 „ „ 8th „ „ „ „
„ 5 Exhaust „ 9th „ „ „ „

On the TB, TC, TD, and 1¼ Litre Model the tappet clearances are set at .019 in. Inlet and Exhaust, when the engine is hot.

While the clearance is being set the tappet of the valve being operated on is exactly on the heel of the cam. Counting from opposite ends the pistons are paired together and move in unison, the valve of each being fully opened with the corresponding valve of the other fully closed. No difficulty should be experienced in ensuring this, since it is only necessary to rotate the engine by the starting handle until the corresponding valve belonging to the other cylinder paired with it is fully opened.

The correct valve adjusting sequence is given herewith.

Adjust No. 1 Rocker with No. 8 Valve fully open
„ „ 3 „ „ „ 6 „ „ „
„ „ 5 „ „ „ 4 „ „ „
„ „ 2 „ „ „ 7 „ „ „
„ „ 8 „ „ „ 1 „ „ „

Adjust No. 6 Rocker with No 3 Valve fully open
„ „ 4 „ „ „ 5 „ „ „
„ „ 7 „ „ „ 2 „ „ „

Turn the engine by the starting handle until No. 8 valve is wide open. No. 1 valve can then be adjusted. Turn engine until No. 6 is open and adjust No. 3 and follow on in sequence.

Valve clearance adjustment is effected by means of the ball ended screws which engage the tops of the push rods. Releasing the lock nut enables the valve clearance to be accurately set by rotating the adjustment screw with a screw driver and checking with a feeler gauge of the correct thickness.

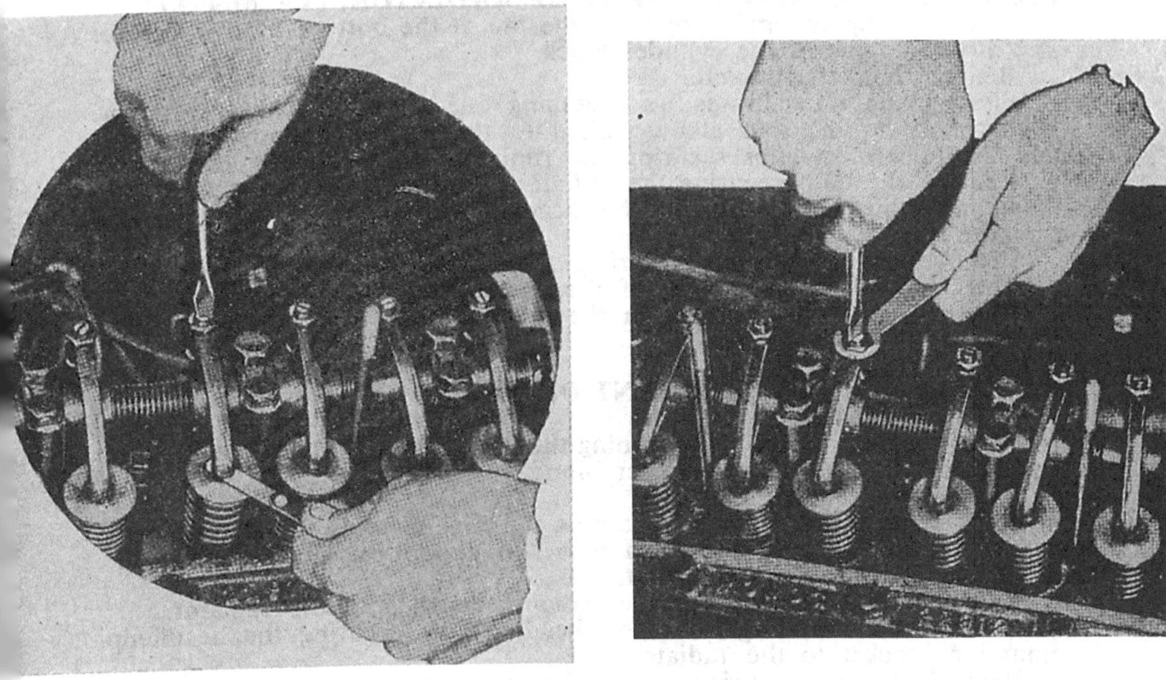

Fig. 68. Adjusting the valves (TA, TB, TC, TD, $1\frac{1}{4}$, $1\frac{1}{2}$, *2 and 2.6 Litre models).*

REMOVAL AND REPLACEMENT OF ROCKER SHAFT AND ROCKERS. (TA, TB, TC, TD, $1\frac{1}{4}$, $1\frac{1}{2}$, **2 and 2.6 Litre.)**

Remove the air cleaner(s) and cylinder head cover.

Remove the push rods using the tool No. T59 on the TA, $1\frac{1}{2}$, 2 and 2.6 Litres.

Knock down the lock plates and remove the 8 bolts, 12 on the 2 and 2.6 Litre Models, securing the rocker shaft brackets to the cylinder head.

NOTE:—The securing bolts should be unscrewed a turn at a time, until all the load has been taken off the rocker shaft.

Remove the rocker shaft complete with brackets and rockers.

Remove the circlip, washer and spring from the end of the shaft and slide the rockers, brackets and springs (spacer tubes where fitted) from the shaft, carefully noting their positions for re-assembling.

Care should be taken not to lose the shaft bracket washers, and note that the centre bracket washers are plain circular washers engaging with key-ways cut in the shaft, the remaining washers being "D" shaped washers.

Replacement is carried out in the reverse manner to that as detailed for removal.

NOTE:—When removable end plugs are fitted to the rocker shaft they should be removed and the rocker shaft oilways thoroughly cleaned.

Fit the rocker shaft oil feed holes uppermost.

Fit the rocker shaft bracket with the oil feed hole drilled up its centre, from the bottom, at the rear of the shaft.

Make sure that the plain circular washers are fitted to the centre bracket and engage correctly with the slots cut in the shaft and that the "D" shaped washers are fitted to the remaining brackets.

Fit new lock plates to holding down bolts.

Adjust valve clearances as previously detailed.

REMOVAL AND REPLACEMENT OF RADIATOR. (TA Model.)

Drain cooling system by opening the tap at the bottom of the radiator and in the off side of the cylinder block.

Remove bonnet complete.

Slacken the water hoses by loosening one of the clips securing the rubber hose to the cast elbow bolted to the rear of the cylinder head and the uppermost clips securing the rubber hose about the thermostat.

Remove the bolts securing the radiator tie rods to the radiator.

Remove the bolts securing the headlamp mounting brackets to the radiator.

Remove the two nuts securing the radiator to the front cross member and lift radiator clear of the chassis.

Replacement is carried out in the reverse manner to that as detailed for removal.

REMOVAL AND REPLACEMENT OF RADIATOR. (TB, TC, 1¼, 2 and 2.6 Litre Models.)

Drain the water system by opening the tap at the bottom of the radiator and in the off side of the cylinder block.

Remove the bonnet complete.

Slacken the water hoses by loosening the clips securing the rubber hoses to the bottom radiator pipe and also the uppermost clips securing the rubber hoses above the thermostat.

Remove the bolts securing the radiator tie rods to the radiator.

Remove the two large hexagon headed bolts securing the headlamp mounting bracket to the radiator.

Remove the two securing nuts under the front cross member and lift the radiator clear of the chassis.

Replacement is carried out in the reverse manner to that detailed for removal.

REMOVAL AND REPLACEMENT OF RADIATOR. (1¼ Litre Model.)

Open the drain tap at the bottom of the radiator and in the off side of the cylinder block.

Remove the bonnet clip at the rear of the bonnet and lift off the bonnet.

Remove both head lamps, pulling the wiring through the bracket and disconnect the left hand lamp bracket from the wing.

Remove the pinch bolt and slide the bracket over the tube.

Remove the right hand pinch bolt and pull the tube out of the right hand lamp bracket.

Remove the two bolts connecting the stay rods to the radiator.

Slacken off the clips on the top and bottom water hoses.

Remove from under the chassis front cross member the four nuts (two locking) securing the radiator in position and lift radiator clear of car.

Replacement is carried out in the reverse manner to that detailed for removal.

REMOVAL AND REPLACEMENT OF RADIATOR. (TD Midget.)

Open the drain tap in the bottom of the radiator and the off side of the cylinder block.

Undo the rear bonnet hinge bracket and lift off bonnet complete.

Detach the two bolts securing the radiator stays to the radiator.

Disconnect the by-pass hose at the thermostat, the hose on the elbow at the pump and the main hose at the top of the thermostat.

Remove the two large shoulder bolts holding the head lamp mounting brackets to the radiator.

Remove the securing nuts and locknuts from the mounting bracket noting the position of the rubber buffers, retaining rings and washers.

Lift radiator complete from car.

Replacement is carried out in the reverse manner to that detailed for removal.

TO DRAIN, CLEANSE, FLUSH AND REFILL COOLING SYSTEM. (All Models.)

The cooling system should periodically be cleansed, particularly where water having a high lime content is used for replenishing the radiator.

Drain the water system as previously detailed.

Clear out the drain tap by means of a piece of stiff wire.

When the water system has drained completely close drain tap.

Fill the radiator with a solution of cleansing compound, several reliable brands of which are available, and run the engine for about fifteen minutes or as directed by the makers of the compound.

Drain off the cleansing compound, and flush out the system with water. Turn off the drain tap and re-fill the radiator to the correct level with preferably soft water.

NOTE:—Avoid splashing the cellulose finish with the flushing compound, as this may have an injurious effect.

Do not leave the compound longer in the radiator than recommended by the makers.

REMOVAL AND REPLACEMENT OF ENGINE COMPLETE WITH GEARBOX. (TA Midget.)

Remove bonnet and radiator complete as previously detailed.

Remove the seats, carpets, rubber gearbox cover or gearbox carpet and floorboards. Undo the rubber muff where the steering column passes through the dash board and slide muff and securing plate down the column.

Disconnect the battery and dynamo cables and uncouple harness cables from the ramp plate.

Remove the screws securing the ramp plate to the dash and lift out the ramp plate.

Remove the two leads from the starter motor terminal and the Bowden wire from the starter motor switch.

Uncouple the clutch withdrawal rod from the lever and leave it hanging on the pedal.

Remove the six nuts securing the gearbox remote control to the gearbox and remove remote control. It is advisable to cover the gearbox with a piece of cardboard to prevent the loss of the selector balls and springs.

Uncouple the propeller shaft flange from the gearbox and suitably mark the flange before uncoupling to ensure that they are re-assembled in the same relative position.

Disconnect the throttle and mixture controls and the petrol pipe at the fuel pump end.

Remove the three nuts securing the exhaust pipe flange to the exhaust manifold and remove exhaust pipe.

Disconnect the oil feed pipe to the oil pressure gauge.

Disconnect the speedometer cable at the gearbox end.

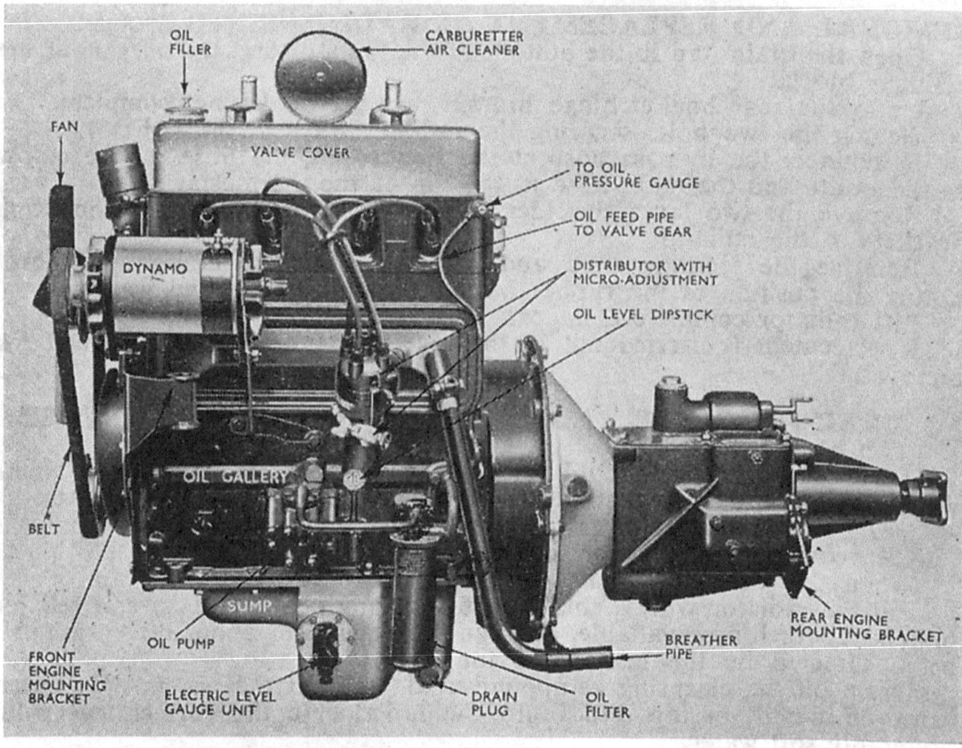

Fig. 69. View of 1½ Litre engine, left hand side.

Disconnect the revolution counter cable from the reduction box at the rear of the dynamo.

Remove the distributor cover and leads to the plugs. (To prevent breakage.)

Take the weight of the gearbox on a jack placed under the gearbox.

Remove the four main rear engine bolts, gearbox to rubber mounting.

Suitably place a sling around the engine and take the weight.

Remove the nuts from the front engine mounting bolts, together with the rubber washers and retaining cups which are also threaded, and uncouple the engine front tie strap.

Unscrew the bolts from the rubber mounting and lift the engine and gearbox complete from the chassis.

Replacement is carried out in the reverse manner to that detailed for removal.

NOTE:—It may be necessary to loosen and turn the starter motor dust cover to clear the steering column as the engine is withdrawn.

REMOVAL AND REPLACEMENT OF ENGINE WITH GEARBOX. (1½ Litre Series VA.)

Remove the bonnet and radiator as previously detailed.

Remove the front seat, front carpets and rubber excluders from the bases of the clutch and brake pedals and slide excluders and retaining plates up the pedals.

Remove the floorboards, disconnect the dipper switch and remove the metal gearbox cover. Disconnect the battery leads and the high tension lead from the coil.

Disconnect the cables connected to the oil gauge sump attachment and to the reverse light switch in the gearbox. (Chassis No. VA1371 and onwards.)

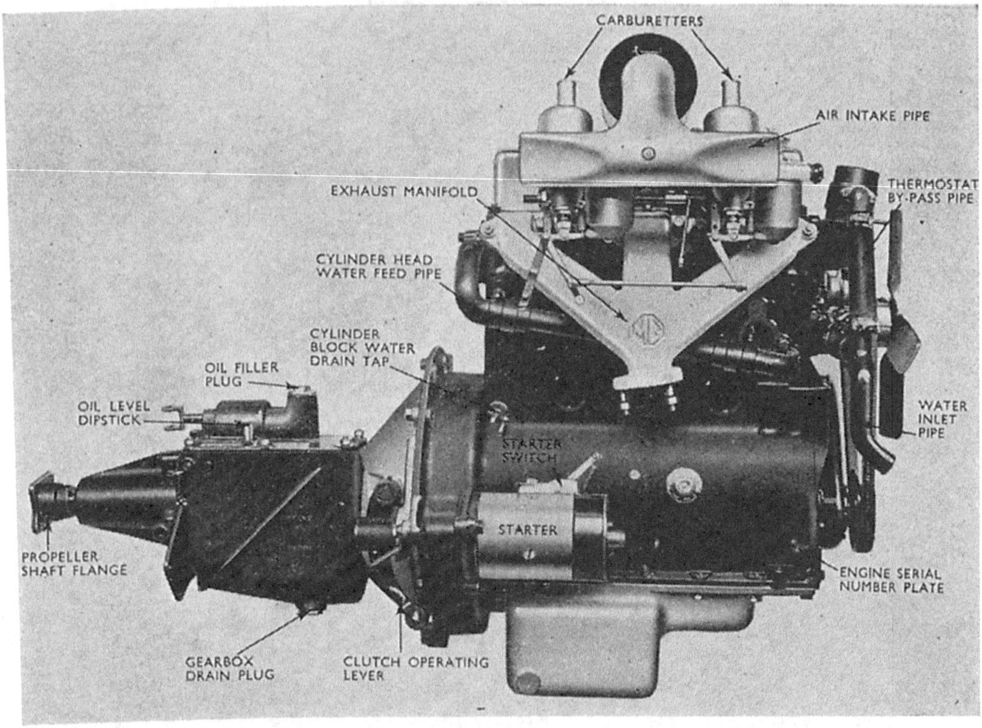

Fig. 70. *View of* 1½ *Litre engine, right-hand side.*

Undo the screws securing the off-side ramp plate to the dash and lift out the panel.

Disconnect the two leads to the starter motor terminal and remove the Bowden wire from the starter motor switch.

Uncouple the clutch withdrawal rods from the lever and leave it hanging on the pedal.

Uncouple the remote control at forward universal joint.

Disconnect the propeller shaft flange from the gearbox, suitably marking prior to removal, to ensure that they are re-assembled in their same relative positions.

Remove speedometer cable at gearbox end.

Remove the revolution counter cable from the reduction box at the rear end of the dynamo.

Disconnect the mixture and throttle controls and disconnect the petrol pipe at the fuel pump end.

Disconnect oil feed pipe to the pressure gauge.

Remove nuts securing the front exhaust pipe flange to the exhaust manifold, and remove front pipe from manifold.

Take the weight of the gearbox on a suitably placed jack and remove the four bolts from the mounting in the rear of the gearbox.

Place a sling around the engine taking the weight.

Uncouple the engine front tie straps. (These are not fitted to later models.)

Remove the front engine mounting bolts and lift engine from frame.

NOTE:—To facilitate removal, the rubber seal between the flywheel housing and the engine shield can be removed. The rubber retainer is bolted to the flywheel housing.

Replacement is carried out in the reverse manner to that detailed for removal.

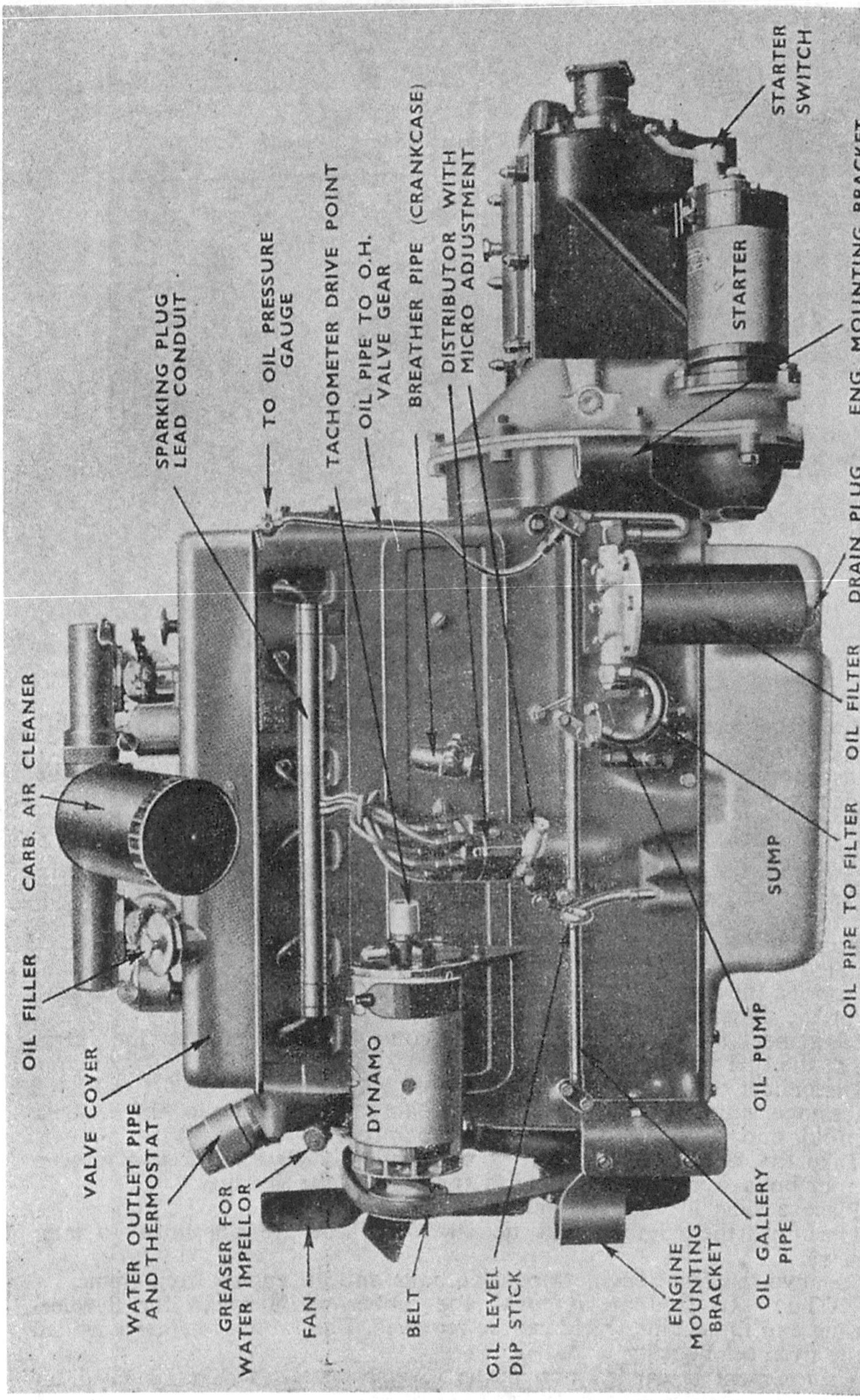

Fig. 71. Left-hand side view, 2 Litre engine.

REMOVAL AND REPLACEMENT OF ENGINE WITH GEARBOX. (2 and 2.6 Litre Models.)

Remove the bonnet complete and radiator as previously detailed.

Remove the front seats and front carpet, undo the rubber muffs at the base of the clutch and brake pedals, and slide muff complete with retainer plates up the pedals.

Undo the rubber muff where the steering column goes through the dash and slide muff and securing plate up the column. (2 Litre Models only.)

Remove the gearbox cover and floorboards, disconnect the high tension cables to the battery.

Disconnect the cables from the reverse light switch on top of the clutch housing and from the dipper switch and pull the cables through to the front side of the bulkhead.

Disconnect the cables from the oil gauge sump attachment.

Remove the high tension lead from the coil.

Remove the gearbox front cover, undo the screws which secure the flywheel housing shroud to the engine shield and lift out shroud.

Remove the engine shield completely. (2.6 Litre Models only.)

Unbolt the Lockheed and Jackall fluid containers and leave them on the ends of their pipes. The accelerator pedal stop which is secured to the inside of the dashboard by two of the bolts for the Lockheed fluid container will come away at the same time.

Undo the screws which secure the ramp plate to the dash and lift out the panel. (2 Litre Models only.)

Remove the two leads from the starter motor terminal; remove the Bowden wire from starter motor switch.

Remove the Jackall fluid tank. (2.6 Litre Models only.)

Uncouple the clutch withdrawal rod from the lever and leave hanging on the pedal. (2 Litre Models only.)

On the 2 Litre Models fitted with the one piece remote control assembly, remove gearbox top cover complete with the remote control. Do not lose the selector springs and balls.

On the 2.6 Litre and on the 2 Litre Models fitted with the two piece remote control assembly, uncouple the remote control at the forward universal joint.

Remove speedometer cable at gearbox end.

Remove revolution counter cable from the reduction box at rear of dynamo.

Uncouple the propeller shaft flange from the gearbox flange, suitably marking before removal to ensure that they are assembled in their same relative positions.

Disconnect the mixture and accelerator controls and petrol pipe at the petrol pump end.

Disconnect the oil pipe to the oil pressure gauge.

Remove the nuts securing the exhaust pipe flange to the exhaust manifold and remove exhaust pipe from flange.

Remove exhaust pipe support bracket from side of the crankcase and let exhaust pipe hang.

Slacken the nuts securing the front end of the gearbox steady blocks and remove gearbox steady cross piece.

Suitably place a sling around the engine and take the weight.

Remove the four main engine holding down nuts and lift out engine, leaving the rubber mountings attached to the chassis. The engine will have to be lifted about 5 in. to clear the mounting bolts.

On the 2.6 Litre Models it will be necessary to remove the nuts securing the engine front steadies.

Replacement is carried out in the reverse manner to that detailed for removal.

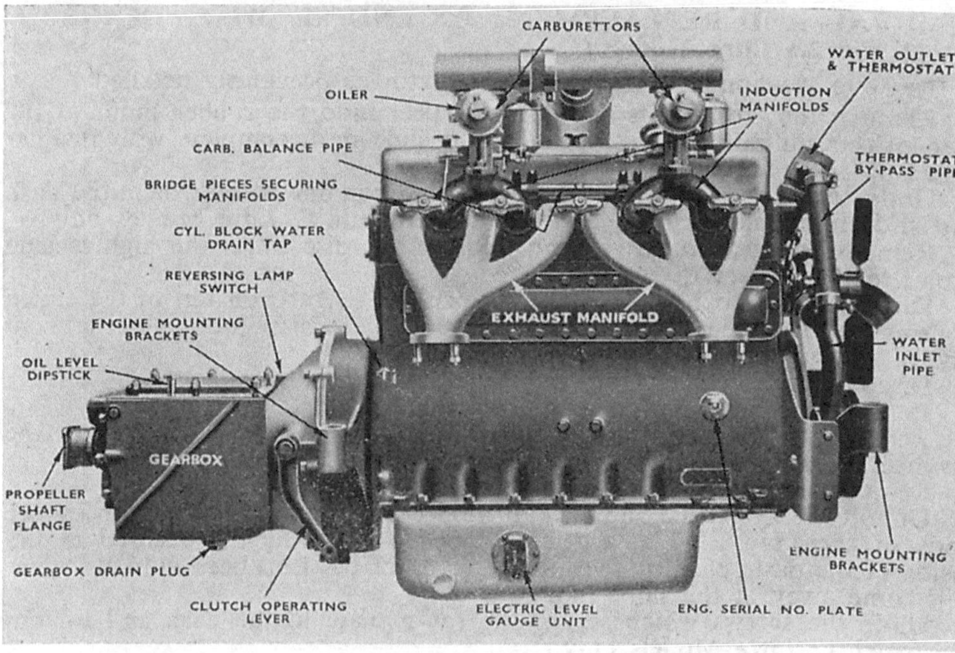

Fig. 72. Right-hand side view, 2 Litre engine.

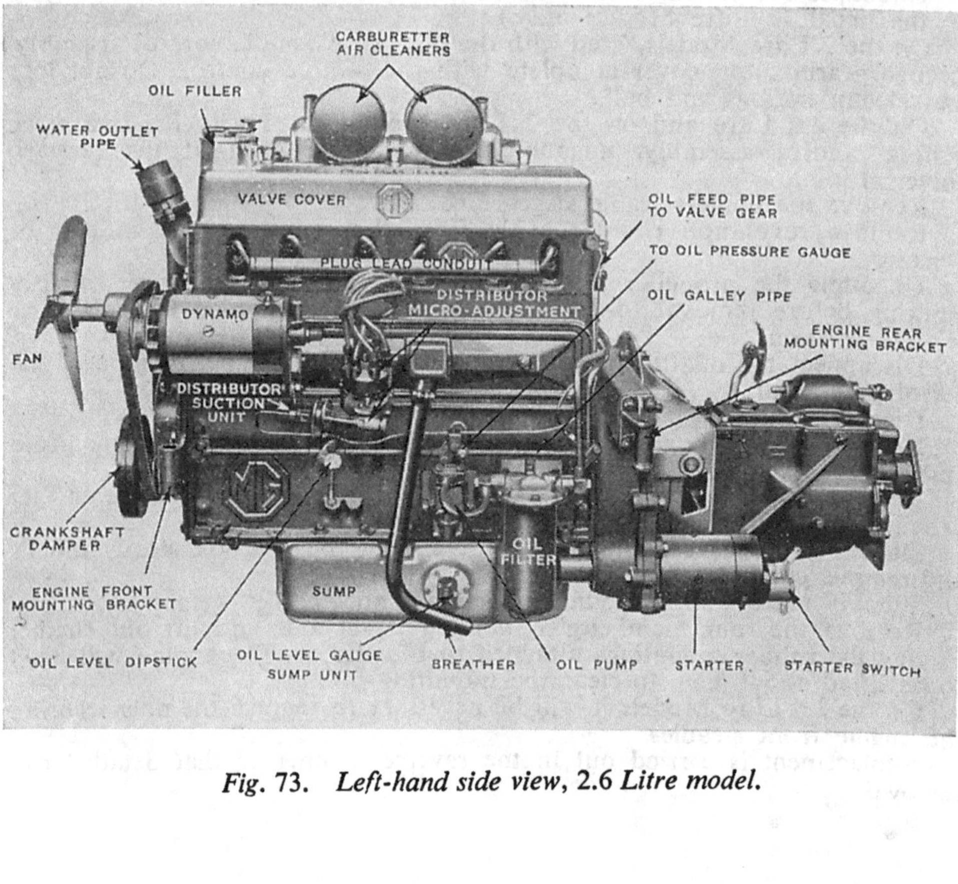

Fig. 73. Left-hand side view, 2.6 Litre model.

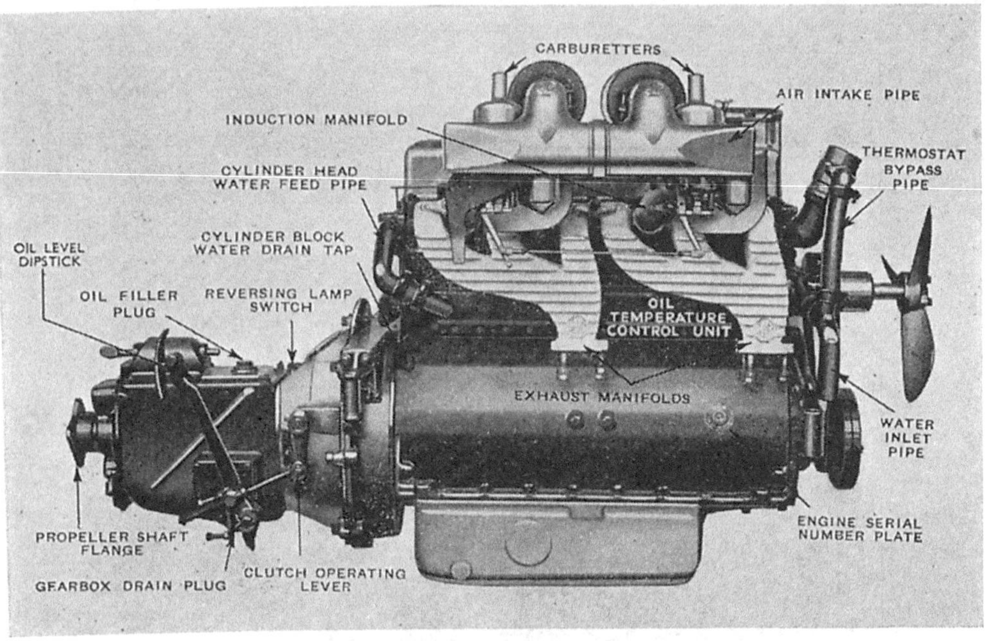

Fig. 74. *Right-hand side view,* 2.6 *Litre model.*

REMOVAL AND REPLACEMENT OF ENGINE WITH GEARBOX. (TB and TC Models.)

Remove the radiator and bonnet complete as previously detailed.

Remove the seats, carpets and rubber gearbox cover and carpet.

Remove the floorboard securing screws and lift out the floorboards.

Slacken off the rubber muff where the steering column passes through the dash and slide muff and securing plate down column.

Disconnect the high tension leads to the battery.

Uncouple the harness cables from the ramp plate.

Disconnect the cables from the dynamo.

Uncouple the clutch operating chain from the pedal.

Extract the split pin at the outer end of the pedal pivot pin, pushing this pin through its bracket.

Remove clutch and brake pedals.

Remove accelerator pedal gear.

Remove the screws securing the ramp plate to the dash and lift out ramp plate.

Remove the two leads from the starter motor terminal and the Bowden wire from the starter motor switch.

Remove the six nuts securing the remote control assembly to the gearbox top and remove the remote control complete, taking care not to lose the selector balls and springs.

It is advisable to cover the top of the gearbox with cardboard to prevent dirt, etc., entering the gearbox.

Remove the mixture and accelerator controls and petrol pipe at the petrol pump end.

Fig. 75. Left-hand view of TC Midget engine.

Disconnect the oil feed pipe to the oil pressure gauge.

Disconnect the speedometer cable at the gearbox end.

Disconnect the revolution counter cable from the reduction box at the rear end of the dynamo.

Remove the three nuts securing the exhaust pipe flange to the exhaust manifold and remove the exhaust pipe.

Take the weight of the gearbox by placing a jack beneath the gearbox and remove the four main rear engine bolts (gearbox to rubber mounting).

Suitably place a sling around the engine and take the weight.

Remove the locknut and sleeve nut from the bottom of the front engine mounting bolt, remove the steel washer and rebound rubber.

Unscrew the engine mounting bolts from rubber mounting bolts and remove engine from chassis.

Replacement is carried out in the reverse manner to that detailed for removal.

REMOVAL AND REPLACEMENT OF ENGINE WITH GEARBOX (1¼ Litre and TD Midget Models.)

Drain the cooling system, remove the bonnet complete and remove the radiator, all as previously detailed.

Remove the battery terminal leads.

Remove the front seats and carpets and gearbox cover.

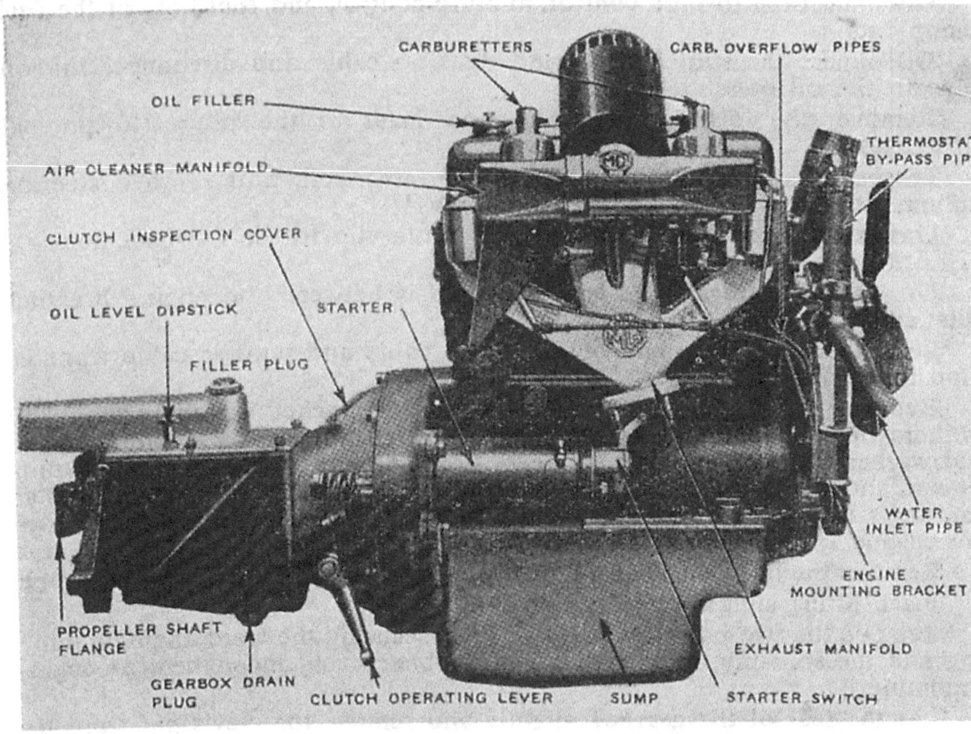

Fig. 76. Right-hand view of TC engine.

Remove the floorboards and footramp.

Remove the gear lever and gearbox extension cover complete, securing a piece of cardboard over the gearbox opening to prevent dirt entering inside of gearbox.

On the 1¼ Litre Model the exhaust system should be removed completely by removing the two bolts, nuts, and spring washers securing rear and intermediate tail pipe supports to their mounting brackets.

Remove the exhaust clip from the exhaust bracket attached to the right hand side of the gearbox (support the system while this work is being carried out).

Remove the three nuts securing the exhaust pipe flange to the exhaust manifold and lower system and withdraw from underneath the car.

On the TD Midget it will be sufficient to remove the exhaust pipe flange from the exhaust manifold, and undo the clip holding the exhaust pipe to the gearbox, allowing the exhaust pipe to be drawn clear of the manifold.

Carefully mark the propeller shaft flange and gearbox driving flange to ensure they are re-assembled in the same relative position, and uncouple the flanges. (Move car backwards and forwards to bring bolts into position for easy removal.)

Disconnect the reverse light switch cables on the gearbox.

Disconnect speedometer cable from the gearbox.

Detach the revolution counter drive from the rear end of the dynamo. (TD Midget only.)

Remove the engine breather pipe by slackening the clip securing it to the inspection cover and unscrewing the bolt securing it to the flywheel housing.

Disconnect the throttle control, mixture control, and fuel hose at the fuel pump end.

Disconnect the starter cable and dynamo cable and disconnect the oil pipe to the oil pressure gauge.

Remove the distributor cover and leads to the plugs (to prevent breakages.)

Disconnect the high tension lead from the coil, and remove steering column complete as detailed in Chapter VIII.

Unscrew the bolt securing the earth cable clip to the flywheel housing. (1¼ Litre Model.)

On the TD Midgets the earthing strip is between the engine breather pipe and the chassis.

Disconnect the front end of the clutch cable and remove cable from its stop bolted to the sump.

Remove the engine control link from its bracket on the chassis by withdrawing the split pins, removing the slotted nuts A & Q, removing the flat washer B and P, cups and rubbers C, D, O, and N, loosen the locknuts H and shorten adjuster J, when control links complete with inner rubbers and cups can be removed. (See Fig. 79.) This will need re-adjusting when the engine is replaced.

Remove the air cleaner and branch pipe and carburetter(s) and fuel pipe.

Place lifting sling around the engine.

Remove the two bolts, nuts and washers securing the front engine mounting and the split pin, nut, washers, and rubber bush under the rear engine mounting.

Lift the rear of the gearbox slightly and remove the clevis pin from the forked bolt. (See Fig. 78.)

Lift engine up and forward clear of the chassis.

Replacement is carried out in the reverse manner to that detailed for removal.

NOTE:—See page 112 for adjustment of engine mounting and pages 113 and 114 for replacement of engine control link.

ENGINE MOUNTING RUBBERS (FRONT). (TA and 1½ Litre Models.)

A point that may not be generally realised is the importance of the correct adjustment of the front engine mountings.

The seating rubbers with bolts secure the engine to the frame, and at the same time allowing flexibility, and the small rubber washers under the support brackets prevent the engine from lifting.

If the small rubber washers are adjusted too hard against the engine brackets, much of the general flexibility is prevented, therefore retaining plates should be fitted so that the rubber washer just makes contact with the support brackets and held in this position while the locknut is tightened.

ADJUSTMENT OF ENGINE MOUNTING RUBBERS (FRONT). (TC Midget Model.)

Adjustment of the front engine mounting is carried out as follows:—

The bolt is threaded through the engine plate into the rubber mounting block and is tightened in the normal way.

Thread on the small rubber block with its small diameter inserted in the hole in the frame bracket.

Fit the washer and sleeve nut which is screwed up through the rubber block so that a slight tension is exerted on the rubber.

Hold the sleeve nut in this position, fit the locknut and properly tighten.

NOTE:—Engine vibration is likely to be felt through the chassis if the stability rubber under the front engine mounting is compressed too much.

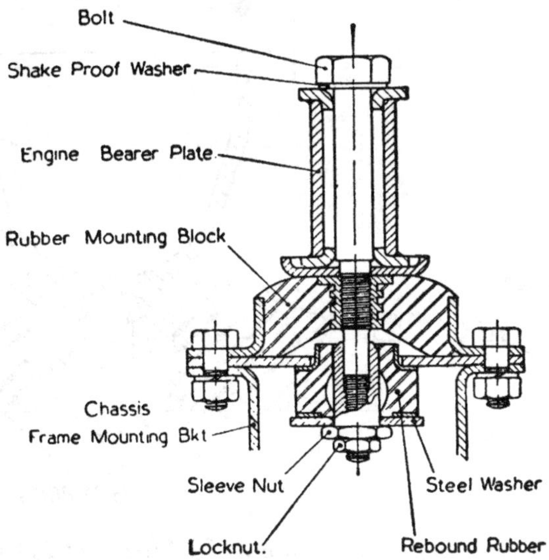

Fig. 77. Front engine mounting rubber. TC Midget.

ENGINE MOUNTINGS. (1¼ Litre and TD Midget Models.)

The power unit, complete with exhaust system, is flexibly mounted to the chassis frame.

A rubber block supports the front of the engine while the rear engine mounting consists of two rubber blocks, upon which the engine rests, and which fit in a cradle on the chassis frame. A rebound rubber is provided to limit the upward movement.

A control link is fitted at the front right hand side of the engine to restrict the torque reaction of the engine.

It is important when replacing the engine complete with gearbox that the rebound rubber is not over tightened. The slotted nut should be tightened only sufficiently to fit the split pin to the forked bolt.

On early 1¼ Litre engines a flat washer was fitted in place of the radius washer " B." The flat washer may be changed for the radiused type and at the same time a $\frac{1}{16}$ in. radius can be added to the edge of the hole in the cradle at the point "A." This will maintain the good condition of the rebound rubber.

The cradle bracket for the rear engine mounting rubbers is slotted where it is bolted to the chassis frame. The slots allow for adjustment so that the front and rear engine rubbers may assume a natural fore and aft position without strain. If the bracket has been disturbed the engine should be lifted slightly by a sling, with the cradle bolts loose and the front and rear mounting rubbers fixed. Allow the mounting rubbers to assume a free position, lower the engine and tighten the rear cradle bolts.

REFITTING THE ENGINE CONTROL LINK. (1¼ Litre and TD Midget Models.)

Fit the adjusting rod " K " with its rubber " M " and " N." cups " L " and " O," flat washer " P " and slotted nut " Q."

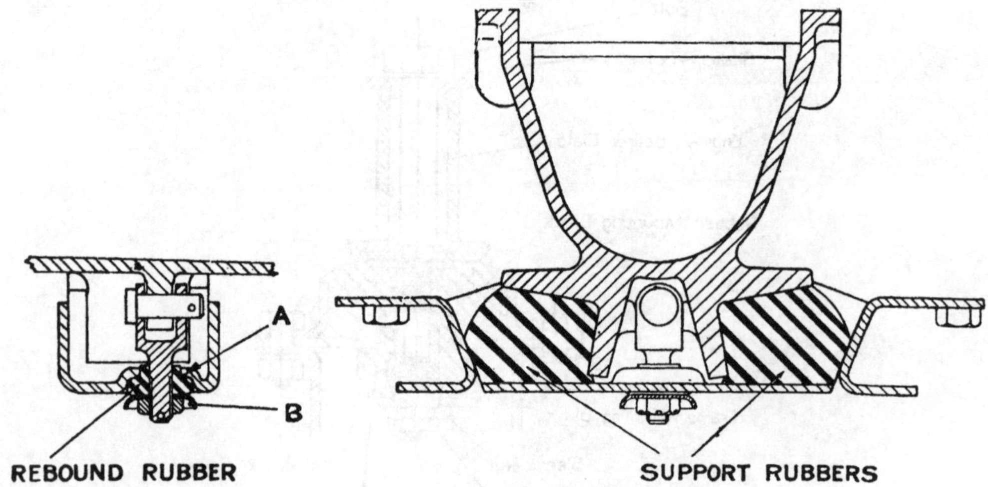

Fig. 78. Rear engine mounting. 1¼ *Litre and TD models.*

The slotted nut should be tightened only sufficiently to allow the split pin to enter.

Fit the outer adjusting nut " G," complete with the adjuster "J," lock-nuts " H " and the inner rubber and cup " E " and " F." Do not fit the outer rubber and cup " C " and " D " at this stage.

Leave the adjuster " J " short and the inner rubber " E " clear of the face of the frame bracket.

Rock the engine until it assumes a natural poise on its mounting rubbers.

To ensure that the engine is quite free on its mountings, it is advisable to disconnect the exhaust system at the manifold and the gearbox bracket clip and when the control link adjustment is set, refit the exhaust system.

If a knock is heard when slightly rocking the engine, suggesting a foul, this should be located and cleared.

It may be the gearbox touching its cover or the body foot ramp plate.

Lengthen the adjuster " J " until the inner rubber " E " is firmly against the frame bracket face, taking care to do this without disturbing the position of the engine.

Fit the outer rubber " D," cup " C," washer " B " and slotted nut "A." Tighten the nut only sufficiently to allow the split pin to enter.

NOTE:—The engine control link is only to control engine movement and must not support weight of the engine. It is important that the control link is adjusted as explained as this will ensure the minimum engine vibration.

REMOVAL OF TIMING CHAIN. (TA, TB, TC, TD, 1¼, 1½, 2, and 2.6 Litre Models.)

Drain the water system, remove bonnet complete and remove radiator.

Slacken off the dynamo securing and adjusting bolts and push dynamo towards engine and remove fan belt.

Remove engine control link as detailed under " Removal of Engine complete with gearbox," page 112, suitably marking position of adjuster so that it may be fitted at the same setting. (1¼ Litre and TD Midget only.)

Remove the set bolts securing the water pump to the crankcase, and withdraw water pump.

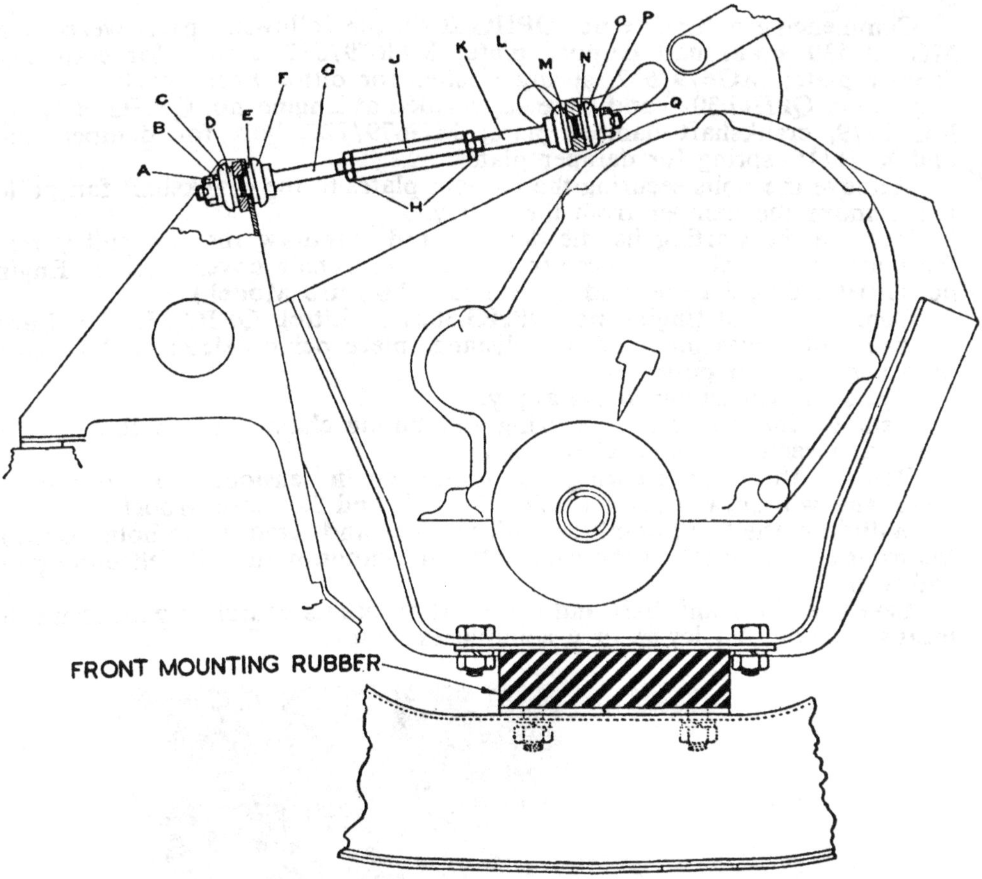

Fig. 79. *Engine Control link and front engine support rubber* : (1¼ *Litre and TD Midget*).

Remove the starting handle dog nut and withdraw crankshaft fan pulley from the end of the crankshaft. (TA, TB, TC, TD, 1¼, 1½ Litre Models only.)

Up to and including Chassis no. SA.1500 the starter dog nut formed part of the crankshaft damper. Remove the three bolts securing the crankshaft damper plate, springs, etc., to the crankshaft fan pulley and withdraw damper plates. Remove nut on end of crankshaft and withdraw fan pulley from shaft. Pulley is keyed on to the shaft so it is essential to ensure that the key is retained.

Commencing at Chassis no. SA.1501, the front crankshaft damper plate was modified, the starting handle dog nut being attached direct to the end of the crankshaft.

Removal instructions as above apply.

It may be of interest to record that several modifications have been introduced on the crankshaft damper plates.

Up to and including Engine no. QPHG.2028, 2-Litre Model, the following parts were used: MG679/154, crankshaft damper plate; X679/17, dowel for damper plate; MG679/156, spring for damper plate.

Commencing at Engine no. QPHG.2029, the following parts were used: MG679/530, crankshaft damper plate; MG679/531, dowel for crankshaft damper plate; MG679/532, spring washer for ditto. These parts ceased at Engine no. QPHG.3020 and were superseded at Engine no. QPHG.3021 by: MG827/9, crankshaft damper plate; MG679/227, bolt for damper plate and X151/15, spring for damper plate.

Remove the bolts securing the damper plates to the crankshaft fan pulley and remove the damper from the pulley.

Remove the starting handle dog nut and withdraw the fan pulley from the shaft and the distance piece from the timing chain cover. (Ceased Engine no. QPHG.3020, 2 Litre, and QPHG.751, 2.6 Litre Model.)

Commencing at Engine no. QPHG.3021 (2 Litre), QPJG.752 (2.6 Litre), the fan pulley was modified, the distance piece being deleted and replaced by a cone for fan pulley.

Removal instructions above apply.

Remove the set screws securing the timing chain case to the cylinder block and remove timing chain cover.

Remove the bolts securing the timing chain tensioner to the cylinder block and withdraw tensioner. (TC, TD, 1¼, and 2.6 Litre Models.)

Withdraw the front crankshaft oil thrower and remove the bolts securing the camshaft sprocket to the camshaft and withdraw the oil deflecting plate and washer.

Lever off the crankshaft and camshaft sprockets complete with chain, by means of short flat levers or a suitable extractor.

Fig. 80. *Damper assembly (up to Engine nos. QPHG* 3020 *and QPJG* 751).

REPLACEMENT OF TIMING CHAIN. (TA, TB, TC, TD, 1¼, 1½, 2, and 2.6 Litre Models.)

The two timing sprockets can only be fitted to the shaft in one position for they are secured to the crankshaft and camshaft by means of single keys.

Each of the sprockets has a tooth marked " T " and each timing chain has two white links.

Between the white links are 13 black links one side of the chain and 17 black links on the other. (TA Midget Model.)

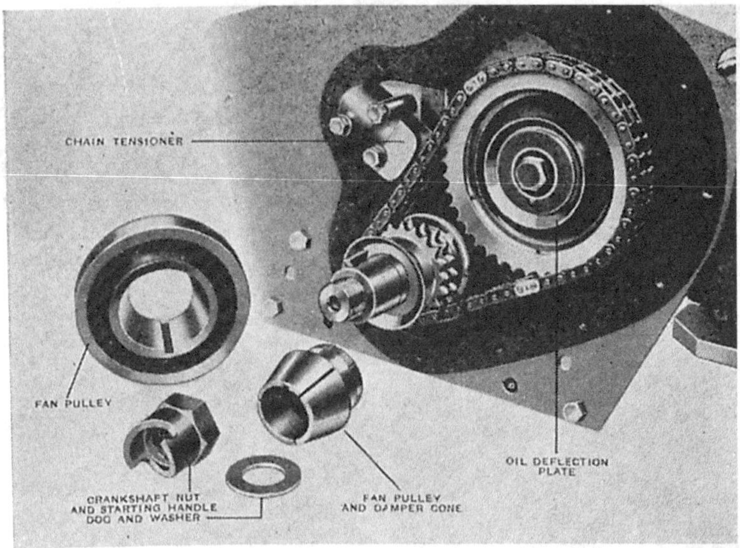

Fig. 81. *Method of mounting damper plates.* (*Cammencing at Engine nos QPHG* 3021 *and QPJG* 752).

Fig. 82. *Timing Chain (TA and TB Midget).*

Between the plated links there are 12 black links on one side of the chain, 15 black links on the other. (1½, 2 and 2.6 Litre Models.)

Fig. 83. *Timing Chain* (1½ *and* 2 *Litre models*).

Fig. 84. *Timing Chain* (2.6 *Litre model*).

Between the white links are 13 black links on one side of the chain and 15 black links on the other. (TB, TC, TD and 1¼ Litre Models.)

With the two white links forward and with the shorter portion of the chain to the left, engage the camshaft sprocket tooth marked "T" with the top white link and the crankshaft sprocket tooth marked "T" with the other white link.

Place the key way of the camshaft and crankshaft in a suitable position and push home the sprockets.

Replace the bolts securing the camshaft sprocket, replace the chain tensioner (where fitted), the crankshaft oil thrower and timing chain case.

Replace the crankshaft fan pulley, starter dog nut and crankshaft damper (where fitted).

Replace the water pump.

Replace the engine control link and readjust if necessary (TD and 1¼ Litre Models only).

Replace fan belt and readjust, refit radiator and bonnet and fill cooling system.

Fig. 85. Timing Chain TC, TD & 1¼ Litre models.

REMOVAL AND REPLACEMENT OF PISTON AND CONNECTING RODS. (ENGINE IN SITU.) (TA, TB, TC, TD, 1¼, 1½, 2 and 2.6 Litre Models.)

Place the car on a hoist or over a pit and drain the sump.

If a hoist or pit are not available, raise the front of the car on a suitable trestle.

Remove the exhaust system. (1¼ Litre and TD Midget Models.)

Remove the bolts securing the sump to the cylinder block and lower sump.

Remove the split pins from the big-end bolts and withdraw the connecting rod bearing cap.

Rotate the crankshaft slightly to give the necessary clearance and draw the piston and connecting rod down the right hand side of the engine.

NOTE:—It is advantageous to replace each bearing in its original position and mark each connecting rod to their corresponding bores.

Replacement is carried out in the reverse manner to that detailed for removal.

IMPORTANT:—The pistons must be replaced in the same bores and the same way round as when removed, that is, with the gudgeon pin pinch bolt towards the right hand side of the engine. The same connecting rod and cap must be fitted to the same crank pin journal from which it is removed.

REMOVING PISTONS FROM CONNECTING RODS. (TA, TB, TC, TD, 1¼, 1½, 2 and 2.6 Litre Models.)

The gudgeon pin is a two-thumb push fit in the piston, being locked in position by the small end clamp bolt, the diameter of which picks up the small groove in the gudgeon pin.

Insert a plug into each hollow end of the gudgeon pin and place in a vice.

Remove the clamp bolt, remove piston from vice, push out gudgeon pin, and separate piston from connecting rods.

Replacement is carried out in the reverse manner to that detailed for removal, and it is important that particular attention be paid to the following points.

1. Ensure that the clamp bolt will screw down readily in the small end of the connecting rod.
2. That the spring washer has sufficient tension to make it effective.
3. Ensure that the gudgeon pin is positioned in the piston and rod so that the groove formed in it is in line with threaded hole for clamp bolt in the connecting rod.
4. Ensure that the clamp bolt, which should be screwed in with the threaded hole for the clamp bolt as in (1) is tightened firmly with a box spanner.

REMOVAL AND REPLACEMENT OF CRANKSHAFT (ENGINE OUT OF CHASSIS). (TA, TB, TC, TD, 1¼, 1½, 2 and 2.6 Litre Models.)

Remove the gearbox and clutch assembly complete and remove starter motor.

Remove the sump, remove the bolts attaching the flywheel to the crankshaft and remove flywheel complete.

Remove the fan driving pulley, timing chain case, timing chain and sprockets as already detailed.

Remove the bolts securing the flywheel housing to the crank case (two having their heads inside the crank chamber), and withdraw flywheel housing. (TA, TB, 1½ and 2.6 Litre Models.)

Remove the pistons and connecting rods as detailed previously (page 119).

Remove the two nuts securing each main bearing cap and withdraw caps and bearings. It is advisable to mark each bearing cap and bearing to ensure correct position for replacement.

Remove crankshaft.

Replacement is carried out in the reverse manner to that detailed for removal, but before doing so thoroughly clean out the oilways in the crankshaft.

All bearings should be thoroughly cleaned before replacement and all parts should be oiled with clean oil.

On the TA Midget, 1½ Litre Models, up to and including Engine no. TPBG.1509, and 2 Litre engines with the oil immersed clutch, the main bearing studs on the left hand side of the engine, that is the camshaft side, should be removed and thoroughly cleaned, as the oil feed passes through the machined undercut on the studs.

REMOVAL AND REPLACEMENT OF CAMSHAFT (ENGINE IN SITU). (TA, TB, TC, TD, 1¼, 1½, 2 and 2.6 Litre Models.)

Remove air cleaner and cylinder head cover and unbolt the breather pipe from the tappet inspection cover.

Remove the rocker gear from the cylinder head and withdraw the push rods (TB, TC, TD and 1¼ Litre Models.)

Using special tool No. T59, gently force rocker to one side and withdraw push rods. (TA, 1½, 2 and 2.6 Litre Models.)

Now remove the dynamo and its support bracket, slacken off the distributor locking screws and withdraw the distributor, and remove the bolts securing the oil pump to the crankcase and withdraw the oil pump.

Remove the tappet inspection cover and lift out the tappets.

In the case of the TB, TC, TD Midgets and 1¼ Litre Saloon, the tappets are located directly in the cylinder block and are easily withdrawn.

In the case of the TA Midget 1½, 2 and 2.6 Litre Models, unbolt the tappet blocks and prise off their dowels. The tappets are restrained from falling out of the tappet blocks by circlips fitted around their upper ends, and will come away with the tappet blocks.

Remove the sump, timing chain case, timing chain and sprockets as previously detailed.

Remove the dowel screw(s) securing the centre camshaft bearings to the engine body and remove the front thrust plate.

Remove the camshaft by drawing it forward from its rear bearing and through the front bearing taking the centre bearing(s) with it. The bearings have grooves turned round their ends, by which they may be conveniently tied or wired to prevent their falling when free of their housings.

When the camshaft has been withdrawn sufficiently to bring the centre bearing(s) free of the housing, they may be untied and lifted out, and the withdrawal of the camshaft completed.

Replacement is carried out in the reverse manner to that detailed for removal, not forgetting to replace and re-wire the dowel bearing screws. When the shaft is in place the bearing(s) may be moved about until their locating holes are opposite the locking screw holes, using a tool such as a blunt-ended scriber to feel that the locking screw will enter directly.

WATER PUMP. (TA, TB, TC, TD, 1¼, 1½, 2 and 2.6 Litre Models.)

The pump is fitted to the front face of the cylinder block and driven by a belt from the crankshaft pulley, this belt also driving the dynamo. A special carbon gland ring is employed which ensures an efficient water seal and eliminating the necessity of lubrication and adjustment.

With the exception of the TA Series Midget the unit complete can be withdrawn from the block with the radiator *in situ*. Should the necessity arise, proceed as follows:—

Drain the radiator and remove radiator assembly complete, as detailed on page 102. (TA Midget only.)

Withdraw the set bolts securing the fan blades to the fan boss, remove the fan blades, slacken off the dynamo and remove the fan belt. Remove the water pump set bolts (or nuts from water pump studs where fitted) and remove water pump by breaking the joint between the impeller housing and the block.

NOTE: As one end of the engine control link is attached to the engine by one of the water pump bolts, it may be necessary to move the inner control link bracket outwards to clear. (1¼ Litre Series Y and YB and TD Midget.)

To dismantle the pump, tap out the taper pin attaching the impeller to the shaft, taking care to see that the pin is knocked out in the right direction. Withdraw the pressure spring and washer, which will give access to the carbon seal and gland washer assembly. Ensure that no damage is done to the carbon ring, this being relatively brittle, the working face of the rubber seal, and do not lose the driving pin for the carbon gland, which is a loose fit in the shaft.

The water pump spindle is carried on two ball races, which should be trouble free, unless they are badly neglected.

To gain access to the bearings, release the impeller and gland, remove the gland drive pin, remove securing nut in the centre of the drive pulley, withdraw the pulley, pulley key, felt sealing ring and retainer. With a pair of long nosed pliers, engage the end lugs of the outer bearing circlip and withdraw circlip.

To remove the outer bearing tap the inner end of the spindle on a piece of wood, and withdraw the distance piece between the bearings.

NOTE: A little paraffin poured into the impeller body around the outer bearing will facilitate removal.

Contract the lugs on the inner circlip securing the remaining bearing and withdraw the bearing and impeller spindle.

Check the felt oil sealing rings and renew if worn. Check the bearing for slackness and renew if necessary. Examine the carbon sealing ring for excessive wear or cracks and renew if necessary. Examine the face of the brass sealing washer for flatness and remove all burrs from the edges, to prevent possibility of damage to the synthetic rubber seal. Fit a new seal if there is any doubt as to the condition of the one removed.

Reassembly is carried out in the reverse manner to that detailed for removal, taking care that the flange jointing washer is in good condition. Partially fill the space between the ball races with grease and liberally soak the felt washers in engine oil or grease before replacement.

Do not overtighten the slotted nut retaining the pulley. If it is just tight it will be satisfactory.

Commencing at engine numbers XPAG/TD/6482 and XPAG/SC/16463. TD and 1¼ Litre respectively, a new water pump with an improved type of water seal was fitted. Except for the gland, the pump is structurally similar and the method of dismantling and reassembling is the same. The sealing for the new seal is against the boss of the impeller vane instead of the pump body.

ENGINE MODIFICATIONS. (1½ Litre Series VA.)

Oil pressure Control Valve.—Reference to Fig. 48, the oil pressure control valve was changed to a slightly different type at Engine no. TPBG.1510. The valve operates exactly the same and is non-adjustable. The correct oil pressure is between 60/80 lbs. per sq. in. at all speeds over 1,500 RPM.

Oil Filter.—Commencing at Engine no. TPBG.1510, the oil filter was slightly modified.

The filter was positioned approximately 3 in. higher allowing the filter body to be connected direct to the oil gallery, provision for the oil outlet from the filter being actually made in the body deleting the copper pipe.

Crankshaft.—A new crankshaft, which differed only in regard to the flywheel flange, was introduced at Engine no. TPBG.1510. The change being due to the fitting of the dry clutch in place of the oil immersed clutch.

The flange is modified to receive an oil thrower to prevent oil from leaking into the clutch pit. Into this thrower the spigot race of the gearbox first motion shaft is fitted.

The correct clearance between oil return threads of the oil thrower and flywheel housing is: .003 in./.006 in.

The flywheel is secured over the oil thrower and both flywheel and oil thrower are secured to the flywheel by six bolts which are locked by wire through the heads.

To remove the flywheel oil thrower, remove flywheel and flywheel housing.

Valve and Ignition Timing.—In addition to the normal method of obtaining T.D.C. (Fig. 57) the flywheel has been marked 1 and 4, and a scribed line which must line up with a pointer fixed to the clutch housing, seen through the clutch inspection cover, indicating T.D.C., of cylinders Nos. 1 and 4.

In addition there are also six scribed marks either side of T.D.C. marks indicating degrees.

This modification was introduced at Engine no. TPBG.1510.

ENGINE MODIFICATIONS. (TD Midget, 1¼ Litre.)

Pistons.—Commencing at Engine No. XPAG/SC/15576 for the 1¼ Litre Model pistons of a slightly modified type were fitted. The earlier type of piston had an oval ground skirt for the full length, but the later type of piston has an oval ground skirt gradually running into a nearly circular form at the extreme bottom edge producing what may be termed an oil scraping effect; and giving only .005 in. ovality when measured at the extreme bottom edge of the skirt. A further modification was the reducing of the top lands to 2.601 in./2.599 in. diameter.

The piston skirt is oval and tapers from top to bottom, the measurement at the bottom of the skirt taken across the thrust faces, i.e., 90° away from the gudgeon pin centre line being .00125 in. diameter larger than at the top of the skirt.

Engine Sump Pick Up.—Commencing at Engine No. XPAG/SC/16729 for the 1¼ Litre and Engine No. XPAG/TD/7576 for the TD Midget, the position of the oil suction filter in the sump was removed to a central position, with a view to giving a more effective oil pick up.

Purolator Oil Filter.—Commencing at Engine No. XPAG/SC/17293 on the 1¼ Litre Engine No. XPAG/TD2/14224 on the TD Midget a combined oil pump and Purolator oil filter was fitted. This type of pump and oil filter cannot be fitted to engines previous to the above Engine Numbers due to a change in the cylinder block itself.

The oil filter body is attached to the oil pump by a retaining bolt through the centre of the filter. By unscrewing the bolt the filter complete can be removed from the pump body, the filter cleaned and the element renewed.

It is recommended that the filter element be replaced every 6,000 miles.

Before replacing the filter check the condition of the rubber washer located in the filter location in the pump body, and if there is any doubt as to its condition, it should be replaced otherwise oil leaks could arise at these points.

from the 'M'
MG
to the 'TF'

Oil Filter.—On engines previous to Engine Nos. XPAG/TD/2985, TD Midget and XPAG/SC/15405 1¼ Litre Saloon, Purolator filter type PTX 346, Part No. 24475 can be used in place of the Wilmot Breeden oil filter, MG.862/39, providing that the narrow straps part Nos. MG.862/394 and MG.862/395 for securing the filter bottle are used.

Sumps.—Commencing at Engine Nos. XPAG/TD2/14948 and XPAG/SC2/17383 TD Midget and 1¼ Litre Saloons respectively, a larger sump was introduced, the sump capacity being increased to 10½ pints.

Rocker Gear.—Revised rocker gear was introduced at Engine Nos. XPAG/TD/9008 TD Midget and XPAG/SC/16831 1¼ Litre Saloon. The long spacer spring fitted in the centre of the rocker shaft has been deleted. The two medium spacer springs originally fitted in the Intermediate position have been reduced to one, this now being fitted in the central position. The two short springs fitted front and rear have been increased in quantity to four off, being now fitted front, rear, and intermediate. The washer for the rocker shaft has been increased to 10 off, these washers being fitted at the end of each spring, interposed between spring and rocker: Rockers Numbers 4 and 8 and 1 and 5 have been modified to suit.

Camshaft.—At Engine No. XPAG/SC/16831 1¼ Litre Series YB, a modified camshaft was fitted giving valve timing as follows:

Inlet Valve Opens	5° B.T.D.C.
Inlet Valve Closes	45° A.B.D.C.
Exhaust Valve Opens	45° B.B.D.C.
Exhaust Valve Closes	5° A.T.D.C.
Valve Lift 8 m/m.	

Push Rods.—Commencing at Engine No. XPAG/TD2/17289 and XPAG/SC2/17432 on the TD and 1¼ Litre respectively, shorter push rods in conjunction with longer rocker adjusting screws were introduced. This does not affect the tappet adjustment.

Cylinder Block, Head and Gasket.—The TD Midget Mark II is now to be fitted with a cylinder block and head with revised water circulating holes.

The standard production engines will, in due course, be modified accordingly.

Great care must be taken therefore with regard to which cylinder head gasket is used.

The early standard gasket has pear shaped water holes and a small rear water hole. Commencing at Engine No. XPAG/TD2/17969 and Engine No. XPAG/SC2/17463 TD Midget and 1¼ Litre Saloon respectively, the new type cylinder block only has been fitted. This necessitates the use of a cylinder head gasket with pear shaped holes and a large rear water hole.

If a new type block is fitted to an old type cylinder head, or a new type cylinder head is fitted to an old type block the gasket with the pear shaped water holes and large rear water hole must be used.

When the new type cylinder head is used in conjunction with the new type cylinder block, the latest gasket with round water holes and a large rear water hole with reinforced bar across to centre, must be used; it is essential therefore that Engine Numbers must be clearly stated when ordering replacement cylinder head gaskets.

Oil Pump.—With a view to improving accessibility of pump priming for use after engine draining, a priming plug has been incorporated in the oil pump cover. This modification was introduced at Engine Nos. XPAG/SC2/17670 and XPAG/TD2/20972 1¼ and TD respectively.

The oil pump cover has been modified to take the plug.

Flywheel Assemblies.—Up to and including Engine No. XPAG/SC/14022 a flywheel assembly with 93 teeth on the starter ring was used, in conjunction with a large starter motor with 10 teeth on the Bendix Pinion.

Commencing at Engine No. XPAG/SC/14023 a modified type of flywheel assembly was fitted incorporating 120 teeth on the starter ring and used in conjunction with a starter motor with 9 teeth on the Bendix Pinion.

The starter motors and flywheel assemblies are not interchangeable.

Commencing at Engine No. XPAG/SC/16916 an 8 in. diameter clutch was introduced, necessitating a further alteration to the flywheel to accommodate the larger diameter clutch, the starter ring remaining unchanged.

LOCATING ENGINE TROUBLES

ENGINE WILL NOT FIRE

Symptom: **The starter will not turn the engine**

Cause : 1. The battery is discharged or defective.
„ 2. Disconnected or broken lead.
„ 3. Faulty starter switch.
„ 4. Faulty starter motor.
„ 5. Starter cables shorting to earth.
„ 6. Battery terminals badly corroded or battery leads loose.

Symptom: **The starter turns engine very slowly**

Cause : 1. Partly discharged battery.
„ 2. Loose terminals or connections.
„ 3. Dirty or corroded connections
„ 4. Faulty insulation on starter cables.
„ 5. Tightness in engine.

Symptom: **The starter turns engine smartly, but engine will not fire**

Cause : 1. Plugs not sparking.
„ 2. Spark at coil ; if coil gives good spark, check :—
(*a*) Plug gaps too wide or close.
(*b*) Plugs oiled up.
(*c*) Plug insulators damaged or excessively dirty.
„ 3. Spark at coil ; if coil gives poor spark, check :—
(*a*) Leads from coil to distributor loose or corroded.
(*b*) Distributor points dirty, worn or out of adjustment.
(*c*) Carbon brush not making contact.
(*d*) Rotor cracked.
(*e*) Faulty coil or condenser (substitute replacement known to be in order).

Cause : 4. Check carburetter for fuel supply; if no fuel in float-chamber, check :—
(*a*) Functioning of the fuel pump.
(*b*) Air-leak in pipe line, indicated by rapid action of the fuel pump.
(*c*) Float-chamber needle sticking.
„ 5. If fuel is reaching float-chamber, check :—
(*a*) For choked jet.
(*b*) Water in the fuel.
(*c*) Dirt in the carburetter.
(*d*) Air-leak in the induction system.
(*e*) Adjustment of carburetter controls.

THE ENGINE STARTS BUT RUNS ERRATICALLY

Cause : 1. Check the following ignition points :—
- (*a*) Loose high-tension leads to sparking plugs, or corroded connections.
- (*b*) Incorrect setting of plug points.
- (*c*) Damaged plugs or moisture on plugs.
- (*d*) Loose connections on battery or in ignition circuit.
- (*e*) Faulty high-tension leads.
- (*f*) Battery charge low.
- (*g*) Battery connections faulty.
- (*h*) Defective contact breaker.
- (*i*) Defective distributor.

" 2. Check the following carburetter points :—
- (*a*) Water in float-chamber.
- (*b*) Choked filter in carburetter or fuel pump, indicated by slow pumping of the fuel pump.
- (*c*) For faulty fuel pump, if action is sluggish.
- (*d*) Jets partially choked.
- (*e*) Carburetter set too rich, indicated by sooty exhaust.
- (*f*) Fuel cap filler vent choked.
- (*g*) Obstruction in fuel feed pipe lines.
- (*h*) Air-leaks in induction system.

" 3. Check the following mechanical points :—
- (*a*) Sticking valves.
- (*b*) Incorrect valve clearance.
- (*c*) Burnt or broken valves.
- (*d*) Broken or weak valve springs.
- (*e*) Incorrect valve timing.
- (*f*) Incorrect ignition timing.
- (*g*) Valve guides worn, causing air-leaks.
- (*h*) Faulty cylinder head gasket.
- (*i*) Back-pressure due to damaged exhaust system.

THE ENGINE STARTS AND STOPS

Cause : 1. Check the following ignition points :—
- (*a*) Loose low-tension leads.
- (*b*) Loose distributor clamp screw.
- (*c*) Faulty ignition switch contact.

" 2. Check the following carburetter points :—
- (*a*) Incorrect setting of carburetter controls.
- (*b*) Blocked fuel line.
- (*c*) Water in float-chamber.
- (*d*) Sticking needle valve.
- (*e*) Fuel pump failing to pump regularly.
- (*f*) Air-leaks in fuel line.
- (*g*) Fuel level low in tank.

ENGINE WILL NOT IDLE OR RUN SLOWLY

Cause : 1. Check the following carburetter points :—
- (*a*) Incorrectly set slow-running adjustment.
- (*b*) Incorrectly set throttle stop screw.
- (*c*) Incorrectly set throttle controls.
- (*d*) Weak mixture or over-rich mixture.
- (*e*) Faulty functioning of fuel pump.

" 2. Check the following mechanical points :—
- (*a*) Sticking valves.
- (*b*) Incorrect valve tappet clearance.

ENGINE WILL NOT IDLE OR RUN SLOWLY *(contd.)*

(*c*) Air-leak in induction system.
(*d*) Burnt or broken valve, indicated by low compression.
(*e*) Broken valve spring.
(*f*) Damaged cylinder head or gasket.

Cause : 3. Check the following ignition points :—
(*a*) Loose high-tension leads.
(*b*) Incorrect setting of plug points.
(*c*) Damaged plugs or moisture on plugs.
(*d*) Loose connections on battery or in ignition circuit.
(*e*) Faulty high-tension leads.
(*f*) Battery charge low.
(*g*) Battery connections loose.
(*h*) Defective contact breaker or burnt points.
(*i*) Defective distributor.

ENGINE FAILS TO GIVE FULL POWER

Cause : 1. Check the following carburetter points :—
(*a*) Faulty or insufficient fuel supply.
(*b*) Air-leaks in induction or fuel pipes.
(*c*) Partly choked jets.

„ 2. Check the following mechanical points :—
(*a*) Incorrect valve tappet clearance.
(*b*) Burnt or badly seating valves.
(*c*) Cylinder head stud nuts not tight.
(*d*) Damaged cylinder head gasket.
(*e*) Valve timing not correct.
(*f*) Broken or weak valve springs.
(*g*) Excessive carbon deposit.
(*h*) Excessively worn pistons and cylinders (no compression).

„ 3. Check the following ignition points :—
(*a*) Ignition retarded too far.
(*b*) High-tension leads shorting or loose.
(*c*) Dirty sparking plugs.
(*d*) Sparking plug points incorrectly set.
(*e*) Contact breaker points incorrectly set.
(*f*) Contact breaker points pitted.
(*g*) Faulty coil.
(*h*) Faulty condenser.
(*i*) Low-tension connections or leads faulty.
(*j*) Battery run down or faulty.

ENGINE KNOCKS

Cause : 1. Check the following :—
(*a*) Ignition timing too far advanced.
(*b*) Excessive carbon deposit.
(*c*) Fuel unsuitable or weak mixture.
(*d*) Loose or worn bearings or pistons.
(*e*) Camshaft bearings loose or worn.
(*f*) Automatic timing contact mechanism defective.
(*g*) Defective or unsuitable plugs.
(*h*) Valve timing or tappet clearance incorrect.

ENGINE BACKFIRES

Cause : 1. Check the following ignition points :—
(*a*) High-tension cables defective or connections loose.
(*b*) High-tension leads incorrectly fitted.

(*c*) Low-tension leads defective or connections loose.
(*d*) Switch contact faulty.
(*e*) Distributor gaps incorrect or points pitted or dirty.
(*f*) Contact breaker arm sticking or defective.
(*g*) Distributor cover cracked or loose.
(*h*) Distributor not correctly timed.
(*i*) Rotor carbon brush pick-up defective or worn.
(*j*) Clearance between rotor arm and distributor stud excessive.
(*k*) Coil defective or wet.
(*l*) Defective condenser.
(*m*) Plugs overheated, unsuitable, or points incorrectly set.

Cause : 2. Check the following carburation points:—
(*a*) Jets choked or restricted.
(*b*) Jets incorrectly set, causing weak mixture.
(*c*) Water in fuel.
(*d*) Choked fuel filters.
(*e*) Inlet manifold joints leaking or manifold cracked.
(*f*) Air cleaner passages choked.
(*g*) Engine running temperature too cold.
(*h*) Throttle not closing completely. Indicated by the engine backfiring when proceeding downhill with throttle shut.

" 3. Check the following mechanical points :—
(*a*) Valve timing incorrect.
(*b*) Valve tappet clearance incorrectly set.
(*c*) Valve sticking.
(*d*) Valve seats pitted or faulty.
(*e*) Valve springs weak or broken.
(*f*) Valve excessively worn, causing air-leaks.
(*g*) Timing chain excessively loose.
(*h*) Excessive carbon deposit.

BEARING MEASUREMENTS, (STANDARD.)

Type	Crankpin Bearing Dia	Crankpin Bearing Width	Crankpin Bearing Radius	Front Main Bearing Dia	Front Main Bearing Width	Front Main Bearing Radius	Inter Bearing Front Dia	Inter Bearing Front Width	Inter Bearing Front Radius	Inter Bearing Rear Dia	Inter Bearing Rear Width	Inter Bearing Rear Radius	Rear Main Bearing Dia	Rear Main Bearing Width	Rear Main Bearing Radius
M	1·5015" MAX. 1·501" MIN.	1·12 +·000" −·001"	·13" MAX. ·12" MIN.	BALL BEARING			—	NIL	—	—	NIL	—	1 3/4" +·0015" −·001"	1 13/16" ±·015"	1/32"
C	1 5/8" +·0015" +·001"	1 1/8" +·000" −·001"	·13" MAX. ·12" MIN.	BALL BEARING			—	NIL	—	—	NIL	—	1 7/8" +·0015" +·001"	1 13/16" ±·015"	1/32" MAX
D	1·5015" MAX 1·501" MIN.	1·12" +·000" −·001"	·13" MAX. ·12" MIN.	BALL BEARING			—	NIL	—	—	NIL	—	1 3/4" +·0015" +·001"	1 13/16" ±·015"	1/32"
F	1·5015" MAX 1·501" MIN.	1·12" +·000" −·001"	1/8" x 45° CHAM.	BALL BEARING			1·501" 1·5014"	1 1/2" ±·01"	1/32" MAX 1/64" MIN.	1·501" 1·5014"	1 1/2" ±·01"	1/32" MAX 1/64" MIN.	1 3/4" +·0015" +·001"	1 13/16" ±·015"	1/32"
J	1·5015" MAX. 1·501" MIN.	1·12" +·000" −·001"	·13" MAX. ·12" MIN.	BALL BEARING			—	NIL	—	—	NIL	—	1 3/4" +·0015 +·001	1 13/16" ±·015"	1/32"
K	1 5/8" +·0015" +·001"	1·12" +·000" −·001"	·13" MAX. ·12" MIN.	1·625" +·0011" +·0004"	1 13/32" +·000" −·05"	5/64" MAX. 1/16" MIN.	1·6264" 1·626"	1 1/2" ±·01"	NIL	1·6264" 1·626"	1 1/2" ±·010"	NIL	1 7/8" +·0015" +·001"	1 13/16" ±·015"	1/32" MAX.
L	1 5/8" +·0015" +·001"	1·12" +·000" −·001"	·13" MAX. ·12" MIN.	1·625" +·0011" +·0004"	1 13/32" +·000" −·05"	5/64" MAX. 1/16" MIN.	1·6264" 1·626"	1 1/2" ±·01"	NIL	1·6264" 1·626"	1 1/2" ±·010"	NIL	1 7/8" +·0015" +·001"	1 13/16" ±·015"	1/32" MAX.
NA	1 5/8" +·0015" +·001"	1·12" +·000" −·001"	·13" MAX. ·12" MIN.	1·625" +·0011" +·0004"	1 13/32" +·000" −·05"	5/64" MAX 1/16" MIN.	1·6264" 1·626"	1 1/2" ±·010"	NIL	1·6264" 1·626"	1 1/2" ±·010"	NIL	1 7/8" +·0015" +·001"	1 13/16" ±·015	1/32"
NE	1 5/8" +·0015" +·001"	1·12" +·000" −·001"	·13" MAX. ·12" MIN	1·625" +·0011" +·0004"	1 13/32" +·000" −·05"	5/64" MAX. 1/16" MIN.	1·6264" 1·626"	1 1/2" ±·010"	NIL	1·6264" 1·626"	1 1/2" ±·010"	NIL	1 7/8" +·0015" +·001"	1 13/16" ±·015"	1/32"
PA	1 5/8" +·0015" +·001"	1·12" ±·000" ·001"	·13" MAX. ·12" MIN.	1·625" +·0011" −·0004"	1 13/32" +·000" −·05"	5/64" MAX. 1/16" MIN.	1·6264" 1·626"	1 1/2" ±·010"	NIL	—	NIL	—	1 7/8" +·0015" +·001"	1 13/16" ±·015"	1/32"
PB	1 5/8" +·0015" +·001"	1·12" ±·000" ·001"	·13" MAX. ·12" MIN.	1·625" ±·0011" ·0004"	1 13/32" +·000" −·05"	5/64" MAX. 1/16" MIN.	1·6264" 1·626"	1 1/2" ±·010"	NIL	—	NIL	—	1 7/8" +·0015" +·001"	1 13/16" ±·015"	1/32"
QA	1 5/8" +·0015" +·001"	1·12" ±·000" ·001"	·13" MAX. ·12" MIN.	1·625" ±·0011" ·0004"	1 13/32" +·000" −·05"	5/64" MAX. 1/16" MIN.	1·6264" 1·626"	1 1/2" ±·010"	NIL	—	NIL	—	1 7/8" +·0015" +·001"	1 13/16" ±·015"	1/32"
RA	1 5/8" +·0015" +·001"	1·12" ±·000" ·001"	·13" MAX. ·12" MIN.	1·625" ±·0011" ·0014"	1 13/32" +·000" −·05"	5/64" MAX. 1/16" MIN.	1·6264" 1·626"	1 1/2" ±·010"	NIL	—	NIL	—	1 7/8" +·0015" +·001"	1 13/16" ±·015"	1/32"
TA	45 M/M +·025 +·005	28 M/M −·07 −·085	2·5 x 2·5 M/M CHAMF.	52 M/M +·025 +·005	35 M/M ±·05	2·5 x 2·5 M/M CHAMF.	52 M/M +·025 +·005	38 M/M −·05 −·075	2·5 x 2·5 M/M CHAMF.	—	NIL	—	52 M/M +·025 +·005	35 M/M ±·05	2·5 x 2·5 M/M CHAMF.
TB TC	THICK-NESS 1·825 M/M 1·831 M/M	22·46 M/M 22·21	CHAMF. ON CON. ROD	52·020 M/M 52·005 M/M	35·12 M/M 34·88	·15 M/M	52·020 M/M 52·005	37·955 M/M 37·925	UNDERCUT ·25/·20 2·75 RAD.	—	NIL	—	52·020 M/M 52·005	35·12 M/M 34·88	·15 M/M
VA	48 M/M +·025 +·005	30 M/M −·07 −·095	2·5 x 2·5 CHAMF.	55 M/M +·025 +·005	38 M/M ±·05	2·5 M/M x 2·5 M/M CHAMF.	55 M/M +·025 +·005	41 M/M −·05 −·075	2·5 M/M x 2·5 M/M CHAMF.	—	NIL	—	55 M/M +·025 +·005	38 M/M ±·05	2·5 M/M x 2·5 M/M CHAMF.
SA	48 M/M +·025 +·005	30 M/M −·07 −·095	2·5 x 2·5 CHAMF.	55 M/M +·025 +·025	38 M/M ±·05	2·5 M/M x 2·5 M/M CHAMF.	55 M/M +·025 +·005	38 M/M ±·05	2·5 M/M x 2·5 M/M CHAMF.	55 M/M +·025 +·005	38 M/M ±·05	2·5 M/M x 2·5 M/M CHAMF.	55 M/M +·025 +·005	41 M/M ±·05	2·5 M/M x 2·5 M/M CHAMF.
WA	THICK-NESS 1·831 M/M 1·825 M/M	24·46 M/M 22·21	CHAMF. ON CON. ROD	55 M/M +·025 +·005	38 M/M ±·05	2·5 M/M x 2·5 M/M CHAMF.	55 M/M +·025 +·005	38 M/M ±·05	2·5 M/M x 2·5 M/M CHAMF.	55 M/M +·025 +·005	38 M/M ±·05	2·5 M/M x 2·5 M/M CHAMF.	55 M/M +·025 +·005	41 M/M ±·05	2·5 M/M x 2·5 M/M CHAMF.

CRANKSHAFT MEASUREMENT. (STANDARD.)

	Crankpin Journal			Front Main Journal			Inter. Journal Front			Inter Journal Rear			Rear Main Journal		
Type	Dia.	Width	Radius	Dia.	Width	Radius	Dia.	Width	Radius	Dia.	Width	Radius	Dia.	Width	Radius
M	$1.5''^{-.0005}_{-.0011}$	$1\frac{1}{8}''^{+.002}_{-.0005}$	·11″ MAX. ·10″ MIN.	BALL BEARING PX		110/25	——	NIL	——	——	NIL	——	$1\frac{3}{4}''^{-.00095}_{-.0011}$	$1\frac{29}{32}''^{+.005}_{-}$	$\frac{1}{16}''$
C	$1\frac{5}{8}''^{-.0005}_{-.0011}$	$1\frac{1}{8}''^{+.002}_{-.0005}$	·11″ MAX. ·10″ MIN.	BALL BEARING PX		110/99	——	NIL	——	——	NIL	——	$1\frac{7}{8}''^{-.0009}_{-.0011}$	$1\frac{29}{32}''^{+.005}_{-}$	$\frac{1}{16}''$
D	$1.5''^{-.0005}_{-.0011}$	$1\frac{1}{8}''^{+.002}_{-.0005}$	·11″ MAX. ·10″ MIN.	BALL BEARING PX		239/107	——	NIL	——	——	NIL	——	$1\frac{3}{4}''^{-.00095}_{-.0011}$	$1\frac{29}{32}''^{+.005}_{-}$	$\frac{1}{16}''$
F	$1.5''^{-.0005}_{-.0011}$	$1\frac{1}{8}''^{+.002}_{-.0005}$	·11″ MAX. ·10″ MIN.	BALL BEARING PX		110/25	$1.5''^{-.0005}_{-.0011}$	$1\frac{3}{4}''^{+.010}_{-}$	·11″ MAX. ·10″ MIN.	$1.5''^{-.0005}_{-.0011}$	$1\frac{3}{4}''^{+.010}_{-}$	·11″ MAX. ·10″ MIN.	$1\frac{3}{4}''^{-.00095}_{-.0011}$	$1\frac{29}{32}''^{+.005}_{-}$	$\frac{1}{16}''$
J	$1.5''^{-.0005}_{-.0011}$	$1\frac{1}{8}''^{+.002}_{-.0005}$	·11″ MAX. ·10″ MIN.	BALL BEARING P		239/107	——	NIL	——	——	NIL	——	$1\frac{3}{4}''^{-.00095}_{-.0011}$	$1\frac{29}{32}''^{+.005}_{-}$	$\frac{1}{16}''$
K	$1.625''^{-.0005}_{-.0011}$	$1\frac{1}{8}''^{+.002}_{-.0005}$	·11″ MAX. ·10″ MIN.	$1.625''^{-.0005}_{-.0011}$	$1\frac{33}{64}''^{-.000}_{+.002}$	$\frac{1}{16}''$ MAX. $\frac{3}{64}''$ MIN.	$1.625''^{-.0005}_{-.0011}$	$1\frac{3}{4}''^{+.010}_{-}$	·11″ MAX. ·10″ MIN.	$1.625''^{-.0005}_{-.0011}$	$1\frac{3}{4}''^{+.010}_{-}$	·11″ MAX. ·10″ MIN.	$1\frac{7}{8}''^{-.00095}_{-.0011}$	$1\frac{29}{32}''^{+.005}_{-}$	$\frac{1}{16}''$
L	$1.625''^{-.0005}_{-.0011}$	$1\frac{1}{8}''^{+.002}_{-.0005}$	·11″ MAX. ·10″ MIN.	$1.625''^{-.0005}_{-.0011}$	$1\frac{33}{64}''^{-.000}_{+.002}$	$\frac{1}{16}''$ MAX. $\frac{3}{64}''$ MIN.	$1.625''^{-.0005}_{-.0011}$	$1\frac{3}{4}''^{+.010}_{-}$	·11″ MAX. ·10″ MIN.	$1.625''^{-.0005}_{-.0011}$	$1\frac{3}{4}''^{+.010}_{-}$	·11″ MAX. ·10″ MIN.	$1\frac{7}{8}''^{-.00095}_{-.0011}$	$1\frac{29}{32}''^{+.005}_{-}$	$\frac{1}{16}''$
NA	$1.625''^{-.0005}_{-.0011}$	$1\frac{1}{8}''^{+.002}_{-.0005}$	·11″ MAX. ·10″ MIN.	$1.625''^{-.0005}_{-.0011}$	$1\frac{33}{64}''^{-.000}_{+.002}$	$\frac{1}{16}''$ MAX. $\frac{3}{64}''$ MIN.	$1.625''^{-.0005}_{-.0011}$	$1\frac{3}{4}''^{+.010}_{-}$	·11″ MAX. ·10″ MIN.	$1.625''^{-.0005}_{-.0011}$	$1\frac{3}{4}''^{+.010}_{-}$	·11″ MAX. ·10″ MIN.	$1\frac{7}{8}''^{-.00095}_{-.0011}$	$1\frac{29}{32}''^{+.005}_{-}$	$\frac{1}{16}''$
NE	$1.625''^{-.0005}_{-.0011}$	$1\frac{1}{8}''^{+.002}_{-.0005}$	·11″ MAX. ·10″ MIN.	$1.625''^{-.0005}_{-.0011}$	$1\frac{33}{64}''^{-.000}_{+.002}$	$\frac{1}{16}''$ MAX. $\frac{3}{64}''$ MIN.	$1.625''^{-.0005}_{-.0011}$	$1\frac{3}{4}''^{+.010}_{-}$	·11″ MAX. ·10″ MIN.	$1.625''^{-.0005}_{-.0011}$	$1\frac{3}{4}''^{+.010}_{-}$	·11″ MAX. ·10″ MIN.	$1\frac{7}{8}''^{-.00095}_{-.0011}$	$1\frac{29}{32}''^{+.005}_{-}$	$\frac{1}{16}''$
PA	$1.625''^{-.0005}_{-.0011}$	$1\frac{1}{8}''^{+.002}_{-.0005}$	·11″ MAX. ·10″ MIN.	$1.625''^{-.0005}_{-.0011}$	$1\frac{33}{64}''^{-.000}_{+.002}$	$\frac{1}{16}''$ MAX. $\frac{3}{64}''$ MIN.	$1.625''^{-.0005}_{-.0011}$	$1\frac{3}{4}''^{+.010}_{-}$	·11″ MAX. ·10″ MIN.	——	NIL	——	$1\frac{7}{8}''^{-.00095}_{-.0011}$	$1\frac{29}{32}''^{+.005}_{-}$	$\frac{1}{16}''$
PB	$1.625''^{-.0005}_{-.0011}$	$1\frac{1}{8}''^{+.002}_{-.0005}$	·11″ MAX. ·10″ MIN.	$1.625''^{-.0005}_{-.0011}$	$1\frac{33}{64}''^{-.000}_{+.002}$	$\frac{1}{16}''$ MAX. $\frac{3}{64}''$ MIN.	$1.625''^{-.0005}_{-.0011}$	$1\frac{3}{4}''^{+.010}_{-}$	·11″ MAX. ·10″ MIN.	——	NIL	——	$1\frac{7}{8}''^{-.00095}_{-.0011}$	$1\frac{29}{32}''^{+.005}_{-}$	$\frac{1}{16}''$
QA	$1.625''^{-.0005}_{-.0011}$	$1\frac{1}{8}''^{+.002}_{-.0005}$	·11 MAX ·10 MIN	$1.625''^{-.0005}_{-.0011}$	$1\frac{33}{64}^{+.000}_{-.002}$	$\frac{1}{16}''$ MAX $\frac{3}{64}''$ MIN	$1.625''^{-.0005}_{-.0011}$	$1\frac{3}{4}''^{+.010}_{-}$	·11″ MAX. ·10″ MIN.	——	NIL	——	$1\frac{7}{8}''^{-.00095}_{-.0011}$	$1\frac{29}{32}''^{+.005}_{-}$	$\frac{1}{16}''$
RA	$1.625''^{-.0005}_{-.0011}$	$1\frac{1}{8}''^{-.002}_{-.0005}$	·11 MAX ·10 MIN	$1.625''^{-.0005}_{-.0011}$	$1\frac{33}{64}^{+.000}_{-.002}$	$\frac{1}{16}''$ MAX $\frac{3}{64}''$ MIN	$1.625''^{-.0005}_{-.0011}$	$1\frac{3}{4}''^{+.010}_{-}$	·11 MAX. ·10 MIN.	——	NIL	——	$1\frac{7}{8}''^{-.00045}_{-.0011}$	$1\frac{29}{32}''^{+.005}_{-}$	$\frac{1}{16}''$
TA	$45\,m/m^{-.015}_{-.035}$	$28\,m/m^{+.015}_{-.01}$	$2.5\,m/m^{+}_{-}.15$	$52\,m/m^{-.015}_{-.035}$	$38\,m/m^{+}_{-}.05$	$2.5\,m/m^{+}_{-}.15$	$52\,m/m^{-.015}_{-.035}$	$38\,m/m^{+.02}_{-.01}$	$2.5\,m/m^{+}_{-}.15$	——	NIL	——	$52\,m/m^{-.015}_{-.035}$	$38\,m/m^{+.1}_{-.0}$	$2.5\,m/m^{+}_{-}.15$
TB.TC	$45\,m/m^{-.015}_{-.035}$	$28\,m/m^{+.015}_{-.01}$	$2.5\,m/m^{+}_{-}.15$	$52\,m/m^{-.015}_{-.035}$	$38\,m/m$	$2.5\,m/m^{+}_{-}.15$	$52\,m/m^{-.035}_{-.015}$	$38\,m/m^{+.02}_{-.01}$	$2.5\,m/m^{+}_{-}.15$	——	NIL	——	$52\,m/m^{-.015}_{-.035}$	$40\,m/m$	$2.5\,m/m^{+}_{-}.15$
VA	$48\,m/m^{-.015}_{-.035}$	$30\,m/m^{+.015}_{-.01}$	$2.5\,m/m^{+}_{-}.15$	$55\,m/m^{-.015}_{-.035}$	$41\,m/m$	$2.5\,m/m^{+}_{-}.15$	$55\,m/m^{-.035}_{-.015}$	$41\,m/m^{+.02}_{-.01}$	$2.5\,m/m^{+}_{-}.15$	——	NIL	——	$55\,m/m^{-.015}_{-.035}$	$41\,m/m^{+.1}_{-.0}$	$2.5\,m/m^{+.15}_{-}$
SA	$48\,m/m^{-.035}_{-.015}$	$30\,m/m^{+.015}_{-.01}$	$2.5\,m/m^{+}_{-}.15$	$55\,m/m^{-.035}_{-.015}$	$41\,m/m$	$2.5\,m/m^{+}_{-}.15$	$55\,m/m^{-.035}_{-.015}$	$41\,m/m^{+.1}_{-.0}$	$2.5\,m/m^{+}_{-}.15$	$55\,m/m^{-.035}_{-.015}$	$41\,m/m^{+.1}_{-.0}$	$2.5\,m/m^{+}_{-}.15$	$55\,m/m^{-.015}_{-.035}$	$41\,m/m^{+.1}_{-.0}$	$2.5\,m/m^{+.15}_{-}$
WA	$48\,m/m^{-.035}_{-.015}$	$30\,m/m^{+.015}_{-.01}$	$2.5\,m/m^{+}_{-}.15$	$55\,m/m^{-.015}_{-.035}$	$41\,m/m$	$2.5\,m/m^{+}_{-}.15$	$55\,m/m^{-.035}_{-.015}$	$41\,m/m^{+.1}_{-.0}$	$2.5\,m/m^{+}_{-}.15$	$55\,m/m^{-.035}_{-.015}$	$4\,m/m^{+.1}_{-.0}$	$2.5\,m/m^{+}_{-}.15$	$55\,m/m^{-.015}_{-.035}$	$41\,m/m^{+.02}_{-.01}$	$2.5\,m/m^{+}_{-}.15$

CHAPTER III

THE CLUTCH

CLUTCH

DESCRIPTION OF CLUTCH (M, D, J, P Type Midgets and F and L Magnas, and K. KN, and N Type Magnettes).

The operation of the clutch is, comparatively speaking, foolproof. That is to say, when the clutch pedal is pushed down, the power from the engine is disconnected, but when pressure on the pedal is released the clutch engages.

There are only two points which require attention, the periodical lubrication of the clutch withdrawal ball race or thrust, and the adjustment of the set screws attached to the withdrawal arms in order to give the necessary clearance between the clutch plates when the clutch is disengaged and equally, to prevent slipping when engaged.

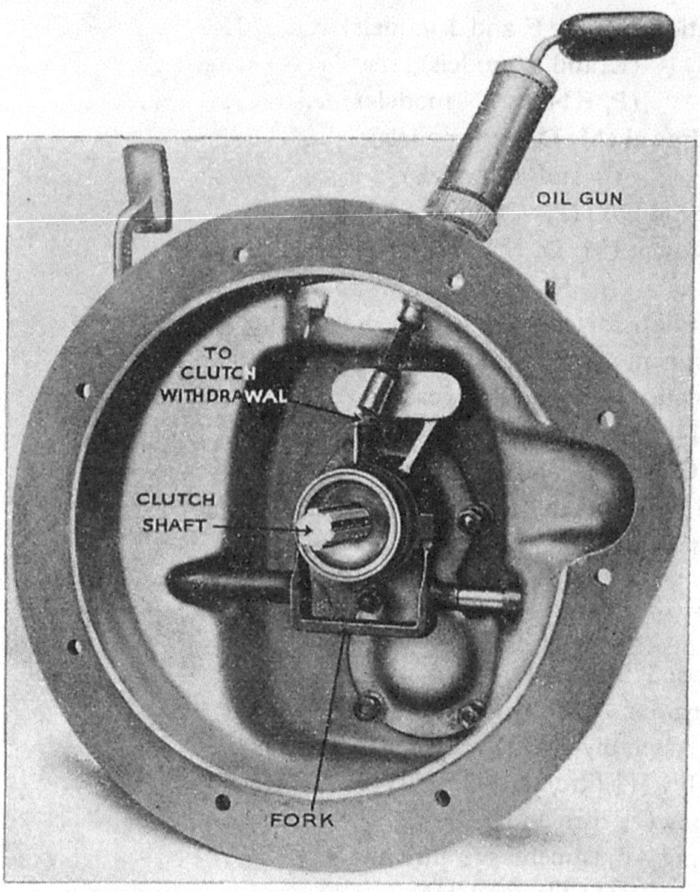

Fig. 1. Showing point of lubrication.

The withdrawal requires only a small quantity of lubricant and too much is worse than too little. Excess lubricant will be flung on to the clutch plates causing slipping. Driving with the foot pressing on the clutch pedal will overheat the withdrawal and cause a slipping clutch.

CLUTCH OPERATION.

For the M, D, and J Type Midgets and the F Type Magna, the assembly of the clutch is as follows: first the flywheel, then the pressure plate, followed by the floating plate and the clutch cover plate.

The clutch cover plate and pressure plate are both fitted with Ferodo discs held by means of countersunk rivets.

Fig. 2. Showing clutch components. (M, D, J, and F models.)

On the P Type Midget and the KN and N Type Magnettes, the assembly of the clutch is as follows: first of all there is the flywheel, then the clutch driven plate, which has fabric facings on either side, and then the pressure plate. All these components are within the clutch cover plate.

On the late PA and all PB Midgets, heat resisting fibre discs are inserted in front of the clutch springs in the pressure plate. These discs are not shown in Fig. 3.

The assembly of the clutch on the L Type Magna and K Type Magnette is as follows: first there is the flywheel, followed by the pressure plate in the rear of which there is a series of cups to locate the clutch springs. Next there is a steel floating plate, a duralumin centre plate with a fabric facing, another steel floating plate and, finally, the clutch cover plate.

The driven plate(s) are made of laminated steel for flexibility and are a sliding fit on the splined first motion shaft of the gearbox. There are six clutch springs (twelve on the P Type Midget and N Type Magnette), which fit into recess cups on the flywheel side of the pressure plate. The flywheel is fitted with six register pins that pass through slots in the pressure plate and when the clutch is bolted up these pins register with holes in the cover plate.

When the clutch is bolted up solid, spring pressure forces the pressure plate into contact with the floating plate and the latter is gripped between the pressure and cover plates. Oil must not be allowed to come into contact with the Ferodo rings and any fierce operation of the clutch may tear the friction faces.

If the clutch is slipped excessively, the centre plate will become very hot and this heat will be transmitted to the clutch surfaces with detrimental results.

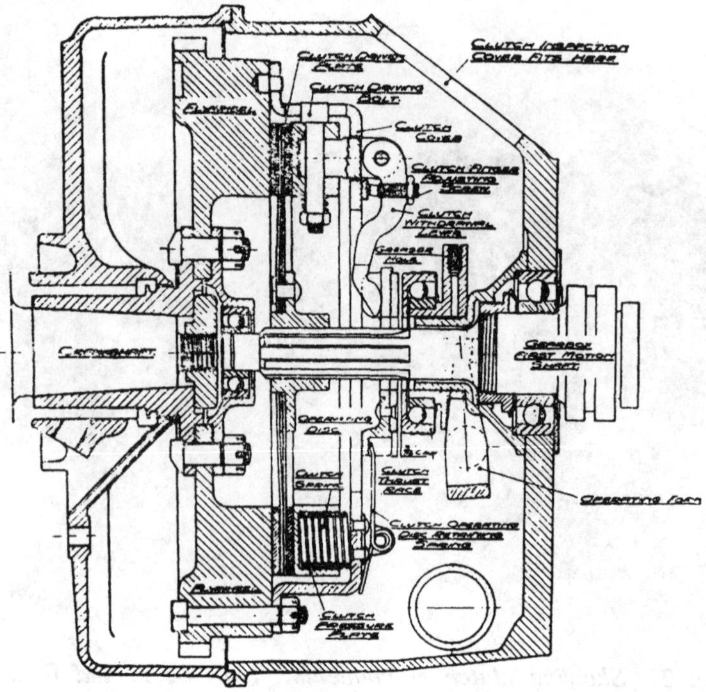

Fig. 3. Diagrammatic view of the clutch. (P, KN, and NA models.)

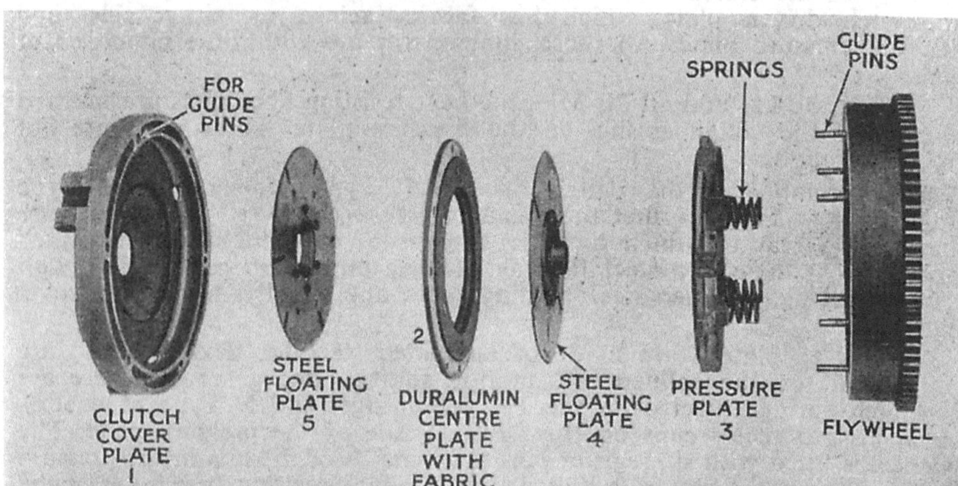

Fig. 4. Clutch components of the L type Magna and K type Magnette.

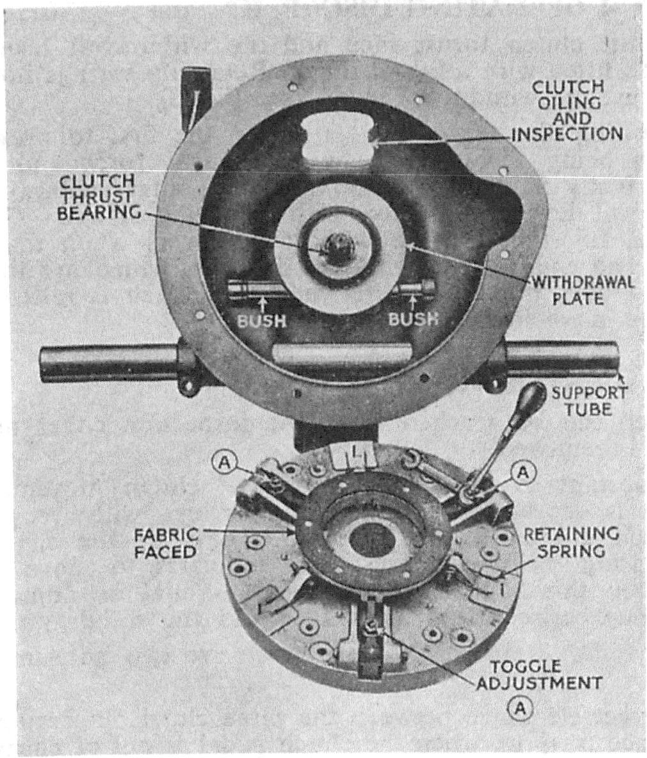

Fig. 5. Clutch cover plate as seen from the exterior of the clutch. The clutch housing and withdrawal mechanism is shown above.

CLUTCH WITHDRAWAL (M, D, J and F Type Models).

In order to disengage the clutch, the pressure plate must be moved out of engagement with the floating plate by compressing the clutch springs. On the inner side of the clutch cover plate are three short plungers marked "P" (Fig. 2), which, when the clutch is assembled, contact the three hardened steel abutments "B" (Fig. 2).

The withdrawal arms are fitted with three adjusting screws which contact these plungers. Pressure on the clutch pedal moves the withdrawal arms forward, thus forcing the adjusting screws against the plungers "P" which, in turn, move the clutch pressure plate towards the flywheel against the pressure of the clutch springs.

CLUTCH WITHDRAWAL (L and K Type Models).

The clutch is disengaged by moving the pressure plate out of engagement with the floating plate against the pressure of the clutch springs. On the inner side of the clutch cover plate there are three holes in which short plungers register. When the clutch is assembled, these plungers contact three hardened steel abutments. These plungers are controlled by adjusting screws carried in the withdrawal arms. Pressure on the clutch pedal moves the withdrawal arms forward, thereby pressing the adjusting screws against the plungers. This movement forces the clutch pressure plate towards the flywheel against the pressure of the clutch springs.

THE CLUTCH OPERATING DISC (P, KN, and NA Models).

Between the clutch thrust race and the withdrawal lever there is an operating disc fitted with a fabric lining. Retaining springs hold the operating disc against the withdrawal levers (see Fig. 3).

The operating disc serves two purposes, the first to prevent the withdrawal lever being thrown out by centrifugal force and the other to ensure a perfectly flat and smooth face for contact with the thrust race. The withdrawal levers are fitted with adjusting screws. These adjusting screws act as fulcrums for the withdrawal levers when the clutch pedal is operated. The ends of the withdrawal levers, which are in contact with the operating disc, move forward when the clutch is released, while the opposite ends move in the reverse direction.

CLUTCH ADJUSTMENT.

The clutch can be reached when the inspection cover on top of the bell housing is removed.

It is important to limit the amount of clutch movement when the clutch pedal is depressed, otherwise the springs will become coilbound. As the clutch gradually wears, the thickness of the fabric rings will decrease, causing the driven and pressure plates to move closer to the flywheel. When this happens the setscrews should be adjusted to give a small clearance between the fabric face and the withdrawal disc.

On the M, D, J, and F Models, there are two adjustments (see Fig. 6):—

1. The correct clearance between the three clutch fingers and the clutch thrust race is $\frac{3}{32}$ in. when the clutch pedal is out of engagement.
2. This adjustment concerns the amount of movement the clutch fingers have between the retaining springs and the clutch operating pins. The duty of the springs is to prevent the fingers fouling the thrust race and they should be set to a clearance of .010 in. between the clutch adjustment screw and the clutch thrust pin.

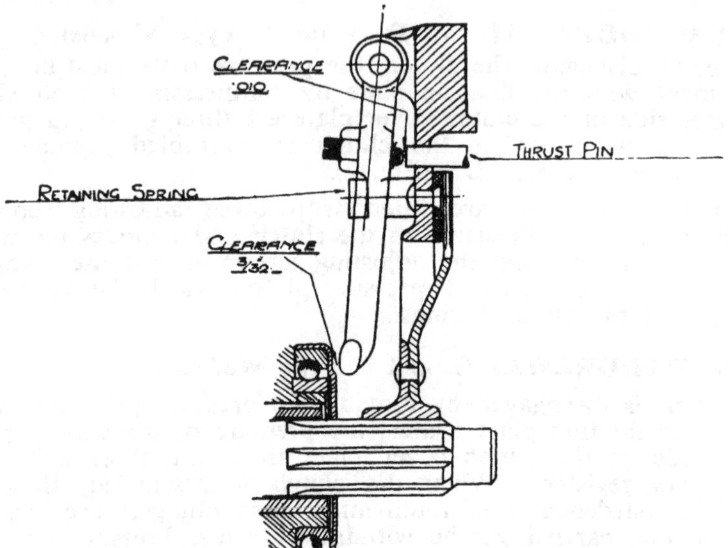

Fig. 6. Correct clutch adjustment for M, D, J and F models.

On the L and K Type Models the clearance between the fabric face and the withdrawal disc should also be $\frac{3}{32}$ in. when the clutch pedal is out of engagement. On the P, KN and N Type Models this clearance should be $\frac{3}{16}$ in.

It is absolutely essential that the clutch withdrawal lever adjusting screws are set so that the clearance between the operating disc and the thrust race opposite each lever is the same. The clearance must be set to a gauge which is inserted opposite the lever which is being adjusted.

After setting the levers, check the clearances again and make quite sure the operating disc is running true when the engine is turning over very slowly.

If the operating disc is not running true, then readjustment and rechecking should be carried out.

On the P, N and KN Models the clutch pedal stop should be adjusted so that there is 1½ in. clearance between the back of the pedal and the stop when the clutch pedal is in its static position.

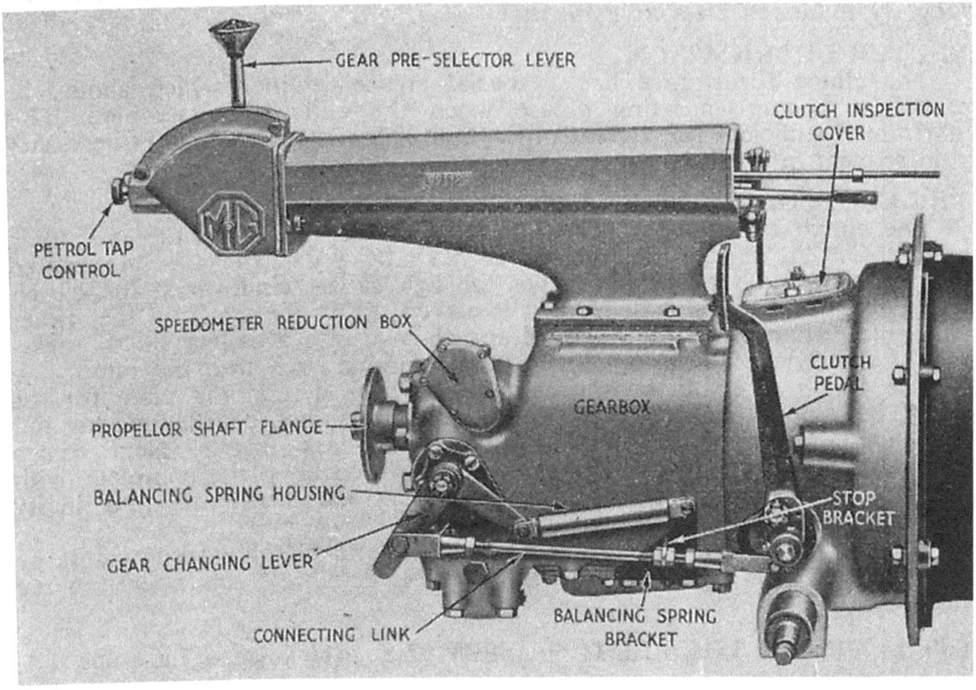

Fig. 7. This shows the clutch connecting link and the correct method of adjustment.

CLUTCH INSTRUCTIONS FOR CARS FITTED WITH THE PRE-SELECTOR GEARBOX.

Before proceeding with adjustment, it is as well to deal with synchronization of the main clutch with the preselector gearbox clutch. The main clutch is connected to the preselector clutch by a link motion which is so arranged that when the clutch pedal is depressed, the first movement disengages the main clutch and further movement changes the gear previously selected. It follows that when the clutch pedal is released the preselector clutch engages first, followed by the main clutch. Adjustment of this link motion is highly skilled and unless the owner is fully aware of the correct procedure, it is advisable to take the car to the nearest expert.

CLUTCH ADJUSTMENT.

There are four withdrawal arms which contact the withdrawal bearing. When the clutch pedal is in the engaged position, clearance between the withdrawal race and the withdrawal fingers should be $\frac{3}{16}$ in. Reference to Fig. 7 which shows the link mechanism between the preselector gearbox and the main clutch will be of assistance.

The adjustment to reduce or increase the clearance between the withdrawal arms and the withdrawal race is effected from the outside of the gearbox. It will be seen that by adjusting the clutch pedal forward, the withdrawal race will be brought closer to the withdrawal levers. Conversely, the withdrawal race can be brought further away from the withdrawal levers. Reference to Fig. 7 will show that the connecting link between the clutch pedal mechanism and the gear change lever is threaded with nuts fitted either side of the stop bracket. This bracket contacts the stop on the balancing spring bracket. If these nuts are slackened and the bracket moved towards the rear, the clutch pedal will be adjusted forward. If the bracket is moved forwards the clutch pedal will be allowed to move further back. Care must be taken to keep the adjustable stop bracket square with its stop, and to see that the nuts are properly tightened after adjustment.

CLUTCH LUBRICATION.

The clutch thrust race has a special grease retainer which should be packed with non-separating grease when the clutch is assembled. This lubricant is sufficient for a great mileage and this attention is only necessary at infrequent intervals.

THE CLUTCH. (1½ Litre, 2 Litre and TA Models.)

The clutch on these models runs in oil which is fed to the unit by a hole drilled in the centre of the crankshaft. When the oil has passed through the clutch plates it escapes through gauze windows in the clutch spring plate into the clutch. Here it is picked up by the teeth on the starter motor ring attached to the flywheel and thrown into a gallery on the near-side of the flywheel housing. From here it drains back into the sump.

The clutch assembly comprises: the flywheel, a retaining plate for the spigot bearing, the gearbox first motion shaft spigot bearing centre and the clutch driven plate which is fitted with cork inserts. Next comes the pressure plate, clutch springs, the clutch cover plate complete with thimbles for the clutch springs, the washer for the thrust race and finally the thrust race and its locking nut.

In order to reach the clutch the gearbox and clutch housing must be removed from the flywheel housing. For full details of this operation see Chapter IV.

DISMANTLING THE CLUTCH ASSEMBLY. (1½ Litre, 2 Litre and TA Models.)

The clutch cover plate is held to the flywheel by a ring of bolts and the complete assembly can be lifted off as a single unit.

With the unit on the bench, it is possible to lift out the driven plate when once the indented C-shaped circlips have been removed from the three studs which project through the cork-faced driven plate.

To dismantle the clutch plates, springs and pressure plate, remove the ring nut from the hub of the pressure plate, when the clutch withdrawal ball race and washer can be taken off. The pressure plate can then be taken out quite easily.

On the 1½ Litre Model, commencing at Engine no. TPBG1510, a dry plate clutch was introduced in place of the oil-immersed clutch. The procedure for access and dismantling for the dry clutch is the same as that already described above.

Fig. 8. *Clutch dismantled showing the various components.* (1½ *Litre and TA models.*)

ADJUSTMENT OF CLUTCH PEDAL. (1½ Litre, 2 Litre and TA Models.)

It is important that the clutch pedal has 1 in. minimum free movement before operating the withdrawal mechanism. This dimension should be taken at the foot pad of the pedal. The object of this dimension is to prevent the clutch thrust race from fouling the operating fork when the cork-lined driven plate wears.

Adjustment takes place at the bottom of the clutch pedal and the maximum pedal travel is controlled by an adjustable stop. The pedal should be adjusted so that the stop comes into contact with the pedal at the same moment as the clutch stops spinning.

CLUTCH MODIFICATIONS. (P and N Types.)

From " PA " Engine no. 2002 and " NA " Engine no. 1001, a lighter design of driven plate with a proportionately thicker pressure plate was introduced. (See Fig. 9.)

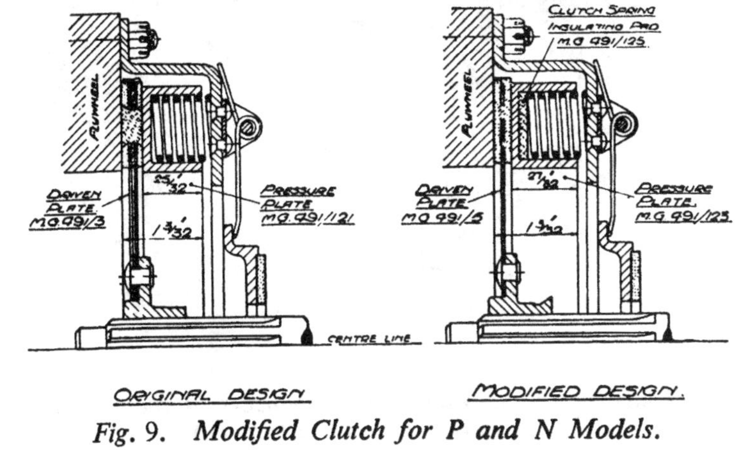

Fig. 9. *Modified Clutch for P and N Models.*

The original design consisted of a driven plate having three discs with the linings secured by means of cork inserts and tabs on the periphery of the disc, turned over to prevent the lining from being flung off by centrifugal force. The pressure plate used with this driven plate is $\frac{25}{32}$ in. thick.

The modified design of clutch consists of a driven plate having a single disc with the lining moulded in position. The pressure plate in this case is $\frac{27}{32}$ in. thick. Insulating pads are inserted in the spring guide holes to protect the clutch springs from heat.

The following are the original part numbers:—

Driven Plate MG491/3
Pressure Plate MG491/121

The new parts superseding the old are as follows:—

Driven Plate MG491/5
Pressure Plate MG491/123
Clutch Spring Insulating Pad (8 per set) MG491/125

The new driven plate, MG491/5, can be fitted to engines produced prior to those given above, providing the pressure plate, MG491/123, and insulating pads, MG491/125, are fitted also.

CLUTCH. (TB and TC Models, also 2.6 Litre Model and 1½ Litre Model commencing at Engine no. TPBG1510.)

A single dry plate clutch is fitted to these models and it consists of a driven plate assembly, a cover assembly and a graphite release bearing assembly.

Access to the clutch unit is obtained by removing the gearbox and clutch housing. (See Chapter IV.)

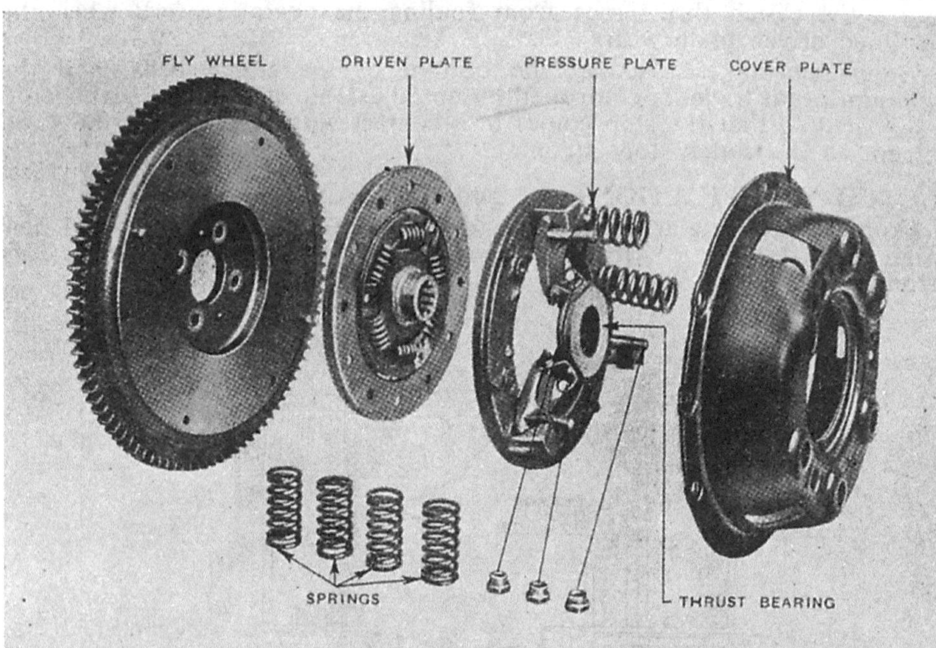

Fig. 10. Clutch dismantled showing the components in their relative positions (TB TC, TD and 1½ Litre models.)

The clutch is removed from the flywheel by undoing a ring of bolts.

To dismantle the clutch unit it is only necessary to remove the three nuts which hold the clutch pressure plate to the clutch cover. These nuts screw on to the fulcrum bolts.

Figs. 10 and 11 show the clutch dismantled, but it should be remembered that this is not a job to be carried out unless the proper equipment for reassembling and balancing is available.

Before tackling this operation, reference should be made to the details of the 1¼ Litre and TD Models shown on page 145.

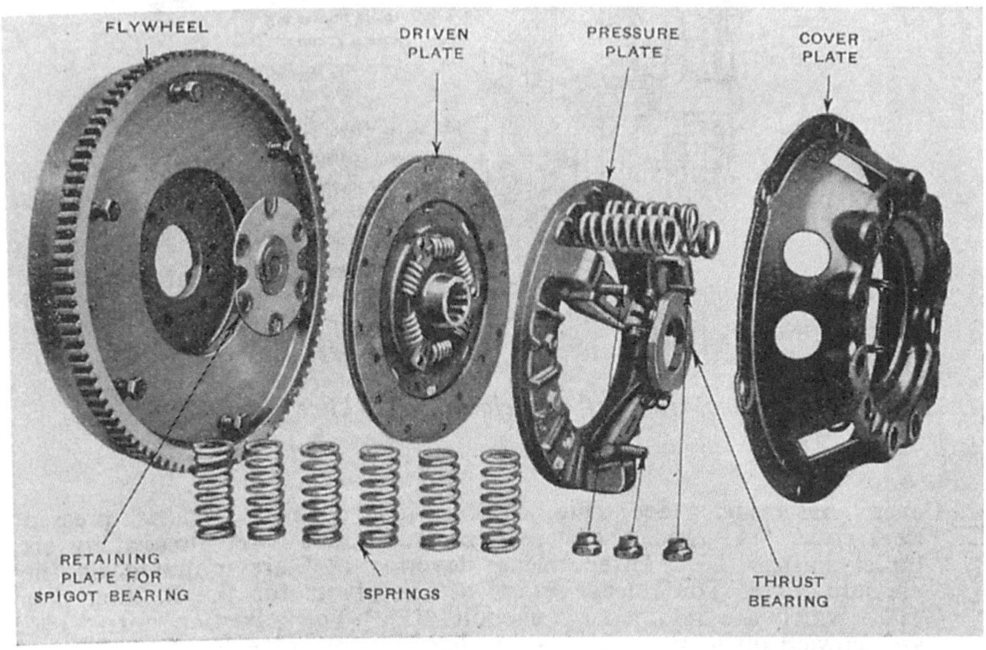

Fig. 11. *The clutch dismantled.* (2.6 *Litre model and later* 1½ *Litre models, except that* 1½ *Litre clutch has six springs only.*)

ADJUSTMENT OF CLUTCH PEDAL. (TB and TC Midgets, 2.6 and later 1½ Litre Models.)

It is important that the clutch pedal has 1 in. minimum free movement before operating the withdrawal mechanism. This dimension should be taken at the foot pad of the pedal and adjustment is by means of the adjuster at the bottom of the clutch pedal. On the 1½ Litre, TB and TC Models, the stop which limits the travel of the clutch pedal should be adjusted to give a movement of 3 in., and on the 2.6 Litre Model a movement of 4 in. These two dimensions are in addition to the 1 in. free movement mentioned earlier.

GENERAL DESCRIPTION OF THE CLUTCH. (1¼ Litre and TD Models.)

The clutch fitted to these cars is a single plate Borg and Beck Model consisting of a driven plate assembly, cover assembly and a graphite release valve assembly.

1. **Driven Plate Assembly.** This has a flexible centre in which the splined hub is indirectly attached to the clutch plate and transmits the power and overrun through a number of coil springs held in position by retaining wires. The clutch plate has a facing on either side held by rivets.

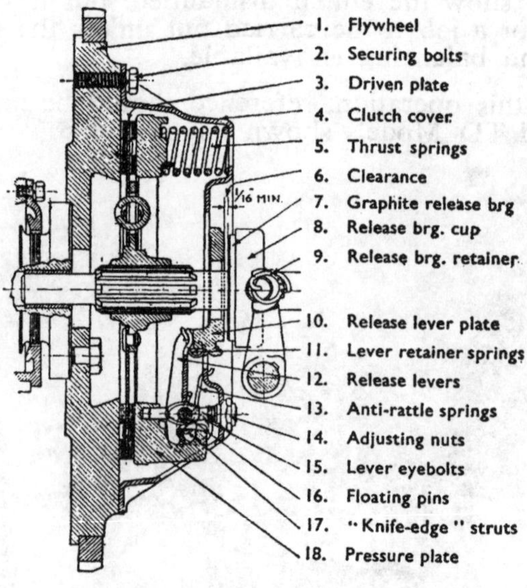

Fig. 12. *Sectional view of the clutch.* (1¼ *Litre and TD Midget.*)

2. **Cover Assembly.** (See Fig. 12). This consists of a pressed steel cover "4" and a cast iron pressure plate "18," loaded by six thrust springs "5." Three release levers "12" are mounted on the pressure plate. The release levers act as pivots for the floating pins "16," which are retained by eyebolts "15." The adjusting nuts "14" are screwed on to the eyebolts and held by split pins. Struts "17" are interposed between the lugs on the pressure plate and the outer ends of the release levers. The release levers are loaded by the anti-rattle springs "13" and retainer springs "11" connect the release lever plate "10" to the release levers.

3. **Release Bearing.** (See Fig. 12). The release bearing consists of a graphite block "7," shrunk into a bearing cup "8." The cup is located by the operating fork and release bearing retainer springs "9."

CLUTCH RUNNING ADJUSTMENTS. (1¼ and TD Midget.)

The only adjustment necessary is to restore free movement at the clutch pedal. This free movement gradually decreases as the driven plate facings wear. This will have a tendency to prevent the clutch from engaging and will also permit too great a movement on withdrawal.

The minimum of clearance at "F" (Fig. 13) between the withdrawal lever plate and the face of the thrust bearing is $\frac{1}{16}$ in. This gives a free pedal movement of ¾ in. and this should always be maintained. When the clutch pedal free movement becomes less than ¾ in., adjustment should take place at the nuts "A" and the locknut "B" at the forward end of the clutch operating cable.

Care must be taken to see that pedal travel is not excessive, because this will place an unnecessary strain on the carbon thrust block.

There should be approximately $\frac{7}{8}$ in. clearance at " H " between the stop nut " C " and its stop " D," when the clutch pedal is held lightly with the carbon block in contact with the thrust ring. To adjust for this dimension, grip the nut " C," slacken off the locknut " J," grip the nut " K " and set the stop nut " C " to the required position.

This adjustment is needed when there is a tendency for the clutch to spin when the pedal is fully depressed.

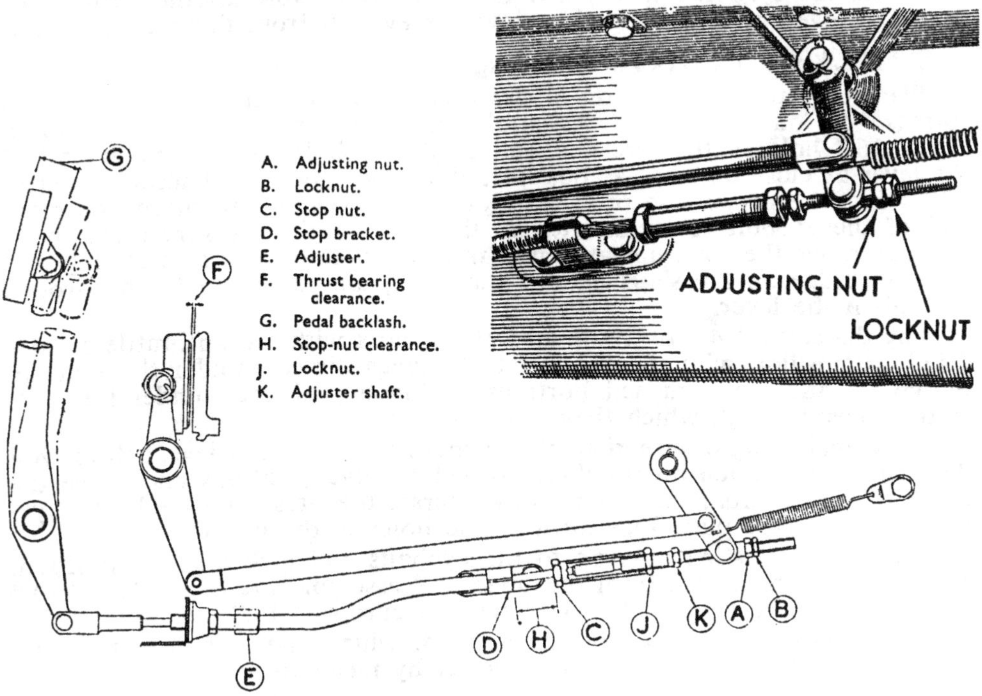

Fig. 13. *Clutch pedal adjustment.* (1¼ *Litre and TD Midget.*)

The adjuster " E " is provided for the initial adjustment of the outer cable to give the correct degree of flexibility between the engine and the pedal box. This is set when the car is built and should not require any further adjustment.

NOTE:—The adjusting nuts " 14 " (Fig. 12) should not be altered unless the clutch has been dismantled and new parts fitted. Unnecessary interference here will throw the pressure plate out of position and cause clutch judder.

REMOVAL OF CLUTCH. (1¼ Litre and TD Models.)

Remove the gearbox as described in Chapter IV. Slacken the securing bolts a turn at a time so that spring pressure is relieved and the clutch assembly is not distorted. The complete assembly can then be lifted away from the flywheel and the driven plate assembly removed.

DISMANTLING THE CLUTCH. (See Fig. 12.)

The cover " 4," pressure plate lugs " 18 " and release levers " 12," should be marked for dismantling so that they can be assembled in the same relative positions in order to preserve balance and adjustment.

Detach the release lever plate " 10 " and the retainer springs " 11." Now place the cover assembly under a press and with the pressure plate " 18 " resting on blocks so arranged that the cover is free to move downwards when pressure is applied via a piece of wood to the top of the cover. The adjuster nuts " 14 " can now be removed. After that release the pressure slowly so that the thrust springs " 5 " do not fly out. Lift off the cover and this will expose all the parts for your inspection.

Remove the release levers "12 " by grasping the lever and eyebolt between the finger and thumb, so that the inner end of the lever and the threaded end of the eyebolt are as near together as possible. At the same time keep the eyebolt pin in position in the lever. Now lift the strut " 17 " over the ridge on the lever and remove the eyebolt from the pressure plate.

ASSEMBLY OF CLUTCH. (See Fig. 12.)

Place the pressure plate on blocks under the press and insert the thrust springs " 5 " on their bosses. Assemble the release levers "12 " and eyebolt pins " 16," holding the threaded end of the eyebolt and the inner end of the lever as close together as possible. With the other hand insert the strut " 17 " in the slots on the pressure plate lug sufficiently to allow the plain end of the eyebolt to be inserted in the hole in the pressure plate. Now you can move the strut upwards into the slot in the pressure plate lug and over the ridge on the short end of the lever, and drop it into the groove formed in the lever.

Lay the cover " 4 " over the assembled parts with the anti-rattle springs " 13 " in position with the tops of the springs directly under the seats in the cover, and the hardened portions of the lugs directly under the slots in the cover through which they have to pass.

Now make quite sure that the components you marked before you dismantled the clutch are in their correct relative positions and place the block of wood across the cover. Now operate the press and guide the eyebolts and pressure plate lugs through the holes in the cover.

Screw the adjusting nuts on to the eyebolts and lock them in position. Next, remove the clutch from the press and assemble the lever plate " 10 " on the tips of the levers " 12 " and retainer springs " 11."

NOTE:—If new parts have been fitted, which may possibly affect the clutch adjustment, the levers should be set by means of a special Borg and Beck gauge plate No. CG.12916.

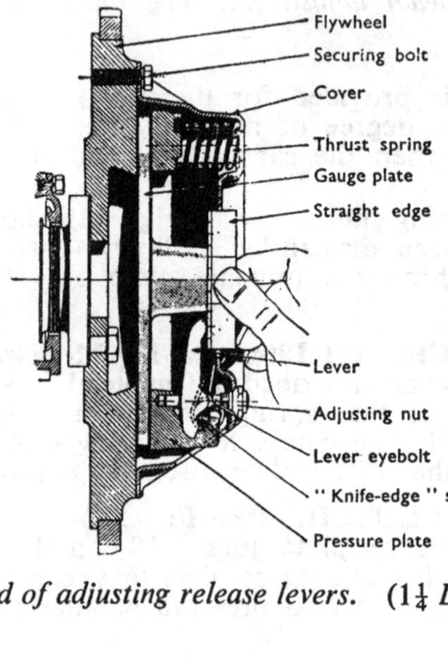

Fig. 14. Method of adjusting release levers. (1¼ Litre and TD Midget.)

ADJUSTING THE RELEASE LEVERS. (See Fig. 14.)

The release levers should only need adjusting if new parts have been fitted, but upon this adjustment depends the satisfactory operation of the clutch.

The maximum difference allowed in height of the levers is .015 in. To set the levers, use the special gauge plate in conjunction with the flywheel.

Fig. 15. *Gauge plate CG* 12916.

Place the gauge plate centrally on the flywheel in place of the driven plate. Now fit the cover to the flywheel by tightening the securing bolts a turn at a time by diagonal selection until fully secured.

Place a straight edge across the gauge plate boss and the tip of one release lever, and adjust by turning the eyebolt until the tip of the lever is level with the top of the boss. Adjust the other levers in a similar manner. You will find that this setting should be within .005 in. if it is carried out carefully.

Now undo the securing bolts, take the clutch from the flywheel, and extract the gauge plate.

REFACING THE DRIVEN PLATE. (1¼ Litre and TD Models.)

Do not punch out the rivets, but drill them out with a $\frac{5}{32}$ in. dia. drill.

Rivet one new face in position, using a blunt ended centre punch or a special tool. Now rivet the second facing to the opposite side of the plate with the clearance holes over the rivet heads already formed.

Mount the plate on a mandrel between centres and check for runout as near as possible to the edge. If the error is more than .015 in., dress over the high spots until the plate is running true.

REFITTING THE CLUTCH. (1¼ Litre and TD. Midget.)

Assemble the driven plate assembly with the larger chamfered splined end of the driven plate hub towards the gearbox. Centralise the driven plate by means of a clutch alignment bar (special tool number T.124).

Now fit the cover assembly to the flywheel, leaving the alignment bar in position until the screws are securely tightened. Remove the alignment bar and refit the withdrawal bearing and gearbox. Make sure that the gearbox is supported so that its weight does not distort the clutch before it is finally bolted to the engine.

INTRODUCTION OF 8 in. DIA. CLUTCH IN PLACE OF 7¼ in. DIA. CLUTCH. (TD and 1¼ Litre.)

Commencing at Engine Nos. XPAG/SC/16916 and XPAG/TD/9408 an 8 in. Dia. Single Plate, type BB.8/70 clutch was introduced with the view to improving the torque carrying capacity. Engines fitted with this type of clutch have the engine description amended from XPAG/SC and XPAG/TD to XPAG/SC/2 and XPAG/TD/2.

The engine and gearbox units separately are not interchangeable, for the clutch thrust race is in a different position relative to the engine bell housing, with the clutch shaft and thrust face position altered in the gearbox bell housing.

The earlier type gearboxes can be identified by the clutch fork shaft with a diameter of ⅝ in. where as the later type of gearbox has a clutch fork shaft of ¾ in. diameter.

CLUTCH ASSEMBLY. (TD and 1¼ Litre Models.)

Commencing at Engine Nos. XPAG/TD/2/15861 and XPAG/SC/2/17392 respectively, clutch plate linings giving a higher friction value were incorporated. The clutch driven plate assembly fitted with the moulded facings was replaced by a clutch plate fitted with wound yarn facings.

CHAPTER IV

GEARBOX

PAGE

GEARBOX

DESCRIPTION (J and P Midgets, K, KN and N Magnettes, and L Type Magna.)

The gearbox is a four-speed unit with double-helical gears for the constant mesh and third-speed. First and second gears have straight teeth. The casing is in three sections, the rear end cover, the centre casing and the clutch pit. The change-speed lever is fitted to an extension of the gearbox lid. The gate is provided with a reverse stop in order to prevent inadvertent engagement of this gear. The gears are operated by a series of forks, coupled to selector rods.

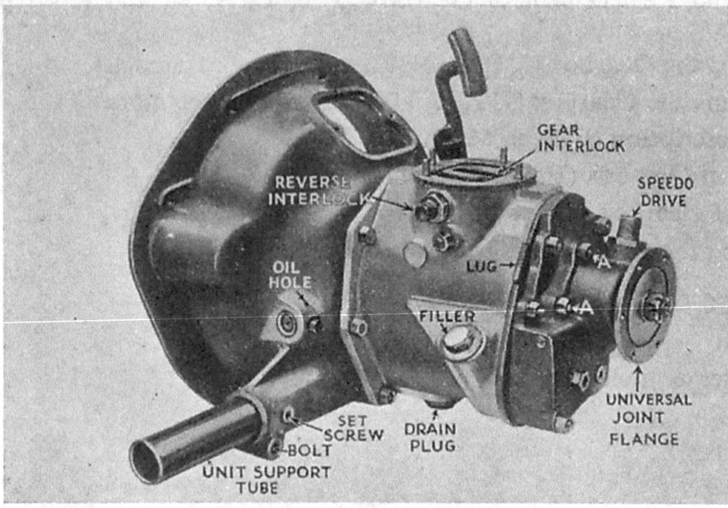

Fig. 1. External view of Gearbox.

Figs. 2 and 3 will give an idea of the manner in which the selector rods are locked in position when the gear lever is placed in any position. The external view of the gearbox shows the oil filler which acts as an automatic high-level indicator. The drain plug from the underside of the box can be removed via the trap door in the undershield.

The rear end of the engine and the forward end of the gearbox are supported in the chassis frame by the tubular member which is located in the bell housing by means of a bolt and setscrew on either side.

The Clutch withdrawal shaft is located in the bell housing and the bushes are held by means of setscrews. There is an oil hole either side for lubrication purposes. When dismantling the gearbox there is no necessity to remove the nut "A" (Fig. 1), and it will be noticed that the rear end housing of the gearbox is provided with lugs to prevent the introduction of a tool likely to damage the faces.

OPERATION OF GEARBOX.

Steel balls register in slots in the selector rods and as the gear lever is moved the balls ride out of one slot and re-engage in another.

Reference to Fig. 2 will give an excellent impression of the gearbox, apart from the gears themselves, for it shows the three principal parts: the front end of the gearbox formed by the bell housing or clutch pit, the centre of the gearbox acting as a casing and support for the reverse shafting, and the rear cover supporting the shafts and their gears. The principal point to note in this illustration is the gear on the layshaft, which is indicated as "the floating third gear." This is always in constant mesh with the third-speed pinion on the mainshaft.

Top gear is obtained by moving the third gear on the main shaft forward so that the splines on the hub of the gear mesh internally with the constant mesh pinion on the first motion shaft. It will be noted that the spigot bearing is of the idling type.

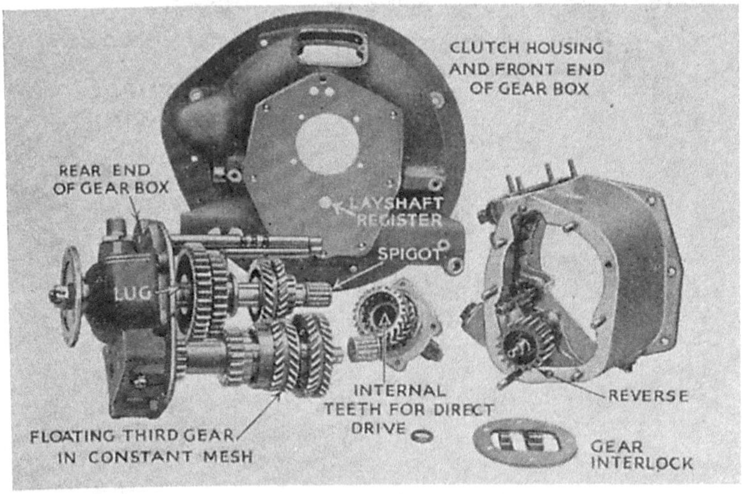

Fig. 2. View of Gearbox partially dismantled.

When, through the movement of the change-speed lever, the third-speed gear on the mainshaft is caused to slide rearwards, the nature of the tooth engagement will automatically cause the third-speed gear on the layshaft to move backwards in unison. It is then that the gear on the layshaft, which has internally cut teeth, engages with the set of teeth on the front of the second-speed pinion. When any other gear is engaged, except third, this third gear on the layshaft simply rotates on its sleeve.

In order to clarify the matter, refer to Fig. 3, which shows the layshaft train of gears locked up on a tube removed from the layshaft itself.

The layshaft is fitted with roller bearings at each end and over these roller bearings the complete layshaft is fitted.

The internal mesh dog engagements of the direct drive on the mainshaft and the third-speed free sliding gear on the layshaft should be examined. It will be noted that the mainshaft protrudes through the gearbox, passing through the gearbox front end cover. The cover will be seen in Fig. 3, and alongside will be noted a steel ring with four bolts passing through it. This steel ring is placed behind the front ball bearing on the mainshaft to act as an oil retaining washer. The ring has a number of lips to act as registers for the bolts which secure the front end of the gearbox cover. A small hole is drilled on the lower portion of the front gearbox cover. This allows a certain amount of oil to pass through so that the clutch thrust is lubricated. In the lower right hand corner of Fig. 3, a gear striker is shown. This operates the gears. There are two rods above the mainshaft, in which grooves are machined. These grooves are intended to register the striker in its proper position, aided by steel balls which fit into a recess in the gear striker.

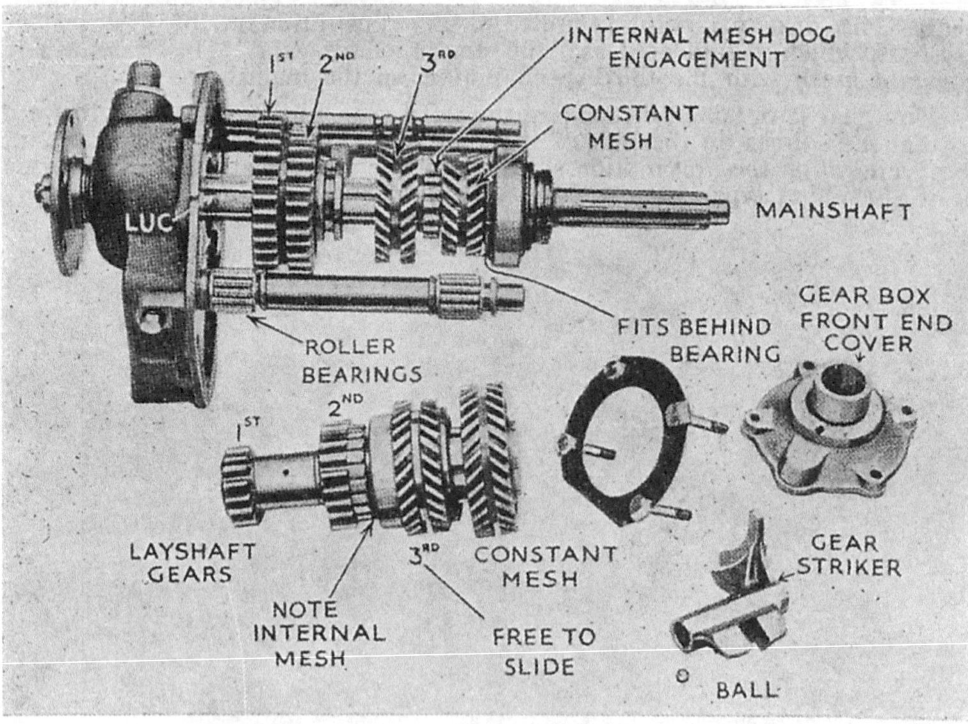

Fig. 3. Another view of gearbox components.

REMOVAL AND REPLACEMENT OF GEARBOX.

Disconnect the mixture and slow running controls just forward of the gear lever. Remove the seats, carpet, and floor boards. Take out the foot ramp.

Take off the four nuts holding the remote control assembly to the centre casing which can then be lifted off. Now cover the aperture in the centre casing with a piece of cardboard. Undo the propeller shaft flange from the gearbox universal joint flange. Mark the flanges for reassembly purposes.

Jack up under the rear of the engine sump, using a piece of wood to spread the load.

NOTE: On the J type Midget this is not necessary because the rear engine mountings are integral with the flywheel housing.

Disconnect the speedometer drive at the gearbox end. Remove the nuts holding the clutch pit to the flywheel housing. (On the J type Midget the gearbox can now be removed.) On the other models, remove the unit support tube fixings and lift the gearbox clear of the car.

NOTE: When removing the gearbox, make quite sure that no load is placed on the drive gear shaft and clutch centre.

DISMANTLING THE GEARBOX.

It is much easier to dismantle the gearbox if it is held in a vice, and this can be done by machining a thread on the end of a square piece of bar to screw into the gearbox drain plug hole.

1. Remove the seven nuts securing the end cover, leaving the three nuts round the mainshaft undisturbed. Take off the $\frac{5}{16}$ in. nut holding the reverse shaft and remove the locking screw holding the layshaft in the rear cover. Leave the universal joint flange attached to the mainshaf'

and remove the rear cover complete. The change-speed rods and reverse shaft remain in the gearbox.

If the rear cover sticks a little on the dowels, a few sharp taps with a lead hammer on the lugs will loosen it. Do not use a chisel or screwdriver to break this joint.

2. Place the reverse gear lever in the engaged position. This holds the spring loaded plunger and the first and second change-speed fork can be removed. Now push the selector fork and the first and second sliding gear as far forward as possible. Now drive out the change-speed rod from the front end (this procedure will save the trouble of reassembling the ball and spring in the fork). The fork, rod and gear can now be withdrawn from the gearbox.
3. The third and fourth-speed selector rod and fork can be removed if you follow the procedure in (2) above.
4. Drive out the layshaft from the front end and withdraw the layshaft gears complete.
5. The rollers in the third-speed layshaft gear can be taken out without withdrawing the layshaft constant mesh gear.
6. Do not remove the constant mesh gear unless it is absolutely necessary. This gear is a press fit on a shallow spline and considerable care would be needed for removal and replacement. It is necessary for a clock reading to be taken in order to make sure that the gear is running true with the sleeve (laterally, this dimension must be within .0015 in.).
7. To remove the speedometer driving gear, or the rear main bearing, the universal joint flange must be taken off. This job is made easier if an extractor is used.

With the flange removed, take off the bearing securing plate by undoing the three bolts on the rear cover. This will allow the mainshaft and the bearing to be taken out.

REASSEMBLY OF THE GEARBOX.

Reassembly is the reverse of the dismantling procedure and the following points should be noted. When the layshaft gears are in position and a straight edge laid across the face of the gearbox, the clearance between the straight edge and the first-speed layshaft gear should be .005 in. (minimum). The layshaft must be free to move endways; it is positioned by the meshing of the double-helical constant mesh gears.

It will be noticed that the spring loaded plunger is carried in an eccentric sleeve. This is to give adjustment for lining up the slot in the reverse lever with the slots in the change-speed forks. This adjustment may be necessary if the plunger sleeve is taken out at any time.

GEARBOX

GENERAL DESCRIPTION. (F type Magna.)

The change speed lever is fitted to an extension of the gearbox lid. The gate is provided with a stop in order to prevent inadvertent engagement of reverse gear. The gears are operated by a series of forks coupled to selector rods. The external view of the gearbox shows the oil filler which acts as an automatic high level indicator. The drain plug on the underside of the gearbox can be removed via the trap door in the undershield.

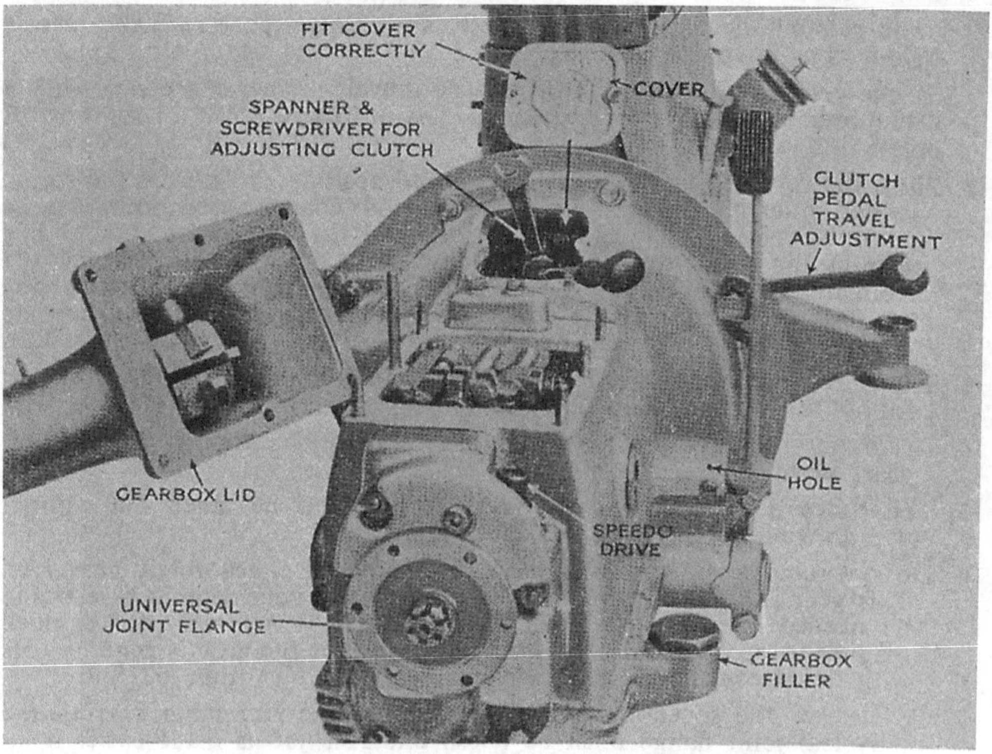

Fig. 4. General view of gearbox and Clutch mounting.

Fig. 5 will give an idea of the manner in which the selector rods are locked in position when the gear lever is placed in any position. Between the gearbox lid and the clutch inspection cover plate is a small aluminium cover, held in position by two bolts containing three springs which make contact with three steel balls. Steel balls register in slots in the selector rods and as the gear lever is moved the balls ride out of one slot and re-engage in another.

OPERATION OF GEARBOX. (F type Magna.)

Reference to the composite Fig. 6A shows the complete set of gears, with the exception of reverse gear, as they would be seen taken from the gearbox. To the left is the clutch extension and to the right the splined extension to which is attached the front universal joint of the propeller shaft. The gears are in second gear and the pair of gears at the front and rear end of the shafts are also in engagement.

Reference to Fig. 6B shows the layshaft which is a chain of gears fitted to a castellated shaft, being separated by three spacing tubes. The whole shaft is locked up solid by ring nuts, which also hold the ball bearings in position.

The mainshaft is built up in two sections, the constant mesh gear or first motion shaft being carried at the forward end, into which fits the tailshaft. The actual bearing surface is a spigot bearing (a number of small steel rollers), or roller bearing. The constant mesh gear rotates, leaving the remainder of the shaft stationary, but always driving the layshaft, being in constant mesh. There is another constant mesh gear fitted to the end of the main shaft, fitted on to the cylindrical portion of the mainshaft between

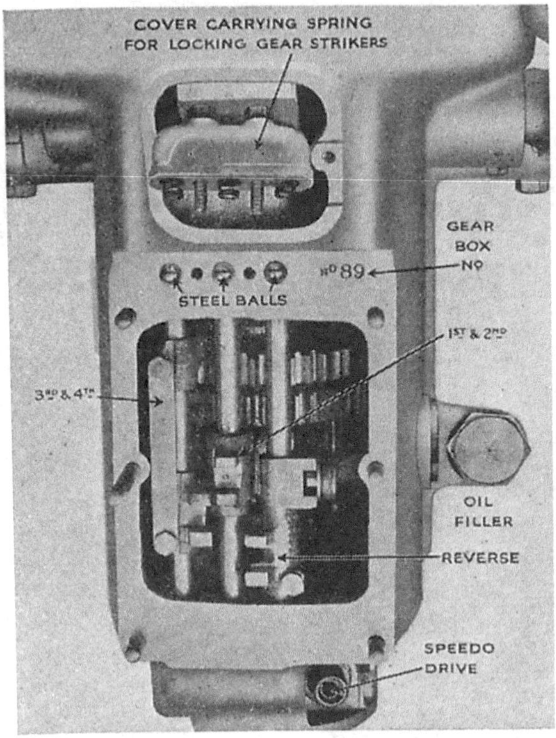

Fig. 5. Gearbox with lid removed.

the large set of splines and the small set at the rear. This gear is also fitted with a needle bearing.

When the two large gears and the two sets of small gears with short teeth fitted to the mainshaft are moved laterally, they will engage with the internal gears on the ends of the mainshaft in the manner of a top direct drive. When first gear on the mainshaft is in engagement with the gear on the layshaft the gears operating are A and B (Fig. 6A), first layshaft gear drives first mainshaft, which in turn drives the propeller shaft. The same applies to second gear except that the second speed on the layshaft drives the second speed gear on the mainshaft.

The third and fourth selector rod carries two separate forks, which engage with the strip tooth gears on the mainshaft. When the gear lever is pulled back, the third speed engaging pinion registers inside the interior of the free running constant mesh gear at the rear end of the mainshaft. The drive then takes place through the front constant gear on the mainshaft to the corresponding gear on the layshaft, through gear " C " of the layshaft, and on to gear " D " on the mainshaft (Fig. 6A), the actual coupling being through the agency of the internal tooth gear. During this time the small internal gear at the front end of the mainshaft is out of engagement with the constant mesh "A." Before top speed is engaged the gear lever is slipped into neutral, disengaging the third speed pinion from the constant mesh wheel.

Fig. 6. A, B & C—Details of gearbox.

The direct drive is engaged when top gear is obtained. The mainshaft is now locked up solid with a direct drive from the engine to the propeller shaft.

The layshaft will continue to rotate on the ball bearings when all gears are running idle, as will gear " D " on its needle bearing.

REMOVAL OF GEARBOX. (F type Magna.)

Remove seats, carpets and floorboards and drain gearbox.

Disconnect mixture control and slow running control just forward of gear change lever.

Remove gearbox extension and lid and cover aperture with piece of cardboard. Remove footramp and disconnect speedometer drive at the gearbox end. Suitably mark the propeller shaft flange and gearbox universal joint flange, for reassembly purposes and disconnect propeller shaft flange from gearbox universal flange. Remove the bolts securing the clutch housing to the flywheel housing, suitably supporting the gearbox during the operation, and lift gearbox clear of car. When removing gearbox, make quite sure that no load is placed on the drive gear shaft and clutch centre.

DISMANTLING THE GEARBOX. (F type Magna.)

Dismantling is facilitated if the gearbox is supported in a vice and this can be done by machining a thread on the end of a small piece of bar to screw into the drain plug hole.

Refer to Figs. 7, 8, 9 and 10 before commencing dismantling. Removal of the front end cover exposes the ends of mainshaft and layshafts. Withdraw the clutch shaft complete with ball bearings and constant mesh gear. Tie a piece of string round the needle bearings before the clutch shaft is completely withdrawn, to prevent dropping out.

The rear end of the gearbox with cover removed is shown in Fig. 9. When dismantling the gearbox, there are a number of shims employed which should be carefully laid aside and marked with a label where they belong.

The only point which might cause difficulty when reassembling the box may be the manner in which the gear wheel and needle bearing is fitted to the rear end of the box. The needles do not run between the gear onto the shaft direct, there being a sleeve bush between the needle and shaft.

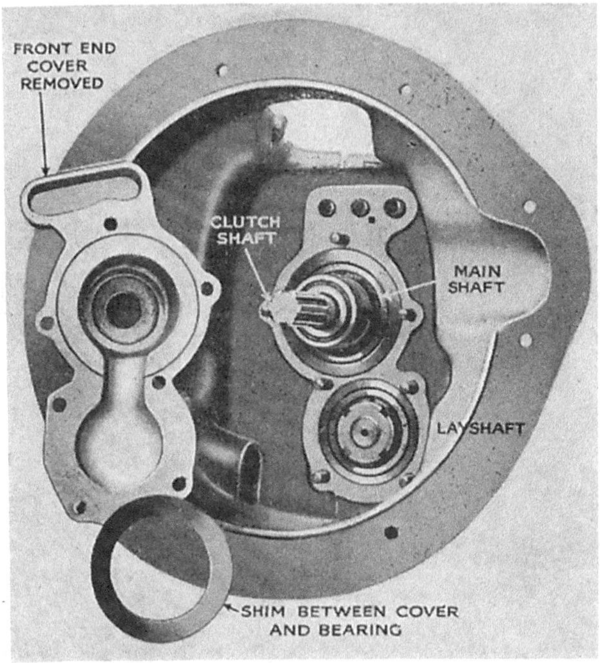

Fig. 7. Clutch housing showing end cover removed.

While withdrawing the mainshaft from the gearbox, through the front end, tap the needle bearing centre sleeve backwards with the gear until it disengages from the shaft.

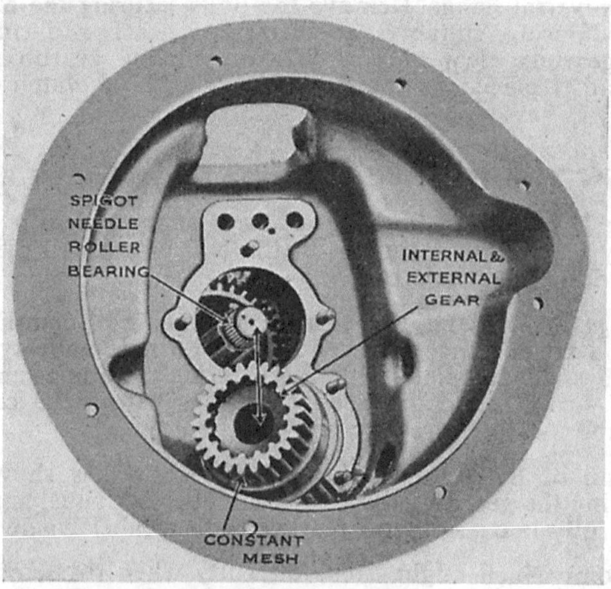

Fig. 8. After front end of cover has been removed it is possible to withdraw front end mainshaft and bearing and front end housing. The remaining gears are removed by extracting shafts and lifting gears through top opening of gear box.

Fig. 9. Rear end of gearbox, cover removed.

Reference to Fig. 10 shows the ends of the reverse and layshafts, the layshaft being locked up by means of tab washers. Removal of one of the nuts enables the shaft to be driven out through the gears from either end. Assembly of the gears is carried out by putting the gears into the box through the top, and drawing the shafts through the gears from the ends.

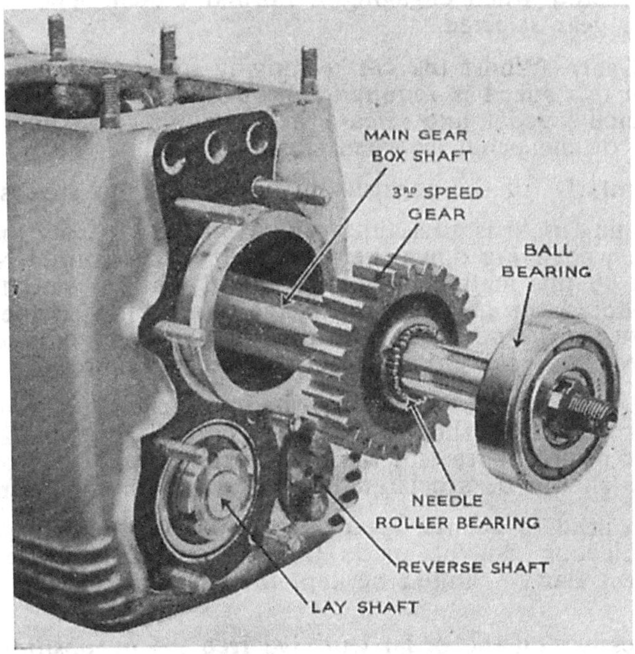

Fig. 10. *Showing how roller bearing fits inside third speed gear on mainshaft. The gear cannot be drawn through the rear housing. Illustration is purely to illustrate the method of construction.*

PRE-SELECTOR GEARBOX

The pre-selector gearbox, manufactured under Wilson patents, fitted to M.G. Magnette gears requires comparatively little instruction. It is as well, however, to know what is taking place inside the box during the operation of " de-clutching " and selecting the gears.

The gears work on what is known as the epicyclic principle, or, as they are sometimes called, planetary gears. The gear wheels are always in constant mesh, and the action of the gears takes place by the contraction of a number of brake bands which grip a series of brake-drums or annuli, depending upon the position of the selector lever; or in other words, the selector lever predetermines which gear will be operated when the " clutch " pedal is pressed down while the engine is running, and then the pedal let up in the ordinary way. The term " clutch " pedal is used in the sense that it is the medium which disengages the power from the engine and, while having the same effect as an ordinary clutch, operates, of course, in an entirely different manner.

HOW TO OPERATE THE CONTROLS.

To Start.—Before starting the engine, see that the hand brake is applied, make sure that the gear is in neutral by moving the pre-selecting lever to the neutral position, and depressing the pedal to its full extent and releasing it. Then start the engine and move the pre-selecting lever to the first gear position (or reverse, should it be required) and depress the pedal to its full extent and release it, controlling the final part of the upward stroke of the pedal as one would when engaging a normal clutch. The car will begin to move in the gear selected.

To Change Gear.—Whilst the car is moving forward in 1st speed, select 2nd, and when this speed is required relieve the accelerator pedal, depress fully the " clutch " pedal and release it, controlling the final part of the upward stroke of the pedal as when starting.

Proceed similarly for the engagement of 3rd and 4th speeds.

When changing up, it is advisable to relieve the foot from the accelerator pedal, and when changing down the accelerator pedal should remain partly depressed, and by this method it will be found that the smoothest changes are effected. Should the gear fail to engage on account of the pre-selector lever being between the positions indicated, move pre-selector to the correct position, depress and release clutch pedal a second time.

It is usual to run on top with 3rd selected and ready for an instantaneous change when required. It should, however, be remembered that complete declutching for any purpose will then cause the engagement of 3rd speed, and the engine should be accelerated to ensure a smooth change.

There is no need to change speed merely because the selector has been moved, and selection can be made long before the change is actually required. " Select Early " should be kept in mind and driving will become much simpler.

A slight pressure on the pedal will give free engine position.

To Reverse.—Although reverse may be selected while the car is moving forward, on no account should it be engaged until the car has been brought to rest. The engagement of reverse is effected by the depression and release of the pedal as explained above.

A safety catch is incorporated on the selector lever to prevent reverse being selected by accident.

To Stop.—Depress and hold the clutch pedal right down before the car comes to rest. If it is wished to proceed immediately, select 1st speed and engage clutch as described in paragraph " To start." If on the other hand a prolonged stop is expected and the driver wishes to release his left foot, neutral must be selected before the pedal is allowed to return. When neutral is engaged the pedal will remain partly depressed.

When changing down care should be exercised to see that the speed of the car is not greater than the equivalent engine speed; for example, if the car is running at 30 miles an hour it would not be advisable to change from top into 1st gear, because this would cause the engine to run at a much greater speed than that for which it is designed, and it would be exactly the same thing as coasting down a steep hill on 1st gear with the clutch out, and after the car has attained a momentum of 30 miles an hour it would be inadvisable to let the clutch in.

Before attempting to follow the operation of the gears it is as well for the owner to study Figs. 11 to 14, which show the exterior of the gearbox from four points of view.

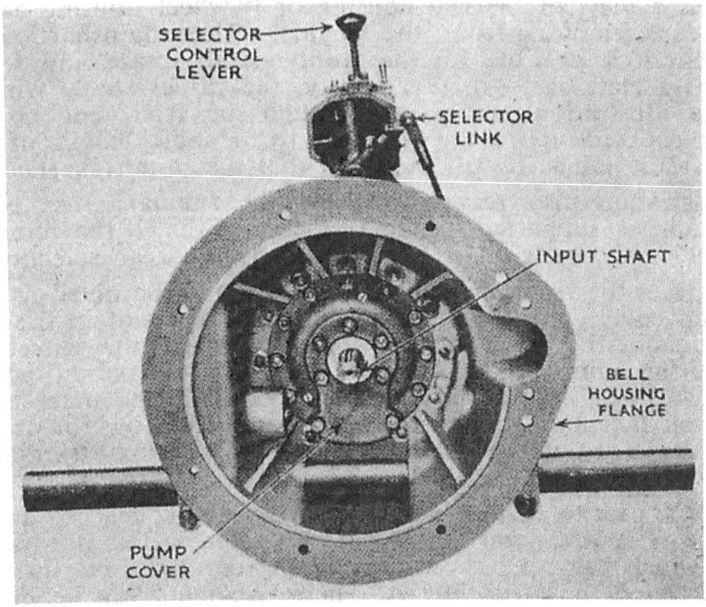

Fig. 11. View of the gearbox units attached to the bell housing. The input shaft is coupled to the flywheel by means of a short coupling shaft having rubberised fabric joints.

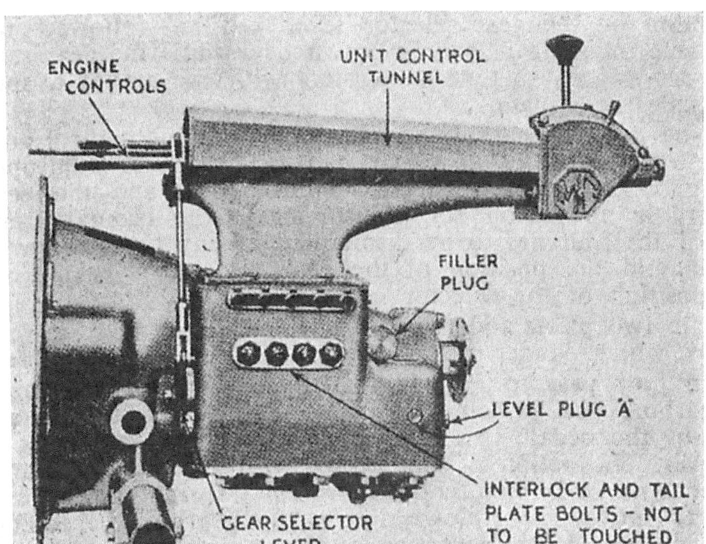

Fig. 12. Near side of the gearbox showing the position of the unit control tunnel and the gear selector lever. The principal points in this illustration which concern the user are the position of the gearbox filler plug and the level plug " A ". Under no circumstances must heavy oil be used in the gearbox, and it should not be overfilled.

Fig. 11 is a view of the bell housing or flywheel housing as seen after this has been removed from the engine. The mainshaft is in two parts, so that it is possible for the front end to rotate and for the rear end to remain stationary—that is to say, the in-put shaft, which can be seen in the illustration protruding through the front end cover of the gearbox, is coupled up to the flywheel by a short length of shaft and rubberised fabric joints. It will be noted that the in-put shaft is splined.

The illustration also serves to show the tubular cross member by which the unit is supported in the frame. In Fig. 11 the remote control tunnel supports the selector link which effects the gear selection.

The shape of the front end gearbox cover should be noted as it contains the oil pump cover. The object of the pump is to maintain the circulation of oil throughout the whole of the working parts of the gearbox, but the operation of the pump will be described later. It will be noticed, however, that there are three rows of nuts in the illustration, the inner circle being used to attach the pump to the housing, the middle row for attaching the housing to the body of the gearbox, and the outer for attaching the bell housing to the gearbox.

We will now pass to Fig. 12, which is a view of the gearbox as seen from the near-side. The first point that should be noted in this illustration is the filler plug, which has been shown removed from the filling orifice. Below this will be found the level plug A. These two items can be more clearly seen by referring to Fig. 14. Under no circumstances should the gearbox be filled without removing the level plug A.

The oil should be changed when the car has completed the first 1,000 miles. It should be replenished every 1,000 miles, and every 3,000 miles the plug at the bottom of the gearbox should be removed and the old oil drained out, and the gearbox subsequently refilled up to the proper level. Under no circumstances should thick oil be used.

The action of the gear selector lever can be followed in Fig. 12. When the selector control lever is drawn backwards the gear selector lever is moved downwards so that it registers with the aid of a spring-loaded ball in recesses in a plate.

It will be noticed that the rod between the unit control and the gear selector lever is threaded at either end to permit of adjustment should same be necessary on the selector control lever and the gear selector lever. There is no reason for adjustment in the ordinary sense of the word, but if the unit has to be dismantled at any time, the adjustment is provided should the position of the gear selector lever not synchronise with the position of the selector control lever.

There are two plates and two rows of nuts to be seen on the near-side of the box which, under no circumstances, require attention.

We will now pass to Fig. 13, which is a view of the opposite side of the gearbox; the principal point to note being the lever which is controlled by the pedal. This lever operates what is known as the busbar rocking lever. The action of the busbar will be explained later, but, in short, it consists of a plate coupled up to a spring; the plate or busbar is caused to move up or down inside the gearbox, depending whether it is being pulled up by the action of the spring or pushed down by the action of the pedal.

The last illustration of the exterior of the gearbox shows what has been previously described as seen from the rear of the box. Fig. 14 shows the filler plug and level plug A; the selector control lever operating in a slot which, besides having the number of the gear marked on it, has stops which make the pre-selection a matter of some simplicity. A little experience in handling the box makes it clear that a slight rocking

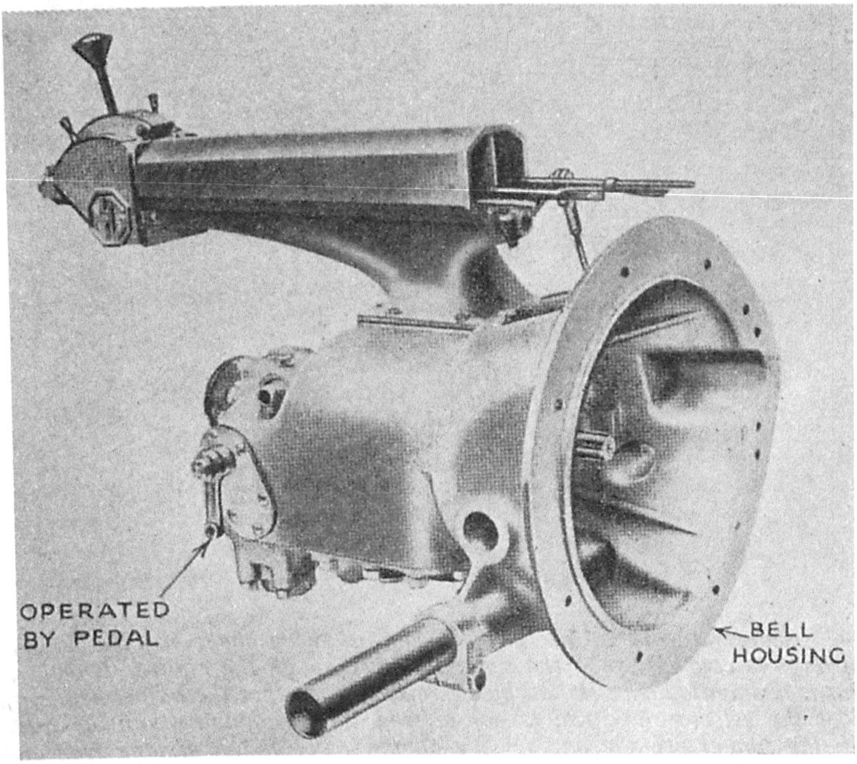

Fig. 13. Offside of the gearbox showing the method of suspension and the short lever which is coupled up to the " clutch " pedal. When the lever is moved it has the same effect as a clutch, although its operation is entirely different.

movement is necessary when the control lever is moved forward and backwards to make a change of gear. The remaining control levers are neatly arranged around the selector control lever for ignition, carburation and shutting off petrol.

There are two things that the owners will probably want to know at the outset, namely, what happens inside the gearbox when the selector lever is moved to and fro, and also what happens when the pedal is pushed down and then let up again. These two questions can be answered as follows. If the engine is not running, or if the engine is running and the pedal is not being pushed out, movement of the control lever simply rotates a camshaft inside the gearbox without causing anything to happen, and secondly when the clutch pedal is operated it causes a lever to rise and fall in the gearbox to either grip or slacken the brake bands which will be referred to later. The operations of these two units will be seen on referring to Fig. 15. Here we see the gearbox upside down after the contents have been removed.

Previous reference has been made to the in-put and out-put shafts. They can be seen in Fig. 15, wherein the in-put shaft registers in a spigot bush inside the out-put shaft. The train of gears, brake bands and drums have been removed so as to show the shafts and to afford a view of those parts of the gearbox which otherwise could not be seen.

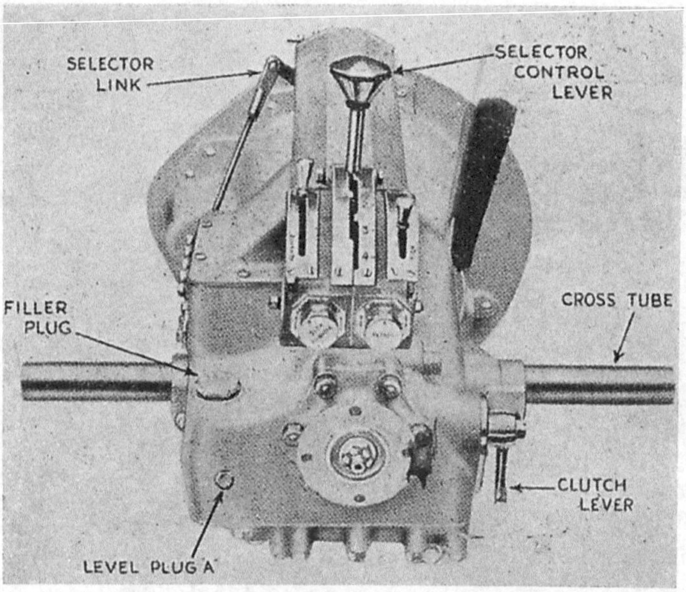

Fig. 14. *Rear end view of the gearbox unit showing the controls. A selector control lever is coupled up to the selector link through the control tunnel, and if ever this is dismantled care should be exercised to see that the adjustment with the length of the selector link rod is not altered. The remaining controls operating through the tunnel are the engine slow-running, petrol tap change-over and the ignition and jet control.*

The first question asked was, What does the camshaft do? The camshaft does not have cams similar to those employed in operating engine valves, but they consist of nicks in circular discs permitting any one of the series of five selector springs to fall into the notch in the cam depending upon the position of the selector lever. When the lever is put into position of first gear selection the camshaft rotates, allowing the notch in No. 1 cam to let the selector spring No. 1 stand proud of all the others, and it is the action of this spring which will cause the brake band of No. 1 train of gears to come into operation through the action of the operating struts. A number of small levers will be seen below the camshaft which act as gear interlocks, making it impossible for two gears to be engaged at once. Their action is entirely automatic, and the manner in which they are worked will be described later.

The second question which was asked was, What is the action of the clutch pedal inside the gearbox? Fig. 15 answers this question in so far that it shows that the "clutch" pedal causes a lever inside the box to move up and down. This lever is known as the busbar rocking lever. It will be noted that at the inner extremity of this lever there is a steel roller let into the fork end of the lever. Before passing to the next illustration it will be remembered that Fig. 15 is the view of the gearbox taken from below, and is really, so far as the reader is concerned, upside down. With this in mind pass to Fig. 16 and first of all look at the part marked 1 Busbar, and then carry the eye along to the point marked 2, which is a raised track on the busbar which engages with the busbar rocking lever shown in Fig. 15.

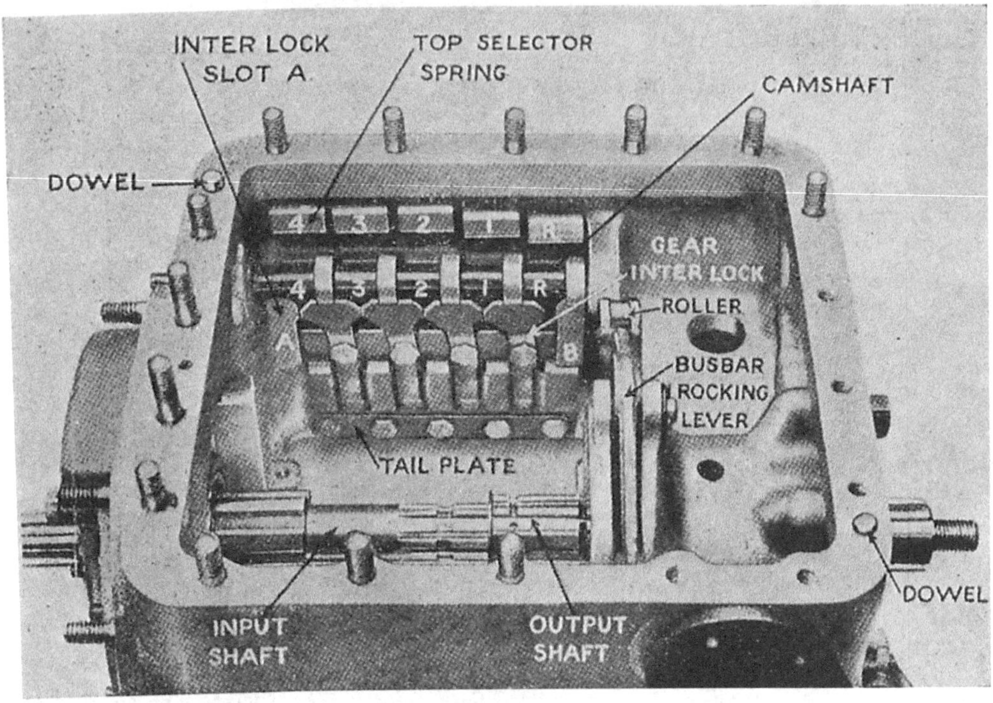

Fig. 15. *Interior of the gearbox, showing the input and output shafts, but without the actual gears and bands.*

To understand the gearbox properly, it is necessary to read the text. The action of the pedal is to cause the busbar rocking lever to rise and fall. On referring to Fig. 16 *the roller on the end of the rocking lever comes into contact with the boss* (2) *on the busbar. Fig.* 12 *shows the external view of the interlock and tail-plate bolts.*

In this illustration the following parts can be seen : the camshaft, the selector springs and the gear locks. These three parts control the actual gear to be engaged. Rotating the camshaft by moving the selector control lever to and fro permits the selector springs engaging with the operating struts shown in Fig. 16, *and the action of the gear interlock device prevents more than one gear being engaged at once.*

When the pedal is pushed down the rocking lever moves downwards, pressing upon the point of the busbar marked 2. The busbar pivots on supports (3), and it will be noted that the point of support has a knife edge. The outer extremity of the busbar, at the point 4, is coupled by means of a link to the spring compression lever 5, which is clearly under the action of the spring 6. It follows, therefore, that when the pedal is pressed down the busbar is forced downwards, thereby compressing the spring, and when the pedal is let up or back, the spring 6 causes the busbar to rise, carrying with it one of the operating struts which through the action of the various linkages at the top will cause one of the brake bands to contract. The busbar has a groove running the complete length, and when the selector lever is in neutral the first and reverse struts are engaged, and owing to the action of the gear interlock neither strut can reach the engaged position and the busbar will remain partly depressed.

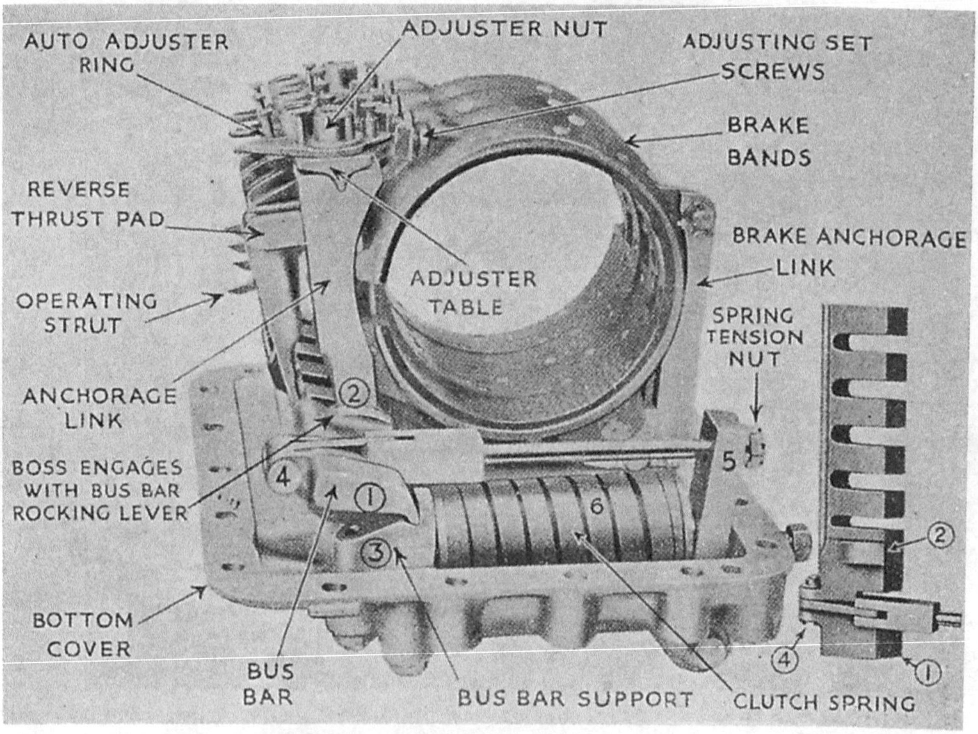

Fig. 16. *This is a view of the base of the gearbox on to which the brake band assembly is mounted on anchorage links. Operating struts are caused to rise and fall by the action of a busbar which is under the control of the spring* (6); *the busbar terminates at* (1), *making knife-edge contact on the bracket* (3) *when the clutch pedal is in its normal position. Expansion of the clutch spring pulls on the point* (4), *causing the busbar to rise with it and contract the selector brake band. The mechanism on the top of the anchorage link consists of an automatic adjusting device, to take care of any wear in the bands and give uniform pressure. When the clutch pedal is pressed down, the busbar rocking lever forces the busbar outwards, thus freeing the brake band until the pedal is let up again.*

A side view of the bands and operating mechanism can be examined after the gearbox casing has been removed from the base plate. In Fig. 17 the busbar, through the action of the spring, is in the up position, and it will be noted that the third speed actuating strut has been engaged with the groove in the busbar, and has been forced upwards while the other operating struts are entirely free of the busbar's action. The whole action of the busbar is to spring load the brake band contracting levers or linkages so as to cause the bands to grip the drum.

The first question that was asked was, What was the action of the camshaft? This action will now be made clear when it is realised that the flat springs inside the box, which are allowed to move forward when they drop into a notch in the cam, come in contact with the operating struts, with the result that when the pedal is pushed out and the busbar is allowed to fall, the spring will force the operating strut sufficiently far forward to engage with the groove in the busbar and allow the

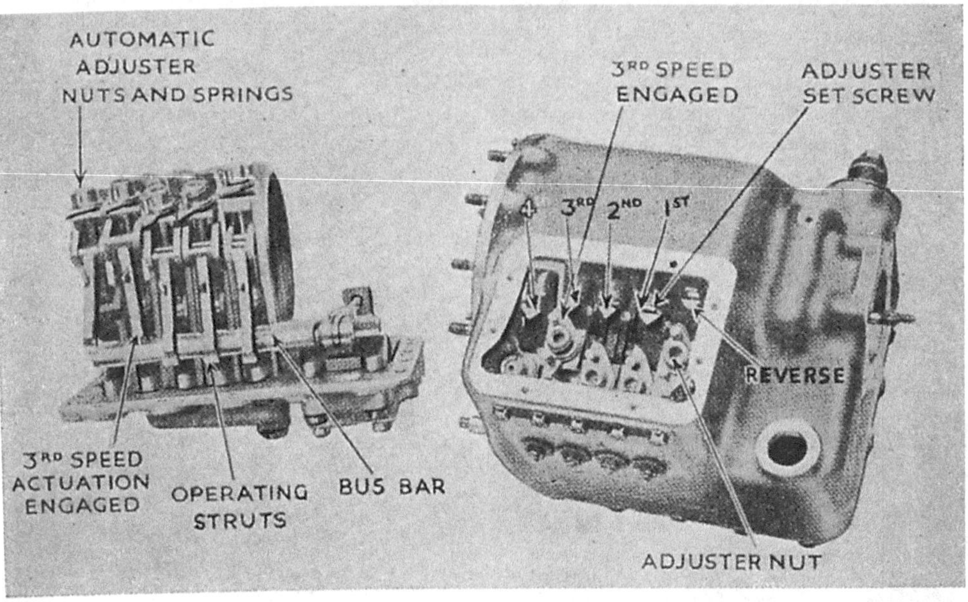

Fig. 17. *The left-hand illustration will make the action of the busbar and operating struts quite clear. It will be seen that the second strut from the left is engaged by the busbar and is forced upwards, while the other struts are free of the busbar and do not engage with it at all. The action of the busbar rising can be seen by referring to the right-hand illustration. It will be noted that the adjuster set screws on the bands are all free of the adjuster nut, except the third gear, which has been forced up against the adjuster set screw by the action of the busbar.*

particular pre-selected gear to come into operation. Carrying this further, the true understanding of the pre-selector gearbox will be grasped. Suppose, for example, that the car is running in third gear, the busbar will be up and the third speed brake bands will be contracted. The action of rotating the camshaft by moving the selector lever will have no other effect than causing the springs to move forward and backward, with the result that they will simply press up against the operating struts which are lying idly in the box. These operating struts cannot engage with the busbar because, on referring to Fig. 17, it will be seen that it is necessary for the busbar to drop considerably first. When the busbar does, however, drop through being forced down by the pedal, and another gear has been pre-selected, the small spring contained in the linkages of the operating struts will throw the strut that has previously been in operation outwards free of the busbar, and then the action of the flat selector springs on the side of the box will cause the next pre-selected gear strut to be forced forward or towards the busbar, so that it in turn will engage with the groove of the busbar and in turn will be forced upwards and cause the particular band of the pre-selector gear to be contracted.

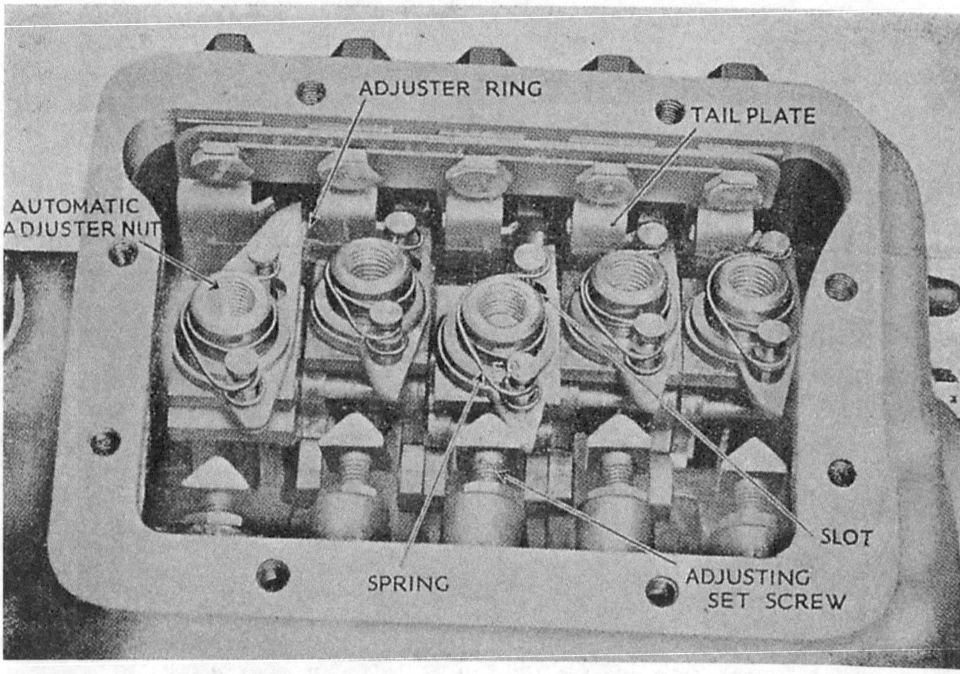

Fig. 18. View that can be obtained after removal of the gear-box cover lid. Follow the arrows to these parts: automatic adjusting nut, adjuster ring, adjusting set screw, tail plate, and spring, then note the slot in the adjuster ring, and the part marked FP, i.e. fixed pin, in the opposite half of the adjuster ring. Unless proper advice is obtained from somebody who knows definitely what the action of the adjusting set screw really is, under no circumstances should these be touched. They have a definite function to perform, but it has nothing to do with the actual contraction of the brake bands; they form stops up against which the adjuster ring abuts, so that when a gear is engaged, the adjuster ring is forced up against the stop and a certain angular motion is imparted, causing the adjuster plate to pivot around the nut, to the limit of the slot. The manner in which the spring is actually around the nut, when the clutch pedal is pushed out, will cause the adjuster ring to move outwards against the tail plate and rock the adjuster ring in the opposite direction, when it comes in contact with the adjuster set screw. This to and fro action causes the spring surrounding the automatic adjuster nut to grip it, which will cause it to rotate slightly on its bolt and so shorten its length of travel and automatically maintain proper adjustment.

If there is any semblance of drag or lack of quick response when the clutch is let in, " pump " clutch pedal, which means pushing it in and out fully several times quickly to take up any slack.

On the right-hand side of Fig. 17 it is possible to examine the interior of the gearbox as is shown when the top cover is removed. The top cover is formed, in the case of the M.G. Magnette, by the unit control tunnel mounting, and in this illustration the third speed engagement that was shown previously is viewed from above. The two further points that can be seen in this illustration are what are known as the adjuster set screws and the adjuster nuts. The object of these is to permit of the wear that takes place on the brake bands being automatically compensated for by a corresponding shortening of the link which is attached to the brake band at the lower extremity and to the adjuster nut at the upper extremity.

Fig. 18 is a close-up view looking inside the gearbox after removal of the cover. It will be noted that the adjuster nuts can be seen in Figs. 16 and 17, screwed on to a link, and beneath the nut there is what is known as an auto adjuster ring and below this an adjuster table. The auto adjuster ring has a special shaped pin let into one side of it and a hole on the other side through which a pin, riveted into the adjuster table, passes. Surrounding each nut there is a spring specially coiled. The object of this spring is to grip the circular adjuster nut. This is effected by the shape and an action of the adjuster ring. It will be noticed that these are tapered at either end. The centre adjuster ring is contacting with what

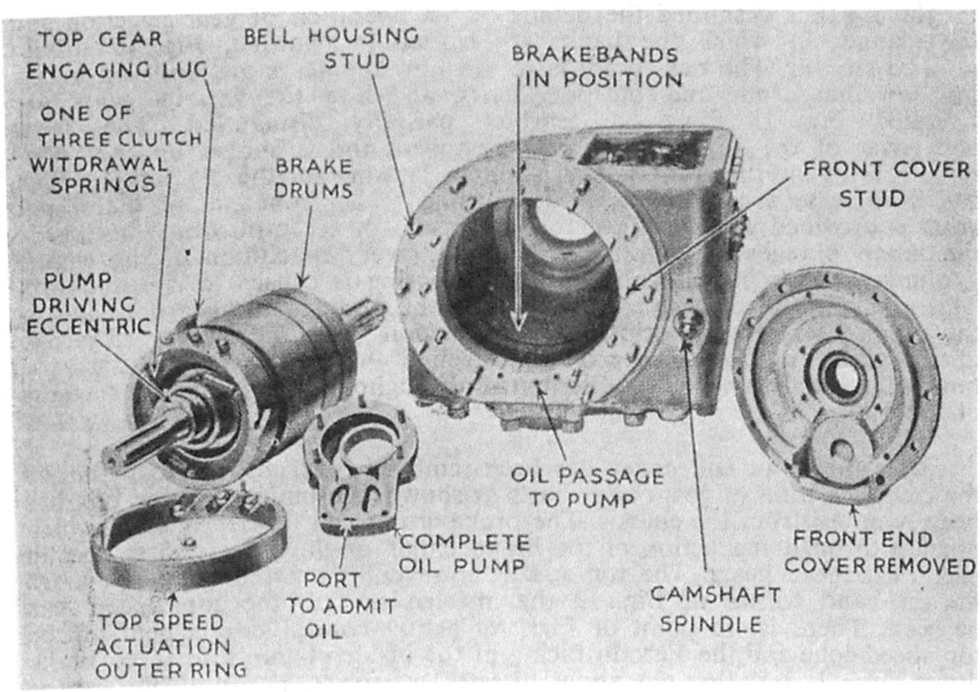

Fig. 19. The front end of the gearbox with the cover removed and complete brake-drum assembly withdrawn to show the brake bands in position. The oil pump is fitted on to the front cover and is driven by an eccentric on the front end input shaft. Oil is drawn from the gearbox and is delivered to the hollow shaft, so that all the gears are constantly lubricated.

is known as the brake band adjuster screw. Reference to Fig. 16 will show these screws and their relationship to the brake band. They are virtually abutments or stops against which the adjuster ring rests when any particular gear is engaged. In Fig. 18 the centre gear is obviously engaged. As wear takes place the two angular faces of the adjuster set screw and adjuster ring at the time of engagement will cause the adjuster ring to pivot slightly. This action, owing to the shape of the spring, will cause it to rotate around the adjuster nut. When another gear is engaged and the whole unit is forced outwards, the opposite end of the adjuster ring will come in contact with what is known as the tail plate and rock the adjuster ring in the opposite direction and tend to tighten it round on the thread of the link. By this means the brake bands are being constantly kept in correct adjustment, and should for any reason a band slip when engaged, it is only necessary to move the pedal fully about twenty times to and fro for the automatic adjuster to take up the slack.

As in every other form of mechanism, excessive wear can be caused by abuse, and it is advisable in changing up to allow the engine speed to be reduced before the next gear is engaged, which is done by lifting the accelerator pedal while the clutch pedal is depressed and not re-accelerating before the bands have had time to grip the drums.

Having thus described the details of the operation of gear selection and the manner in which the bands are caused to contract, Fig. 19 should now be studied. The next point is to see how the gears are built up into a unit on the in-put and out-put shafts, and how the various gears are operated. Fig. 19 shows a gearbox partially dismantled. The front end cover of the gearbox has been removed, and is shown on the right-hand side of the illustration. This forms a housing for the oil pump, which can also be seen complete with its housing. The front end of the in-put shaft is provided with a cam-shaped extension or eccentric which fits inside the pump plunger extension, so that with every revolution of the engine a plunger which fits into an oscillating housing is caused to rise and fall. This sucks up oil through a small hole in the body of the gearbox, and as the pump oscillates and closes up the suction hole, the downward stroke of the pump plunger forces oil up through the hollow plunger into the hollow gearshaft, and is so distributed throughout the whole inner works of the box.

After the front end cover has been removed it is possible to withdraw the complete train of gears and shafts as shown, presuming that the box has been removed from the chassis. The brake-drums can be clearly seen, which operate through the action of the brake bands of the reverse, first, second and third speed gears. The top speed, however, is differently operated. On the left-hand corner of Fig. 19 the mechanism of the top speed can be seen. There is, in point of fact, an actual conical engagement of the top speed cone and the Ferodo facing of the inside of the third speed brake drum. This is not brought about directly by spring pressure, but by the action of the top speed actuating outer ring, which is held stationary up against the front end cover by means of three set screws and the steel balls which are positioned in helical slots.

Reference to Fig. 20 shows the train of gears, brake-drums or annuli after the drums have been slightly drawn apart to show the inner planetary gears. This illustration also shows the top speed cone and the helical ball tracks which, when top speed is engaged, cause the cone to travel forward on the shaft and engage with the Ferodo facing on the inside of the third speed brake-drum. When top gear is engaged, the whole train of gears and drums rotate as a single unit, there being no gear movement at all. It will be seen from this illustration that the shaft is supported by radial and thrust bearings, the rear end bearing, however, only being visible. The rear end of the out-put shaft is provided with a keyway for fitting a gear wheel to drive the speedometer.

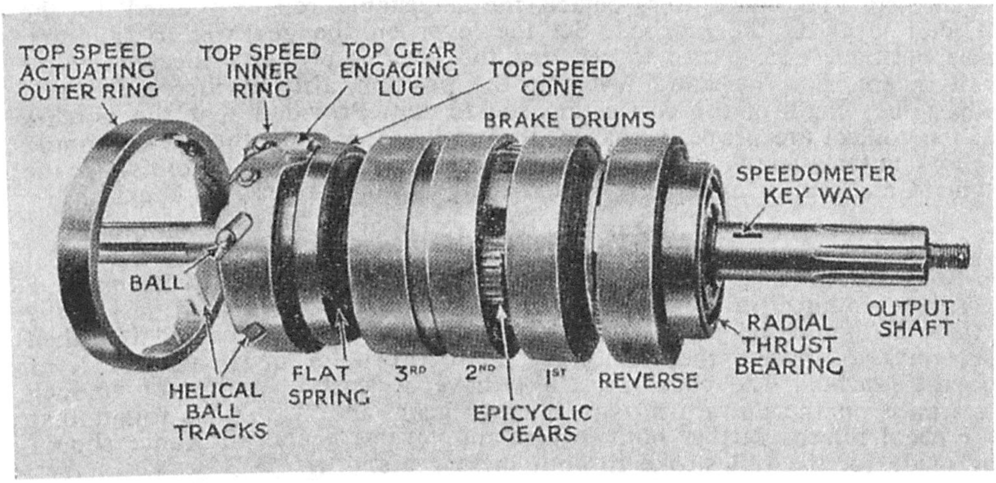

Fig. 20. *Complete set of brake-drums and epicyclic gears. The parts have been slightly strung out to show the internal gears, the top speed cone engagement with the third speed drum, and the manner in which the top speed is actuated by its steel balls, forcing the top speed inner ring forward under the action of the helical ball tracks, and the top speed actuating outer ring.*

The following points should be carefully observed by the Driver:—

1. Always feel that the pre-selector lever is well in its notch opposite the desired gear before actuating the pedal.
2. Always depress the pedal to its full extent when making a change of gear, and do not rest your foot on the pedal after the change is made.
3. Always remember that it is the end of the upward stroke of the pedal which decides the smooth engagement of the gear. Therefore, control this part of the stroke of the pedal carefully with your foot.
4. Before leaving the car always make sure that neutral is not only selected but engaged by fully depressing the clutch pedal with neutral selected.
5. IMPORTANT.—K. Type cars fitted with the pre-selector gearbox have an oil pump fitted inside, which operates only when the car is in gear. These cars therefore must not be towed, unless the propeller shaft is disconnected. Failure to do this will cause the gearbox to seize, as when the gear lever is in the neutral position the oil pump does not function and seizure takes place.

ADJUSTMENTS TO PRE-SELECTOR GEARBOXES

NO ADJUSTMENTS SHOULD BE ATTEMPTED WITH THE ENGINE RUNNING

Checking the Selection. Should any gear persistently fail to engage it will probably be caused through a derangement of the selector control mechanism. To check this, detach the adjustable rod connected to the selector lever on the gearbox. Set the lever on the gearbox to say, first gear position, taking care to see that the lever is properly located by the ball detent. Set the hand lever in the position also to first speed and adjust the length of the connecting rod to suit. Provided that the notches of the control are properly spaced, the positions of the other gears should follow automatically. It is also advisable to go round all joints on the control mechanism making sure that there is no looseness or backlash.

Check Over Pedal Clearance. It is essential that there should be a definite clearance between the pedal and the floorboards when the pedal is fully depressed and that the lever on the box is moved to its extreme position, coming up as it should do against the stop situated inside the gearbox. Should the pedal touch the floorboards on depression, it is then necessary to shorten the connecting rod between the pedal and the lever on the gearbox. The pedal must also have at least a quarter of an inch clearance on the upward stroke in every gear, and it will be found that the pedal returns further both in first and reverse gears, allowance should be made for the full stroke in both these gears.

Adjustment to Overcome Fierceness

1. Remove top cover.
2. Screw the 'V' shaped adjuster setscrew on the band requiring adjustment one turn out and lock with locknut. Select the gear and with the engine stationary actuate the pedal a number of times, say 20 times, to allow the automatic adjustment to take up the new setting.

Adjustment to Overcome Slipping

Slipping could occur for three reasons:—

1. That the automatic adjustment is failing to take up. This can be determined by removing the top inspection cover of the gearbox and selecting and engaging the speed required. If the wedge-shaped plate on top of the control mechanism strikes the 'V' shaped set screw on the band and is moved an appreciable amount, then the automatic adjuster should work by actuation of the pedal.

   To test this, make a pencil mark on the thimble nut, actuate the pedal briskly 20 times or so, and note if it has moved at all. If the adjuster ring is being moved a reasonable amount with each stroke of the pedal and the mark on the nut does not turn, change the adjuster nut spring and repeat the "pumping action."

THE AUTOMATIC ADJUSTER SHOULD ONLY WORK IF THE WEDGE-SHAPED PLATE IS TURNED AN APPRECIABLE AMOUNT ON STRIKING THE SETSCREW.

2. Insufficient toggle action.

   This is shown in Fig. 18, page 168, and the remedy is to screw IN the 'V' shaped adjuster screw on the band of the speed to be adjusted to the extent of about one turn.

   Next unhook the top loop of the automatic adjuster spring, unscrew the thimble nut by means of a B.S.F. Bolt and Lock nut about a quarter turn, replace the automatic adjuster spring exactly as before, select the speed required and after marking the thimble nut, actuate the pedal briskly until the nut comes to rest.

The amount to which the 'V' shaped set-screw is to be put in can best be found by experiment. Whenever the set-screw is put in, remember to unscrew the thimble nut a certain amount to enable the toggle action to come up to the set-screw, and always allow the automatic adjustment to bring the thimble nut to its correct position.
NOTE:—It is dangerous to screw in the 'V' shaped set-screw too far as when the busbar exceeds 45 degrees as shown in Fig. 19, the busbar reaches a stop and no power will be applied to the brake.

3. If the pedal fouls the floorboards on return the gear cannot be fully engaged—consequently the length of the pedal rod should be increased.

Sticking in Top Gear

Should the top gear fail to disengage immediately the pedal is depressed this is probably caused by high spots in the fabric lining and the cone should be rebedded.

This is effected by selecting Top speed with the car stationary and the hand brake on, the engine should be run at approximately 1,500 r.p.m. The pedal should be let back to reduce the engine speed to say, 500 r.p.m. but not far enough to stall the engine. This treatment should not be carried out too long but should be repeated several times with short intervals to prevent overheating.

Drawing the oil and refilling may also be tried.

DISMANTLING '75' TYPE PRE-SELECTOR GEARBOX

1. Remove top cover and undo each automatic adjuster nut three or four turns. To do this, remove the large eye of the automatic adjuster spring from its pin, place a $\frac{7}{16}$ in. B.S.F. set-screw with locknut into adjuster nut, lock up locknut and unscrew the required number of turns.
2 Remove front cover. Undo the eleven $\frac{1}{4}$ in. B.S.F. nuts on the outer diameter and five smaller nuts on the inner diameter, also the two cheese-headed screws, tap off the front cover preferably with two sharp wedges. While doing this tap gently on the five studs to keep them coming away with the cover. Now remove the pump plunger and oscillating cylinder, next, the pump housing. Note the position of the three coil springs for assembly. Now remove the two countersunk screws seen at "one o'clock" on the actuating ring and remove this ring. If tight it can be extracted by using a plate and the two holes used by the cheese-headed screws of the front cover, but first remove the input shaft, taking care of the steel balls which will come away when this ring is removed.
3. Remove gearing. The running gear can now be removed piece by piece, noting carefully the location of the planetary trains, bushes, shims, etc. To remove reverse and last annulus this should be driven out from the rear end after having first removed the output shaft from the rear end.
4. To remove gear case from bottom cover. Remove all the nuts fixing bottom cover to case, also extract all the long studs, numbering eight. Now remove the plug immediately below the busbar actuating lever bracket and insert here a $\frac{7}{16}$ in. x $1\frac{1}{2}$ in. Whitworth stud screwed right up to the head and screw this in until solid. This depresses the busbar and enables the case to be removed. All the bands and busbar can now be removed from the bottom cover. Before removing the spring tension nut from the busbar note carefully the number of washers between the end of the nut and split pin hole. The correct quantity should be replaced when renewing the busbar.
5. To re-assemble. The case is placed in position over the finished bottom cover and as no paper joints are fitted a good jointing compound should be used. See that the two dowels locate, then tap the case into position with a hide mallet, refit the eight long studs, replace all the 18 nuts and tighten up.

6. To replace running gear. This is best assembled complete on the bench after having first removed the output shaft rear end ball race. The whole mass can now be pushed through the case, steadying the rear end with the output shaft, and a steady pressure put on the input shaft to ensure that the ball race fitted to the rear annulus goes well home into its housing. It is now best to tip up the box so that the front cover can be put on working from the top, taking care that the output shaft does not slip out of place. Should there be any difficulty in replacing the running gear, remove the speedo pinion as this might not mesh with the wheel on the output shaft. The outer actuating ring can now be replaced, locating the two dowels on the 4th speed actuating support, the two countersunk screws replaced and tightened up. Make sure that the lip on the inner actuating ring engages correctly with the hook on the top speed pull rod. Now drop in the steel balls into their respective slots, replace the pump housing and see that the three springs are hooked on correctly and return the pump plunger and cylinder.
7. The front cover can then be replaced using jointing compound. Tighten up the eleven outer nuts, the five nuts on the pump housing studs and the two cheese-headed set-screws. All that is now required is to re-adjust the automatic adjuster nuts to within one turn of the correct adjustment, the remainder should be pedalled up to ensure the correct working of the automatic adjuster springs.
   NOTE:—Do not forget to remove the $\frac{7}{16}$ in. Whitworth stud after assembly and replace plug.

To Refit New Bands

Great care should be taken in the fitting of new bands to ensure the centralizers take up their correct position.

REVERSE BAND. See that the centralizer is fitted to band lug correctly, i.e., that the small tit on centralizer is to the top. Next fit band to bottom cover and contract the band on to one of the annuli, using the strut gear and busbar. When band is fully engaged force the centralizer round so that it comes into contact with the facing on the bottom cover. This gives the correct position when the band is off.

FORWARD BAND. See that the correct off and near side centralizers are fitted, then proceed to contract the band as in the reverse. When this is done force the near side centralizer up to the busbar support, and the off side down to the facing on the bottom cover. Do this with all three forward bands, and they will then be centralized properly in their off position.

Inspection of Parts and Gearbox when Dismantled

BUSBAR. If it is found that the operating pad on the busbar is worn, replace with spare busbar supplied.

BRAKE DRUMS. If grooves on brake drums are seriously damaged, we should recommend you to clean up the grooves with a pointed tool.

NOTE:—These grooves are not screwed but are parallel annular grooves.

Lubrication

As the question of lubrication is of the utmost importance in obtaining a sweet pick-up, it is possible that in hot weather slightly heavier grade of oil can be used to advantage. Price's Motorine D, de Luxe oil is most satisfactory. Alternatively Castrol XL could be used. Please see that the oil is changed regularly.

GEARBOX. (1½ Litre, 2 Litre, 2.6 Litre, 1¼ Litre, TA, TB, TC and TD Models).

The gearbox has four forward and one reverse gear. On the 1½ Litre, 2 Litre, 2.6 Litre and TA Midget, synchromesh was incorporated on third and top gears. Synchromesh is incorporated on second, third and top gears on the TB Midget and all post-war models.

Top gear is direct, third and second are in constant mesh, while first and reverse are obtained through sliding gears.

REMOVAL AND REPLACEMENT OF GEARBOX. (1½ Litre, 2 Litre and 2.6 Litre Models).

Remove the front seats, front carpets and the rubber draught excluders from the bases of the clutch and brake pedals. Remove the floorboards and the metal cover of the gearbox (the dipper switch must be disconnected so that it may be withdrawn).

Undo the screws holding the offside ramp plate to the dash and lift out the panel. Drain the gearbox and remove the rubber seal and its retainer from the flywheel housing.

Remove the gearbox lid (after uncoupling the isolated remote control on cars so fitted). Remove the forward section of the propeller shaft tunnel (complete with isolated remote control on cars so fitted). Series SA 2 Litre only.

Remove the starter motor. Uncouple the gear lever remote control at the forward universal joint and the clutch control at the front end. Let the control hang on the clutch pedal. Disconnect the reverse lamp switch and undo the speedometer drive at the gearbox end.

Uncouple the propeller shaft flange from the gearbox flange. Mark these parts for subsequent assembly.

Remove the clutch operating lever from its cross shaft through the clutch housing (1½ Litre Series VA). As the front engine bolts will have to be slackened off to allow slight engine movement, first of all remove the rubber washers and retainers from underneath and undo the bolts. Jack up the rear of the engine and remove the bolts holding the gearbox rear bearer plate to the rear mounting blocks. Also undo the bolts holding the clutch housing to the flywheel housing. The rear of the engine must now be lifted until the gearbox rear bearer plate is clear of the frame brackets, when the gearbox can be withdrawn (1½ Litre Series VA).

On cars fitted with the divided propeller shaft, it is necessary to remove the bolts holding the centre bearing support bracket to the frame brace, but under no circumstances must the centre coupling be dismantled (2.6 Litre Series WA).

Slacken the forward nuts of the gearbox steady blocks and remove the gearbox steady crosspiece downwards (2 Litre and 2.6 Litre Series SA and WA).

Remove the bolts holding the clutch housing to the flywheel housing. Now extract the two studs from the flywheel housing and rotate the gearbox counter-clockwise to give the clutch lever sufficient clearance as the gearbox is being withdrawn (2 Litre and 2.6 Litre Series SA and WA).

NOTE: On the 2 Litre Model Series SA, the studs referred to above apply to early cars only. Later they were changed to bolts.

Replacement is carried out in the reverse manner to that detailed for removal.

REMOVAL AND REPLACEMENT OF GEARBOX. (TA, TB and TC Models.)

Remove the seats, carpets, rubber gearbox cover and floorboards. Undo the rubber muff where the steering column passes through the dash and slide it and its securing plate down the column.

Disconnect the battery, uncouple the harness cable from the ramp plate and also the cables from the dynamo. Remove the ramp plate and undo the electric cables at the starter motor. Disconnect the Bowden wire from the starter motor switch.

Drain the gearbox. Take off the propeller shaft tunnel and the gearbox lid. (On earlier TA Models the rear flange is bolted behind the heelboard of the body so that it cannot be removed completely from the car without lifting the body. This is unnecessary if both front and rear propeller shaft flanges are disconnected and all details securing the shroud removed. The shroud can then be moved to one side at the front end to give sufficient clearance for gearbox removal.)

Remove the starter motor, uncouple the clutch control rod at the forward end and let it hang on the clutch pedal. Disconnect the speedometer drive and uncouple the propeller shaft at the front end, taking care to mark the flanges for subsequent reassembly.

Unbolt the steering box from the frame and also the exhaust pipe between the manifold and the silencer. Place this portion of the exhaust system on one side.

Jack up under the rear of the engine and remove the four bolts holding the gearbox to the rear engine mounting rubbers.

Remove the clutch operating lever from the withdrawal shaft by knocking out the taper pin. Undo the bolts holding the clutch housing to the flywheel housing and withdraw the gearbox complete.

Replacement is carried out in the reverse manner to that detailed for removal.

REMOVAL AND REPLACEMENT OF GEARBOX. (1¼ Litre Model).

Remove front seats from their runners. Unfasten and roll back the carpets and remove carpet from over the gearbox cover.

Remove the four bolts securing the cover to the ramp plate and the nine setbolts securing the cover to the floorboards and remove the gearbox cover. Remove the setscrews remaining and lift out the front floorboards.

Remove the ramp plate, uncouple the speedometer cable, disconnect the reverse light switch, remove the setbolts securing the exhaust pipe bracket to the gearbox and slacken off the nuts securing the exhaust pipe at the manifold flange. Remove the split pin, nut, washer and rubber bush underneath the rear engine mounting.

Raise rear of gearbox and extract clevis pin which is in the forked bolt. Remove the clutch operating rod from the operating lever. Uncouple propeller shaft flange from the gearbox flange, marking the flanges for reassembly in the same relative positions.

Jack up the rear of the engine, remove the ten setscrews securing the gearbox to the flywheel housing, the two bolts securing the inspection cover, and lift out the gearbox.

Replacement is carried out in the reverse manner to that detailed for removal.

Fig. 21. *Gearbox with cover removed showing selector mechanism* ($1\frac{1}{2}$ *Litre series VA*).

REMOVAL AND REPLACEMENT OF GEARBOX. (TD Midget).

Remove floor mats and floorboards. Remove gearbox cowl; disconnect propeller shaft at its forward end, marking flanges for reassembly. Jack up the engine unit under the rear of the sump, using a large piece of wood between the jack and the sump to spread the load.

Disconnect the speedometer drive at the gearbox end; disconnect the clutch operating lever from its connecting rod. Release the rear engine mounting by removing the nuts and bolts from the rubber mounting.

Undo all the retaining screws holding the bell housing to the crankcase and lift out the gearbox, taking care not to place any load on the drive gear shaft or the clutch centre.

Replacement is carried out in the reverse manner to that detailed for removal.

DISMANTLING THE GEARBOX. (TB and TC Midgets, $1\frac{1}{2}$ Litre and 2.6 Litre Models).

The dismantling of the gearbox is a comparatively simple operation, but due to the fact that a second gear synchromesh unit is incorporated, the operation is somewhat unorthodox.

1. Remove the gearbox top. Care must be taken that the selector ball springs are not lost.

2. Remove the propeller shaft flange after suitably marking the flange and shaft so that it can be replaced in exactly the same position.
3. Remove the rear cover, the rearmost ball bearing, the distance tube and the speedometer drive gear.

NOTE: When the divided propeller shaft is fitted (2.6 Litre Series WA) the gearbox rear cover is shorter and the rearmost ball bearing and distance tube referred to above are not fitted.

4. Remove the clutch housing from the gearbox casing.
5. Remove the layshaft locking screw which is underneath the gearbox casing and withdraw the layshaft, allowing the gear unit to fall into the bottom of the box.
6. Remove the selection operating gear, taking care not to lose the two interlocking balls between the selector shafts at the rear end.
7. Engage second gear fully to give the necessary clearance for removing the mainshaft.
8. The first motion shaft can now be removed complete with ball race.
9. The mainshaft is then pushed back sufficiently to draw its ball race

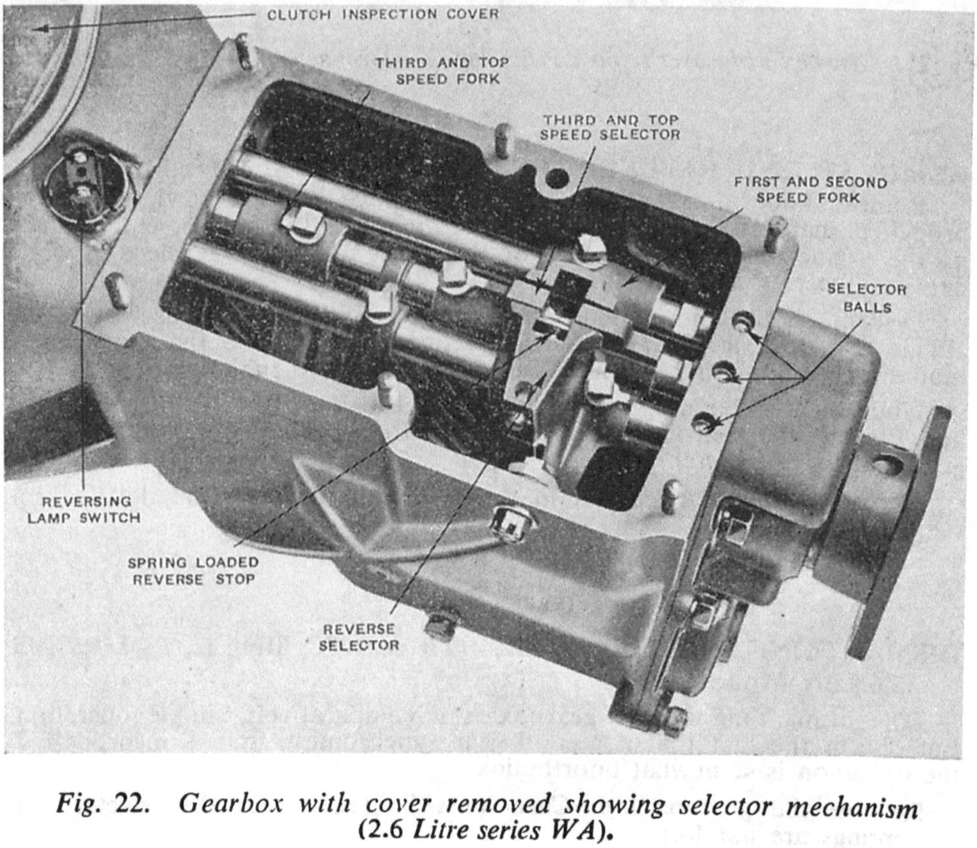

Fig. 22. Gearbox with cover removed showing selector mechanism (2.6 Litre series WA).

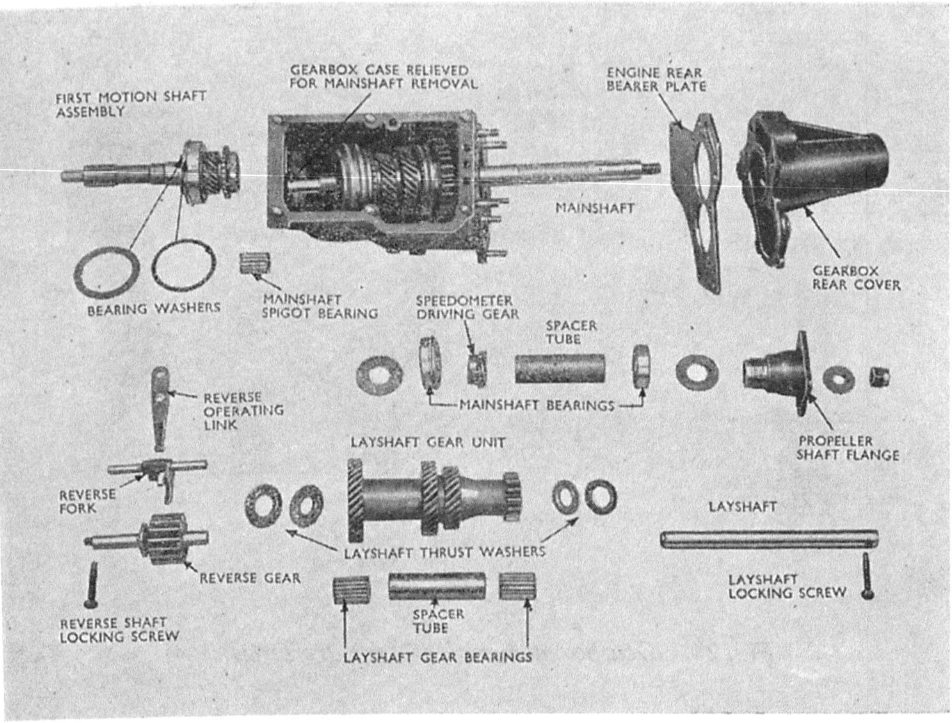

Fig. 23. *Gearbox dismantled* (1½ *Litre series VA*).

from the gearbox casing, when the ball race can be withdrawn from the mainshaft.

10. The mainshaft can be withdrawn from the gearbox casing as shown in Fig. 24.

The reverse gear and its operating mechanism are now the only parts left in the casing and it remains only to remove the locking screw securing the reverse gear shaft from the side of the box and the bolt securing the actuating lever, for the whole mechanism to be removed.

The layshaft bearings and spacer tubes are retained inside the layshaft gears by steel dowels which fit the internal bore, over which are pressed steel washers which make surfaces for the bronze thrust washers. As these washers are of different sizes, they cannot very well be confused with one another.

DISMANTLING THE GEARBOX. (TA Midget and 2 Litre Models).

Remove the propeller shaft flange, marking the flange in relation to the gearbox primary shaft for reassembly.

Remove the gearbox rear cover and the speedometer driven gear. Lift out the rear ball race packing and withdraw the various gearbox components backwards as follows: Remove the grubscrews securing the striking forks to their rods; draw the selector rods out one at a time, taking care at the last stage of withdrawal to catch the detent and interlocking balls as they are freed. Lift out selector forks, press down the spring loaded plunger and turn the splined retaining plate at the rear face of the third-speed double-helical gear until the gear is free to move on the mainshaft.

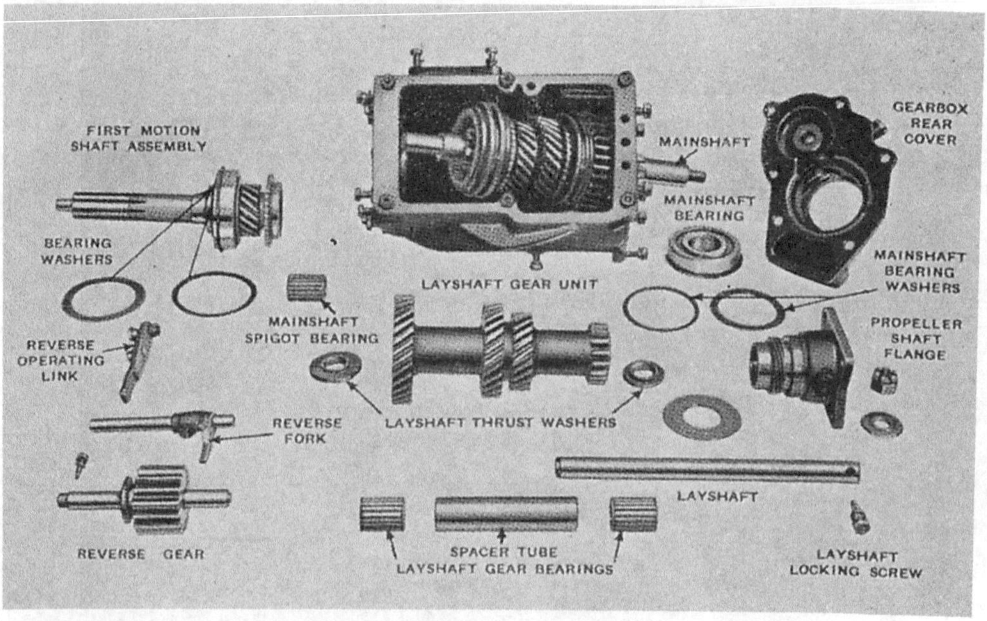

Fig. 24. Gearbox dismantled (2.6 Litre series WA).

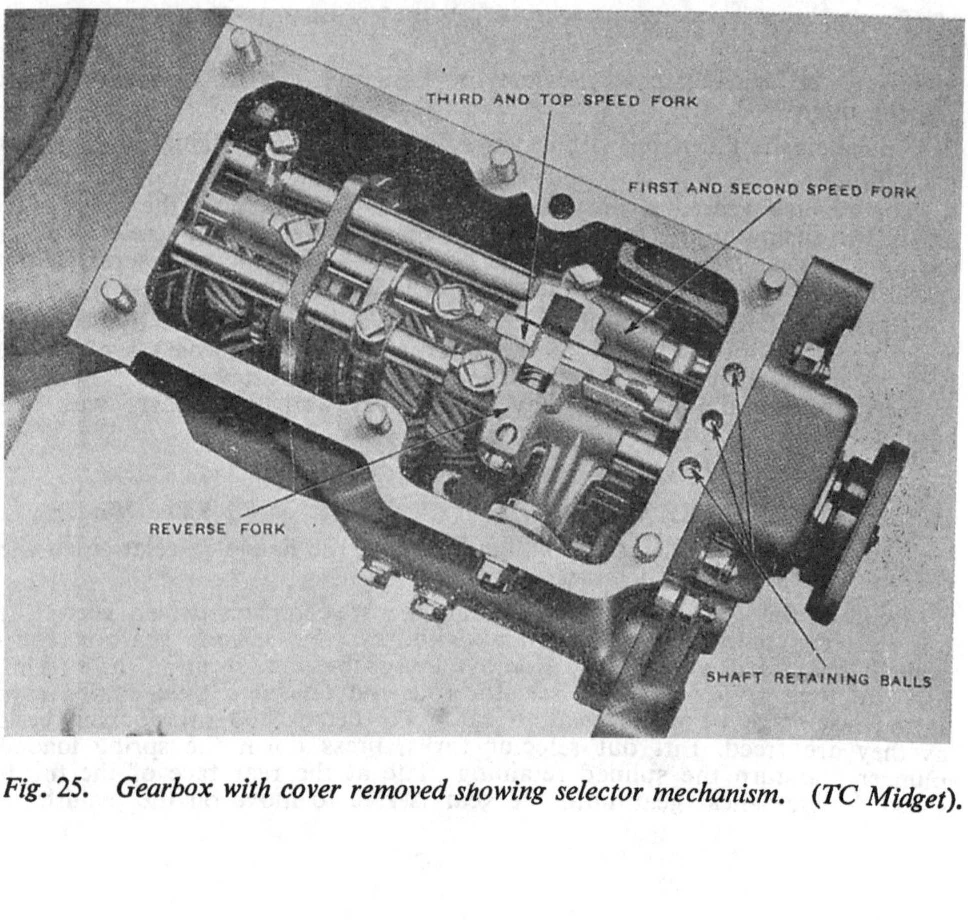

Fig. 25. Gearbox with cover removed showing selector mechanism. (TC Midget).

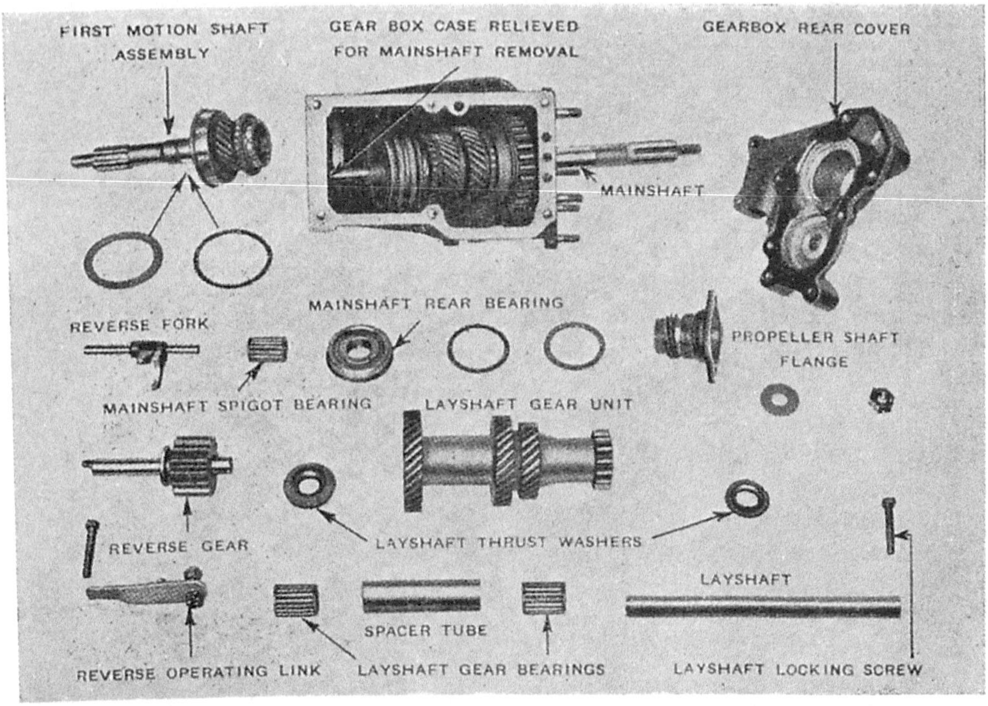

Fig. 26. Gearbox dismantled (TC Midget).

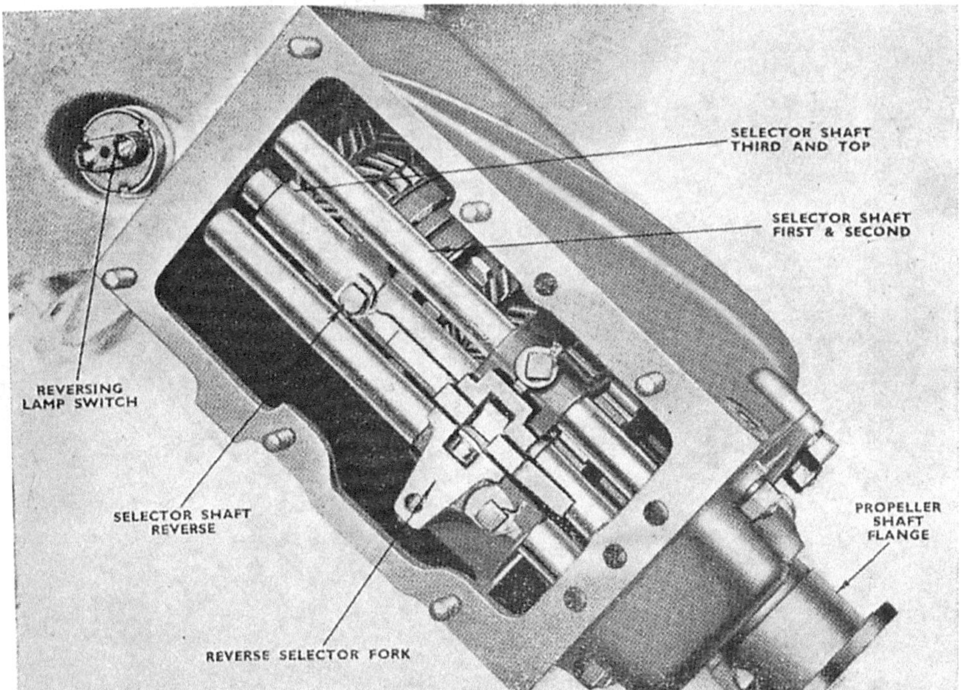

Fig. 27. View of Gearbox used when " one piece " remote control is fitted. (2 Litre series SA).

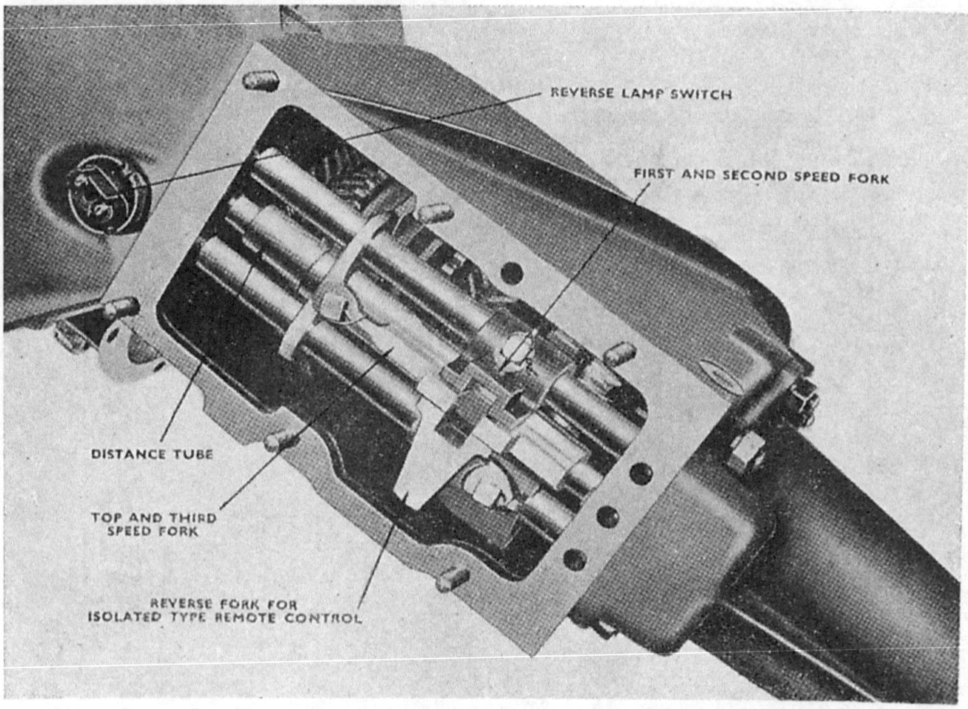

Fig. 28. Gearbox lid removed showing selector with " two-piece " remote control fitted (2 Litre series SA).

Fig. 29. View of gearbox with cover removed. (TA Midget.)

Draw out the mainshaft complete with its rear ball race and chip excluding washer, lifting out each gear as it is freed. Remove the countershaft locking grubscrew and withdraw layshaft, allowing layshaft cluster complete with roller races and spacing sleeve to drop to the bottom of the box. Remove the reverse shaft locking screw and withdraw shaft, lifting out the reverse gear cluster as soon as it is freed. Remove the clutch housing complete with front oil seal and lift out the first motion shaft ball bearing packing washer.

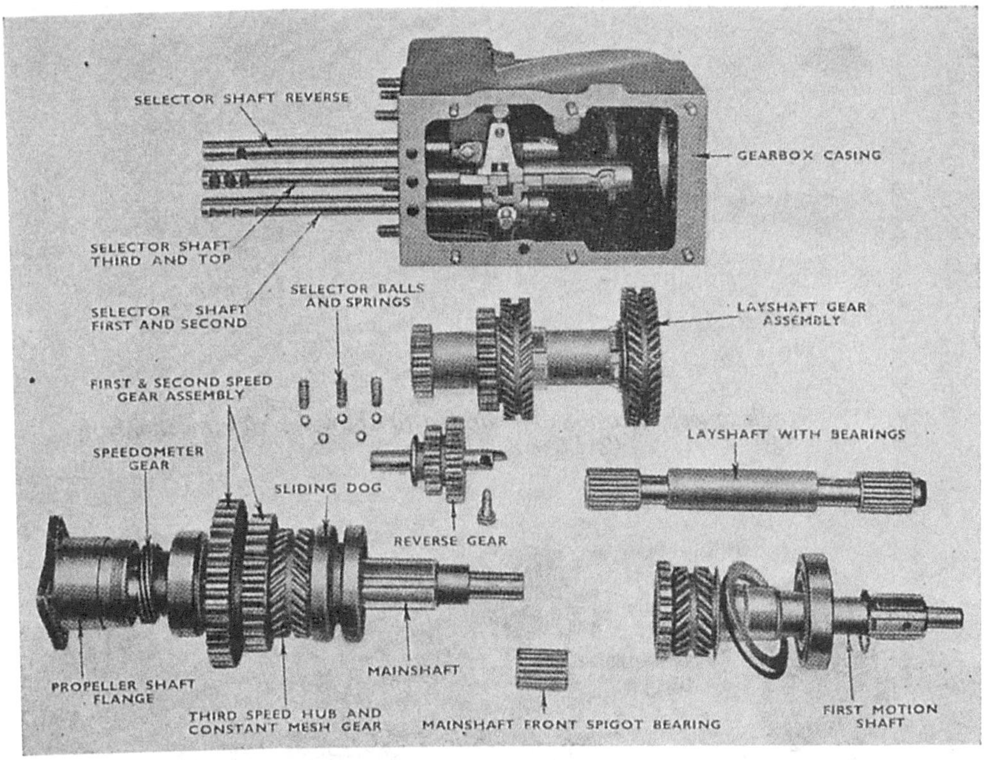

Fig. 30. *Components of crash type gearbox.* (2 *Litre series SA*).

Remove the first motion shaft retaining circlip in front of the ball bearing and press out the first motion shaft. The layshaft gear cluster can now be lifted out from the bottom of the gearbox.

As the construction of the synchromesh gearbox fitted to the 2 Litre Series SA Model is identical to the crash type gearbox, instructions for dismantling are as above with the exception of those for dismantling " third and top " mechanism.

DISMANTLING THE GEARBOX. (1¼ Litre and TD Midget.)

Support the gearbox in a vice by means of a steel bar approximately 1½ in. sq. by 5 in. long suitably machined and threaded at one end to enable it to be screwed into the gearbox drain plug hole. (See Fig. 33.) Remove dipstick and drain off the oil. Release the clutch housing from the gearbox.

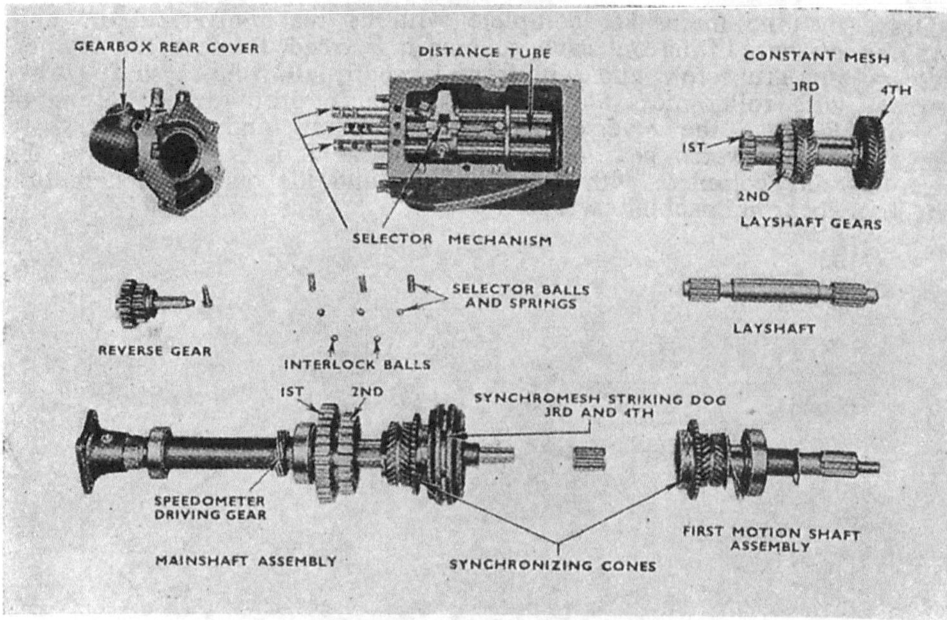

Fig. 31. *Synchromesh gearbox dismantled to show interior mechanism.* (2 *Litre series SA*).

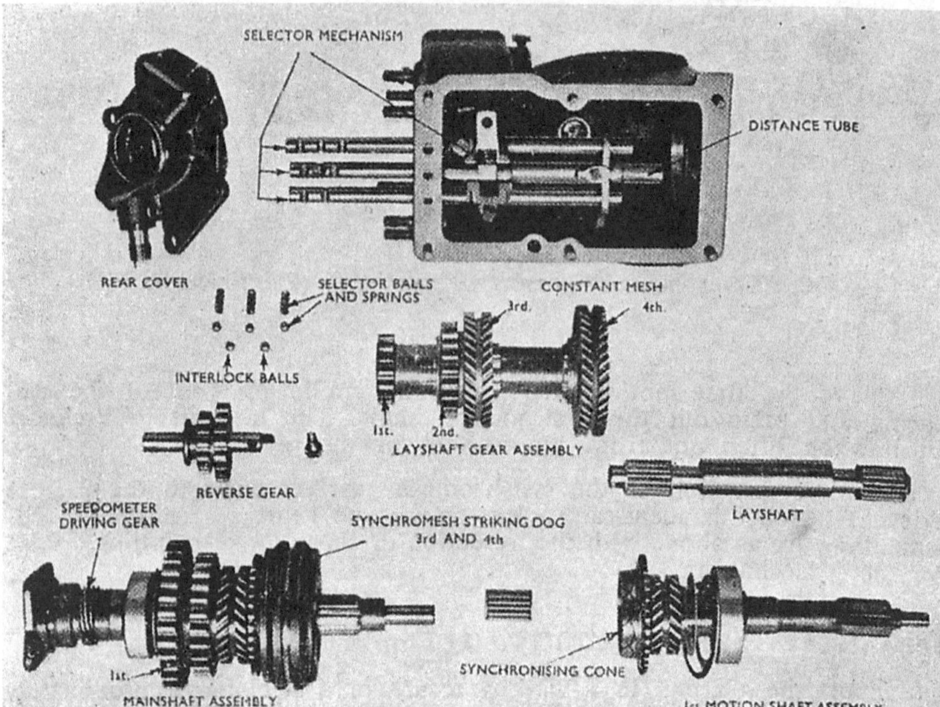

Fig. 32. *Gearbox dismantled* (*TA Midget*).

Fig. 33. (*above*) *Method of supporting Gearbox in vice.*

Fig. 34 (*right*) *Showing method of withdrawing mainshaft assembly.*

Extract the split pin from the nut retaining the drive flange at the rear of the gearbox sliding shaft and remove nut and plain washer. Remove nuts securing top cover assembly and bolts securing gear lever cover assembly to gearbox extension.

Remove the three selector springs. Withdraw propeller shaft flange using tool number T108 to prevent distortion, after suitably marking flange and shaft for reassembly.

Detach speedometer drive housing taking care not to damage the paper gasket on the joint face of the housing.

Extract the lock-wire from the eight square-headed bolts locking the gear shifters and stops to the selector shafts and remove screws. Slacken nuts and setbolts securing the gearbox rear casing to the gearbox and withdraw sufficiently to allow the gear shifters to be removed from the ends of the selector shafts.

On earlier type gearboxes, withdraw the selector shafts one at a time, taking care not to lose the selector lock balls in the process. The later type gearboxes have a third and top gear selector shaft extended at its front end and fitted with a circlip to prevent its accidental withdrawal and loss of synchromesh balls. In this case, the circlip must be removed before the shaft can be withdrawn.

Remove the layshaft spindle locating screw from the rear of the gearbox and extract the layshaft by tapping at the forward end with a suitable copper or brass drift. Remove the drive gear with its journal bearing by tapping the mainshaft towards the rear of the gearbox, using a suitable copper drift.

Using a suitable drift, extract the journal from its housing, when the mainshaft can be removed. Extract the layshaft gear unit, observing that the tabs on the thrust pads line up with the slots cut in the boss at the front and rear walls of the gearbox.

Removing Third-speed Constant Mesh Gear from Mainshaft. (1¼ Litre, 1½ Litre, 2.6 Litre, TB, TC and TD Models.)

This gear is retained on the mainshaft by a splined ring locked in position by a spring and plunger. (See Fig. 35.)

There are two slots cut in the outer face of the ring to enable the plunger to be depressed. Rotate the locking ring until its male splines match with the female splines of the mainshaft and withdraw.

NOTE: Commencing at Engine no. TPBG1510 (1½ Litre), the number of roller bearings was increased from 32 to 36.

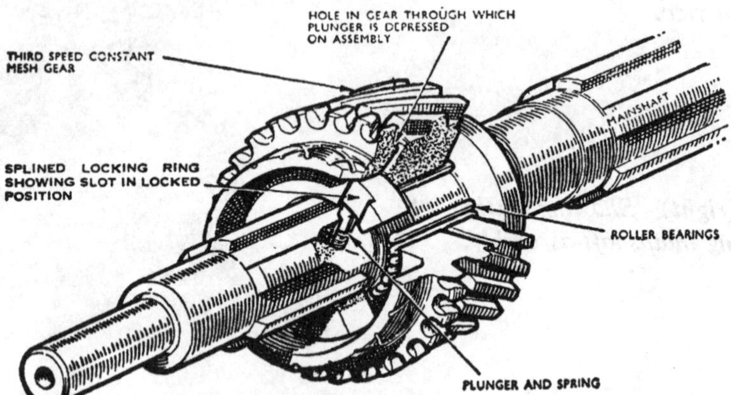

Fig. 35. Showing method of securing third-speed constant mesh gear to mainshaft.

Removing Second-speed Constant Mesh Gear from Mainshaft. (1¼ Litre, 1½ Litre and 2.6 Litre Models and TB, TC and TD Midgets.)

The retaining arrangement for this gear is slightly different from that of the third-speed constant mesh (see Fig. 36), although it is also retained by a spring and plunger. Two holes in the coned portion of the gear give access to the plunger. By inserting a split pin in the hole in the gear and the hole in the splined ring, the plunger can be depressed, which will allow the ring to be rotated until its female splines match up with the male splines of the mainshaft and the ring withdrawn. The plunger can then be removed, allowing the needle roller washer (two halves) and the gear wheel and rollers to be removed.

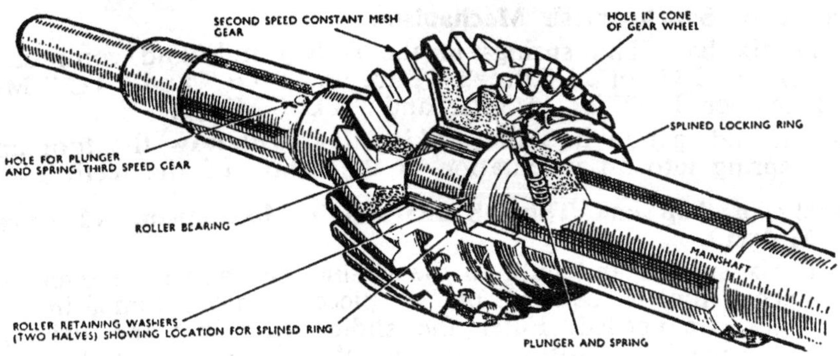

Fig. 36. *Showing method of securing second-speed constant mesh gear to main shaft.*

Dismantling Synchromesh Mechanism. (1¼ Litre, 1½ Litre, 2 Litre, 2.6 Litre, TA, TB, TC and TD Models.)

The striking dogs for top and third and second-speed gears are retained in sliding hubs. Apply sufficient pressure to overcome the springs and the sliding hub can be pushed out from the striking dog.

Commencing at Engine no. XPAG/TD/3114 (TD Midget) and XPAG/SC/15499 (1¼ Litre), the synchro balls are peened into the synchromesh hub.

Commencing at Engine no. XPAG/TD/1653 (TD Midget) and XPAG/SC/16431 (1¼ Litre), one of the spring holes of the top and third synchromesh hub has been bored right through and an additional ball added to maintain it positively in the correct position. This ball locates in an indent in the gearbox mainshaft.

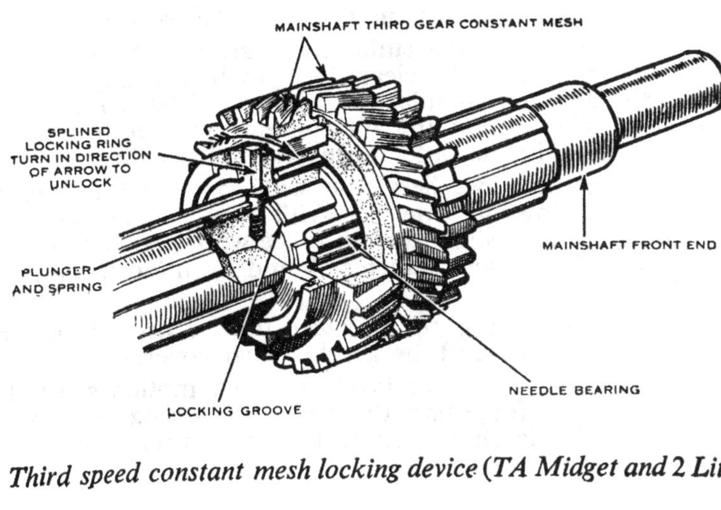

Fig. 37. *Third speed constant mesh locking device (TA Midget and 2 Litre Saloon).*

Reassembly of Synchromesh Mechanism.

Fit the six balls and springs in the sliding hub and compress by a special tool No. T83 for 1½ Litre, 2.6 Litre and "TB" and "TC" Midgets, and tool number T109 for 1¼ Litre and "TD" Midget.

Push the sliding hub into the striking dog, withdraw the tool and the balls will spring into an indentation in the centre of the teeth.

Reassembly of Top and Third Synchromesh Mechanism. (2 Litre and TA Midget.)

Fit the six springs and balls in the sliding hub and slide over striking dog. No special tool is required, but six pieces of steel shim ⅜ in. wide × .015 in. thick are needed. Enter the sliding hub into the striking dog and fit one spring into position, and with the aid of one of the pieces of shim, the ball is retained in the sliding hub.

Push the shim down between the hub and striking dog. Holding the six balls and springs in this manner, push the sliding hub right home. when the balls will click into position in the centre of the teeth.

REASSEMBLY OF THE GEARBOX.

Reassembly of the Gearbox is carried out in the reverse manner to that detailed for dismantling, but the following points should be noted:—

1½ Litre, 2.6 Litre, TB and TC Midgets.—Before placing the layshaft gears in the bottom of the gearbox, insert roller bearings, spacer tube, steel and bronze thrust washer assemblies. The layshaft should not be replaced until the mainshaft and first motion shaft have been fitted.

The locking plunger and ring for the third-speed constant mesh gear must be placed in the shaft, followed by the splined locking ring. The two slots cut on the outer face of the ring must line up with the plunger when in the locked position. When sliding on the ring, the plunger can be depressed through a hole between mesh gear and striking dog mating gear.

The assembly procedure for the second-speed constant mesh on the mainshaft differs slightly. Apply a little grease to hold the collar in position. Place the locking plunger spring in position and slide on the second-speed gear. Insert the two steel half washers inside the gear next to the rollers so that their protrusions line up between two splines. Depress the plunger and place the locking ring in position with the slots cut on the inner face in line with the protrusions on the two half washers.

Engage second gear allowing sufficient clearance for the spigot on the front end of the mainshaft to clear the gearbox casing and enter the mainshaft into the gearbox complete with gears. The mainshaft rear ball race can be fitted after fitting the oil spinning washer between the race and the second gear, with its outer edge clear of the ball race.

Fit the two large steel washers, the wide one next to the ball race and the thin one following. The "Belville" type thin steel washer is designed to eliminate any slight variation between the ball race and the recess in the cover. The inner diameter should be fitted towards the large flat steel washer.

Fit the speedometer driving gear, spacer tube, rearmost ball race, the engine rear support bracket, and the gearbox rear cover.

Insert the mainshaft roller race inside the first motion shaft. Insert the selector mechanism, not forgetting the two interlocking balls between the selector shafts. Fit the layshaft and lock into position from underneath the gearbox.

When refitting the clutch housing, do not forget to replace the two steel washers, the wider one next to the ball race and the thinner one following it. The same remarks apply to these washers as to those fitted to the mainshaft rear bearing.

Before replacing the gearbox top, do not forget to insert the three locking balls and springs.

TA Midget and $1\frac{1}{2}$ Litre.—Fit roller bearings and distance tubes into the layshaft cluster, secure the reverse gear shaft, also layshaft, by its grub screw, fit plunger and spring into the mainshaft, lock the third-speed double-helical gear by rotating the loose locking plate in its groove in the mainshaft, until the plunger engages, fit three balls and springs underneath the three selector shafts, also one ball between each shaft, and replace propeller shaft flange on the mainshaft in its original position.

TD Midget and $1\frac{1}{4}$ Litre.—When fitting the layshaft, ensure that the tags on the thrust washers at each end fit into the bosses of the gearbox. For easy assembly of the layshaft with its bearings, it is recommended that a dummy shaft $\frac{9}{16}$ in. dia. × $6\frac{11}{32}$ in. long be used.

Care must also be taken in the case of the later gearboxes, to replace the circlip on the forward end of the third and top selector shaft.

Gear Lever Remote Control. (2 Litre.)

When the one-piece remote control is fitted, the reverse gear selector is fitted with a plunger, ball and spring to prevent the gear from entering reverse inadvertently (see Fig. 27). When the isolated gear remote control is fitted, these parts are not required (see Fig. 28), as the gear lever turret is fitted with a catch which prevents the gear lever from entering the reverse gear gate. As both the crash type and synchromesh type gearbox are fitted with both types of gear remote controls, it is important that the aforementioned details are checked before fitting a replacement gearbox.

GEARBOX MODIFICATIONS. (TD and $1\frac{1}{4}$ Litre Models.)

With the introduction of the 8 in. diameter clutch at Engine Nos. XPAG/SC2/16916 and XPAG/TD2/9408, $1\frac{1}{4}$ Litre and TD Midget respectively, the clutch fork shaft of the gearbox has been modified. The earlier clutch fork shaft was $\frac{5}{8}$ in. diameter where as the later type has a diameter of $\frac{3}{4}$ in.

3rd and 4th Gear Shifter Shaft (TD Midget).—The gearboxes fitted to the later TD Midget Models were fitted with a longer 3rd and 4th gear shifter shaft, the shaft being supported by a supporting cap fitted to the speedo gear housing.

This modification was introduced at Gearbox No. TV.840.

Speedometer Gear (TD and $1\frac{1}{4}$ Litre Models).—On the above Models commencing at Gearbox Nos. SJ/337 and TW/396, $1\frac{1}{4}$ Litre Saloon, Series YB and TD Midgets respectively, the speedometer gear has been keyed on to the mainshaft.

CHAPTER V

REAR AXLE

REAR AXLE

GENERAL DESCRIPTION. (All models except TD and YB.)

The drive for the rear axle is through the medium of the propeller shaft from the gearbox. The drive shafts transmit only the driving torque, and do not carry the main load.

The rear hub bearings are mounted on extensions of the rear axle casing. The integral rear hub flanges and driving shafts are easily withdrawn.

Spiral bevel drive gears are employed. The bearings are assembled into the housing with an initial pinch or pre-load.

Preloading of bearings prevents vibration and assists in maintaining a quiet running axle.

The crown wheel is adjustable from side to side by two adjusting nuts on the outer side of the differential ball bearings.

On all models except the 1¼ and TD, the bevel pinion is adjusted to or from the crown wheel by shims between the pinion housing and differential carrier.

On the 1¼ and TD models, no provision is made for the adjustment of the drive pinion, the correct position being held by accurately controlling, in production, the position of the shoulder (dimension "A") and the width of taper roller bearing " B " (Fig. 7).

MAINTENANCE. (All models.)

Check and tighten if necessary, (1) the nuts securing the axle drive gear housing to the centre gear casing; (2) the nuts securing the propeller shaft to the axle and gearbox driving flanges; (3) the nuts securing the axle " U " bolts.

REMOVAL OF REAR HUB. (All models except TD and YB.)

Jack up rear axle and remove road wheel. Remove the set screws or nuts securing the brake drum to the hub, release the brake shoe adjustment and withdraw brake drum. Replace the wheel nuts, and drive out the axle half shaft and hubs complete, by hitting the nut " ears " with a copper hammer (D, J, F, L, P, K, KN, N, TA, TB, TC, 1½, 2 and 2.6 litre models.) There are two threaded holes in the axle drive shaft flange. Screw into the holes two of the countersunk screws which secure the brake drums, using them as extractors for removing the axle drive shaft and hub. (1¼ litre.) Tap back the tabs of the locking washer and remove the nut. Using a suitable claw extractor, withdraw the hub complete, with bearing and oil seal.

REPLACEMENT OF REAR HUB. (All models except TD and YB.)

The tip of the sealing edge of the seal should face towards the bearing. Replace the hub complete with bearing and oil seal. The bearing should be gently tapped into position for it is a tight driving fit on the axle extension. Use a piece of tube large enough to slide over the centre tube of the axle. Fit the lock washer and tighten the nut, tapping over the tabs of the lock washer.

NOTE:—The hub retaining nuts have right and left hand threads; right hand side are right hand threads and left hand side, left hand threads.

Replace the axle shaft and hub and brake drum, resetting the brake shoe adjustment.

REMOVAL OF THE DIFFERENTIAL ASSEMBLY. (All models except TD and YB.)

This operation can be carried out with the axle in position. In models where the batteries are carried in carriers between the chassis side members, it will be necessary to disconnect and remove the batteries. Drain the oil and jack up the rear of the car until the wheels are clear of the ground. Remove the wheels, brake drums and axle shafts. Uncouple propeller shaft at the rear axle flange, after suitably marking for reassembly. Now remove the nuts securing the differential assembly to the axle casing and withdraw assembly complete.

DISMANTLING DIFFERENTIAL ASSEMBLY. (M, D, F, J, K, L, N, P, KN, TA, TB, TC, 1½, 2 and 2.6 Litre models.)

Using lead clamps, support the differential assembly in a vice. Withdraw the lock plates after removing the bolt and spring washers.

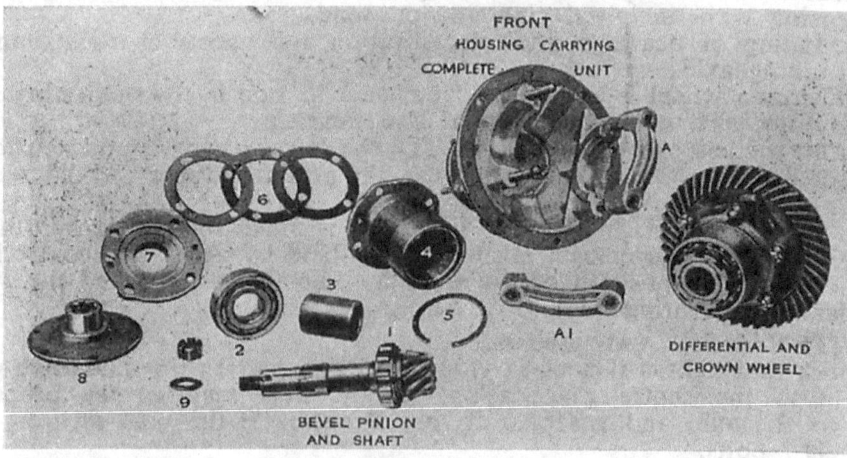

Fig. 1. Differential dismantled (M, D, F, J, K, KN, L, N & P models).

1. Bevel pinion roller bearing and shaft. 2. Bevel pinion bearing. 3. Distance piece for bevel pinion bearing. 4. Bevel pinion housing. 5. Spring ring. 6. Shims for bevel pinion. 7. Bevel pinion cap. 8. Propeller shaft flange. 9. Slotted nut and washer for propeller-shaft flange. A and A1. Differential bearing caps.

Remove the nuts securing the pinion bearing assembly to the differential carrier and withdraw the assembly. On the 2 litre the bevel pinion and its three bearings can be withdrawn as a unit after removing the pinch bolt and locking tab from the axle nose piece and screwing out the adjusting sleeve.

Remove the slotted nut and washer from the end of the pinion and withdraw the propeller shaft flange, using a suitable puller to avoid distortion.

Extract the screws securing the bevel pinion housing end cap, and withdraw the ball race with housing, distance piece and roller race from the pinion. Now remove the rear retaining circlip when the outer track of the ball race can be withdrawn from the differential carrier. On the 2.6 litre model remove the end cover, pick out the oil seal and press the pinion shaft from the adjusting sleeve. Between the bearing distance piece and the inner race of the outer ball bearing are a number of shims. The correct amount is the smallest that will enable the pinion to be just free in the adjusting sleeve, when the propeller shaft flange is fully tightened.

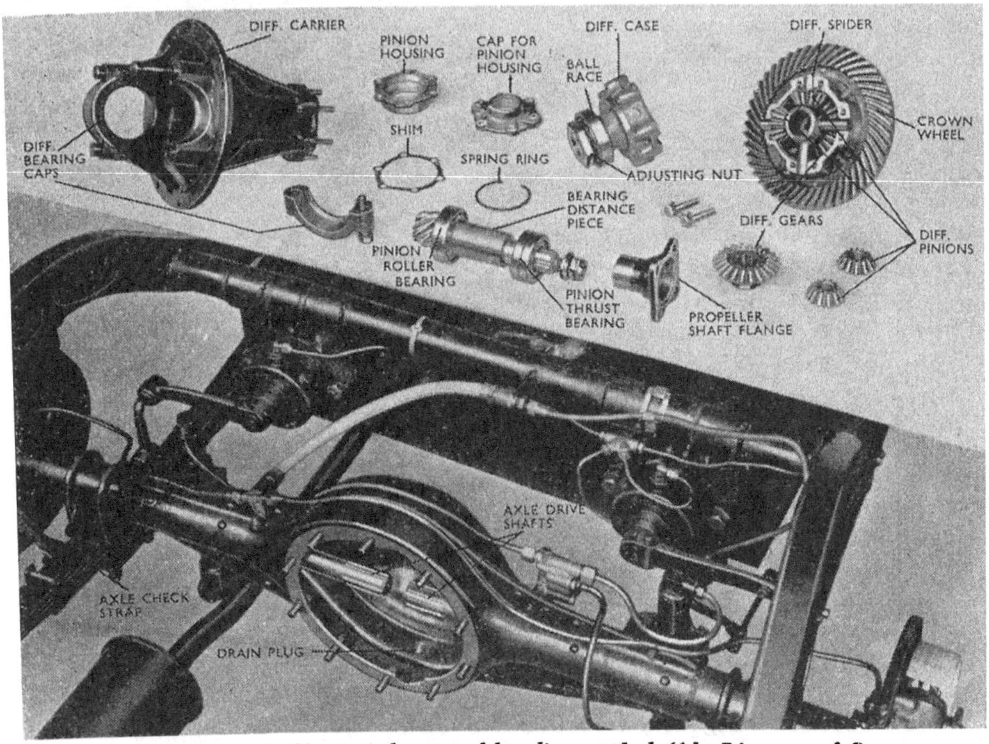

Fig. 2. *Differential assembly dismantled* (1½ *Litre model*).

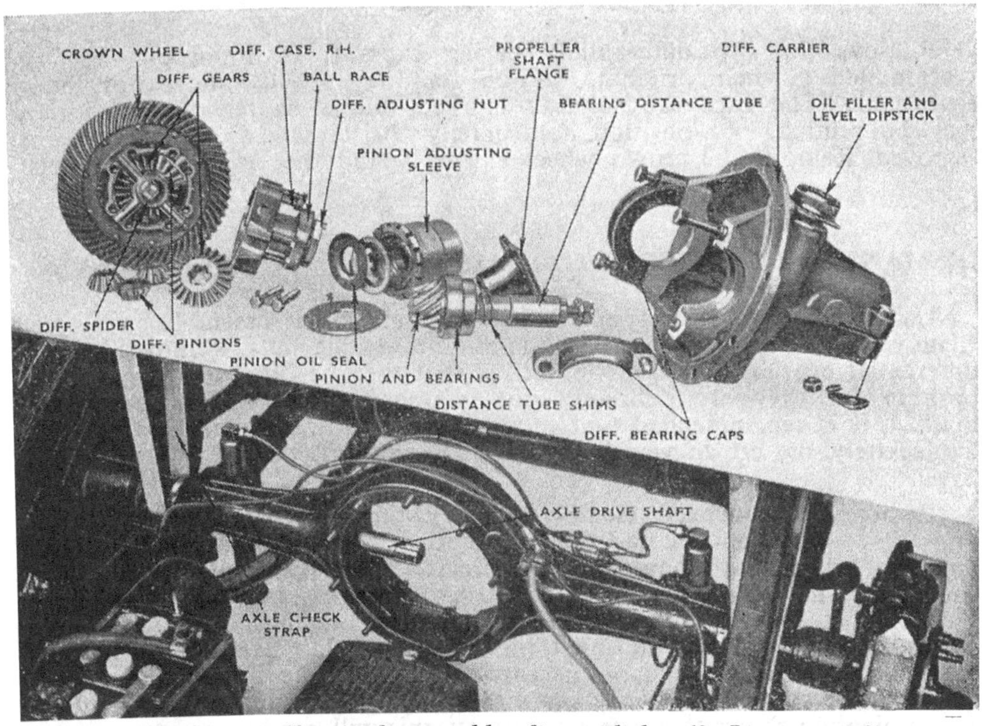

Fig. 3. *Differential assembly dismantled.* (2 *Litre model.*)

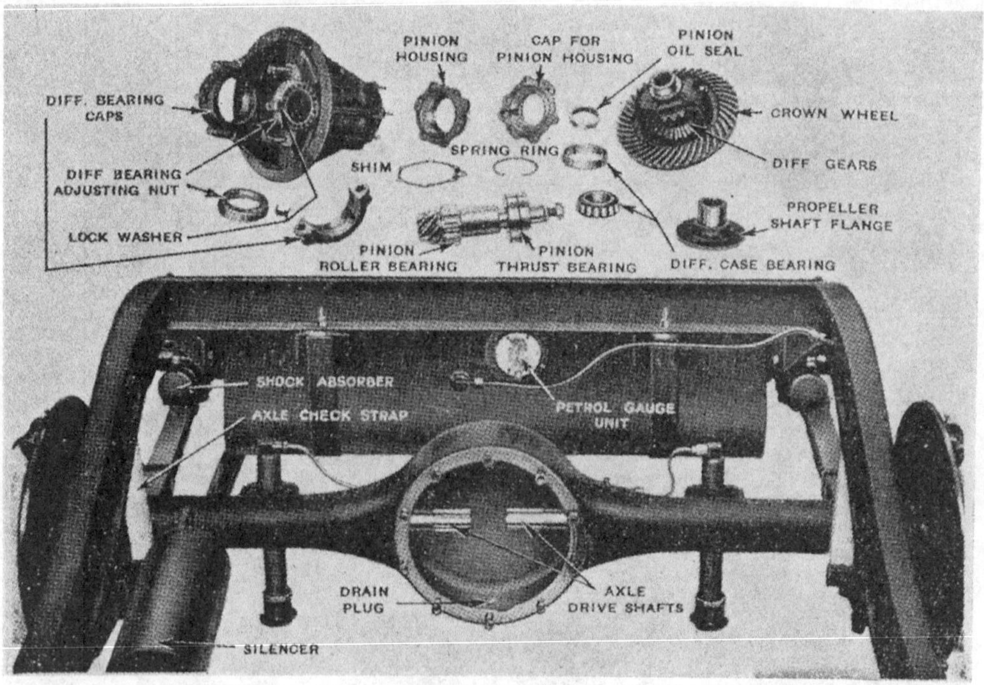

Fig. 4. Differential unit dismantled. (2.6 *Litre model.*)

Remove the two differential bearing caps. It is important they be reassembled in their original position for they are definitely not interchangeable. The caps are suitably marked, but it is recommended that they be refitted into position immediately the crown wheel assembly is removed. Remove the crown wheel assembly and the two adjusting nuts.

DISMANTLING DIFFERENTIAL ASSEMBLY. (1¼ Litre, Series Y.)

Using lead clamps, support the differential carrier assembly in a vice. Remove the lock plates for the differential adjusting nuts. Remove the two differential bearing caps. It is important that they be fitted in their original position for they are definitely not interchangeable. Although suitably marked, it is recommended that they be replaced and bolted into position immediately the crown wheel assembly is removed.

Remove crown wheel assembly and adjusting nuts. Remove slotted nut and flat washer from the end of the pinion shaft after extracting the cotter pin. The pinion shaft should be now pressed out towards the crown wheel end of the shaft. The inner race of the large bearing will come away with the pinion, the outer race and the smaller race and the oil seal remaining in the carrier. Press out the inner race of the smaller bearing and oil seal from the crown wheel end of the carrier. Through the slots in the carrier, drive out the outer races of the bearings, using a steel drift ⅜ in. in diameter. If examination shows they are satisfactory, this will not be necessary.

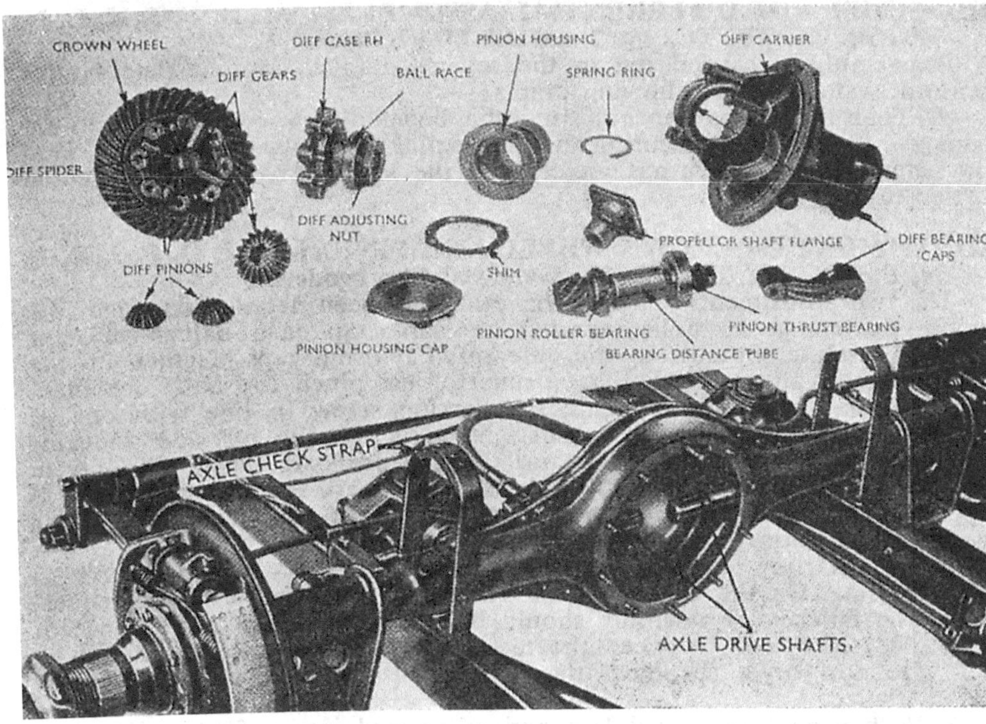

Fig. 5. Differential assembly dismantled. (TA, TB and TC models.)

Fig. 6. Differential carrier supported in a vice.

REASSEMBLY OF DIFFERENTIAL ASSEMBLY. (M, D, F, J, K, L, N, KN, P, TA, TB, TC, 1½, 2 and 2.6 litre models.)

Reassembly is carried out in the reverse manner to that detailed for removal, watching the following points:—

(1) That the shims for adjusting the bevel pinion are clean and undamaged. (2) The bolts securing the differential assembly are tight. (3) That the differential ball bearings are against the shoulders in the differential carrier caps.

ADJUSTMENT OF CROWN WHEEL AND PINION. (M, D, F, J, K, L, N, P, KN, TA, TB, TC, 1½, 2 and 2.6 litre models.)

The two adjustments provided have already been detailed on page 192. Before tightening the bolts securing the caps, tap each ball race outer track against the shoulder in the differential housing caps. Tighten up the cap bolts and adjust the ball race nuts to just pinch the ball race inner track, at the same time bringing the locking screw in line with one of the serrations in the locking washer. Now tighten the off side adjusting nut one further serration to obtain a slight pre-load. This will prevent slackness in the races when the axle is under load. Each crown wheel is lapped with its pinion to ensure perfect tooth contact, necessitating meshing the gears in the same relative position.

1. As a starting point, insert .010 thickness of shims behind the pinion housing. On the TC model .007 thickness of shims is recommended.
2. The differential assembly should be adjusted to give a clearance of .007 to .010 in. between the teeth of the crown wheel and pinion. Do not forget to adjust the differential thrust races as previously detailed.
3. With a mixture of red lead and oil, mixed into a fairly stiff paste, cover the teeth of the crown wheel using a stiff brush to ensure even distribution.
4. Now revolve the differential in both directions using a handle made from round bar and with two collars to fit into either side of the differential casing. One collar should be secured to the handle, the other one must be loose on the shaft the whole being locked up with a nut. After the gears have been removed, the tooth mating markings can be seen on the crown wheel. The markings should be identical on each side of the tooth and in the centre of each tooth. The tooth markings should only extend half the length of the tooth, extending the full length under load. Very narrow tooth markings is an indication of too much backlash between the gears and tooth rooting or bottoming, insufficient backlash. If further adjustment is necessary, shims can be removed from or additional shims inserted behind the pinion housing not exceeding .004 in. If the adjustment proves unsatisfactory before condemning the gears, all bearings should be inspected for slackness, particular attention being paid to the pinion rear roller race. It should be renewed if a feeler .015 in. thick can be inserted between the outer track and the roller.

REASSEMBLING DIFFERENTIAL ASSEMBLY. (1¼ litre Series Y.)

The bearings of both crown wheel and pinion are pre-loaded, the pinion bearings being supplied in preselected sets comprising the large and small bearings "H" and "J" and distance piece "G"—(Fig. 7). They should assemble into any axle, giving the correct amount of pre-load.

Remove any burrs from the shoulders machined in the carrier which prevent the possibility of the correct pre-loading being obtained and correct engagement of the crown wheel and pinion.

Press into the carrier the outer races of the pinion shaft bearings; the larger race from the crown wheel side and the smaller race from the driving flange side. Press the large taper bearing complete with rollers on to the pinion shaft making sure it is tight home against the shoulder on the pinion. Now fit the distance piece on the pinion shaft and place the shaft in position in the carrier. The smaller inner race complete with rollers may now be placed in position on the pinion shaft and the assembly pressed together. Fit the driving flange, washer and nut, omitting the oil seal at this stage, and tighten the nut. Now check the pre-load using the special torque recording spanner (T.110) attached to the rear axle universal joint flange. When the pinion is rotated, the correct pre-load is 5-6 lbs./ins. If the readings are incorrect check all fittings and test again. If still not correct, replace bearings and distance piece. When the correct pre-load is obtained, remove the nut, washer and driving flange, fit the oil seal with the knife edge of the central leather portion towards the bearing, replace the driving flange, washer and nut; tighten up securely and fit the cotter pin.

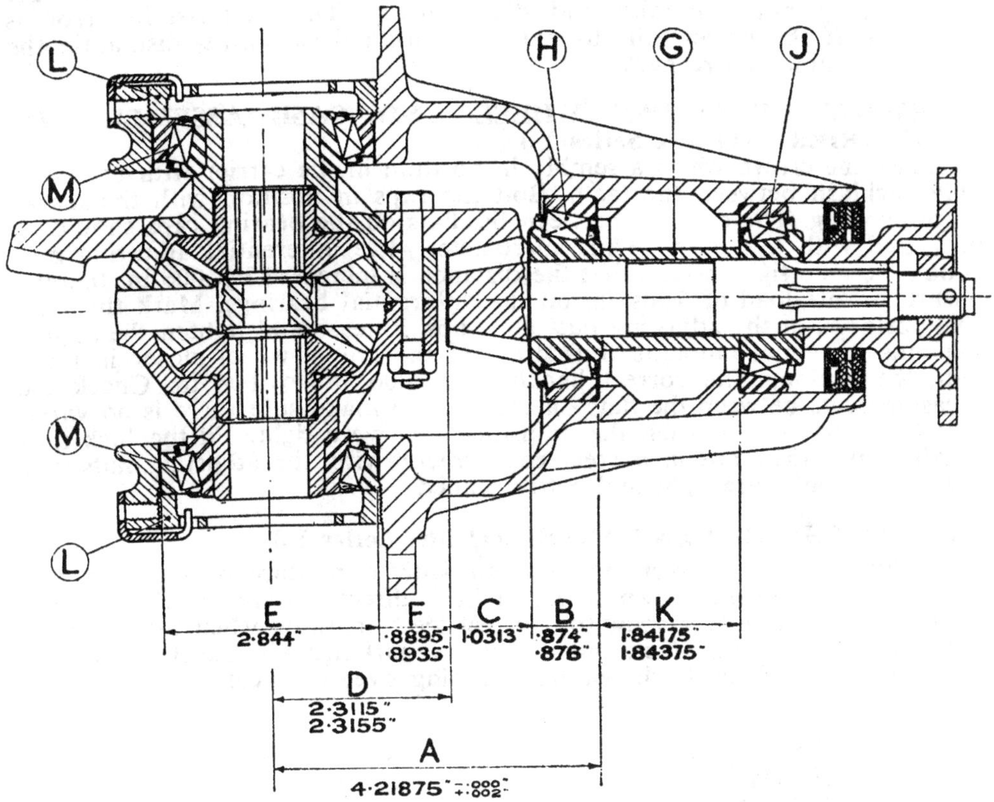

Fig. 7. Sectional view of drive gear, showing dimensions controlling pinion position.

Now check the distance between the end face of the pinion and the centre line of the differential bearings, the distance from the pinion measured with slip gauges being within the limits .8895 in. and .8935 in. as at "F." (Fig. 7.) If not within these limits the races are not home against the shoulders on the pinion and carrier, which will mean the removal of the pinion and the repeating of the operations.

ASSEMBLY OF THE DIFFERENTIAL CASE AND CROWN WHEEL. (1¼ litre Series Y.)

Absolute cleanliness of the joint faces is essential. The shafts of the differential gears shouid be smeared with grease and the gears inserted into each half of the case. Lightly lubricate the differential pinion spider and thread on the differential pinions with their teeth facing inwards; inserting the assembly into the half of the case to which is attached the crown wheel and with the pinion engaging the gears correctly. Place the two halves of the case together with the identification marks corresponding and engage the crown wheel with the register machined in the case. Now bolt the crown wheel and both halves of the case together making sure the differential assembly rotates freely, and lock the nuts with cotter pins. The roller bearings should now be mounted on the spigots in the case, with the smaller diameter of the inner ring facing outwards from the case. Temporarily assemble the crown wheel in the carrier and check the crown wheel for true running both laterally and diametrically. The permissible error is ±.003 in. If the wheel fails to run true within these limits, dismantle the case, reassemble and recheck.

ASSEMBLY OF CROWN WHEEL AND CASE ASSEMBLY TO CARRIER. (1¼ litre Series Y.)

Place the crown wheel assembly in position in the carrier with considerable backlash between the teeth. Bolt the caps in position with the marks corresponding. The distance between the flats on the bearing caps should be measured with a 5 in. to 6 in. micrometer. The differential adjusting nuts should now be tightened so that the distance is increased by .006 in. thereby ensuring a pre-load of 5 lbs/in. on the differential bearings. Mark the relative positions of the adjusting nuts and caps. Release and tighten the adjusting nuts by equal amounts thus drawing the crown wheel closer in mesh with the pinion. The correct backlash is .006 in to .008 in. Check the measurements between the flats on the caps to make sure there is no variation from that taken when the adjusting nuts were tight. Fit the lock nuts making sure they are in correct engagement with the adjusting nuts and see that all bolts are tight and cotter pinned.

GEAR MESHING—REAR AXLE. (1¼ litre Series Y.)

Lightly coat the crown wheel teeth with a mixture of red lead and engine oil. Rotate the crown wheel in both directions and check marking. Reference to Fig. 8 will show the correct teeth marking when the assembly is tested without load. When under load the markings will travel up towards the full length of the teeth without creeping over the heel.

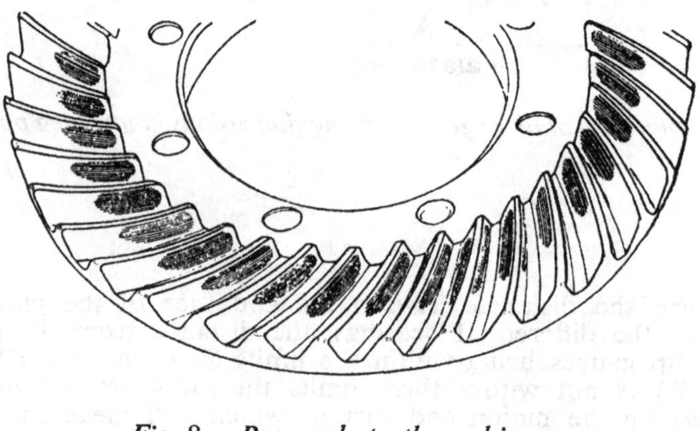

Fig. 8. Rear axle teeth markings.

TO COMPLETE ASSEMBLY OF AXLE. (1¼ litre Series Y.)

Clean all joint faces and fit new gaskets in all places where gaskets are required. Place the differential assembly in position taking care to avoid damaging the threads of the securing bolts. Fit the axle drive shafts and hubs, and brake drums and reset brake shoe adjustment. Couple up the propeller shaft flange, making sure that the dismantling marks correspond. Check all nuts to see they are tightened and cotter pinned and fill axle to correct level, with the recommended oil.

THE REAR AXLE

GENERAL DESCRIPTION. (TD and 1¼ litre Series YB.)

The rear axle is of the semi-floating type employing the use of Hypoid gears, the axle housing being divided close to its centre and the pinion assembly being mounted in the right hand half or axle casing. It is of unit construction, no repairs or adjustments, apart from those connected with the half shafts, rear wheel bearings, brake drums and shoe mechanism, can be carried out without removing the unit complete from the car. The wheel hubs are attached to the axle shafts by splines and a tapered split collar, and can only be withdrawn after removing the wheel, wheel hub, and brake drum, brake back plate assembly and bearing housing. The brake drum is integral with the hub or permanently attached to the wheel hub flanges by countersunk screws, which must not be disturbed. The crown wheel and pinion are supplied in pairs, the adjustment of the position of both the crown wheel and pinion in the axle being effected by initially selected distance pieces. There is no other provision for adjustment.

LUBRICATION.

It is absolutely essential that only the approved hypoid oils be employed and the level of the oil not be raised above the lip of the filling aperture.

REMOVAL AND REPLACEMENT OF BRAKE DRUM AND HUB.

Jack up the axle until the wheel is clear of the ground and release the hand brake. Withdraw the road wheel, remove the split pin from the axle nut and unscrew the nut. The axle shafts are interchangeable for they are both right hand threaded. In addition to the driving splines a split collar is also employed to lock the wheel hub to the axle shaft. Some resistance will be evident when the extractor is used to free the hub from the taper. When replacing the rear hub it is essential that the tapered split collar be

Fig. 9. Removal of brake plate assembly from axle flange.

right home against the inner race of the wheel bearing. A hide hammer can be used to tap the collar tightly into position. Check the oil seal for damage and make sure that the parallel portion of the collar engaging the oil seal is free of blemishes before replacing.

REMOVAL AND REPLACEMENT OF BRAKE PLATE ASSEMBLY.

Jack up the axle, release hand brake and remove wheel and brake drum and hub. Four bolts with the nuts fitted on the inner side of the axle flange secure the brake plate in position. Remove these four nuts, withdraw the bolts and lift back plate clear of axle flange.

Replacement is carried out in the reverse order to removal.

REMOVAL AND REPLACEMENT OF AXLE SHAFT.

Jack up the rear axle and remove the road wheel, hub and brake drum assembly. See that the hand brake is fully released. Fit the special tool to the end of the axle half shaft and using the sliding impact weights, release the shaft complete with bearing, housing and oil seal. Now press out the half shaft from the bearing.

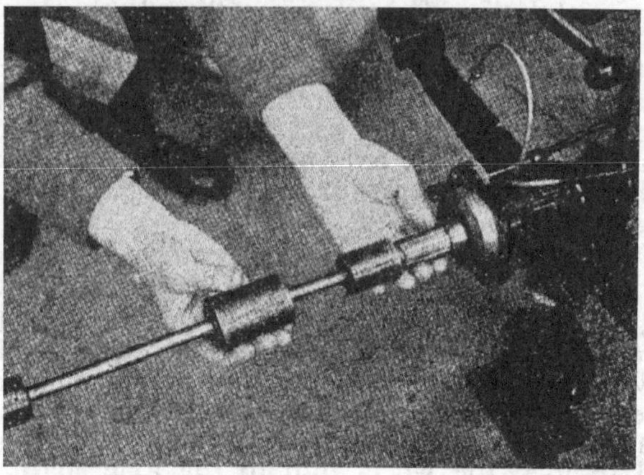

Fig. 10. *Removing half shaft, using special extractor.* (*Part no.* 68823.)

REMOVAL OF REAR AXLE. (TD model.)

Raise the rear of the car by the use of a suitable sling and support chassis just forward of the rear axle spring front mounting. Release handbrake and remove both road wheels. The Lockheed flexible pipe should now be removed at its junction to the bracket on the chassis. Disconnect the brake cable casings from their anchorage to the spring brackets. Disconnect the brake cables from the brake shoe actuating lever on the brake plate. Disconnect lower ends of shock absorber arms. Uncouple the propeller shaft flange after suitably marking for reassembly purposes, and support the tail end of the propeller shaft through the aperture in the rear seat. Undo the spring " U " bolts and rest the axle on the rebound straps. Using jacks or axle stands, take the weight of the axle. Remove rebound straps, lower exhaust pipe and wihdraw axle sideways. Reassembly is carried out in the reverse manner to that detailed for removal, remembering to bleed the brakes after coupling up the flexible hose.

IMPORTANT POINTS CONCERNING AXLE ATTENTION.

As the dismantling and reassembly of the rear axle necessitates the use of special tools, including checking gauges and a full range of distance pieces and spacers from which to select the required new sizes, it is recommended that a replacement axle, when possible, be fitted. Dismantling for

Fig. 11. Axle ready for removal.

examination and cleaning is permissible provided care is taken to refit the distance pieces and spacers in exactly the same position. In the accepted sense there are no adjustments provided. The crown wheel and pinion are set in their correct relation to each other by means of distance pieces and spacers selected to provide the correct location of the components on initial assembly. Should it be necessary to dismantle the rear axle, carefully observe their relative positions and suitably mark each component to enable them to be reassembled correctly in the original positions. There are various components which can be replaced by correctly combining the markings on the original components against those on the new parts. The position of the crown wheel in relation to the centre line of the pinion is controlled by spacers between the outer races of the differential bearings and faces of the machined recesses in the axle casing and cover. Varying the thickness of the pinion washer controls the adjustment position of the pinion and crown wheel adjustment is carried out by the varying thickness of the differential bearing spacers. Various operations can be carried out without the use of special tools. (1) To replace the crown wheel and pinion with a pair carrying markings which are identical to the originals. (2) To replace a crown wheel bearing alone, providing genuine replacement parts are used. (3) To replace an axle cover which carries marking identical to the original. These operations merely call for the fitting of new parts in the positions occupied by the old.

DISMANTLING THE AXLE AND REMOVING DIFFERENTIAL ASSEMBLY.

Remove axle from car, remove hub and brake drum assembly and brake back plates and withdraw axle shafts. Remove the bolts joining the axle casing and cover. To avoid damage to the differential assembly suitably support both halves of the axle before carefully parting the casing and cover. The withdrawal of the axle cover from the casing releases the differential and crown wheel assembly, which can now be withdrawn. Carefully note the positions of the spacers fitted between the differential bearings and the bearing housings, for they control the position of the differential assembly in the axle, and it is absolutely imperative that they be assembled in their original locations. The original spacers are marked O/S and N/S.

DISMANTLING THE DIFFERENTIAL AND CROWN WHEEL ASSEMBLY.

When the differential assembly has been removed from the axle casing, bend back the tab of its locking plate of the bolt locating the differential pinion shaft, withdraw the bolt and remove the shaft. The differential pinions with their thrust plates should now be swung round until they register with the openings in the differential cage, through which they can be removed. Now through the openings and from inside the differential, withdraw the cage gears, together with their thrust washers. Tap back the tabs of the locking plates on crown wheel bearing bolts, remove the bolts and withdraw the crown wheel from the differential cage.

NOTE:—The crown wheels are marked on their back faces with one of the following figures: +2. +1. ZERO (or no marking) −1. −2.

Fig. 12. The differential and crown wheel assembly with ball races in position on differential cage.

ASSEMBLING THE DIFFERENTIAL AND CROWN WHEEL.

Insert the differential gears with their thrust washers in position, inside the differential cage. Insert the pinions with the distance pieces and thrust washers through the openings of the cage and rotate the pinions until they register with the holes for the cage in the shaft. The pinion spindle, which should be a light push fit in the cage, is now inserted, taking care to line up the locking bolt holes.

NOTE:—The slot in the shaft can be used as a guide.

Fit the locking bolt and turn up the tab of the locking washer. Check the mating faces of the differential cage for cleanliness and fit the crown wheel into position. Check the crown wheel for truth by spinning the assembly on a roller fixture with a dial gauge registering against the outer cage of the crown wheel. The maximum permissible error is .001 in. If the figure registered is in excess of these limits the crown wheel should be removed from the cage and the flange of the cage checked for truth. If the flange is within the permissible error, clean all parts carefully and reassemble the crown wheel to the cage in a different position to that in which it was first assembled and checked, and recheck. The differential ball races can then be pressed on.

REMOVAL OF THE PINION.

Remove the cotter pin securing the slotted nut on the forward end of the pinion shaft and remove the nut and flat washer. Gently tap out the pinion shaft towards the crown wheel end, using a soft drift of a suitable diameter that will pass through the flange bore, so that the flange will be removed at the same time. The inner race of the large bearing will come away with the pinion leaving the outer race in the pinion housing, together with the complete smaller bearing and oil seal. Now withdraw the smaller bearing and oil seal with a suitable extractor and tap out the outer race of the large bearing towards the crown wheel end. Press off the pinion, the inner race of the larger bearing and withdraw the special washer.

REPLACEMENT OF THE PINION.

Fit the pinion outer bearing races in the pinion housing, making sure they are tight up to the shoulders machined in the housing. The special washer should be threaded on to the shaft, bevelled side against the pinion, and press on the rear roller bearing inner race with the protecting side against the washer. Now insert the sub-assembly into the casing through the axle cover opening and locate in position in the pinion housing. Fit the distance piece and forward roller bearing inner race on to the pinion shaft, with the protecting side of the inner race facing forward. Fit the pinion flange with the retaining washer and nut, and tighten up the nut.

NOTE:—The oil seal should be omitted at this stage. Rotate the pinion using special tool, part No. 68839. to ascertain the correct degree of pre-load which should be 11-13 lbs./ins. If the pre-load is correct, remove the nut, retaining washer and flange, fit the oil seal (sharp edge of the bore toward bearing) replace the flange, retaining washer and nut. Tighten up the nut and fit the cotter pin. If the pre-load is not correct, the distance piece and bearings will have to be replaced by selector process until the correct pre-load is obtained.

NOTE:—These instructions apply only to the replacement of material that bears the same markings as that of the original components. If examination shows that the material required is to bear different markings to that of the original, then it is recommended that a replacement axle be fitted.

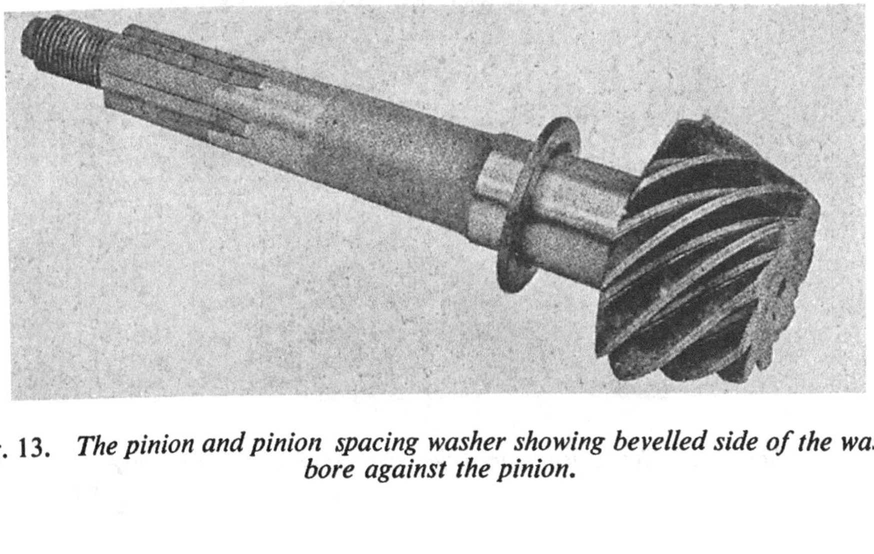

Fig. 13. *The pinion and pinion spacing washer showing bevelled side of the washer bore against the pinion.*

REASSEMBLING THE AXLE.

Providing no replacement parts have been fitted the assembly of the axle is quite straightforward, for assembly of the differential and crown wheel and assembly of pinion housing have been dealt with. Place the spacers in position in the axle casing and cover, making sure that they are in their original positions. A new gasket must be used between the axle cover and axle casing, and it is imperative that only a genuine M.G. replacement is used. The calculation made for adjustment provides for the thickness of the gasket which is .005 in. when compressed. Place the differential assembly in position in the axle casing, making sure that the bearing in the axle casing is right home and the gasket in position on

Fig. 14. *Showing bearing spacer, indicated by arrow, in position in axle casing.*

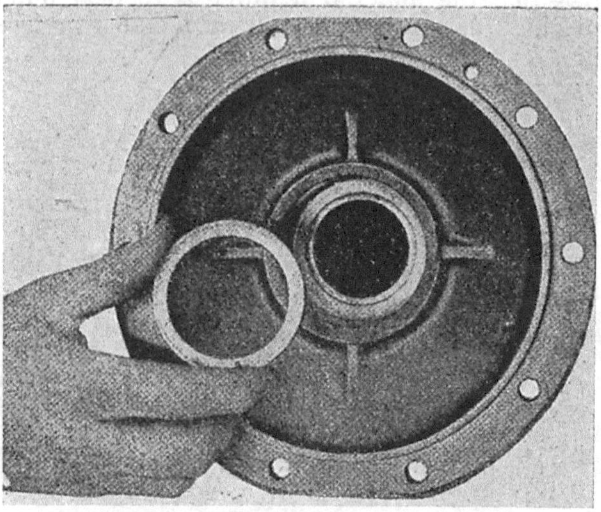

Fig. 15. *The spacer for axle cover and recess in cover*

the joint surface. Place the axle cover in position over the axle casing and carefully push home. Lightly screw into position the ten nuts securing the halves of the axle housing and then tighten up a quarter of a turn at a time in a diagonal sequence to ensure even tightening and absence of distortion.

The replacement of the brake plates and axle half shafts and hub and brake drum assemblies are detailed on pages 199 and 200. Make sure that the pinion differential and axle half shafts are free from undue restriction before replacing the axle in the car.

REFITTING THE REAR AXLE. (TD model.)

This is largely a reversal of the removal procedure, and after lifting the axle over the springs it must be carefully positioned before the "U" bolts are tightened.

Connect the propeller shaft and make sure the self locking nuts are quite tight.

Reconnect the rebound straps and hand brake cables; fit shock absorbers to their brackets; connect up brake hoses, finally bleeding the braking system is detailed in Chapter VIII.

REMOVAL AND REPLACEMENT OF REAR AXLE. (1¼ litre Series YB.)

With hypoid rear axle no repairs, such as the fitting of a replacement crown wheel and pinion, etc., can be carried out with the axle in position in the chassis. When repairs become necessary the axle complete must be removed from the chassis. Proceed as follows:—

Remove the drain plug and drain oil from the axle. This is best carried out after a run when the oil is warm. Lift the rear of the car by the use of a suitable sling and support the chassis just forward of the rear spring front mounting. Release the hand brake and remove the road wheels. Remove the right hand side brake drum and hub assembly as detailed on page 199. Uncouple the propeller shaft flange after suitably marking for reassembly and release the Lockheed flexible hose from its junction on the chassis, and release the rubber Jackall hose from its junction on the chassis also: Remove the nuts securing the hydraulic damper lower links: Remove the two bolts securing each axle check strap to the chassis: Disconnect the brake cables from their anchorage on the chassis: Disconnect the brake cables from the brake shoe actuating lever on the brake back plate: Disconnect the brake pipe from the back of the wheel cylinder R.H. Back plate only and remove back plate as detailed on page 200.

Remove the nuts and locknuts off the spring "U" Clips and withdraw "U" clips from the axle.

Tilt the axle to allow the hydraulic jacks to clear the springs and chassis and withdraw axle through left hand side wheel arch.

Replacement is carried out in the reverse manner to that detailed for removal, remembering to bleed and adjust the brakes after coupling up the flexible hose.

from the 'M'
MG
to the 'TF'

CHAPTER VI

FUEL SYSTEM

CARBURETTORS

With the exception of the "M" and "D" Midgets and the One and a Quarter Litre Model, all types are fitted with twin carburettors. On all models except the MG Two-Litre which is fitted with downdraught carburettors, the carburettors are semi-downdraught.

CONSTRUCTION. (All Models.)

Reference to the illustrations showing section views of the carburettor and jet assembly will prove valuable if studied before dismantling.

Functioning is as follows: the level of petrol in the float chamber is controlled by the float lifting the needle lever and consequently, the needle on to its seating. This shuts off the petrol supply at a predetermined level. There is a mesh filter situated at the top of the float chamber.

Petrol leaving the float chamber passes through ducts into the jet assembly and is drawn up the jet, around the taper needle and into the mixing chamber. Here, between the piston and the top of the jet, it is mixed with air and the mixture passes into the engine.

The jet is free to move until the jet head abuts against the jet adjusting nut. At the top end of the jet is an accurately calibrated hole in which the jet needle moves. The jet needle is held in the carburettor piston by means of a grub screw.

On the engine side of the piston and jet assembly is the throttle valve which is controlled by the accelerator pedal. When the accelerator pedal is depressed and the throttle valve opened, the suction created by the engine pistons draws the mixture into the combustion chambers where it is ignited.

The top of the carburettor piston is formed into what is termed the suction disc which is an accurate fit in its housing. When the throttle valve is opened, engine suction, which is transferred to the piston housing, causes the piston and needle to rise. This movement is in proportion to the speed of the engine. This means that the size of the choke and the position of the needle is varied automatically, to give a correct mixture at all engine speeds.

On late TC Models and on all 1¼ Litre and TD Models, there is a small hydraulic damper fitted to the hollow piston rod. This damper consists of a plunger with a one-way valve, and its function is to produce a slightly rich mixture on acceleration, by preventing the piston from rising too quickly.

If this damper is removed at any time, it should be tested for truth. If the wire rod holding the damper plunger to the screwed cap of the suction chamber is bent, it will prevent the piston from rising and cause spitting in the carburettor.

MAINTENANCE AND ADJUSTMENTS.

The piston guide rod should be lubricated with bicycle oil. Three drops through the small lubricator on the side every 1,000 miles will be sufficient.

On carburettors fitted with the hydraulic damper, the brass cap must be unscrewed and the damper plunger removed, when the reservoir should be filled with Duckham's "Carburettor Piston Oil." This attention is necessary once each month.

It is essential that the recommended oil be used, otherwise the damping will be upset and this in turn will affect acceleration. Never lubricate the suction disc of the piston.

The jet adjusting nut is there for setting the mixture strength, in order to obtain good slow running. Screwing the jet adjusting nut upwards, inwards in the case of the Two-Litre Model, and subsequently pushing in the jet until the jet head abuts against the nut, weakens the mixture. This

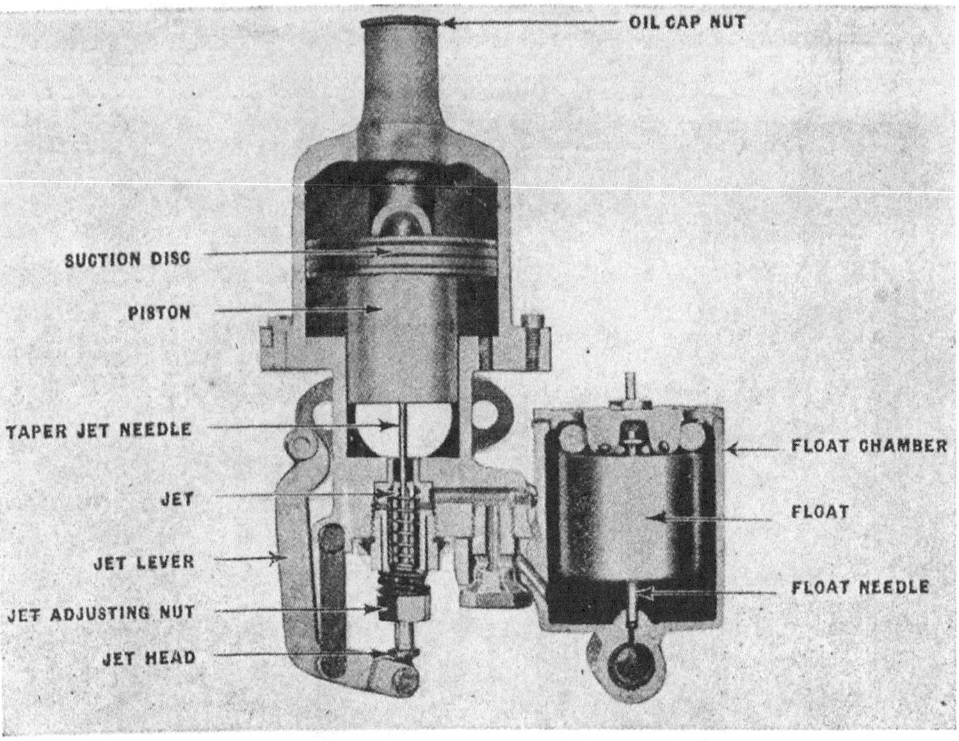

Fig. 1. *Sectional view of carburettor M, D and F Models.*

jet should be adjusted so that when the engine is thoroughly warm the weakest setting is provided which is consistent with good slow running. When the jet is adjusted in this manner, provided the correct needle is fitted, the performance of the carburettor in other respects should be correct. If a richer or weaker setting is desired, either to improve the performance or obtain more economical running, a different needle should be fitted.

It must be emphasised here that adjustment of the jet position will not achieve a richer or weaker setting, except at the expense of upsetting the idling.

The use of the rich mixture control is intended for cold starting only, and should be used as little as possible. The unnecessary use of this control will cause surplus petrol to reach the cylinder bores with consequent rapid wear.

When two carburettors are fitted, the coupling rod for the jet control levers should be disconnected before any jet adjustment is made. When the jets are finally set, the coupling rod should be lengthened or shortened to fit the two levers when the jets are hard up against their adjusting nuts.

Occasionally, the suction chamber should be removed together with the piston so that these parts may be cleaned. Make sure the suction chamber is replaced in the position from which it was removed. On the twin carburettor installation, do not change a suction chamber or piston from one carburettor to another.

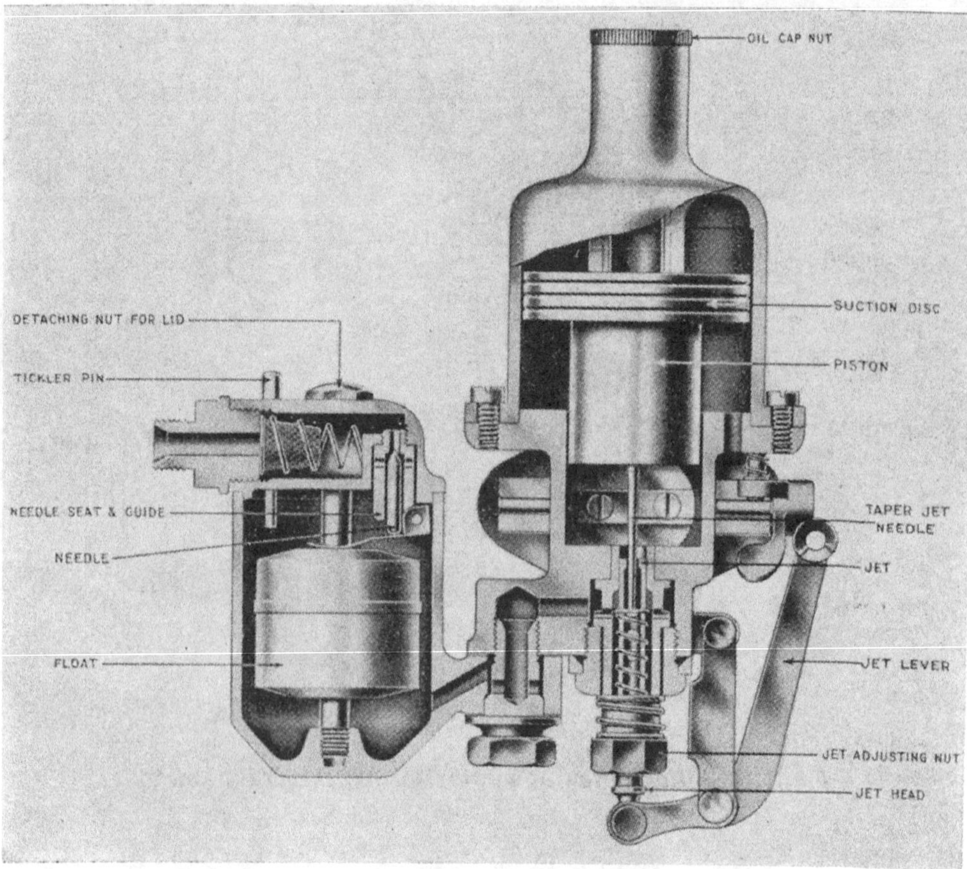

Fig. 2. Sectional view of carburettor as fitted to L, J, P, N, K, KN, TA, TB, 1½ *Litre and* 2.6 *Litre models.*

CARBURETTOR TROUBLES.

1. **Sticking piston.** This should be checked by lifting the piston with a screwdriver and allowing it to fall to its original position. If the action is sluggish or erratic, the assembly should be removed and carefully cleaned. A small amount of thin oil should then be applied to the *piston rod only,* and the carburettor reassembled.

2. **Water or dirt in the carburettor.** When this condition is suspected, the following routine should be employed: (a) lift the piston; (b) depress the tickler to flood the carburettor; (c) watch the jet; (d) observe the flow of petrol through the jet and if this flow is poor then the jet is blocked, so:—(i) start the engine; (ii) open the throttle; (iii) close the air intake with the hand; (iv) keep the throttle open until the engine speeds up. If the above procedure does not cure the trouble, the jet should be removed and cleaned. Removal and refitting of the jet is not difficult, but correct assembly is extremely important. If this is not carried out correctly, the carburettor will give a poor performance. The jet can be removed by undoing the jet holding screw and as the jet is only slightly larger than the needle, it will be obvious that great care must be taken in refitting, particularly as the jet and needle must be concentric.

The following is the sequence for reassembling:—(a) screw the jet adjusting nut to its top position; (b) move the jet up until its head is against the adjusting nut; (c) refit the complete jet; (d) check the piston for freedom of movement; (e) if the piston is not free, the jet screw must be slackened and the procedure repeated. This may have to be done several times; (f) bring the adjusting nut to its original position.

3. **Flooding of the Float Chamber.** This is generally due to dirt on the guide of the needle valve and can usually be rectified by depressing the tickler. This allows a flood of petrol to flow through the valve.

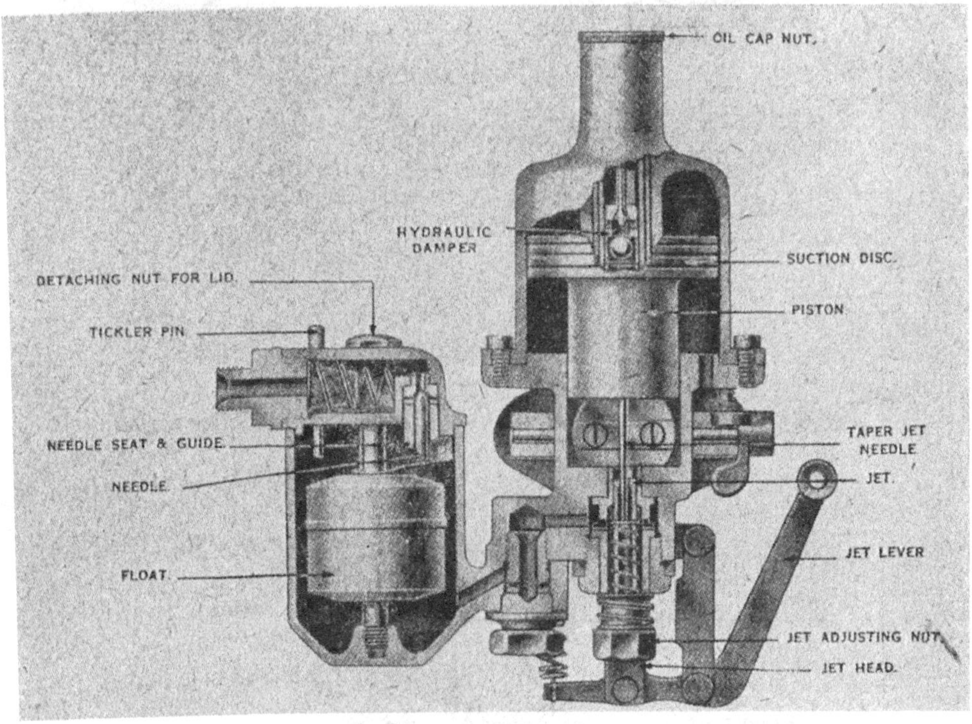

Fig. 3. *Showing section through carburettor TC, TD, and* $1\frac{1}{4}$ *Litre Models.*

4. **Float Needle Sticking.** This will cause lack of fuel at the carburettor and the needle together with its seating should be removed and cleaned.

A sticking float needle might indicate that there is a considerable amount of foreign matter in the petrol, and it might be a wise plan to clean out the whole system.

ADJUSTING THE CARBURETTOR. (M and D Models).

With the engine at its normal running temperature, screw the jet adjusting nut up as far as it will go and disconnect the mixture control rod. Now push the brass actuating lever towards the body of the carburettor as far as it will go and then move it slowly away from the carburettor (thereby weakening the mixture), until the engine idles evenly.

The jet adjusting nut should now be unscrewed until its head comes into contact with the head of the jet. This will now be the normal slow running position when the engine is hot. The mixture control rod should now be reconnected.

Provided the jet needle is of the correct type, then the performance of the carburettor on the road should be satisfactory. The correct position of the taper needle is with its shoulder flush with the piston.

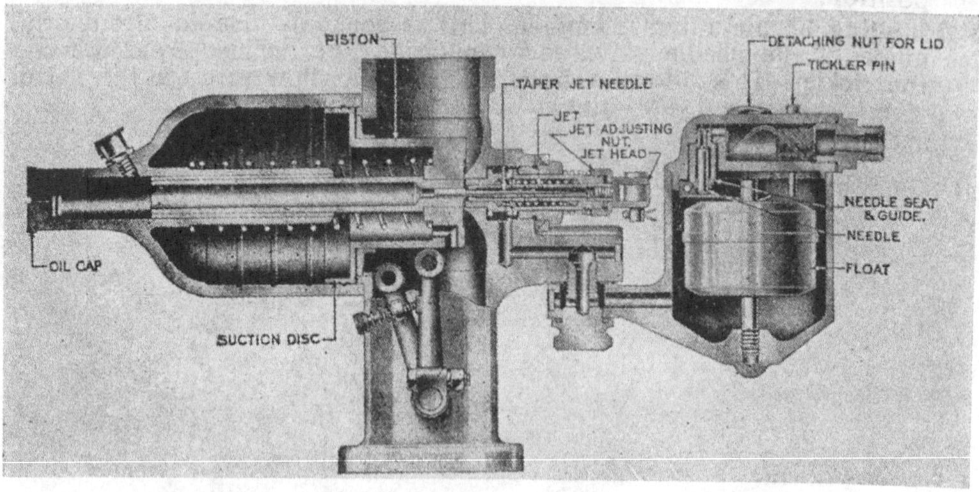

Fig. 4. Sectional view of downdraught carburettor 2 Litre model.

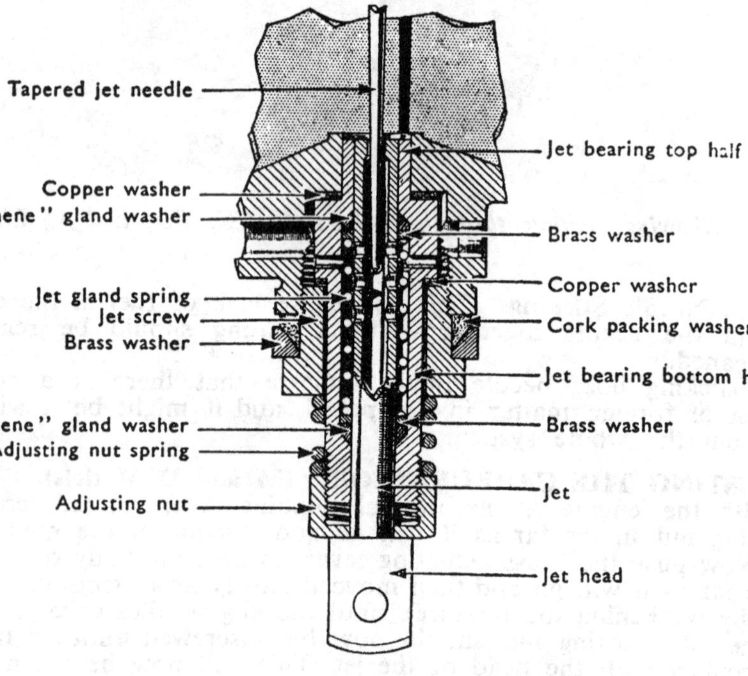

Fig. 5. Section through jet assembly. All models.

SYNCHRONISATION OF THE CARBURETTORS. (F, L, J, P, K, KN, N, 1½ Litre, 2 Litre, 2.6 Litre, TA, TB, TC, and TD Models.)

Before attempting to set the carburettors, it is advisable to check over the following points:—Ignition timing, Distributor points, Plug points, and Valve clearances. Details of these items are given on the data sheets at the end of the book.

Remove the dashpot covers and dashpots. Disconnect the jet coupling rod and screw the adjusting nuts up as far as they will go. Now push the jets up to the weak position and check for correct petrol level, which should be 1 mm. below the tops of the jets. If this is not so, and assuming the floats are in good condition, the levels can be adjusted by bending the guides that control the float needles. These guides should be bent up to raise the petrol level and down to lower it.

See that the jet needles have their shoulders flush with the bottom face of their respective pistons. Now refit the dashpots and covers to the carburettors, making quite sure that the pistons are perfectly free in their suction chambers when the latter are screwed firmly in position.

Now screw the two jet adjusting nuts down two complete turns and screw off the slow running rod.

Next screw down the butterfly adjusting screws so that they are not in use. Then slacken one butterfly flexible couple bolt; make sure that both butterflies are shut and retighten the bolt, making sure that neither butterfly moves during this operation. (F, J and K Models.)

Now set the butterfly adjusting screws so that the engine will run at something like 300 or 400 r.p.m. with the ignition fully retarded. (F, J, K and L Models.)

NOTE:—On cars fitted with a suction type windscreen wiper, make quite sure that the wiper valve screw is left open when adjusting the carburettors. This applies to F, J, K and L Type Models.

Now slacken one of the flexible coupling bolts on the coupling connecting the throttle spindles. Start the engine, screwing down the jet adjusting nuts further if necessary, in order to enable it to run. The slow running screws on the throttle spindles should now be screwed up and eventually adjusted so that the engine is idling fairly slowly. Place the ear to the mouth of the two carburettors to determine that there is an equal flow of air through them. Adjust the slow running screws until this occurs, i.e., until there is an equal hiss at the mouth of each carburettor. Now retighten the coupling bolt (P, KN, N, 1½ Litre, 2 and 2.6 Litre, TA, TB, TC and TD Models).

Allow the engine to attain its normal running temperature before attempting final adjustments. When the engine is warm, the mixture may be judged by the exhaust note. If the engine is running with a regular rhythm, but suggesting that more work is being done by one cylinder than the others, the mixture is too rich. This is accompanied by black smoke at the exhaust. In this case, screw up one of the adjusting nuts, pushing up the jet head to meet it at the same time. If no improvement is made, return this adjusting nut to its original position and adopt the same procedure with the other.

If, on the other hand, the exhaust note is regular without any perceptible rhythm, it is probable that the mixture is too weak. In this case first one and then the other of the jet adjusting nuts should be screwed down.

Turn the slow running adjustment screws counter-clockwise until the slowest possible tickover is obtained consistent with even firing.

A good check on mixture strength can be obtained by letting the engine idle fairly rapidly and lifting each of the pistons in turn, with a pencil, by about $\frac{1}{32}$ in. This should cause a slight increase in engine speed, with no upset in the evenness of firing. If the engine stops, then it is an indica-

Fig. 6. Adjusting the slow running by means of the throttle stop screw.

tion that the setting of the carburettor is too weak. If an appreciable increase in engine speed occurs and continues to occur as the piston is lifted by another $\frac{1}{32}$ in., then the indication is that the mixture on this carburettor is too rich and the jet adjusting nut should be screwed up.

On the TA, TB, TC and TD Midgets, $1\frac{1}{2}$ Litre, 2 Litre, 2.6 Litre and $1\frac{1}{4}$ Litre Models, it is recommended that the carburettor air cleaner and branch pipe be removed before attempting to synchronise the twin carburettors.

ADJUSTING THE SLOW RUNNING. (One and a Quarter Litre.)

Run the engine until it attains its normal working temperature and then set the slow running screw so that the engine runs at a fast idling speed.

Disconnect the mixture control wire from the end of the jet lever and screw the jet adjusting nut outwards. The engine will now run with a constant uneven beat known as "hunting." This is caused by the mixture being too rich. NOTE:—The jet lever must be kept in contact with the head of the jet during the whole time that the slow running is being set.

Now screw the jet adjusting nut inwards and this will gradually weaken the mixture until the engine runs evenly. Screw in the jet so far as possible to obtain the weakest setting without causing the engine to run in an irregular manner. This is the normal slow running position and if the jet needle is of the correct size, then the general performance will be satisfactory.

To test for rich mixture when the engine is idling, raise the piston by $\frac{1}{32}$ in., and if the engine speeds up then the mixture is too rich.

If the mixture is too rich when the adjusting nut is screwed up tight, with its spring compressed as far as it will go, then remove the suction chamber, take out the piston and set the jet needle so that it protrudes $\frac{1}{64}$ in. from the recommended position. (The normal position for the needle is with its shoulder just flush with the bottom face of the piston.) Now reset the carburettor as described above. If the mixture is still too rich, the needle or jet may be worn and the cure is new parts.

With the carburettor set correctly, reconnect the mixture control wire to the jet lever and make quite sure that the control rod has ample clearance when the jet is in contact with the adjusting nut.

Finally, set the carburettor throttle stop screw until gentle running is attained and make quite sure that there is a small clearance between the mixture and throttle interconnecting lever and its spring-loaded abutment screw.

THE FUEL PUMP

(Late J and F type Models, All P, TA, TB, TC, TD, F, L, K, KN and N Models, 1¼ Litre, 1½ Litre, 2 Litre and 2.6 Litre Models.)

The S.U. 12v. Type " L " Electric Pump is constructed to give prolonged service with the minimum of attention. The filter must be removed occasionally and cleaned with a stiff brush.

REMOVAL AND REPLACEMENT OF FUEL PUMP.

Disconnect the battery. Undo the pipe from the pump to the carburettor and the pipe from the pump to the tank. Remove the securing bolts that hold the pump to the dash wall or battery box and withdraw the pump.

CONSTRUCTION OF THE FUEL PUMP. (See Fig. 7.)

The pump consists of three main assemblies—the body, the magnet assembly, and the contact breaker. The body is composed of two aluminium die castings " A " (the older type of pump had a body made from a brass stamping); into the bottom of the larger one the filter " B " is screwed. The inlet union " C " is screwed in at an angle on one side. The outlet union " D " is screwed into the top. This tightens down onto the delivery valve cage " E," which is clamped between two fibre washers " F " and " G." In the top of the cage is the delivery valve, a thin brass disc " H " held in position by a spring clip " I," the suction valve " K " being a similar disc resting on a seating machined in the body. Holes connect the space between the valves to the pumping chamber, which is a shallow depression on the forward face of the smaller body casting. This space is closed by a diaphragm assembly " L," which is clamped at the outside between the magnet housing " M " and the body. A bronze rod " P " is screwed through the centre of the armature to which the diaphragm is attached and passes through the magnet core to the contact breaker which is located at the far end. A spring " S1 " is interposed between the armature and the end plate of the coil. There is a fabric joint washer between the larger and smaller body castings.

The magnet consists of a cast-iron pot, having an iron core " Q," on which is wound a coil of copper wire which energises the magnet. Between the magnet housing and the armature are fitted eleven spherical-edged brass rollers " S."' These locate the armature centrally within the magnet and allow absolute freedom of movement in a longitudinal direction.

THE CONTACT BREAKER MECHANISM.

The contact breaker consists of a small bakelite moulding carrying two rockers " U " and " U1," which are both hinged to the moulding at one end and are connected together at the top end by two small springs arranged to give a " throw over " action. A trunnion is fitted into the centre of the inner rocker, and the bronze rod " P " connected to the armature is screwed into this. The outer rocker " U1 " is fitted with a tungsten point which makes contact with a further tungsten point on the spring blade " V." This spring blade is connected to one end of the coil and the other coil end is connected to the terminal " W."

A volute spring is interposed between the armature and the diaphragm. A short length of flexible wire connects the outer rocker to one of the screws which hold the bakelite moulding onto the magnet housing, in order to ensure a good earth. Two fibre rollers are fitted to one of the spindles of the " throw-over " mechanism in order to silence the operation of the contact breaker.

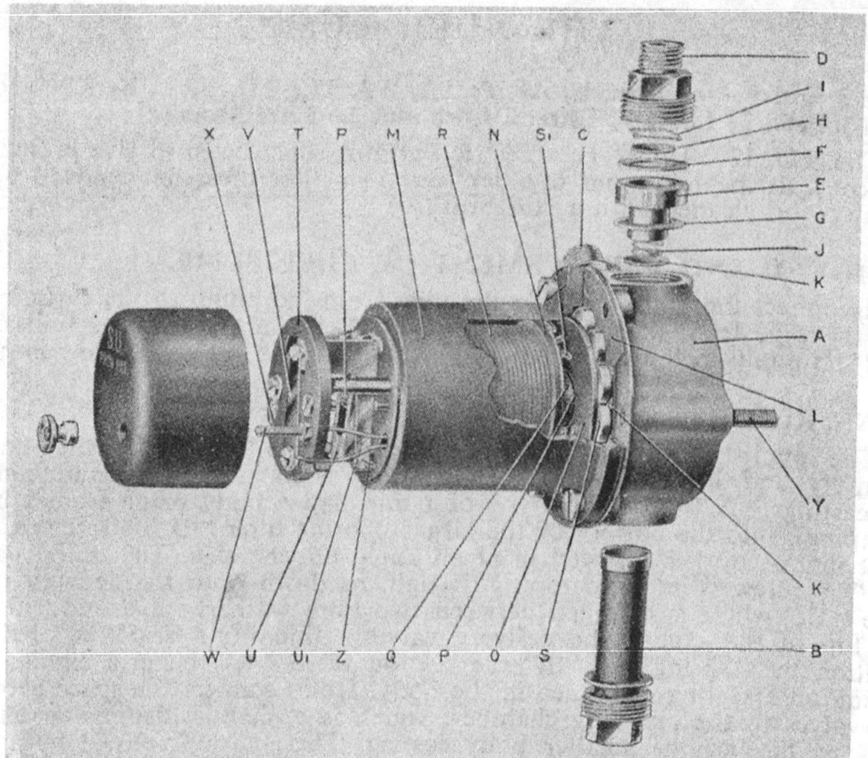

Fig. 7. The S.U. electrical Fuel Pump.

FUEL PUMP ACTION.

The action of the pump is as follows:—When the pump is at rest the outer rocker lies in the outer position and the tungsten points are in contact. The current passes from the terminal, through the coil, back to the blade, through the points and to earth, thus energising the magnet and attracting the armature. This comes forward, bringing the diaphragm with it, and sucking petrol through the suction valve into the pumping chamber. When the armature has advanced nearly to the end of its stroke, the " throw over " mechanism operates, and the outer rocker flies back, separating the points and breaking the circuit. The spring "S1" then pushes the armature and the diaphragm back, forcing petrol through the delivery valve at a rate determined by the requirements of the engine. As soon as the armature gets near the end of this stroke, the " throw over " mechanism again operates, the points again make contact, and the cycle of operations is repeated.

The spring blade rests against a small projection on the bakelite moulding, and it should be set so that when the points are in contact it is deflected back from the moulding. The width of the gap at the points is approximately .030 in.

If the magnet is removed from the body for any reason, care should be taken that the rollers " S " do not drop out. If the armature and centre rod have been unscrewed it will be necessary to reset these. IN ORDER TO DO THIS THE SPRING BLADE WHICH CARRIES THE CONTACT MUST BE SWUNG TO ONE SIDE. The armature should be screwed in as far

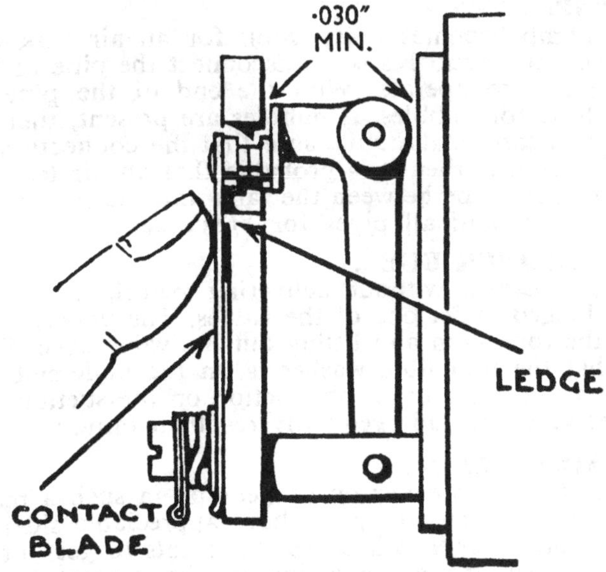

Fig. 8. *Showing the correct setting for the Contact Breaker Points.*

as possible and should then be screwed back gradually and pressed in and out until it is found that when it is pushed in slowly but firmly the " throw over " mechanism operates. IT SHOULD THEN BE UNSCREWED A FURTHER TWO-THIRDS OF A TURN. The six screws which hold the magnet to the body may now be screwed into place, BUT BEFORE TIGHTENING THESE DOWN IT IS ESSENTIAL THAT THE DIAPHRAGM SHOULD BE STRETCHED TO THE OUTERMOST POSITION. This is most easily done by inserting a matchstick behind one of the white fibre rollers on the outer rocker and thus holding the points in contact. If a current is then passed through the pump the magnet will be energised and will pull the armature forward, and while it is in this position the screws should be tightened. For correct adjustment see page 218.

TRACING TROUBLES.

In the event of trouble, first disconnect the petrol delivery pipe from the pump. If the latter then works, the most likely cause of the trouble is a sticking needle in the float-chamber of the carburettor. Should the pump not work, disconnect the lead from the terminal and strike against the body of the pump to see if it sparks and therefore if any current is available in the wire. If the current is there, remove the bakelite cover and touch the terminal with the lead. If the pump does not operate and the points are in contact and a spark cannot be struck off the terminal, it is probable there is some dirt on the points. These may be cleaned by inserting a piece of card between them, pinching them together and sliding the card backwards and forwards. If, when the wire is connected to the terminal and the tickler of the carburettor is depressed, the points fail to break, it is possible that there is either an obstruction in the suction pipe, which should be cleaned by blowing down it with a tyre pump, or something in the pump itself preventing a correct movement. This may be due either to the diaphragm having stiffened or may be due to foreign matter in the roller assembly which supports the diaphragm and, therefore, the diaphragm should be removed, the whole assembly cleaned and reassembled in accordance with the instructions given in the previous paragraphs.

A NOISY PUMP.

If the fuel pump becomes noisy, look for an air leak on the suction side. The best way to check this is to disconnect the pipe at the carburettor and allow the pump to operate. With the end of the pipe submerged in petrol in a can look for bubbles. If bubbles are present, then there must be an air leak. In this case, make quite sure that the connections at the pump are in order. If they are, then it is probable that an air leak has developed somewhere in the feed pipe between the tank and the pump. Check all connections for looseness and all pipes for breakage.

FAILURE TO DELIVER FUEL.

If the pump operates without delivering petrol, it is possible that a piece of dirt is lodged under one of the valves. The valves can be removed by unscrewing the top union and lifting out the valve cage. When replacing the cage, see that the thin fibre washer is on the underside and the thick one on top. A choked filter or an obstruction on the suction side will cause the pump to get very hot and eventually result in failure.

FUEL PUMP ADJUSTMENTS.

Assemble the contact breaker on its pedestal in such a manner that the rockers are free in their mountings without appreciable sideplay. Excessive sideplay on the outer rocker will cause the points to get out of line while excessive tightness will make the action of the contact breaker sluggish. To obtain the required freedom in case of tightness, it may be necessary to straighten the outer rocker with a pair of pliers. The hinge pin is case hardened and it is not advisable to use an ordinary piece of wire as a replacement.

The contact breaker blade should always bear directly against the bakelite pedestal, that is, underneath the tags.

When the blade is fitted correctly it should rest against the ledge below the opening in the pedestal for the contact breaker points when the points are separated. The blade should not be so stiff as to prevent the outer rocker from coming right forward when the points are in contact. The points should make contact when the rocker is in its mid-position. A simple way to check this is to hold the blade in contact with the pedestal, taking care not to press on the overhanging portion and see that a feeler gauge .030 in. thick can be inserted between the white rollers and the body of the pump. The gap at the points will then be approximately .030 in. If necessary, the tip of the blade may be set to give the correct clearance.

It is important that the contact points are set in the correct vertical position relative to each other. As the rocker moves the points have a sliding motion and they should be set to give an equal amount of movement each way. (See Fig. 9.) This position can be obtained by slackening the screw that holds the contact blade. The blade can then be moved up or down, whichever is required.

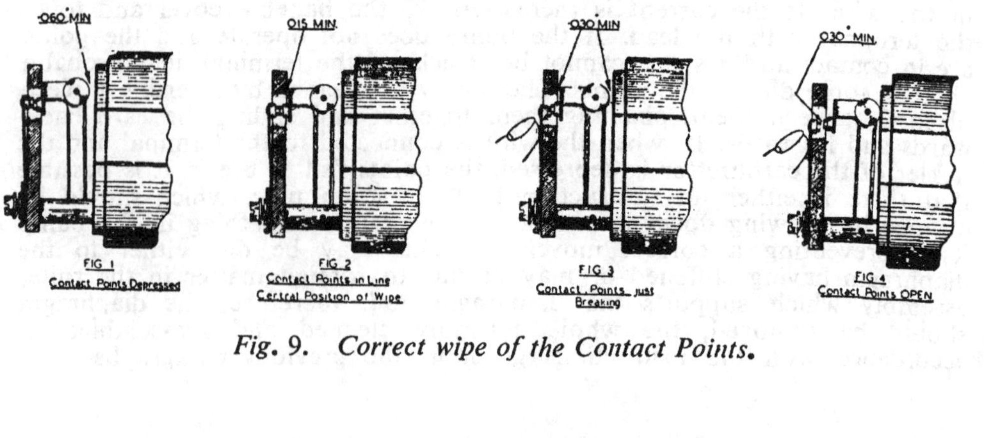

Fig. 9. Correct wipe of the Contact Points.

NOTE.—The spring washer on the 2BA screw to which the earth connection is made should be fitted between the tag and the pedestal. The spring washer is not a reliable conductor and the brass tag must bear directly against the head of the screw.

The two ends of the earthing tag and the two ends of the coil should be soldered. The coil end leading to the terminal should be soldered to its tag and not to the retaining nut. Similar conditions apply to the terminal screw holding the bakelite cover in position. The assembly is as follows: spring washer, wiring tag, lead washer and recessed nut. The spring washer must not be omitted or the assembly shortened, because this may lead to breakage of the pedestal when the cover retaining nut is tightened.

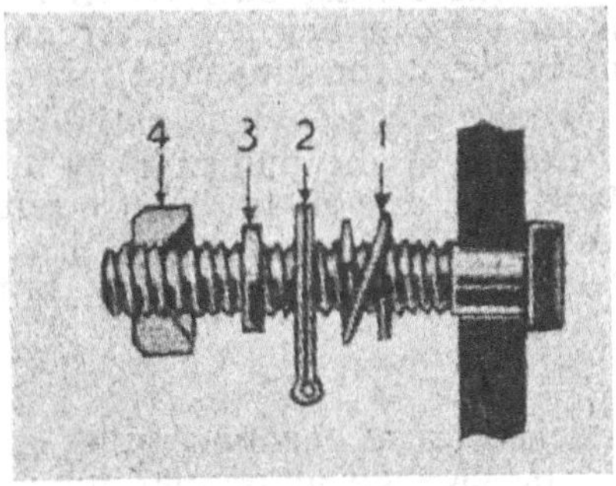

Fig. 10. *Correct assembly of the connecting components of the Terminal Screw.*

ADJUSTING THE MAGNET ASSEMBLY.

The core of the magnet is pressed into position with a special tool and should not be disturbed. Fit the armature return spring with its larger diameter towards the coil and its smaller diameter resting against the armature. Do not stretch or interfere with the spring or the action of the pump will be impaired.

The correct adjustment for the armature is as follows: (i) swing the contact blade on the pedestal to one side; (ii) fit the impact washer in the recess of the armature; (iii) screw the armature into position; (iv) place the eleven guide rollers in position around the armature; (v) hold the magnet assembly in a horizontal position; (vi) push the armature in with the thumb using a firm and steady action. The armature should be screwed in until the contact breaker ceases to throw over. Now unscrew one sixth of a turn at a time until a position is found where the contact breaker will just throw over. Care must be taken to avoid jerking the armature. Now unscrew two thirds of a turn, i.e., four holes. The armature setting is now correct. At this point it is important to remember that the setting must be carried out when the points are not in contact. If a new diaphragm is fitted it is quite probable that considerable pressure will be required to push the armature right home. If any doubt arises about the point at which the contact breaker throws over, come back one sixth of a turn, (vii) place the cast-iron body in position on the brass body, making sure that the drainhole in the cast-iron part is at the bottom and in line with the filter plug of the brass body. Make sure also that the rollers are in their correct positions. If a roller drops out, it will become trapped between the two parts and it will cut a hole in the diaphragm. Make sure that the cast-iron body seats correctly on the brass part and insert the five coupling screws and the earth terminal screw. Do not tighten these screws completely at this stage, because it is essential that the diaphragm is stretched to its uttermost position. A special forked wedge will keep the armature in this position. The trunnion in the centre of the inner rocker will be lifted as far as it will go if the wedge is inserted between the white rollers of the outer rocker and pressed under the tips of the inner rocker. The diaphragm can be stretched with the points in contact by inserting a matchstick under one of the white fibre rollers and passing an electric current through the pump. This will excite the magnet, actuate the armature, and so stretch the diaphragm. With the diaphragm in this position, the five screws and the earth terminal screw may be fully tightened; (viii) the pump should now be tested and the action of the contact breaker can be observed if a cutaway cover is used to prevent the hinge pin from falling out. When testing, mount the pump three feet above a tank and use either petrol or paraffin and a glass tube with rubber connections between the tank and pump.

NOTE:—Three important points, which should not be overlooked, can impair the correct functioning of the pump. They are as follows: 1. Keep the contact breaker blade out of contact whilst setting the diaphragm. 2. Press firmly and steadily on the armature whilst obtaining the correct setting. 3. Keep the diaphragm stretched to the limit of its stroke whilst tightening the body screws.

MAINTENANCE.

There is a filter attached to a hexagon plug on the underside of the pump. This filter should be cleaned at regular intervals with a stiff brush.

Always keep the contact breaker points clean.

NOTE:—Dual petrol pumps are fitted to the Two-Litre model Series "SA." A "clacking" noise from the pumps usually indicates lack of petrol. This is not always the case, however, and the noise can be due to one or both of the following defects: (a) an air leak on the suction side or dirty suction valves; (b) should both pumps make this noise, it is probable that the defect lies in the pipeline. In this case, inspect the pipes and unions. If the noise occurs on one pump only, it is an indication of a dirty suction valve.

When "ticks" are heard at fairly long intervals it is an indication that the pumps are operating satisfactorily. It is only when a continuous noise is heard that a defect is to be suspected.

THE S.U. PETROLIFT

(M, D, F and Early J Models.)

This instrument is constructionally very simple and it is improbable that it will give trouble. Should it cease to function, the trouble will probably be due to

1. The pump plungers (C or K) sticking, due to dirt or grit getting between the pump plungers and the body. Often a blow on the pump with the fist is sufficient to get it working, when the dirt will pass right through. Should it not do so, the remedy is to remove the filter bowl (U) and foot valve (Y), also the top cap of the pump and the cork float, when it will be possible to push the plunger (C) through the bottom after which a clean rag can be drawn through the bore of the pump. Note when assembling the plunger of the pump that valve (E) is on top.

If the above is found to be in order

2. First of all remove the cap (V) from the pump to see if the float-chamber contains petrol. If it does then the trouble is not due to the pump.

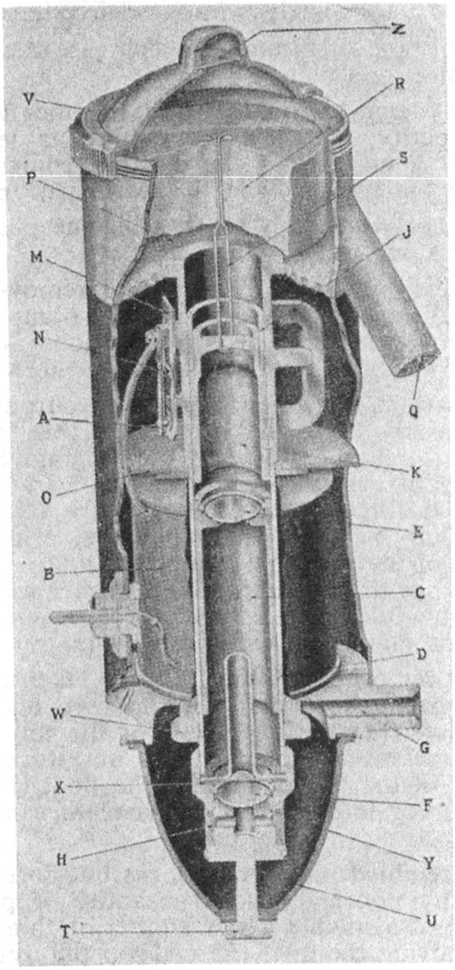

Fig. 11. *The S.U. Petrolift.*

3. If the pump continues to make a pumping noise without delivering petrol, it is due to one of the following causes:—

(a) Lack of petrol in the back tank.

(b) Air leak, which may be due to

(1) a bad joint between the filter bowl (U) and the casing, in which event tightening up will generally correct. If it does not do so a new washer will have to be fitted

or

(2) a loose petrol union on the suction pipe, that is to say any point between the bottom union of the pump and the back tank. The washer between the filter bowl and its bolt (T) should also be inspected.

(c) Foot valve (F) held up. This is a very rare source of trouble. To rectify remove the filter bowl (U), filter (H) and foot valve (Y) by means of a tommy bar through one of the holes. The foot valve can then be cleaned. A second filter (X) will be found in the foot valve underneath the priming tube.

4. If the pump works very slowly without delivering petrol, it is due to
   (a) Blocked petrol pipe or filters, in which case the filters or pipe must be cleaned out.
   (b) Batteries run down, in which case fill the float-chamber of the pump with petrol. This will probably enable the engine to be started up by hand, and as the dynamo comes into action it will boost up the batteries sufficiently to run the pump.
5. Should the pump not work at all, providing the plunger has not stuck, the trouble will be due to
   (a) A bad electrical connection. To test this remove the terminal from the pump and flash the wire across the pump body. If there is a bright flash this is in order. If not, the trouble is due to the batteries being run down or bad connections somewhere in the system.

The electrical apparatus is to all intents and purposes absolutely foolproof. Practically the only thing that can cause this to cease to function is a broken wire. If reference is made to the diagram the connections will be seen exactly. To gain access to the electrical part of the pump it will be necessary to remove filter bowl (U), foot valve (Y), unscrew the large hexagon nut (W) holding the inlet ring, when the casing can be drawn off and the internal parts of the electrical equipment and connections inspected. Care must be taken to see that the cork gland washer which makes a petrol-tight joint between the inlet ring and electrical equipment is in perfect condition. A new cork gland washer is advisable.

When the casing is removed care must also be taken to see that the wires are not broken, and particularly that the top wire does not come across the rocking contact plate (M). A simple test for the contacts being in working order, providing the bottom plunger has not stuck, is to remove the cap (V) from the top of the pump and lift the float (R) up and down its full stroke. If listened for intently the rocker-plate can be heard to click as it breaks the contact.

If, after being reassembled, the pump works but does not deliver petrol, it should be primed by pouring a small quantity of petrol into the top chamber. If petrol is not available a few squirts of thin oil down the tube of the pump after removing the float (R) and top plunger (K) will have the same effect. Please note the oil must be thin.

NOTE:—In cases where trouble is due to sodden floats, this can be put right in the following manner: A sodden float causes the valve assembly to become inoperative and to remedy this the cork assembly should be baked in a slow oven. When the cork is dry, paint it with several layers of quick-drying lacquer varnish, paying particular care to those parts where the metal fittings pierce the cork.

Modifications to Fuel Supply—Petrolift Pump. (J1 and J2 Models.)
On the J1 and J2 Models fitted with a Petrolift, there was a tendency for the fuel supply to limit the maximum speed of the car and cause what is commonly known as "drying-up." The difficulty can normally be rectified as follows:

1. The Petrolift should be fitted 1¼ in. higher. No alteration is necessary to the fuel pipes to effect this change. Commencing at Chassis No. JO300 on the J1 and Chassis No. J2300 on the J2, this modification was incorporated.
2. The existing float chamber cover fixing screws should be replaced by screws having slots cut in them to provide larger vents to the float chambers.

3. The float chamber needles required are $\frac{5}{32}$ in. in diameter with seats to correspond.

4. The screws holding the float chamber to the carburettor bodies must be tightly screwed up, as tests show that if the screws are loose, drying-up may occur.

5. When reassembling the carburettors, care must be taken to see that the toggles are not bent. With the main jets in the full rich position the petrol level should be just below the top of the jet. If this is not so, the toggles should be bent slightly as required.

6. In the case of the J type cars fitted with a pressure petrol pump, it has been found that a $\frac{1}{8}$ in. diameter needle and seating will give satisfactory results.

NOTE:—As a great many J type models have been modified on these lines, in the event of any defect in the fuel supply, careful check should be made for each of the various points mentioned, before it is assumed that the modification has been made.

CHAPTER VII

ELECTRICAL AND IGNITION EQUIPMENT

M MODEL

D MODEL

J MODEL

1¼ LITRE MODEL

1½ AND 2.6 LITRE MODELS

2 LITRE MODEL

MODIFICATIONS

ELECTRICAL EQUIPMENT
(M Model)

COIL IGNITION EQUIPMENT.

Very little attention is needed to keep the ignition equipment in first class condition. We advise that it is inspected occasionally, and the following instructions on lubrication, cleaning and adjustment should be carried out.

DISTRIBUTOR.

Occasionally remove the distributor moulding by pushing aside its two securing springs. See that the electrodes are clean and free from deposit. If necessary, wipe out the distributor with a dry duster and clean the electrode with a cloth moistened with petrol. Clean the outside of the moulding, particularly the spaces between the terminals. Next examine the contact breaker; it is important that the contacts are kept free from any grease or oil. If they are burned or blackened, they may be cleaned with very fine emery cloth and afterwards with a cloth moistened with petrol. Care must be taken that all particles of dirt and metal dust are wiped away. Misfiring may be caused if the contacts are not kept clean.

Provided that the cam is kept clean and that the instructions on cam lubrication are carried out, the contact breaker gap will only need adjustment at very long intervals. It is not advisable to alter the setting unless the gap varies considerably from the gauge. If adjustment is necessary, proceed as follows: When the contacts are fully opened, slacken the locking nut on the stationary contact screw, and rotate it by its hexagon head until the gap is set to the thickness of the gauge. After making the adjustment, care must be taken to tighten the locking nut.

LUBRICATION.

(1) **Distributor Shaft.**—The greaser on distributor shaft should be given one turn about every 500 miles.

Repack the greaser with a good quality high melting point grease when necessary.

(2) **Cam.**—About every 3,000 miles, give the cam the slightest smear of vaseline.

Coil.—The coil unit is not adjustable in any way, and requires no attention beyond seeing that the terminal connections are kept tight, and the moulded coil top is kept clean.

Warning Lamp.—A warning lamp is provided in the instrument panel, which gives a red light when the ignition is " ON " and the car is stationary. The warning lamp will also light when the engine is running very slowly, due to the fact that the dynamo is not running at sufficient speed to generate a high enough voltage to actuate the cut-out.

THE DETECTION AND REMEDY OF IGNITION FAULTS.

If a failure of ignition or misfiring occurs, unless the cause is at once apparent the owner is strongly recommended to proceed in accordance with the following routine, which should quickly enable him to locate the trouble.

Before proceeding with the examination, make sure that the trouble is not due to defects in the engine, carburetter, petrol supply, sparking plugs, etc.

Engine will not Fire.—Switch on the ignition, turn the engine and observe the ammeter reading. The engine should be turned by hand if it is known that the battery is in a low state of charge.

If an ammeter reading is given which rises and falls with the closing and opening of the contacts, then the low tension wiring is in order. If the reading does not fluctuate in this way, a short in the low tension wiring is indicated, or the contacts are remaining closed. When no reading is given, a broken or loose connection in the low tension wiring is indicated, or the battery may be exhausted.

Examine the high tension cables, i.e., cables from the coil to the distributor, and from the distributor to the plugs. If the rubber shows signs of deterioration or cracking, the cable should be renewed. Remove the distributor moulding and examine the contacts; if necessary, clean them as described on page 231. Turn the engine over by hand and see that the contacts come together.

If a fault is indicated in the low tension wiring, examine the cables from the switch or junction box to coil, and from coil to distributor. See that the battery terminals are tight and that the cables from the switch-box to the battery are secure. The battery may be dismissed as the cause of the trouble if the lamps will light.

Test the coil independently of the distributor as follows: Remove the cable from the centre distributor terminal, and hold it about ¼ in. from some metal part of the chassis and turn the engine. The sparking should be strong and regular if the coil is functioning correctly.

Misfiring and Bad Starting.—Examine the high tension cables and the plugs. If necessary, adjust the gaps to the correct setting (about 20 thousandths of an inch). Sooty or oiled plugs may be dismantled and washed with petrol.

The plugs and high tension cables may be tested by removing the plugs in turn and allowing them to rest on the cylinder head and observing whether a spark occurs at the points when the engine is turned by hand. It should, however, be noted that this is only a rough test, since it is possible that a spark may not take place when the plug is under compression.

Remove the distributor moulding and see that the electrodes and contacts are clean. If necessary, clean them as described on page 231. See that the contact gap setting is correct.

If after carrying out the examination suggested, the trouble cannot be found, we advise that the equipment should be examined by the nearest Lucas Service Depôt.

DYNAMO.

The dynamo is mounted on a platform at the front end of the engine, and forms part of the overhead-camshaft drive.

The dynamo is a singularly efficient shunt wound machine, with third brush control, and requires but little attention; there are a few points, however, which should be inspected occasionally to ensure satisfactory results. These parts are readily accessible by slackening the screw and springing the cover off the dynamo.

Brushes.—It is very important to make sure that the three brushes work freely in their holders. This can easily be ascertained by holding back the spring and gently pulling each flexible lead, when the brush should move without the slightest suggestion of sluggishness. The brushes should be clean and should " bed " over the whole of their working surface; that is to say, the face in contact with the commutator should appear uniformly polished. Dirty brushes may be cleaned with a cloth moistened with petrol.

If any of the brushes become so badly worn that it is necessary to replace them, this is accomplished by releasing the brush lead eyelet by removal of the screw, then, while holding the spring lever back out of the way, withdrawing the brush from its holder. The new brush can then be fitted by reversing the operation.

NOTE:—When ordering brush replacements state whether they are main or control brushes, and for what type of machinery they are required.

The brush springs should be inspected occasionally to see that they have sufficient tension to keep the brushes firmly pressed against the commutator when the machine is running. It is particularly necessary to keep this in mind when the brushes have been in use for a long time and are very much worn down.

Owners are cautioned that it is unwise to insert brushes of a grade other than that supplied with the machine, or to change the tension springs.

The arrangement provided has been made only after many years' experience, and will be found to give the best results.

Commutator. The surface of the commutator should be kept clean and free from oil and brush dust, etc.; neglect of this precaution will result in the commutator becoming blackened, causing sparking to occur at the brushes, and consequent shortening of the life of the machine. The best way to clean the commutator is to insert a fine duster, held by means of a suitably shaped piece of wood, against the commutator surface, slowly rotating the armature by the starting handle at the same time.

If the commutator has been neglected for long periods, it may need cleaning with fine glass paper, but this is more difficult to do, and should not be necessary if it has received regular attention. The grooves between the commutator segments should be examined occasionally and any deposits of copper or carbon dust may be cleaned out by means of a thin saw blade or similar article.

Lubrication.—As the bearings are packed with grease before leaving the works, very little attention is needed. A few drops of oil, however, may be added through the lubricators provided, say, every 1,000 miles. The reader is cautioned that far more trouble has been caused by excessive oiling than by too little.

After the car has run about 10,000 miles the dynamo should be cleaned, adjusted and the bearings re-packed with grease. This should be entrusted to the nearest Lucas Service Depôt.

Dynamo Field Fuse.—A fuse is provided in the dynamo field circuit to protect the machine in the event of anything being wrong in the charging circuit, e.g., a loose or broken battery connection. If the dynamo fails to charge the battery at any time (indicated by the ammeter giving a discharge reading during daytime running), inspect the fuse and if it has blown, replace it. If the new fuse blows after starting up, the cause of the trouble must be found, and we advise that the equipment is examined by a Lucas Service Depôt. Replacement fuses must be of the same size as those originally fitted.

With the Two Seater Model, the fuse is carried in the instrument panel. The holder can be withdrawn from the front of the panel.

In the case of the Coupé Model, the fuse is one of the four cartridge type fitted in the CJF type junction box on the engine side of the dash, and is marked F1. The size of the fuse is marked on a coloured paper slip which can be seen inside the fuse.

Removing the Dynamo.—We recommend that this should only be undertaken by an expert but should occasion arise that it is absolutely essential for the owner to remove the dynamo carefully adhering to the following instructions. In order to obtain easy access to the dynamo for removal it is necessary to take off the radiator. Drain the radiator by opening the tap in the lower water tank under the bonnet, and while the water is running away take off the bonnet by removing the two retaining screws which attach the bracket at the rear end of the bonnet rod to the scuttle apron. When the radiator is empty, disconnect the top water connection by removing the two screws which attach it to the front face of the cylinder head. This will enable the top water connection to be withdrawn with the radiator. Disconnect the rear clip on the lower hosepipe and loosen the hose from the water inlet pipe. Remove the two bolts fastening the radiator to the chassis frame, slacken the nuts at the rear end of the bonnet rod and release it from its bracket. This will permit the removal of the radiator complete with bonnet rod.

Between the projecting portion of the cylinder head and the top of the dynamo will be found a circular flexible coupling. Remove the nuts on each of the four coupling bolts in turn, leaving the bolts in position, so that the coupling can be rotated by use of the starting handle to bring each bolt into a position where the nut may easily be reached.

Having removed all four nuts, take off the clutch inspection cover and turn the engine by the starting handle until the timing marks on the flywheel are in the centre of the aperture. This will bring the driving yoke on the cylinder head across the engine, and the driving yoke on the dynamo parallel to the engine centre line. The bolts themselves may now be withdrawn, care being taken not to lose the distance washers, which must be replaced in the same position. Removal of the bolts enables the flexible coupling to be withdrawn.

Detach the two cables on the distributor side of the dynamo, noting from which terminal they are removed.

Unscrew the four set-screws which attach the dynamo to its platform at the front of the engine, thus releasing the dynamo. Lift the dynamo approximately a quarter of an inch and tilt it towards the near-side of the car until the driving yoke on the dynamo is just clear of that on the cylinder head. The dynamo can then be tilted forwards and easily withdrawn.

Replacement of the dynamo is effected in the reverse way, but it is necessary to make sure that the engine timing has not been disturbed while the dynamo was removed. Removal of the rectangular cover-plate in the clutch housing—just in front of the gear lever—should reveal the timing mark on the flywheel for Nos. 1 and 4 cylinders, exactly in the centre of the opening. Place the brass packing pieces which fit under the dynamo base in position on the dynamo platform, making sure that you replace the same number that you took off. If for any reason the dynamo is replaced by another, it may be necessary to readjust the mesh of the driving gears to obtain silent running by suitable selection of the packing shims used. Turn the dynamo spindle until the timing mark on the dynamo drive gear coincides with the centre line of the dynamo and is at the rear—that is, ready for engagement with the correspondingly marked teeth of the drive gear on the crankshaft. The holes in the dynamo coupling yoke will then be parallel with the engine centre line. Tilting the dynamo towards the near-side of the car and holding the dynamo coupling yoke in this position, insert the drive gear into the opening of the dynamo platform and swing the dynamo backwards and downwards into position. The gears can be felt to be meshing properly if the coupling is slightly oscillated as the dynamo is replaced, but do not overdo the oscillations or you may engage the wrong teeth. The correctness of adjustment may be determined after having bolted down the dynamo on its platform by checking the degree of movement obtainable on the coupling yoke on the top of the dynamo.

This movement should be very slight but nevertheless existent, otherwise the driving pinions will hum and consequent uneven wear set in.

See that the bolts in the dynamo base are coinciding with the holes in the dynamo platform, and then observe if the dynamo coupling yoke is exactly parallel to the engine centre line. If it is not, withdraw the dynamo, reset the coupling yoke and re-insert the dynamo. No difficulty should be experienced in getting the dynamo in position with the correct teeth in mesh, as the distance between one tooth and the next is sufficient to make an appreciable difference to the position of the dynamo coupling yoke, a difference which is immediately discernible.

Having satisfied yourself that the correct gear teeth are in mesh, replace the screws in the dynamo base, taking particular care to tighten them up evenly a partial turn at a time until all are quite tight.

Now make a final test. The flywheel mark 1/4 should show exactly in the centre of the inspection cover opening with the distributor rotating arm pointing towards No. 1 cylinder. (This can easily be found by tracing the high-tension lead from No. 1 sparking plug to its junction on the distributor. Removal of the distributor cover should show the distributor arm directly beneath it.) The dynamo coupling bolt holes should be exactly fore and aft and at right angles to the coupling yoke on the cylinder head, and the timing marks on the camshaft driving gears should be coinciding.

If all the foregoing are correct, replace the flexible coupling and coupling bolts, taking care to replace the distance washers in exactly the same position as they were originally. Tighten up the nuts firmly and rotate the engine slowly by hand. If the distance washers are in the correct position, the flexible coupling should run absolutely true. If it does not do so, note where the error is and adjust the distance washers accordingly. Replace the valve cover, attach the dynamo cables on to their correct terminals.

Replace the radiator and bolt the upper water connection to the forward face of the cylinder head, not forgetting to replace the jointing washer. Engage the rear end of the bonnet rod in its bracket and tighten up the nuts. You may now connect up the bottom hose and replace the bonnet, thus completing the reassembly.

Starting Switch.—The switch is extremely simple in construction, and as there are no working parts liable to get out of order, the contacts are permanently enclosed. Operate the switch firmly and quickly, both in switching on and releasing.

Starter Motor.—The armature spindles of these machines are fitted with a pinion which, on rotation, runs into engagement with the geared ring on the flywheel. Immediately the engine begins to fire, the pinion is automatically thrown out of mesh.

If, for any reason, the pinion wheel on the motor does not engage with the flywheel teeth, examine the screwed sleeve on the armature spindle to see that it is free from dust; if necessary, wash over with paraffin. Occasionally give it a few drops of thin machine oil.

As in the case of the dynamo, the surface of the commutator must be kept clean and free from oil, brush dust, etc.

The starter is designed for starting the engine under normal conditions, but any unnecessary or additional loading will considerably diminish the life of the machine and battery. In order to facilitate starting in cold weather, it is advisable t› flood the carburetter, and, before using the electric starter, crank the engine over slowly by the starting handle for two or three revolutions; this will break the oil film and considerably diminish the load for starting.

In the event of the engine refusing to fire after being turned by the starter, make sure that the ignition switch is " on."

Battery.—It is of the utmost importance that the battery should receive regular attention, as upon its good condition depends the satisfactory functioning of the ignition, starting motor, and the lamps.

At least once a fortnight the vent plugs in the top of the battery should be removed, and the level of the acid solution examined. If necessary, distilled water (which can be obtained at all chemists and most garages) should be added to bring the level well above the plates. If, however, acid solution has been spilled, it should be replaced by a diluted sulphuric acid solution of specific gravity 1.320. **It is important, when examining the cells, that naked lights should not be held near the vents, on account of the possible danger of igniting the gas which is generated by the plates.** It is advisable to complete the inspection by measuring the specific gravity of the acid, as this gives a very good indication of the state of charge of the battery. An instrument known as a hydrometer is employed for this purpose; these can be bought from your dealer or from any of the Lucas Service Depôts.

For the battery fitted to the M.G. Midget, the specific gravity readings will be 1.285—1.300 for a fully charged battery, about 1.210 when half discharged and about 1.150 when fully discharged.

If one cell gives a reading very different from the rest, it may be that electrolyte has been spilled or has leaked from this cell, or there may be a " short " between the plates. In the latter case, the battery should be examined as soon as possible by a Lucas Service Depôt.

Finally, see that the tops of the cells are clean and dry, and that the terminals are tight and smeared with vaseline.

If the equipment is laid by for several months, the battery must be given a small charge from a separate source of electrical energy about once a fortnight, in order to obviate any permanent injury to the plates.

Under no circumstances must the acid be removed from the battery and the plates allowed to dry, as certain changes take place which result in loss of efficiency.

The battery must never be left in a fully discharged condition, and, unless some long runs are to be taken, it is advisable to have the battery removed from the car periodically and charged up from an independent electrical supply.

SWITCHES.

Coupé Only.—The S.A.1 Switch is a combined Ignition and Lighting Switch. The Ignition Switch is key controlled in the centre of the unit.

The Lighting Switch has three positions, which, from left to right, are:—

"OFF"—Lights off. Dynamo on.

"SIDE"—Sidelamps and Tail on. Dynamo on.

"HEAD"—All Lamps on. Dynamo on.

Two-Seater Model.—The Ignition Switch is of the key type, positioned between the speedometer and the ammeter. The Lamp Switch is immediately above the ammeter and has three positions marked D, S and H, which gives similar control to the type S.A.1 switch but working from right to left.

Ammeter.—The centre-zero ammeter which is incorporated in the instrument panel indicates the actual current flowing into or out of the battery. For instance, suppose two amperes are consumed when the side- and tail-lamps are switched on, and the ignition coil takes one ampere, then if the dynamo is generating at seven amperes the meter will show four amperes on the charge side of the scale. This is the current in excess of the lamp and ignition load which is available for battery charging purposes.

COMBINED JUNCTION BOX, CUT-OUT AND FUSE UNIT, TYPE CJF1.

Coupé Model.—This unit is mounted on the engine side of the dash, and it houses the cut-out and fuses on its one side, while its other side forms a junction box. All terminals are marked with letters on the moulding, and the ends of the cables are given on the wiring diagram so that the wiring can easily be traced.

Terminals are provided in the Junction box for wiring up any extra accessories the owner may wish to fit. These terminals are readily accessible when the cover is removed, the one marked "AUX" is the positive feed, and the other marked "E" is the negative or earth return. When wiring up accessories it is advisable to remove the fuse while wiring is being carried out. To make efficient connections to terminals proceed as follows:—

Bare about ⅜ in. of the cable, twist the wire strands together and turn back about ⅛ in. so as to form a small ball. Remove the grub screw from the appropriate terminal and insert the wire so that the ball fits in the terminal post. Now replace and tighten the grub screw, this will compress the ball to make a good electrical connection.

Fuses.—There are four fuses, of the cartridge type, which protect the equipment in the event of short circuits. One, marked "H," protects the headlamps; another, marked "S & T," the side and tail lamps; the third, marked "AUX," the circuits of the auxiliary accessories, which include the horn, electric windscreen wiper, etc., and the fourth "F1" the dynamo.

The separate fusing of the various circuits ensures that a short in any one does not affect the rest of the electrical equipment. This is particularly important when coil ignition is fitted. It will be noticed that the fusing of the lamps is such that there is no risk of the driver being plunged into total darkness. If both the headlamps, or the side and tail lamps, or all the units connected to the auxiliary accessory terminal fail to function, examine the particular fuse protecting them. Remove the fuse from its holder and see whether there is a break in the fuse wire. Before replacing the fuse, inspect the units that have failed, for evidence of short circuits or other faults that may have caused the fuse to blow.

If a fuse blows repeatedly, and the cause cannot be traced, we advise that the equipment is examined by a Lucas Service Station.

The Cut-out.—The function of the cut-out is to close the charging circuit, as the increasing engine speed when the car is starting, causes the dynamo voltage to rise above that of the battery. When the engine slows down, the dynamo voltage falls below that of the battery, and the reverse action takes place, i.e., the cut-out opens and thereby prevents the battery from discharging itself through the dynamo.

The question is sometimes asked whether the operation of the cut-out in any way depends upon the state of charge of the battery. There is no such relation between the two; the sole function of the cut-out is to switch on the dynamo with rising engine speed and to disconnect it when the engine slows down to below a certain speed. The cut-out is accurately set before leaving the works and does not need any adjustment, and therefore the cover protecting it is sealed.

Two-Seater Model.—The Cut-out is a separate unit and no Junction Box is provided.

The Dynamo shunt field fuse is a detachable unit on the F.T.55/3 facia plate on the dash.

Headlamps.—The headlamps are provided with a patented universally adjustable mounting which allows the beam of light to be set to the best advantage. This adjustment is obtained by slackening the hexagon locking nut, turning the lamp to the desired position and then locking it by tightening up the nut.

The near side Headlamp is set to throw its beam to the near side of the road and the off side (set straight) can be turned off or on independently.

Removing the Front and Reflector.—The Headlamps are constructed with detachable fronts, parabolic reflectors and focusing devices. To remove the front the round headed screw at the top of the lamp should be unscrewed and the front turned to the left as indicated by the arrow stamped on the rim. The reflector is held in position by three supports from which it can be readily removed when necessary to re-focus the bulb. Lamps are correctly focused when sent out from the works, but if a bulb is replaced it may be necessary to adjust the focus of the lamp. Accurate focusing is imperative if the maximum results are to be attained. The method of focusing is as follows:—

Remove the front as above, remove the screws or spring ring holding reflector and then carefully draw forward the reflector until the lamp holder is exposed. This holder is fixed by a screw and may be moved backwards or forwards when the screw is loosened. Each lamp must be focused separately, care being taken that lamps are properly set in line, not pointing up or down. Correct focusing adjustment is obtained by reflecting lamps on to a wall or, preferably, a white object at a distance of about 100 yards, sliding holder backwards or forwards until the light reaches its highest point of brilliancy with total absence of shadows. Then tighten screw and replace the reflector and front.

With some lamps the lamp holder is provided with three notches. By trying the bulb in the alternative positions the best results can be obtained.

Wiring Headlamps.—Remove the front and reflector as described in a previous paragraph. Then depress the insulating washer against the spring until the hole is exposed. Thread the cable end through the hole, release the pressure on washer when the cable will be securely held in position and good contact made. Cables should not be pushed too far through the terminal, as there is a danger of shorting on to the reflector or body of the lamp.

Side-lamps.—The fronts of the wing lamps are secured by small grub screws; when these are removed the fronts can be pulled out.

Tail-lamp.—The tail-lamp is fixed on to the car by means of a flanged base. To replace a bulb turn the front portion of the lamp to the left and withdraw it from its base.

When replacing see that the studs locate with the slots in the lamp front, then push it home to lock it in position. Should it be necessary to re-wire the tail-lamp, unscrew the coupling nut, when the cable covering shell and the cable plug can be withdrawn from the lamp; pass the cable through the shell and the rubber washer. Thread the bared end of the lead into the terminal socket and secure by the screw. Replace the cable plug in its holder and secure by tightening the coupling nut.

Replacement of Bulbs.—When the replacement of any bulb is necessary we strongly advise that Lucas bulbs are used. The filaments are arranged to be in focus and give the best results with the lamps fitted. The following are the correct bulbs to use:—

Headlamps	B.A.S. No. 2S,	6 volt, 18 watt.
Side and Tail	B.A.S. No. 8S,	6 volt, 3 watt.

Cleaning Lamps.—The reflectors are protected by a transparent and colourless covering, which enables any accidental finger marks to be removed with chamois leather or a soft cloth without affecting the surface of the reflector. Do not use metal polishes on Lucas Reflectors. Ebony black lamps can be cleaned with a good car polish. Chromium plated lamps will not tarnish and only need wiping over with a damp cloth to remove dust or dirt.

Lucas Altette Horn (Type HF311).—All electric horns, before being passed out of the Works, are adjusted to give the best performance, and they will give long periods of service without any attention.

The adjustment of a horn does not alter the characteristic of the note as is sometimes supposed, but takes up the wear of vibrating parts. Do not alter the adjustment unnecessarily. If the horn becomes uncertain in its action, giving only a choking sound or does not vibrate, it does not follow that the horn has broken down. First ascertain that the trouble is not due to some outside source, e.g., a discharged battery, a loose connection or short circuit in the wiring of the horn, or in some cases a blown fuse. It is also possible that the performance of a horn may be upset by the horn becoming loose on its mounting. Also see that the front or grille is secure.

The horn is adjusted by moving the screw in the centre of the grille. This screw is notched on its underside. Do not turn it through more than one or two notches at a time. The best adjustment can be found by trial. If, after the adjustment, the tone of the horn is still unsatisfactory, the owner is urged to return it to a Lucas Service Depôt to have other adjustments made.

Electric Windscreen Wiper (Type MT1).—To start the wiper, pull out the curved handle and swing it aside so as to move the cleaning arm into position on the screen. Then pull out the switch knob and give it a spin. To stop the wiper, push in the knob. Then pull out the curved handle to disengage the wiper blade from its drive, and turn the end of the handle over the switch knob. This locks the cleaning arm out of the line of vision of the driver and ensures that the wiper is switched off.

Unless the wiper is switched off when not in use, it may be irretrievably damaged.

The wiper requires absolutely no attention; all moving parts are packed with grease during assembly, and no lubrication is necessary.

When cleaning the windscreen, the wiper arm can be easily lifted from the glass, but care must be taken that it is not moved from side to side.

If the rubber squeegee becomes worn or perished, it can be easily replaced at very small cost.

ELECTRICAL EQUIPMENT
(D Model)

The wiring diagram (see Contents) may appear a little complicated to the average car user, and is really only intended to act as a guide in the case of dismantlement of all the wiring system. It shows, nevertheless, the run of the wires from the contact breaker to the coil, and from the coil to the junction box. It will also be seen from this illustration that the wiring is carried out on the earth return system, that the negative terminal of the battery is earthed, and there is an automatic earth from the lamps, starter and dynamo. There are, however, two separate earths that have to be made, namely, from the horn and the junction box.

The principal points connected with the electrical equipment on which the owner will require instruction are the instrument board, the fuses and the wiring.

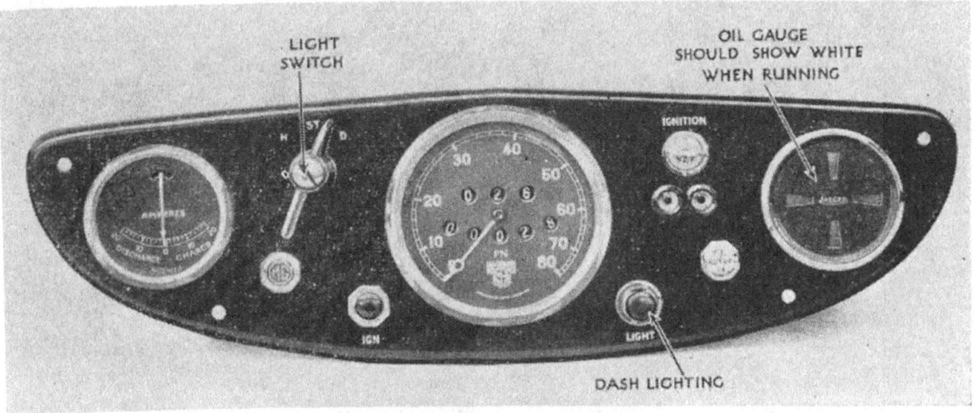

Fig. 1. *View of instrument board with instruments suitably lettered. (D, J, and F models).*

A front and rear view of the instrument board are as shown in Figs. 1 and 2. The instrument board contains, reading from left to right, an ammeter, a switch controlling the side and tail and head lamp illumination. The speedometer (which is driven by the gear box) is placed in the centre of the instrument panel on the right of which will be seen towards the upper edge of the panel, the red ignition tell-tale light and beneath this a socket for inspection lamp, and on the extreme right an oil circulation indicator. This indicator has four slots which when the oil is not circulating show red, but as soon as the oil pressure reaches a predetermined pressure the four slots show white through them. On the lower side of the panel there is the ignition switch and on the right hand side a switch to illuminate the panel from behind. The rear view of the panel shows the central speedometer which has a small arm attached to it for resetting the trip. There are a number of connections that, of course, have to be made on the junction box to the switchboard and these are indicated by colour and description on the wiring diagram. It will be noticed that the two lamps that illuminate the dials and also the ignition tell-tale lamp are fitted at the back of the panel into clips, each having its own individual socket holder. By this means a bulb could easily be changed and should ever the ignition tell-tale bulb prove defective one of the other bulbs can be temporarily used in place of it. The wiring of the car is carried in armoured tubing at all essential points and the connections are made behind the combined junction and cut-out. This will be seen on reference to Fig. 3 of the engine side of the dash-board.

A better view of this is shown in Fig. 4. The junction box is fitted with a number of glass barrel fuses, and holes are provided in the body of the box to carry spares. The fuses control the side and tail lamps, separate fuses for the head lamps, another fuse for auxiliaries such as the horn and a further fuse marked " B " which is for the dynamo field. The cut-out which is mounted on the same base as the fuses, is provided with a cover and under normal circumstances need never be touched, its operations being entirely automatic.

A small circular junction box is placed beneath the cut-out which forms a convenient connection point for wires to the horn and off side head lamp, in order to be able to control the latter through the dipper switch. It will be seen on referring to the wiring diagram that the lower terminal of the junction box is coupled to the horn and switch, and the upper terminal to the dipper switch and off side head lamp.

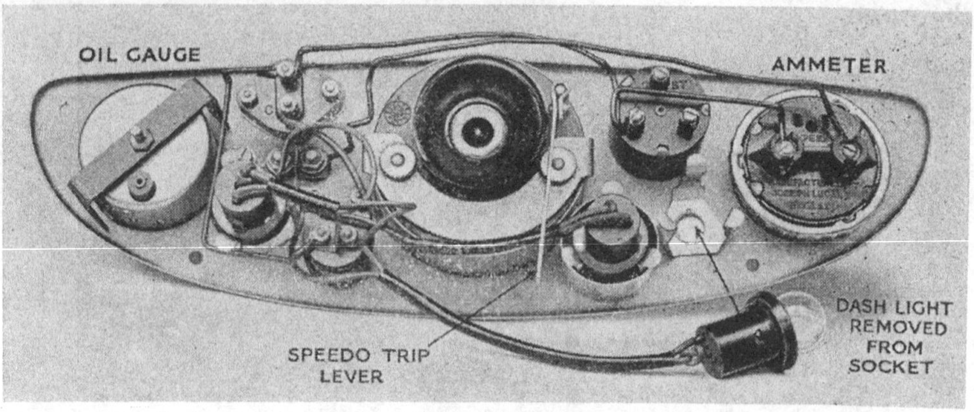

Fig. 2. Rear view of instrument board. (D, J, and F models).

DYNAMO.

The dynamo is mounted on a platform at the front end of the engine and forms part of the overhead camshaft drive. Two views of the dynamo can be seen in Fig. 5. On the left hand side it will be seen that the pinion is held on to a shaft by a bolt having a very wide head and also a tab washer which registers into the key way in the armature spindle. The two holes in the pinion are drilled and tapped ¼ in. B.S.F. in order to facilitate removal at any time with the aid of two bolts, one of which is shown in the illustration. The dynamo requires but little attention, the brush gear can be examined by slackening off the screw holding the cover in position. The cover is shown in the illustration above the dynamo.

Brushes.—It is very important to make sure that the three brushes work freely in their holders. This can easily be ascertained by holding back the spring and gently pulling each flexible lead, when the brush should move without the slightest suggestion of sluggishness. The brushes should be clean and should " bed " over the whole of their working surface; that is to say, the face in contact with the commutator should appear uniformly polished. Dirty brushes may be cleaned with a cloth moistened with petrol.

If any of the brushes become so badly worn that it is necessary to replace them, this is accomplished by releasing the brush lead eyelet by removal of the screw then, while holding the spring lever back out of the way, withdrawing the brush from its holder. The new brush can then be fitted by reversing the operation.

When ordering new brush replacements state whether they are main or control brushes, and for what type of machine they are required.

The brush springs should be inspected occasionally to see that they have sufficient tension to keep the brushes firmly pressed against the commutator when the machine is running. It is particularly necessary to keep this in mind when the brushes have been in use for a long time and are very much worn down.

Readers are cautioned that it is unwise to insert brushes of a grade other than that supplied with the machine, or to change the tension springs. The arrangement provided has been made only after many years' experience, and will be found to give the best results.

Commutator.—The surface of the commutator should be kept clean and free from oil and brush dust, etc.; neglect of this precaution will result in the commutator becoming blackened, causing sparking to occur at the brushes, and consequent shortening of the life of the machine. The best way to clean the commutator is to insert a fine duster, held by means of a suitably shaped piece of wood against the commutator surface, slowly rotating the armature by the starting handle at the same time.

If the commutator has been neglected for long periods, it may need cleaning with fine glass paper, but this is more difficult to do, and should not be necessary if it has received regular attention. The grooves between the commutator segments should be examined occasionally and any deposit of copper or carbon dust may be cleaned out by means of a thin saw blade or similar article.

Lubrication.—As the bearings are packed with grease before leaving the Works, very little attention is needed. A few drops of oil, however, may be added through the lubricators provided, say, every 1,000 miles. The reader is cautioned that far more trouble has been caused by excessive oiling than by too little. (After the car has run about 10,000 miles the dynamo should be removed, cleaned and adjusted and the bearings re-packed with grease. This should be entrusted to the nearest Lucas Service Depôt.)

Never attempt to make any alteration to the electrical equipment without first disconnecting the battery lead, on replacement watch ammeter for any undue discharge.

Dynamo Field Fuse.—A fuse is provided in the dynamo field circuit to protect the machine in the event of anything being wrong in the charging circuit, e.g., a loose or broken battery connection. If the dynamo fails to charge the battery at any time (indicated by the ammeter giving a discharge reading during daytime running), inspect the fuse and if it has blown, replace it. If the new fuse blows after starting up, the cause of the trouble must be found, and we advise that the equipment is examined by a Lucas Service Station. Replacement fuses must be of the same size as those originally fitted, i.e., 5 amps.

In the case of the 8/33 M.G. Midget model, the fuse is one of the four cartridge type fitted in the junction box on the engine side of the dash and is marked "E." The size of the fuse is marked on a coloured paper slip which can be seen inside the fuse.

Removing the Dynamo.—In order to obtain easy access to the dynamo for removal it is necessary to take off the radiator. Remove the two bolts fastening the radiator to the chassis frame, slacken the nuts at the rear end of the bonnet rod and release it from its bracket. Remove water return pipe from cylinder head and loosen clips holding bottom water hose.

N.B.—It is unnecessary to interfere with any of the oiling system to remove dynamo.

Between the projecting portion of the cylinder head and the top of the dynamo will be found a circular flexible coupling. Remove the nuts on each of the four coupling bolts in turn, leaving the bolts in position, so that the

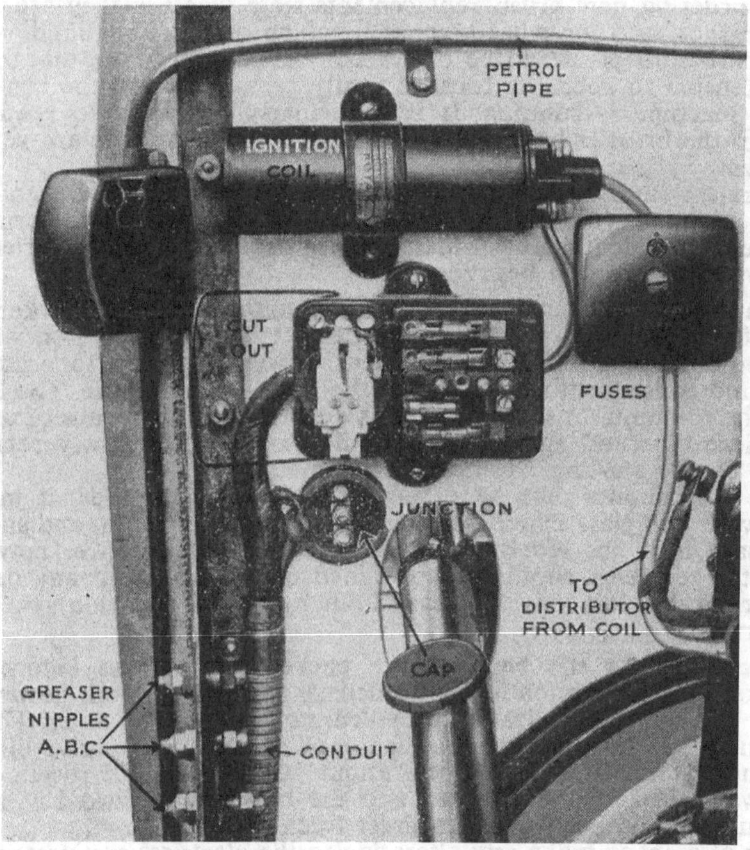

Fig. 3. Close-up view of the dashboard fittings. (D and F models).

coupling can be rotated by use of the starting handle to bring each bolt into a position where the nut may easily be reached.

Having removed all four nuts, take off the valve gear cover and turn the engine by the starting handle until the timing marks on the spiral bevel gears coincide. These will bring the driving yoke on the cylinder head across the engine, and the driving yoke on the dynamo parallel to the engine centre line. The bolts themselves may now be withdrawn, care being taken not to lose the distance washers, which must be replaced in the same position. Removal of the bolts enables the flexible coupling to be withdrawn.

Detach the two cables on the distributor side of the dynamo noting from which terminal they are removed.

Now unscrew the four set screws which attach the dynamo to its platform at the front of the engine thus releasing the dynamo. Lift the dynamo approximately a quarter of an inch and tilt it towards the near side of the car until the driving yoke on the dynamo is just clear of that on the cylinder head. The dynamo can then be tilted forwards and easily withdrawn.

Replacement of the dynamo is effected in the reverse way, but it is necessary to make sure that the engine timing has not been disturbed while the dynamo was removed. Removal of the rectangular cover plate in the clutch housing should reveal the timing mark on the flywheel for nos. 1 and 4 cylinders, exactly in the centre of the opening. Place the brass packing pieces which fit under the dynamo base in position on the dynamo platform, making sure that you replace the same number that you took off. If for any reason the dynamo is replaced by another, it may be necessary to re-adjust the mesh of the driving gears to obtain silent running by suitable selection of the packing shims used. Turn the dynamo spindle until the timing mark on the dynamo drive gear coincides with the centre line of the dynamo and is at the rear, that is, ready for engagement with the correspondingly marked teeth of the drive gear on the crankshaft. The holes in the dynamo coupling yoke will then be parallel with the engine centre line. Tilting the dynamo towards the near side of the car and holding the dynamo coupling yoke in this position, insert the drive gear into the opening of the dynamo platform and swing the dynamo backwards and downwards into position. The gears can be felt to be meshing properly if the coupling is slightly oscillated as the dynamo is replaced, but do not overdo the oscillations or you may engage the wrong teeth.

See that the bolts in the dynamo base are coinciding with the holes in the dynamo platform, and then observe if the dynamo coupling yoke is exactly parallel to the engine centre line. If it is not, withdraw the dynamo, reset the coupling yoke and re-insert the dynamo. No difficulty should be experienced in getting the dynamo in position with the correct teeth in mesh, as the distance between one tooth and the next is sufficient to make an appreciable difference to the position of the dynamo coupling yoke, a difference which is immediately discernible.

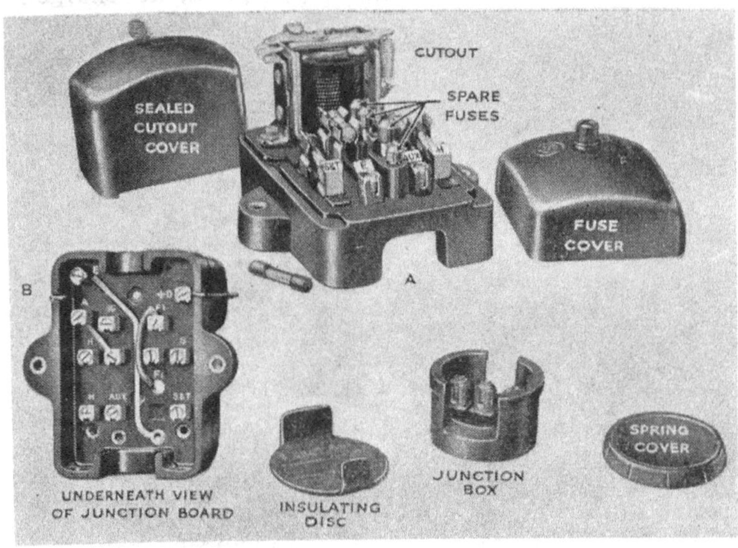

Fig. 4. *Cut-out and junction board.* (*D, J, F, L and K models*).

Having satisfied yourself that the correct gear teeth are in mesh, replace the screws in the dynamo base, taking particular care to tighten them up evenly a partial turn at a time until all are quite tight. Now make a final test. The flywheel mark 1-4 should show exactly in the centre of the inspection cover opening with the distributor rotating arm pointing towards No. 1 cylinder. (This can easily be found by tracing the high tension lead from No. 1 sparking plug to its junction on the distributor. Removal of the distributor cover should show the distributor arm directly beneath it.) The dynamo coupling bolt holes should be exactly fore and aft and at right angles to the coupling yoke on the cylinder head, and the timing marks on the camshaft driving gears should be coinciding.

If all the foregoing are correct, replace the flexible coupling and coupling bolts, taking care to replace the distance washers in exactly the same position as they were originally. Tighten up the nuts firmly and rotate the engine slowly by hand. If the distance washers are in the correct position, the flexible coupling should run absolutely true. If it does not do so, note where the error is and adjust the distance washers accordingly. Then replace the valve cover and attach the dynamo cables on to their correct terminals.

Third Brush Regulator.—The output of the dynamo, that is to say its rate of charge, is controlled by a Third Brush. It is inadvisable for anybody who is not an electrician to tamper with this, but in case the ammeter reads more than 2 amps. when all the head lamps are on, or in the event of its reading less than 8 amps. when no lights are alight, the Third Brush can be moved relative to the armature by slackening off the locking nut and turning the brush gear either towards the direction of rotation of the armature to increase the rate of charge or in the opposite direction to decrease the charge rate. **The indicated charging rate must not exceed 8 amps. with all lights off or severe damage will be done to the dynamo and cut-out.**

The Third Brush is to be found on the near side of the car and on the same side as the small ball covered lubricator. Never omit to tighten up the lock screw after any adjustment has been effected.

When replacing the brush cover care must be taken that the cover does not make contact with dynamo terminal (symptom, no charge).

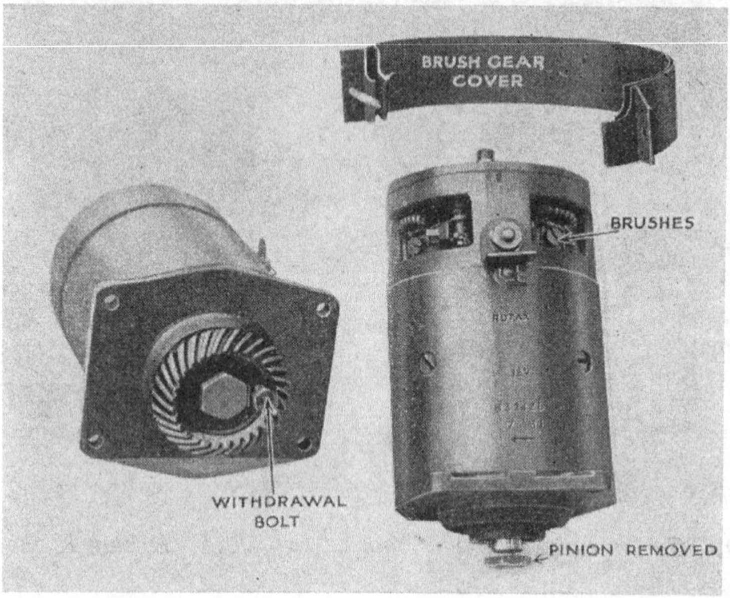

Fig. 5. Two views of the dynamo. (D, F, J, L, K, P, KN and N models)

Starter Motor.—The armature spindles of these machines are fitted with a pinion which, on rotation, runs into engagement with the geared ring on the flywheel. Immediately the engine begins to fire, the pinion is automatically thrown out of mesh.

If, for any reason, the pinion wheel on the motor does not engage with the flywheel teeth, examine the screwed sleeve on the armature spindle to see that it is free from dust; if necessary wash over with paraffin. Occasionally, give it a few drops of thin machine oil.

As in the case of the dynamo, the surface of the commutator must be kept clean and free from oil, brush dust, etc.

The starter is designed for starting the engine under normal conditions, but any unnecessary or additional loading will considerably diminish the life of the machine and battery. In order to facilitate starting in cold weather, it is advisable to flood the carburetter, and, before using the electric starter, crank the engine over slowly by the starting handle for two or three revolutions; this will break the oil film and considerably diminish the load for starting.

In the event of the engine refusing to fire after being turned by the starter, make sure that the ignition switch is " on."

It is not advisable to use the self starter with the Ignition Lever in the full advance position. It is better to put the Ignition Lever half advance and if necessary increase the amount of advance while the Starter Motor is spinning the engine. With coil ignition this is not usually necessary.

Battery.—It is of the utmost importance that the battery should receive regular attention, as upon its good condition depends the satisfactory functioning of the ignition, starter motor, and the lamps.

At least once a fortnight the vent plugs in the top of the battery should be removed, and the level of the acid solution examined. If necessary, distilled water (which can be obtained at all chemists and most garages) should be added to bring the level well above the plates. If, however, acid solution has been spilled, it should be replaced by a diluted sulphuric acid solution of specific gravity 1.320. It is important, when examining the cells, that naked lights should not be held near the vents, on account of the possible danger of igniting the gas which is generated by the plates. It is advisable to complete the inspection by measuring the specific gravity of the acid, as this gives a very good indication of the state of charge of the battery. An instrument known as a hydrometer is employed for this purpose; these can be bought from your Dealer or from any of the Lucas Service Stations.

For the battery fitted to the 8/33 M.G. Midget the specific gravity readings will be 1.285—1.300 for a fully charged battery, about 1.210 when half discharged and about 1.150 when fully discharged.

If one cell gives a reading very different from the rest, it may be that electrolyte has been spilled or has leaked from this cell, or there may be a " short " between the plates. In the latter case, the battery should be examined as soon as possible by a Lucas Service Station.

Finally, see that the tops of the cells are clean and dry, and that the terminals are tight and smeared with vaseline.

If the equipment is laid by for several months, the battery must be given a small charge from a separate source of electrical energy about once a fortnight, in order to obviate any permanent injury to the plates.

Under no circumstances must the acid be removed from the battery and the plates allowed to dry, as certain changes take place which result in loss of efficiency.

The battery must never be left in a fully discharged condition, and, unless some long runs are to be taken, it is advisable to have the battery removed from the car periodically and charged up from an independent electrical supply.

Ammeter.—The centre-zero ammeter which is incorporated in the instrument panel indicates the actual current flowing into or out of the battery. For instance, suppose two amperes are consumed when the side and tail lamps are switched on, and the ignition coil takes one ampere, then if the dynamo is generating at seven amperes the meter will show four amps. on the charge side of the scale. This is the current in excess of the lamp and ignition load which is available for battery charging purposes.

Fuses.—The separate fusing of the various circuits ensures that a short in any one does not affect the rest of the electrical equipment. This is particularly important when coil ignition is fitted. It will be noticed that the fusing of the lamps is such that there is no risk of the driver being plunged into total darkness. If both the headlamps, or the side and tail lamps, or all the units connected to the auxiliary accessory terminal fail to function, examine the particular fuse protecting them.

Remove the fuse from its holder and see whether there is a break in the fuse wire. Before replacing the fuse, inspect the units that have failed, for evidence of short circuits or other faults that may have caused the fuse to blow.

If a fuse blows repeatedly, and the cause cannot be traced we advise that the equipment is examined by a Lucas Service Station.

The Cut-out.—The function of the cut-out is to close the charging circuit, as the increased engine speed when the car is starting causes the dynamo voltage to rise above that of the battery. When the engine slows down, the dynamo voltage falls below that of the battery, and the reverse action takes place, i.e., the cut-out opens and thereby prevents the battery from discharging itself through the dynamo.

The question is sometimes asked whether the operation of the cut-out in any way depends upon the state of charge of the battery. There is no such relation between the two; the sole function of the cut-out is to switch on the dynamo with rising engine speed and to disconnect it when the engine slows down to below a certain speed. The cut-out is accurately set before leaving the Works and does not need any adjustment, and therefore the cover protecting it is sealed.

Very little attention is needed to keep the ignition equipment in first class condition. We advise that it is inspected occasionally, and the following instructions on lubrication, cleaning and adjustment should be carried out.

Distributor.—Occasionally remove the distributor moulding by pushing aside its two securing springs. See that the electrodes are clean and free from deposit. If necessary, wipe out the distributor with a dry duster and clean the electrodes with a cloth moistened with petrol. See that the carbon brush is clean and moves freely in its holder. Clean the outside of the moulding, particularly the spaces between the terminals. Next examine the contact breaker; it is important that the contacts are kept free from any grease or oil. If they are burned or blackened, they may be cleaned with very fine emery cloth and afterwards with a cloth moistened with petrol. Care must be taken that all particles of dirt and metal dust are wiped away. Misfiring may be caused if the contacts are not kept clean.

The contact breaker gap is carefully set before leaving the Works, and a gauge is provided on the spanner despatched with each distributor. Provided that the cam is kept clean and that the instructions on cam lubrication are carried out, the contact breaker gap will only need adjustment at very long intervals. It is not advisable to alter the setting unless the gap varies considerably from the gauge. If adjustment is necessary, proceed as follows:— When the contacts are fully opened, slacken the locking nut on the stationary contact screw, and rotate it by its hexagon head until the gap is set to the thickness of the gauge. After making the adjustment, care must be taken to tighten the locking nut.

LUBRICATION.

(1) **Distributor Shaft.**—The greaser on distributor shaft should be given one turn about every 500 miles.

Repack the greaser with a good quality high melting point grease when necessary.

(2) **Cam.**—About every 3,000 miles, give the cam the slightest smear of vaseline.

(3) **Contact Breaker Pivot.**—Every 5,000 miles, place a single drop of oil on the pivot on which the contact breaker works.

Coil.—The coil unit is not adjustable in any way, and requires no attention beyond seeing that the terminal connections are kept tight, and the moulded coil top is kept clean.

Warning Lamp.—A warning lamp is provided in the instrument panel, which gives a red light when the ignition is "ON" and the car is stationary. The warning lamp will also light when the engine is running very slowly, due to the fact that the dynamo is not running at sufficient speed to generate a high enough voltage to actuate the cut-out.

THE DETECTION AND REMEDY OF IGNITION FAULTS.

If a failure of ignition or misfiring occurs, unless the cause is at once apparent the owner is strongly recommended to proceed in accordance with the following routine which should quickly enable him to locate the trouble.

Before proceeding with the examination, make sure that the trouble is not due to defects in the engine, carburetter, petrol supply, sparking plugs, etc.

Engine will not Fire.—Switch on the ignition, turn the engine and observe the ammeter reading. The engine should be turned by hand if it is known that the battery is in a low state of charge.

If an ammeter reading is given which rises and falls with the closing and opening of the contacts, then the low tension wiring is in order. If the reading does not fluctuate in this way, a short in the low tension wiring is indicated, or the contacts are remaining closed. When no reading is given, a broken or loose connection in the low tension wiring is indicated, or the battery may be exhausted.

Examine the high tension cables, *i.e.*, cables from the coil to the distributor, and from the distributor to the plugs. If the rubber shows signs of deterioration or cracking, the cable should be renewed. Remove the distributor moulding and examine the contacts; if necessary, clean them. Turn the engine over by hand, and see that the contacts come together.

If a fault is indicated in the low tension wiring, examine the cables from the switch or junction box to coil, and from coil to distributor. See that the battery terminals are tight and that the cables from the switch-box to the battery are secure. The battery may be dismissed as the cause of the trouble if the lamps will light.

Test the coil independently of the distributor as follows:—Remove the cable from the centre distributor terminal, and hold it about ¼ in. from some metal part of the chassis and turn the engine. The sparking should be strong and regular if the coil is functioning correctly.

Misfiring and Bad Starting.—Examine the high tension cables and the plugs. If necessary, adjust the gaps to the correct setting (about 20 thousandths of an inch). Sooty or oiled plugs may be dismantled and washed out with petrol.

The plugs and high tension cables may be tested by removing the plugs in turn and allowing them to rest on the cylinder head and observing whether a spark occurs at the points when the engine is turned by hand. It should, however, be noted that this is only a rough test, since it is possible that a spark may not take place when the plug is under compression.

Remove the distributor moulding and see that the electrodes and contacts are clean. If necessary, clean them. See that the contact gap setting is correct.

If after carrying out the examination suggested, the trouble cannot be found, we advise that the equipment should be examined by the nearest Lucas Service Depôt.

Headlamps. The headlamps are provided with a patented universally adjustable mounting which allows the beam of light to be set to the best advantage. This adjustment is obtained by slacking the hexagon locking nut turning the lamp to the desired position and then locking it by tightening up the nut.

The near side headlamp is set to throw its beam to the near side of the road and the off side (set straight) can be turned off or on independently.

Removing the Front and Reflector.—The headlamps are constructed with detachable fronts, parabolic reflectors and focusing devices. To remove a lamp front, give the screw head at the top of the lamp about a quarter of a turn—a coin will serve as a screw driver. This movement presses a cam against the body and pushes the front away from the body, enabling it to be removed. To replace, turn back the screw head and locate the cam between the two lugs on the body, then push the front into place. The reflector is secured by means of four fixing screws. If a bulb is replaced it may be necessary to adjust the focus of the lamp. Accurate focusing is imperative if the maximum results are to be attained. The method of focusing is as follows:—

Remove the front as above, remove the screws holding reflector and then carefully draw forward the reflector until the lamp holder is exposed. This holder is fixed by a screw and may be moved backwards or forwards when the screw is loosened. Each lamp must be focused separately, care being taken that lamps are properly set in line, not pointing up or down. Correct focusing adjustment is obtained by reflecting lamps on to a wall, or, preferably, a white object at a distance of about 100 yards, sliding the holder backwards or forwards until the light reaches its highest point of brilliancy with total absence of shadows. Then tighten screw and replace the reflector and front.

Wiring Headlamps.—Remove the front and reflector as described in a previous paragraph. Then depress the insulating washer on the bulb holder terminal against the spring until the cable hole is exposed. Thread the cable end through the hole, release the pressure on washer, when the cable will be securely held in position and good contact made. Cables should not be pushed too far through the terminal, as there is a danger of shorting on to the reflector or body of the lamp.

Side lamps.—The fronts of the wing lamps are secured by small grub screws; when these are removed the fronts can be withdrawn by first pulling the top forward.

Tail-lamp.—The tail-lamp is fixed on to the car by means of a flanged base. To replace a bulb turn the front portion of the lamp to the left and withdraw it from its base.

When replacing see that the studs locate with the slots in the lamp front, then push it home to lock it in position. Should it be necessary to re-wire the tail-lamp, unscrew the coupling nut, when the cable covering shell and the cable plug can be withdrawn from the lamp; pass the cable through the shell and the rubber washer. Thread the bared end of the lead into the terminal socket and secure by the screw. Replace the cable plug in its holder and secure by tightening the coupling nut.

Replacement of Bulbs.—When the replacement of any bulb is necessary we strongly advise that Lucas bulbs are used. The filaments are arranged to be in focus and give the best results with the lamps fitted. The following are the correct bulbs to use:—Head lamps, B.A.S. No. 35; Side and tail lamps, B.A.S. No. 10S.

Cleaning Lamps.—The reflectors are protected by a transparent and colourless covering, which enables any accidental finger-marks to be removed with chamois leather or a soft cloth without affecting the surface of the reflector. Do not use metal polishes on Lucas Reflectors. Ebony black lamps can be cleaned with a good car polish. Chromium plated lamps will not tarnish and only need wiping over with a damp cloth to remove dust or dirt.

Lucas Altette Horn (Type HF318).—All electric horns, before being passed out of the Works, are adjusted to give the best performance, and they will give long periods of service without any attention; no subsequent adjustment is required.

If the horn becomes uncertain in its action, giving only a choking sound, or does not vibrate, it does not follow that the horn has broken down. First ascertain that the trouble is not due to some outside source, e.g. a discharged battery, a loose connection or short circuit in the wiring of the horn, or a blown fuse. It is also possible that the performance of a horn may be upset by the horn becoming loose on its mounting. If the cause of the trouble cannot be located, do not attempt to dismantle the horn, but return it to a Lucas Service Depôt for examination.

ELECTRICAL EQUIPMENT
(J Model)

The Electrical Equipment on the range of J Models is the direct result of many years of experiment with various types of installations, and we believe that its simplicity of design, together with the incorporation of high efficiency components which have proved capable of meeting the exacting demands of a high speed motor car, will give lasting and trouble free service.

The wiring diagrams (see Contents) may appear a little complicated to the average car user, and are really only intended to act as a guide in the case of dismantlement of all the wiring system. They show, nevertheless, the run of the wires from the contact breaker to the coil, and from the coil to the junction box. It will also be seen from these illustrations that the wiring is carried out on the earth return system, that the negative terminal of the battery is earthed, and there is an automatic earth from the lamps, starter and dynamo. There are, however, three separate earths that have to be made, namely from the tail lamp, horn switch and junction box.

The principal points connected with the electrical equipment on which the owner will require instruction are the instrument board, the fuses and the various components.

The front and rear views of the instrument board as fitted to J1 Models are as shown in Figs. 1 and 2. The instrument board contains, reading from left to right, an ammeter, a switch controlling the side and tail and head lamp illumination; the speedometer (which is driven by the gear box) placed in the centre of the instrument panel, on the right of which will be seen towards the upper edge of the panel, the ignition switch and beneath this is a socket for inspection lamp, and on the extreme right an oil circulation indicator. This indicator has four slots which when the oil is not circulating show red, but as soon as the oil pressure reaches a predetermined pressure the four slots show white through them. On the lower side of the panel there is the ignition warning lamp and on the right hand side a switch to illuminate the panel from behind. The rear view of the panel shows the central speedometer which has a small arm attached to it for resetting the trip. There are a number of connections that, of course, have to be made from the junction box to the switchboard and these are indicated on the wiring diagram. It will be noticed that the two lamps that illuminate the dials and also the ignition tell-tale lamp are fitted at the back of the panel into clips, each having its own individual socket holder. By this means a bulb could easily be changed and should ever the ignition tell-tale bulb prove defective one of the other bulbs can be temporarily used in place of it. The instrument panel fitted to the J2 Models, while including all the essential instruments as described in the preceding paragraph, has been designed to give a strikingly neat and compact effect. The principal points of difference are in the Ignition Warning Lamp which in this case is a simple 2½ volt S.S.C. bulb, the "On" and "Off" positions for the dynamo in the PLC Switch and the Oil Gauge which gives a direct reading in lbs. per sq. in. In the centre of the facia board the combined Horn and Dipping switch is fitted, together with a small but efficient panoramic mirror. The Speedometer, located in the right hand side of the facia board, is a composite instrument, recording not only m.p.h. but also engine r.p.m. in 3rd and top gears. The wiring of the car is carried in armoured tubing at all essential points and the connections are made behind the combined junction and cut-out. This will be seen on reference to Fig. 6 of the engine side of the dash-board.

A good view of the junction box is shown in Fig. 4. It is fitted with a number of glass barrel fuses, and holes are provided in the body of the box to carry spares. The fuses control the side and tail lamps, the head lamps, auxiliaries such as the horn and one marked "B" which is for the dynamo field. The cut-out which is mounted on the same base as the fuses, is provided with a cover and under normal circumstances need never be touched, its operations being entirely automatic.

A small circular junction box is placed beneath the cut-out which forms a convenient connection point for wires to the horn and off side head lamp, in order to be able to control the latter through the dipper switch. It will be seen on referring to the wiring diagram that the lower terminal of the junction box is coupled to the horn and switch, and the upper terminal to the dipper switch and off side head lamp.

DYNAMO.

The dynamo is mounted on a platform at the front end of the engine and forms part of the overhead camshaft drive. Two views of the dynamo can be seen in Fig 5. On the left hand side it will be seen that the pinion is held on to a shaft by a bolt having a very wide head and also a tab washer which registers into the withdrawal holes of the bevel pinion. The two holes in this pinion are drilled and tapped ¼ in. B.S.F. in order to facilitate removal at any time with the aid of two bolts, one of which is shown in the illustration. The dynamo has been specially designed to meet the requirements of high engine revolutions, and is particularly robust in constructional and electrical details, and beyond an occasional inspection of brushes should require but little attention. The brushes and attendant gear are readily accessible on removing the cover as shown in Fig. 5.

Brushes.—It is very important to make sure that the three brushes work freely in their holders. This can easily be ascertained by holding back the spring and gently pulling each flexible lead, when the brush should move without the slightest suggestion of sluggishness. The brushes should be clean and should "bed" over the whole of their working surface; that is to say, the face in contact with the commutator should appear uniformly polished. Dirty brushes may be cleaned with a cloth moistened with petrol.

If any of the brushes become so badly worn that it is necessary to replace them, this is accomplished by releasing the brush lead eyelet by removal of the screw, then, while holding the spring lever back out of the way, withdrawing the brush from its holder. The new brush can then be fitted by reversing the operation.

When ordering new brush replacements state whether they are main or control brushes, and for what type of machine they are required.

The brush springs should be inspected occasionally to see that they have sufficient tension to keep the brushes firmly pressed against the commutator when the machine is running. It is particularly necessary to keep this in mind when the brushes have been in use for a long time and are very much worn down.

Owners are cautioned that it is unwise to insert brushes of a grade other than that supplied with the machine, or to change the tension springs. The arrangement provided has been made only after many years' experience, and will be found to give the best results.

Commutator.—The surface of the commutator should be kept clean and free from oil and brush dust, etc.; neglect of this precaution will result in the commutator becoming blackened, causing sparking to occur at the brushes, and consequent shortening of the life of the machine. The best way to clean the commutator is to insert a fine duster, held by means of a suitably shaped piece of wood against the commutator surface, slowly rotating the armature by the starting handle at the same time.

Fig. 6. General view of the layout of electrical equipment on the J model.

If the commutator has been neglected for long periods, it may need cleaning with fine glass paper, but this is more difficult to do, and should not be necessary if it has received regular attention. The grooves between the commutator segments should be examined occasionally and any deposit of copper or carbon dust may be cleaned out by means of a thin saw blade or similar article. It is strongly recommended, however, that this operation should be entrusted to a competent electrical engineer if ever necessary.

Lubrication.—As the bearings are packed with grease before leaving the Works, very little attention is needed. A few drops of oil, however, may be added through the lubricators provided, say, every 1,000 miles. The reader is cautioned that far more trouble has been caused by excessive oiling than by too little. Every 10,000 miles the dynamo should be removed, cleaned and adjusted and the bearings re-packed with grease. This should be entrusted to the nearest Lucas Service Depôt.

Never attempt to make any alteration to the electrical equipment without first disconnecting the battery lead, on replacement watch ammeter for any undue discharge.

Dynamo Field Fuse.—A fuse is provided in the dynamo field circuit to protect the machine in the event of anything being wrong in the charging circuit, e.g., a loose or broken battery connection. If the dynamo fails to charge the battery at any time (indicated by the ammeter giving a discharge reading during daytime running), inspect the fuse and if it has blown, replace it. If the new fuse blows after starting up, the cause of the trouble must be found, and we advise that the equipment is examined by a Lucas Service Station. Replacement fuses must be of the same size as those originally fitted, i.e. 4.5 amps.

The fuse is one of the four cartridge type fitted in the junction box on the engine side of the dash (see Fig. 4), and is marked "E." The size of the fuse is marked on a coloured paper slip which can be seen inside the fuse.

Removing the Dynamo.—In order to obtain easy access to the dynamo for removal it is necessary to take off the radiator. Remove the two bolts fastening the radiator to the chassis frame, slacken the nuts at the rear end of the bonnet rod and release it from its bracket. Remove water return pipe from cylinder head and loosen clips holding bottom water hose.

N.B.—It is unnecessary to interfere with any of the oiling system to remove dynamo.

Between the projecting portion of the cylinder head and the top of the dynamo will be found a circular flexible coupling. Remove the nuts on each of the four coupling bolts in turn, leaving the bolts in position, so that the coupling can be rotated by use of the starting handle to bring each bolt into a position where the nut may easily be reached.

Having removed all four nuts, take off the valve gear cover and turn the engine by the starting handle until the timing marks on the spiral bevel gears coincide. These will bring the driving yoke on the cylinder head across the engine, and the driving yoke on the dynamo parallel to the engine centre line. The bolts themselves may now be withdrawn, care being taken not to lose the distance washers, which must be replaced in the same position. Removal of the bolts enables the flexible coupling to be withdrawn.

Detach the two cables on the distributor side of the dynamo noting from which terminal they are removed.

Now unscrew the four set screws which attach the dynamo to its platform at the front of the engine thus releasing the dynamo. Lift the dynamo approximately a quarter of an inch and tilt it towards the near side of the car until the driving yoke on the dynamo is just clear of that on the cylinder head. The dynamo can then be tilted forwards and easily withdrawn.

Replacement of the dynamo is effected in the reverse way, but it is necessary to make sure that the engine timing has not been disturbed while the dynamo was removed. Removal of the rectangular cover plate in the clutch housing should reveal the timing mark on the flywheel for Nos. 1 and 4 cylinders, exactly in the centre of the opening. Place the brass packing pieces which fit under the dynamo base in position on the dynamo platform, making sure that you replace the same number that you took off. If for any reason the dynamo is replaced by another, it may be necessary to re-adjust the mesh of the driving gears to obtain silent running by suitable selection of the packing shims used. Turn the dynamo spindle until the timing mark on the dynamo drive gear coincides with the centre line of the dynamo and is at the rear, that is, ready for engagement with the correspondingly marked teeth of the drive gear on the crankshaft. The holes in the dynamo coupling yoke will then be parallel with the engine centre line. Tilting the dynamo towards the near side of the car and holding the dynamo coupling yoke in this position, insert the drive gear into the opening of the dynamo platform and swing the dynamo backwards and downwards into position. The gears can be felt to be meshing properly if the coupling is slightly oscillated as the dynamo is replaced, but do not overdo the oscillations or you may engage the wrong teeth.

See that the bolts in the dynamo base are coinciding with the holes in the dynamo platform, and then observe if the dynamo coupling yoke is exactly parallel to the engine centre line. If it is not, withdraw the dynamo, reset the coupling yoke and re-insert the dynamo. No difficulty should be experienced in getting the dynamo in position with the correct teeth in mesh, as the distance between one tooth and the next is sufficient to make an appreciable difference to the position of the dynamo coupling yoke, a difference which is immediately discernible.

Having satisfied yourself that the correct gear teeth are in mesh, replace the screws in the dynamo base, taking particular care to tighten them up evenly a partial turn at a time until all are quite tight. Now make a final test. The flywheel mark 1-4 should show exactly in the centre of the inspection cover opening with the distributor rotating arm pointing towards No. 1 cylinder. (This can be readily found by tracing the high tension lead from No. 1 sparking plug to its junction on the distributor. Removal of the distributor cover should show the distributor arm directly beneath it). The dynamo coupling bolt holes should be exactly fore and aft and at right angles to the coupling yoke on the cylinder head, and the timing marks on the camshaft driving gears should be coinciding.

If all the foregoing are correct, replace the flexible coupling and coupling bolts, taking care to replace the distance washers in exactly the same position as they were originally. Tighten up the nuts firmly and rotate the engine slowly by hand. If the distance washers are in the correct position, the flexible coupling should run absolutely true. If it does not do so, note where the error is and adjust the distance washers accordingly. Then replace the valve cover and attach the dynamo cables on to their correct terminals.

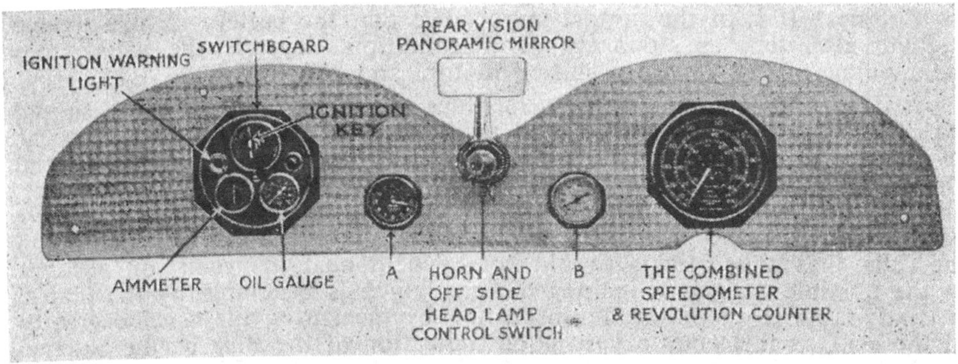

Fig. 7. Instrument panel on J2 models : equipment marked A and B on de-luxe models only.

Third Brush Regulator.—The output of the dynamo, that is to say, its rate of charge, is controlled by a Third Brush. It is inadvisable for anybody who is not an electrician to tamper with this, but in case the ammeter reads more than 2 amps. when all the head lamps are on, or in the event of its reading less than 6 amps. when no lights are alight, the Third Brush can be moved relative to the armature by slackening off locking nut and turning the brush gear either towards the direction of rotation of the armature to increase the rate of charge or in the opposite direction to decrease the charge rate. **It is very important that the increased charging rate must not exceed 8 amps. with all lights off or severe damage will be done to the dynamo and cut-out.**

The Third Brush is to be found on the near side of the car and on the same side as the small ball covered lubricator. Never omit to tighten up the lock screw after any adjustment has been effected.

When replacing the brush cover care must be taken that the cover does not make contact with dynamo terminal (symptom, no charge).

Starter Motor.—The armature spindles of these machines are fitted with a pinion which, on rotation, runs into engagement with the geared ring on the flywheel. Immediately the engine begins to fire, the pinion is automatically thrown out of mesh.

If, for any reason, the pinion wheel on the motor does not engage with the flywheel teeth, examine the screwed sleeve on the armature spindle to see that it is free from dust; if necessary wash over with paraffin. Occasionally, give it a few drops of thin machine oil.

As in the case of the dynamo, the surface of the commutator must be kept clean and free from oil, brush dust, etc.

The starter is designed for starting the engine under normal conditions. but any unnecessary or additional loading will considerably diminish the life of the machine and battery. In order to facilitate starting in cold weather, it is advisable to flood the carburetter, and, before using the electric starter, crank the engine over slowly by the starting handle for two or three revolutions; this will break the oil film and considerably diminish the load for starting.

In the event of the engine refusing to fire after being turned by the starter, make sure that the ignition switch is "on."

It is not advisable to use the self starter with the Ignition Lever in the full advance position. It is better to put the Ignition Lever half advance and if necessary increase the amount of advance while the Starter Motor is spinning the engine. With coil ignition this is not usually necessary.

Battery.—It is of the utmost importance that the battery should receive regular attention, as upon its good condition depends the satisfactory functioning of the ignition, starter motor, and the lamps.

At least once a fortnight the vent plugs in the top of the battery should be removed, and the level of the acid solution examined. If necessary, distilled water (which can be obtained at all chemists and most garages) should be added to bring the level well above the plates. If, however, acid solution has been spilled, it should be replaced by a diluted sulphuric acid solution of specific gravity 1.320. It is important, when examining the cells, that naked lights should not be held near the vents, on account of the possible danger of igniting the gas which is generated by the plates. It is advisable to complete the inspection by measuring the specific gravity of the acid, as this gives a very good indication of the state of the battery. An instrument known as a hydrometer is employed for this purpose; these can be bought from your dealer or from any of the Lucas Service Stations.

For the battery fitted to the J models the specific gravity readings will be 1.285—1.300 for a fully charged battery, about 1.210 when half discharged and about 1.150 when fully discharged.

Should the quantity of distilled water required to top up be excessive, overcharging should be suspected. Additional symptoms will be, furious "gassing" of the cells when the dynamo is charging, a gradual but distinct increase in the intensity of the illumination of the headlamps as the dynamo output builds up, and a noticeable heating up of the battery after, say, a two hours' daylight run.

The remedy is either to reduce dynamo output, withdraw field fuse in the junction box for a while, or, in the case of the J2 Models, put P.L.C. switch in "Off" position for a period of daylight running. It should always be remembered that overcharging can and ultimately will appreciably reduce the useful life of a battery.

The symptoms of an undercharged battery are too well known and need not be dealt with in these pages except to emphasise the necessity to maintain the battery in a *normal* condition at all times.

If one cell gives a reading very different from the rest, it may be that electrolyte has been spilled or has leaked from this cell, or there may be a "short" between the plates. In the latter case, the battery should be examined as soon as possible by a Lucas Service Station.

Finally, see that the tops of the cells are clean and dry, and that the battery posts and cable lugs are thoroughly impregnated with vaseline before replacing and tightening up. More batteries are damaged by neglect of this simple precaution than by any other agency, often with attendant damage to all the other components of the installation.

If the equipment is laid by for several months, the battery must be given a small charge from a separate source of electrical energy about once a fortnight, in order to obviate any permanent injury to the plates.

Under no circumstances must the acid be removed from the battery and the plates allowed to dry, as certain chemical changes take place which result in loss of efficiency.

The battery must never be left in a fully discharged condition, and unless some long runs are to be taken, it is advisable to have the battery removed from the car periodically and inspected, when it should be charged from an independent source of supply if found necessary.

Ammeter.—The centre-zero ammeter which is incorporated in the instrument panel indicates the actual current flowing into or out of the battery. For instance, suppose two amperes are consumed when the side and tail lamps are switched on, and the ignition coil takes one ampere, then if the dynamo is generating at seven amperes the meter will show four amps. on the charge side of the scale. This is the current in excess of the lamp and ignition load which is available for battery charging purposes.

Fuses.—The separate fusing of the various circuits ensures that a short circuit in any one does not affect the rest of the electrical equipment. This is particularly important when coil ignition is fitted. It will be noticed that the fusing of the lamps is such that there is no risk of the driver being plunged into total darkness. If both the headlamps, or the side and tail lamps, or all the units connected to the auxiliary accessory terminal fail to function, examine the particular fuse protecting them.

Remove the fuse from its holder and see whether there is a break in the fuse wire. Before replacing the fuse, inspect the units that have failed, for evidence of short circuits or other faults that may have caused the fuse to blow.

If a fuse blows repeatedly, and the cause cannot be traced we advise that the equipment be examined by a Lucas Service Station.

The Cut-out.—The function of the cut-out is to close the charging circuit, as the increased engine speed when the car is starting causes the dynamo voltage to rise above that of the battery. When the engine slows down, the dynamo voltage falls below that of the battery, and the reverse action takes place, i.e., the cut-out opens and thereby prevents the battery from discharging itself through the dynamo.

The question is sometimes asked whether the operation of the cut-out in any way depends upon the state of charge of the battery. There is no such relation between the two; the sole function of the cut-out is to switch on the dynamo with rising engine speed and to disconnect it when the engine slows down to below a certain speed. The cut-out is accurately set before leaving the Works and does not need any adjustment, and therefore the cover protecting it is sealed.

Very little attention is needed to keep the ignition equipment in first class condition. We advise that it is inspected occasionally, and the following instructions on lubrication, cleaning and adjustment should be carried out.

Distributor.—Occasionally remove the distributor moulding by pushing aside its two securing springs. See that the electrodes are clean and free from deposit. If necessary, wipe out the distributor with a dry duster and clean the electrodes with a cloth moistened with petrol. See that the carbon brush is clean and moves freely in its holder. Clean the outside of the moulding, particularly the spaces between the terminals. Next examine the contact breaker; it is important that the contacts are kept free from any grease or oil. If they are burned or blackened, they may be cleaned with very fine emery cloth and afterwards with a cloth moistened with petrol. Care must be taken that all particles of dirt and metal dust are wiped away. Misfiring may be caused if the contacts are not kept clean.

The contact breaker gap should be between 15 and 20 thousandths and is carefully set before leaving the Works. Provided that the cam is kept clean and that the instructions and cam lubrication are carried out, the contact breaker gap will only need adjustment at very long intervals. It is not advisable to alter the setting unless the gap varies considerably. If adjustment is necessary, proceed as follows:—When the contacts are fully opened, slacken the locking nut on the stationary contact screw, and rotate it by its hexagon head until the gap is set to the thickness of the gauge. After making the adjustment, care must be taken to tighten the locking nut.

LUBRICATION.

(1) **Distributor Shaft.**—The greaser on distributor shaft should be given one turn about every 500 miles.

Repack the greaser with a good quality high melting point grease when necessary.

(2) **Cam.**—About every 3,000 miles, give the cam the slightest smear of vaseline.

(3) **Contact Breaker Pivot.**—Every 5,000 miles, place a single drop of oil on the pivot on which the contact breaker works.

Coil.—The coil unit is not adjustable in any way, and requires no attention beyond seeing that the terminal connections are kept tight and the moulded coil top is kept clean.

Warning Lamp.—A warning lamp is provided in the instrument panel which gives a red light when the ignition is "ON" and the car is stationary. The warning lamp will also light when the engine is running very slowly, due to the fact that the dynamo is not running at sufficient speed to generate a high enough voltage to actuate the cut-out.

THE DETECTION AND REMEDY OF IGNITION FAULTS.

If a failure of ignition or misfiring occurs, unless the cause is at once apparent the owner is strongly recommended to proceed in accordance with the following routine which should quickly enable him to locate the trouble.

Before proceeding with the examination, make sure that the trouble is not due to defects in the engine, carburetter, petrol supply, sparking plugs, etc.

Engine will not Fire.—Switch on the ignition, turn the engine and observe the ammeter reading. The engine should be turned by hand if it is known that the battery is in a low state of charge.

If an ammeter reading is given which rises and falls with the closing and opening of the contacts, then the low tension wiring is in order. If the reading does not fluctuate in this way, a short in the low tension wiring is indicated, or the contacts are remaining closed. When no reading is given, a broken or loose connection in the low tension wiring is indicated, or the battery may be exhausted.

Examine the high tension cables, i.e., cables from the coil to the distributor, and from the distributor to the plugs. If the rubber shows signs of deterioration or cracking, the cable should be renewed. Remove the distributor moulding and examine the contacts; and clean them if

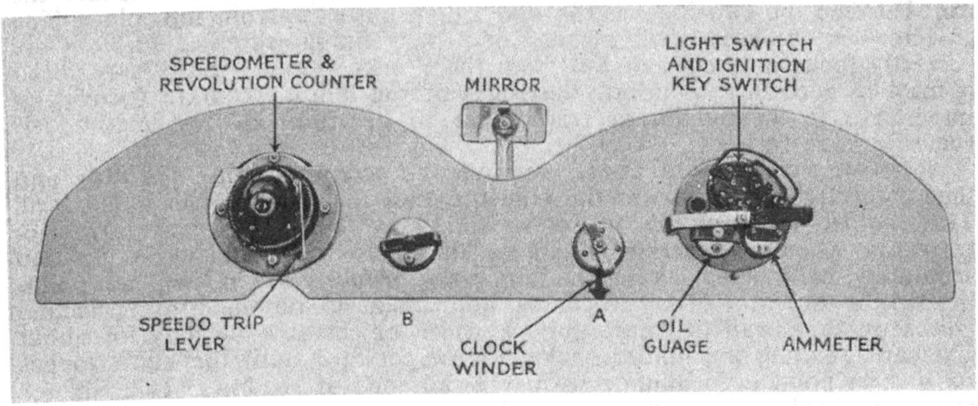

Fig. 8. *Behind the instrument panel of the J2 Midget.*

necessary. Turn the engine over by hand, and see that the contacts come together.

If the fault is indicated in the low tension wiring, examine the cables from the switch or junction box to coil, and from coil to distributor. See that the battery terminals are tight and that the cables from the switch-box to the battery are secure. The battery may be dismissed as the cause of the trouble if the lamps will light.

Test the coil independently of the distributor as follows.—Remove the cable from the centre distributor terminal, and hold it about ¼ in. from some metal part of the chassis and turn the engine. The sparking should be strong and regular if the coil is functioning correctly.

Misfiring and Bad Starting.—Examine the high tension cables and the plugs. If necessary, adjust the gaps to the correct setting (about 20 thousandths of an inch). Sooty or oiled plugs may be dismantled and washed out with petrol.

The plugs and high tension cables may be tested by removing the plugs in turn and allowing them to rest on the cylinder head and observing whether a spark occurs at the points when the engine is turned by hand.

It should, however, be noted that this is only a rough test, since it is possible that a spark may not take place when the plug is under compression.

Remove the distributor moulding and see that the electrodes and contacts are clean. See that the contact gap setting is correct.

If after carrying out the examination suggested, the trouble cannot be found, we advise that the equipment should be examined by the nearest Lucas Service Depôt.

Headlamps.—The headlamps are provided with a patented universally adjustable mounting which allows the beam of light to be set to the best advantage. This adjustment is obtained by slacking the hexagon locking nut, turning the lamp to the desired position and then locking it by tightening up the nut.

The near side headlamp is set to throw its beam to the near side of the road and the off-side (set straight) can be turned off or on independently.

Removing the Front and Reflector.—The headlamps are constructed with detachable fronts parabolic reflectors and focusing devices. To remove a lamp front, give the screw head at the top of the lamp about a quarter of a turn—a coin will serve as a screwdriver. This movement presses a cam against the body and pushes the front away from the body, enabling it to be removed. To replace, turn back the screw head and locate the cam between the two lugs on the body, then push the front into place. The reflector is secured by means of four fixing screws. Lamps are correctly focused when sent out from the Works, but if a bulb is replaced it may be necessary to adjust the focus of the lamp. Accurate focusing is imperative if the maximum results are to be attained. The method of focusing is as follows:—

Remove the front as above, remove the screws holding reflector and then carefully draw forward the reflector until the lamp holder is exposed. This holder is fixed by a screw and may be moved backwards or forwards when the screw is loosened. Each lamp must be focused separately, care being taken that lamps are properly set in line, not pointing up or down. Correct focusing adjustment is obtained by reflecting lamps on to a wall or, preferably, a white object at a distance of about 100 yards, sliding the holder backwards or forward until the light reaches its highest point of brilliancy with total absence of shadows. Then tighten screw and replace the reflector and front.

Wiring Headlamps.—Remove the front and reflector as described in a previous paragraph. Then depress the washer on the bulb holder terminal against the spring until the cable hole is exposed. Thread the cable end through the hole, release the pressure on washer, when the cable will be securely held in position and good contact made. Cables should not be pushed too far through the terminal, as there is a danger of shorting on to the reflector or body of the lamp.

Side-Lamps.—The fronts of the wing lamps are secured by small grub screws; when these are removed the fronts can be withdrawn by first pulling the top forward.

Tail-lamp.—The tail-lamp is fixed on to the car by means of a flanged base. To replace a bulb turn the front portion of the lamp to the left and withdraw it from its base.

When replacing see that the studs locate with the slots in the lamp front, then push it home to lock it in position. Should it be necessary to rewire the tail-lamp, unscrew the coupling nut when the cable covering shell and the cable plug can be withdrawn from the lamp; pass the cable through the shell and the rubber washer. Thread the bared end of the lead into the terminal socket and secure by the screw. Replace the cable plug in its holder and secure by tightening the coupling nut.

Replacement of Bulbs.—When the replacement of any bulb is necessary we strongly advise that Lucas bulbs are used. The filaments are arranged to be in focus and give the best results with the lamps fitted. The following are the correct bulbs to use:—Head lamps, B.A.S. No. 35; side and tail lamps, B.A.S. No. 10S.

Cleaning Lamps.—The reflectors are protected by a transparent and colourless covering, which enables any accidental finger marks to be removed with chamois leather or a soft cloth without affecting the surface of the reflector. Do not use metal polishes on Lucas Reflectors. Ebony black lamps can be cleaned with a good car polish. Chromium plated lamps will not tarnish and only need wiping over with a damp cloth to remove dust or dirt.

Horn.—All electric horns, before being passed out of the Works, are adjusted to give the best performance, and they will give long periods of service without any attention; no subsequent adjustment is required.

If the horn becomes uncertain in its action, giving only a choking sound, or does not vibrate, it does not follow that the horn has broken down. First ascertain that the trouble is not due to some outside source, e.g. a discharged battery, a loose connection or short circuit in the wiring of the horn, or a blown fuse. It is also possible that the performance of a horn may be upset by the horn becoming loose on its mounting. If the cause of the trouble cannot be located, do not attempt to dismantle the horn, but return it to a Service Depot for examination.

ELECTRICAL EQUIPMENT
(P Model)

IGNITION.

Very little attention is needed to keep the ignition equipment in first class condition. We advise that it is inspected occasionally, and the following instructions on lubrication, cleaning and adjustment are carried out.

Lubrication.—The High Tension distributor is very accessible. It is provided with an oiler to lubricate the spindle, a few drops every thousand miles being sufficient.

About every 3,000 miles, give the cam the slightest smear of vaseline. Also withdraw the rotor from the top of the spindle by lifting it off, and add a few drops of thin machine oil. Do not remove the screw exposed to view, as there is a clearance between the screw and the inner face of the spindle through which the oil passes to lubricate the automatic timing control.

Every 5,000 miles, place a single drop of oil on the contact breaker pivot.

The contacts can be inspected by unclipping the two springs when the cover carrying the High Tension Leads can be removed. The points may require adjustment from time to time, and the clearance between the points should be 15-18/1,000 of an inch when they are fully open.

Provided that the cam is kept clean and that the instructions on cam lubrication are carried out, the contact breaker gap will only need adjustment at very long intervals. It is not advisable to alter the setting unless the gap varies considerably. If adjustment is necessary, proceed as follows.—When the contacts are fully opened, slacken the locking nut on the stationary contact screw, and rotate it by its hexagon head until the gap is set to the thickness of the gauge. After making the adjustment, care must be taken to tighten the locking nut.

The centre spindle of the High Tension distributor carries a "Bakelite" Arm called the "Rotor." It can only fit on the spindle one way. If this part is removed, it affords an excellent thief proof device, but care should be exercised to see that it does not become chipped or in any way damaged, not to mention lost, after it has been removed. The ends of the High Tension Wires are secured to brass discs in order to make good contact with the terminals inside the cover. See that the terminal screws are always tight.

Occasionally remove the distributor moulding (by pushing aside its two securing springs). See that the electrodes are clean and free from deposit. If necessary, wipe out the distributor with a dry duster and clean the electrodes with a cloth moistened with petrol. See that the carbon brush is clean and moves freely in its holder. Clean the outside of the moulding, particularly the spaces between the terminals. Next examine the contact breaker; it is important that the contacts are kept free from any grease or oil. If they are burned or blackened, they may be cleaned with very fine emery cloth and afterwards with a cloth moistened with petrol. Care must be taken that all particles of dirt and metal dust are wiped away. Misfiring may be caused by any of the following:—

1. Dirty sparking plug or plugs.
2. Bad connection to High Tension Leads.
3. Bad connection from High Tension Distributor to Coil.
4. Improper adjustment of or burnt make and break points.
5. Dirt between make and break points.
6. Defective Coil.
7. Faulty condenser, causing repeated burning of points.
8. Temporary petrol starvation.

There is one other remote cause of electrical failure, namely, bad earth contacts from the battery to the frame. Other causes of misfiring can be attributed to carburetter and incorrect valve adjustment. Separate instructions are given on these last two items elsewhere.

Distributor Timing.—It has been found on test that the ignition timing is fairly critical and that on the average the amount of advance required is equal to exactly 1½ inches measured on the flywheel in advance of the top dead centre mark. It will be found that there is a line on the flywheel marked "Ign"—the engine should be turned by means of the starting handle until the mark is in the centre of the clutch inspection aperture, the points of the make and break in the distributor should then be on the point of opening.

In order that the timing is set correctly it is advised that the setting should always be done using as a guide the line of the flywheel.

To actually time the distributor slacken the nut on the fixing screw and turn the distributor body in a clockwise direction to retard and anti-clockwise to advance.

Coil.—The coil unit is not adjustable in any way, and requires no attention beyond seeing that the terminal connections are kept tight, and the moulded coil top is kept clean.

Warning Lamp.—A warning lamp is provided in the instrument panel, which gives a red light when the ignition is "ON" and the car is stationary. The warning lamp will also light when the engine is running very slowly, due to the fact that the dynamo is not running at sufficient speed to generate a high enough voltage to actuate the cut-out.

THE DETECTION AND REMEDY OF IGNITION FAULTS.

If a failure of ignition or misfiring occurs, unless the cause is at once apparent the owner is strongly recommended to proceed in accordance with the following routine, which should quickly enable him to locate the trouble.

Before proceeding with the examination, make sure that the trouble is not due to defects in the engine, carburetter, petrol supply, sparking plugs, etc.

Engine will not Fire.—Switch on the ignition, turn the engine and observe the ammeter reading. The engine should be turned by hand if it is known that the battery is in a low state of charge.

If an ammeter reading is given which rises and falls with the closing and opening of the contacts, then the low tension wiring is in order. If the reading does not fluctuate in this way, a short in the low tension wiring is indicated, or the contacts are remaining closed. When no reading is given, a broken or loose connection in the low tension wiring is indicated, or the battery may be exhausted.

Examine the high tension cables, i.e., cables from the coil to the distributor, and from the distributor to the plugs. If the rubber shows signs of deterioration or cracking, the cable should be renewed. Remove the distributor moulding and examine the contacts, and clean if necessary. Turn the engine over by hand, and see that the contacts come together.

If a fault is indicated in the low tension wiring, examine the cables from the switch or junction box to coil, and from coil to distributor. See that the battery terminals are tight and that the cables from the switch-box to the battery are secure. The battery may be dismissed as the cause of the trouble if the lamps will light.

Test the coil independently of the distributor as follows: Remove the cable from the centre distributor terminal, and hold it about ¼ in. from some metal part of the chassis and turn the engine. The sparking should be strong and regular if the coil is functioning correctly.

Misfiring and Bad Starting.—Examine the high tension cables and the plugs. If necessary, adjust the gaps to the correct setting (about 18 thousandths of an inch). Sooty or oiled plugs may be dismantled and washed with petrol.

The plugs and high tension cables may be tested by removing the plugs in turn and allowing them to rest on the cylinder head and observing whether a spark occurs at the points when the engine is turned by hand. It should, however, be noted that this is only a rough test, since it is possible that a spark may not take place when the plug is under compression.

If after carrying out the examination suggested, the trouble cannot be found, we advise that the equipment should be examined by the nearest Lucas Service Depôt.

SPARKING PLUGS.

The correct gap is .018 in., and no advantage will be gained by either a wider or a closer gap. A definite mileage when cleaning will be necessary cannot be specified, as there are a number of governing factors. It is, however, wise to leave well alone if there is no indication of trouble in that direction.

Plugs are easily ruined by inexperienced handling. The only satisfactory way of ascertaining whether a plug is sparking correctly is to test it under compression. Most service stations have a "plug tester," and we recommend that M.G. owners have their plugs checked in this way periodically.

A good plan is to keep a spare set of plugs and if plug trouble is suspected to change over to the spares. This allows plenty of time to have the others attended to as suggested.

Electrical Equipment.—The electrical equipment of the modern motor-car is often and quite erroneously regarded as something very complex and difficult. Yet in the present day motor-car simple application of thoroughly tried-out ideas has made possible easy adjustment and maintenance of practically all the electrically operated parts. In this section an endeavour has been made to explain the various instruments and the best way to set about adjustments.

The wiring diagram may at first appear to be a little complicated, but it shows clearly exactly where each individual cable goes and the manner in which they are carried from point to point.

Wiring is arranged on the earth return system, and particular care has been taken to ensure that good continuity obtains at all points of the circuit, by the inclusion, where necessary, of independent earthing cables. It will also be noticed that cables at all essential points are encased in flexible metallic armouring. This not only protects the cables from possible abrasion, but excludes road dirt and water, prevents premature wear and makes for long and trouble-free service.

Instrument Panel.—The panel—see Fig. 9—contains the control switch for the dynamo, ignition and lights, a centre zero ammeter and various other items, all of which are clearly marked. A word of explanation on the dual function of the ignition warning light may prove of assistance. Its obvious object is to warn the owner that the ignition switch is "on" and that the battery is being discharged, but, being in effect "shunted" across the cut-out points, it will allow only a very small current of electricity to flow through the armature and field coils of the dynamo, thus giving the dynamo an initial excitation. A resistance is always included in the warning lamp circuit to prevent the voltage rise of the dynamo burning out the bulb, and in the light under consideration the resistance is of sufficient value to permit the use of a small 2½ volt bulb of considerably longer life than the earlier heavy consumption type.

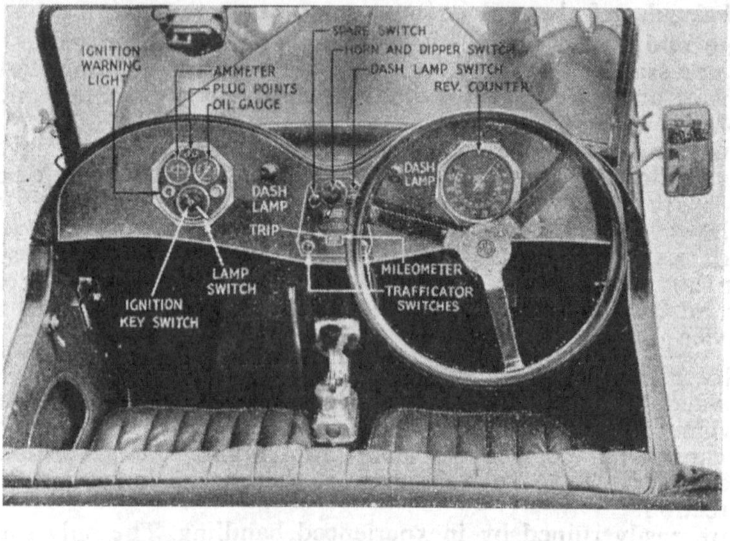

Fig. 9. The facia board of the P models.

Cut-out and Junction Box.—A good view of this component can be had by referring to Fig. 10. All cables are taken to terminal posts in a recessed space at the back of the box, and it is here that the two main harnessed cables—the panel harness associated with the control panel and the junction box harness which constitutes the cables going to the various points on the chassis—are coupled up. The fuses, reading from left to right, are auxiliary, near-side head, off-side head, side and tail, and horn. Above will be seen the dynamo field fuse. These are so arranged that should a short-circuit develop in any lamp or other components, the fuse associated with that component only will blow and the remainder of the installation will remain normal. The operation of tracing the fault is simplified enormously as it is at once apparent which circuit is at fault. It should be noted that the field fuse is only of 4·5 amperes.

On the left of the fuse block the spare fuse carrier is provided with one complete replacement set. At the side of the fuse block a pair of terminal posts are available for coupling up any other instruments fitted subsequently. This pair is fused in the auxiliary circuit and is intended for additional horns, fog lamps, etc. It will perhaps be helpful to note here that there are many accessories on the market, some of which take a very high operating current, and the owner should avoid fitting any of these unless approved by a competent electrical engineer. In any case a heavy current accessory must be connected directly to the battery and not through the junction box or ammeter.

Fuses.—It will be noticed that the fusing of the lamps is such that there is no risk of the driver being plunged into total darkness. If both the headlamps, or the side and tail-lamps, or all the units connected to the auxiliary accessory terminal fail to function, examine the particular fuse protecting them.

Remove the fuse from its holder and see whether there is a break in the fuse wire. Before replacing the fuse, inspect the units that have failed, for evidence of short circuits or other faults that may have caused the fuse to blow.

If a fuse blows repeatedly, and the cause cannot be traced, we advise that the equipment is examined by a Lucas Service Station.

The Cut-out.—The function of the cut-out is to close the charging circuit, as the increased engine speed when the car is starting causes the dynamo voltage to rise above that of the battery. When the engine slows down, the dynamo voltage falls below that of the battery, and the reverse action takes place, i.e., the cut-out opens and thereby prevents the battery from discharging itself through the dynamo.

The question is sometimes asked whether the operation of the cut-out in any way depends upon the state of charge of the battery. There is no such relation between the two; the sole function of the cut-out is to switch on the dynamo with rising engine speed and to disconnect it when the engine slows down to below a certain speed. The cut-out, which is housed under a separate cover at the top of the junction-box, see Fig. 10. is entirely self-operating and is accurately adjusted by the makers before fitting. It should operate for a considerable period without attention, and because considerable damage to the entire installation could be done by anyone attempting to vary the setting it is sealed by the manufacturers.

Ammeter.—The centre-zero ammeter which is incorporated in the instrument panel indicates the actual current flowing into or out of the battery. For instance, suppose two amperes are consumed when the side- and tail-lamps are switched on, and the ignition coil takes one ampere, then if the dynamo is generating at seven amperes the meter will show four amperes on the charge side of the scale. This is the current in excess of the lamp and ignition load which is available for battery charging purposes.

Dynamo.—The dynamo is mounted on a platform at the front end of the engine and forms part of the overhead camshaft drive. Two views of the dynamo can be seen in Fig. 5. On the left-hand side it will be seen that the pinion is held on to a shaft by a bolt having a very wide head and also a tab washer which registers in one of the two holes in the pinion. These two holes are drilled and tapped ¼ in. B.S.F. in order to facilitate removal at any time with the aid of two bolts, one of which is shown in the illustration. The dynamo is what is known as a simple shunt machine and has been designed to stand up to the very high revolutions per minute and acceleration movement. Beyond an occasional inspection of the brush gear it requires but little attention. The brush gear can be examined by slackening off the screw holding the cover in position. The cover is shown in Fig. 5 above the dynamo.

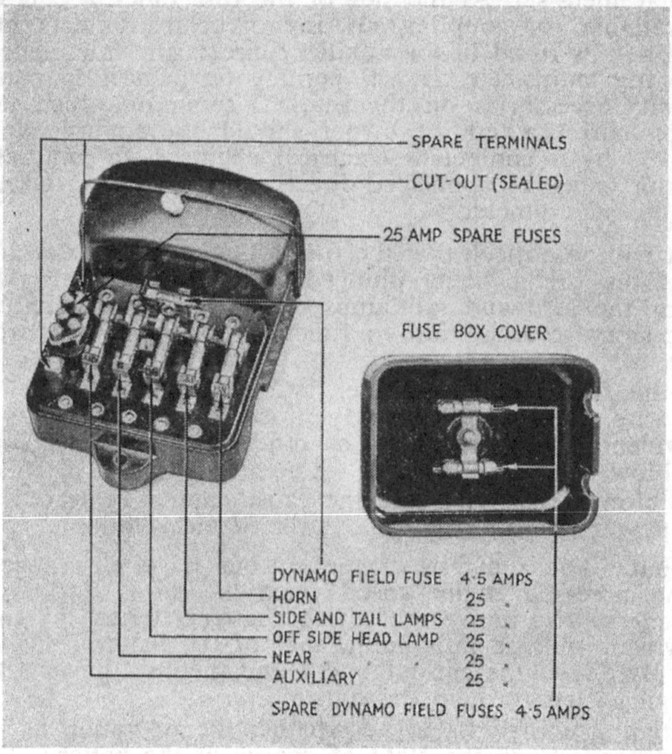

Fig. 10. *The fuse box and cut out with details of the various fuses.* (*P, N and KN models*).

Brushes.—It is very important to make sure that the three brushes work freely in their holders. This can easily be ascertained by holding back the spring and gently pulling each flexible lead, when the brush should move without the slightest suggestion of sluggishness. The brushes should be clean and should " bed " over the whole of their working surface; that is to say, the face of contact with the commutator should appear uniformly polished. Dirty brushes may be cleaned with a cloth moistened with petrol.

If any of the brushes become so badly worn that it is necessary to replace them, this is accomplished by releasing the brush lead eyelet by removal of the screw then, while holding the spring lever back out of the way, withdrawing the brush from its holder. The new brush can then be fitted by reversing the operation.

When ordering new brush replacements state whether they are main or control brushes, and for what type of machine they are required, also chassis and engine numbers.

The brush springs should be inspected occasionally to see that they have sufficient tension to keep the brushes firmly pressed against the commutator when the machine is running. It is particularly necessary to keep this in mind when the brushes have been in use for a long time and are very much worn down.

Owners are cautioned that it is unwise to insert brushes of a grade other than that supplied with the machine, or to change the tension springs. The arrangement provided has been made only after many years' experience, and will be found to give the best results.

Commutator.—The surface of the commutator should be kept clean and free from oil and brush dust, etc.; neglect of this precaution will result in the commutator becoming blackened, causing sparking to occur at the brushes, and consequent shortening of the life of the machine. The best way to clean the commutator is to insert a fine duster, held by means of a suitably shaped piece of wood, against the commutator surface, slowly rotating the armature by the starting handle at the same time.

If the commutator has been neglected for long periods, it may need cleaning with fine glass paper; but this is more difficult to do, and should not be necessary if it has received regular attention. The grooves between the commutator segments should be examined occasionally and any deposit of copper or carbon dust may be cleaned out by means of a thin saw blade or similar article. The owner is strongly advised to entrust this operation to a fully qualified electrician, as serious damage can unconsciously be done by any one not familiar with this class of work.

Lubrication.—As the bearings are packed with grease before leaving the Works, very little attention is needed. A few drops of oil, however, may be added through the lubricators provided, say, every 1,000 miles. The reader is cautioned that far more trouble has been caused by excessive oiling than by too little. After the car has run about 20,000 miles the dynamo should be removed, cleaned and adjusted and the bearings repacked with grease. This should be entrusted to the nearest Lucas Service Depôt.

Dynamo Field Fuse.—A fuse is provided in the dynamo field circuit to protect the machine in the event of anything being wrong in the charging circuit, e.g., a loose or broken battery connection. If the dynamo fails to charge the battery at any time (indicated by the ammeter giving a discharge reading during daytime running), inspect the fuse and, if it has blown, replace it. If the new fuse blows after starting up, the cause of the trouble must be found, and we advise that the equipment is examined by a Lucas Service Station. Replacement fuses must be of the same size as those originally fitted.

Removing the Dynamo.—In order to obtain easy access to the dynamo for removal it is desirable, although not absolutely necessary, to take off the radiator. Remove the lock nuts and nuts from the radiator bolts carried through the engine nose piece extension, and the bolts retaining the yokes of the bonnet tie rods. Also, after emptying the water system, remove the top and bottom water connections. It is unnecessary to interfere with any of the oiling system to remove dynamo.

Between the projecting portion of the cylinder head and the top of the dynamo will be found a circular flexible coupling. Remove the nuts on each of the four coupling bolts in turn, leaving the bolts in position, so that the coupling can be rotated by use of the starting handle to bring each bolt into a position where the nut may easily be reached.

Having removed all four nuts, take off the clutch inspection cover and turn the engine by the starting handle until the timing marks on the flywheel are in the centre of the aperture. This will bring the driving yoke on the cylinder head across the engine, and the driving yoke on the dynamo parallel to the engine centre line. The bolts themselves may now be withdrawn, care being taken not to lose the distance washers, which must be replaced in the same position. Removal of the bolts enables the flexible coupling to be withdrawn.

Fig. 11. *Facia board on the PB models.*

Detach the two cables on the distributor side of the dynamo, noting from which terminal they are removed.

Now unscrew the four set screws which attach the dynamo to its platform at the front of the engine, thus releasing the dynamo. Lift the dynamo approximately a quarter of an inch and tilt it towards the near-side of the car until the driving yoke on the dynamo is just clear of that on the cylinder head. The dynamo can then be tilted forwards and easily withdrawn.

Replacement of the dynamo is effected in the reverse way, but it is necessary to make sure that the engine timing has not been disturbed while the dynamo was removed. Removal of the rectangular cover-plate in the clutch housing should reveal the timing mark on the flywheel for Nos. 1 and 4 cylinders, exactly in the centre of the opening. Place the brass packing shims which fit under the dynamo base in position on the dynamo platform, making sure that you replace the same number that you took off. If for any reason the dynamo is replaced by another, it may be necessary to readjust the mesh of the driven gears to obtain silent running by suitable selection of the packing shims used. Turn the dynamo spindle until the timing mark on the dynamo drive gear coincides with the centre line of the dynamo and is at the rear—that is, ready for engagement with the correspondingly marked teeth of the drive gear on the crankshaft. The holes in the dynamo coupling yoke will then be parallel with the engine centre line. Tilting the dynamo towards the near-side of the car and holding the dynamo coupling yoke in this position, insert the drive gear into the opening of the dynamo platform and swing the dynamo backwards and downwards into position. The gears can be felt to be meshing properly if the coupling is slightly oscillated as the dynamo is replaced, but do not overdo the oscillations or you may engage the wrong teeth.

See that the bolts in the dynamo base are coinciding with the holes in the dynamo platform, and then observe if the dynamo coupling yoke is exactly parallel to the engine centre line. If it is not, withdraw the dynamo, reset the coupling yoke and re-insert the dynamo. No difficulty should be experienced in getting the dynamo in position with the correct teeth in mesh, as the distance between one tooth and the next is sufficient to make an appreciable difference to the position of the dynamo coupling yoke, a difference which is immediately discernible.

Having satisfied yourself that the correct gear teeth are in mesh, replace the screws in the dynamo base, taking particular care to tighten them up evenly a partial turn at a time until all are quite tight. Now make a final test. The flywheel mark 1—4 should show exactly in the centre of the inspection cover opening with the distributor rotating arm pointing towards No. 1 cylinder. (This can easily be found by tracing the high-tension lead from No. 1 sparking plug to its junction on the distributor. Removal of the distributor cover should show the distributor arm directly beneath it.) The dynamo coupling bolt holes should be exactly fore and aft and at right angles to the coupling yoke on the cylinder head, and the timing marks on the camshaft driving gears should be coinciding.

If all the foregoing are correct, replace the flexible coupling and coupling bolts, taking care to replace the distance washers in exactly the same position as they were originally. Tighten up the nuts firmly and rotate the engine slowly by hand. If the distance washers are in the correct position, the flexible coupling should run absolutely true. If it does not do so, note where the error is and adjust the distance washers accordingly. Then replace the valve cover and attach the dynamo cables on to their correct terminals.

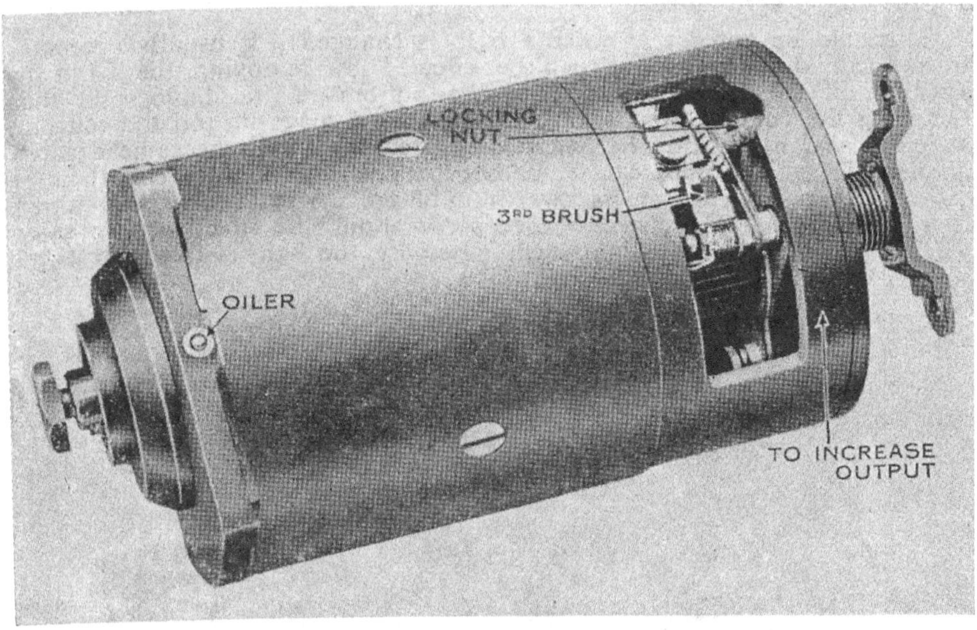

Fig. 12. *Dynamo showing* 3rd *regulating brush and greaser.* (*M, D, J, F, P, L, K models*).

Dynamo Adjustment.—The output of the dynamo—that is to say, its rate of charge—is controlled by a third brush which is shown quite clearly in Fig. 12. All machines are adjusted to an output of 9 amperes before leaving the factory, and any marked variation from this standard usually indicates some other factor at fault which must be cleared before normal conditions are possible for the dynamo. Should, however, the indicated charging rate increase to a value above 10 amperes **with all lights out,** it is necessary to reduce this at once, as a continuation of high output will rapidly destroy the machine. It is inadvisable for anybody who is not an electrician to tamper with this. To reduce the output the third brush or control brush must be moved in a direction opposite to that of rotation. To move the brush it is necessary to slacken the stud marked " Locking Nut " in Fig. 12, and when the adjustment has been satisfactorily effected it is important to make sure that the stud is tightened.

The third brush is to be found on the near-side of the car and on the same side as the small ball covered lubricator. Never omit to tighten up the lock screw after any adjustment has been effected.

Headlamps.—The **L.B.D.140** headlamps manufactured by Messrs. Joseph Lucas, Ltd., are specially designed to give the highest illuminating factor without unduly overloading the electrical installation. The near-side lamp is equipped with a permanently deflected reflector and is always in circuit when the headlamp switch is on. The off-side lamp has a straight focus reflector, and while primarily wired in the same circuit as the other, can also be controlled by the dipping switch. The lamps are carefully focused before leaving the factory so that the near-side lamp will throw a slightly open beam and thus effectively illuminate the road and kerb up to about 40 yards, while the off-side beam is relatively closely focused to throw a penetrating light ahead.

It should be noted that when a bulb is changed it is usually necessary to re-focus the beam. This can be effected by removing the front by slacking off the retaining screw at the bottom of the lamp, and then removing the cork washer, taking care not to damage it, and the reflector which is held in position by one screw, when the small cheese-head screw holding the bulb holder can be loosened. The bulb holder can then be moved either forwards or backwards until the correct beam is obtained. It is essential that the cheese-headed screw should be effectively tightened when the correct focus is obtained, otherwise the light will be erratic.

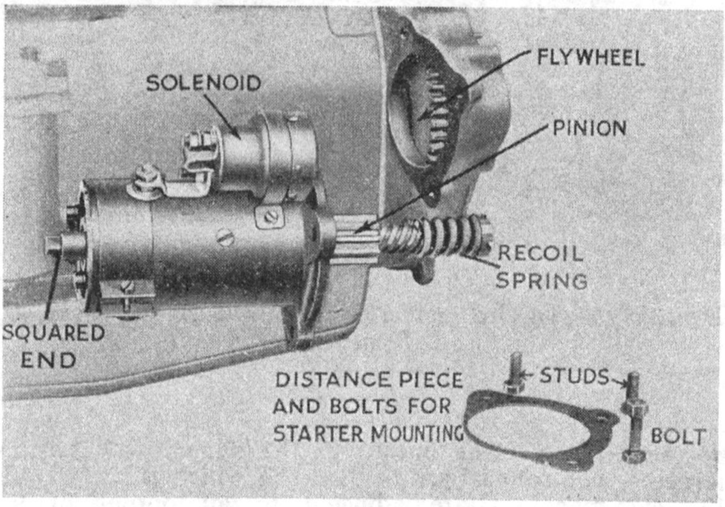

Fig. 13. Starter motor removed from engine.

ELECTRICAL EQUIPMENT

(F Model)

The accompanying wiring diagram may appear a little complicated to the average car user, and is really only intended to act as a guide in the case of dismantlement of all the wiring system. It shows, nevertheless, the run of the wires from the contact breaker to the coil, and from the coil to the junction box. It will also be seen from this illustration that the wiring is carried out on the earth return system, that the negative terminal of the battery is earthed, and there is an automatic earth from the lamps, starter and dynamo. There are, however, two separate earths that have to be made, namely from the horn and the junction box.

The principal joints connected with the electrical equipment on which the owner will require instruction are the instrument board, the fuses and the wiring.

A front and rear view of the instrument board are as shown in Figs. 1 and 2. The instrument board contains, reading from left to right, an ammeter, a switch controlling the dynamo charge, side and tail and head lamp illumination. The speedometer (which is driven by the gear box) is placed in the centre of the instrument panel on the right of which will be seen towards the upper edge of the panel, the red ignition tell-tale light beneath this a socket for inspection lamp, and on the extreme right an oil circulation indicator. This indicator has four slots which when the oil is not circulating show red, but as soon as the oil pressure reaches a predetermined pressure the four slots show white through them. On the lower side

Fig. 14. General view of dashboard from engine side showing junction fusebox F, cut-out C, accessory junction box J.

of the panel there is the ignition switch and on the right hand side a switch to illuminate the panel from behind. The rear view of the panel shows the central speedometer which has a small arm attached to it for reseting the trip. There are a number of connections that, of course, have to be made on the junction box to the switchboard and these are indicated by colour and description on the wiring diagram. It will be noticed that the two lamps that illuminate the dials and also the ignition tell-tale lamp are fitted at the back of the panel into clips, each having its own individual socket holder. By this means a bulb could easily be changed and should ever the ignition tell-tale bulb prove defective one of the other bulbs can be temporarily used in place of it. The wiring of the car is carried in armoured tubing at all essential points and the connections are made behind the combined junction and cut-out. This will be seen on reference to Fig. 3 of the engine side of the dash-board.

A better view of this is shown in Fig. 4. The junction box is fitted with a number of glass barrel fuses, and holes are provided in the body of the box to carry spares. The fuses control the side and tail lamps, separate fuses for the head lamps, another fuse for auxiliaries such as the horn and a further fuse marked " B " which is for the dynamo field. The cut-out which is mounted on the same base as the fuses, is provided with a cover and under normal circumstances need never be touched, its operations being entirely automatic.

A small circular junction box is placed beneath the cut-out which forms a convenient connection point for wires to the horn and off-side head lamp, in order to be able to control the latter through the dipper switch. It will be seen on referring to the wiring diagram that the lower terminal of the junction box is coupled to the horn and switch, and the upper terminal to the dipper switch and off-side head lamp.

DYNAMO.

The dynamo is mounted on a platform at the front end of the engine and forms part of the overhead camshaft drive. Two views of the dynamo can be seen in Fig. 5. On the left hand side it will be seen that the pinion is held on to a shaft by a bolt having a very wide head and also a tab washer which registers into the key way in the armature spindle. The two holes in the pinion are drilled and tapped ¼ in. B.S.F. in order to facilitate removal at any time with the aid of two bolts, one of which is shown in the illustration. The dynamo requires but little attention, the brush gear can be examined by slackening off the screw holding the cover in position. The cover is shown in the illustration above the dynamo.

Brushes.—It is very important to make sure that the three brushes work freely in their holders. This can easily be ascertained by holding back the spring and gently pulling each flexible lead, when the brush should move without the slightest suggestion of sluggishness. The brushes should be clean and should " bed " over the whole of their working surface; that is to say, the face in contact with the commutator should appear uniformly polished. Dirty brushes may be cleaned with a cloth moistened with petrol.

If any of the brushes become so badly worn that it is necessary to replace them, this is accomplished by releasing the brush lead eyelet by removal of the screw then, while holding the spring lever back out of the way, withdrawing the brush from its holder. The new brush can then be fitted by reversing the operation.

When ordering new brush replacements state whether they are main or control brushes, and for what type of machine they are required.

The brush springs should be inspected occasionally to see that they have sufficient tension to keep the brushes firmly pressed against the commutator when the machine is running. It is particularly necessary to keep this in mind when the brushes have been in use for a long time and are very much worn down.

Readers are cautioned that it is unwise to insert brushes of a grade other than that supplied with the machine, or to change the tension springs. The arrangement provided has been made only after many years' experience, and will be found to give the best results.

Commutator.—The surface of the commutator should be kept clean and free from oil and brush dust, etc.; neglect of this precaution will result in the commutator becoming blackened, causing sparking to occur at the brushes, and consequent shortening of the life of the machine. The best way to clean the commutator is to insert a fine duster, held by means of a suitably shaped piece of wood against the commutator surface, slowly rotating the armature by the starting handle at the same time.

If the commutator has been neglected for long periods, it may need cleaning with fine glass paper, but this is more difficult to do, and should not be necessary if it has received regular attention. The grooves between the commutator segments should be examined occasionally and any deposit of copper or carbon dust may be cleaned out by means of a thin saw blade or similar article.

Lubrication.—As the bearings are packed with grease before leaving the Works, very little attention is needed. A few drops of oil, however, may be added through the lubricators provided, say, every 1,000 miles. The reader is cautioned that far more trouble has been caused by excessive oiling than by too little. (After the car has run about 10,000 miles the dynamo should be removed, cleaned and adjusted and the bearings re-packed with grease. This should be entrusted to the nearest Lucas Service Depôt.)

Dynamo Field Fuse.—A fuse is provided in the dynamo field circuit to protect the machine in the event of anything being wrong in the charging circuit, e.g., a loose or broken battery connection. If the dynamo fails to charge the battery at any time (indicated by the ammeter giving a discharge reading during daytime running), inspect the fuse and if it has blown, replace it. If the new fuse blows after starting up, the cause of the trouble must be found, and we advise that the equipment is examined by a Lucas Service Station. Replacement fuses must be of the same size as those originally fitted.

In the case of the 12/70 M.G. Magna model, the fuse is one of the four cartridge type fitted in the junction box on the engine side of the dash, and is marked " E." The size of the fuse is marked on a coloured paper slip which can be seen inside the fuse.

Removing the Dynamo.—In order to obtain easy access to the dynamo for removal it is necessary to take off the radiator. Remove the two bolts fastening the radiator to the chassis frame, slacken the nuts at the rear end of the bonnet rod and release it from its bracket. Remove water return pipe from cylinder head together with fan, and loosen clips holding bottom water hose.

N.B.—It is unnecessary to interfere with any of the oiling system to remove dynamo.

Between the projecting portion of the cylinder head and the top of the dynamo will be found a circular flexible coupling. Remove the nuts on each of the four coupling bolts in turn, leaving the bolts in position, so that the coupling can be rotated by use of the starting handle to bring each bolt into a position where the nut may easily be reached.

Having removed all four nuts, take off the valve gear cover and turn the engine by the starting handle until the timing marks on the spiral bevel gears coincide. These will bring the driving yoke on the cylinder head across the engine, and the driving yoke on the dynamo parallel to the engine centre line. The bolts themselves may now be withdrawn, care being taken not to lose the distance washers, which must be replaced in the same position. Removal of the bolts enables the flexible coupling to be withdrawn.

Detach the two cables on the distributor side of the dynamo noting from which terminal they are removed.

Now unscrew the four set screws which attach the dynamo to its platform at the front of the engine thus releasing the dynamo. Lift the dynamo approximately a quarter of an inch and tilt it towards the near side of the car until the driving yoke on the dynamo is just clear of that on the cylinder head. The dynamo can then be tilted forwards and easily withdrawn.

Replacement of the dynamo is effected in the reverse way, but it is necessary to make sure that the engine timing has not been disturbed while the dynamo was removed. Removal of the rectangular cover plate in the clutch housing should reveal the timing mark on the flywheel for nos. 1 and 6 cylinders, exactly in the centre of the opening. Place the brass packing pieces which fit under the dynamo base in position on the dynamo platform, making sure that you replace the same number that you took off. If for any reason the dynamo is replaced by another, it may be necessary to re-adjust the mesh of the driving gears to obtain silent running by suitable selection of the packing shims used. Turn the dynamo spindle until the timing mark on the dynamo drive gear coincides with the centre line of the dynamo and is at the rear, that is, ready for engagement with the correspondingly marked teeth of the drive gear on the crankshaft. The holes in the dynamo coupling yoke will then be parallel with the engine centre line. Tilting the dynamo towards the near side of the car and holding the dynamo coupling yoke in this position, insert the drive gear into the opening of the dynamo platform and swing the dynamo backwards and downwards into position. The gears can be felt to be meshing properly if the coupling is slightly oscillated as the dynamo is replaced, but do not overdo the oscillations or you may engage the wrong teeth.

See that the bolts in the dynamo base are coinciding with the holes in the dynamo platform, and then observe if the dynamo coupling yoke is exactly parallel to the engine centre line. If it is not, withdraw the dynamo, reset the coupling yoke and re-insert the dynamo. No difficulty should be experienced in getting the dynamo in position with the correct teeth in mesh, as the distance between one tooth and the next is sufficient to make an appreciable difference to the position of the dynamo coupling yoke, a difference which is immediately discernible.

Having satisfied yourself that the correct gear teeth are in mesh, replace the screws in the dynamo base, taking particular care to tighten them up evenly a partial turn at a time until all are quite tight. Now make a final test. The flywheel mark 1-6 should show exactly in the centre of the inspection cover opening with the distributor rotating arm pointing towards No. 1 cylinder. (This can easily be found by tracing the high tension lead from no. 1 sparking plug to its junction on the distributor. Removal of the distributor cover should show the distributor arm directly beneath it.) The dynamo coupling bolt holes should be exactly fore and aft and at right angles to the coupling yoke on the cylinder head, and the timing marks on the camshaft driving gears should be coinciding.

If all the foregoing are correct, replace the flexible coupling and coupling bolts, taking care to replace the distance washers in exactly the same position as they were originally. Tighten up the nuts firmly and rotate the engine slowly by hand. If the distance washers are in the correct position, the flexible coupling should run absolutely true. If it does not do so, note where the error is and adjust the distance washers accordingly. Then replace the valve cover and attach the dynamo cables on to their correct terminals.

Third Brush Regulator.—The output of the dynamo, that is to say its rate of charge, is controlled by a Third Brush which is shown quite clearly in Fig. 12. It is inadvisable for anybody who is not an electrician to tamper with this, but in case the Amp. Meter reads more than 2 amps. when all the head lamps are on, or in the event of its reading less than

8 amps. when no lights are alight, the Third Brush can be moved relative to the armature by slackening off the nut marked " locking nut " and turning the brush gear either towards the direction of rotation of the armature to increase the rate of charge or in the opposite direction to decrease the charge rate.

The Third Brush is to be found on the near side of the car and on the same side as the small ball covered lubricator. Never omit to tighten up the lock screw after any adjustment has been effected.

Self Starter.—The following instructions may be of some assistance in case the Starter is not functioning properly.

Starter Motor.—The armature spindles of these machines are fitted with a pinion which, on rotation, runs into engagement with the geared ring on the flywheel. Immediately the engine begins to fire, the pinion is automatically thrown out of mesh.

If, for any reason, the pinion wheel on the motor does not engage with the flywheel teeth, examine the screwed sleeve on the armature spindle to see that it is free from dust; if necessary wash over with paraffin. Occasionally, give it a few drops of thin machine oil.

As in the case of the dynamo, the surface of the commutator must be kept clean and free from oil, brush dust, etc.

The starter is designed for starting the engine under normal conditions, but any unnecessary or additional loading will considerably diminish the life of the machine and battery. In order to facilitate starting in cold weather, it is advisable to flood the carburetter, and, before using the electric starter, crank the engine over slowly by the starting handle for two or three revolutions; this will break the oil film and considerably diminish the load for starting.

In the event of the engine refusing to fire after being turned by the starter, make sure that the ignition switch is " on."

It is not advisable to use the self starter with the Ignition Lever in the full advance position. It is better to put the Ignition Lever half advance and if necessary increase the amount of advance while the Starter Motor is spinning the engine. With coil ignition this is not usually necessary.

Battery.—It is of the utmost importance that the battery should receive regular attention, as upon its good condition depends the satisfactory functioning of the ignition, starting motor, and the lamps.

At least once a fortnight the vent plugs in the top of the battery should be removed, and the level of the acid solution examined. If necessary, distilled water (which can be obtained at all chemists and most garages) should be added to bring the level well above the plates. If, however, acid solution has been spilled, it should be replaced by a diluted sulphuric acid solution of specific gravity 1.320. It is important, when examining the cells, that naked lights should not be held near the vents, on account of the possible danger of igniting the gas which is generated by the plates. It is advisable to complete the inspection by measuring the specific gravity of the acid, as this gives a very good indication of the state of charge of the battery. An instrument known as a hydrometer is employed for this purpose; these can be bought from your Dealer or from any of the Lucas Service Stations.

For the battery fitted to the 12/70 M.G. Magna the specific gravity readings will be 1.285—1.300 for a fully charged battery, about 1.210 when half discharged and about 1.150 when fully discharged.

If one cell gives a reading very different from the rest, it may be that electrolyte has been spilled or has leaked from this cell, or there may be a " short " between the plates. In the latter case, the battery should be examined as soon as possible by a Lucas Service Station.

Finally, see that the tops of the cells are clean and dry, and that the terminals are tight and smeared with vaseline.

If the equipment is laid by for several months, the battery must be given a small charge from a separate source of electrical energy about once a fortnight, in order to obviate any permanent injury to the plates.

Under no circumstances must the acid be removed from the battery and the plates allowed to dry, as certain changes take place which result in loss of efficiency.

The battery must never be left in a fully discharged condition, and, unless some long runs are to be taken, it is advisable to have the battery removed from the car periodically and charged up from an independent electrical supply.

Ammeter.—The centre-zero ammeter which is incorporated in the instrument panel indicates the actual current flowing into or out of the battery. For instance, suppose two amperes are consumed when the side and tail lamps are switched on, and the ignition coil takes one ampere, then if the dynamo is generating at seven amperes the meter will show four amps. on the charge side of the scale. This is the current in excess of the lamp and ignition load which is available for battery charging purposes.

Fuses.—The separate fusing of the various circuits ensures that a short in any one does not affect the rest of the electrical equipment. This is particularly important when coil ignition is fitted. It will be noticed that the fusing of the lamps is such that there is no risk of the driver being plunged into total darkness. If both the headlamps, or the side and tail lamps, or all the units connected to the auxiliary accessory terminal fail to function, examine the particular fuse protecting them.

Remove the fuse from its holder and see whether there is a break in the fuse wire. Before replacing the fuse, inspect the units that have failed, for evidence of short circuits or other faults that may have caused the fuse to blow.

If a fuse blows repeatedly, and the cause cannot be traced we advise that the equipment is examined by a Lucas Service Station.

The Cut-out.—The function of the cut-out is to close the charging circuit, as the increased engine speed when the car is starting causes the dynamo voltage to rise above that of the battery. When the engine slows down, the dynamo voltage falls below that of the battery, and the reverse action takes place, i.e., the cut-out opens and thereby prevents the battery from discharging itself through the dynamo.

The question is sometimes asked whether the operation of the cut-out in any way depends upon the state of charge of the battery. There is no such relation between the two; the sole function of the cut-out is to switch on the dynamo with rising engine speed and to disconnect it when the engine slows down to below a certain speed. The cut-out is accurately set before leaving the Works and does not need any adjustment, and therefore the cover protecting it is sealed.

Very little attention is needed to keep the ignition equipment in first class condition. We advise that it is inspected occasionally, and the following instructions on lubrication, cleaning and adjustment should be carried out.

Distributor.—Occasionally remove the distributor moulding by pushing aside its two securing springs. See that the electrodes are clean and free from deposit. If necessary, wipe out the distributor with a dry duster and clean the electrodes with a cloth moistened with petrol. See that the carbon brush is clean and moves freely in its holder. Clean the outside of the moulding, particularly the spaces between the terminals. Next examine the contact breaker; it is important that the contacts are kept free from any grease or oil. If they are burned or blackened, they may be cleaned with very fine emery cloth and afterwards with a cloth moistened with petrol. Care must be taken that all particles of dirt and metal dust are wiped away. Misfiring may be caused if the contacts are not kept clean.

Provided that the cam is kept clean and that the instructions on cam lubrication are carried out, the contact breaker gap will only need adjustment at very long intervals. It is not advisable to alter the setting unless the gap varies considerably from the gauge. If adjustment is necessary, proceed as follows:—When the contacts are fully opened, slacken the locking nut on the stationary contact screw, and rotate it by its hexagon head until the gap is set to the thickness of the gauge. After making the adjustment, care must be taken to tighten the locking nut.

LUBRICATION.

(1) **Distributor Shaft.**—The greaser on distributor shaft should be given one turn about every 500 miles.

Repack the greaser with a good quality high melting point grease when necessary.

(2) **Cam.**—About every 3,000 miles, give the cam the slightest smear of vaseline.

(3) **Contact Breaker Pivot.**—Every 5,000 miles, place a single drop of oil on the pivot on which the contact breaker works.

Coil.—The coil unit is not adjustable in any way, and requires no attention beyond seeing that the terminal connections are kept tight, and the moulded coil top is kept clean.

Warning Lamp.—A warning lamp is provided in the instrument panel, which gives a red light when the ignition is "ON" and the car is stationary. The warning lamp will also light when the engine is running very slowly, due to the fact that the dynamo is not running at sufficient speed to generate a high enough voltage to actuate the cut-out.

THE DETECTION AND REMEDY OF IGNITION FAULTS.

If a failure of ignition or misfiring occurs, unless the cause is at once apparent the owner is strongly recommended to proceed in accordance with the following routine which should quickly enable him to locate the trouble.

Before proceeding with the examination, make sure that the trouble is not due to defects in the engine, carburetter, petrol supply, sparking plugs. etc.

Engine will not Fire.—Switch on the ignition, turn the engine and observe the ammeter reading. The engine should be turned by hand if it is known that the battery is in a low state of charge.

If an ammeter reading is given which rises and falls with the closing and opening of the contacts, then the low tension wiring is in order. If the reading does not fluctuate in this way, a short in the low tension wiring is indicated, or the contacts are remaining closed. When no reading is given, a broken or loose connection in the low tension wiring is indicated, or the battery may be exhausted.

Examine the high tension cables, i.e., cables from the coil to the distributor, and from the distributor to the plugs. If the rubber shows signs of deterioration or cracking, the cable should be renewed. Remove the distributor moulding and examine the contacts, and clean if necessary. Turn the engine over by hand, and see that the contacts come together.

If a fault is indicated in the low tension wiring, examine the cables from the switch or junction box to coil, and from coil to distributor. See that the battery terminals are tight and that the cables from the switch-box to the battery are secure. The battery may be dismissed as the cause of the trouble if the lamps will light.

Test the coil independently of the distributor as follows:—Remove the cable from the centre distributor terminal, and hold it about ¼ in. from some metal part of the chassis and turn the engine. The sparking should be strong and regular if the coil is functioning correctly.

Misfiring and Bad Starting.—Examine the high tension cables and the plugs. If necessary, adjust the gaps to the correct setting (about 20 thousandths of an inch). Sooty or oiled plugs may be dismantled and washed out with petrol.

The plugs and high tension cables may be tested by removing the plugs in turn and allowing them to rest on the cylinder head and observing whether a spark occurs at the points when the engine is turned by hand. It should, however, be noted that this is only a rough test, since it is possible that a spark may not take place when the plug is under compression.

Remove the distributor moulding and see that the electrodes and contacts are clean. If necessary, clean them as previously detailed. See that the contact gap setting is correct.

If after carrying out the examination suggested, the trouble cannot be found, we advise that the equipment should be examined by the nearest Lucas Service Depôt, the addresses of which are given later.

Headlamps.—The headlamps are provided with a patented universally adjustable mounting which allows the beam of light to be set to the best advantage. This adjustment is obtained by slacking the hexagon locking nut, turning the lamp to the desired position and then locking it by tightening up the nut.

The near side headlamp is set to throw its beam to the near side of the road and the off side (set straight) can be turned off or on independently.

Removing the Front and Reflector.—The headlamps are constructed with detachable fronts, parabolic reflectors and focusing devices. To remove a lamp front, give the screw head at the top of the lamp about a quarter of a turn—a coin will serve as a screw driver. This movement presses a cam against the body and pushes the front away from the body, enabling it to be removed. To replace, turn back the screw head and locate the cam between the two lugs on the body, then push the front into place. The reflector is secured by means of four fixing screws. Lamps are correctly focused when sent out from the Works, but if a bulb is replaced it may be necessary to adjust the focus of the lamp. Accurate focusing is imperative if the maximum results are to be attained. The method of focusing is as follows:—

Remove the front as above, remove the screws holding reflector and then carefully draw forward the reflector until the lamp holder is exposed. This holder is fixed by a screw and may be moved backwards or forwards when the screw is loosened. Each lamp must be focused separately, care being taken that lamps are properly set in line, not pointing up or down. Correct focusing adjustment is obtained by reflecting lamps on to a wall, or, preferably, a white object at a distance of about 100 yards, sliding the holder backwards or forwards until the light reaches its highest point of brilliancy with total absence of shadows. Then tighten screw and replace the reflector and front.

Wiring Headlamps.—Remove the front and reflector as described in a previous paragraph. Then depress the insulating washer on the bulb holder terminal against the spring until the cable hole is exposed. Thread the cable end through the hole, release the pressure on washer, when the cable will be securely held in position and good contact made. Cables should not be pushed too far through the terminal, as there is a danger of shorting on to the reflector or body of the lamp.

Side-lamps.—The fronts of the wing lamps are secured by small grub screws; when these are removed the fronts can be withdrawn by first pulling the top forward.

Tail-lamp.—The tail-lamp is fixed on to the car by means of a flanged base. To replace a bulb turn the front portion of the lamp to the left and withdraw it from its base.

When replacing see that the studs locate with the slots in the lamp front, then push it home to lock it in position. Should it be necessary to re-wire the tail-lamp. unscrew the coupling nut, when the cable covering shell and the cable plug can be withdrawn from the lamp; pass the cable through the shell and the rubber washer. Thread the bared end of the lead into the terminal socket and secure by the screw. Replace the cable plugs in its holder and secure by tightening the coupling nut.

Replacement of Bulbs.—When the replacement of any bulb is necessary we strongly advise that Lucas bulbs are used. The filaments are arranged to be in focus and give the best results with the lamps fitted. The following are the correct bulbs to use:—Headlamps, B.A.S. No. 35; Side and tail lamps, B.A.S. No. 10S.

Cleaning Lamps.—The reflectors are protected by a transparent and colourless covering, which enables any accidental finger marks to be removed with chamois leather or a soft cloth without affecting the surface of the reflector. Do not use metal polishes on Lucas Reflectors. Ebony black lamps can be cleaned with a good car polish. Chromium plated lamps will not tarnish and only need wiping over with a damp cloth to remove dust or dirt.

Lucas Altette Horn (Type HF318).—All electric horns, before being passed out of the Works, are adjusted to give the best performance, and they will give long periods of service without any attention; no subsequent adjustment is required.

If the horn becomes uncertain in its action, giving only a choking sound, or does not vibrate, it does not follow that the horn has broken down. First ascertain that the trouble is not due to some outside source, e.g. a discharged battery, a loose connection or short circuit in the wiring of the horn, or a blown fuse. It is also possible that the performance of a horn may be upset by the horn becoming loose on its mounting. If the cause of the trouble cannot be located, do not attempt to dismantle the horn, but return it to a Lucas Service Depôt for examination.

ELECTRICAL EQUIPMENT
(L Model)

The electrical equipment of the modern motorcar is often and quite erroneously regarded as something very complex and difficult. Yet in the present-day motorcar simple application of thoroughly tried out ideas has made possible easy adjustment and maintenance of practically all the electrically operated " bits." In this section an endeavour has been made to explain the various instruments and the best way to set about adjustments.

The wiring diagram may at first appear to be a little complicated, but it shows clearly exactly where each individual cable goes and the manner in which they are carried from point to point.

Wiring is arranged on the earth return system, and particular care has been taken to ensure that good continuity obtains at all points of the circuit, by the inclusion, where necessary, of independent earthing cables. It will also be noticed that cables at all essential points are encased in flexible metallic armouring. This not only protects the cables from possible abrasion, but excludes road dirt and water, prevents premature wear and makes for long and trouble-free service.

INSTRUMENT PANEL.

The panel—see Fig. 15, which shows the types in use on the Two-seater and Four-seater and the Saloon and Continental Coupé models—contains the control switch for the dynamo, ignition and lights (shown at " C," a centre zero ammeter " A," two-pin plug sockets " E," oil pressure gauge " B," and the ignition warning light " D.") A word of explanation on the dual function of the ignition warning light may prove of assistance. Its obvious object is to warn the owner that the ignition switch is "on" and the battery is being discharged, but, being in effect "shunted" across the cut-out points, it will allow only a very small current of electricity to flow through the armature and field coils of the dynamo, thus giving the dynamo an initial excitation. A resistance is always included in the warning lamp circuit to prevent the voltage rise of the dynamo burning out the bulb, and in the light under consideration the resistance is of sufficient value to permit the use of a small 2½-volt bulb of considerably longer life than the earlier heavy consumption type.

Views of the back of the instrument panels can be seen in Fig. 16, which is intended primarily for reference to the electrical equipment, but shows at the same time the general layout of the instruments seen from behind. The electrical panel is suitably lettered, and reference to the wiring diagram shows clearly the points of connection in relation to the other electrical components, such as the cut-out, junction box, ammeter, etc.

Fig. 15. *Facia boards on the "L" Magna ; Top :* 2 *and* 4 *seater. Bottom : saloon and continental coupe. Following letters indicate items. A. Ammeter. B. Oil Pressure Gauge. C. Lighting and Ignition switch. D. Ignition warning light. E. Plug sockets. F. Dashlamp switch. G. Horn and Dipper switch. H. Dashlamps.*

Cut-out and Junction Box.—A good view of this component can be had by referring to Fig. 4. It will be seen that all cables are taken to terminal posts in a recessed space at the back of the box, and it is here that the two main harnessed cables—the panel harness associated with the control panel and the junction box harness which constitutes the cables going to the various points on the chassis—are coupled up. On the front of the junction box, and immediately opposite the junction terminal posts, are the fuses. Reading from left to right, these are "Side and Tail," "S" or Field, "Auxiliary," and "Head." These are so arranged that should a short-circuit develop in any lamp or other component, the fuse associated with that component only will blow and the remainder of the installation will remain normal. The operation of tracing the fault is simplified enormously, as it is at once apparent which circuit is at fault. It should be noted that the "Field" or "S" fuse is only 4.5 amperes, and this must never be replaced by one of higher value or serious damage to the dynamo and cut-out is inevitable.

In the centre of the fuse block the spare fuse carrier is provided with one complete replacement set. At the bottom of the fuse block a pair of terminal posts are available for coupling up any other instruments fitted subsequently. This pair is fused in the auxiliary circuit and is intended for additional horns, foglamps, etc. It will perhaps be helpful to note here that there are many accessories on the market, some of which take a very high operating current, and the owner should avoid fitting any of these unless approved by a competent electrical engineer. In any case a heavy current accessory must be connected directly to the battery and not through the junction box or ammeter.

A small circular four-terminal junction box is placed close to the cut-out which forms a convenient connection point for wires, details of which will be seen on referring to the wiring diagram. There is in addition a further junction box on the Continental Coupé model.

Dynamo.—The dynamo is mounted on a platform at the front end of the engine and forms part of the overhead camshaft drive. Two views of the dynamo can be seen in Fig. 5. On the left-hand side it will be seen that the pinion is held on to a shaft by a bolt having a very wide head and also a tab washer which registers in one of the two holes in the pinion. These two holes are drilled and tapped ¼ in. B.S.F. in order to facilitate removal at any time with the aid of two bolts, one of which is shown in the illustration. The dynamo is what is known as a simple shunt machine and has been designed to stand up to the very high revolutions per minute and acceleration movement. Beyond an occasional inspection of the brush gear it requires but little attention. The brush gear can be examined by slackening off the screw holding the cover in position. The cover is shown in Fig. 5 above the dynamo.

Brushes.—It is very important to make sure that the three brushes work freely in their holders. This can easily be ascertained by holding back the spring and gently pulling each flexible lead, when the brush should move without the slightest suggestion of sluggishness. The brushes should be clean and should "bed" over the whole of their working surface; that is to say, the face in contact with the commutator should appear uniformly polished. Dirty brushes may be cleaned with a cloth moistened with petrol.

If any of the brushes become so badly worn that it is necessary to replace them, this is accomplished by releasing the brush lead eyelet by removal of the screw, then, while holding the spring lever back out of the way, withdrawing the brush from its holder. The new brush can then be fitted by reversing the operation.

When ordering new brush replacements state whether they are main or control brushes, and for what type of machine they are required.

The brush springs should be inspected occasionally to see that they have sufficient tension to keep the brushes firmly pressed against the commutator when the machine is running. It is particularly necessary to keep this in mind when the brushes have been in use for a long time and are very much worn down.

Owners are cautioned that it is unwise to insert brushes of a grade other than that supplied with the machine, or to change the tension springs. The arrangement provided has been made only after many years' experience, and will be found to give the best results.

Commutator.—The surface of the commutator should be kept clean and free from oil and brush dust, etc.; neglect of this precaution will result in the commutator becoming blackened, causing sparking to occur at the brushes, and consequent shortening of the life of the machine. The best way to clean the commutator is to insert a fine duster, held by means of a suitably shaped piece of wood, against the commutator surface, slowly rotating the armature by the starting handle at the same time.

If the commutator has been neglected for long periods, it may need cleaning with fine glass paper, but this is more difficult to do, and should not be necessary if it has received regular attention. The grooves between the commutator segments should be examined occasionally and any deposit of copper or carbon dust may be cleaned out by means of a thin saw blade or similar article. The owner is strongly advised to entrust this operation to a fully qualified electrician, as serious damage can unconsciously be done by anyone not familiar with this class of work.

Lubrication.—As the bearings are packed with grease before leaving the Works, very little attention is needed. A few drops of oil, however, may be added through the lubricators provided, say, every 1,000 miles. The reader is cautioned that far more trouble has been caused by excessive oiling than by too little. (Every 10,000 miles the dynamo should be removed, cleaned and adjusted and the bearings re-packed with grease. This should be entrusted to the nearest Lucas Service Depôt.)

Dynamo Field Fuse.—A fuse is provided in the dynamo field circuit to protect the machine in the event of anything being wrong in the charging circuit, e.g., a loose or broken battery connection. If the dynamo fails to charge the battery at any time (indicated by the ammeter giving a discharge reading during daytime running), inspect the fuse and, if it has blown, replace it. If the new fuse blows after starting up, the cause of the trouble must be found, and we advise that the equipment is examined by a Lucas Service Station. Replacement fuses must be of the same size as those originally fitted.

As explained elsewhere in this section, the fuse is one of the four cartridge type fitted in the junction box on the engine side of the dash. The size of the fuse is marked on a coloured paper slip which can be seen inside the fuse.

Removing the Dynamo.—In order to obtain easy access to the dynamo for removal it is desirable, although not absolutely necessary, to take off the radiator. Remove the lock nuts and nuts from the radiator bolts carried through the engine nose piece extension, and the bolts retaining the yokes of the bonnet tie rods. Also, after emptying the water system, remove the water return pipe from cylinder head and loosen clips holding bottom water hose. It is unnecessary to interfere with any of the oiling system to remove dynamo.

Between the projecting portion of the cylinder head and the top of the dynamo will be found a circular flexible coupling. Remove the nuts on each of the four coupling bolts in turn, leaving the bolts in position, so that the coupling can be rotated by use of the starting handle to bring each bolt into a position where the nut may easily be reached.

Having removed all four nuts, take off the valve gear cover and turn the engine by starting handle until the timing marks on the spiral bevel gears coincide. These will bring the driving yoke on the cylinder head across the engine, and the driving yoke on the dynamo parallel to the engine centre line. The bolts themselves may now be withdrawn, care being taken not to lose the distance washers, which must be replaced in the same position. Removal of the bolts enables the flexible coupling to be withdrawn.

Detach the two cables on the distributor side of the dynamo, noting from which terminal they are removed.

Now unscrew the four set screws which attach the dynamo to its platform at the front of the engine, thus releasing the dynamo. Lift the dynamo approximately a quarter of an inch and tilt it towards the near-side of the car until the driving yoke on the dynamo is just clear of that on the cylinder head. The dynamo can then be tilted forwards and easily withdrawn.

Replacement of the dynamo is effected in the reverse way, but it is necessary to make sure that the engine timing has not been disturbed while the dynamo was removed. Removal of the rectangular cover-plate in the clutch housing should reveal the timing mark on the flywheel for Nos. 1 and 6 cylinders, exactly in the centre of the opening. Place the brass packing pieces which fit under the dynamo base in position on the dynamo platform, making sure that you replace the same number that you took off. If for any reason the dynamo is replaced by another, it may be necessary to readjust the mesh of the driving gears to obtain silent running by suitable selection of the packing shims used. Turn the dynamo spindle until the timing mark on the dynamo drive gear coincides with the centre line of the dynamo and is at the rear—that is, ready for engagement with the correspondingly marked teeth of the drive gear on the crankshaft. The holes in the dynamo coupling yoke will then be parallel with the engine centre line. Tilting the dynamo towards the near-side of the car and holding the dynamo coupling yoke in this position, insert the drive gear into the opening of the dynamo platform and swing the dynamo backwards and downwards into position. The gears can be felt to be meshing properly if the coupling is slightly oscillated as the dynamo is replaced, but do not overdo the oscillations or you may engage the wrong teeth.

See that the bolts in the dynamo base are coinciding with the holes in the dynamo platform, and then observe if the dynamo coupling yoke is exactly parallel to the engine centre line. If it is not, withdraw the dynamo, reset the coupling yoke and re-insert the dynamo. No difficulty should be experienced in getting the dynamo in position with the correct teeth in mesh, as the distance between one tooth and the next is sufficient to make an appreciable difference to the position of the dynamo coupling yoke, a difference which is immediately discernible.

Having satisfied yourself that the correct gear teeth are in mesh, replace the screws in the dynamo base, taking particular care to tighten them up evenly a partial turn at a time until all are quite tight. Now make a final test. The flywheel mark 1—6 should show exactly in the centre of the inspection cover opening with the distributor rotating arm pointing towards No. 1 cylinder. This can easily be found by tracing the high-tension lead from No. 1 sparking plug to its junction on the distributor. Removal of the distributor cover should show the distributor arm directly beneath it. The dynamo coupling bolt holes should be exactly fore and aft and at right angles to the coupling yoke on the cylinder head, and the timing marks on the camshaft driving gears should be coinciding.

If all the foregoing are correct, replace the flexible coupling and coupling bolts, taking care to replace the distance washers in exactly the same position as they were originally. Tighten up the nuts firmly and rotate the engine slowly by hand. If the distance washers are in the correct position, the flexible coupling should run absolutely true. If it does not do so, note where the error is and adjust the distance washers accordingly. Then replace the valve cover and attach the dynamo cables on to their correct terminals.

Third Brush Regulator.—The output of the dynamo—that is to say, its rate of charge—is controlled by a third brush which is shown quite clearly in Fig. 12. All machines should be adjusted to an output of 9 amperes, and any marked variation from this standard usually indicates some other factor at fault which must be cleared before normal conditions are possible for the dynamo. Should, however, the indicated charging rate increase to a value above 10 amperes **with all lights out,** it is necessary to reduce this at once, as a continuation of high output will rapidly destroy the machine. It is inadvisable for anybody who is not an electrician to tamper with this. To reduce the output the third brush or control brush must be moved in a direction opposite to that of rotation. To move the brush it is necessary to slacken the stud marked "Locking Nut" in Fig. 12, and when the adjustment has been satisfactorily effected it is important to make sure that the stud is tightened.

The third brush is to be found on the near-side of the car and on the same side as the small ball covered lubricator. Never omit to tighten up the lock screw after any adjustment has been effected.

Starter Motor.—The armature spindles of these machines are fitted with a pinion which, on rotation, runs into engagement with the geared ring on the flywheel. Immediately the engine begins to fire, the pinion is automatically thrown out of mesh.

If, for any reason, the pinion wheel on the motor does not engage with the flywheel teeth, examine the screwed sleeve on the armature spindle to see that it is free from dust; if necessary wash over with paraffin. Occasionally, give it a few drops of thin machine oil.

As in the case of the dynamo, the surface of the commutator must be kept clean and free from oil, brush dust, etc.

The starter is designed for starting the engine under normal conditions, but any unnecessary or additional loading will considerably diminish the life of the machine and battery. In order to facilitate starting in cold weather, it is advisable to flood the carburetter, and, before using the electric starter, crank the engine over slowly by the starting handle for two or three revolutions; this will break the oil film and considerably diminish the load for starting.

In the event of the engine refusing to fire after being turned by the starter, make sure that the ignition switch is "on."

Battery.—It is of the utmost importance that the battery should receive regular attention, as upon its good condition depends the satisfactory functioning of the ignition, starting motor, and the lamps.

At least once a fortnight the vent plugs in the top of the battery should be removed, and the level of the acid solution examined. If necessary, distilled water (which can be obtained at all chemists and most garages) should be added to bring the level well above the plates. If, however, acid solution has been spilled, it should be replaced by a diluted sulphuric acid solution of specific gravity 1.320. It is important, when examining the cells, that naked lights should not be held near the vents, on account of the possible danger of igniting the gas which is generated by the plates. It is advisable to complete the inspection by measuring the specific gravity of the acid, as this gives a very good indication of the state of charge of the battery. An instrument known as a hydrometer is employed for this purpose; these can be bought from your dealer or from any of the Lucas Service Stations.

For the 12-volt 53 ampere battery fitted to the M.G. Magna the specific gravity readings will be 1.285—1.300 for a fully charged battery, about 1.210 when half discharged and about 1.150 when fully discharged.

If one cell gives a reading very different from the rest, it may be that electrolyte has been spilled or has leaked from this cell, or there may be a "short" between the plates. In the latter case, the battery should be examined as soon as possible by a Lucas Service Station.

Finally, see that the tops of the cells are clean and dry, and that the terminals are tight and smeared with vaseline.

If the equipment is laid by for several months, the battery must be given a small charge from a separate source of electrical energy about once a fortnight, in order to obviate any permanent injury to the plates.

Under no circumstances must the acid be removed from the battery and the plates allowed to dry, as certain changes take place which result in loss of efficiency.

The battery must never be left in a fully discharged condition, and, unless some long runs are to be taken, it is advisable to have the battery removed from the car periodically and charged up from an independent electrical supply.

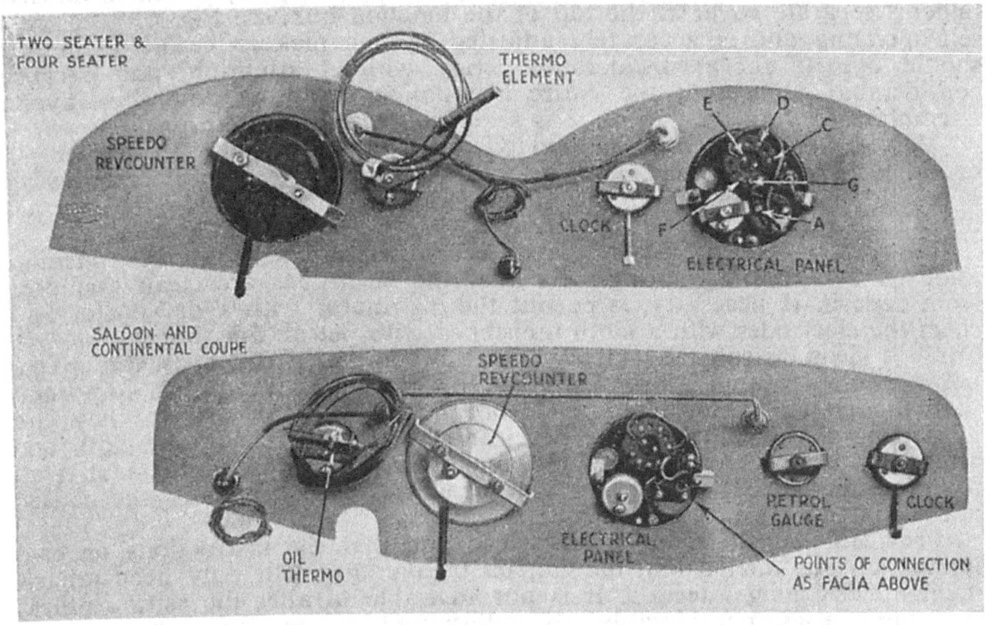

Fig. 16. *Rear view of the facia boards.*

Ammeter.—The centre-zero ammeter which is incorporated in the instrument panel indicates the actual current flowing into or out of the battery. For instance, suppose two amperes are consumed when the side and tail-lamps are switched on, and the ignition coil takes one ampere, then if the dynamo is generating at seven amperes the meter will show four amps. on the charge side of the scale. This is the current in excess of the lamp and ignition load which is available for battery charging purposes.

Fuses.—The separate fusing of the various circuits ensures that a short in any one does not affect the rest of the electrical equipment. This is particularly important when coil ignition is fitted. It will be noticed that the fusing of the lamps is such that there is no risk of the driver being plunged into total darkness. If both the headlamps, or the side and tail-lamps, or all the units connected to the auxiliary accessory terminal fail to function, examine the particular fuse protecting them.

Remove the fuse from its holder and see whether there is a break in the fuse wire. Before replacing the fuse, inspect the units that have failed, for evidence of short circuits or other faults that may have caused the fuse to blow.

If a fuse blows repeatedly, and the cause cannot be traced, we advise that the equipment is examined by a Lucas Service Station.

The Cut-out.—The function of the cut-out is to close the charging circuit, as the increased engine speed when the car is starting causes the dynamo voltage to rise above that of the battery. When the engine slows down, the dynamo voltage falls below that of the battery, and the reverse action takes place, i.e., the cut-out opens and thereby prevents the battery from discharging itself through the dynamo.

The question is sometimes asked whether the operation of the cut-out in any way depends upon the state of charge of the battery. There is no such relation between the two; the sole function of the cut-out is to switch on the dynamo with rising engine speed and to disconnect it when the engine slows down to below a certain speed. The cut-out, which is housed under a separate cover at the top of the junction box, see Fig. 4, is entirely self-operating and is accurately adjusted by the makers before fitting. It should operate for a considerable period without attention, and because considerable damage to the entire installation could be done by anyone attempting to vary the setting it is sealed by the manufacturers.

Very little attention is needed to keep the distributor in first class condition. We advise that it is inspected occasionally, and the following instructions on lubrication, cleaning and adjustment should be carried out.

Distributor.—Occasionally remove the distributor moulding by pushing aside its two securing springs. See that the electrodes are clean and free from deposit. If necessary, wipe out the distributor with a dry duster and clean the electrodes with a cloth moistened with petrol. See that the carbon brush is clean. Clean the outside of the moulding, particularly the spaces between the terminals. Next examine the contact breaker; it is important that the contacts are kept free from any grease or oil. If they are burned or blackened, they may be cleaned with very fine emery cloth and afterwards with a cloth moistened with petrol. Care must be taken that all particles of dirt and metal dust are wiped away. Misfiring may be caused if the contacts are not kept clean.

Provided that the cam is kept clean and that the instructions on cam lubrication are carried out, the contact breaker gap will only need adjustment at very long intervals. It is not advisable to alter the setting unless the gap varies considerably. If adjustment is necessary, proceed as follows:—When the contacts are fully opened, slacken the locking nut on the stationary contact screw, and rotate it by its hexagon head until the gap is set to the thickness of the gauge. After making the adjustment, care must be taken to tighten the locking nut.

LUBRICATION.

(1) **Distributor Shaft.**—Add one or two drops of thin machine oil through the oiler provided about every 1,000 miles.

(2) **Cam.**—About every 3,000 miles, give the cam the slightest smear of vaseline.

(3) **Automatic Timing Control.**—About every 3,000 miles withdraw the rotating arm from the top of the spindle by lifting it off, and add a few drops of thin machine oil. Do not remove the screw exposed to view, as there is a clearance between the screw and the inner face of the spindle through which the oil passes to lubricate the automatic timing control.

Coil.—The coil unit is not adjustable in any way, and requires no attention beyond seeing that the terminal connections are kept tight, and the moulded coil top is kept clean.

Warning Lamp.—A warning lamp is provided in the instrument panel, which gives a red light when the ignition is " ON " and the car is stationary. The warning lamp will also light when the engine is running very slowly, due to the fact that the dynamo is not running at sufficient speed to generate a high enough voltage to actuate the cut-out.

THE DETECTION AND REMEDY OF IGNITION FAULTS.

If a failure of ignition or misfiring occurs, unless the cause is at once apparent the owner is strongly recommended to proceed in accordance with the following routine, which should quickly enable him to locate the trouble.

Before proceeding with the examination, make sure that the trouble is not due to defects in the engine, carburetter, petrol supply, sparking plugs, etc.

Engine will not Fire.—Switch on the ignition, turn the engine and observe the ammeter reading. The engine should be turned by hand if is known that the battery is in a low state of charge.

If an ammeter reading is given which rises and falls with the closing and opening of the contacts, then the low-tension wiring is in order. If the reading does not fluctuate in this way, a short in the low-tension wiring is indicated, or the contacts are remaining closed. When no reading is given, a broken or loose connection in the low-tension wiring is indicated, or the battery may be exhausted.

Examine the high-tension cables, i.e., cables from the coil to the distributor, and from the distributor to the plugs. If the rubber shows signs of deterioration or cracking, the cable should be renewed. Remove the distributor moulding and examine the contacts; and if necessary, clean them. Turn the engine over by hand, and see that the contacts come together.

If a fault is indicated in the low-tension wiring, examine the cables from the switch or junction box to coil, and from coil to distributor. See that the battery terminals are tight and that the cables from the switchbox to the battery are secure. The battery may be dismissed as the cause of the trouble if the lamps will light.

Test the coil independently of the distributor as follows:—Remove the cable from the centre distributor terminal, and hold it about $\frac{1}{4}$ in. from some metal part of the chassis and turn the engine. The sparking should be strong and regular if the coil is functioning correctly.

Misfiring and Bad Starting.—Examine the high-tension cables and the plugs. If necessary, adjust the gaps to the correct setting (about 20 thousandths of an inch). Sooty or oiled plugs may be dismantled and washed out with petrol.

The plugs and high-tension cables may be tested by removing the plugs in turn and allowing them to rest on the cylinder head and observing whether a spark occurs at the points when the engine is turned by hand. It should, however, be noted that this is only a rough test, since it is possible that a spark may not take place when the plug is under compression.

Remove the distributor moulding and see that the electrodes and contacts are clean. See that the contact gap setting is correct.

If after carrying out the examination suggested, the trouble cannot be found, we advise that the equipment should be examined by the nearest Lucas Service Depôt.

Headlamps.—The headlamps are provided with a patented universally adjustable mounting which allows the beam of light to be set to the best advantage. This adjustment is obtained by slacking the hexagon locking nut, turning the lamp to the desired position and then locking it by tightening up the nut.

The near-side headlamp is set to throw its beam to the near-side of the road and the off-side (set straight) can be turned off or on independently.

Removing the Front and Reflector.—The headlamps are constructed with detachable fronts, parobolic reflectors and focusing devices. To remove a lamp front for fitting a new lamp bulb, give the screw head at the top of the lamp about a quarter of a turn—a coin will serve as a screwdriver. This movement presses a cam against the body and pushes the front away from the body, enabling it to be removed. To replace, turn back the screw head and locate the cam between the two lugs on the body, then push the front into place. The reflector is secured by means of four fixing screws. Lamps are correctly focused when sent out from the Works, but if a bulb is replaced it may be necessary to adjust the focus of the lamp.

Accurate focusing is imperative if the maximum results are to be attained. The method of focusing is as follows:—

Remove the front as above, remove the screws holding reflector and then carefully draw forward the reflector until the lamp holder is exposed. This holder is fixed by a screw and may be moved backwards or forwards when the screw is loosened. Each lamp must be focused separately, care being taken that lamps are properly set in line, not pointing up or down. Correct focusing adjustment is obtained by reflecting lamps on to a wall or, preferably, a white object at a distance of about 100 yards, sliding the holder backwards or forwards until the light reaches its highest point of brilliancy with total absence of shadows. Then tighten screw and replace the reflector and front.

Wiring Headlamps.—Remove the front and reflector as described in a previous paragraph. Then depress the insulating washer on the bulbholder terminal against the spring until the cable hole is exposed. Thread the cable end through the hole, release the pressure on washer, when the cable will be securely held in position and good contact made. Cables should not be pushed too far through the terminal, as there is a danger of shorting on to the reflector or body of the lamp.

Sidelamps.—The fronts of the wing lamps are secured by small grub screws; when these are removed the fronts can be withdrawn by first pulling the top forward.

Stop and Tail-lamp.—By virtue of its extreme simplicity there is practically nothing which can go wrong with this instrument. In the event of a bulb burning out it can be replaced by removing the front of the lamp. This is done by removing the screw carried through the front of the lamp and gently prising the front away from the body at the screw side. With the front removed either of the bulbs are easily accessible.

The stop light is operated by a switch spring coupled to the brake cross shaft on the off-side of the car. This switch is wired into the ignition circuit so that with the ignition switched "off" it is not possible to leave the light on. It is perhaps as well to note here that this is the only light point on the car which is not fused.

Replacement of Bulbs.—When the replacement of any bulb is necessary we strongly advise that Lucas bulbs are used. The filaments are arranged to be in focus and give the best results with the lamps fitted. The following are the correct bulbs to use:—Headlamps, B.A.S. No. 35; side and tail-lamps, B.A.S. No. 10S.

Cleaning Lamps.—The reflectors are protected by a transparent and colourless covering, which enables any accidental finger marks to be removed with chamois leather or a soft cloth without affecting the surface of the reflector. Do not use metal polishes on Lucas reflectors. Ebony black lamps can be cleaned with a good car polish. Chromium plated lamps will not tarnish and only need wiping over with a damp cloth to remove dust or dirt.

Lucas Altette Horn (Type HF318).—All electric horns are adjusted to give the best performance, and they will give long periods of service without any attention; no subsequent adjustment is required.

If the horn becomes uncertain in its action, giving only a choking sound, or does not vibrate, it does not follow that the horn has broken down. First ascertain that the trouble is not due to some outside source, e.g., a discharged battery, a loose connection or short circuit in the wiring of the horn, or a blown fuse. It is also possible that the performance of a horn may be upset by the horn becoming loose on its mounting. If the cause of the trouble cannot be located, do not attempt to dismantle the horn, but return it to a Lucas Service Depôt for examination.

ELECTRICAL EQUIPMENT
(K Model)

A view of a facia board will be seen on referring to Fig. 17 and it will be seen that there are several electrical fittings in addition to the instrument panel. The form of the facia board and the positions of the controls vary in different models, but the instruments remain fundamentally the same in each case. The instrument panel contains a centre zero ammeter, ignition lock switch, dynamo and lights switch, dashlamp switch, dashlamps and a remote control solenoid switch for the starter motor. The ignition locking switch is so arranged that when in the "off" position it is impossible to start the car or alter the position of the light switch. This is particularly useful when leaving the car unattended in a public place. Front and rear views of the panel are given in Figs. 18 and 19, while the wiring can easily be traced on the wiring diagram.

Fig. 17. Showing layout of facia board.

DYNAMO.

The dynamo is mounted on a platform at the front end of the engine and forms part of the overhead camshaft drive. Two views of the dynamo can be seen in Fig. 5. On the left-hand side it will be seen that the pinion is held on to a shaft by a bolt having a very wide head and also a tab washer which registers into the key way in the armature spindle. The two holes in the pinion are drilled and tapped $\frac{1}{4}$ in. B.S.F. in order to facilitate removal at any time with the aid of two bolts, one of which is shown in the illustration. The dynamo requires but little attention, the brush gear can be examined by slackening off the screw holding the cover in position. The cover is shown in the illustration above the dynamo.

Brushes.—It is very important to make sure that the three brushes work freely in their holders. This can easily be ascertained by holding back the spring and gently pulling each flexible lead, when the brush should move without the slightest suggestion of sluggishness. The brushes should be clean and should "bed" over the whole of their working surface; that is to say, the face in contact with the commutator should appear uniformly polished. Dirty brushes may be cleaned with a cloth moistened with petrol.

If any of the brushes become so badly worn that it is necessary to replace them, this is accomplished by releasing the brush lead eyelet by removal of the screw, then, while holding the spring lever back out of the way, withdrawing the brush from its holder. The new brush can then be fitted by reversing the operation.

When ordering new brush replacements state whether they are main or control brushes, and for what type of machine they are required.

The brush springs should be inspected occasionally to see that they have sufficient tension to keep the brushes firmly pressed against the commutator when the machine is running. It is particularly necessary to keep this in mind when the brushes have been in use for a long time and are very much worn down.

Readers are cautioned that it is unwise to insert brushes of a grade other than that supplied with the machine, or to change the tension springs. The arrangement provided has been made only after many years' experience, and will be found to give the best results.

Commutator.—The surface of the commutator should be kept clean and free from oil and brush dust, etc.; neglect of this precaution will result in the commutator becoming blackened, causing sparking to occur at the brushes, and consequent shortening of the life of the machine. The best way to clean the commutator is to insert a fine duster, held by means of a suitably shaped piece of wood against the commutator surface, slowly rotating the armature by the starting handle at the same time.

If the commutator has been neglected for long periods, it may need cleaning with fine glass paper, but this is more difficult to do, and should not be necessary if it has received regular attention. The grooves between the commutator segments should be examined occasionally and any deposit of copper or carbon dust may be cleaned out by means of a thin saw blade or similar article. Generally speaking, this should not be undertaken by anyone other than a qualified electrician, as extensive damage can be unconsciously done by anyone not familiar with the operation.

Dynamo Field Fuse.—A fuse is provided in the dynamo field circuit to protect the machine in the event of anything being wrong in the charging circuit, e.g., a loose or broken battery connection. If the dynamo fails to charge the battery at any time (indicated by the ammeter giving a discharge reading during daytime running), inspect the fuse and if it has blown, replace it. If the new fuse blows after starting up, the cause of the trouble must be found, and we advise that the equipment is examined by a Lucas Service Station. Replacement fuses must be of the same size as those originally fitted, viz., 4.5 amps.

In the case of the M.G. Magnette model the fuse is one of the four cartridge type fitted in the junction box on the engine side of the dash (see Fig. 4), and is marked E. The size of the fuse is marked on a coloured paper slip which can be seen inside the fuse.

Removing the Dynamo.—In order to obtain easy access to the dynamo for removal it is advisable to take off the radiator. Remove the lock nuts and nuts from the radiator bolts carried through the engine nose piece extension, and the bolts retaining the yokes of the bonnet tie-rods. Remove water return pipe from cylinder head, and loose clips holding bottom water hose.

N.B.—It is unnecessary to interfere with any of the oiling system to remove dynamo.

Between the projecting portion of the cylinder head and the top of the dynamo will be found a circular flexible coupling. Remove the nuts on each of the four coupling bolts in turn, leaving the bolts in position, so that the coupling can be rotated by use of the starting handle to bring each bolt into a position where the nut may easily be reached.

Having removed all four nuts, take off the valve gear cover and turn the engine by the starting handle until the timing marks on the spiral bevel gears coincide. These will bring the driving yoke on the cylinder head across the engine, and the driving yoke on the dynamo parallel to the engine centre line. The bolts themselves may now be withdrawn, care being taken not to lose the distance washers, which must be replaced in the same position. Removal of the bolts enables the flexible coupling to be withdrawn.

Detach the two cables on the distributor side of the dynamo, noting from which terminal they are removed.

Now unscrew the four set screws which attach the dynamo to its platform at the front of the engine, thus releasing the dynamo. Lift the dynamo approximately a quarter of an inch and tilt it towards the near-side of the car until the driving yoke on the dynamo is just clear of that on the cylinder head. The dynamo can then be tilted forwards and easily withdrawn.

Replacement of the dynamo is effected in the reverse way, but it is necessary to make sure that the engine timing has not been disturbed while the dynamo was removed. Removal of the rectangular cover-plate in the clutch housing should reveal the timing mark on the flywheel for Nos. 1 and 6 cylinders, exactly in the centre of the opening. Place the brass packing shims which fit under the dynamo base in position on the dynamo platform, making sure that you replace the same number that you took off. If for any reason the dynamo is replaced by another, it may be necessary to readjust the mesh of the driving gears to obtain silent running by suitable selection of the packing shims used. Turn the dynamo spindle until the timing mark on the dynamo drive gear coincides with the centre line of the dynamo and is at the rear—that is, ready for engagement with the correspondingly marked teeth of the drive gear on the crankshaft. The holes in the dynamo coupling yoke will then be parallel with the engine centre line. Tilting the dynamo towards the near-side of the car and holding the dynamo coupling yoke in this position, insert the drive gear into the opening of the dynamo platform and swing the dynamo backwards and downwards into position. The gears can be felt to be meshing properly if the coupling is slightly oscillated as the dynamo is replaced, but do not overdo the oscillations or you may engage the wrong teeth.

See that the bolts in the dynamo base are coinciding with the holes in the dynamo platform, and then observe if the dynamo coupling yoke is exactly parallel to the engine centre line. If it is not, withdraw the dynamo, reset the coupling yoke and re-insert the dynamo. No difficulty should be experienced in getting the dynamo in position with the correct teeth in mesh, as the distance between one tooth and the next is sufficient to make an appreciable difference to the position of the dynamo coupling yoke, a difference which is immediately discernible.

Having satisfied yourself that the correct gear teeth are in mesh, replace the screws in the dynamo base, taking particular care to tighten them up evenly a partial turn at a time until all are quite tight. Now make a final test. The flywheel mark 1—6 should show exactly in the centre of the inspection cover opening with the magneto high-tension rotor pointing towards No. 1 cylinder lead. (This can be easily found by tracing the high-tension lead from No. 1 sparking plug to its junction on the distributor. Removal of the distributor cover should show the high-tension rotor directly beneath it.) The dynamo coupling bolt holes should be exactly fore and aft and at right angles to the coupling yoke on the cylinder head, and the timing marks on the camshaft driving gears should be coinciding.

If all the foregoing are correct, replace the flexible coupling and coupling bolts, taking care to replace the distance washers in exactly the same position as they were originally. Tighten up the nuts firmly and rotate the engine slowly by hand. If the distance washers are in the correct position, the flexible coupling should run absolutely true. If it does not do so, note where the error is and adjust the distance washers accordingly. Then replace the valve cover and attach the dynamo cables on to their correct terminals.

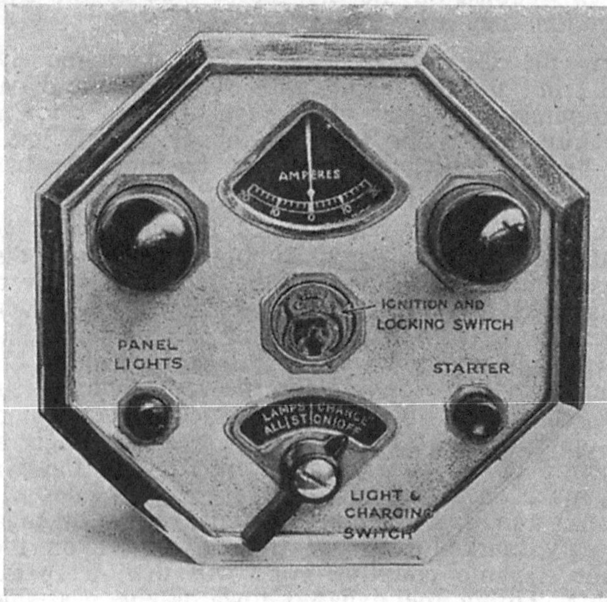

Fig. 18. Front view of switchboard and instrument panel.

Third Brush Regulator.—The output of the dynamo—that is to say, its rate of charge—is controlled by a third brush which is shown quite clearly in Fig. 12. It is inadvisable for anybody who is not an electrician to tamper with this, as all machines are adjusted to an output of 9 amperes, and any marked variation from this standard usually indicates some other factor at fault, which must be cleared before normal conditions are possible for the dynamo. Should, however, the indicated rate of charge increase to a value above 10 amperes **with all lights out,** it is necessary to reduce this at once, as a continuation of high output will rapidly destroy the machine. This may be done by moving the third brush or control brush in a direction opposite to that of rotation. To move the brush it is necessary to slacken the stud marked " Locking Nut " in Fig. 12, and when the adjustment has been effected, it is important to make sure that the stud is tightened.

Lubrication.—A few drops of oil may be added through the lubricators provided, say, every 2,000 miles. The reader is cautioned that far more trouble has been caused by excessive oiling than by too little. (Every 10,000 miles the dynamo should be removed, cleaned and adjusted and the bearings re-packed with grease. This should be entrusted to the nearest M.G. or Lucas Service Depôt.)

Ammeter.—The centre-zero ammeter which is incorporated in the instrument panel indicates the actual current flowing into or out of the battery. For instance, suppose two amperes are consumed when the side and tail-lamps are switched on, then if the dynamo is generating at seven amperes the meter will show five amps. on the charge side of the scale. This is the current in excess of the lamp and other load which is available for battery charging purposes.

A flickering of the instrument needle is caused by the petrol pump, but should cease when the pump is switched on and the engine is not running after the pump has filled up. Shortage of petrol or air leak on the suction side will cause the petrol pump to keep on operating.

Fuses.—The separate fusing of the various circuits ensures that a short in any one does not affect the rest of the electrical equipment. It will be noticed that the fusing of the lamps is such that there is no risk of the driver being plunged into total darkness. If both the headlamps, or the side and tail lamps, or all the units connected to the auxiliary accessory terminal fail to function, examine the particular fuse protecting them.

Remove the fuse from its holder and see whether there is a break in the fuse wire. Before replacing the fuse, inspect the units that have failed, for evidence of short circuits or other faults that may have caused the fuse to blow.

If a fuse blows repeatedly and the cause cannot be traced, we advise that the equipment is examined by a Lucas Service Station.

The Cut-out.—The function of the cut-out is to close the charging circuit as the increased engine speed when the car is starting causes the dynamo voltage to rise above that of the battery. When the engine slows down, the dynamo voltage falls below that of the battery, and the reverse action takes place, i.e., the cut-out opens and thereby prevents the battery from discharging itself through the dynamo.

The question is sometimes asked whether the operation of the cut-out in any way depends upon the state of charge of the battery. There is no such relation between the two; the sole function of the cut-out is to switch on the dynamo with rising engine speed and to disconnect it when the engine slows down to below a certain speed. The cut-out is accurately set before leaving the Works and does not need any adjustment, and therefore the cover protecting it is sealed.

Junction Box.—This is fitted with a number of glass barrel fuses, and holes are provided in the body of the box to carry spares. The fuses control the side and tail-lamps, separate fuses for the headlamps, another fuse for auxiliaries such as the horn and a further fuse marked E, which is for the dynamo field. The cut-out, which is mounted on the same base as the fuses, is provided with a cover and under normal circumstances need never be touched, its operations being entirely automatic.

A small circular junction box is placed beneath the fusebox and cut-out, which forms a convenient connection point for wires to the horn, off-side headlamp, direction indicators and main battery positive pick-up. The scheme of connections in this unit can be best understood by referring to the wiring diagram.

Battery.—It is of the utmost importance that the battery should receive regular attention, as upon its good condition depends the satisfactory functioning of the ignition, starting motor, and the lamps.

At least once a fortnight the vent plugs in the top of the battery should be removed, and the level of the acid solution examined. If necessary, distilled water (which can be obtained at all chemists and most garages) should be added to bring the level well above the plates. If, however, acid solution has been spilled, it should be replaced by a diluted sulphuric acid solution of specific gravity 1.320. It is important, when examining the cells, that naked lights should not be held near the vents, on account of the possible danger of igniting the gas which is generated by the plates. It is advisable to complete the inspection by measuring the specific gravity of the acid, as this gives a very good indication of the state of charge of the battery. An instrument known as a hydrometer is employed for this purpose; these can be bought from your dealer or from any of the Lucas Service Stations.

For the battery fitted to the M.G. Magnette the specific gravity readings will be 1.285—1.300 for a fully charged battery, about 1.210 when half discharged and about 1.150 when fully discharged.

If one cell gives a reading very different from the rest, it may be that electrolyte has been spilled or has leaked from this cell, or there may be a "short" between the plates. In the latter case, the battery should be examined as soon as possible by a Lucas Service Station.

Finally, see that the tops of the cells are clean and dry, and that the terminals are tight and smeared with vaseline.

If the equipment is laid by for several months, the battery must be given a small charge from a separate source of electrical energy about once a month, in order to obviate any permanent injury to the plates.

Under no circumstances must the acid be removed from the battery and the plates allowed to dry, as certain changes take place which result in loss of efficiency.

The battery must never be left in a fully discharged condition, and, unless some long runs are to be taken, it is advisable to have the battery removed from the car periodically and charged up from an independent electrical supply.

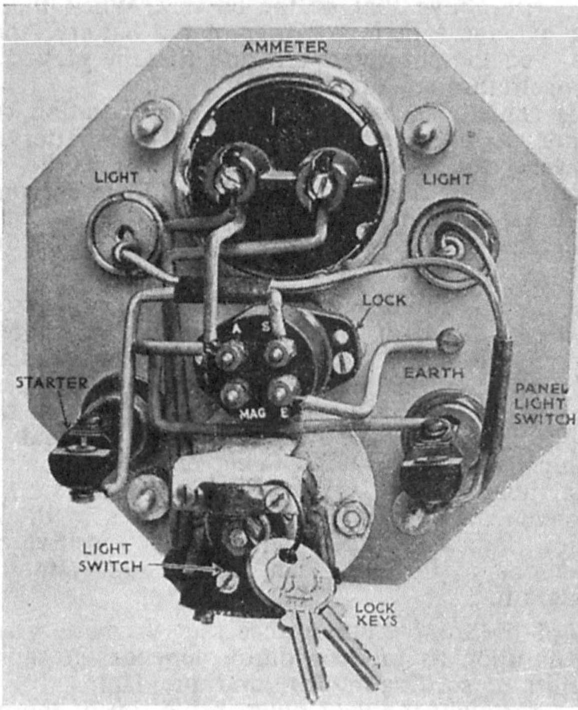

Fig. 19. *Rear view of instrument panel.*

Headlamps.—The **L.150** headlamps manufactured by Messrs. Joseph Lucas, Ltd., are specially designed to give the highest illuminating factor without unduly overloading the electrical installation. The near-side lamp is equipped with a permanently deflected reflector and is always in circuit when the headlamp switch is on. The off-side lamp has a straight focus reflector, and while primarily wired in the same circuit as the other, can also be controlled by the dipping switch. The lamps are carefully focused before leaving the factory so that the near-side lamp will throw a slightly open beam and thus effectively illuminate the road and kerb up to about 40 yards, while the off-side beam is relatively closely focused to throw a penetrating light straight ahead. By a careful association of the Biflex reflector principle and diffusing glass in the front, the "black spots" are entirely eliminated while good side lighting and a clear and steady beam ahead are ensured.

It should be noted that when the bulb is changed it is usually necessary to re-focus the beam. This can be effected by removing the front by slacking off the retaining screw at the bottom of the lamp, and then removing the reflector which is held in position by three small screws, when the small cheese-head screw holding the bulb holder can be loosened. The bulb holder can then be moved either forwards or backwards until the correct beam is obtained. It is essential that the cheese-headed screw should be effectively tightened when the correct focus is obtained, otherwise the light will be erratic.

Wiring Headlamps.—Remove the front and reflector. Then depress the insulating washer on the bulb holder terminal against the spring until the cable hole is exposed. Thread the cable end through the hole, release the pressure on washer, when the cable will be securely held in position and good contact made. Cables should not be pushed too far through the terminal, as there is a danger of shorting on to the reflector or body of the lamp.

Sidelamps.—The fronts of the wing lamps are secured by small grub screws; when these are removed the fronts can be withdrawn by first pulling the top forward.

Replacement of Bulbs.—When the replacement of any bulb is necessary we strongly advise that Lucas bulbs are used. The filaments are arranged to be in focus and give the best results with the lamps fitted.

Cleaning Lamps.—The reflectors are protected by a transparent and colourless covering, which enables any accidental finger marks to be removed with chamois leather or a soft cloth without affecting the surface of the reflector. Do not use metal polishes on Lucas reflectors. Ebony black lamps can be cleaned with a good car polish. Chromium plated lamps will not tarnish and only need wiping over with a damp cloth to remove dust or dirt.

Direction Indicator.—The Saloon model is equipped with a combined direction indicator and stop lamp. The stop lamp is wired in the usual manner and is controlled by the brake pedal. The direction indicator is so connected that a red warning lamp situated in the centre of the switch is brought into circuit when either of the indicator lamps are operated. Thus it is impossible to unconsciously leave either of the indicators "on."

Bulb replacement is effected by removing the case of the lamp when the lamp holders are easily accessible.

Starter Motor.—The starter motor is a specially designed four-brush machine with very high start and run torque. It is brought into operation by the solenoid or relay fixed to its yoke, which in turn is controlled by the small push switch in the instrument panel referred to elsewhere. Beyond an occasional inspection of the brushes and the usual cleaning instructions, the starter should require very little attention. It is, however, important that the engine should never be started with the ignition advanced, as, owing to the high starting torque of the starter, a back-fire would possibly damage the spindle.

The armature spindles of these machines are fitted with a pinion which, on rotation, runs into engagement with the geared ring on the flywheel

Immediately the engine begins to fire, the pinion is automatically thrown out of mesh.

If, for any reason, the pinion wheel on the motor does not engage with the flywheel teeth, examine the screwed sleeve on the armature spindle to see that it is free from dust; if necessary wash over with paraffin. Occasionally, give it a few drops of thin machine oil.

As in the case of the dynamo, the surface of the commutator must be kept clean and free from oil, brush dust, etc.

The starter is designed for starting the engine under normal conditions, but any unnecessary or additional loading will considerably diminish the life of the machine and battery. In order to facilitate starting in cold weather, it is advisable to flood the carburetter, and, before using the electric starter, crank the engine over slowly by the starting handle for two or three revolutions; this will break the oil film and considerably diminish the load for starting.

In the event of the engine refusing to fire after being turned by the starter, make sure that the ignition switch is "on."

Magneto.—Beyond strict observance of the makers' instructions with regard to lubrication, the magneto should require very little attention. It is advisable to periodically remove the distributor moulding and thoroughly clean with a soft cloth moistened with petrol. Providing that the correct lubrication instructions are observed the contact should not require any attention for a long period. Should the points become foul through any foreign matter getting on them, they should be removed and cleaned and on replacement the gap should be set to recommended clearance.

It is particularly important to see that the switch wire terminal on the magneto is at all times tight, as this not only anchors the switch wire but also the link cable between the low-tension coil and the contact breaker.

The Detection and Remedy of Ignition Faults.—If a failure of ignition or misfiring occurs, unless the cause is at once apparent the owner is strongly recommended to proceed in accordance with the following routine which should quickly enable him to locate the trouble.

Before proceeding with the examination, make sure that the trouble is not due to defects in the engine, carburetter, petrol supply, sparking plugs, etc.

Examine the high-tension cables, i.e., cables from the magneto to the plugs. If the rubber shows signs of deterioration or cracking, the cable should be renewed. Remove the distributor moulding and examine the contacts; if necessary, clean them as previously described. Turn the engine over by hand, and see that the contacts come together.

Should all appear to be in order, carefully check the switch cable from the magneto to the magneto switch for evidence of earthing out. If there is still no spark the fault will probably be in the magneto itself and must be investigated by an electrician.

Misfiring and Bad Starting.—Examine the sparking plugs. If necessary, adjust the gaps to the correct setting (about 18 thousandths of an inch). Sooty or oiled plugs may be dismantled and washed out with petrol.

The plugs and high tension cables may be tested by removing the plugs in turn and allowing them to rest on the cylinder head and observing whether a spark occurs at the points when the engine is turned by hand. It should, however, be noted that this is only a rough test, since it is possible that a spark may not take place when the plug is under compression.

Lucas Altette Horns (Type HF317).—All electric horns are adjusted to give the best performance, and they will give long periods of service without any attention; no subsequent adjustment is required.

If the horns become uncertain in their action, giving only a choking sound, or do not vibrate, it does not follow that they have broken down. First ascertain that the trouble is not due to some outside source, e.g., a discharged battery, a loose connection or short circuit in the wiring of the horn, or a blown fuse. It is also possible that the performance of a horn may be upset by the horn becoming loose on its mounting. If the cause of the trouble cannot be located, do not attempt to dismantle the horn, but return it to a Lucas Service Depôt for examination.

COIL IGNITION INSTRUCTIONS.

The high-tension distributor with automatic advance mechanism is very accessible. It is provided with an oiler to lubricate the spindle, and only a few drops of thin machine oil should be put in every 1,000 miles.

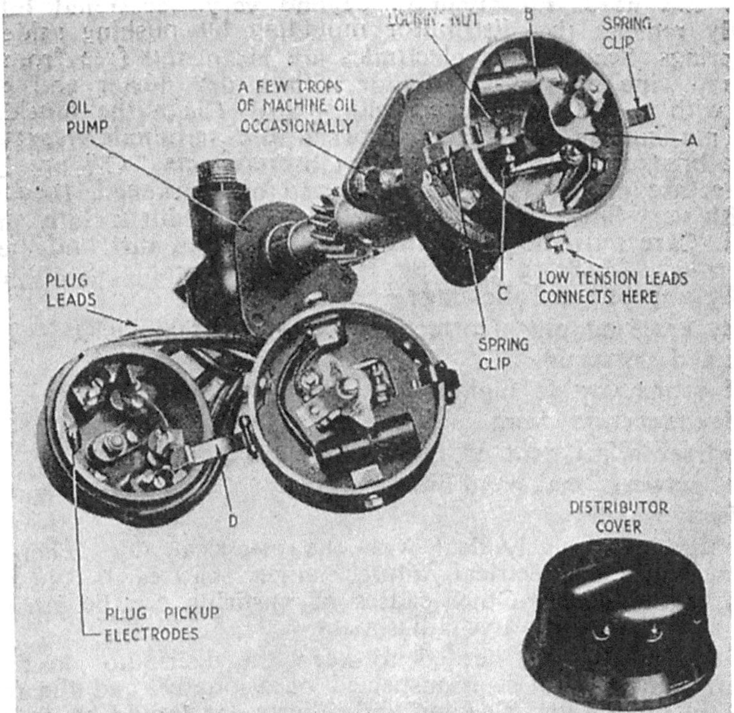

Fig. 20. *Distributor and oil pump.*

The contacts can be inspected by unclipping the two springs, when the cover carrying the high-tension leads can be removed. Should for any reason it be necessary to check or adjust, the ignition timing should be set so that No. 1 piston is on top dead centre, measured on the flywheel when the 1/6 marking is in the centre of the clutch inspection aperture and the inlet valve of No. 6 is opening and the contact breaker arm pointing towards No. 1 plug segment. Then, after slackening the securing bolt, rotate the body of the distributor and set the contact breaker arm so that the fibre heel on the contact breaker arm is on the apex of the cam on the rotor. The clearance between the points should be 15—20/1,000 of an inch, they should be carefully set. It is not advisable to alter the setting unless the gap varies considerably. If adjustment is necessary, proceed as follows:—When the contacts are fully opened, slacken the locking nut on the stationary contact screw, and rotate it by its hexagon head until the gap is set to the thickness of the gauge. After making the adjustment, care must be taken to tighten the locking nut.

The centre spindle of the high-tension distributor carries a "Bakelite" arm called the "rotor." It can only fit on the spindle one way. If this part is removed (it simply pulls off) it affords an excellent thiefproof device, but care should be exercised to see that it does not become chipped or in any way damaged, not to mention lost, after it has been removed. The ends of the high-tension cables are secured by means of pointed contact screws which pierce the rubber insulation to make good contact with the cable strands, at the same time securing the cables tightly in the terminals. The screw heads are located on the inside of the distributor moulding.

The automatic device, which consists of a centrifugal governor, is housed in the distributor body. It is packed with grease before leaving the Works and needs no attention beyond very occasional lubrication. Occasionally remove the distributor moulding by pushing aside its two securing springs. See that the electrodes are clean and free from deposit. If necessary, wipe out the distributor with a dry duster and clean the electrodes with a cloth moistened with petrol. Clean the outside of the moulding, particularly the spaces between the terminals. Next examine the contact breaker; it is important that the contacts "C" are kept free from any grease or oil. If they are burned or blackened, they must be cleaned with very fine emery cloth and afterwards with a cloth, moistened with petrol. Care must be taken that all particles of dirt and metal dust are wiped away. Misfiring may be caused if the contacts are not kept clean, or by any of the following:—

1. Dirty sparking plug (correct plug gap 20/1,000 inch).
2. Cracked porcelain.
3. Bad connection to high-tension leads.
4. Bad connection from high-tension distributor to coil.
5. Improper adjustment of make-and-break points.
6. Dirt between make-and-break points.
7. Defective coil.

These causes exclusively deal with the electrical side. There is one other remote cause of electrical failure, namely, bad earth contacts from the battery to the frame. Other causes of misfiring can be attributed to carburetter and improper valve adjustment.

Very little attention is needed to keep the distributor in first class condition. We advise that it is inspected occasionally, and the following instructions on lubrication, cleaning and adjustment should be carried out.

Distributor.—Occasionally remove the distributor moulding by pushing aside its two securing springs. See that the electrodes are clean and free from deposit. If necessary, wipe out the distributor with a dry duster and clean the electrodes with a cloth moistened with petrol. See that the carbon brush is clean. Clean the outside of the moulding, particularly the spaces between the terminals. Next examine the contact breaker; it is important that the contacts are kept free from any grease or oil. If they are burned or blackened, they may be cleaned with very fine emery cloth and afterwards with a cloth moistened with petrol. Care must be taken that all particles of dirt and metal dust are wiped away. Misfiring may be caused if the contacts are not kept clean.

Provided that the cam is kept clean and that the instructions on cam lubrication are carried out, the contact breaker gap will only need adjustment at very long intervals. It is not advisable to alter the setting unless the gap varies considerably from the gauge. If adjustment is necessary, proceed as follows:—When the contacts are fully opened, slacken the locking nut on the stationary contact screw, and rotate it by its hexagon head until the gap is set to the thickness of the gauge. After making the adjustment, care must be taken to tighten the locking nut.

LUBRICATION.

(1) Distributor Shaft.—Add one or two drops of thin machine oil through the oiler provided about every 1,000 miles.

(2) Cam.—About every 3,000 miles, give the cam the slightest smear of vaseline.

(3) Automatic Timing Control.—About every 3,000 miles withdraw the rotating arm from the top of the spindle by lifting it off, and add a few drops of thin machine oil. Do not remove the screw exposed to view, as there is a clearance between the screw and the inner face of the spindle through which the oil passes to lubricate the automatic timing control.

Coil.—The coil unit is not adjustable in any way, and requires no attention beyond seeing that the terminal connections are kept tight, and the moulded coil top is kept clean.

Warning Lamp.—A warning lamp is provided in the instrument panel, which gives a red light when the ignition is "ON" and the car is stationary. The warning lamp will also light when the engine is running very slowly, due to the fact that the dynamo is not running at sufficient speed to generate a high enough voltage to actuate the cut-out.

THE DETECTION AND REMEDY OF IGNITION FAULTS.

If a failure of ignition or misfiring occurs, unless the cause is at once apparent the owner is strongly recommended to proceed in accordance with the following routine, which should quickly enable him to locate the trouble.

Before proceeding with the examination, make sure that the trouble is not due to defects in the engine, carburetter, petrol supply, sparking plugs, etc.

Engine will not Fire.—Switch on the ignition, turn the engine and observe the ammeter reading. The engine should be turned by hand if it is known that the battery is in a low state of charge.

If an ammeter reading is given which rises and falls with the closing and opening of the contacts, then the low-tension wiring is in order. If the reading does not fluctuate in this way, a short in the low-tension wiring is indicated, or the contacts are remaining closed. When no reading is given, a broken or loose connection in the low-tension wiring is indicated, or the battery may be exhausted.

Examine the high-tension cables, i.e., cables from the coil to the distributor, and from the distributor to the plugs. If the rubber shows signs of deterioration or cracking, the cable should be renewed. Remove the distributor moulding and examine the contacts; clean if necessary. Turn the engine over by hand, and see that the contacts come together.

If a fault is indicated in the low tension wiring, examine the cables from the switch or junction box to coil, and from coil to distributor. See that the battery terminals are tight and that the cables from the switch-box to the battery are secure. The battery may be dismissed as the cause of the trouble if the lamps will light.

Test the coil independently of the distributor as follows: Remove the cable from the centre distributor terminal, and hold it about ¼ in. from some metal part of the chassis and turn the engine. The sparking should be strong and regular if the coil is functioning correctly.

Misfiring and Bad Starting.—Examine the high-tension cables and the plugs. If necessary, adjust the gaps to the correct setting (about 20 thousandths of an inch). Sooty or oiled plugs may be dismantled and washed out with petrol.

The plugs and high-tension cables may be tested by removing the plugs in turn and allowing them to rest on the cylinder head and observing whether a spark occurs at the points when the engine is turned by hand. It should, however, be noted that this is only a rough test, since it is possible that a spark may not take place when the plug is under compression.

Remove the distributor moulding and see that the electrodes and contacts are clean. See that the contact gap setting is correct.

If after carrying out the examination suggested, the trouble cannot be found, we advise that the equipment should be examined by the nearest Lucas Service Depôt.

ELECTRICAL EQUIPMENT
(K N & N Model)

The electrical equipment of the modern motor-car is often and quite erroneously regarded as something very complex and difficult. Yet in the present day motor-car simple application of thoroughly tried out ideas has made possible easy adjustment and maintenance of practically all the electrically operated parts. In this section an endeavour has been made to explain the various instruments and the best way to set about adjustments.

The wiring diagram may at first appear to be a little complicated, but it shows clearly exactly where each individual cable goes and the manner in which they are carried from point to point.

Wiring is arranged on the earth return system, and particular care has been taken to ensure that good continuity obtains at all points of the circuit, by the inclusion, where necessary, of independent earthing cables. It will also be noticed that cables at all essential points are encased in flexible metallic armouring. This not only protects the cables from possible abrasion, but excludes road dirt and water, prevents premature wear and makes for long and trouble-free service.

INSTRUMENT PANEL.

The panel—see Fig. 21—contains the control switch for the dynamo, ignition and lights, a centre zero ammeter and various other items, all of which are clearly marked. A word of explanation on the dual function of the ignition warning light may prove of assistance. Its obvious object is to warn the owner that the ignition switch is " on " and the battery is being discharged, but, being in effect " shunted " across the cut-out points, it will allow only a very small current of electricity to flow through the armature and field coils of the dynamo, thus giving the dynamo an initial excitation. A resistance is always included in the warning lamp circuit to prevent the voltage rise of the dynamo burning out the bulb, and in the light under consideration the resistance is of sufficient value to permit the use of a small 2½-volt bulb of considerably longer life than the earlier heavy consumption type.

Cut-out and Junction Box.—A good view of this component can be had by referring to Fig. 10. All cables are taken to terminal posts in a recessed space at the back of the box, and it is here that the two main harnessed cables—the panel harness associated with the control panel and the junction box harness which constitutes the cables going to the various points on the chassis—are coupled up. The fuses, reading from left to right, are auxiliary, near-side head, off-side head, side and tail, and horn. Above will be seen the dynamo field fuse. These are so arranged that should

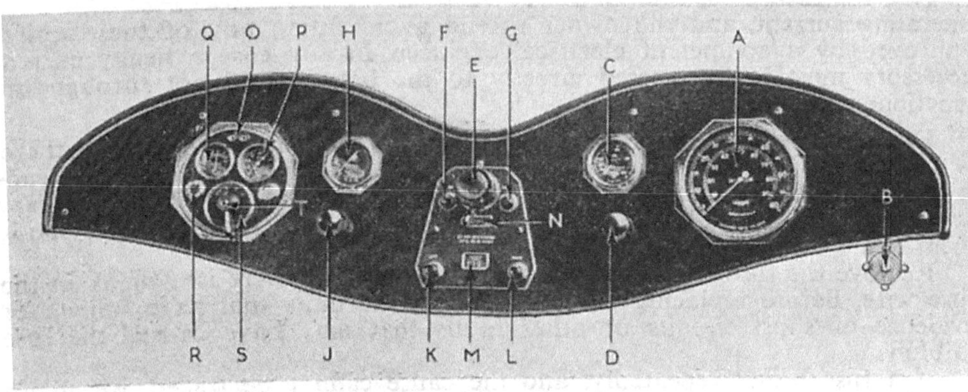

Fig. 21. *The facia board, N Magnette. (a) revolution counter. (b) petrol tap. (c) oil thermo. (d) dashlamp. (e) horn arm and dipper switch. (f) foglamp switch (spare). (g) dashlamp switch. (h) petrol gauge. (j) dashlamp. (k) trafficator n/s. (l) trafficator o/s. (m) mileometer (trip). (n) mileometer. (o) plug points. (p) oil gauge. (q) ammeter. (r) ignition warning light. (s) lamp and dynamo switch. (t) ignition switch.*

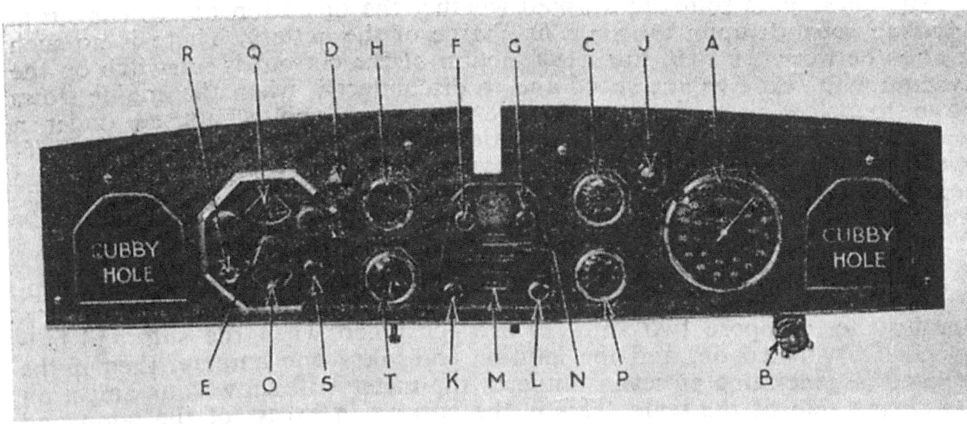

Fig. 21a. *Facia board. KN Magnette (a) revolution counter. (b) petrol tap. (c) oil thermo. (d) dashlamp. (e) ignition and locking switch. (f) foglamp switch. (g) dashlamp switch. (h) petrol gauge. (j) dashlamp switch. (k and l) trafficator n/s., & o/s. (m) mileometer (trip). (n) mileometer. (o) light and charging switch. (p) oil gauge. (q) ammeter. (r) ignition warning light. (s) starter. (t) clock.*

a short-circuit develop in any lamp or other component, the fuse associated with that component only will blow and the remainder of the installation will remain normal. The operation of tracing the fault is simplified enormously as it is at once apparent which circuit is at fault. It should be noted that the field fuse is only of 4.5 amperes.

On the left of the fuse block the spare fuse carrier is provided with one complete replacement set. At the side of the fuse block a pair of terminal posts are available for coupling up any other instruments fitted subsequently. This pair is fused in the auxiliary circuit and is intended for additional horns, foglamps, etc. It will perhaps be helpful to note here that there are many accessories on the market, some of which take a very high operating current, and the owner should avoid fitting any of these unless approved by a competent electrical engineer. In any case a heavy current accessory must be connected directly to the battery and not through the junction box or ammeter.

Fuses.—It will be noticed that the fusing of the lamps is such that there is no risk of the driver being plunged into total darkness. If both the head-lamps, or the side and tail-lamps, or all the units connected to the auxiliary accessory terminal fail to function, examine the particular fuse protecting them.

Remove the fuse from its holder and see whether there is a break in the fuse wire. Before replacing the fuse, inspect the units that have failed, for evidence of short circuits or other faults that may have caused the fuse to blow.

If a fuse blows repeatedly, and the cause cannot be traced, we advise that the equipment is examined by a Lucas Service Station.

The Cut-out.—The function of the cut-out is to close the charging circuit, as the increased engine speed when the car is starting causes the dynamo voltage to rise above that of the battery. When the engine slows down, the dynamo voltage falls below that of the battery, and the reverse action takes place, i.e., the cut-out opens and thereby prevents the battery from discharging itself through the dynamo.

The question is sometimes asked whether the operation of the cut-out in any way depends upon the state of charge of the battery. There is no such relation between the two; the sole function of the cut-out is to switch on the dynamo with rising engine speed and to disconnect it when the engine slows down to below a certain speed. The cut-out, which is housed under a separate cover at the top of the junction box, see **Fig. 10**, is entirely self-operating and is accurately adjusted by the makers before fitting. It should operate for a considerable period without attention, and because considerable damage to the entire installation could be done by anyone attempting to vary the setting it is sealed by the manufacturers.

Ammeter.—The centre-zero ammeter which is incorporated in the instrument panel indicates the actual current flowing into or out of the battery. For instance, suppose two amperes are consumed when the side and tail-lamps are switched on, and the ignition coil takes one ampere, then if the dynamo is generating at seven amperes the meter will show four amps. on the charge side of the scale. This is the current in excess of the lamp and ignition load which is available for battery charging purposes.

DYNAMO:—The dynamo is mounted on a platform at the front end of the engine and forms part of the overhead camshaft drive. Two views of the dynamo can be seen in Fig. 5. On the left-hand side it will be seen that the pinion is held on to a shaft by a nut and tab washer which registers in the hole in the pinion. The dynamo is what is known as a simple shunt machine and has been designed to stand up to the very high revolutions per minute and acceleration movement. Beyond an occasional inspection of the brush gear it requires but little attention. The brush gear can be examined by slackening off the screw holding the cover in position. The cover is shown in Fig. 5 above the dynamo.

Brushes.—It is very important to make sure that the three brushes work freely in their holders. This can easily be ascertained by holding back the spring and gently pulling each flexible lead, when the brush should move without the slightest suggestion of sluggishness. The brushes should be clean and should " bed " over the whole of their working surface; that is to say, the face of contact with the commutator should appear uniformly polished Dirty brushes may be cleaned with a cloth moistened with petrol.

If any of the brushes become so badly worn that it is necessary to replace them, this is accomplished by releasing the brush lead eyelet by removal of the screw, then, while holding the spring lever back out of the way, withdrawing the brush from its holder. The new brush can then be fitted by reversing the operation.

When ordering new brush replacements state whether they are main or control brushes, and for what type of machine they are required, also chassis and engine numbers.

The brush springs should be inspected occasionally to see that they have sufficient tension to keep the brushes firmly pressed against the commutator when the machine is running. It is particularly necessary to keep this in mind when the brushes have been in use for a long time and are very much worn down.

Owners are cautioned that it is unwise to insert brushes of a grade other than that supplied with the machine, or to change the tension springs. The arrangement provided has been made only after many years' experience, and will be found to give the best results.

Commutator.—The surface of the commutator should be kept clean and free from oil and brush dust, etc.; neglect of this precaution will result in the commutator becoming blackened, causing sparking to occur at the brushes, and consequent shortening of the life of the machine. The best way to clean the commutator is to insert a fine duster, held by means of a suitably shaped piece of wood, against the commutator surface, slowly rotating the armature by the starting handle at the same time.

If the commutator has been neglected for long periods, it may need cleaning with fine glass paper, but this is more difficult to do, and should not be necessary if it has received regular attention. The grooves between the commutator segments should be examined occasionally and any deposit of copper or carbon dust may be cleaned out by means of a thin saw blade or similar article. The owner is strongly advised to entrust this operation to a fully qualified electrician, as serious damage can unconsciously be done by anyone not familiar with this class of work.

Lubrication.—As the bearings are packed with grease before leaving the Works, very little attention is needed. A few drops of oil, however, may be added through the lubricators provided, say, every 1,000 miles. The reader is cautioned that far more trouble has been caused by excessive oiling than by too little. (Every 20,000 miles the dynamo should be removed, cleaned and adjusted and the bearings repacked with grease. This should be entrusted to the nearest Lucas Service Depôt.)

Dynamo Field Fuse.—A fuse is provided in the dynamo field circuit to protect the machine in the event of anything being wrong in the charging circuit, e.g., a loose or broken battery connection. If the dynamo fails to charge the battery at any time (indicated by the ammeter giving a discharge reading during daytime running), inspect the fuse and, if it has blown, replace it. If the new fuse blows after starting up, the cause of the trouble must be found, and we advise that the equipment is examined by a Lucas Service Station. Replacement fuses must be of the same size as those originally fitted.

Removing the Dynamo.—In order to obtain easy access to the dynamo for removal it is desirable, although not absolutely necessary, to take off the radiator. Remove the lock nuts and nuts from the radiator bolts carried through the engine nose piece extension, and the bolts retaining the yokes of the bonnet tie rods. Also, after emptying the water system, remove the top and bottom water connections. It is unnecessary to interfere with any of the oiling system to remove dynamo.

Between the projecting portion of the cylinder head and the top of the dynamo will be found a circular flexible coupling. Remove the nuts on each of the four coupling bolts in turn, leaving the bolts in position, so that the coupling can be rotated by use of the starting handle to bring each bolt into a position where the nut may easily be reached.

Fig. 22. View of dynamo showing withdrawal gear.

Having removed all four nuts, take off the clutch inspection cover and turn the engine by the starting handle until the timing marks on the flywheel are in the centre of the aperture. This will bring the driving yoke on the cylinder head across the engine, and the driving yoke on the dynamo parallel to the engine centre line. The bolts themselves may now be withdrawn, care being taken not to lose the distance washers, which must be replaced in the same position. Removal of the bolts enables the flexible coupling to be withdrawn.

Detach the two cables on the distributor side of the dynamo, noting from which terminal they are removed.

Now unscrew the four set screws which attach the dynamo to its platform at the front of the engine, thus releasing the dynamo. Lift the dynamo approximately a quarter of an inch and tilt it towards the near-side of the car until the driving yoke on the dynamo is just clear of that on the cylinder head. The dynamo can then be tilted forwards and easily withdrawn.

Replacement of the dynamo is effected in the reverse way, but it is necessary to make sure that the engine timing has not been disturbed while the dynamo was removed. Removal of the rectangular cover-plate in the clutch housing should reveal the timing mark on the flywheel for nos. 1 and 6 cylinders, exactly in the centre of the opening. Place the brass packing shims which fit under the dynamo base in position on the dynamo platform, making sure that you replace the same number that you took off. If for any reason the dynamo is replaced by another, it may be necessary to readjust the mesh of the driving gears to obtain silent running by suitable selection of the packing shims used. Turn the dynamo spindle until the timing mark on the dynamo drive gear coincides with the centre line of the dynamo and is at the rear—that is, ready for engagement with the correspondingly marked teeth of the drive gear on the crankshaft. The holes in the dynamo coupling yoke will then be parallel with the engine centre line. Tilting the dynamo towards the near-side of the car and holding the dynamo coupling yoke in this position, insert the drive gear into the opening of the dynamo platform and swing the dynamo backwards and downwards into position. The gears can be felt to be meshing properly if the coupling is slightly oscillated as the dynamo is replaced, but do not overdo the oscillations or you may engage the wrong teeth.

See that the bolts in the dynamo base are coinciding with the holes in the dynamo platform, and then observe if the dynamo coupling yoke is exactly parallel to the engine centre line. If it is not, withdraw the dynamo, reset the coupling yoke and re-insert the dynamo. No difficulty should be experienced in getting the dynamo in position with the correct teeth in mesh, as the distance between one tooth and the next is sufficient to make an appreciable difference to the position of the dynamo coupling yoke, a difference which is immediately discernible.

Having satisfied yourself that the correct gear teeth are in mesh, replace the screws in the dynamo base, taking particular care to tighten them up evenly a partial turn at a time until all are quite tight. Now make a final test. The flywheel mark 1—6 should show exactly in the centre of the inspection cover opening with the distributor rotating arm pointing towards No. 1 cylinder. (This can easily be found by tracing the high-tension lead from No. 1 sparking plug to its junction on the distributor. Removal of the distributor cover should show the distributor arm directly beneath it). The dynamo coupling bolt holes should be exactly fore and aft and at right angles to the coupling yoke on the cylinder head, and No. 1 inlet and exhaust valves should be closed.

If all the foregoing are correct, replace the flexible coupling and coupling bolts, taking care to replace the distance washers in exactly the same position as they were originally. Tighten up the nuts firmly and rotate the engine slowly by hand. If the distance washers are in the correct position, the flexible coupling should run absolutely true. If it does not do so, note where the error is and adjust the distance washers accordingly. Then replace the valve cover and attach the dynamo cables on to their correct terminals.

Dynamo Adjustment.—The output of the dynamo—that is to say, its rate of charge—is controlled by a third brush which is shown quite clearly in Fig. 23. All machines are adjusted to an output of 9 amperes, and any marked variation from this standard usually indicates some other factor at fault which must be cleared before normal conditions are possible for the dynamo. Should, however, the indicated charging rate increase to a value above 10 amperes **with all lights out,** it is necessary to reduce this at once, as a continuation of high output will rapidly destroy the machine. It is inadvisable for anybody who is not an electrician to tamper with this. To reduce the output the third brush or control brush must be moved in a direction opposite to that of rotation. To move the brush it is necessary to slacken the stud marked "Locking Nut" in Fig. 23, and when the adjustment has been satisfactorily effected it is important to make sure that the stud is tightened.

The third brush is to be found on the near-side of the car and on the same side as the small ball covered lubricator. Never omit to tighten up the lock screw after any adjustment has been effected, and don't try to undo or tighten the lock screw with pliers.

Headlamps.—The **L.B.D.150** headlamps manufactured by Messrs. Joseph Lucas, Ltd., are specially designed to give the highest illuminating factor without unduly overloading the electrical installation. The near-side lamp is equipped with a permanently deflected reflector and is always in circuit when the headlamp switch is on. The off-side lamp has a straight focus reflector, and while primarily wired in the same circuit as the other, can also be controlled by the dipping switch. The lamps are carefully focused before leaving the factory so that the near-side lamp will throw a slightly open beam and thus effectively illuminate the road and kerb up to about 40 yards, while the off-side beam is relatively closely focused to throw a penetrating light ahead.

It should be noted that when a bulb is changed it is usually necessary to re-focus the beam. This can be effected by removing the front by slacking

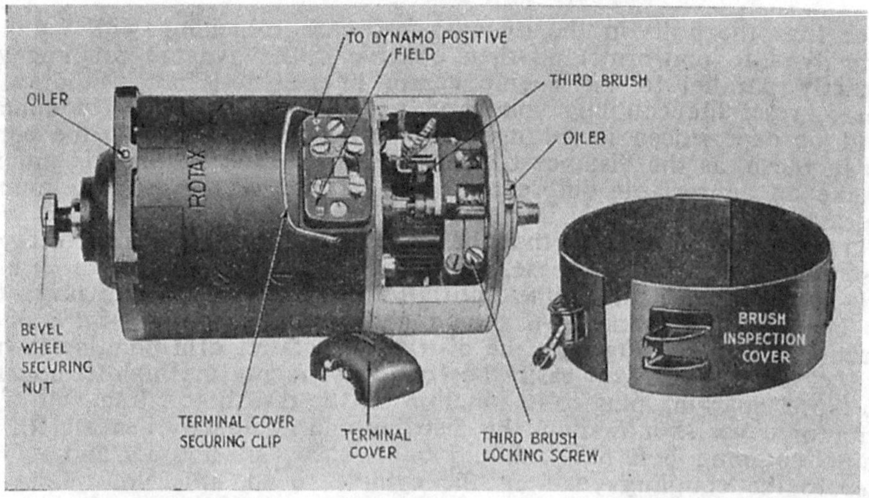

Fig. 23. View of dynamo showing position of third regulating brush and locking screw, and lubricating hole.

off the retaining screw at the bottom of the lamp, and then removing the cork washer, taking care not to damage it, and the reflector which is held in position by one screw, when the small cheese-head screw holding the bulb holder can be loosened. The bulb holder can then be moved either forwards or backwards until the correct beam is obtained. It is essential that the cheese-headed screw should be effectively tightened when the correct focus is obtained, otherwise the light will be erratic.

Wiring Headlamps.—Remove the front and reflector. Then depress the insulating washer on the bulb holder terminal against the spring until the cable hole is exposed. Thread the cable end through the hole, release the pressure on washer, when the cable will be securely held in position and good contact made. Cables should not be pushed too far through the terminal, as there is a danger of shorting on to the reflector or body of the lamp.

Sidelamps.—The fronts of the wing lamps are secured by small grub screws; when these are removed the fronts can be withdrawn by first pulling the top forward.

Replacement of Bulbs.—When the replacement of any bulb is necessary genuine Lucas replacement bulbs only should be used.

Cleaning Lamps.—The reflectors are protected by a transparent and colourless covering, which enables any accidental finger marks to be removed with chamois leather or a soft cloth without affecting the surface of the reflector. Do not use metal polishes on Lucas reflectors. Chromium plated lamps will not tarnish and only need wiping over with a damp cloth to remove dust or dirt.

Electric Starter.—The engagement consists of a coarse threaded shaft on which a pinion is fitted. When the starter button is pressed the starter motor rotates quickly, causing the pinion to be rotated, but owing to the form of this pinion it has a tendency to travel along the starter extension shaft, and thus engage with the teeth of the flywheel, the shock being taken up by a recoil spring.

The moment the engine starts up and the self-starter button is released the flywheel rotation causes the pinion to travel back along the starter shaft out of engagement.

The front end of the self-starter shaft is provided with a square end, and it is possible in the case of only a slight jam to rotate this spindle with a spanner and thus disengage the starter pinion from the flywheel. The above explanation will be made clearer on examining Fig. 13, which shows the starter removed from the engine. It is held in position by two studs and one bolt, and there is a distance plate between the starter housing and the flywheel housing. The flywheel teeth can be seen and also the spiral thread on the extension of the starter shaft, along which the engaging pinion travels to and fro; the recoil or buffer spring is clearly visible.

If, for any reason, the pinion wheel on the motor does not engage with the flywheel teeth, examine the screwed sleeve on the armature spindle to see that it is free from dust; if necessary wash over with paraffin. Occasionally, give it a few drops of thin machine oil.

As in the case of the dynamo, the surface of the commutator must be kept clean and free from oil, brush dust, etc.

The starter is designed for starting the engine under normal conditions, but any unnecessary or additional loading will considerably diminish the life of the machine and battery. In order to facilitate starting in cold weather, it is advisable, before using the electric starter, to crank the engine over slowly by the starting handle for two or three revolutions; this will break the oil film and considerably diminish the load for starting. By pushing out the clutch when using the starter the load of the gearbox is taken off the starter and the engine turns faster and more easily.

In the event of the engine refusing to fire after being turned by the starter, make sure that the ignition switch is " on."

Battery.—It is of the utmost importance that the battery should receive regular attention, as upon its good condition depends the satisfactory functioning of the ignition, starting motor, and the lamps.

At least once a month the vent plugs in the top of the battery should be removed, and the level of the acid solution examined. If necessary, distilled water (which can be obtained at all chemists and most garages) should be added to bring the level well above the plates. If, however, acid solution has been spilled, it should be replaced by a diluted sulphuric acid solution of specific gravity 1.320. It is important, when examining the cells, that naked lights should not be held near the vents, on account of the possible danger of igniting the gas which is generated by the plates. It is advisable to complete the inspection by measuring the specific gravity of the acid, as this gives a very good indication of the state of charge of the battery. An instrument known as a hydrometer is employed for this purpose; these can be bought from your dealer or from any of the Lucas Service Stations.

For the battery the specific gravity readings will be 1.285—1.300 for a fully charged battery, about 1.210 when half discharged, and about 1.150 when fully discharged.

If one cell gives a reading very different from the rest, it may be that electrolyte has been spilled or has leaked from this cell, or there may be a " short " between the plates. In the latter case, the battery should be examined as soon as possible by a Lucas Service Station.

Finally, see that the tops of the cells are clean and dry, and that the terminals are tight and smeared with vaseline.

If the equipment is laid by for several months, the battery must be given a small charge from a separate source of electrical energy about once a month, in order to obviate any permanent injury to the plates.

Under no circumstances must the acid be removed from the battery and the plates allowed to dry, as certain changes take place which result in loss of efficiency.

The battery must never be left in a fully discharged condition, and, unless some long runs are to be taken, it is advisable to have the battery removed from the car periodically and charged up from an independent electrical supply.

Ignition.—Very little attention is needed to keep the ignition equipment in first class condition. We advise that it is inspected occasionally, and the following instructions on lubrication, cleaning and adjustment are carried out.

Lubrication.—The High Tension distributor is very accessible. It is provided with an oiler to lubricate the spindle, a few drops every thousand miles being sufficient.

About every 3,000 miles, give the cam the slightest smear of vaseline. Also withdraw the rotor from the top of the spindle by lifting it off, and add a few drops of thin machine oil. Do not remove the screw exposed to view, as there is a clearance between the screw and the inner face of the spindle through which the oil passes to lubricate the automatic timing control.

Every 5,000 miles, place a single drop of oil on the contract breaker pivot.

The contacts can be inspected by unclipping the two springs when the cover carrying the High Tension Leads can be removed. The points may require adjustment from time to time and the clearance between the points should be 15-18/1,000 of an inch when they are fully open.

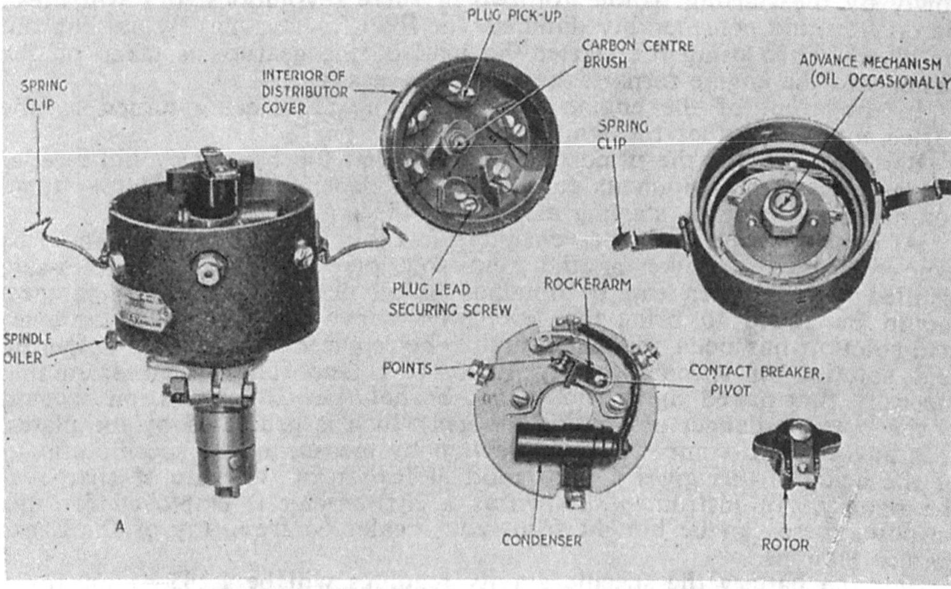

Fig. 24. View of distributor with cover removed.

Provided that the cam is kept clean and that the instructions on cam lubrication are carried out, the contact breaker gap will only need adjustment at very long intervals. It is not advisable to alter the setting unless the gap varies considerably from the gauge supplied. If adjustment is necessary, proceed as follows: When the contacts are fully opened, slacken the locking nut on the stationary contact screw, and rotate it by its hexagon head until the gap is set to the thickness of the gauge. After making the adjustment, care must be taken to tighten the locking nut.

The centre spindle of the High Tension distributor carries a "Bakelite" Arm called the "Rotor." It can only fit on the spindle one way. If this part is removed, it affords an excellent thief proof device, but care should be exercised to see that it does not become chipped or in any way damaged, not to mention lost, after it has been removed. The ends of the high-tension wires are held in position by securing screws from inside the cover. See that the screws are always tight.

Occasionally remove the distributor moulding (by pushing aside its two securing springs). See that the electrodes are clean and free from deposit. If necessary, wipe out the distributor with a dry duster and clean the electrodes with a cloth moistened with petrol. See that the carbon brush is clean and moves freely in its holder. Clean the outside of the moulding, particularly the spaces between the terminals. Next examine the contact breaker; it is important that the contacts are kept free from any grease or oil. If they are burned or blackened, they may be cleaned with very fine emery cloth and afterwards with a cloth moistened with petrol. Care must be taken that all particles of dirt and metal dust are wiped away. Misfiring may be caused by any of the following:—

1. Dirty sparking plug or plugs.
2. Bad connection to High Tension Leads.
3. Bad connection from High Tension Distributor to Coil.
4. Improper adjustment of or burnt make and break points.
5. Dirt between make and break points.
6. Defective Coil.
7. Faulty condenser, causing repeated burning of points.
8. Temporary petrol starvation.

There is one other remote cause of electrical failure, namely, bad earth contacts from the battery to the frame. Other causes of misfiring can be attributed to Carburetter and incorrect valve adjustment. Separate instructions are given on these last two items elsewhere.

Distributor Timing.—It has been found on test that the ignition timing is fairly critical and that on the average the amount of advance required is equal to exactly 1½ inches measured on the flywheel in advance of the top dead centre mark. It will be found that there is a line on the flywheel marked " Ign "—the engine should be turned by means of the starting handle until the mark is in the centre of the clutch inspection aperture, the points of the make and break in the distributor should then be on the point of opening.

In order that the timing is set correctly it is advised that the setting should always be done using as a guide the line of the flywheel.

Coil.—The coil unit is not adjustable in any way, and requires no attention beyond seeing that the terminal connections are kept tight, and the moulded coil top is kept clean.

Warning Lamp.—A warning lamp is provided in the instrument panel, which gives a red light when the ignition is " ON " and the car is stationary. The warning lamp will also light when the engine is running very slowly, due to the fact that the dynamo is not running at sufficient speed to generate a high enough voltage to actuate the cut-out.

THE DETECTION AND REMEDY OF IGNITION FAULTS.

If a failure of ignition or misfiring occurs, unless the cause is at once apparent the owner is strongly recommended to proceed in accordance with the following routine which should quickly enable him to locate the trouble.

Before proceeding with the examination, make sure that the trouble is not due to defects in the engine, carburetter, petrol supply, sparking plugs, etc.

Engine will not Fire.—Switch on the ignition, turn the engine and observe the ammeter reading. The engine should be turned by hand if it is known that the battery is in a low state of charge.

If an ammeter reading is given which rises and falls with the closing and opening of the contacts, then the low-tension wiring is in order. If the reading does not fluctuate in this way, a short in the low-tension wiring is indicated, or the contacts are remaining closed. When no reading is given, a broken or loose connection in the low-tension wiring is indicated, or the battery may be exhausted.

Examine the high-tension cables, i.e., cables from the coil to the distributor, and from the distributor to the plugs. If the rubber shows signs of deterioration or cracking, the cable should be renewed. Remove the distributor moulding and examine the contacts, cleaning if necessary. Turn the engine over by hand and see that the contacts come together.

If a fault is indicated in the low tension wiring, examine the cables from the switch or junction box to coil, and from coil to distributor. See that the battery terminals are tight and that the cables from the switch-box to the battery are secure. The battery may be dismissed as the cause of the trouble if the lamps will light.

Test the coil independently of the distributor as follows: Remove the cable from the centre distributor terminal, and hold it about ¼ in. from some metal part of the chassis and turn the engine. The sparking should be strong and regular if the coil is functioning correctly.

Misfiring and Bad Starting.—Examine the high tension cables and the plugs. If necessary, adjust the gaps to the correct setting (about 18 thousandths of an inch). Sooty or oiled plugs may be dismantled and washed out with petrol.

The plugs and high tension cables may be tested by removing the plugs in turn and allowing them to rest on the cylinder head and observing whether a spark occurs at the points when the engine is turned by hand. It should, however, be noted that this is only a rough test, since it is possible that a spark may not take place when the plug is under compression.

If after carrying out the examination suggested, the trouble cannot be found, we advise that the equipment should be examined by the nearest Lucas Service Depôt.

SPARKING PLUGS.

The correct gap is .018 in., and no advantage will be gained by either a wider or a closer gap. A definite mileage when cleaning will be necessary but cannot be specified, as there are a number of governing factors. It is, however. wise to leave well alone if there is no indication of trouble in that direction.

Plugs are easily ruined by inexperienced handling. The only satisfactory way of ascertaining whether a plug is sparking correctly is to test it under compression. Most service stations have a "plug tester," and we recommend that M.G. owners have their plugs checked in this way periodically.

A good plan is to keep a spare set of plugs and if plug trouble is suspected to change over to the spares. This allows plenty of time to have the others attended to as suggested.

ELECTRICAL EQUIPMENT
(TA Model)

THE BATTERIES.

About once a month unscrew the filler caps and pour a small quantity of distilled water into each of the cells to bring the acid just level with the tops of the separators. This is done to replace acid solution which has been lost by evaporation. If any of the acid solution is spilled it must be replaced by a dilute sulphuric acid solution of the same specific gravity as the electrolyte in the cell to which it is to be added. The batteries must not be allowed to remain with the plates uncovered as they will suffer a chemical change and be irretrievably damaged.

Keep the Terminals and Battery Fixing Bolts Clean and Tight and Smeared with Vaseline.—A smearing of vaseline protects the terminals and fixing bolts from the corrosive action of the acid, which, if allowed to continue, will eventually result in trouble. When inspecting the batteries, put a spanner on the bolt securing the cable end to the chassis in order to make sure that it is quite tight.

Keep the Outside of the Batteries Clean and Dry, particularly the Tops of the Cells.—Water and dirt form a conductor of electricity and if such a path is allowed to form between the positive and negative terminals of the battery or between the negative battery terminal and the chassis, there will be a leakage of current which will eventually cause the battery to run down. Give the cell tops a wipe over regularly and you will avoid this.

Occasionally make a point of Examining the Health of your Batteries by taking Hydrometer Readings.—The hydrometer allows you to take a sample of the acid solution. It contains a graduated float which indicates the specific gravity of the acid solution in the cell from which the sample was taken.

The readings and their indications are as follows:—

1.250—1.300 ... ...
Battery fully charged.
1.150—1.250 ... ...
Battery about half discharged.
Below 1.150 ... ...
Battery fully discharged.

The readings for each of the cells should be approximately the same. If one cell gives a very different reading from the rest it may be that acid has been spilled or has leaked from this particular cell, or there may be a short circuit between the plates. In this case we advise the owner to have his battery examined by a Lucas Service Depôt—neglect may mean costly repairs later on.

It must be remembered that you cannot get more energy out of a battery than is put in. If the car is left parked at night with the lights on, or if the starter motor is used excessively with very little daytime running, then the battery may get into a low state of charge—particularly in winter. This may be remedied either by running the car for longer periods during daytime or by economising in the use of the headlamps and other accessories.

Never leave the Batteries in a Discharged Condition for any length of time.—If the car is to be out of use for any length of time, see that the batteries are fully charged and about every fortnight give them a short refreshing charge to prevent any tendency to permanent sulphation of the plates.

THE DYNAMO (LUBRICATION).

All dynamos are sent out from the works with the bearings packed with grease. This lasts until the car is taken down for a general overhaul, when it is advisable to have the machine dismantled, preferably by a Lucas Service Depôt, for cleaning and adjustment and repacking the bearings with grease.

The dynamo is provided with oilers to enable you to give it a little additional lubrication. The oilers should be given several drops of good quality engine oil every 1,000 miles. It is of the utmost importance that the oiler on top of the lubricator shown in Fig. 25 receives special attention or this bearing will run dry.

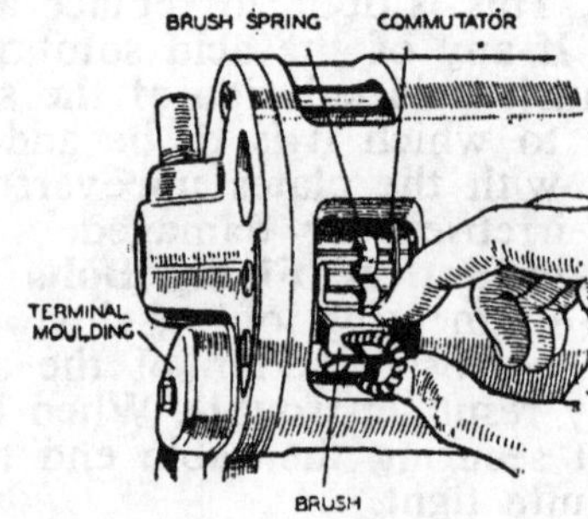

Fig. 25. Removing the Brushes (TA, TB, TC, 1½, 2 and 2.6 Litre Models).

Inspection of Commutator and Brushgear.—About once a season remove the metal cover from the dynamo for inspection of the commutator and the carbon contacts, or brushes as they are called. Take care not to lose the nut when removing the cover fixing screw, as the cover is liable to spring open when the screw is released.

It is essential that the brushes make good firm contact with the commutator. The brushes are held in boxes by means of springs. Hold back the spring, and at the same time move the brush to see that it is free to slide in its holder. If there are any signs of sticking, remove it from its box and clean it with a cloth moistened with petrol.

After removing brushes for cleaning or any other purpose, care must be taken to replace them in their original positions, otherwise they will not "bed" properly on the commutator.

If, after long service, the brushes have become worn to such an extent that they will not "bed" properly on the commutator, they should be replaced. Always fit genuine Lucas brushes, as these are made specifically for work on our machines and will give by far the best results and the longest life. We advise you to have the brushes fitted at a Lucas Service Depôt, so that they can be properly "bedded" to the commutator.

Next examine the commutator. It should be clean and free from any trace of oil or dirt, and should have a highly polished appearance. The best way to clean a dirty or blackened commutator is by pressing against it a fine dry duster and getting someone to turn the engine over slowly by hand. If the commutator is very dirty, the duster may be moistened with petrol.

Dynamo Field Fuse.—A fuse is provided in the dynamo field circuit and will blow in the event of anything being wrong in the charging circuit, e.g., the battery becoming disconnected from the dynamo due to broken wiring or a loose connection. The fuse is located in the cut-out and fuse unit on the engine side of the dash.

DYNAMO OUTPUT.

The dynamo output is accurately set before leaving the Works to suit the requirements of the equipment fitted on your car, and in normal service the battery will be kept in good condition. If very special running conditions are encountered, however, you may find that the battery is not kept in a charged condition, or is being excessively overcharged; in which case we advise you to consult your nearest Lucas Service Depot, where any necessary adjustments can be made.

AMMETER READINGS.

The ammeter, which forms one of the instruments on the dash, indicates the amount of current passing into or out of the battery. For instance, suppose the dynamo is generating 7 amperes at a particular speed and that some of the lamps are in use, taking, together with the ignition coil, say 5 amperes, then 2 amperes are left for charging the battery—this is the reading given on the ammeter.

THE CUT-OUT—AN AUTOMATIC DYNAMO SWITCH.

It will be noticed from the ammeter readings that the dynamo does not charge at very low engine speeds. This is because it is not rotating fast enough to generate sufficient energy to charge the battery.

Connected between the dynamo and the battery is the cut-out—an automatic switch which acts as a "valve" allowing the flow of current from the dynamo to the battery only. It closes when the dynamo is running fast enough to charge the battery, and opens when the speed is low or the engine is stationary, thus preventing current from flowing from the battery through the dynamo windings.

The cut-out is located along with the fuses on the engine side of the dash.

THE STARTER.

When Starting.—Observe the following points when starting the engine:—

(i) Operate the starter switch firmly, and, of course, release it as soon as the engine fires.

(ii) Never operate the starter when the engine is running. If the engine will not fire at once, allow it to come to rest before pressing the switch again.

(iii) Do not run the battery down by keeping the starter on when the engine will not start.

(iv) In cold weather, starting will be facilitated by giving the engine a few turns by the starting handle before using the starter.

Attention Needed by the Starter.—Give the commutator and brush gear similar attention to that described for the dynamo.

Occasionally, if dirty, wash over the screwed sleeve on the armature spindle with paraffin and afterwards give it the merest trace of thin machine oil.

INSTRUMENT PANEL.

The instrument panel houses the combined lighting and ignition switch, ammeter, starter switch control, etc.

HOW TO USE THE CHARGING SWITCH.

This switch has two positions—" High Charge " and " Low Charge." The dynamo is arranged to give its maximum output whenever the head-lamps are switched on. For cars running under average conditions, keep the switch in the " High Charge " position during winter, and in the " Low Charge " position during summer.

In some cases, where exceptional use is made of the lamps and starter in the summer, causing the battery to be in a low state of charge (hydrometer readings of 1.200 or under), run with the charging switch in the " High Charge " position. On the other hand, if in winter the car is run regularly during the day, with practically no night running, and the hydrometer readings are always found to be about 1.285, and if the acid level gets unusually low, then it is probable that the battery is being overcharged. In these circumstances, move the charging switch to the " Low Charge " position.

What to do if the Battery does not keep in a Charged Condition.—If the battery is found to be in a half discharged or low state of charge, the charging switch should, if possible, be left in the " High Charge " position for longer periods of running. It should be remembered that the battery will be helped to regain its normal condition if its load is temporarily lessened, as, for instance, by using the side instead of the headlamps. If the condition of the battery does not improve in a reasonable time, it is advisable to have the battery inspected at a Lucas Service Depôt.

The battery must never be left in a fully discharged condition, and unless some long runs are to be taken, it is advisable to have the battery charged up from an independent electrical supply.

CUT-OUT AND FUSE UNIT.

This unit houses the cut-out and six fuses, and forms a junction box for distributing the cables to the various accessories.

The fuses incorporated in this unit protect the dynamo, the head, side and tail-lamps, the electric horn and the auxiliary accessories (such as the wiper).

A blown fuse will be indicated by the failure of all the units protected by it, and is confirmed by examination of the fuse. If it has blown, the broken ends of the wire will be visible inside the glass tube. Before replacing a blown fuse, inspect the wiring of the units that have failed for evidence of short circuits or other faults that may have caused the fuse to blow, and remedy.

Spare fuses are provided, and it is important to use only the correct replacement fuse—the fusing value is marked on a coloured paper slip inside the glass tube.

Fig. 26. *Cut-out, Fuse and Junction box* (*TA*, *TB*, 1½, 2 *and* 2.6 *Litre commencing at chassis no. S.A.* 903).

LAMPS. HEADLAMPS.

Removing Lamp Front and Reflector.—To remove the lamp front, slacken the single securing screw and move it downwards from the slot in which it fits. When replacing the front locate the top of the rim first, then press on the bottom and tighten the fixing screw. See Fig. 27.

To remove the reflector, turn back the ends of the cork washer at the top of the rim and remove the screw opposite the medallion in the top of the lamp. Turn reflector until markings " O " stamped on the reflector rim and lamp body coincide. The reflector can then be withdrawn. When replacing reflector engage it with the lamp body, then turn it until the screw hole in its rim is opposite to the screw hole on top of lamp body. Secure reflector by means of screw.

Checking the Alignment.—The simplest way of checking the adjustment of the lamps is to take the car on a straight level stretch of road at night and examine the direction of the beams. They should be parallel with the road and with each other. If one appears to be out of alignment, adjust it as follows:—Slacken the single fixing nut at the base of the lamp under the wing and move the lamp on its universal mounting to the required position, finally locking the adjustment by tightening the fixing nut.

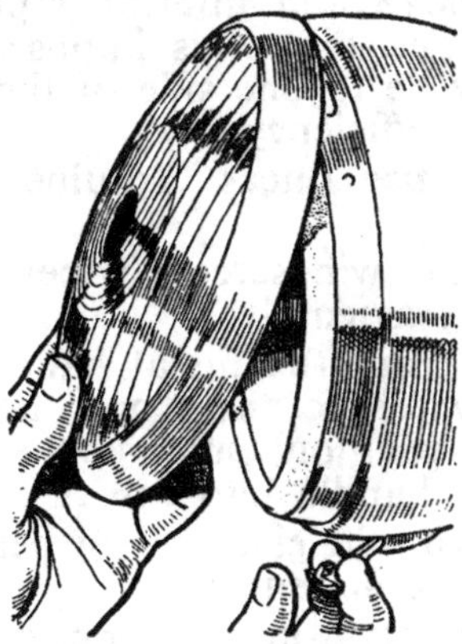

Fig. 27. *Showing method of removing headlamp front* (*TA, TB, TC,* 1½, 2 *and* 2.6 *Litre models*).

Focusing.—For the lamps to give a parallel beam, the filament of the bulb must coincide as nearly as possible with the focal point of the reflector. If the filament of the bulb is behind the focal point of the reflector, the beam will be divergent, while on the other hand, if the filament is in front of the focal point, the beam will be convergent, with a dark area in the centre of the beam. In either case, the lamps will have a poor range and will cause unnecessary dazzle to approaching traffic.

Before lamps are dispatched from the Works, the bulbs are carefully focused to give the best results. Provided that official Lucas bulbs are fitted as replacements, it should not be necessary to disturb the setting. If for any reason these are not obtainable, and ordinary bulbs have to be fitted, it may be necessary to refocus by moving the bulb backwards or forwards along the axis of the reflector until the best lighting is obtained.

Cover up one lamp while testing the other. If the lamp does not give a uniform long-range beam without any dark centre, the bulb needs adjusting. To do this remove the front and reflector, and slacken the clamping clip at the back of the reflector.

After each adjustment, note the effect with the reflector and front refitted.

When the best position for the bulb holder has been found, see that the clamping screw is tightened.

SIDELAMPS.

The front of the lamp is removed for bulb replacement when the screw at the top of the lamp is withdrawn. When refitting the front, locate the bottom and secure with the screw.

STOP TAIL-LAMP.

The lamp front can be removed when the securing screw is withdrawn

FOGLAMP.

To remove the lamp front withdraw the screw from the fixing band, which can then be sprung off the lamp. Remove rubber band and lift off lamp glass. To withdraw the bulb, it is first necessary to remove the shield. The lamp can be focused in the same manner as that described for the headlamps.

REPLACEMENT OF BULBS.

When the replacement of a bulb is necessary, it is important not only that the same size bulb is fitted, but that it has a high efficiency and will focus in the reflector. Cheap and inferior replacement bulbs often have the filament of such a shape that it is impossible to focus correctly; for example, the filament may be to one side of the axis of the bulb, resulting in loss of range and light efficiency.

It always pays you to use Lucas Genuine Spare Bulbs as then these problems will not arise.

Windscreen Wiper: The windscreen wiper is carefully adjusted and packed with grease during assembly.

To start the wiper pull out the metal handle and turn to disengage it from the switch. Move switch to the " on " position. To stop the wiper, move switch to the " off " position, pull out the handle to disengage wiper blade from gears and turn handle into top of switch control.

Electric Horn: The horn will give a long period of service without any attention and no adjustment is required.

If the horn becomes uncertain in its action, giving only a choking sound, or ceases to vibrate, ensure first that the trouble is not due to a discharged battery, a loose connection, a short-circuit in the wiring of the horn, or a blown fuse. The performance of the horn can be upset by the horn becoming loose on its mounting. If the trouble cannot be traced return the instrument to a Lucas Service Depôt for examination.

Coil Ignition Equipment: The attention to ensure absolutely reliable service is very small.

Periodically check the contact breaker points of the distributor and if necessary reset to .012 in. and smear the cam with clean engine oil.

Lubrication: The following parts of the distributor require lubrication every 3,000 miles:—

1. **Cam.**—Give the cam a light smear of engine oil.
2. **Cam Bearing.**—Remove the moulded cap, pull off the moulded rotating arm from the top of the spindle and add a few drops of oil through the lubricator hole provided in the top of the spindle. Ensure the arm is refitted correctly, pushing on the shaft as far as possible to prevent possibility of tracking and burning of the moulded cap.
3. **Contact Breaker Pivot.**—Place a spot of oil on the pivot on which the contact breaker rocker-arm works.
4. **Automatic Advance Mechanism.**—Add a few drops of thin machine oil through the hole in the contact breaker base through which the cam passes.

Cleaning: Keep the outside of the distributor clean, particularly spaces between high-tension terminals. Occasionally remove the moulding, wipe the inside clean with a dry cloth and see that the carbon brush is quite free in its holder. Clean the metal electrodes inside the moulding and also the rotating electrode on the distributor arm, using, if necessary, a cloth moistened with a drop of petrol.

Next, examine the contact breaker; you must keep the contacts free from any grease or oil. If they are burned or blackened, clean them with fine carborundum stone, or, if this is not available, you can use very fine emery cloth. Finish off with a cloth moistened with petrol, and remove all traces of dirt and metal dust. Misfiring is sometimes caused by dirty contacts.

CHECKING AND ADJUSTING THE CONTACTS.

After the contacts are reset at the end of 500 miles' running they require only occasional adjustment. The chief cause of variation in the gap is wear of the heel of the contact rocker-arm which bears upon the actuating cam. Provided you keep the cam smeared with engine oil, however (see Lubrication Instructions on previous page), the wear on the heel will be negligible and the contact gap setting should only require adjustments at infrequent intervals.

To check the setting, turn the engine by hand until the contacts are fully opened. Now insert the gauge, provided on the ignition screwdriver, between the contacts. The gauge has a thickness of about 12 thousandths of an inch and it should be a sliding fit between the contacts when the gap is correct. We do not advise you to alter the setting unless there is quite an appreciable variation from the gauge. To make the adjustment, keep the engine in the position to give maximum opening of the contacts and slacken the two screws securing the contact plate. Then move the plate until the gap is set to the thickness of the gauge. After making the adjustment, care must be taken to tighten the locking screws.

RENEWING THE HIGH-TENSION CABLE.

The high-tension cables are those connecting the coil to the distributor and the distributor to the sparking plugs. When these cables show signs of perishing or cracking they must be replaced by 7 mm. rubber-covered ignition cable.

The method of connecting the cable is to thread the knurled moulded nut over the lead, bare the end of the cable for about ¼ in., thread the wire through the brass washer provided and bend back the strands. Finally screw the nut into its respective terminal.

THE COIL.

The coil requires no attention whatever beyond keeping its exterior clean, particularly between the terminals, and occasionally checking that the terminal connections are quite tight.

IGNITION SWITCH AND WARNING LIGHT.

The ignition switch, besides forming a means of stopping the engine, is provided for the purpose of preventing the battery being discharged by the current flowing through the coil windings when the engine is stopped. A warning lamp is usually provided in the instrument panel which gives a red light when the ignition is switched on and the car is running very slowly or is stationary, thus reminding you to switch off.

Should the warning lamp bulb burn out, this will not in any way affect the ignition system, but you should replace it as soon as possible in order to safeguard your battery.

The bulb used is a No. 252MES.

Micrometer Timing Adjustment.—In order to obtain very fine timing of the ignition to the engine and to allow for altered engine conditions, e.g., state of carbonisation of engine, change of fuel, etc., a micrometer adjustment is provided which allows fine adjustment to be made simply by the movement of a knurled knob.

With a clean engine, and using first grade fuel, the micrometer scale should be at "0" before timing.

The final setting can be made by use of the micrometer adjustment after running the engine. If the firing is found to be slightly too early or too late, adjust the knurled knob until the best engine performance is obtained. The adjustment should not be altered more than one distributor degree at a time (one division on the scale is equivalent to two distributor degrees or 4° on the crankshaft).

ELECTRICAL EQUIPMENT
(TB & TC Models)

THE BATTERY.

About once a month unscrew the filler caps and pour a small quantity of distilled water into each of the cells to bring the acid just level with the tops of the separators. This is done to replace acid solution which has been lost by evaporation. If any of the acid solution is spilled it must be replaced by a dilute sulphuric acid solution of the same specific gravity as the electrolyte in the cell to which it is to be added. The batteries must not be allowed to remain with the plates uncovered as they will suffer a chemical change and be irretrievably damaged.

Keep the Outside of the Battery Clean and Dry, particularly the Tops of the Cells.—Water and dirt form a conductor of electricity and if such a path is allowed to form between the positive and negative terminals of the battery or between the negative battery terminal and the chassis, there will be a leakage of current which will eventually cause the battery to run down. Give the cell tops a wipe over regularly and you will avoid this.

Occasionally make a point of Examining the Health of your Battery by taking Hydrometer Readings.—The hydrometer allows you to take a sample of the acid solution. It contains a graduated float which indicates the specific gravity of the acid solution in the cell from which the sample was taken.

The readings and their indications are as follows:—

1.280—1.300 ... ... Battery fully charged.

About 1.210 ... ... Battery about half discharged.

Below 1.150 ... ... Battery fully discharged.

The readings for each of the cells should be approximately the same. If one cell gives a very different reading from the rest it may be that acid has been spilled or has leaked from this particular cell, or there may be a short circuit between the plates. In this case we advise the owner to have his battery examined by a Lucas Service Depôt—neglect may mean costly repairs later on.

Never leave the Battery in a Discharged Condition for any length of time.—If the car is to be out of use for any length of time, see that the battery is fully charged and about every fortnight give it a short refreshing charge to prevent any tendency to permanent sulphation of the plates.

THE DYNAMO (LUBRICATION).

All dynamos are sent out from the works with the bearings packed with grease. This lasts until the car is taken down for a general overhaul, when it is advisable to have the machine dismantled, preferably by a Lucas Service Depôt, for cleaning and adjustment and repacking with grease.

About every 12,000 miles or 20,000 km. unscrew the cap of the lubricator at the commutator end of the dynamo. If the lubricating wick is dry, refill the cap with recommended grease.

Inspection of Commutator and Brushgear.—About once a season remove the metal cover from the dynamo for inspection of the commutator and brushes. Take care not to lose the nut when removing the cover fixing screw, as the cover is liable to spring open when the screw is released.

It is essential that the brushes make good firm contact with the commutator. The brushes are held in boxes by means of springs. Hold back the spring, and at the same time move the brush to see that it is free to slide in its holder. If there are any signs of sticking, remove it from its box and clean it with a cloth moistened with petrol.

After removing brushes for cleaning or any other purpose, care must be taken to replace them in their original positions, otherwise they will not "bed" properly on the commutator.

If, after long service, the brushes have become worn to such an extent that they will not "bed" properly on the commutator, they should be replaced. Always fit genuine Lucas brushes, as these are made specifically for work on M.G. machines and will give by far the best results and the longest life. It is recommended the brushes be fitted at a Lucas Service Depôt.

Next examine the commutator. It should be clean and free from any trace of oil or dirt, and should have a highly polished appearance. The best way to clean a dirty or blackened commutator is by pressing against it a fine dry cloth and getting someone to turn the engine over slowly by hand. If the commutator is very dirty, the cloth may be moistened with petrol.

DYNAMO OUTPUT.

The dynamo automatically keeps the battery in good condition. It is of the compensated voltage control type and works in conjunction with a regulator unit which is housed along with the cut-out. The regulator causes the dynamo to give an output which varies according to the load on the battery and its state of charge. When the battery is discharged, the dynamo gives a high output so that the battery receives a quick re-charge which brings it back to its normal state in the minimum time. On the other hand, if the battery is fully charged, the dynamo is arranged to give only a trickle charge which is sufficient to keep it in good condition without any possibility of causing damage to the battery by overcharging.

In addition to controlling the output of the dynamo according to the condition of the battery, the regulator provides for an increase of output to balance the current taken by the lamps or other accessories whenever they are switched on.

A further refinement causes the charging current to vary to suit all climatic conditions and in addition causes the dynamo to give an increased charge to the battery during the first half-hour after the car is started. This quickly restores to the battery the energy that has been used in the operation of starting.

The regulator unit is accurately set before leaving the works to suit the requirements of the equipment fitted to the car, and in normal service the battery will be kept in good condition. If, however, you should find that the battery is not being kept in a charged condition, or if it is being excessively overcharged, we advise you to consult your nearest Lucas Service Depôt, where any necessary adjustments can be made.

AMMETER READINGS.

Normally, during daytime running, when the battery is in good condition, the dynamo gives only a trickle charge, so that the charge reading will seldom be more than a few amperes. A discharge reading may be shown sometimes immediately after switching on the lamps. This usually happens after a long run, when the voltage of the battery is high. After a short time, the voltage of the battery will fall and the regulator will respond, causing the dynamo output to balance the load. When starting, the driver will notice the rise of current, which reaches a maximum at about 20 m.p.h. (32 km.p.h.). The current then falls slowly until after about 30 minutes' running it reaches a steady charge which is most suitable for the particular conditions of the battery.

THE CUT-OUT—AN AUTOMATIC DYNAMO SWITCH.

It will be noticed from the ammeter readings that the dynamo does not charge at very low engine speeds. This is because it is not rotating fast enough to generate sufficient energy to charge the battery.

Connected between the dynamo and the battery is the cut-out—an automatic switch which allows the flow of current from the dynamo to the battery only. It closes when the dynamo is running fast enough to charge the battery, and opens when the speed is low or the engine is stationary, thus preventing current from flowing from the battery through the dynamo windings.

The cut-out is located along with the dynamo regulator unit and fuses on the engine side of the dash.

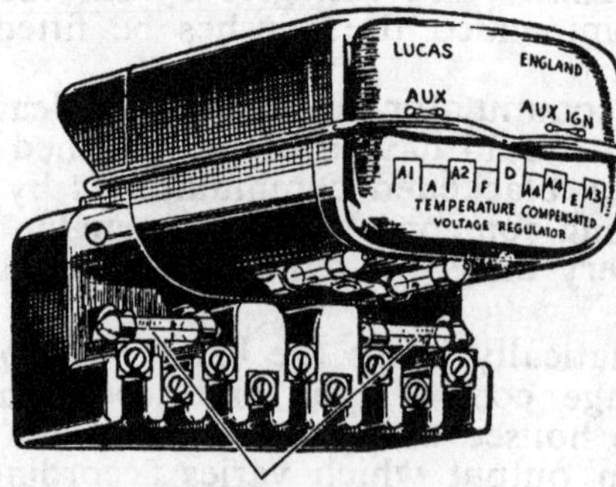

Fig. 28. The cut-out, fuse and junction box.

THE STARTER.

When Starting.—Observe the following points when starting the engine:—

(i) See that the controls are properly set.

(ii) Operate the starter switch firmly, and, of course, release it as soon as the engine fires.

(iii) Never operate the starter when the engine is running. If the engine will not fire at once, allow it to come to rest before pulling the switch again.

(iv) Do not run the battery down by keeping the starter on when the engine will not start.

Attention Needed by the Starter.—Give the commutator and brush gear similar attention to that described for the dynamo.

Occasionally, if dirty, wash over the screwed sleeve on the armature spindle with paraffin.

INSTRUMENT PANEL.

The instrument panel houses the combined lighting and ignition switch, ammeter, starter switch control, etc.

REGULATOR, CUT-OUT AND FUSE UNIT.

This unit houses the voltage regulator cut-out and two fuses, and forms a junction box for distributing the cables to the various accessories. The fuses are accessible without removing the cover protecting the regulator and cut-out. One fuse protects those accessories which are operative only when the ignition is switched on, while the second protects the accessories which can be operated irrespective of whether the ignition is on or off.

A blown fuse will be indicated by the failure of all units protected by it, and is confirmed by examination of the fuse. If it has blown, the broken ends of the wire will be visible inside the glass tube. Before replacing a blown fuse, inspect the wiring of the units that have failed for evidence of short circuits or other faults that may have caused the fuse to blow. A spare fuse is provided, and it is important to use only the correct replacement fuse —the fusing value is marked on a coloured paper slip inside the glass tube.

Both the regulator and the cut-out are accurately set before leaving the works and do not need any adjustment; the cover protecting them is therefore sealed.

NOTE:—Some cars will be found fitted with an alternative type control box RF95/2, on which the fuses are external: in this case there is no need to remove the cover.

LAMPS. HEADLAMPS.

Removing Lamp Front and Reflector.—To remove the lamp front, pull forward the fixing clip and swing it aside from the location. The front can then be withdrawn from the lamp body. When replacing the front, locate the top of the rim first, then press on the bottom and fit the fixing clip in its location. (See Fig. 27.)

The reflector is secured to the lamp body by a rubber bead and can be withdrawn when the bead is removed. When replacing the reflector, the projection on the rim must fit in the location at the top of the lamp body. When refitting the rubber bead, locate its thinner edge between the reflector rim and the edge of the lamp body.

SETTING AND FOCUSING

Lucas Beam Setter.—All Lucas Service Depôts and many Service Agents now include among their special testing facilities an apparatus known as the Lucas Beam Setter. This is a specially designed instrument by means of which headlamps and foglamps can be set and focused with extreme accuracy, thereby ensuring the maximum efficiency from the lamps.

Owners are strongly advised to make use of this service. If, however, it is not possible to have the lamps adjusted by this method, the procedure for focusing and setting as per instructions below.

SETTING

Headlamps.—The headlamps must be set so that the main beams of light are parallel with the road and with each other. They must also be arranged so that they give a light which is incapable of dazzling any person standing on the same horizontal plane as the car at a greater distance than 25 ft. from the lamp and whose eye-level is not less than 3 ft. 6 in. above that plane. To make the adjustment slacken the single fixing nut at the base of the lamp and move the lamp on its adjustable mounting to the required position. Finally tighten the locknut.

Foglamp.—The foglamp must be set so that the beam does not rise above the horizontal when the car is standing on level ground. To ensure this, dip the lamp very slightly to compensate for road inequalities or an extra heavy load in the rear of the car, and also tilt the lamp to the right to allow for road camber. In addition the lamp can be swung slightly to the left in order to give additional illumination on the near-side of the road. To make the adjustment slacken the single fixing nut at the base of the lamp and move the lamp on its adjustable mounting to the required position. Finally tighten the locknut.

FOCUSING

Headlamps.—For the lamps to give a parallel beam, the filament of the bulb must coincide as nearly as possible with the focal point of the reflector. If the filament of the bulb is behind the focal point of the reflector, the beam will be divergent, while on the other hand, if the filament is in front of the focal point, the beam will be convergent, with a dark area in the centre of the beam. In either case the lamps will have a poor range and will cause unnecessary dazzle to approaching traffic.

Before lamps are dispatched from the works, the bulbs are carefully focused to give the best results. Provided that official Lucas bulbs are fitted as replacements, it should not be necessary to disturb the setting. If for any reason these are not obtainable, and ordinary bulbs have to be fitted, it may be necessary to re-focus by moving the bulb backwards or forwards along the axis of the reflector until the best lighting is obtained.

Cover up one lamp while testing the other. If the lamp does not give a uniform long-range beam without any dark centre, the bulb needs adjusting.

To do this, remove the front and reflector, and slacken the clamping clip at the back of the reflector.

After each adjustment, note the effect with the reflector and front refitted. When the best position for the bulb-holder has been found, see that the clamping screw is tightened.

Foglamps.—The foglamp is fitted with a special pre-focus bulb which ensures that the filament of the bulb is always at the focal point of the reflector. No focusing is therefore necessary.

BULB REPLACEMENT

Headlamps.—To gain access to the bulb remove the lamp front and reflector as described on page 319.

Foglamp.—To gain access to the bulb, slacken the screw at the top of the lamp and remove the front rim and light unit assembly. Twist the back shell in an anti-clockwise direction and pull it off. The bulb can now be removed from the rear of the reflector.

Place the correct replacement bulb (see below) in the holder and engage the projections of the back shell with the slots in the bulb-holder, press on and secure by twisting to the left.

Sidelamps.—The front of the lamp is removed for bulb replacement when the screw at the top of the lamp is withdrawn. When refitting the front, locate the bottom and secure with the screw.

Stop and Tail-lamp.—The lamp front can be removed when the securing screw is withdrawn.

Replacement of Bulbs.—When the replacement of a bulb is necessary, it is important not only that the same size bulb is fitted, but that it has a high efficiency and will focus in the reflector. Cheap and inferior replacement bulbs often have the filament of such a shape that it is impossible to focus correctly; for example, the filament may be to one side of the axis of the bulb, resulting in loss of range and light efficiency.

It always pays you to use Lucas Genuine Spare Bulbs, as then these problems will not arise.

Bulbs Ftted.—

Headlamp, R.H. only	...	Lucas No. 54	12-v. 36 watt
„ L.H. only	...	Lucas No. 171	12-v. 36-36 watt
Sidelamps	...	Lucas No. 207	12-v. 6 watt
Stop and tail-lamp	...	Lucas No. 207	12-v. 6 watt
Foglamp SET 462	...	Lucas No. 162	12-v. 36 watt
Panel illumination bulb	...	Lucas No. 987	12-v. 2.2 watt
Map lamp	...	Lucas No. 207	12-v. 6 watt
30 m.p.h. warning light	...	Lucas No. 207	12-v. 6 watt
Ignition warning light	...	Lucas No. 970	2.5-v. .5 watt

Windscreen Wiper.—The windscreen wiper is carefully adjusted and packed with grease during assembly, and needs no attention.

To start the wiper, pull out the metal handle and turn to disengage it from the switch. Then move the switch to the " on " position. To stop the wiper move the switch to the " off " position, pull out the handle to disengage wiper blade from the gears, and turn end of handle into top of switch control.

Electric Horn.—The horn will give a long period of service without any attention; no adjustment is required.

Should the horn become uncertain in its action, giving only a choking sound, or cease to vibrate, it does not follow that the horn has broken down. First ascertain that the trouble is not due to some outside source, e.g., a discharged battery a loose connection or short circuit in the wiring of the horn, or a blown fuse. It is also possible that the performance of a horn may be upset by the horn becoming loose on its mounting.

If the cause of the trouble cannot be found, do not attempt to dismantle the horn, but return it to a Lucas Service Depôt for examination.

Coil Ignition Equipment.—Considering the exacting duties coil ignition has to perform, the attention needed to ensure absolutely reliable service is very small.

Periodically the contact breaker gap of the distributor should be checked and, if necessary, reset to .012 in. (.30 mm.) maximum opening, and the cam must be given a smear of clean engine oil.

LUBRICATION.

The following parts of the distributor require lubrication about every 3,000 miles or 5,000 km.

1. **Cam:** Give the cam a light smear of clean engine oil.

2. **Cam Bearing:** Take off the moulded cap, withdraw the moulded rotating arm from the top of the spindle by pulling it off, and add a few drops of thin oil through the lubricator hole provided in the top of the spindle. Take care to refit the arm correctly and to push it on the shaft as far as possible, otherwise there is a risk of tracking and burning of the moulded cap.

3. **Contact Breaker Pivot:** Place a spot of oil on the pivot on which the contact breaker rocker-arm works.

4. **Automatic Advance Mechanism:** Lubricate the automatic advance mechanism by adding a few drops of thin oil through the hole in the contact breaker base through which the cam passes.

Cleaning.—Keep the outside of the distributor clean, particularly the spaces between the high-tension terminals. Very occasionally remove the moulding by springing aside its two securing spring clips. Wipe the inside clean with a dry cloth, and see that the carbon brush is quite free in its holder. Clean the metal electrodes inside the moulding and also the rotating electrode on the distributor arm; if necessary, use a cloth moistened with a drop of petrol for this.

Next, examine the contact breaker; you must keep the contacts free from any grease or oil. If they are burned or blackened, clean them with fine carborundum stone, or, if this is not available, you can use very fine emery cloth. Finish off with a cloth moistened with petrol, and remove all traces of dirt and metal dust. Misfiring is sometimes caused by dirty contacts.

Checking and Adjusting the Contacts.—After the contacts are reset at the end of 500 miles or 800 km. they require only occasional adjustment. The chief cause of variation in the gap is wear of the heel of the contact rocker-arm which bears upon the actuating cam. Provided you keep the cam smeared with engine oil, however (see Lubrication Instructions above), the wear on the heel will be negligible and the contact gap setting should only require adjustment at infrequent intervals.

To check the setting, turn the engine by hand until the contacts are fully opened. Now insert the gauge, provided on the ignition screwdriver, between the contacts. The gauge has a thickness of about 12 thousandths of an inch (.30 mm.), and it should be a sliding fit between the contacts when the gap is correct. Do NOT alter the setting unless there is quite an appreciable variation from the gauge. To make the adjustment, keep the engine in the position to give maximum opening of the contacts and slacken the two screws securing the contact plate. Then move the plate until the gap is set to the thickness of the gauge. After making the adjustment, care must be taken to tighten the locking screws.

Renewing the High-tension Cable.—The high-tension cables are those connecting the coil to the distributor and the distributor to the sparking plugs. When these cables show signs of perishing or cracking they must be replaced by 7 mm. rubber-covered ignition cable.

The method of connecting the cable is to thread the knurled moulded nut over the lead, bare the end of the cable for about $\frac{1}{4}$ in. (6 mm.), thread the wire through the brass washer provided and bend back the strands. Finally screw the nut into its respective terminal.

The Coil.—The coil requires no attention whatever beyond keeping its exterior clean, particularly between the terminals, and occasionally checking that the terminal connections are quite tight.

Ignition Switch and Warning Light.—The ignition switch, besides forming a means of stopping the engine, is provided for the purpose of preventing the battery being discharged by the current flowing through the coil windings when the engine is stopped. A warning lamp is usually provided in the instrument panel which gives a red light when the ignition is switched on and the car is running very slowly or is stationary, thus reminding you to switch off.

Should the warning lamp bulb burn out, this will not in any way affect the ignition system, but you should replace it as soon as possible in order to safeguard your battery.

Micrometer Timing Adjustment.—In order to obtain very fine timing of the ignition to the engine and to allow for altered engine conditions, e.g., state of carbonisation of engine, change of fuel, etc., a micrometer adjustment is provided which allows fine adjustment to be made simply by the movement of a knurled knob.

With a clean engine, and using first grade fuel, the micrometer scale should be at " 0 " before timing.

The final setting can be made by use of the micrometer adjustment after running the engine. If the firing is found to be slightly too early or too late, adjust the knurled knob until the best engine performance is obtained. The adjustment should not be altered by more than one distributor degree at a time (one division on the scale is equivalent to two distributor degrees or 4° on the crankshaft).

The Wiring System.—Apart from the starter motor, which is a heavy gauge circuit complete in itself, there are twenty-two places where current, either from the battery or from the dynamo, is used. These may be divided into four groups:—

1. Those which can be used independently of the ignition switch. They are:
   - The two headlamps.
   - The two sidelamps.
   - The tail-lamp.
   - The map reading lamp.
   - The horn.
2. Those which are automatically switched on with the ignition switch and have no fuse protection. They are:
   - The ignition.
   - The ignition warning light.
   - The petrol pump.
   - The petrol reserve indicator.
3. Those which have separate controls, but which can only be used when the ignition is switched on. They are all protected by a fuse and are:
   - The windscreen wiper.
   - The foglamp.
   - The stop light.

4. The inspection posts are a distinct circuit of their own, as they (apart from the starter motor) constitute the only circuit through which the current can flow without passing through the ammeter. The electric clock is connected to the inspection posts.

   The six panel lamps can only be used when the sidelamps are switched on.

The current leaves the battery negative terminal by the starter wire and returns to the positive terminal by the frame (earth) and earth connection.

These two wires are the only battery connections, and the negative wire to the starter carries all current so far. From there it is picked up by a brown wire which goes to the ammeter.

These two wires and the ammeter are the main current connection, along which the current flows one way when the battery is supplying it, and the other way when the dynamo is charging the battery.

Diagrams of the electrical system are given on pages 515-516 which show the complete car wiring.

N.B.—All the wires in the circuits are identified by colours; where these wires are braided the braiding is woven in colours.

THE IGNITION AND ELECTRICAL EQUIPMENT (TD Model)

IGNITION EQUIPMENT—GENERAL DESCRIPTION.

The coil ignition equipment is provided with an automatic advance mechanism which relieves the driver of the necessity of adjusting the timing. The advantages are particularly evident when accelerating and during hill climbing, since the danger of knocking or pinking through excessive advance is very much reduced.

The automatic advance device is housed in the distributor unit, and consists of a centrifugally operated mechanism by means of which the ignition is advanced in proportion to the engine speed.

Like the rest of the electrical equipment, it is wired on the " positive earth " system, which results in longer sparking plug life.

Distributor Type.

The distributor is a Lucas Model DKY4A, Service No. 40162. These identification marks are stamped on the side of the distributor. When ordering replacements, always quote these numbers.

Ignition Coil Type.

The coil is a Lucas Model Q12, Service No. 45020. These identification marks are stamped on the base of the ignition coil. When ordering, always quote these numbers.

Sparking Plugs.

The standard sparking plugs for the M.G. " TD " Midget are Champion L.10s, 14 mm.

LOCATING THE CAUSE OF UNEVEN FIRING.

To Test with Sparking Plugs in Position.

(*a*) Start the engine and set it to run at a fairly fast idling speed.

(*b*) Short-circuit each plug in turn by placing a hammer head or the blade of a screwdriver with a wooden or insulated handle between the terminal and the cylinder head. No difference in the engine performance will be noted when short-circuiting the plug in the defective cylinder. Shorting the other plugs will make uneven running more pronounced.

(*c*) Having located the cylinder which is at fault, stop the engine and remove the cable from the terminal of the sparking plug. Restart the engine and hold the end of the cable about $\frac{3}{16}$ in. (4.8 mm.) from the cylinder head.

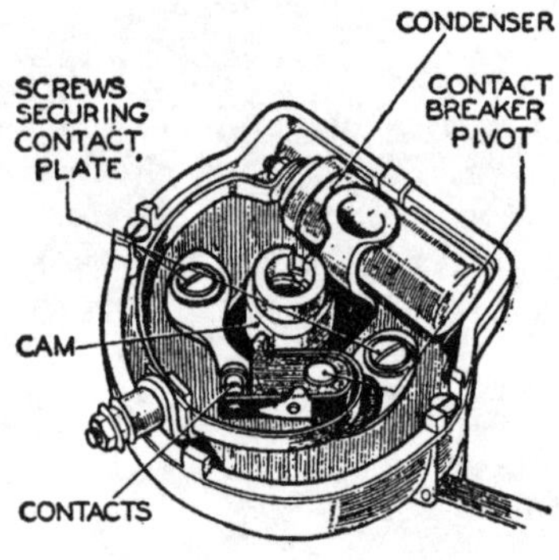

Fig. 29. The distributor with cover and arm removed.

(*d*) If the sparking is strong and regular, the fault probably lies in the sparking plug. Remove the plug, clean, and adjust the gap to the correct setting or alternatively fit a replacement plug.

(*e*) If there is no spark or if it is weak and irregular, examine the cable from the sparking plug to the distributor. After a long period of service the rubber insulation may be cracked or perished, in which case the cable should be replaced. Finally, examine the distributor moulded cap, wipe the inside and outside with a clean dry cloth, see that the carbon brush moves freely in its holder and examine the moulding closely for signs of breakdown. After long service, it may have become tracked, that is, a conducting path may have formed between two or more of the electrodes or between one of the electrodes and some part of the distributor in contact with the cap. Evidence of a tracked cap is shown by the presence of a thin black line in the places indicated. A replacement distributor cap must be fitted in place of one that has become tracked.

TESTING LOW-TENSION CIRCUIT.

Low-tension Circuit. Testing in Position.

(*a*) Spring back the securing clips on the distributor and remove the moulded cap and rotor. If the rotor is a tight fit, it can be levered off carefully with a screwdriver.

(*b*) Check that the contacts are clean and free from pits, burns, oil or grease. Turn the engine and check that the contacts are opening and closing correctly and that the clearance when the contacts are fully opened is between .010 in. and .012 in. (.25 mm.—.30 mm.). Correct the gap if necessary.

(*c*) Disconnect the cable at the contact breaker terminal of the coil (C.B.) and at the low-tension terminal of the distributor, and connect a test lamp between these terminals. If the lamp lights when the contacts close and goes out when the contacts open, the low-tension circuit is in order.

Low-tension Circuit—to Locate Fault.

(*a*) Having determined, by testing as previously described, that the fault lies in the low-tension circuit, switch on the ignition, and turn the engine until the contact breaker points are fully opened.

(*b*) Refer to the wiring diagram and check the circuit with a voltmeter (0—20 volts) as follows:—

NOTE.—If the circuit is in order the reading on the voltmeter should be approximately 12 volts.

(*c*) **Battery to starter switch terminal.** Connect a voltmeter between the starter switch terminal and a good earthing point. No reading indicates a damaged cable or loose connections.

(*d*) **Starter switch to ammeter** (brown lead). Connect a voltmeter to ammeter terminal and earth. No reading indicates faulty cable or loose connections

(*e*) **Ammeter.** Connect voltmeter to other ammeter terminal and earth. No reading indicates fault in ammeter, which must be replaced.

(*f*) **Ammeter to control box terminal " A "** (brown with white lead). Connect voltmeter to control box terminal " A " and earth. No reading indicates faulty cable or loose connection.

(*g*) **Control box terminal** " AI." Connect voltmeter to control box terminal " AI " and earth. No reading indicates fault in series winding of control box.

(*h*) **Control box terminal " AI " to lighting and ignition switch terminal " A "** (brown with blue lead). Connect voltmeter to " A " terminal on switch and earth. No reading indicates faulty cable or loose connections.

(*j*) **Lighting and ignition switch terminal** " IG." Connect voltmeter to " IG " terminal on switch and earth. No reading indicates fault in switch, which must be replaced.

(*k*) " IG " **terminal to control box terminal** " A3 " (white lead). Connect voltmeter to "A3 " terminal on control box and earth. No reading indicates faulty cable or loose connections.

(*l*) **Control box terminal** " A3 " **to ignition coil terminal** " SW " (white lead). Connect a voltmeter to the ignition coil terminal " SW " and to earth. No reading indicates a damaged cable or loose connections.

(*m*) **Ignition coil.** Disconnect the cable from the " CB " terminal of the ignition coil and connect a voltmeter to this terminal and to earth. No reading indicates a fault in the primary winding of the coil and a replacement coil must be fitted. If the correct reading is given, remake the connections to the coil terminal.

(*n*) **Ignition coil to distributor** (white with black lead). Disconnect the cable from the low-tension terminal on the distributor and connect the voltmeter to the end of this cable and to earth. No reading indicates a damaged cable or loose connections.

(*o*) **Contact breaker and condenser.** Connect the voltmeter across the contact breaker points. No reading indicates a fault in the condenser.

HIGH-TENSION CABLES.

(*a*) The high-tension cables must be examined carefully and any which have the insulation cracked, perished or damaged in any way, must be replaced by 7 mm. rubber-covered ignition cable.

(*b*) To fit the cables, thread the knurled moulded terminal nut over the lead, bare the end of the cable for about ¼ in. (6 mm.), thread the wire through the brass washer removed from the original cable and bend back the strands over the washer. Finally screw the nut into its terminal.

(*c*) The cables from the distributor to the sparking plugs must be connected up in the correct firing order, which is 1, 3, 4, 2.

ATTENTION TO SPARKING PLUGS.

To obtain the best engine performance and the most economical running, the sparking plugs must be kept clean and correctly adjusted. They should be removed and cleaned after the first 500 miles (800 km.) of use in a new engine. This is advisable since the slower engine speeds and the conditions of the " running-in " period have a tendency to cause fouling of the plugs.

Plugs should subsequently be removed for inspection, cleaning and adjustment after each period of 3,000 miles (5,000 km.).

When sparking plugs are removed from the engine their gaskets should be removed with them and replaced on the plugs, which should be placed in a suitable holder. It is advisable to identify each plug with the number of the cylinder from which it was removed so that any faults can be traced back to the cylinder concerned.

When examining plugs place a new plug of the same type beside the others to afford a ready comparison of the relative condition of the used plugs.

When examining plugs note the condition of their gaskets. A large proportion of the heat from the insulator is dissipated to the cylinder head by means of the copper gasket between the plug and cylinder head. Plugs not screwed down tight become over-heated, causing pre-ignition, short plug life and " pinking."

On the other hand it is unnecessary and unwise to tighten up the plugs too much. What is required is a reasonably good seal between the plug and the cylinder head.

If the gaskets are at all questionable they should be replaced bv new ones without hesitation.

Plug Inspection.

After removal of the plug, the condition of the electrodes and deposits on the insulator and plug body should be examined.

1. If the insulator is brown in colour, the electrodes grey and the plug body dry or covered with a thin layer of soot, the engine condition and mixture strength are satisfactory.
2. A dry, greyish-yellow or brown insulator with a thin layer of light fawn powder deposit indicates the use of a leaded fuel or a rich mixture.
3. When the insulator is dry and fawn or white in colour, and the electrodes are corroded and burnt at the tips, the plug temperature is too high. This is caused either through the use of an unsuitable plug; by a weak mixture; or by high combustion temperatures.
4. Soot deposits, forming a black velvety coating on the insulator and plug body, show that the plug does not reach a self-cleansing temperature. This may be due to a mixture which is too rich, but if the deposit is wet it indicates that oil is also reaching the combustion space in excessive quantities. Correct operation may be restored by adjusting the mixture, but an overhaul of the engine is necessary to reduce the amount of oil passing the piston.
5. After cleaning, examine the plugs for cracked insulators and the lower end for wear produced through previous cleaning.

Whenever possible sparking plugs should be cleaned in a special plug cleaner of the type supplied by the plug manufacturer. Oily plugs should be washed with petrol first. A compressed air jet should then be used to remove any abrasive from the interior of the plug body and the insulator. If a plug cleaner is not available, a wire brush is the best substitute. This should also be used to clean any accumulation of carbon from the threads. The thread portion of the plug body is often neglected when cleaning owing to the fact that it is not generally realised that, like the gaskets, the threads are an important means of heat dissipation and that when they are coated with carbon it retards the flow of heat from the plug and leads to overheating.

Cleaning the thread will also facilitate refitting of the plug and avoid the use of unnecessary force on the plug spanner.

Having ensured that the plug is thoroughly clean and still serviceable, the electrodes should be reset. A combination gauge and setting tool produced by the makers of Champion sparking plugs greatly facilitates the correct and easy setting of the sparking plug points, but care should be taken to avoid a false reading through distortion of the points by burning.

When resetting the points, the side electrode only should be adjusted to give the correct clearance.

Never Bend the Centre Electrode.

Champion L.10.S plugs are fitted as standard and their correct spark gap should be set between .020 in. and .022 in. (.50 mm. and .56 mm.).

Remember that electrode corrosion and the development of oxides at the gap area vitally affects the sparking efficiency. The special plug cleaner can remove oxides and deposits from the insulator, but the cleaner stream does not always reach this area with full effect owing to its location, and cannot necessarily deal with corrosion effectively as this sometimes requires too strong a blast for proper removal.

When plugs appear worthy of further use it is good practice to dress the gap area on both centre and side electrodes with a small file before resetting them to the correct gap. The intense heat, pressures, explosion shock, electrical and chemical action to which the plugs are submitted during miles of service are so intense that the molecular structure of the metal of the points is eventually affected. Plugs then reach a worn out condition where resetting of the points no longer serves a useful purpose and where plug replacement is called for. Every 12,000 miles (20,000 km.) new plugs should, therefore, be fitted.

Before replacing a used plug in the engine, test it for correct functioning under air pressure in a plug tester, following out the instructions issued by the makers of the tester. Generally speaking, a plug may be considered satisfactory for further service if it sparks continuously under a pressure of 100 lb./sq. in. (7 kg.cm.²) with the gap between the points set at .022 in. (.56 mm.). It is essential that the plug points then be reset to the smaller gap of .020 in. (.5 mm.) before the plug is refitted to the engine.

While the plug is under pressure in the tester it should be inspected for leakage by applying oil round the terminal and insulator. Leakage is indicated by the production of air bubbles, the intensity of which give an indication of the degree of leakage. The leaking gases have a "blow-torch" effect when the engine is running which rapidly raises the temperature of the plug to above its heat range, thus producing overheating, pre-ignition, and rapid electrode destruction.

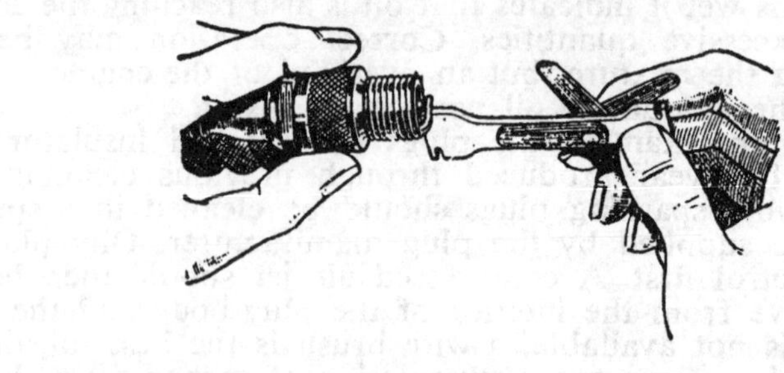

Fig. 30. Adjusting sparking plug gap using setting tool.

The top half of the insulator is also frequently responsible for poor plug performance due to the following faults: splashes, accumulation of dirt or dust, cracked insulators, caused by a slipping spanner; overtightness of the terminals. Examine for a cracked insulator at the shoulder and the terminal post and remove any accumulations of dirt and dust.

Since each engine design has its own particular working temperature and pressure inside the cylinder, it is essential that only sparking plugs recommended by The M.G. Car Company, Ltd., be used. A plug designed for a hot dry engine will not function satisfactorily in relatively cool oily engines, as it will constantly oil up and cause trouble. On the other hand, a plug suitable for the oily engine will not function in the hot type engine as the points will overheat and cause pre-ignition.

The threaded portion or "reach" of the plug is also important since it determines the position of the points in the combustion chamber and may produce pre-ignition if the threads on the plug body protrude beyond the cylinder head.

CONTACT BREAKER MECHANISM.

After the first 500 miles (800 km.) and subsequently every 3,000 miles (5,000 km.) check the contact breaker as follows:—

(*a*) Turn the engine until the contact breaker points are fully opened, and check the gap with a gauge having a thickness of from .010 in. to .012 in. (.25 mm. to .30 mm.). If the gap is correct, the gauge should be a sliding fit. Do not alter the setting unless the gap varies considerably from the gauge thickness.

To adjust the setting, keep the engine in the position which gives maximum opening of the contacts and then slacken the two screws securing the fixed contact plate. Adjust the position of the plate until the gap is set to the thickness of the gauge and then tighten the two locking screws.

Remember that the cam only keeps the contact points fully open over 10° and that care must be taken to ensure that the points are in the fully open position.

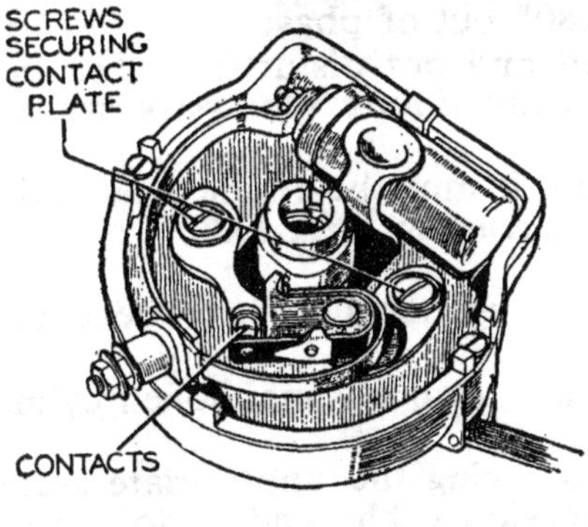

Fig. 31. *The contact breaker, showing the adjusting screws.*

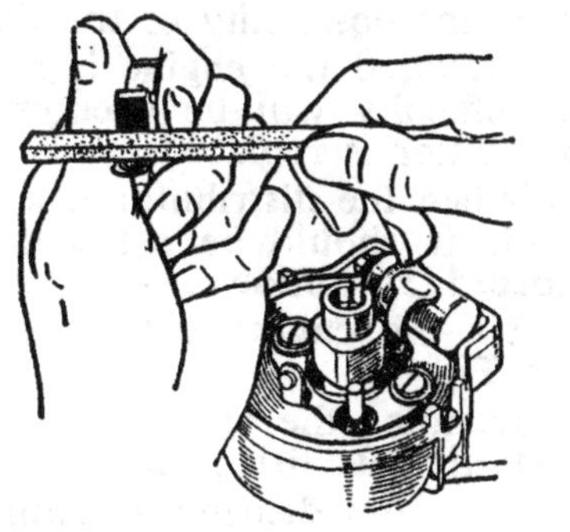

Fig. 32. *Cleaning the distributor points.*

(*b*) If the contacts are dirty or pitted, they must be cleaned by polishing them with a fine carborundum stone, and afterwards wiping them with a petrol-moistened cloth. The moving contact can be removed from its mounting in order to assist cleaning (see Fig. 32). Check and adjust the contact breaker setting after cleaning the contacts.

(*c*) Check that the moving arm moves freely on its pivot. If it is sluggish, remove the moving arm and polish the pivot pin with a strip of fine emery cloth. Afterwards clean off all trace of emery dust and apply a spot of clean engine oil to the top of the pivot.

The contact breaker spring tension should be between 20 and 24 oz. (567-680 gm.) measured at the contacts.

DISTRIBUTOR LUBRICATION.

To be carried out after servicing the distributor and at intervals of about 3,000 miles (5,000 km.).

(*a*) Give the cam a light smear of grease and apply a slight trace of oil to the top of the contact breaker lever pivot pin.

(*b*) Lift the rotor arm off the top of the spindle and add a few drops of thin machine oil through the lubricating passage provided in the spindle to lubricate the cam bearing and distributor shaft. (Do not remove the screw in the top of the spindle as an oilway is provided.) Refit the rotor correctly, and push it on to the shaft as far as it will go.

(*c*) Add a few drops of thin machine oil through the hole in the contact breaker base through which the cam passes, in order to lubricate the automatic timing control. Do not allow any oil to get on or near the contacts.

REMOVAL OF THE DISTRIBUTOR.

Disconnect the low-tension lead from the terminal on the side of the distributor body. Spring back the two retaining clips and lift off the distributor head which can be lodged on the cylinder block. To facilitate replacement turn the engine over until the rotor arm is pointing to the segment in the cover for No. 1 cylinder plug lead.

Remove the locking wire from the dowel bolt holding the clamp plate to the cylinder block and take out the bolt. Disconnect the bonding wire from the cylinder block to the clamp and then the distributor can be lifted straight up.

It should be noted that the drive tongue on the end of the distributor shaft is offset. This is to ensure its correct replacement on reassembly and avoid the possibility of the timing being 180° out of phase.

Provided the engine is set as indicated and not disturbed subsequently no difficulty will be encountered in correctly timing the spark when the distributor is replaced.

When the distributor has been removed, it must be retimed on replacement. It should be set with points just breaking at T.D.C. To do this proceed as follows:—

Set the engine with the Nos. 1 and 4 cylinders on top dead centre (see Fig. 33).

Examine the valves to see which of the previously mentioned cylinders is starting its firing stroke.

Turn the distributor until the rotor is facing the appropriate segment (i.e., the segment connected to the high-tension cable leading to the same cylinder), and insert the distributor in its housing, "feeling" it in so that the nearest tooth is engaged. Turn the body about until the locking screw will enter, and lock it.

Set the contact points to .012 in. (.3 mm.) and check that the hole in the crankshaft pulley still coincides with the arrow on the timing cover. The contact breaker points should now just be commencing to open. Should this not be the case, release the clamping bolt at the base of the distributor, turn the distributor, anti-clockwise until the points are fully closed, and then turn carefully clockwise until the contact points just commence to open.

Securely tighten the clamp bolt.

Re-check timing to make sure that tightening the clamp bolt has not altered the setting.

Fig. 33. *T.D.C. mark.*

NOTE.—Before setting the timing make sure that the automatic advance and retard mechanism is working properly and is in its fully retarded position while the timing is being set.

IMPORTANT.—To obtain an accurate setting an electrical method should be used to determine the actual position at which the points break, and the following method can be used:—

With the low-tension lead connected to the distributor, turn on the ignition switch and connect a 12-volt lamp in parallel with the contact breaker point (i.e., one lead from the distributor low-tension terminal and the other to earth) and turn the distributor until the lamp lights, which indicates that the points have just opened.

OR

Get a second operator to watch the ammeter and then turn the distributor body, when it will be observed that the small reading recorded on the ammeter when the points are closed " flicks " back to zero immediately the points open.

DISMANTLING THE DISTRIBUTOR.

Before dismantling, carefully note the positions in which the various components are fitted so that they can be replaced correctly, then:—

(*a*) Spring back the securing clips and remove the moulded cap.

(*b*) Lift the rotor off the top of the spindle. If it is a tight fit it should be levered off carefully with a screwdriver.

(*c*) Slacken off the nut on the terminal post and lift off the end of the contact breaker spring which is slotted to assist removal. Lift the contact breaker lever off its pivot pin and remove the insulating washer. Take out the two screws, complete with spring washers and flat steel washers, from the plate carrying the fixed contact and remove the plate.

(*d*) Undo the two screws fitted at the edge of the contact breaker base and lift them out together with the spring washers. The contact breaker base can then be removed from the body of the distributor.

(*e*) Unscrew the condenser terminal nut, lift off the spring washer and remove the connector strip. Soften the solder securing the condenser in its clip, with a hot iron, and remove the condenser by applying pressure at one end.

NOTE.—The condenser should not be removed unless absolutely necessary.

(*f*) Drive out the parallel driving pin passing through the collar of the driving tongue member at the lower end of the spindle, and withdraw the driving tongue from the spindle. Note that the driving tongue itself is offset and that the small offset is towards the front of the engine when the slot for the rotating arm faces the condenser in the distributor body.

(*g*) Lift the cam, automatic timing control and shaft assembly from the distributor. Take out the screw from inside the top of the cam spindle and lift the cam off. The automatic timing control is then accessible.

THE CONDENSER.

The best method of testing the condenser is by substitution. Disconnect the original condenser and connect a new one between the low-tension terminal of the distributor and earth.

Should a new condenser be necessary, it is advisable to fit a complete condenser and contact breaker plate assembly, but should a condenser only be available, use a hot iron to soften the solder securing the defective condenser in the clip and remove the condenser by applying pressure at one end. Care must be taken not to overheat the new condenser when soldering it in position.

The capacity of the condenser is 0.2 microfarads.

FITTING NEW DISTRIBUTOR BUSHES.

In order to ensure easy running of the distributor shaft when the shank has been rebushed, the new bushes must be fitted so that they are in correct alignment. The bushes must be fitted by means of a vertical drilling machine or hand press, using a mandrel and a packing block.

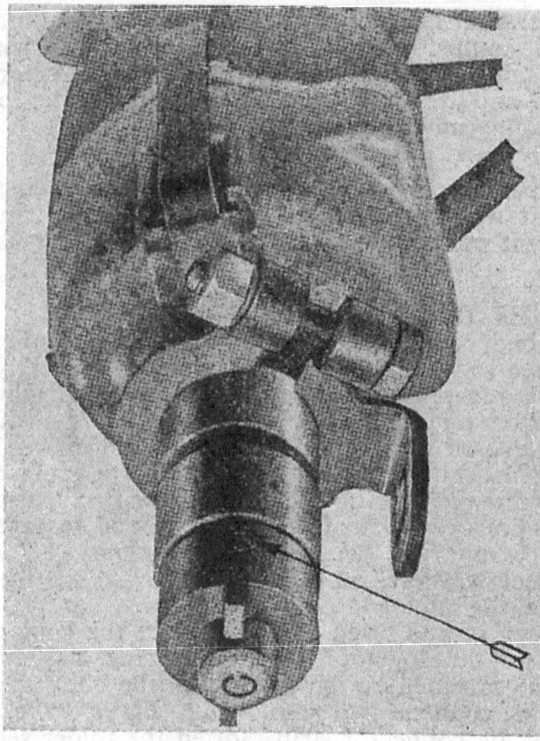

Fig. 34. Showing pin locating the driving collar to the contact breaker spindle.

(*a*) Fit the mandrel in the drilling machine or hand press and place the distributor body in an inverted position on the table below it.

(*b*) To remove the bushes, a sleeve must be fitted over the mandrel to build it up to the required size. With this sleeve fitted in position, force the old bushes out of the shank by applying a steady pressure. Before new bushes are fitted they should be allowed to soak for twenty-four hours in thin engine oil.

(*c*) Take the sleeve off the mandrel. Place one of the longer bushes on the mandrel, then the distributor body in an inverted position and finally one of the smaller bushes.

(*d*) Locate the end of the mandrel through the packing piece and press the mandrel downwards, taking care that both bushes enter the distributor shank squarely. Continue forcing the bushes into the shank until the mandrel reaches the end of its travel.

(*e*) After fitting, the bushes must not be opened out by reamering or any other means, as this would tend to impair the porosity of the bushes, and so prevent effective lubrication from being obtained.

REASSEMBLING THE DISTRIBUTOR.

NOTE.—Before assembly, the automatic advance mechanism, distributor shaft, and the portion of the shaft on which the cam fits, must be lubricated with thin, clean engine oil.

(*a*) Assemble the automatic timing control, taking care that the parts are fitted in their original positions and that the control springs are not stretched. Two holes are provided in each toggle; the springs must be fitted to the inner hole in each case. Place the cam on its spindle and secure by tightening the locking screw.

(*b*) Fit the shank in its bearings and replace the driving member. Remembering that the small offset of the driving tongue lies towards the front of the engine when the slot for the rotating arm in the cam faces towards the centre of the engine (or towards the condenser in the distributor body), fit the driving pin and burr over the collar each side, to retain it in position, with a suitable punch.

(*c*) Place the contact breaker base in position on the distributor body and secure it by replacing the two screws. A spring washer must be fitted under each of the screw heads, and the screws must be fully tightened.

(*d*) Place the end of the connector strip over the condenser terminal post, refit the spring washer and secure it by tightening the terminal nut.

(*e*) Position the plate carrying the fixed contact on the contact breaker base and secure it by replacing and lightly tightening the two screws, placing a spring washer and flat steel washer under the heads of each of the screws. Place the insulating washer over the contact breaker pivot pin and position the contact breaker lever over the pivot pin. Locate the slotted end of the contact breaker spring under the head of the terminal screw and tighten the nut to lock the spring in position. Adjust the contact breaker setting to give a maximum opening of from .010 in. to .012 in. (.25 mm. to .30 mm.).

NOTE.—If it becomes necessary to renew the contacts a replacement set comprising fixed and moving contacts must be fitted.

(*f*) Place the rotor on the top of the spindle, locating the register correctly, and push it fully home.

(*g*) Fit the distributor moulding and secure it by means of the spring clips.

ELECTRICAL EQUIPMENT—GENERAL DESCRIPTION.

The electrical equipment fitted to the M.G. Midget Series "TD" is of the 12-volt type incorporating constant voltage control for the charging circuit. The positive earth system of wiring is employed.

The battery is mounted on the dash under the bonnet and is readily accessible for examination and maintenance attention.

The dynamo is mounted on the left of the cylinder block and driven by endless belt from the engine crankshaft. A hinged mounting enables the belt tension to be adjusted.

The control box is sealed and should not normally need attention. The fuses are carried in external holders, as are the spare fuses, so that there is no need to remove the control box cover to gain access to them.

The starter motor is mounted on the flywheel housing on the right-hand side of the engine unit and operates on the flywheel through the usual sliding pinion device.

The headlamps are of the "dip and switch" type, in which the near-side lamp beam is dipped and the off-side lamp extinguished on operation of the dipping switch. On some exported models the lamps are fitted with double filament bulbs in order to comply with the regulations existing in the country concerned.

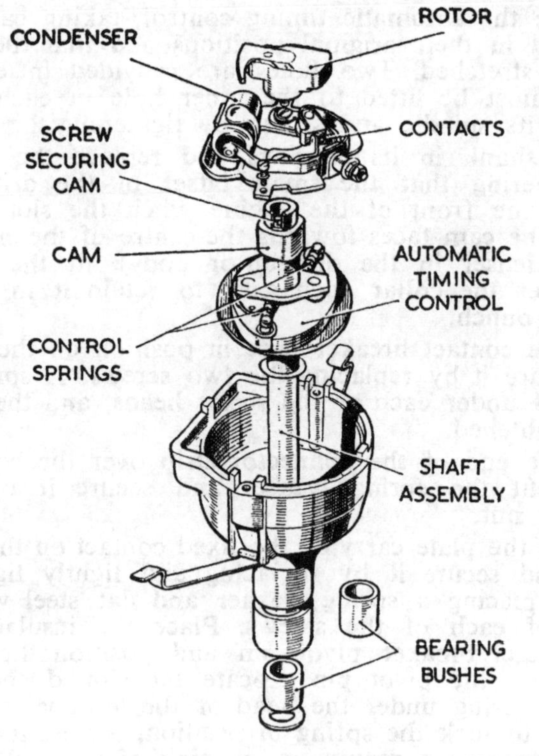

Fig. 35. Component parts of the distributor.

THE BATTERY. Model GTW9A. Capacity (10 hour rate) 51 ampere hours.

ROUTINE MAINTENANCE.

In order to keep the battery in good condition, a periodical inspection should be made and the following carried out:—

(i) Topping Up.

About once a month (more often in warmer climates), remove the vent plugs from the top of each of the cells and examine the level of the electrolyte. If necessary, add distilled water **until the top edges of the separators are just covered. Do not fill above this level, otherwise the excess electrolyte will be thrown out of the cell.** An hydrometer will be found useful for topping up, as it prevents distilled water from being spilled on the top of the battery.

NOTE.—In very cold weather it is essential that the car be used immediately after topping up the battery to ensure that the distilled water is thoroughly mixed with the electrolyte. Neglect of this precaution may result in the distilled water freezing, with consequent damage to the battery.

When examining the cells, do not hold naked lights near the vent holes, as there is a danger of igniting the gas coming from the plates.

(ii) Testing the Condition of the Battery.

Occasionally examine the condition of the battery by taking hydrometer readings. There is no better way of ascertaining the state of charge of the battery. The hydrometer contains a graduated float on which is indicated the specific gravity of the acid in the cell from which the sample is taken.

The specific gravity readings and their indications are as follows:—

1.280—1.300 Battery fully charged.
About 1.210 Battery about half discharged.
Below 1.150 Battery fully discharged.

These figures are given assuming an electrolyte temperature of 60° F. (16° C.). If the temperature of the electrolyte exceeds this, .002 must be added to hydrometer readings for each 5° F. (2.8° C.) rise to give the true specific gravity. Similarly .002 must be subtracted from the hydrometer readings for every 5° F. (2.8° C.) below 60° F. (16° C.).

The readings for each of the cells should be approximately the same. If one cell gives a reading very different from the rest it may be that the electrolyte has been spilled or has leaked from one of the cells, or there may be an internal fault. In this case it is advisable to have the battery examined by a battery specialist. Should the battery be in a low state of charge, it should be recharged by taking the car for a long daytime run or by charging from an external source of D.C. supply at a current rate of 5 amperes until the cells are gassing freely.

After examining the battery, check the vent plugs, making sure that the air passages are clear, and screw the plugs into position. Wipe the top of the battery to remove all dirt and moisture.

STORAGE.

If a battery is to be out of use for any length of time, it should first be fully charged and then given a freshening charge about every fortnight.

A battery must never be allowed to remain in a discharged condition, as this will cause the plates to become sulphated.

INITIAL FILLING AND CHARGING.

Usually the battery will have been filled and initially charged. If, however, it should be found necessary to prepare a new battery, supplied dry, proceed as follows:—

(a) Preparation of Electrolyte.

The specific gravity of the electrolyte necessary to fill the new battery, and the specific gravity at the end of the charge, are as follows:—

Climate.	S.G. of Filling Acid (corrected to 60° F. (16° C.))	S.G. at End of Charge (corrected to 60° F. (16° C.))
Ordinarily below 80° F. (27° C.)	1.350	1.280—1.300
Between 80°—100° F.	1.320	1.250—1.270
Over 100° F. (38° C.)	1.300	1.220—1.240

The electrolyte is prepared by mixing distilled water and concentrated sulphuric acid 1.835 S.G. The mixing must be carried out in a lead-lined tank or a suitable glass or earthenware vessel. Steel or iron containers must **not** be used. The acid must be added slowly to the water, while the mixture is stirred with a glass rod. **Never add the water to the acid,** as the resulting chemical reaction may have dangerous consequences.

To produce electrolyte of the correct specific gravity, use the following proportions of acid and distilled water:—

To obtain Specific Gravity (corrected to 60° F. (16° C.))	Add 1 part by volume of 1.835 S.G. acid to distilled water by volume as below
1.350	1.8 parts
1.320	2.2 „
1.300	2.5 „

Heat is produced by the mixture of acid and water, and it should, therefore, be allowed to cool before pouring it into the battery, otherwise the plates, separators and moulded container may become damaged.

(b) Filling in and Soaking.

The temperature of the filling-in acid, battery and charging room should be above 32° F. (0° C.).

Carefully break the seals in the filling holes and half fill each cell in the battery with dilute sulphuric acid solution of the appropriate specific gravity (according to temperature) (see table in (*a*) above). The quantity of electrolyte required to half fill a two-volt cell is ½ pint.

(c) Duration and Rate of Initial Charge.

Charge at a constant current of 3.5 amps. until the voltage and temperature-corrected specific gravity readings show no increase over five successive hourly readings. This period is dependent upon the length of time the battery has been stored since manufacture, and will be from forty to eighty hours, but usually not more than sixty.

Throughout the charge the acid must be kept level with the tops of the separators in each cell by the addition of acid solution of the same specific gravity as the original filling-in acid.

If, during charge, the temperature of the acid in any cell of the battery reaches the maximum permissible temperature of 120° F. (49° C.), the charge must be interrupted and the battery temperature allowed to fall at least 10° F. (5.5° C.) before charging is resumed.

At the end of the first charge, i.e., when specific gravity and voltage measurements remain substantially constant, carefully check the specific gravity in each cell to ensure that it lies within the limits specified. If any cell requires adjustment, some electrolyte must be syphoned off, and replaced with either acid of the strength used for the original filling in or distilled water, according to whether the specific gravity is too low or too high respectively. After such adjustment the gassing charge should be continued for one or two hours to ensure adequate mixing of the electrolyte. Re-check, if necessary, repeating the procedure until the desired result is obtained. Finally adjust electrolyte to correct level.

THE DYNAMO.

Type.

The dynamo is a Lucas Model C39PV, Service No. 22257A. These identification marks are stamped on the yoke. When ordering replacements always quote these numbers.

To Test on Vehicle.

The cutting-in speed is from 1,050 to 1,200 r.p.m. at 13 dynamo volts.

The output is 17 amps. at 1,800—2,000 r.p.m. at 13.5 dynamo volts taken on a resistance load of 0.8 ohm without regulator.

(*a*) Make sure that the driving belt is not slipping, and it should be capable of being deflected approximately ½ in. (12.7 mm.) at the centre of its run between the pulleys with moderate hand pressure. If the belt is too slack, tightening is effected by slackening the two bolts attaching the dynamo end plate extensions to the cylinder head, loosening the bolt attaching it to the slotted adjustment link and gently pulling the dynamo outwards by hand until the correct tension is obtained. The slotted link bolt must then be tightened, followed by the two upper bolts.

(*b*) Check that the dynamo and control box are connected correctly. The dynamo terminal " D " should be connected to the control box terminal " D " and the dynamo terminal " F " connected to control box terminal " F."

(*c*) After switching off all lights and accessories, disconnect the cables from the terminals of dynamo marked " D " and " F " respectively.

(*d*) Connect the two terminals with a short length of wire.

(*e*) Start the engine and set it to run at normal idling speed.

(*f*) Clip the negative lead of a moving coil type voltmeter, calibrated 0—20 volts, to one dynamo terminal and the other lead to a good earthing point on the dynamo yoke.

(*g*) Gradually increase the engine speed, when the voltmeter reading should rise rapidly and without fluctuation. Do not allow the voltmeter reading to reach 20 volts. Do not race the engine in an attempt to increase the voltage. It is sufficient to run the dynamo up to a speed of 1,000 r.p.m.

If there is no reading—check the brush gear.

If the reading is low (approximately 1 volt) the field winding may be faulty.

If the reading is approximately 5 volts the armature winding may be faulty.

(*h*) Remove the dynamo cover band and examine the brushes and commutator. Hold back each of the brush springs and move the brush by pulling gently on its flexible connector. If the movement is sluggish, remove the brush from its holder and ease the sides by lightly polishing on a smooth file. Always replace brushes in their original positions. If the brushes are worn so that they no longer bear on the commutator, or if the brush flexible has become exposed on the running face, new brushes must be fitted. If the commutator is blackened or dirty, clean it by holding a petrol-moistened cloth against it while the engine is turned slowly by hand cranking. Re-test the dynamo; if there is still no reading on the voltmeter, there is an internal fault and the complete unit should be replaced, if a spare is obtainable.

If the dynamo is in good order, leave the temporary link in position between the terminals and restore the original connections, taking care to connect the dynamo terminal "D" to the control box terminal "D," and the dynamo terminal "F" to the control box terminal "F." Remove the lead from the "D" terminal on the control box and connect the voltmeter between this cable and a good earthing point on the vehicle. Run the engine as before. The reading should be the same as that measured directly at the dynamo. No reading on the voltmeter indicates a break in the cable to the dynamo. Carry out the same procedure for the "F" terminal connecting the voltmeter between cable and earth. Finally remove link from the dynamo.

If the reading is correct test the control box.

REMOVAL AND. REPLACEMENT OF THE DYNAMO. (TA, TB, TC AND TD.)

Remove the set screw securing the insulator pad to the rear face of the dynamo.

Disconnect the dynamo leads from the dynamo terminals.

Disconnect the revolution counter drive from the dynamo.

To remove the dynamo, slacken its three attachment bolts and swing the dynamo towards the engine to give maximum belt slackness. Carefully free the belt from the dynamo pulley.

Supporting the dynamo, completely remove its attachment bolts, enabling it to be lifted from the engine.

Replacement of the dynamo is carried out in the reverse manner to that detailed for removal.

DISMANTLING THE DYNAMO. (TA, TB, TC AND TD.)

Take off the dynamo pulley.

Remove the cover band, hold back the brush springs and remove the brushes from their holders.

Unscrew the locking nuts from the through bolts at the commutator end.

Withdraw the two through bolts from the driving end.

Remove the nut, spring washer and flat washer from the smaller terminal (i.e., field terminal) on the commutator end bracket and remove the bracket from the dynamo yoke.

The driving end bracket which, on removal from the yoke, has withdrawn with it the armature and armature shaft ball bearing, need not be separated from the shaft unless the bearing is suspected and requires examination; in which event the armature should be removed from the end bracket by means of a hand press.

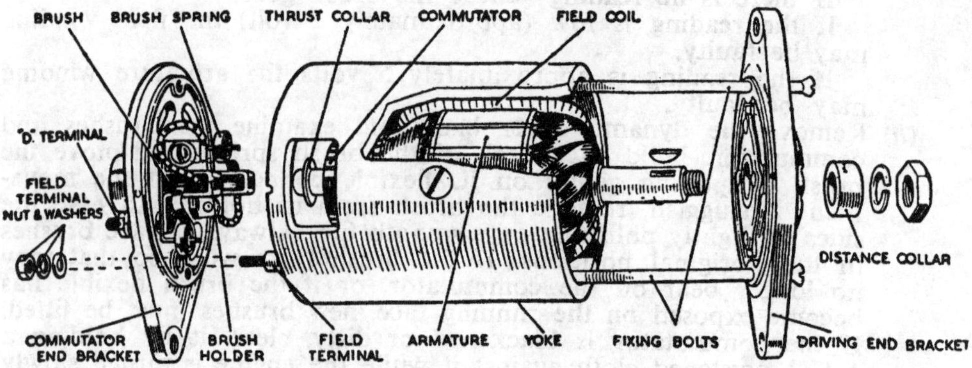

Fig. 36. *The Dynamo components.*

Brushes.

Test if the brushes are sticking. Clean them with petrol and, if necessary, ease the sides by lightly polishing with a smooth file. Replace the brushes in their original positions.

Test the brush spring tension with a spring scale if available. The correct tension is 20/25 oz. (567/709 grams). Fit a new spring if the tension is low.

If the brushes are worn so that the flexible is exposed on the running face, new brushes *must* be fitted. Brushes are pre-formed so that bedding to the commutator is unnecessary.

Commutator.

A commutator in good condition will be smooth and free from pits or burned spots. Clean the commutator with a petrol-moistened cloth. If this is ineffective, carefully polish with a strip of fine glasspaper while rotating the armature. To remedy a badly worn commutator, mount the armature (with or without the drive end bracket) in a lathe, rotate at high speed and take a light cut with a very sharp tool. Do not remove more metal than is necessary. Polish the commutator with very fine glass-paper. Undercut the insulators between the segments to a depth of $\frac{1}{32}$ in. with a hack-saw blade ground down to the thickness of the insulator.

Field Coils.

Test the field coils, without removing them from the dynamo yoke, by means of an ohmmeter. The reading on the ohmmeter should be between 6.0 and 6.3 ohms. If this is not available, connect a 12-volt D.C. supply with an ammeter in series between the field terminal and dynamo yoke The ammeter reading should be approximately 2 amps. If no reading is indicated the field coils are open-circuited and must be replaced. To test for earthed field coils, unsolder the end of the field winding from the earth terminal on the dynamo yoke and, with a test lamp connected from supply mains, test across the field terminal and earth. If the lamp lights the field coils are earthed and must be replaced.

When replacing field coils, carry out the procedure outlined below, using an expander and wheel-operated screwdriver.

(*a*) Remove the insulation piece which is provided to prevent the junction of the field coils from contacting the yoke.

(*b*) Mark the yoke and pole shoes in order that they can be fitted in their original positions.

(*c*) Unscrew the two pole shoe retaining screws by means of the wheel-operated screwdriver.

(*d*) Draw the pole shoes and coils out of the dynamo yoke and lift off the coils.

(*e*) Fit the new field coils over the pole shoes and place them in position inside the yoke. Take care to ensure that the taping of the field coils is not trapped between the pole shoes and the yoke.

(*f*) Locate the pole shoes and field coils by lightly tightening the fixing screw.

(*g*) Insert the pole shoe expander, open it to the fullest extent and tighten the screws.

(*h*) Finally tighten the screws by means of the wheel-operated screwdriver and lock them by caulking.

(*i*) Replace the insulation piece between the field coil connections and the yoke.

Armature.

The testing of the armature winding requires the use of a voltage drop test and growler. If these are not available, the armature should be checked by substitution. No attempt should be made to machine the armature core or to true a distorted armature shaft.

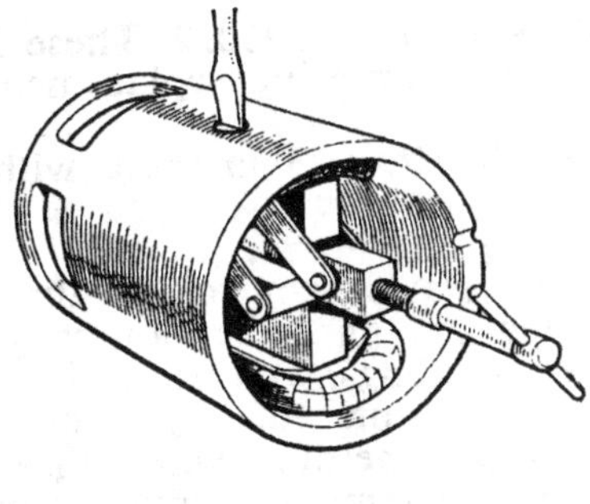

Fig. 37. *Fitting pole shoes using expander.*

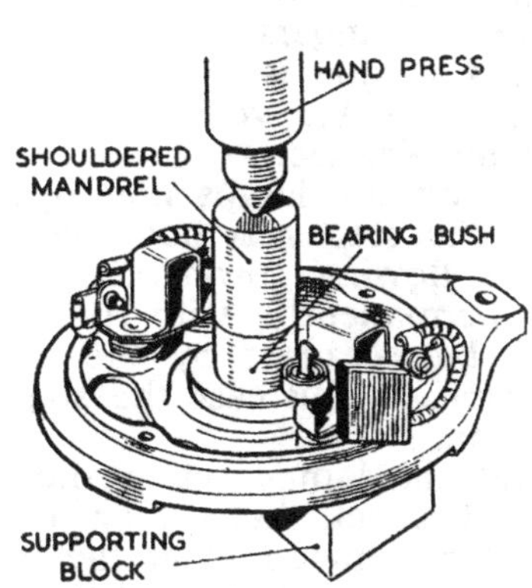

Fig. 38. *Correct method of fitting the Bronze bearing bush on the dynamo.*

Bearings.

Bearings which are worn to such an extent that they will allow side movement of the armature shaft must be replaced.

To replace the bearing bush at the commutator end, proceed as follows:—

(*a*) Press the bearing bush out of the commutator end bracket.

(*b*) Press the new bearing bush into the end bracket, using a shouldered mandrel of the same diameter as the shaft which is to fit in the bearing.

NOTE.—Before fitting the new bearing bush it should be allowed to stand completely immersed for twenty-four hours in thin engine oil. This will allow the pores of the bush to be filled with lubricant.

The ball bearing at the driving end is replaced as follows:—

(*a*) Knock out the rivets which secure the bearing retaining plate to the end bracket and remove the plate.

(*b*) Press the bearing out of the end bracket and remove the corrugated washer, felt washer and oil-retaining washer.

(*c*) Before fitting the replacement bearing see that it is clean and pack it with a high-melting-point grease.

(*d*) Place the oil-retaining washer, felt washer and corrugated washer in the bearing housing in the end bracket.

(*e*) Locate the bearing in the housing and press it home by means of a hand press.

(*f*) Fit the bearing retaining plate. Insert the new rivets from the inside of the end bracket and open the rivets by means of a punch to secure the plate rigidly in position.

Reassembly.

In the main the reassembly of the dynamo is a reversal of the dismantling operations.

Before refitting the dynamo to the vehicle, unscrew the lubricator from the commutator end bracket, lift out the felt wick and spring and refill the cap with high-melting-point grease. Replace spring and wick and screw the lubricator in position in the end bracket.

THE STARTER.

Type.

The starter is a Lucas Model M35G, Service No. 25022. These identification marks are stamped on the yoke. When ordering replacements always quote these numbers.

Its lock torque is approximately 9.3 lb./ft. (1.3 m./kg.), with 300-350 amps. and 7.5—8.0 volts.

To Test on Vehicle.

Switch on the lamps and operate the starter control. If the lights go dim, but the starter is not heard to operate, an indication is given that current is flowing through the starter windings, but that the starter is meshed permanently with the geared ring on the flywheel. This has probably been caused by the starter being operated while the engine is still running. In this case the starter must be removed from the engine for examination.

Should the lamps retain their full brilliance when the starter switch is operated, check that the switch is functioning. If the switch is in order, examine the connections at the battery, starter switch and starter, and also check the wiring between these units. Continued failure of the starter to operate indicates an internal fault in the starter, and the starter must be removed from the engine for examination.

Sluggish or slow action of the starter is usually caused by a poor connection in the wiring which produces a high resistance in the starter circuit. Check as described above.

Damage to the starter drive is indicated if the starter is heard to operate but does not crank the engine.

REMOVAL AND REPLACEMENT OF STARTER.

Remove the two nuts, bolts and spring washers securing the rear exhaust bracket to the rear cross-member.

Remove the two nuts, bolts and spring washers securing the exhaust bracket to the third cross-member.

Remove the nut, bolt and spring washer securing the exhaust pipe clip to the exhaust bracket attached to the gearbox.

Remove the three nuts and spring washers attaching the exhaust pipe to the exhaust manifold and lower the exhaust system complete to the ground.

(The exhaust system should be supported at intervals during these operations.)

Remove the set screw securing the cable from the starter switch to the starter motor at the commutator end.

Unscrew the three set bolts securing the starter motor to the flywheel housing and withdraw the starter.

Replacement is carried out in the reverse manner to that detailed for removal.

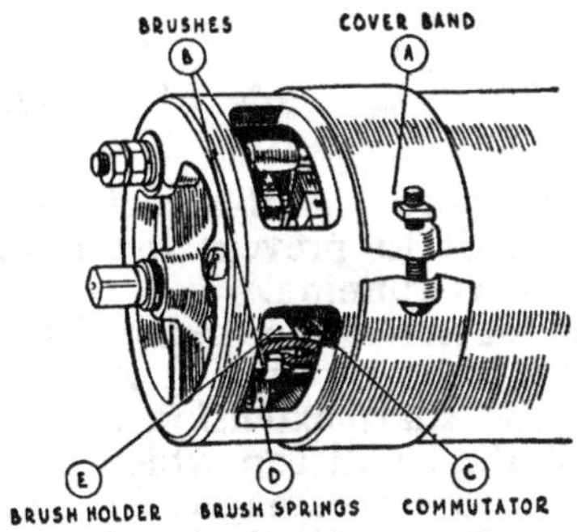

Fig. 39. *Starter brush gear.*

STARTER SWITCH.

The starter switch is fitted on the engine side of the bulkhead and is operated by pulling the control knob in the instrument panel. The knob is connected to the switch by means of a cable. At the back of the starter switch are two terminals, from one of which two cables are connected, the heavy one to the battery and the other to the ammeter. From the second terminal a heavy cable is taken to the terminal on the commutator end bracket of the starter.

THE STARTER COMMUTATOR AND BRUSH GEAR.

Inspect the commutator and brush gear in the manner described for the dynamo on pages 338 and 339. The brush spring tension should be 30—40 oz. (850—1,134 grams). This operation should be carried out every 12,000 miles or 20,000 kilometres.

In the unlikely event of brush replacement being necessary, this must be carried out by a Lucas Service Agent, as the operation entails dismantling of the starter.

THE STARTER DRIVE.

It is recommended that the starter drive be examined every 12,000 miles or 20,000 kilometres, taking particular care to ensure that the pinion moves freely on the screwed sleeve. Any dirt or other foreign matter on the sleeve must be washed away with paraffin.

When the starter is operated, the rotation of the armature screws the pinion along the sleeve to mesh with the flywheel teeth. As soon as the engine starts firing, it will drive the flywheel faster than it is being driven by the starter. This will cause the pinion to screw back along the sleeve

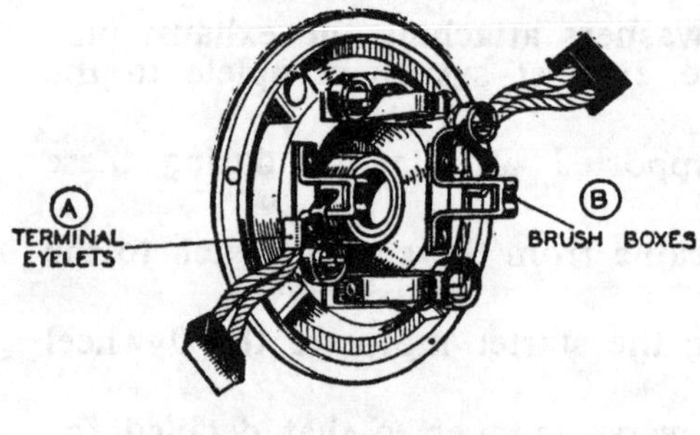

Fig. 40. *Starter end cover.*

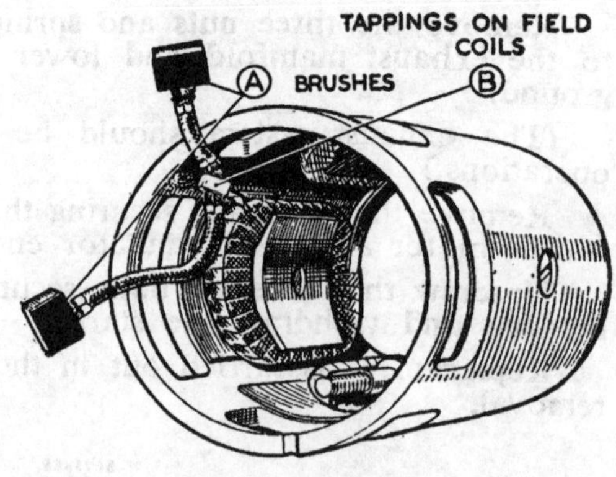

Fig. 41. *Brush connections for the starter.*

and so draw out of mesh with the flywheel teeth. A pinion restraining spring is fitted over the shaft to prevent the pinion from being vibrated into mesh when the engine is running.

In the event of the starter drive being jammed in mesh with the flywheel, it can usually be freed by turning the starter armature by means of a spanner applied to the shaft extension at the commutator end. This is accessible by taking off the small cap which is secured by two screws.

If the pinion is found to be damaged or worn, it must be replaced together with the screwed sleeve. Similarly, if the main spring is broken, or the restraining spring weak, or broken, replacement must be made.

SERVICING THE STARTER.

Examination of Commutator and Brush Gear.

Remove the starter cover band " A " (Fig. 39) and examine the brushes " B " and the commutator " C." Hold back each of the brush springs " D " and move the brush by pulling gently on its flexible connector. If the movement is sluggish, remove the brush from its holder " E " and ease the sides by lightly polishing on a smooth file. Always replace brushes in their original positions. If the brushes are worn so that they no longer bear on the commutator or if the brush flexible has become exposed on the running face they must be replaced.

If the commutator is blackened or dirty, clean it by holding a petrol-moistened cloth against it while the armature is rotated.

Dismantling.

Take off the cover band " A " (Fig. 39) at the commutator end, hold back the brush springs " D " and take out the brushes " B " from their holders.

Withdraw the two through bolts and remove the armature complete with driving end bracket.

Remove the terminal nuts and washers from the terminal post at the commutator end bracket and remove the commutator end bracket.

Brushes.

(*a*) Test the brush springs with a spring scale. The correct tension is 30—40 oz. (850-1,134 grams). Fit a new spring if the tension is low.

(*b*) If the brushes are worn so that they no longer bear on the commutator, or if the flexible connector has become exposed on the running face, they must be replaced. Two of the brushes are connected to terminal eyelets " A " (Fig. 40) attached to the brush boxes " B " on the commutator end bracket, and two " A "

(Fig. 41) are connected to a tapping "B" on the field coils.

The flexible connectors must be removed by unsoldering and the connectors of the new brushes secured in their place by soldering. The brushes are pre-formed so that bedding of their working face to the commutator is unnecessary.

Drive.

(*a*) If the pinion is tight on the screwed sleeve, wash away any dirt with paraffin.

(*b*) If any parts are worn or damaged they must be replaced.

(*c*) Remove the split pin "H" (Fig. 42) from the shaft nut at the end of the starter drive. Hold the squared end of the starter shaft at the commutator end by means of a spanner and unscrew the shaft nut "J."

(*d*) Lift off the main spring "K" (Fig. 42) and remove the retaining ring "L."

(*e*) The control nut "M," sleeve "N" and restraining spring "P" will now slide off.

(*f*) Withdraw the splined washer from the armature shaft and remove the pinion and barrel assembly.

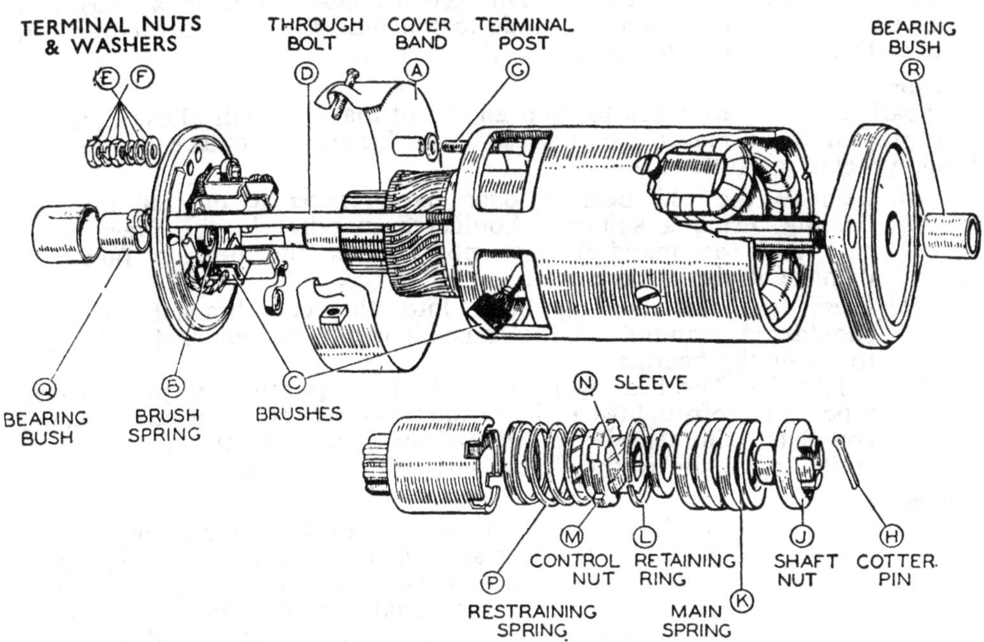

Fig. 42. *The component parts of the starter.*

Commutator.

A commutator in good condition will be smooth and free from pits and burned spots. Clean the commutator with a petrol-moistened cloth. If this is ineffective, carefully polish with a strip of fine glass-paper, while rotating the armature. To remedy a badly worn commutator, dismantle the starter drive as previously described and remove the armature from the end bracket. Now mount the armature in a lathe, rotate it at a high speed and take a light cut with a very sharp tool. Do not remove any more metal than is absolutely necessary, and finally polish with very fine glass-paper.

The insulator between the commutator segments **must not be undercut.**

Field Coils.

The field coils can be tested for an open circuit by connecting a 12-volt battery, with a 12-volt bulb in one of the leads, to the tapping point of the field coils to which the brushes are connected, and the field terminal post. If the lamp does not light, there is an open circuit in the wiring of the field coils.

Lighting of the lamp does not necessarily mean that the field coils are in order, as it is possible that one of them may be earthed to a pole shoe or to the yoke. This may be checked by removing the lead from the brush connector and holding it on a clean part of the starter yoke. Should the bulb now light it indicates that the field coils are earthed.

Should the above tests indicate that the fault lies in the field coils, they must be replaced. When replacing field coils carry out the procedure detailed for the Dynamo on page 337.

Armature.

Examination of the armature will in many cases reveal the cause of failure, e.g., conductors lifted from the commutator due to the starter being engaged while the engine is running and causing the armature to be rotated at an excessive speed. A damaged armature must in all cases be replaced—no attempt should be made to machine the armature core or to true a distorted armature shaft.

Bearings.

Bearings which are worn to such an extent that they will allow excessive side play of the armature shaft must be replaced. To replace the bearing bush, proceed as follows:—

(*a*) Remove the old bearing bush by pressing it out of the end bearing, using a suitable shouldered mandrel in the press. Take care that the mandrel does not contact the bearing face and damage it during removal.

(*b*) Press the new bearing bush into the end bracket, using a shouldered mandrel of the same diameter as the shaft which is to fit in the bearing.

NOTE.—The bearing bushes are of the porous phosphor-bronze type and before fitting them they should be allowed to stand completely immersed for twenty-four hours in thin engine oil in order to fill the pores of the bush with lubricant.

Reassembly.

The reassembly of the starter is a reversal of the dismantling.

NOTE.—Secure the body of the starter in a vice and test by connecting it with heavy gauge cables to a battery of the correct voltage. One cable must be connected to the starter terminal and the other held against the starter body or end bracket. Under these light load conditions the starter should run at a very high speed.

THE CONTROL BOX.

Regulator Adjustment.

The regulator is carefully set before leaving the Works to suit the normal requirements of the standard equipment, and in general it should not be necessary to alter it. If, however, the battery does not keep in a charged condition, or if the dynamo output does not fall when the battery is fully charged, it may be advisable to check the setting and if necessary to readjust it.

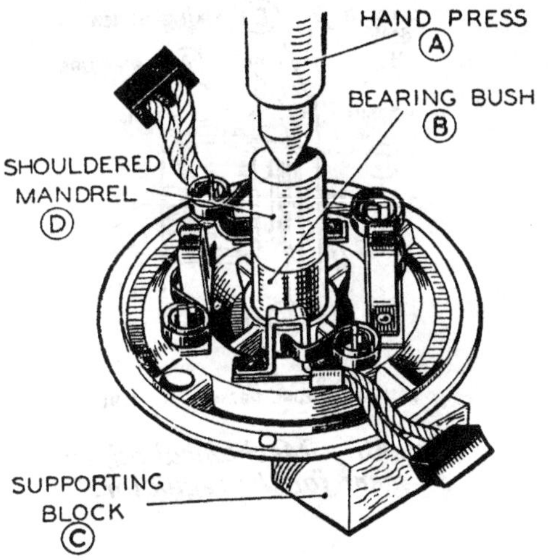

Fig. 43. *Method of inserting a new bush in the starter end cover.*

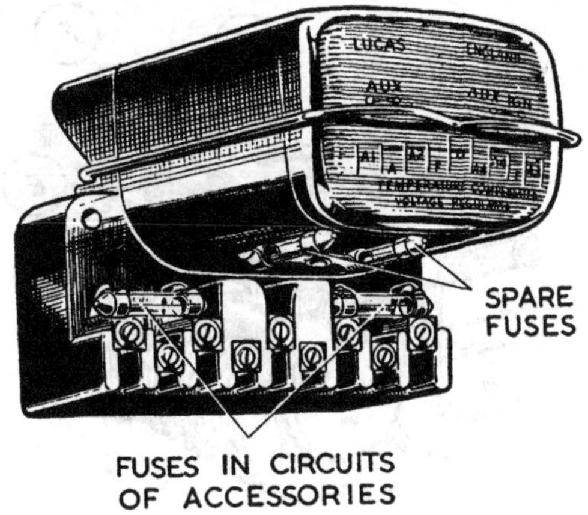

Fig. 44. *The control box showing location of fuses.*

It is important, before altering the regulator setting, when the battery is in a low state of charge, to check that its condition is not due to a battery defect or to the dynamo belt slipping.

How to Check and Adjust Electrical Setting.

The regulator setting can be checked without removing the cover on the control box.

Withdraw the cables from the terminals marked "A" and "A.I" at the control box and join them together. Connect the negative lead of a moving coil voltmeter (0-20 volts full scale reading) to the "D" terminal on the dynamo and connect the other lead from the meter to a convenient chassis earth.

Slowly increase the speed of the engine until the voltmeter needle "flicks" and then steadies; this should occur at a voltmeter reading between the limits given in the next column for the appropriate temperature of the regulator.

Setting at	10° C. (50° F.)	16.1—16.7 volts
" "	20° C. (68° F.)	15.8—16.4 "
" "	30° C. (86° F.)	15.6—16.2 "
" "	40° C. (104° F.)	15.3—15.9 "

If the voltage at which the reading becomes steady occurs outside these limits, the regulator must be adjusted.

Shut off the engine, remove the control box cover, release the locknut "A" (Fig. 45) holding the adjusting screw "B." The screw turns in a clockwise direction to raise the setting or in an anti-clockwise direction to lower the setting. Turn the adjusting screw a fraction of a turn in the required direction and then tighten the locknut.

When the dynamo is run at a high speed on open circuit it builds up a high voltage. When adjusting the regulator do not run the engine up to more than half throttle or a false voltmeter reading will be obtained.

Mechanical Setting.

The mechanical setting of the regulator is accurately adjusted before leaving the Works, and provided that the armature carrying the moving contact is not removed, the regulator will not require mechanical adjustment. If, however, the armature has been removed from the regulator for any reason, the contacts will have to be reset. To do this, proceed as follows:—

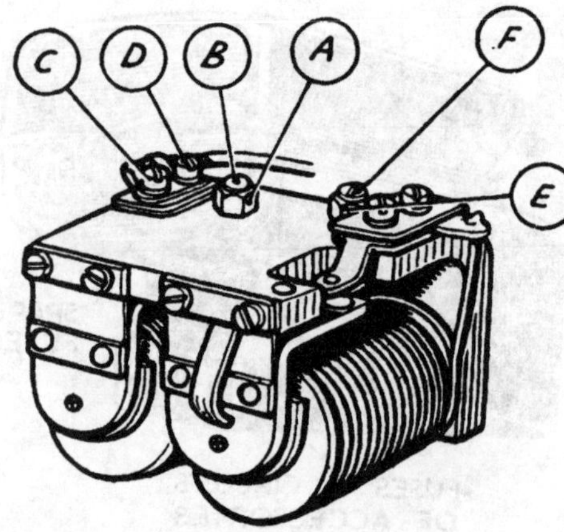

Fig. 45. The cut-out and regulator assembly.

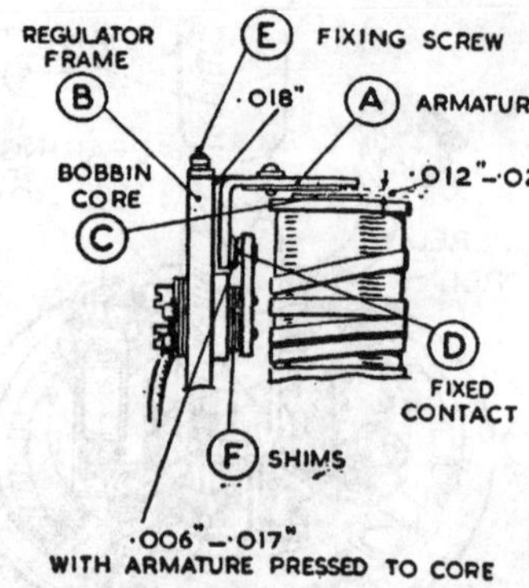

Fig. 46. Mechanical adjustment for the regulator.

(i) Slacken the two armature fixing screws "E" (Fig. 46). Insert a .018 in. (.46 mm.) feeler gauge between the back of the armature "A" and the regulator frame.

(ii) Press back the armature against the regulator frame and down on to the top of the bobbin core with the gauge in position, and lock the armature by tightening the two fixing screws.

(iii) Check the gap between the underside of the arm and the top of the bobbin core. This must be .012—.020 in. (.30—.50 mm.). If the gap is outside these limits correct by adding or removing shims "F" at the back of the fixed contact "D."

(iv) Remove the gauge and press the armature down, when the gap between the contacts should be between .006 in. (.15 mm.) and .017 in. (.43 mm.).

Cleaning Contacts.

To render the regulator contacts accessible for cleaning, slacken the screws securing the plate carrying the fixed contact. It will be necessary to slacken the upper screw "C" (Fig. 45) a little more than the lower "D" so that the contact plate can be swung outwards. Clean the contacts by means of fine carborundum stone or fine emery cloth. Carefully wipe away all traces of dirt or other foreign matter. Finally tighten the securing screws.

CUT-OUT.

Adjustment.

If it is suspected that the cutting-in speed of the dynamo is too high, connect a voltmeter between the terminals marked "D" and "E" at the control box and slowly raise the engine speed. When the voltmeter reading rises to between 12.7 and 13.3 the cut-out contact should close.

If the cut-out has become out of adjustment and operates at a voltage outside these limits it must be reset. To make the adjustment, slacken the locknut "E" (Fig. 45), turn the adjusting screw "F" a fraction of a turn in a clockwise direction to raise the operating voltage or in an anti-clockwise direction to lower the voltage. **Tighten locknut after making the adjustment.**

Cleaning.

To clean the contacts remove the cover, place a strip of fine glass-paper between the contacts and then, closing the contacts by hand, draw the paper through. This should be done two or three times, with the rough side towards each contact.

Radio Suppression.

When it is desired to fit suppressors for radio equipment, make sure that this is done only in accordance with recommended practice. Suppressors and capacitors wrongly fitted may cause damage to the electrical equipment.

WINDSCREEN WIPER—TYPE CW.

The motor is a three-pole wound armature type running in a two-pole field. A celeron plate carries the brush gear for two carbon brushes which bear on the commutator.

If the wiper fails to operate or operates unsatisfactorily, proceed as follows:—

Ascertain if the battery is supplying current to the wiper by switching on and noting if the ammeter responds.

Examine the fuse protecting the wiper. If it has blown, examine the wiring for evidence of short circuits or chafed leads. If, on replacing the fuse, it blows again, remove the wiper from the car. Momentarily connect wiper to a battery and see if the wiper then operates satisfactorily.

1. Switch Setting out of Adjustment or Switch Contacts Dirty.

If the cover of the wiper has been damaged, or if the switch has been tampered with, the contacts may be remaining permanently open. Remove the cover and see that the switch control moves freely—when the cover is removed the switch contacts should remain closed. Examine the contacts and if necessary clean them with fine emery cloth.

2. Loose or Broken Connections.

Remove the wiper cover and examine the wiring of the motor. If necessary, re-solder any connections which are loose or broken.

3. Brush Worn Out or not Bearing on Commutator.

The brushes are carried in small boxes and are pressed against the commutator by means of springs. See that the springs press firmly on the brushes and that they do not foul the sides of the brush boxes. The brushes must be clean and move freely in their holders. To remove the brushes for examination, hold back the springs by means of a bent piece of wire and gently lever the brushes from their holders.

If the brushes are dirty or greasy, causing them to stick in their holders, clean them with a cloth moistened with petrol.

When refitting the brushes, replace them in the same boxes and in the same position as originally fitted, in order to preserve the brush bedding.

If the brushes have become worn to such an extent that they no longer make good contact with the commutator, they must be replaced. Replacement brushes can be obtained from any Lucas Agent.

4. Armature Binding or Bearings Stiff.

Turn armature by hand for several revolutions—if it is consistently tight the wiper probably needs lubrication. If the armature is only tight occasionally, the stiffness is probably in the gearbox or is caused by dirt or other foreign matter in the air gap between the armature and the pole shoes. Remove the wiper back plate and examine the gears and links for evidence of stiffness.

5. Commutator Dirty.

Clean the commutator by means of a cloth moistened with petrol. Carefully remove any carbon dust from between the segments of the commutator.

6. Brush Gear Short-circuited.

This may be caused either by damage to the wiper cover or to a stray conductor making contact with the brush gear. Rectify the damage to the wiper cover or fit a new cover and remedy any other damage.

7. Armature Damaged.

If, after following the preceding examination, the wiper still does not operate or its performance is unsatisfactory, the fault may be due to the armature being damaged. Fit a replacement armature.

8. Lubrication.

If the rotor is consistently stiff when turned by hand, add a few drops of thin machine oil to the lubricator provided. When examining the gearbox, if necessary, lightly pack the gears with a good quality high-melting-point zinc-oxide grease, such as Duckham's Keenol KG.25.

THE ELECTRIC HORNS—MODEL WT614.

All horns before being passed out of the Works are adjusted to give their best performance and will give a long period of service without any attention; no subsequent adjustment is required.

If one of the horns fails or becomes uncertain in its action, it does not follow that the horn has broken down. First ascertain that the trouble is not due to some outside source, e.g., a discharged battery, or loose connection or short circuit in the wiring of the horn; a short circuit in the horn wiring will cause the fuse to blow. If both horns fail or become uncertain in action, the trouble is probably due to a discharged battery or blown fuse. If the fuse has blown, examine the wiring for the fault and replace with the spare fuse provided.

It is also possible that the performance of a horn may be upset by the fixing bolt working loose, or by some component near the horn becoming loose. If after carrying out the above examination the trouble is not rectified, the horn may need adjustment, but this should not be necessary until the horns have been in service for a long period.

Adjustment does not alter the pitch of the note, it merely takes up wear of moving parts. When adjusting the horns, short-circuit the fuse, otherwise it is liable to blow. Again, if the horns do not sound on adjustment, release the push instantly.

When making adjustments to a horn, always disconnect the supply lead of the other horn, taking care to ensure that it does not come into contact with any part of the chassis and so cause a short circuit.

Adjustment.

Remove the horn cover after withdrawing the fixing screw and detach the cover securing bracket by springing it from its fixing.

Slacken the locknut on the fixed contact and rotate the adjusting nut until the contacts are just separated (indicated by horns failing to sound). Turn the adjusting nut half a turn in the opposite direction and secure in this position by tightening the locknut. Finally, if the note is still unsatisfactory, do not dismantle the horn but return it to a Lucas Service Depôt or Service Agent for examination.

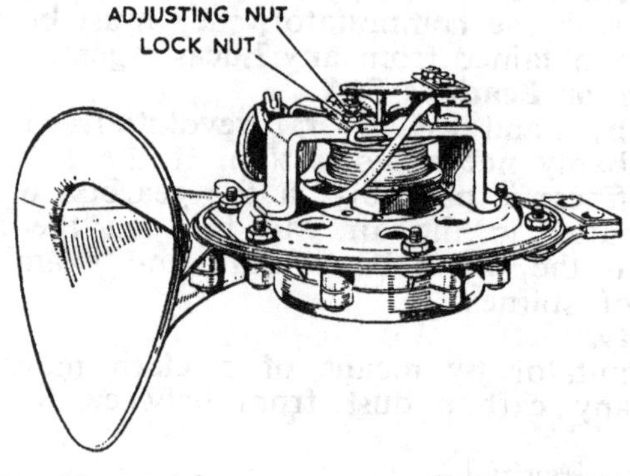

Fig. 47. Horn with cover removed showing adjustment

THE FUSES.

The fuses are accessible without removing the cover over the cut-out and regulator. The cover is therefore sealed.

Fuse Marked " AUX."

This fuse protects the accessories which are connected so that they operate irrespective of whether the ignition switch is " on " or " off."

It is of 35 amp. rating.

Fuse Marked " AUX IGN."

This fuse protects the accessories which are connected so that they operate only when the ignition is switched on.

It is also of 35 amp. rating.

Units Protected.

The units which are protected by each fuse can readily be identified by referring to the wiring diagram.

Blown Fuses.

A blown fuse is indicated by the failure of all the units protected by it, and is confirmed by examination of the fuse, which can easily be withdrawn from the spring clips in which it fits. If it has blown, the fused state of the wire will be visible inside the glass tube. Before replacing a blown fuse, inspect the wiring of the units that have failed for evidence of a short circuit or other faults which may have caused the fuse to blow, and remedy the cause of the trouble. This is essential or the fuse is liable to blow again on replacement. Two spare fuses are carried in the clips on the underside of the regulator cover.

THE HEADLAMPS.

Each headlamp incorporates a Lucas light unit, which consists essentially of a reflector and front glass assembly. The bulb, which is a Lucas " pre-focus " type, is located accurately in the reflector and is secured by a bayonet-fixed backshell which also provides the contact to the bulb. The design of the bulb and of its holder is such that the bulb is correctly positioned in relation to the reflector and no focusing is required when a replacement bulb is fitted.

The Anti-dazzle Device.

Home models are equipped with a double-filament bulb in the left-hand headlamp, controlled by the dipping switch. Operation of the dip switch extinguishes the right-hand lamp and simultaneously deflects the left-hand headlamp beam downwards and to the left to provide good illumination of the left-hand kerb.

On export models both headlamps are fitted with double-filament bulbs. Operation of the dip switch extinguishes the main driving beam in each headlamp and brings the dipping beams into action.

Bulb Replacement.

To gain access to the bulb, slacken the screw at the top of the lamp and remove the front rim and light unit assembly. Twist the backshell in an anti-clockwise direction and pull it off. The bulb can now be removed from the rear of the reflector. Place the replacement bulb in the holder and engage the projections of the backshell with the slots in the bulb holder, press on and secure by twisting to the left.

SETTING THE HEADLAMPS.

Lucas Beam Setter.

All Lucas Service Depôts and many service agents now include among their special testing facilities an apparatus known as the Lucas Beam Setter. This is a specially designed instrument by means of which headlamps can be set with extreme accuracy, thereby ensuring the maximum efficiency from the lamps.

You are strongly advised to make use of this service. If, however, it is not possible to have the lamps adjusted by this method, the procedure for setting is as follows:—

Each lamp must be set so that the main driving beam is parallel with the road surface, or in accordance with the local regulations. To make the adjustments, slacken the single fixing nut at the base of the lamp and move the lamp on its adjustable mounting to the required position. Finally tighten the locknut.

THE SIDELAMPS.

To gain access to the bulb slacken the screw at the top of the lamp and withdraw the front rim and glass assembly.

THE TAIL- AND STOP-LAMPS.

Twin lamps are fitted to all models. Each lamp is fitted with a double-filament bulb (6/24 watt). The small wattage filament is connected to the tail-lamp circuit and the large wattage filament to the stop-lamp circuit.

Two types of tail-lamp bulb holders have been employed, the normal symmetrical type and, more recently, holders with offset bayonet slots.

The latter type have been introduced to ensure that the bulbs cannot be fitted the wrong way round, thus causing the high wattage filament to be in circuit when the tail-lamp switch is on, and the low wattage filament to be operated by the stop switch.

When fitting replacement bulbs no difficulty should be experienced in fitting them correctly, as Lucas No. 189 bulbs (symmetrical type) are marked with the word "TOP" on the cap and Lucas No. 353 bulbs (non-reversible type) are designed so that the locating pegs are offset and therefore can only be fitted into the bulb holder in the correct position.

To gain access to the bulb, slacken the two screws held by nuts and spring washers under the wing, and withdraw the rim and glass assembly.

Fig. 48. Tail-lamp dismantled.

THE NUMBER-PLATE ILLUMINATION LAMP.

To gain access to the bulbs, slacken the single central securing screw for the plated cover and withdraw the cover and glass assembly.

The 12-volt lamp bulbs are of the miniature bayonet type and bear the Lucas Part No. 989.

THE PANEL LAMPS.

The instruments are illuminated by small screw-in type bulbs located behind the instrument panel and access to them can be obtained through a hand-hole cut in the inner panel close to the steering column.

More comfortable access can be achieved by removing the facia board undershield panel.

REMOVAL OF THE INSTRUMENT PANEL.

Release the large facia panel undershield and lower the steering column. This will give access to the rear of the instrument panel which is held in place by six nuts. Disconnect the starter and mixture controls, remove the panel attachment nuts and withdraw the panel.

THE IGNITION WARNING LAMP.

Access to the bulb of the ignition warning lamp is obtained by unscrewing the slotted plated rim surrounding the light. Removal of the rim releases the bulb from its spring holder and no difficulty arises in replacement.

The ignition warning bulb is a Lucas No. 970 2.5-volt, .5 watt.

REPLACEMENT BULBS.

	Lucas No.	Volts	Watts
Headlamps (Main), Home (L/H) ...	300 (dip left) ...	12	36/36
Headlamps (Main), Home (R/H) ...	162	12	36
Headlamps, Export (R.H.D.) ...	300 (dip left) ...	12	36/36
Headlamps, Export (L.H.D.) ...	301 (dip right) ...	12	36/36
Headlamps, Export (Europe and N. Africa)	Philips Duplo No. 12721	12	35/35
Sidelamps	207	12	6
Stop/tail-lamps (reversible) ...	189	12	6/24
Stop/tail-lamps (non-reversible) ...	353	12	6/24
Number-plate illumination lamp ...	989	12	6
Ignition warning light	970	2.5	.5

LOCATION AND REMEDY OF FAULTS.

Although every precaution is taken to eliminate possible causes of trouble, failure may occasionally develop through lack of attention to the equipment, or damage to the wiring. The following pages set out the recommended procedure for a systematic examination to locate and remedy the causes of some of the more usual faults encountered.

The sources of trouble are by no means always obvious, and in some cases a considerable amount of deduction from the symptoms is needed before the cause is disclosed.

For instance, the engine might not respond to the starter switch; a hasty inference would be that the starter motor is at fault. However, as the motor is dependent on the battery, it may be that the battery is exhausted.

This, in turn, may be due to the dynamo failing to charge the battery, and the final cause of the trouble may be, perhaps, a loose connection in some part of the charging circuit.

If, after carrying out an examination, the cause of the trouble is not found, the equipment should be checked by the nearest Lucas Service Depôt or Agent.

Much needless damage can be done to items of the equipment by attempting to correct faults without the necessary equipment and experience.

THE CHARGING CIRCUIT.

1. Battery in Low State of Charge.

(*a*) This state will be shown by lack of power when starting, poor light from the lamps, and hydrometer readings below 1.200. It may be due to the dynamo not charging or giving low or intermittent output. The ignition warning light will not go out if the dynamo fails to charge, or will flicker on and off in the event of intermittent output.

(*b*) Examine the charging and field circuit wiring, tightening any loose connections or replacing broken cables. Pay particular attention to the battery connections.

(*c*) Examine the dynamo driving belt; take up any undue slackness by swinging the dynamo outwards on its mounting after slackening the attachment bolts.

(*d*) Check the regulator setting and adjust if necessary.

(*e*) If, after carrying out the above, the trouble is still not cured, have the equipment examined by a Lucas Service Depôt or Agent.

2. Battery Overcharged.

(*a*) This will be indicated by burnt-out bulbs, very frequent need for topping up the battery, and high hydrometer readings. Check the charging current when the engine is running steadily with a fully charged battery and no lights or accessories in use; the charge reading should be of the order of only 3—4 amperes.

If the charge reading is in excess of this value, it is advisable to check the regulator setting and adjust if necessary.

THE STARTER MOTOR.

1. Starter Motor Lacks Power or Fails to Turn Engine.

(*a*) See if the engine can be turned over by hand. If not, the cause of the stiffness in the engine must be located and remedied.

(*b*) If the engine can be turned by hand, first check that the trouble is not due to a discharged battery.

(*c*) Examine the connections to the battery, starter and starter switch, making sure that they are tight and that the cables connecting these units are not damaged.

(*d*) It is also possible that the starter pinion may have jammed in mesh with the flywheel, although this is by no means a common occurrence. To disengage the pinion, rotate the squared end of the starter shaft by means of a spanner.

2. Starter Operates but Does not Crank Engine.

This fault will occur if the pinion of the starter drive is not allowed to move along the screwed sleeve into engagement with the flywheel, due to dirt having collected on the screwed sleeve. Remove the starter and clean the sleeve carefully with paraffin.

3. Starter Pinion will not Disengage from Flywheel when Engine is Running.

Stop the engine and see if the starter pinion is jammed in mesh with the flywheel, releasing it if necessary by rotation of the squared end of the starter shaft. If the pinion persists in sticking in mesh, have the equipment examined at a Service Depôt. Serious damage may result to the starter if it is driven by the flywheel.

THE LIGHTING CIRCUITS.

1. Lamps Give Insufficient Illumination.

(*a*) Test the state of charge of the battery, recharging it if necessary from an independent electrical supply.

(*b*) Check the setting of the lamps.

(*c*) If the bulbs are discoloured as the result of long service, they should be replaced.

2. Lamps Light When Switched On But Gradually Fade Out.

As paragraph 1 (*a*).

3. Brilliance Varies With Speed of Car.

(*a*) As paragraph 1 (*a*).

(*b*) Examine the battery connections, making sure that they are tight, and replace any faulty cables.

KEY TO CABLE COLOURS.

1 Blue
2 Blue *with* Red
3 Blue *with* Yellow
4 Blue *with* White
5 Blue *with* Green
6 Blue *with* Purple
7 Blue *with* Brown
8 Blue *with* Black
9 White
10 White *with* Red
11 White *with* Yellow
12 White *with* Blue
13 White *with* Green
14 White *with* Purple
15 White *with* Brown
16 White *with* Black
17 Green
18 Green *with* Red
19 Green *with* Yellow
20 Green *with* Blue
21 Green *with* White
22 Green *with* Purple
23 Green *with* Brown
24 Green *with* Black
25 Yellow
26 Yellow *with* Red
27 Yellow *with* Blue
28 Yellow *with* White
29 Yellow *with* Green
30 Yellow *with* Purple
31 Yellow *with* Brown
32 Yellow *with* Black
33 Brown
34 Brown *with* Red
35 Brown *with* Yellow
36 Brown *with* Blue
37 Brown *with* White
38 Brown *with* Green
39 Brown *with* Purple
40 Brown *with* Black
41 Red
42 Red *with* Yellow
43 Red *with* Blue
44 Red *with* White
45 Red *with* Green
46 Red *with* Purple
47 Red *with* Brown
48 Red *with* Black
49 Purple
50 Purple *with* Red
51 Purple *with* Yellow
52 Purple *with* Blue
53 Purple *with* White
54 Purple *with* Green
55 Purple *with* Brown
56 Purple *with* Black
57 Black
58 Black *with* Red
59 Black *with* Yellow
60 Black *with* Blue
61 Black *with* White
62 Black *with* Green
63 Black *with* Purple
64 Black *with* Brown

ELECTRICAL EQUIPMENT
(1¼ Litre Model)

FOREWORD.

The information given below is intended to furnish complete maintenance instructions for the electrical equipment on the car, and also the procedure for carrying out the more usual replacements, such as dynamo brushes, etc. A limitation is necessarily imposed on the scope of servicing, since for detailed examination and adjustments special test fixtures, instruments and tools are required. To meet the need for servicing of a more advanced nature the Lucas Service Organization, with agencies in all parts of the world, is at the disposal of the motorist, dealer or distributor.

A 12-volt electrical system is provided, which is wired on the earth return system. The battery is automatically charged according to its load by a dynamo operating in conjunction with a voltage control unit. No charging switch is therefore provided.

DYNAMO.

Type.

The dynamo is a Lucas Model C.45Y-V3, Type L/1, Service Number 228139. These identification marks are stamped on the yoke. When ordering replacements always quote these numbers.

Commencing engine Types XPAG/SC (R.H.D.) and XPAG/SC/IHY (L.H.D.), each at No. 14023, a smaller dynamo, Lucas Model CP.39.PV, Part No. 162078, is fitted. It is important that control box Lucas Model RF.95/2, Type 6, is used in conjunction with this type of dynamo.

Removal and Replacement of Dynamo.

Remove the set screw securing the insulator pad to the rear of the dynamo and disconnect the cables.

Slacken off the adjuster bolt and the two bolts securing the dynamo to the two brackets.

Move the dynamo towards the engine until the fan belt is loose and remove the fan belt from the dynamo pulley.

Remove the adjuster bolt and the two bolts and nuts securing the dynamo to the brackets and withdraw the dynamo from the car.

Replacement is carried out in the reverse manner to that detailed for removal.

Lubrication.

The dynamo is provided with a wick-type lubricator at the commutator end. Every 12,000 miles or 20,000 kilometres unscrew the cap of the lubricator, lift out the felt pad and spring, and about half-fill the lubricator with grease or petroleum jelly. Replace the spring and felt pad and screw the lubricator in position.

Cleaning the Commutator.

Remove the cover band, which will expose " windows " in the dynamo yoke giving access to the brushgear and commutator. The commutator should be cleaned, and have a highly polished appearance. If it is dirty, clean it by pressing a fine dry duster against it while the armature is slowly rotated; this can be done by turning the engine by hand.

Brush Spring Tension.

Check the tension of the brush springs by means of a spring scale, if available. The correct spring tension is 15-25 oz. (.4-.7 kilogram). If the tension is low, fit a replacement spring.

Brushes.

Check that the brushes move freely in their holders by holding back the brush springs and pulling gently on the flexible connectors. If a brush is inclined to stick, remove it from its holder and clean its sides with a petrol-moistened rag.

Be careful to replace the brushes in their original positions so as to retain the " bedding." Brushes which have worn so that they will not bed correctly on the commutator, or on which the flexible lead is exposed on the running face, must be renewed. To do this proceed as follows:—

Remove the screw securing the eyelet on the end of the brush lead. Hold back the brush spring and remove the brush from its holder. Place the new brush in the holder and lower the spring gently. Secure the eyelet on the end of the brush lead in the original position.

Brushes are pre-formed but it is advisable also to " bed " them to the commutator. To do this, pass a thin strip of very fine glass paper between the commutator and each of the brushes so that the abrasive surface is towards the brush faces. Turn the armature in its normal direction of rotation for a few minutes and then remove the glass paper, afterwards wiping away carbon or glass paper dust. After several days' running the brush gear should again be examined and any dust wiped away.

STARTER.

Type.

The starter motor is a Lucas Model M.418G, Type L/O, Service Number 255378. These identification marks are stamped on the yoke. When ordering replacements always quote these numbers.

Commencing engine Type XPAG/SC No. 14023 on R.H.D. and L.H.D. models, the starter motor fitted is Lucas Model M.35.G1, Part No. X.31401. This type of starter, which is smaller, must only be used in conjunction with flywheel starter ring Part No. X.24418.

Removal and Replacement of Starter.

Remove the two nuts, bolts and spring washers securing the rear exhaust bracket to the rear cross-member.

Remove the two nuts, bolts and spring washers securing the exhaust bracket to the third cross-member.

Remove the nut, bolt and spring washer securing the exhaust pipe clip to the exhaust bracket attached to the gearbox.

Remove the three nuts and spring washers attaching the exhaust pipe to the exhaust manifold and lower the exhaust system complete to the ground.

(The exhaust system should be supported at intervals during these operations.)

Remove completely the tie-bar between the body bulkhead and chassis frame, also the tie-rod between the body and radiator top tank.

Remove the set screw securing the cable from the starter switch to the starter motor at the commutator end.

Unscrew the three set bolts securing the starter motor to the flywheel housing.

Withdraw the starter about 4 in. (10 cm.) and rotate until the terminal is at the bottom. Draw the starter over towards the right up to the Jackall hydraulic jack. The pinion end will now lift upwards clear of the flywheel housing. Slide the starter over, passing it under the steering column so that it lies in the gap between engine and chassis frame. Lift the pinion end up and bring outwards, over and around the flywheel housing. The whole starter will now come out vertically, pinion end first, in the space between the steering column, body bulkhead and wing valance.

Replacement is carried out in the reverse manner to that detailed for removal.

Starter Switch.

The starter switch is fitted on the engine side of the bulkhead and is operated by pressing the control knob in the instrument panel. The knob is connected to the switch by means of a flexible rod. At the back of the starter switch are two terminals, from one of which two cables are connected, the heavy one to the battery and the other (yellow and black) to the ammeter. From the second terminal a heavy cable is taken to the terminal on the commutator end bracket of the starter.

Two contacts are fitted on the commutator end bracket of the starter which on some cars form part of the starter switch. In this case, however, a separate switch is provided and an adaptor plate is fitted to the starter which permanently bridges the two contacts.

Commutator and Brush Gear.

Inspect the commutator and brush gear in the manner described for the dynamo. The brush spring tension should be 30-40 oz. (.85-1.13 kilograms). This operation should be carried out every 12,000 miles or 20,000 kilometres.

In the unlikely event of brush renewal being necessary, this must be carried out by a Lucas Service Agent, as the operation entails dismantling of the starter.

Drive.

It is recommended that the starter drive be examined every 12,000 miles or 20,000 kilometres, taking particular care to ensure that the pinion moves freely on the screwed sleeve. Any dirt or other foreign matter on the sleeve must be washed away with paraffin.

Starter Drive.

When the starter is operated, the rotation of the armature screws the pinion along the sleeve to mesh with the flywheel teeth. As soon as the engine starts firing, it will drive the flywheel faster than it is being driven by the starter. This will cause the pinion to screw back along the sleeve and so draw out of mesh with the flywheel teeth. A pinion restraining spring is fitted over the shaft to prevent the pinion from being vibrated into mesh when the engine is running.

In the event of the starter drive being jammed in mesh with the flywheel, it can usually be freed by turning the starter armature by means of a spanner applied to the shaft extension at the commutator end. This is accessible by pulling off the small cap which is secured by two screws.

If the pinion is found to be damaged or worn, it must be renewed together with the screwed sleeve. Similarly, if the main spring is broken, or the restraining spring weak or broken, a renewal must be fitted.

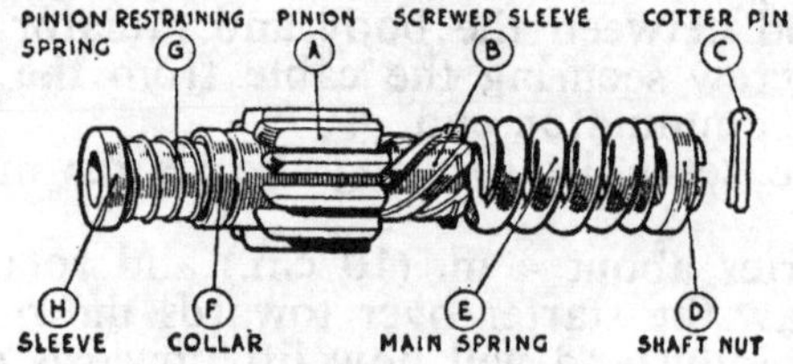

Fig. 49. Starter drive.

Dismantling the Starter Drive.

Remove the starter from the engine as detailed on page 354. Remove the cotter pin from the shaft nut at the end of the starter drive. Hold the squared end of the starter shaft at the commutator end by means of a spanner, and unscrew the square shaft nut. Lift off the main spring, screwed sleeve with pinion, collar, pinion restraining spring and sleeve on which the restraining spring fits. The drive should be reassembled with the new components by reversing the order of dismantling.

CONTROL BOX.

Type.

The control box is a Lucas Model RF.91.L.34, 12-volt, Service No. 37031A or Lucas Model RF.95/2, Type L, 12-volt, Service No. 37057A. These identification marks are stamped on the base of the control box. When ordering replacements always quote these numbers.

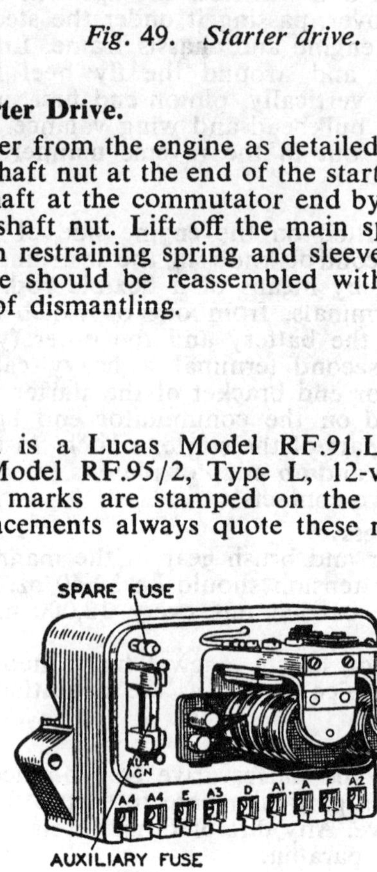

Fig. 50. Control box with cover removed.

Removal and Replacement of the Control Box.

Disconnect the cables from the terminals on the underside of the control box.

Remove the two set screws securing the control box to the engine bulkhead and the control box can then be withdrawn.

Replacement is carried out in the reverse manner to that detailed for removal, the correct connection of the various cables being determined by the colour code.

Description of Control Box.

This unit contains the cut-out and the voltage regulator. Both these units are accurately adjusted during manufacture and further adjustment is unnecessary.

Also mounted under the cover of the control box are two fuses connected in the circuits of the accessories. One of these fuses protects those accessories which are operative only when the ignition is switched on, namely, the windscreen wiper, fuel gauge, stop lamp and trafficators, and is marked AUX IGN.

The other fuse, marked AUX, protects accessories which can be operated irrespective of whether the ignition is on or off, namely, the horn and interior light.

NOTE:—Some cars will be found fitted with an alternative type control box, on which the fuses are external, obviating the necessity of removing the cover to replace a blown fuse.

The fuse will blow in the event of a short circuit occurring in the wiring of any of the components connected through it, or a fault in one of the units. Before replacing the fuse, which is of 25 amperes fusing value, carefully check the wiring of the accessories concerned, and tape any lead which is badly worn or chafed. If it is suspected that the fault lies in the accessory itself, the unit must be examined by a Lucas Service Agent.

Care must be taken when replacing the cover not to close the cut-out contacts, or damage to the equipment may result. If accidentally closed, the contacts must be carefully pulled apart.

BATTERY.

Topping Up.

The electrolyte in each cell should be level with the tops of the separators. Distilled water, which can conveniently be inserted by means of a clean hydrometer, should be used to bring the fluid up to this level. Do not use tap-water, and do not use a naked light when examining the condition of the cells.

The outside of the battery must be clean and dry, particularly the tops of the cells. An accumulation of dirt and moisture will cause a leakage path between the battery terminals which may slowly cause the battery to become discharged. See that the ventilating holes in the vent plugs are quite clean—any dirt can be removed by means of a piece of wire.

Clean away any corrosion from the battery terminals and smear them with petroleum jelly. Also examine the connections, making sure that the terminals are tight.

Measurement of Specific Gravity of Electrolyte.

The state of charge of the battery is checked by measuring the specific gravity of the electrolyte in each cell by means of a hydrometer. The specific gravity of the electrolyte in a cell of serviceable condition will rise during a charging period and fall during discharging. Specific gravity readings and their indications are:—

1.285—1.300.	Battery fully charged.
About 1.210.	Battery about half-discharged.
Below 1.150.	Battery fully discharged.

These figures are given assuming an electrolyte temperature of 60° F. (15.6° C.). If the electrolyte temperature exceeds this, .002 must be added to hydrometer readings for each 5° F. (2.8° C.) rise to give the true specific gravity of 60° F. Similarly, .002 must be subtracted from hydrometer readings for every 5° F. below 60° F.

The readings of the cells should be approximately the same. If one cell gives a reading very different from the rest, it may be that acid has been spilled or has leaked from this particular cell, or there may be a short circuit between the plates. In this case the battery should be examined by a Lucas Service Agent.

When taking specific gravity readings, examine the condition of the electrolyte in the hydrometer. It should be fairly clear. If it is very dirty, it is possible that the plates are in a bad condition.

Never leave the battery in a discharged condition for any length of time. Have it fully charged and every fortnight give it a short refresher charge to prevent any tendency for the plates to become permanently sulphated.

LAMPS.

Setting.

The setting of the lamps is checked by placing the car in front of a blank wall at the greatest possible distance, taking care that the surface on which the car is standing is not sloping relative to the wall.

Headlamps.

The headlamps must be set so that the main beams of light are parallel with the road and with each other. If adjustment is necessary, slacken the fixing nut at the base of the lamp and move the lamp on its adjustable mounting to the required position. Finally tighten the locking nut.

Fog-Lamp.

The lamp must be set so that the beam does not rise above the horizontal when the car is standing on level ground. To ensure this, dip the lamp very slightly to compensate for road inequalities or an extra load in the rear of the car, and also tilt the lamp to the right to allow for road camber. In addition, the lamp can be swung slightly to the left to give additional illumination to the left-hand side of the road. If adjustment is necessary, slacken the single fixing nut and move the lamp on its adjustable mounting to the required position. Finally tighten the locking nut.

Focusing.

To enable the best light to be obtained from the lamps, the bulb filament must be as near as possible to the focal point of the reflector. If a bulb is out of focus the lamp will have a poor range and will cause dazzle to approaching traffic.

Before lamps leave the works, the bulbs are focused to give the best results and, provided that genuine Lucas spare bulbs are fitted as replacements, it should not be necessary to alter the setting. If, however, an ordinary bulb has to be fitted, it may be necessary to re-focus by moving the bulb backwards or forwards until the best light is obtained.

The bulb and its holder can be adjusted when the lamp front and reflector are removed and the clamping clip on the bulb-holder is slackened.

When focusing the fog-lamp, adjust the position of the bulb until any top light is removed and until the semicircular beam of light is of the greatest concentration, i.e., the area of the beam must be as small as possible.

When focusing the headlamps, it is advantageous to cover one lamp while testing the other. If the lamp does not give a uniform long-range beam without a dark centre, the bulb needs adjusting. Tighten the clamping clip after the best position for the bulb has been found.

Headlamp Anti-Dazzle Device.

The left-hand headlamp is fitted with a double-filament bulb and the right-hand headlamp with a single-filament bulb. On operation of the foot dipper switch, the left-hand lamp gives a beam which points downwards for anti-dazzle purposes, while the right-hand lamp is switched off. In countries where the standard arrangement of headlamps is not permitted both lamps are fitted with double-filament bulbs.

REMOVING LAMP FRONTS AND REFLECTORS.

Headlamps.

To remove the lamp front pull forward the fixing clip at the bottom of the lamp and swing it downwards. Remove the front from the bottom of the lamp first. When replacing, the projection on the rim must fit into the left-hand location at the top of the lamp body so that the thinner lip of the bead is in the groove at the front of the lamp body. On later models fitted with prefocused bulbs and light unit the headlamp front is removed by swinging the top outwards after slackening the fixing screw which is inserted through the top of the lamp.

Fog-Lamp.

To remove the lamp front and reflector pull out the clip at the bottom of the lamp and swing it out of its location; the front and the reflector can then be lifted off.

The bulb holder can be pulled out of the back of the reflector. When replacing the front and reflector locate the top of the rim first, then press on at the bottom and secure by springing the clip into its location.

Sidelamps.

To remove the front for bulb renewal undo the fixing screw at the top of the lamp body and withdraw the front. The bulb is then accessible in the lamp body. When refitting the front locate at the bottom first and secure with the screw.

Stop, Tail and Reverse Lamps.

The fronts can be swung open when the fixing screws are removed.

Cleaning.

Care must be taken when handling the reflectors to prevent them from becoming finger-marked. If they become marked, a transparent and colourless protective covering enables any finger-marks to be removed by polishing with a chamois leather or a very soft cloth. **Do not use metal polish on reflectors.**

Metal polish must not be used for cleaning chromium-plated lamp bodies. They must be washed with plenty of water, and when the dirt has been completely removed the lamp bodies must be polished with a chamois leather or a soft dry cloth.

RENEWAL OF BULBS.

When renewing a bulb it is important not only that the same size bulb is fitted, but also that the bulb has a high efficiency and will focus in the reflector. Cheap and inferior bulbs often have the filament of such a shape that it is impossible to focus properly. This will cause dazzle and will result in the loss of range and efficiency.

Lucas genuine spare bulbs are sold by most garages and are specially tested to ensure that the filament is in the correct position to give the best results with Lucas lamps. Lucas bulbs have a number marked on the metal cap; when fitting a replacement see that it is of the same number as the original bulb.

It is advisable to replace the bulbs after long service before they actually burn out, as often the filaments sag, making it impossible for them to be focused properly.

Replacement Bulbs.

	Voltage	*Wattage*	*Lucas No.*
Headlamps			
Left-hand lamp (original type) Home ...	12	36–36	52
Left-hand lamp (pre-focus type) Home	12	36–36	300
Right-hand lamp (original type) Home	12	36	171
Right-hand lamp (pre-focus type) Home	12	36	162
Export (R.H.D.) (dip left) original type	12	36–36	52
Export (R.H.D.) (dip left) pre-focus ...	12	36–36	300
Export (L.H.D.) (dip right) original type	12	36–36	52
Export (L.H.D.) (dip right) pre-focus ...	12	36–36	301
Sidelamps	12	6	207
Stop, tail-lamp	12	6	207
Reverse lamp	12	24	1
Interior lamp	12	6	207
Panel illumination bulbs	12	2.2	986-987

HORN.

All horns are adjusted during manufacture to give their best performance and will give a long period of service without any attention; no subsequent adjustment is required.

If the horn fails or becomes uncertain in its action, it does not follow that it has broken down. First ascertain that the trouble is not due to some outside source, e.g., a discharged battery, a loose connection or short circuit in the wiring of the horn: a short circuit in the horn will cause the fuse to blow.

If the fuse has blown, examine the wiring for the fault and replace with the spare fuse provided.

It is also possible that the performance of the horn may be upset by the fixing bolt working loose, or by some component near the horn being loose. This can be ascertained by removing the horn from its mounting, holding it firmly in the hand and pressing the push.

TRAFFICATORS.

Lubrication.

When the arms become stiff, apply by means of a brush or other suitable article a drop of thin machine oil, such as sewing machine oil, to the catch pin between the arm and the operating mechanism. Use only the slightest trace as any excess may affect the operating mechanism.

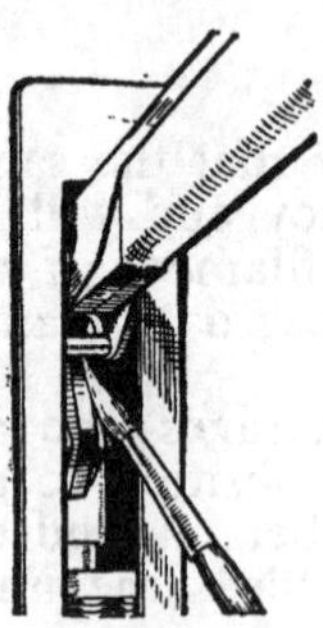

Fig. 51. *Lubricating the catch pin.*

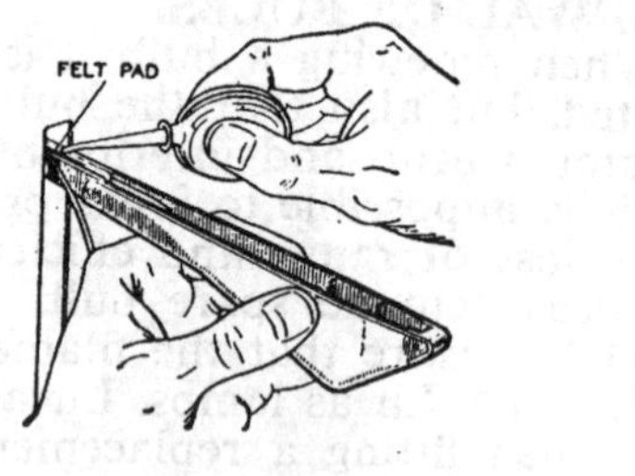

Fig. 52. *Lubricating arm pivot bearing.*

Also withdraw the screw on the under side of the arm and slide off the metal arm cover. Place the connecting wire to the bulb on one side and apply a drop of thin oil to the lubricating pad at the top of the arm. To replace the arm cover, slide it in an upwards direction so that the side plates engage with the slots in the under side of the spindle bearing, and secure with the screw.

Bulb Renewal.

Withdraw the screw on the under side of the arm and slide off the metal plate; the burnt-out bulb may then be renewed.

To replace the metal cover, slide it in an upwards direction so that the side plates engage with the slots on the under side of the spindle bearing. Finally secure the plate by means of the fixing screw.

The replacement bulb is a Lucas No. 256, 12-volt, 3-watt, festoon type.

Removal and Replacement of Trafficator Assembly.

The trafficator assemblies are fitted inside the right- and left-hand door pillars. If it is necessary to remove, proceed as follows:—

Remove the pillar casing by unscrewing the eight self-tapping screws.

Remove the trimming covering the trafficator.

Remove the two set bolts securing the trafficators in the pillars, and also the two wire connections.

The trafficator assemblies can then be withdrawn for further examination.

Replacement of the trafficators is carried out in the reverse manner to that detailed for removal.

Trafficator Switch.

The steering column control which is fitted in the hub of the steering wheel incorporates a horn push and a trafficator switch.

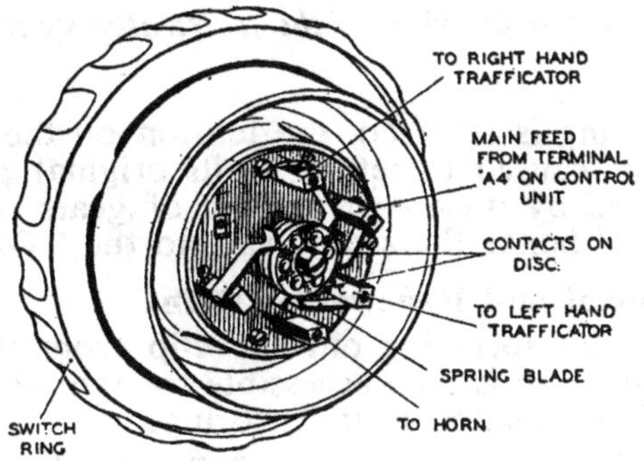

Fig. 53. *The Trafficator switch.*

The trafficator switch is operated by a moulded ring which is moved to either the right or left in order to switch on the right-hand or left-hand trafficator. The action of switching on the trafficator brings into operation a clockwork timing mechanism which holds the trafficator in the " On " position for a period long enough for the driver to turn a corner or overtake. After this the trafficator is automatically switched off and the switch ring returned to its original position.

If for any reason the trafficators need to be switched off whilst the automatic timing mechanism is working, the switch ring can be brought back manually to its original position without damage to the mechanism.

Operation of the Timing Mechanism.

The moulded operating ring of the trafficator switch is carried on three spokes, at the centre of which is a boss. The boss has a hexagon-shaped recess which fits over the spindle of the timing control mechanism, being secured to it by means of a screw.

At the bottom of the switch is the contactor mechanism which completes the circuit to the trafficator when the switch ring is moved to the appropriate position. This consists essentially of three spring contact blades, each secured at one end to a terminal post, the free end bearing on the edge of the disc attached to the switch spindle. This disc is provided with metal contacts through which the circuit to the trafficators can be completed.

Attached to the spindle and fitted just below the boss on the switch ring is a cam which is arranged so that when the switch ring is turned to the left or right it moves a lever against the tension of the operating spring, thereby setting the timing mechanism in motion.

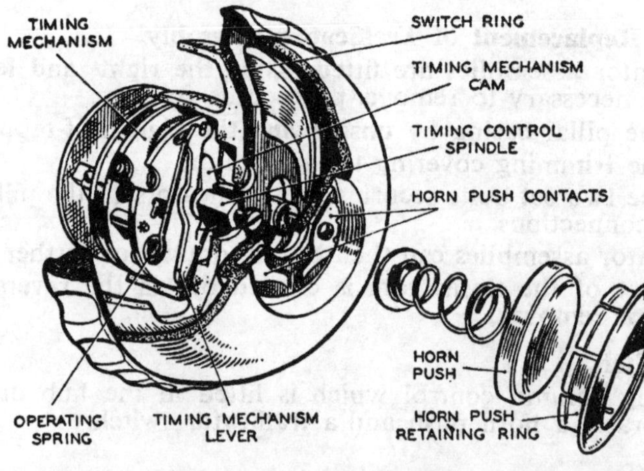

Fig. 54. Exploded view of the trafficator switch and horn push.

When the switch ring is released, the tension on the operating spring is relieved and the lever caused to return to its original position, the rate of return being controlled by means of a train of gears. The movement also returns the spindle and hence the switch ring to the " Off " position.

Interior Light—Removal and Renewal of Bulb.

Remove the glass by springing out the clip from the two lugs on the body of the lamp. The bulb is then accessible for renewal. Replace the glass, and spring the clip into position in the two lugs.

The replacement bulb is a Lucas No. 207, 12-volt, 6-watt.

Windscreen Wiper.

No adjustment or lubrication is necessary as the gears are fully lubricated before leaving the works.

Replacement of Arm and Blade Assembly.

To remove the arm and blade assembly, slacken the fixing nut and tap it sharply to release the collet which clamps the arm on to the spindle. When fitting a replacement arm and blade, slacken the securing nut and push the arm-fixing bush over the end of the spindle as far as it will go. Secure by tightening the nut.

Replacement of Blade.

Take out the rubber bush securing the blade to the arm and remove the blade. Insert the tongue on the replacement blade through the slot in the arm, and secure it by fitting the rubber bush through the hole in the tongue.

Steering Column Slip-Ring Unit.

As there is no outlet at the bottom of the steering gearbox for the cables from the trafficator switch and horn push to pass to the various accessories, a slip-ring unit is incorporated and fitted near the base of the steering column.

The cables are taken down the centre of the steering column and are connected to four spring contacts in a moulded ring clamped around the inner sleeve of the steering column by means of jump rings.

Another moulding, which is secured by means of three screws to a bracket on the outer sleeve, fits over these contacts. This moulding is fitted with four slip-rings and terminals for connecting cables to the appropriate accessories.

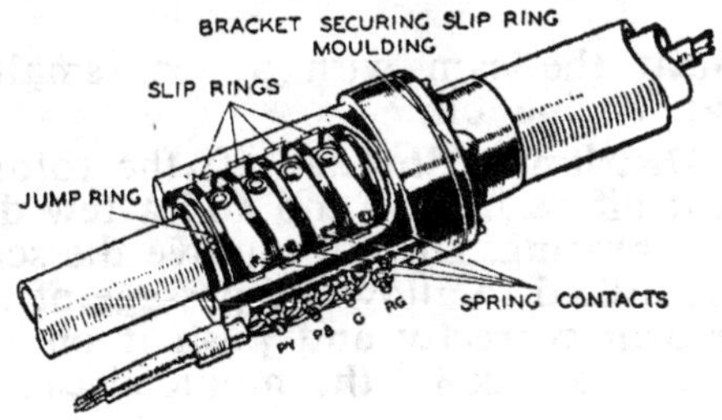

Fig. 55. Steering column slip-ring unit.

The terminals and connections are as follows:—

R.G.—Red and green cable to the right-hand trafficator.

G.—Green cable to the left-hand trafficator.

P.B.—Purple and black cable to terminal A4 on control unit.

P.Y.—Purple and yellow cable to horn.

When the steering column is turned, the inner moulding moves together with the spring contacts which rub on the slip-rings.

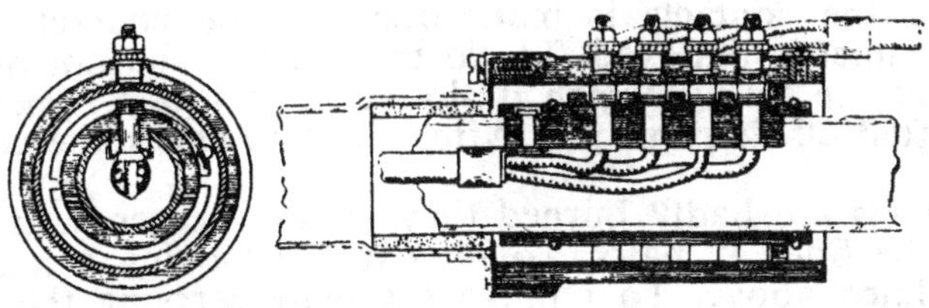

Fig. 56. Sectional view of the steering column slip-ring unit.

Lubrication.

When necessary smear the slip-ring with petroleum jelly which has a melting-point of 43° C. (109° F.). To gain access to the slip-ring remove the three screws securing the outer moulding to its bracket and then slide off.

DISTRIBUTOR.

The distributor is a Lucas Model DKYH.4A, Type DA.36. The identification is stamped on the side of the distributor.

Adjustment.

Remove the moulded distributor cap and turn the engine over by hand until the contacts in the distributor are fully opened. Check the gap with a gauge of thickness .012 in. (.30 mm.), and if the setting is correct the gauge should be a sliding fit. If the gap varies appreciably from the gauge the contact breaker should be adjusted. To adjust, keep the engine in the position to give the maximum opening of the contacts and slacken the two screws which secure the contact plate. Move the plate until the gap is set to the thickness of the gauge and fully tighten the locking screws. Re-check the gap.

Lubrication.

Cam.—Lightly smear the cam with a very small amount of grease, or, alternatively, clean engine oil.

Cam Bearing and Distributor Shaft.—Lift the rotor arm off the top of the spindle by pulling it off vertically, and add a few drops of thin machine oil to lubricate the cam bearing. Do not remove the screw which is exposed to view as the screw is drilled to allow the passage of oil.

Replace the rotor arm correctly and push it on the shaft as far as it will go, otherwise there is a risk of the moulded cap becoming burned or tracked.

Automatic Timing Control.—Carefully add a few drops of thin machine oil through the hole in the contact breaker base through which the cam passes. Do not allow oil or grease to get near the contacts.

Contact Breaker Pivot.—Place a small amount of grease, or, alternatively, clean engine oil on the pivot on which the contact breaker lever works. Do not allow oil or grease to get on the contacts.

Wipe the inside and the outside of the moulded distributor cap with a soft dry cloth, paying particular attention to the space between the terminals. See that the small carbon brush on the inside moulding works freely in its holder.

Examine the contact breaker. The contacts must be free from grease or oil. If they are burned or blackened, clean them with a fine carborundum stone or with very fine emery cloth. Afterwards wipe away any trace of dirt or metal dust with a petrol-moistened cloth.

Cleaning of the contacts is made easier if the contact breaker lever carrying the contact is removed. To do this slacken the nut on the terminal post and lift off the spring, which is slotted to facilitate removal. The lever can then be lifted off its pivot pin. After cleaning check the contact breaker setting.

If the contacts are badly burned they should be renewed. Replacement contacts must be fitted in pairs. To remove the moving contact, follow the procedure outlined above. To remove the plate carrying the fixed contact, take out the two set screws, complete with spring washers and flat steel washers. The replacement set of contacts can now be fitted by a reversal of the above.

Check the adjustment of the contact breaker gap as described on page 363.

IGNITION COIL.

The coil is a Lucas Model Q.12-8, Type L/O, 12-volt, Service Number 45020A. The identification numbers are stamped on the top of the coil.

Adjustment.

The ignition coil requires no attention beyond seeing that the terminals are tight and that the exterior is kept clean, particularly between the terminals.

RENEWAL OF HIGH-TENSION CABLES.

High-tension cables which have the insulation cracked, perished or damaged in any way must be renewed.

The method of connecting the cable to the coil is to thread the knurled moulded nut over the cable, bare the ends of the cable for about ¼ in. (6 mm.), thread the wire through the washer removed from the end of the original cable and bend back the wire strands. Screw the nut into its terminal.

The method of connecting the cable to the distributor is to unscrew the pointed fixing screws on the inside of the moulding, push the cables, which should not be bared but cut off flush to the required length, well home into their respective terminals and tighten the pointed fixing screws.

SPARKING PLUGS.

The sparking plugs are Champion L.10.S. (14 mm.).

The gaps should be set to between .020 in. and .022 in. (.51 mm. to .56 mm.).

IMPORTANT:—Always adjust the side electrode. Never attempt to bend the central electrode.

It is recommended that the plugs be renewed every 12,000 miles or 20,000 kilometres.

LOCATION AND REMEDY OF FAULTS.

Although every precaution is taken to eliminate all possible causes of trouble, failure may occasionally develop through lack of attention to the wiring. The most probable faults are tabulated, according to the symptoms displayed, in the fault-finding tables on the following pages.

It is recommended that a systematic examination is made by following the suggestions in the fault-finding tables, as the sources of many troubles are by no means obvious. In some cases a considerable amount of deduction from the symptoms is needed before the cause of the trouble is disclosed.

For instance, the engine might not respond to the starter switch; a hasty reference would be that the starter motor is at fault. However, as the motor is dependent on the battery, it may be that the battery is exhausted. This, in turn, may be due to the dynamo failing to charge, and the final cause of the trouble may be, perhaps, a loose terminal nut either at the battery or elsewhere in the wiring circuit.

If, after carrying out the examination, the cause of the trouble is not to be found, get in touch with the nearest Lucas Service Depôt or Agent.

HOW TO LOCATE AND REMEDY COIL IGNITION TROUBLE

ENGINE WILL NOT FIRE

Possible Cause : **Battery Discharged.** Starter will not turn engine and lamps do not give a good light.

Remedy : Start engine by hand. Battery should be recharged by running car for a long period during daytime. Alternatively, recharge from an independent electrical supply.

Possible Cause : **Controls not set correctly for starting.**

Remedy : See that ignition is switched on and everything is in order for starting.

Possible Cause : **Test if Coil Sparks** by removing lead from centre distributor terminal and holding it about ¼ in. (6 mm.) away from some metal part of the chassis while engine is turned over.

If the sparks jump gap regularly, the coil and distributor are functioning correctly.

Remedy : Examine the sparking plugs, and if these are clean and the gaps correct, the trouble is due to carburetter, fuel supply, etc.

Possible Cause : **If the Coil does not spark,** the trouble may be due to any of the following causes :—

Fault in low-tension wiring. Indicated by (1) no ammeter reading when engine is slowly turned and ignition is switched on ; or (2) no spark occurs between the contacts when quickly separated by the fingers when the ignition is switched on.

Remedy : Examine all cables in ignition circuit and see that all connections are tight. See that battery terminals are secure.

Possible Cause : **Dirty or Pitted Contacts.**

Remedy : Clean contacts with fine carborundum stone or fine emery cloth and afterwards with a cloth moistened with petrol.

ENGINE MISFIRES

Possible Cause : **Contact breaker out of adjustment.** Turn engine until contacts are fully opened and test gap with gauge.

Remedy : Adjust gap to gauge.

Possible Cause : **Remove each Sparking Plug in turn,** and observe whether a spark occurs at the points when the engine is turned. Irregular sparking may be due to dirty plugs or defective high-tension cables. If the sparking is regular at all plugs, the trouble is probably due to engine defects.

Remedy : Clean plugs and adjust gaps to correct setting. Renew any lead if the insulation shows signs of deterioration or cracking. Examine carburetter, fuel supply, etc.

HOW TO LOCATE AND REMEDY LIGHTING TROUBLES

LAMPS GIVE INSUFFICIENT ILLUMINATION

Probable Fault : **Battery discharged.**

Remedy : Charge battery either by a long period of daytime running or from independent electrical supply

Probable Fault : **Lamps out of alignment or bulbs out of focus.**

Remedy : Align lamps and focus bulbs.

Probable Fault : **Bulbs discoloured through use, or reflectors dirty.**

Remedy : Fit new bulbs or clean reflectors.

LAMPS LIGHT WHEN SWITCHED ON BUT GRADUALLY FADE OUT

Probable Fault : **Battery discharged.**

Remedy : As above.

BRILLIANCE VARIES WITH SPEED OF CAR

Probable Fault : **Battery discharged.**
Remedy : As above.
Probable Fault : **Battery connections loose or broken.**
Remedy : Tighten connections or renew faulty cables.

LIGHTS FLICKER

Probable Fault : **Loose connection.**
Remedy : Locate loose connection and tighten.

FAILURE OF LIGHTS

Probable Fault : **Faulty cable or connection.**
Remedy : Examine wire for faulty cables or connections and remedy.
Probable Fault : **Battery Discharged.**
Remedy : As above.
Probable Fault : **Loose or broken connection.**
Remedy : Locate and tighten loose connection, or re-make broken connection.

HOW TO LOCATE AND REMEDY TROUBLE WITH VOLTAGE DYNAMO EQUIPMENT

BATTERY IN LOW STATE OF CHARGE, SHOWN BY LACK OF POWER WHEN STARTING. (Hydrometer readings less than 1.200.)

Possible Cause : **Dynamo not charging, indicated by ammeter not showing charge reading when running at about 20 m.p.h. (32 k.p.h.) with no light in use.** Due to :—
Broken or loose connection in dynamo circuit, or regulator not functioning properly.
Remedy : Examine charging and field circuit wiring. Tighten loose connection or renew broken lead. Particularly examine battery connections. Return regulator to Lucas Service Depot or Agent for attention.
Possible Cause : **Commutator greasy or dirty.**
Remedy : Clean with soft rag moistened with petrol.
Possible Cause : **Dynamo giving low or intermittent output, indicated by ammeter giving low or intermittent reading when car is running steadily in top gear.** Due to :—
Loose or broken connections in dynamo circuit.
Remedy : Examine charging and field circuits wiring. Tighten loose connections or renew broken lead. Particularly examine battery connections.
Possible Cause : **Brushes Greasy or dirty.**
Remedy : Clean with soft rag moistened with petrol.
Possible Cause : **Brushes worn or not fitted correctly.**
Remedy : Renew worn brushes. See that the brushes " bed " properly.
Possible Cause : **Regulator not functioning correctly.**
Remedy : Have equipment examined by a Lucas Service Depot or Agent.

BATTERY OVERCHARGED, SHOWN BY BURNT-OUT BULBS AND VERY FREQUENT NEED FOR " TOPPING " (Hydrometer readings high.)

Possible Cause : **Dynamo giving high output, indicated by ammeter giving high charge reading.** Due to :—
Regulator not functioning properly.
Remedy : Return regulator to Lucas Service Depot or Agent for attention.

HOW TO LOCATE AND REMEDY STARTER MOTOR TROUBLE

STARTER MOTOR LACKS POWER OR FAILS TO TURN ENGINE

Possible Cause : **Stiff engine, indicated by inability to turn by hand.**
Remedy : Locate and remedy cause of stiffness.

Possible Cause : **If engine can be turned by hand, then trouble may be due to :— Battery discharged.**
Remedy : Start by hand. Charge battery by either long period of daytime running or from independent electrical supply.

Possible Cause : **Broken or loose connection in starter circuit.**
Remedy : See that connections to battery, starter and starter switch are tight, and that cables connecting these units are not damaged.

Possible Cause : **Starter commutator or brushes dirty.**
Remedy : Clean.

Possible Cause : **Brushes worn or not fitted correctly.**
Remedy : See that brushes " bed " correctly. Have worn brushes renewed.

Possible Cause : **Starter pinion jammed in mesh with flywheel.**
Remedy : Rotate squared end of starter with spanner.

STARTER OPERATES BUT DOES NOT CRANK ENGINE

Possible Cause : **Pinion of starter drive does not engage with flywheel,** due to dirt on screwed sleeve.
Remedy : Clean sleeve with paraffin.

STARTER PINION WILL NOT DISENGAGE FROM FLYWHEEL WHEN ENGINE IS RUNNING.

Possible Cause : **Starter pinion jammed in mesh with flywheel.**
Remedy : Rotate squared end of starter shaft with spanner

ELECTRICAL EQUIPMENT

(1½ & 2.6 Litre Models)

THE BATTERY.

About once a month unscrew the filler caps and pour a small quantity of distilled water into each of the cells to bring the acid ¼ in. above the tops of the separators. This is done to replace water which has been lost by evaporation. If any of the acid solution is spilled it must be replaced by a dilute sulphuric acid solution of the same specific gravity as the electrolyte in the cell to which it is to be added. The battery must not be allowed to remain with the plates uncovered, as they will suffer a chemical change and be damaged beyond repair.

Keep the Terminals and Battery Fixing Bolts Clean and Tight and Smeared with Vaseline.—A smearing of vaseline protects the terminals and fixing bolts from the corrosive action of the acid, which if allowed to continue will eventually result in trouble. When inspecting the battery, put a spanner on the bolt securing the cable end to the chassis in order to make sure that it is quite tight. When corrosion is evident, it must be completely removed before vaseline is applied.

Keep the Outside of the Battery Clean and Dry, particularly the Tops of the Cells.—Water and dirt form a conductor of electricity and if such a path is allowed to form between the positive and negative terminals of the battery or between the negative battery terminal and the chassis, there will be a leakage of current which will eventually cause the battery to run down. Give the cell tops a wipe over regularly and you will avoid this.

Occasionally make a point of Examining the Health of your Battery by taking Hydrometer Readings.—The hydrometer allows you to take a sample of the acid solution. It contains a graduated float which indicates the specific gravity of the acid solution in the cell from which the sample was taken. The readings and their indications are as follows:—

1.250—1.300	...	Battery fully charged.
1.150—1.250	...	Battery about half discharged.
Below 1.150	...	Battery fully discharged.

The readings for each of the cells should be approximately the same. If one cell gives a very different reading from the rest it may be that acid has been spilled or has leaked from this particular cell, or there may be a short circuit between the plates. In this case we advise the owner to have his battery examined by a Lucas Service Depôt—neglect may mean costly repairs later on.

It must be remembered that you cannot get more energy out of a battery than is put in. If the car is left parked at night with the lights on, or if the starter motor is used excessively with very little daytime running, then the battery may get into a low state of charge—particularly in winter. This may be remedied either by running the car for longer periods during daytime or by economising in the use of the headlamps and other accessories.

Never leave the Battery in a Discharged Condition for any length of time.—If the car is to be out of use for any length of time, see that the battery is fully charged and about every fortnight give it a short refreshing charge to prevent any tendency to permanent sulphation of the plates.

THE DYNAMO.

All dynamos are sent out from the works with the bearings packed with grease. This lasts until the car is taken down for a general overhaul, when it is advisable to have the machine dismantled, preferably by a Lucas Service Depôt, for cleaning and adjustment and repacking the bearings with grease.

The dynamo is provided with a lubrication wick at the commutator end of the machine. Once every year, unscrew the cap of the lubricator, and if the wick is dry, refill the cap with vaseline. (See Fig. 25.)

Inspection of Commutator and Brush Gear.—About once a season, remove the metal cover from the dynamo for inspection of the commutator and the carbon contacts or brushes as they are called. Take care not to lose the dished nut when removing the cover fixing screw, as the cover is liable to spring open when the screw is released.

It is essential that the brushes make good firm contact with the commutator. The brushes are held in boxes by springs. Hold back the springs and at the same time move the brush to see that it is free to slide in its holder. If there are any signs of sticking, remove it from its box and clean it with a cloth moistened with petrol.

After removing brushes for cleaning or any other purpose, care must be taken to replace them in their original positions, otherwise they will not "bed" properly on the commutator.

If, after long service, the brushes have become worn to such an extent that they will not bear properly on the commutator, they should be replaced. Always fit genuine Lucas brushes, as these are made specifically for work on these machines and will give by far the best results and the longest life. We advise you to have the brushes fitted at a Lucas Service Depôt, so that they can be properly "bedded" to the commutator.

Next examine the commutator. It should be clean and free from any trace of oil or dirt and should have a highly polished appearance. The best way to clean a dirty or blackened commutator is by pressing against it a fine dry duster and getting someone to turn the engine over slowly by hand. If the commutator is very dirty, the duster may be moistened with petrol.

DYNAMO OUTPUT.

The dynamo automatically keeps the battery in good condition.

The dynamo is of the compensated voltage controlled type. These machines work in conjunction with a regulator unit which is housed along with the cut-out.

What the Regulator does—a completely Automatic Control.—The regulator causes the dynamo to give an output which varies according to the lamp load and state of charge of the battery. When the battery is discharged the dynamo gives a high output, so that the battery receives a quick recharge which brings it back to its normal state in the minimum possible time. On the other hand, if the battery is fully charged, the dynamo is arranged to give only a trickle charge which is sufficient to keep it in good condition without any possibility of causing damage to the battery by overcharging.

In addition to controlling the output of the dynamo according to the condition of the battery, the regulator provides for an increase of output to balance the current taken by the lamps or other accessories whenever they are switched on.

The regulator unit is accurately set to suit the requirements of the equipment fitted on the car, and in normal service the battery will be kept in good condition. If, however, you should find that the battery is not kept in a charged condition, or is being excessively overcharged, we advise you to consult your nearest Lucas Service Depôt, where any necessary adjustments can be made. Owners should not attempt the adjustment themselves.

Ammeter Readings.—It must be remembered, when noting ammeter readings, that, normally, during daytime running, when the battery is in good condition, the dynamo gives only a trickle charge, so that the charge reading will seldom be more than a few amperes.

When starting from cold the driver will notice the rise of charging current until it reaches a steady maximum at a speed of, say, twenty miles per hour, after which it will remain fairly high for perhaps ten minutes or so, then fall to a steady charge which is most suitable for the particular condition of the battery.

A discharge reading may be shown sometimes immediately after switching on the lamps. This usually happens after a long run when the voltage of the battery is high. After a short time, the voltage of the battery will fall and the regulator will respond, causing the dynamo output to balance the load.

The Cut-out—an Automatic Dynamo Switch.—It will be noticed from the ammeter readings that the dynamo does not charge at very low engine speeds. This is because it is not rotating fast enough to generate sufficient energy to charge the battery.

Connected between the dynamo and the battery is the cut-out—an automatic switch which acts as a " valve," allowing the flow of current from the dynamo to the battery only. It closes when the dynamo is running fast enough to charge the battery, and opens when the speed is low or the engine is stationary, thus preventing current from flowing through the dynamo windings from the battery.

The cut-out is located together with the fuses on the engine shield.

Cut-out, Regulator and Fuse Unit.—This unit houses the cut-out, regulator and fuses and forms a junction box for distributing the cables to the various units and accessories. The cut-out and regulator are accurately set before leaving the works, and need no further adjustment. The cover protecting them is therefore sealed.

This unit also incorporates five fuses. Two protect the auxiliary accessories which are operative only when the ignition is switched on, one protects those accessories which can be operated irrespective of whether the ignition is on or off, while the remaining two protect the head, side and tail-lamps.

THE STARTER.

Observe the following points when starting:—

1. Operate the starter switch control firmly, and of course release it as soon as the engine is running.
2. Never operate the starter when the engine is running. If the engine will not fire at once, allow it to come to rest before operating the switch again.
3. Do not run the battery down by keeping the starter on when the engine will not start.
4. In cold weather starting will be facilitated by giving the engine a few turns by the starting handle before using the starter.

Attention Needed by the Starter.—Give the starter commutator and brush gear similar attention to that described for the dynamo.

Occasionally, if dirty, wash over the screwed sleeve on the armature spindle with paraffin and afterwards give it the merest trace of thin machine oil.

Instrument Panel and Switch Panel.—The instrument panel houses the ammeter, speedometer, revolution counter, clock, etc. The instruments are illuminated by means of bulbs mounted behind the panel. The switch panel houses the lighting and ignition switch, starter switch, warning lights, etc., and is also illuminated by a bulb behind the instrument panel.

HEADLAMPS.

Removing Lamp Front and Reflector.—To remove the lamp front pull forward the spring loaded catch when the rim is released. When replacing the front, locate the top of the rim first, then press on the bottom.

To remove the reflector, turn back the ends of the cork washer at the top of the rim and remove the screw opposite the medallion in the top of the lamp. Turn reflector until markings " O " stamped on reflector rim and lamp body coincide. The reflector can then be withdrawn. When replacing reflector, engage it with the lamp body, then turn it until the screw hole in its rim is opposite to the screw hole on top of lamp body. Secure reflector by means of screw.

Checking the Alignment.—The simplest way of checking the adjustment of the lamps is to take the car on a straight level stretch of road at night and examine the direction of the beams. They should be parallel with the road and with each other. If one appears to be out of alignment, adjust it as follows: Slacken the single fixing nut at the base of the lamp and move the lamp on its universal mounting to the required position, finally locking the adjustment by tightening the fixing nut.

Focusing.—For the lamps to give a parallel beam, the filament of the bulb must coincide as nearly as possible with the focal point of the reflector. If the filament of the bulb is behind the focal point of the reflector, the beam will be divergent, while on the other hand, if the filament is in front of the focal point, the beam will be convergent, with a dark area in the centre of the beam. In either case, the lamps will have a poor range and will cause unnecessary dazzle to approaching traffic.

Before lamps are dispatched from the works, the bulbs are carefully focused to give the best results. Provided that official Lucas bulbs are fitted as replacements, it should not be necessary to disturb the setting. If for any reason these are not obtainable, and ordinary bulbs have to be fitted, it may be necessary to re-focus by moving the bulb backwards or forwards along the axis of the reflector until the best lighting is obtained.

Cover up one lamp while testing the other. If the lamp does not give a uniform long-range beam without any dark centre, the bulb needs adjusting. To do this, remove the front and reflector, and slacken the clamping clip at the back of the reflector.

After each adjustment. note the effect with the reflector and front refitted.

When the best position for the bulb holder has been found, see that the clamping screw is tightened.

" Pass-light " (2.6 Litre Models only).—To remove the front and reflector pull out the spring clip at the bottom of the lamp and swing it out from its location. The reflector front can then be withdrawn. When replacing, locate top of rim first, then press home the front and secure by the clip.

Focusing is carried out in the same manner as described in the previous paragraphs.

Sidelamps.—The front of the lamp is removed for a bulb replacement when the screw at the top of the lamp is withdrawn. When refitting the front, locate the bottom and secure with the screw.

" Stop," Tail and Reversing Lamps.—The fronts of these lamps can be opened when the fixing screw is unscrewed.

Cleaning.—All Lucas reflectors are protected by a fine transparent and colourless covering which enables any accidental finger marks to be removed with a soft cloth without affecting the surface of the reflector. Never use metal polishes on Lucas reflectors. A light polish with a soft cloth is all that is necessary. Chromium-plated lamps will not tarnish and only need wiping over with a damp cloth to remove dust or dirt.

Lucas Genuine Spare Bulbs.—Lucas genuine spare bulbs are sold by any reputable garage and are specially tested to check that the filament is in the correct position to give the best result with Lucas lamps. To assist in identification, Lucas bulbs are marked on the metal cap with a number. When fitting a replacement, see that it is the same number as the original bulb.

We advise you to replace bulbs after long service before they actually burn out, as very often the filaments sag, making it impossible for them to be focused correctly.

Bulbs Fitted.—

Headlamps	Lucas No. 54	12 v.	36 watt
Sidelamps	Lucas No. 207	12 v.	6 watt
Pass-lamp	Lucas No. 87	12 v.	60 watt
" Stop " and Tail-lamps ...	Lucas No. 207	12 v.	6 watt
Reversing Lamp	Lucas No. 1	12 v.	24 watt
Panel Lights	Lucas No. 1224M	12 v.	2.4 watt
30 M.P.H. Warning Lamp ...	Lucas No. C252A	2.5 v.	.5 watt
Ignition Warning Light ...	Lucas No. C252A	2.5 v.	.5 watt
" Trafficators " (Festoon) ...	Lucas No. 256	12 v.	3 watt
Interior Light	Lucas No. 207	12 v.	3 watt

LUCAS " MELLOTONE HORNS."

The horns are adjusted to give their best performance before being sent out of the works, and will give long periods of service without any attention.

If one of the horns fails or becomes uncertain in its action, first ascertain that the trouble is not due to some outside source, e.g., a loose connection or short circuit in the wiring of the horn. If both horns fail or become uncertain in their action, the trouble is probably due to a discharged battery or blown fuse. If the fuse has blown, examine the wiring for the fault and replace with spare fuse provided, a 35 ampere size.

It is also possible that the performance of a horn may become uncertain, due to its fixing bolt working loose. If after carrying out the above examination the trouble is not rectified, the horn may need adjustment; but this should not be necessary until the horns have been in service for a long period.

Adjustment does not alter the pitch of the note, it merely takes up wear of moving parts. When adjusting the horns, short-circuit the fuse, otherwise it is liable to blow. Again, if the horns do not blow on adjustment, release the push instantly.

Adjustment of Horns.—Remove the horn cover after withdrawing the fixing screw and detach the cover securing bracket by springing it from its fixing.

Adjustment must be made by the Centre Nut only.—Slacken the lock nut and turn the adjustment nut underneath, a few degrees at a time, in an anti-clockwise direction. It must be remembered that the adjustment is a very sensitive one. Tighten the locknut before testing the horns on each occasion.

When making adjustments to a horn, always disconnect the supply lead of the other horn, taking precautions that it does not come into contact with any part of the chassis and so cause a short circuit.

Finally, if the note is still unsatisfactory, do not dismantle the horn, but return it to a Lucas Service Depôt for examination.

" TRAFFICATORS."

Every two-three months or if the arms become stiff at any time, raise each arm, and by means of a brush, match-stick or similar article, apply a drop of thin machine oil, such as sewing machine oil, to the catch pin between the arm and the operating mechanism. Use only the merest drop of oil—any excess may affect the working of the operating mechanism. The " Trafficators " are kept in the closed position by means of a spring. The arms can be pulled out by hand. If any difficulty is experienced, switch the " Trafficator " on and then, supporting the arm in a horizontal position, move the switch to the " off " position.

If at any time the arm fails to light up when in operation, raise the arm in the manner previously described and examine the bulb, replacing it if necessary by a Lucas No. 256 (3 watt festoon type) bulb.

To remove the bulb, withdraw the screw on the underside of the arm and slide off the metal plate; the bulb can then be replaced. To replace the metal plate, slide it on in an upwards direction so that the side plates engage with the slots on the underside of the spindle bearing. Finally secure the plate by means of its fixing screw.

Do not remove bulb while the " Trafficator " is switched on as this may cause a short circuit.

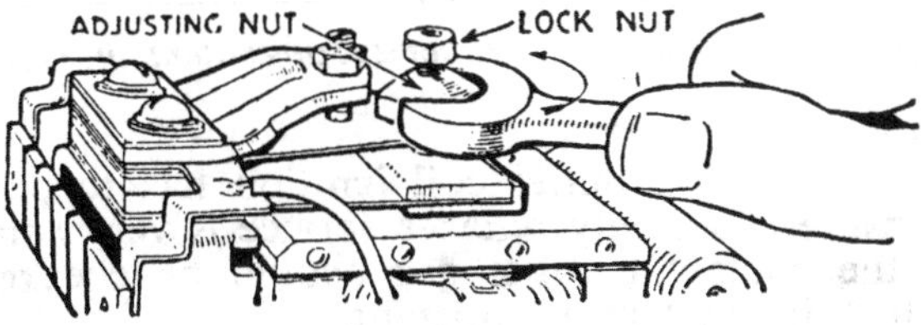

Fig. 57. *Mellotone horn adjustment* (1½, 2 *and* 2.6 *Litre models*).

WINDSCREEN WIPER (CLOSED MODELS).

The motor is mounted on the metal bulkhead under the scuttle. Incorporated in the motor is a reduction gearbox from which the drive is transmitted through a continuously revolving shaft. A crank on the end of the shaft is connected by links to two clutch boxes which transmit motion to the wiper spindles.

The starting switch is combined with the clutch on the driver's side and the action of rotating the knob in an anti-clockwise direction starts the motor and engages the clutch when the wiper comes into operation. To bring the arm on the passenger's side into operation, rotate the knob in a clockwise direction.

To stop the wiper motor push in the knob on the driver's side and turn as indicated to park the arm on the scuttle. The blade on the near-side is parked in a similar manner, but does not control the wiper motor.

If it is found necessary to dismantle the driving links shown in Fig. 58, the position of the pen steel, felt and spring washers should be noted so that they may be reassembled in their correct order.

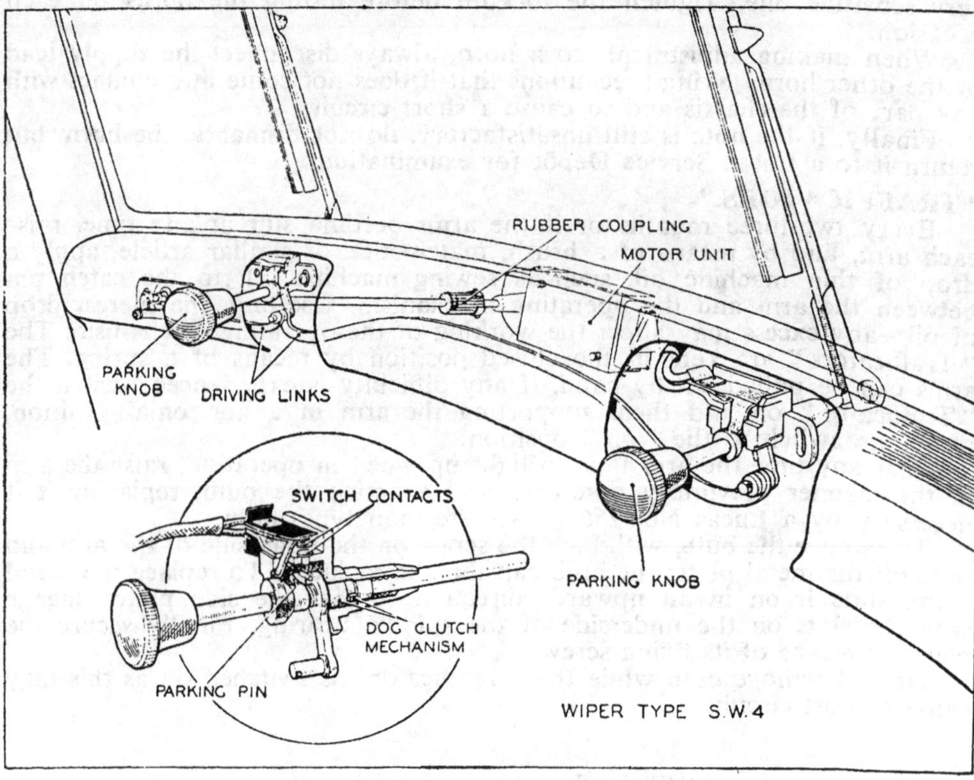

Fig. 58. The windscreen wiper arrangement on closed models.

COIL IGNITION EQUIPMENT.

Considering the exacting duties coil ignition has to perform, the attention needed to ensure absolutely reliable service is very small.

Periodically the contact breaker gap should be checked and, if necessary, reset to .012 in. maximum opening.

LUBRICATION.

The following parts of the distributor require lubrication:—

1. Remove the cover and fill the well with medium grade engine oil every 2,500 miles.
2. *Cam.*—Give the cam a smear of light grease every 5,000 miles.
3. *Cam Bearing.*—About every 3,000 miles, withdraw the moulded rotating arm from the top of the spindle by pulling it off, and add a few drops of thin machine oil. Do not remove the screw exposed to view, as there is a clearance between the screw and the inner face of the spindle through which the oil passes to lubricate the cam bearing. Take care to refit the arm correctly and to push it on to the shaft as far as possible, otherwise there is a risk of tracking and burning of the moulding.
4. *Contact Breaker Pivot.*—Every 5,000 miles smear a little light grease on the pivot on which the contact breaker rocker-arm works.

Cleaning.—Keep the outside of the distributor clean, particularly the spaces between the high-tension terminals. Very occasionally, remove the moulding by springing aside its two securing spring clips. Wipe the inside clean with a dry cloth, and see that the carbon brush is quite free in its holder. Clean the metal electrodes inside the moulding and also the rotating electrode on the distributor arm; if necessary, use a cloth moistened with petrol for this.

Next examine the contact breaker; you must keep the contacts free from any grease or oil. If they are burned or blackened, clean them with fine carborundum stone, or if this is not available, you can use very fine emery cloth. Finish off with a cloth moistened with petrol, and remove all traces of dirt and metal dust. Misfiring is sometimes caused by dirty contacts.

Checking and Adjusting the Contacts.—The contacts require only occasional adjustment. The chief cause of variation in the gap is wear of the heel of the contact rocker-arm which bears upon the actuating cam. Provided you keep the cam smeared with grease, however (see lubrication instructions), the wear on the heel will be negligible and the contact gap setting should only require adjustment at infrequent intervals.

To check the setting, turn the engine by hand until the contacts are fully opened. Now insert the gauge provided on the ignition screwdriver between the contacts. The gauge has a thickness of about 12 thousandths of an inch and it should be a sliding fit between the contacts when the gap is correct. We do not advise altering the setting unless there is quite an appreciable variation from the gauge. To make the adjustment, keep the engine in the position to give maximum opening of the contacts and slacken the two screws securing the contact plate. Then move the plate until the gap is set to the thickness of the gauge. After making the adjustment, care must be taken to tighten the locking screws.

Micrometer Timing Adjustment.—In order to obtain very fine timing of the ignition to the engine and to allow for altered engine conditions, e.g., state of carbonisation of engine, change of fuel, etc., a micrometer adjustment is provided which allows fine adjustment to be made simply by the movement of a knurled knob.

With a clean engine, and using first grade fuel, the micrometer scale should be at 0, before timing.

The final setting can be made by use of the micrometer adjustment after running the engine. If the firing is found to be slightly too early or too late, adjust the knurled knob until the best engine performance is obtained. The adjustment should not be altered by more than one distributor degree at a time (one division on the scale is equivalent to two distributor degrees or four crankshaft degrees).

Renewing the High-tension Cable.—The high-tension cables are those connecting the coil to the distributor and the distributor to the sparking plugs. When these cables show signs of perishing or cracking, they must be replaced by 7 mm. rubber-covered ignition cable.

The method of connecting the cable is to thread the knurled moulded nut over the lead, bare the end of the cable for about ¼ in., thread the wire through the brass washer provided, and bend back the strands. Finally. screw the nut into its respective terminal.

The Coil.—The coil requiries no attention whatever beyond keeping its exterior clean, particularly between the terminals, and occasionally checking that the terminal connections are quite tight.

Ignition Switch and Warning Light.—The ignition switch, besides forming a means of stopping the engine, is provided for the purpose of preventing the battery being discharged by the current flowing through the coil windings when the engine is stopped. A warning lamp is provided in the switch panel which gives a red light when the ignition is switched on and the car is running very slowly or is stationary, thus reminding you to switch off.

Should the warning lamp bulb burn out, this will not in any way affect the ignition system, but you should replace it as soon as possible in order to safeguard your battery.

The bulb used is a Lucas No. C252A.

ELECTRICAL EQUIPMENT

(2 Litre Model)

THE BATTERY.

About once a month unscrew the filler caps and pour a small quantity of distilled water into each of the cells to bring the acid just level with the tops of the separators. This is done to replace water which has been lost by evaporation. If any of the acid solution is spilled it must be replaced by a dilute sulphuric acid solution of the same specific gravity as the electrolyte in the cell to which it is to be added. The battery must not be allowed to remain with the plates uncovered, as they will suffer a chemical change and be irretrievably damaged.

Keep the Terminals and Battery Fixing Bolts Clean and Tight and Smeared with Vaseline.—A smearing of vaseline protects the terminals and fixing bolts from the corrosive action of the acid, which if allowed to continue will eventually result in trouble. When inspecting the battery, put a spanner on the bolt securing the cable end to the chassis in order to make sure that it is quite tight.

Keep the Outside of the Battery Clean and Dry, particularly the Tops of the Cells.—Water and dirt form a conductor of electricity, and if such a path is allowed to form between the positive and negative terminals of the battery or between the negative battery terminal and the chassis, there will be a leakage of current which will eventually cause the battery to run down. Give the cell tops a wipe over regularly and you will avoid this.

Occasionally make a Point of Examining the Health of your Battery by taking Hydrometer Readings.—The hydrometer allows you to take a sample of the acid solution. It contains a graduated float which indicates the specific gravity of the acid solution in the cell from which the sample was taken. The readings and their indications are as follows:—

1.250—1.300	...	Battery fully charged.
1.150—1.250	...	Battery about half discharged.
Below 1.150	...	Battery fully discharged.

The readings for each of the cells should be approximately the same. If one cell gives a very different reading from the rest it may be that acid has been spilled or has leaked from this particular cell, or there may be a short circuit between the plates. In this case we advise the owner to have his battery examined by a Lucas Service Depôt—neglect may mean costly repairs later on.

It must be remembered that you cannot get more energy out of a battery than is put in. If the car is left parked at night with the lights on, or if the starter motor is used excessively with very little daytime running, then the battery may get into a low state of charge—particularly in winter. This may be remedied either by running the car for longer periods during daytime or by economising in the use of the headlamps and other accessories.

Never Leave the Battery in a Discharged Condition for any Length of Time.—If the car is to be out of use for any length of time, see that the battery is fully charged, and about every fortnight give it a short refreshing charge to prevent any tendency to permanent sulphation of the plates.

THE DYNAMO.

All dynamos are sent out from the works with the bearings packed with grease. This lasts until the car is taken down for a general overhaul, when it is advisable to have the machine dismantled, preferably by a Lucas Service Depôt, for cleaning and adjustment and repacking the bearings with grease.

The dynamo is provided with oilers to enable you to give it a little additional lubrication. The oilers should be given several drops of good quality engine oil every 1,000 miles. It is of the utmost importance that the oiler on top of the lubricator shown in Fig. 25 receives special attention or this bearing will run dry.

Inspection of Commutator and Brushgear.—About once a season remove the metal cover from the dynamo for inspection of the commutator and the carbon contacts, or brushes as they are called. Take care not to lose the nut when removing the cover fixing screw, as the cover is liable to spring open when the screw is released.

It is essential that the brushes make good firm contact with the commutator. The brushes are held in boxes by means of springs. Hold back the spring, and at the same time move the brush to see that it is free to slide in its holder. If there are any signs of sticking, remove it from its box and clean it with a cloth moistened with petrol.

After removing brushes for cleaning or any other purpose, care must be taken to replace them in their original positions, otherwise they will not "bed" properly on the commutator.

If, after long service, the brushes have become worn to such an extent that they will not "bed" properly on the commutator, they should be replaced. Always fit genuine Lucas brushes, as these are made specifically for work on our machines and will give by far the best results and the longest life. We advise you to have the brushes fitted at a Lucas Service Depôt, so that they can be properly "bedded" to the commutator.

Next examine the commutator. It should be clean and free from any trace of oil or dirt, and should have a highly polished appearance. The best way to clean a dirty or blackened commutator is by pressing against it a fine dry duster and getting someone to turn the engine over slowly by hand. If the commutator is very dirty, the duster may be moistened with petrol.

DYNAMO OUTPUT.

The dynamo automatically keeps the battery in good condition.

The dynamo is of the compensated voltage controlled type. These machines work in conjunction with a regulator unit which is housed with the cut-out.

What the Regulator Does—a Completely Automatic Control.—The regulator causes the dynamo to give an output which varies according to the lamp load and state of charge of the battery. When a battery is discharged the dynamo gives a high output, so that the battery receives a quick recharge which brings it back to its normal state in the minimum possible time. On the other hand, if the battery is fully charged, the dynamo is arranged to give only a trickle charge which is sufficient to keep it in good condition without any possibility of causing damage to the battery by overcharging.

In addition to controlling the output of the dynamo according to the condition of the battery, the regulator provides for an increase of output to balance the current taken by the lamps or other accessories whenever they are switched on.

The regulator unit is accurately set to suit the requirements of the equipment fitted on the car, and in normal service the battery will be kept in good condition. If, however, you should find that the battery is not kept in a charged condition, or is being excessively overcharged, we advise you to consult your nearest Lucas Service Depôt, where any necessary adjustments can be made. Owners should not attempt the adjustment themselves

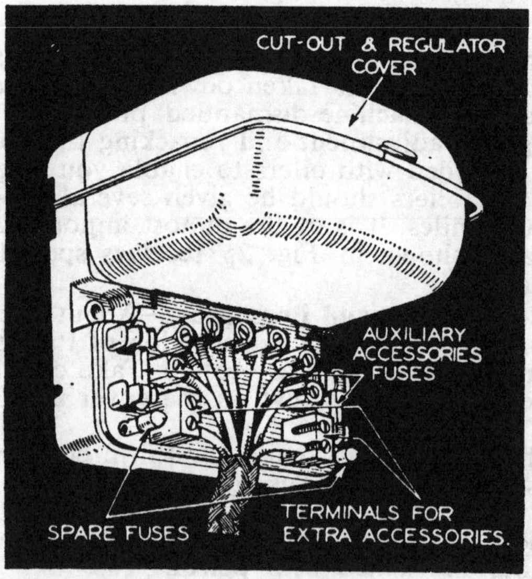

Fig. 59. *Cut-out, regulator fuse and junction boxes* (*ceased Chassis no. SA* 0902).

Ammeter Readings.—It must be remembered, when noting ammeter readings, that, normally, during daytime running, when the battery is in good condition, the dynamo only gives a trickle charge, so that the charge reading will seldom be more than a few amperes.

When starting from cold, the driver will notice the rise of charging current until it reaches a steady maximum at a speed of, say, 20 miles per hour; after which it will remain fairly high for perhaps 10 minutes or so, then fall to a steady charge which is most suitable for the particular condition of the battery.

A discharge reading may be shown sometimes immediately after switching on the lamps. This usually happens after a long run when the voltage of the battery is high. After a short time, the voltage of the battery will fall and the regulator will respond, causing the dynamo output to balance the load.

The Cut-out—an Automatic Dynamo Switch.—It will be noticed from the ammeter readings that the dynamo does not charge at very low engine speeds. This is because it is not rotating fast enough to generate sufficient energy to charge the battery.

Connected between the dynamo and the battery is the cut-out—an automatic switch which acts as a "valve," allowing the flow of current from the dynamo to the battery only. It closes when the dynamo is running fast enough to charge the battery, and opens when the speed is low or the engine is stationary, thus preventing current from flowing through the dynamo windings from the battery.

The cut-out is located together with the fuses on the engine side of the dash.

THE STARTER.

Observe the following points when starting:—

1. Operate the starter switch firmly, and of course release it as soon as the engine fires.
2. Never operate the starter when the engine is running. If the engine will not fire at once, allow it to come to rest before operating the switch again.

3. Do not run the battery down by keeping the starter on when the engine will not start.
4. In cold weather, starting will be facilitated by giving the engine a few turns by the starting handle before using the starter.

Attention Needed by the Starter.—Give the commutator and brushgear similar attention to that described for the dynamo.

Occasionally, if dirty, wash over the screwed sleeve on the armature spindle with paraffin and afterwards give it the merest trace of thin machine oil.

INSTRUMENT PANEL.

The instrument panel houses the combined lighting and ignition switch, ammeter, starter switch control, etc., together with the speedometer, clock, etc. The instruments are illuminated by means of bulbs mounted on the back of the panel.

CUT-OUT, REGULATOR AND FUSE UNIT (Ceasing at Chassis No. SA.0902).

This unit houses the cut-out, regulator and fuses, and forms a junction box for distributing the cables to the various accessories.

The cut-out and regulator are accurately set before leaving the works, and need no further adjustment. The cover protecting them is, therefore, sealed.

This unit also incorporates two fuses. One fuse protects the auxiliary accessories which are operative only when the ignition is switched on, while the other protects those accessories which can be operated irrespective of whether the ignition is on or off.

FUSE UNIT (Commencing at Chassis No. SA.0903) is somewhat similar to the previous type except that it contains five fuses, two protecting the auxiliary accessories which operate through the ignition switch, one the accessories, working independently on the ignition circuit and two protect the head, side and tail lamps.

LAMPS. HEADLAMPS.

Removing Lamp Front and Reflector.—To remove the lamp front, slacken the single securing screw and move it downwards from the slot in which it fits. When replacing the front locate the top of the rim first, then press on the bottom and tighten the fixing screw.

To remove the reflector, turn back the ends of the cork washer at the top of the rim and remove the screw opposite the medallion in the top of the lamp. Turn reflector until markings " O " stamped on reflector rim and lamp body coincide. The reflector can then be withdrawn. When replacing reflector engage it with the lamp body, then turn it until the screw hole in its rim is opposite to the screw hole on top of lamp body. Secure reflector by means of screw.

Checking the Alignment.—The simplest way of checking the adjustment of the lamps is to take the car on a straight level stretch of road at night and examine the direction of the beams. They should be parallel with the road and with each other. If one appears to be out of alignment, adjust it as follows: Slacken the single fixing nut at the base of the lamp and move the lamp on its universal mounting to the required position, finally locking the adjustment by tightening the fixing nut.

Focusing.—For the lamps to give a parellel beam, the filament of the bulb must coincide as nearly as possible with the focal point of the reflector. If the filament of the bulb is behind the focal point of the reflector, the beam will be divergent; while on the other hand, if the filament is in front of the focal point, the beam will be convergent, with a dark area in the centre of the beam. In either case, the lamps will have a poor range and will cause unnecessary dazzle to approaching traffic.

Before lamps are despatched from the works, the bulbs are carefully focused to give the best results. Provided that official Lucas bulbs are fitted as replacements, it should not be necessary to disturb the setting. If for any reason these are not obtainable, and ordinary bulbs have to be fitted, it may be necessary to re-focus by moving the bulb backwards or forwards along the axis of the reflector until the best lighting is obtained.

Cover up one lamp while testing the other. If the lamp does not give a uniform long-range beam without any dark centre, the bulb needs adjusting. To do this, remove the front and reflector, and slacken the clamping clip at the back of the reflector.

After each adjustment, note the effect with the reflector and front refitted.

When the best position for the bulb holder has been found, see that the clamping screw is tightened.

" PASS-LIGHTS " (Ceasing Chassis No. SA.0902).

Removing Lamp Front and Reflector.—Slacken the fixing screw at the bottom of the lamp and press down out of locating slot. The front and reflector can then be withdrawn. When replacing, locate top of rim first, then press front home and secure by tightening fixing screw.

The bulb holder may be withdrawn for bulb replacement when the two securing springs at the rear of the reflector are moved aside.

" PASS-LIGHT " (Commencing Chassis No. SA.0903).

To remove the front and reflector pull out the spring clip at the bottom of the lamp and swing it out from its location. The reflector front can then be withdrawn. When replacing, locate top of rim first, then press home the front and secure by the clip.

Focusing is carried out in the same manner as described in the previous paragraphs.

SIDELAMPS.

The front of the lamp may be removed for a bulb replacement after the screw at the top of the lamp is withdrawn. When refitting the front, locate the bottom and secure with the screw.

" STOP " TAIL LAMP AND REVERSING LAMP.

The lamp front can be removed after the securing screw is withdrawn.

Cleaning.—All Lucas reflectors are protected by a fine transparent and colourless covering, which enables any accidental finger marks to be removed with a soft cloth without affecting the surface of the reflector. Never use metal polishes on Lucas reflectors. A light polish with a soft cloth is all that is necessary. Chromium plated lamps will not tarnish and only need wiping over with a damp cloth to remove dust or dirt.

REPLACEMENT OF BULBS.

When the replacement of a bulb is necessary, it is important not only that the same size bulb is fitted, but that it has a high efficiency and will focus in the reflector. Cheap and inferior replacement bulbs often have the filament of such a shape that it is impossible to focus correctly; for example, the filament may be to one side of the axis of the bulb, resulting in loss of range and light efficiency.

It always pays you to use Lucas Genuine Spare Bulbs as then these problems will not arise.

Bulbs Fitted.—

Headlamps	Lucas No. 54	12 v. 36 watt
Sidelamps	Lucas No. 207	12 v. 6 watt
"Pass-lamps" (ceasing SA.0902) ...	Lucas 57	12 v. 36 watt
"Pass-lamp" commencing SA.0903)	Lucas 87	12 v. 60 watt
"Stop" Tail Lamp	Lucas No. 207	12 v. 6 watt
Reversing Lamp	Lucas No. 1	12 v. 24 watt
Panel Lights	No. 1224 M.E.S.	12 v. 2.4 watt
30 M.P.H.—Warning Lamp	No. 1224 M.E.S.	12 v. 2.4 watt
Ignition Warning Lamp	No. 252 M.E.S.	2.5 v —
Trafficators (Festoon)	Lucas No. 256	12 v. 3 watt
Interior Light	Lucas No. 207	12 v. 6 watt

NEW ALTO HORNS.

The horns require no adjustment. Should one or both of the horns fail or become uncertain in action, first ascertain that the trouble is not due to faulty wiring, a blown fuse or loose mounting. If the cause of the trouble cannot be found, do not attempt to dismantle the horn, but return it to a Lucas Service Depôt for examination.

WINDSCREEN WIPER (Ceasing Chassis No. SA.0902).

The windscreen wiper is carefully adjusted and packed with grease during assembly, and needs no attention.

To start the wiper, pull out the handle and turn to disengage it from the switch. Then move switch to "ON" position. To stop the wiper, move switch to "OFF" position, pull out handle to disengage wiper blades from the gears, and turn end of handle into top of switch control.

WINDSCREEN WIPER (Commencing Chassis No. SA.0903).

To start the wiper, pull out the knob on the driver's side and turn it until the drive is engaged. In this position the wiper is automatically switched on. To bring the arm on the passenger's side into operation, pull out its knob and turn it to bring the arm on to the screen.

To stop the wiper, pull out the knob on the driver's side to its full extent, turn it as far as the concealed stop and then release it. The arm on the passenger's side must be parked by pulling out the knob and turning it in an anti-clockwise direction.

TRAFFICATORS.

Every two to three months or if the arms become stiff at any time, raise each arm, and by means of a brush, matchstick or similar article apply a drop of thin machine oil, such as sewing machine oil, to the catch pin between the arm and the operating mechanism. Use only the merest drop of oil—any excess may affect the working of the operating mechanism. The "Trafficators" are kept in the closed position by means of a spring. The arms can be pulled out by hand. If any difficulty is experienced, switch the "Trafficator" on and then, supporting the arm in a horizontal position, move the switch to the "off" position.

If at any time the arm fails to light up when in operation, raise the arm in the manner previously described and examine the bulb. If necessary, replace the bulb by a Lucas No. 256 (3 watt festoon type) bulb.

To remove the bulb, withdraw the screw on the underside of the arm and slide off the metal plate; the metal plate can then be replaced. To replace the metal plate, slide it on in an upwards direction so that the side plates engage with the slots on the underside of the spindle bearing. Finally, secure the plate by means of its fixing screw.

Do not remove bulb while the "Trafficator" is switched on, as this may cause a short circuit.

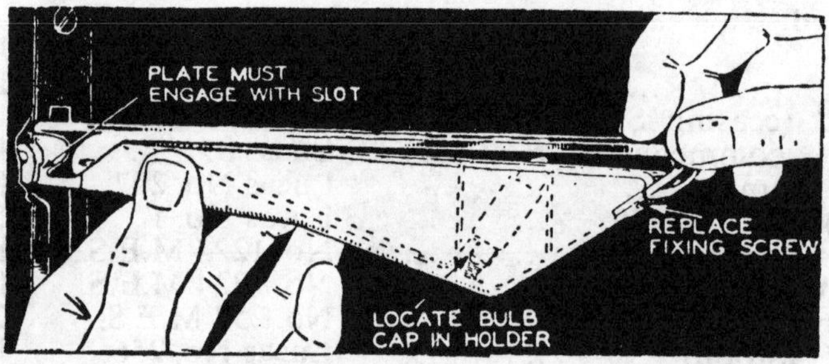

Fig. 60. Replacing trafficator bulb.

COIL IGNITION EQUIPMENT.

Considering the exacting duties coil ignition has to perform, the attention needed to ensure absolutely reliable service is very small.

The contact breaker gap should periodically be checked and, if necessary, reset to .012 in. maximum opening, and the cam given a smear of engine oil.

LUBRICATION.

The following parts of the distributor require lubrication:—

1. **Distributor Shaft.**—Add a few drops of thin machine oil through oiler provided about every 1,000 miles.

2. **Cam.**—Give the cam a smear of engine oil every 3,000 miles.

3. **Cam Bearing.**—About every 3,000 miles, withdraw the moulded rotating arm from the top of the spindle by pulling it off, and add a few drops of thin machine oil. Do not remove the screw exposed to view, as there is a clearance between the screw and the inner face of the spindle through which the oil passes to lubricate the cam bearing. Take care to refit the arm correctly and to push it on to the shaft as far as possible, otherwise there is a risk of tracking and burning of the moulding.

4. **Contact Breaker Pivot.**—Every 5,000 miles, place a spot of oil on the pivot on which the contact breaker rocker arm works.

5. **Automatic Advance Mechanism.**—About once every year the moving parts of the automatic timing control must be lubricated with a good grade engine oil. To render the control accessible, remove the distributor moulding and lift off the rotating distributor arm, then remove the contact breaker moulding by withdrawing its two securing screws.

Cleaning.—Keep the outside of the distributor clean, particularly the spaces between the high tension terminals. Very occasionally, remove the moulding by springing aside its two securing spring clips. Wipe the inside clean with a dry cloth, and see that the carbon brush is quite free in its holder. Clean the metal electrodes inside the moulding and also the rotating electrode on the distributor arm; if necessary, use a cloth moistened with a drop of petrol for this.

Next examine the contact breaker; you must keep the contacts free from any grease or oil. If they are burned or blackened, clean them with fine carborundum stone; or if this is not available, you can use very fine emery cloth. Finish off with a cloth moistened with petrol, and remove all traces of dirt and metal dust. Misfiring is sometimes caused by dirty contacts.

Checking and Adjusting the Contacts.—The contacts require only occasional adjustment. The chief cause of variation in the gap is wear of the heel of the contact rocker arm which bears upon the actuating cam. Provided you keep the cam smeared with engine oil, however (see Lubrication Instructions), the wear on the heel will be negligible, and the contact gap setting should only require adjustment at infrequent intervals.

To check the setting, turn the engine by hand until the contacts are fully opened. Now insert the gauge provided on the ignition screwdriver between the contacts. The gauge has a thickness of about 12 thousandths of an inch and it should be a sliding fit between the contacts when the gap is correct. We do not advise you to alter the setting unless there is quite an appreciable variation from the gauge. To make the adjustment keep the engine in the position to give maximum opening of the contacts and slacken the two screws securing the contact plate. Then move the plate until the gap is set to the thickness of the gauge. After making the adjustment, care must be taken to tighten the locking screws.

Renewing the High Tension Cable.—The high tension cables are those connecting the coil to the distributor and the distributor to the sparking plugs. When these cables show signs of perishing or cracking, they must be replaced by 7 mm. rubber-covered ignition cable.

The method of connecting the cable is to thread the knurled moulded nut over the lead, bare the end of the cable for about ¼ in., thread the wire through the brass washer provided, and bend back the strands. Finally, screw the nut into its respective terminal.

The Coil.—The coil requires no attention whatever beyond keeping its exterior clean, particularly between the terminals, and occasionally checking that the terminal connections are quite tight.

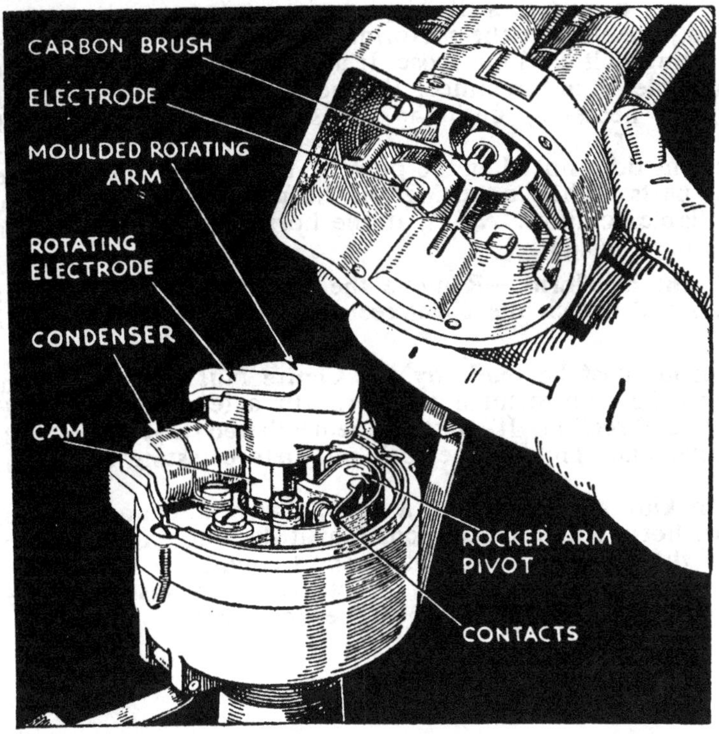

Fig. 61. Distributor with cover removed.

Ignition Switch and Warning Light.—The ignition switch, besides forming a means of stopping the engine, is provided for the purpose of preventing the battery being discharged by the current flowing through the coil windings when the engine is stopped. A warning lamp is provided in the instrument panel which gives a red light when the ignition is switched on and the car is running very slowly or is stationary, thus reminding you to switch off.

Should the warning lamp bulb burn out, this will not in any way affect the ignition system; but you should replace it as soon as possible in order to safeguard your battery.

The bulb used is a No. 252 M.E.S.

Micrometer Timing Adjustment.—In order to obtain very fine timing of the ignition to the engine and to allow for altered engine conditions—e.g., state of carbonisation of engine, change of fuel, etc.—a micrometer adjustment is provided which allows fine adjustment to be made simply by the movement of a knurled knob.

The final setting can be made by use of the micrometer adjustment after running the engine. If the firing is found to be slightly too early or too late, adjust the knurled knob until the best engine performance is obtained. The adjustment should not be altered by more than one distributor degree at a time (one division on the scale is equivalent to two distributor degrees).

Lucas "Mellotone Horns" (when fitted).—The horns are adjusted to give their best performance before being sent out of the works, and will give long periods of service without any attention.

If one of the horns fails or becomes uncertain in its action, first ascertain that the trouble is not due to some outside source, e.g., a loose connection or short circuit in the wiring of the horn. If both horns fail or become uncertain in their action, the trouble is probably due to a discharged battery or blown fuse. If the fuse has blown, examine the wiring for the fault and replace with spare fuse provided, a 35 ampere size.

It is also possible that the performance of a horn may become uncertain, due to its fixing bolt working loose. If after carrying out the above examination the trouble is not rectified, the horn may need adjustment; but this should not be necessary until the horns have been in service for a long period.

Adjustment does not alter the pitch of the note, it merely takes up wear of moving parts. When adjusting the horns, short-circuit the fuse, otherwise it is liable to blow. Again, if the horns do not blow on adjustment, release the push instantly.

Adjustment of Horns.—Remove the horn cover after withdrawing the fixing screw and detach the cover securing bracket by springing it from its fixing.

Adjustment must be made by the Centre Nut only.—Slacken the lock nut and turn the adjustment nut underneath, a few degrees at a time, in an anti-clockwise direction. It must be remembered that the adjustment is a very sensitive one. Tighten the lock nut before testing the horns on each occasion.

When making adjustments to a horn, always disconnect the supply lead of the other horn, taking precautions that it does not come into contact with any part of the chassis and so cause a short circuit.

Finally, if the note is still unsatisfactory, do not dismantle the horn, but return it to a Lucas Service Depôt for examination.

MODIFICATIONS

FOOT OPERATED DIPPER-SWITCH. (TD Midget.)

The later TD Midget Models are now being fitted with a foot operated dipper switch. The earlier Models can be fitted with this equipment, and the necessary conversion material can be obtained through a recognised M.G. Distributor or Dealer or on direct application to the Service Parts Department, The M.G. Car Company Limited, Abingdon-on-Thames.

Fitting details are as follows:—

Midget TD. L.H. Drive.—Fig. 62, gives full details of the position and method of fixing the dipper switch bracket to the left hand side of the

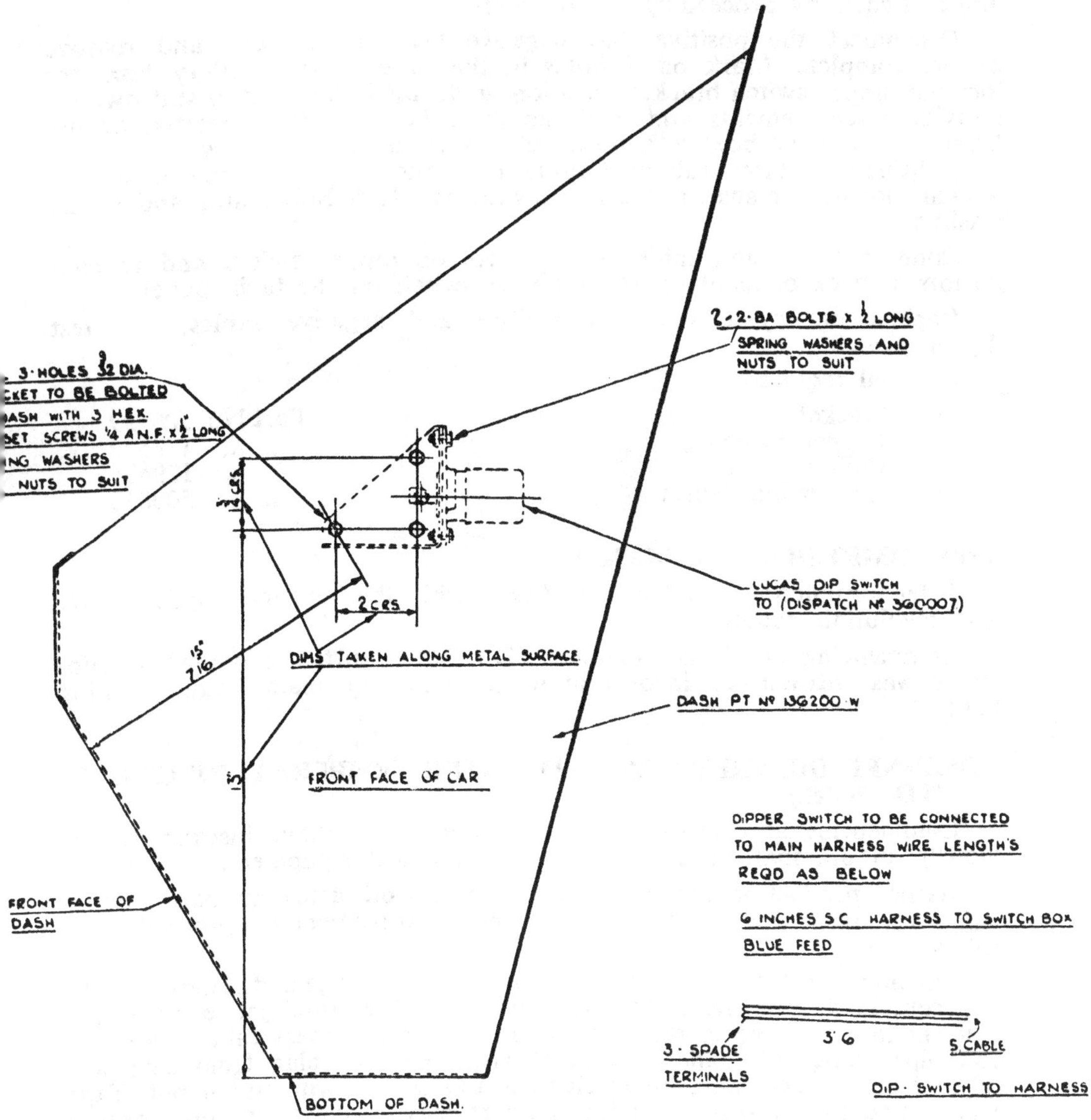

Fig. 62. Showing dipper switch fixing bracket for L.H. Drive TD Midget. Foot Operated.

dash-panel. The cables from the dipper switch contacts are connected to the connectors at the back of the hand operated dipper switch on the facia panel.

Materials available:—

1 off Bracket	Part No. 500804
1 off Dipper Switch	„ „ 138599
1 off Conversion set of Cables	„ „ 500825

Midget TD. R.H. Drive.—In the case of the R.H. Drive Model, the foot operated dipper switch is fitted in almost a central position in the car, attached to the base of the battery box, slightly forward of the clutch pedal and slightly over on the right hand side of the centre line of the gearbox cover. Earlier models can be converted to the foot operated dipper switch by proceeding as follows:—

Disconnect the positive and negative leads to battery and remove battery complete. Mark off 3 holes in the base of the battery box, for locating dipper switch bracket, position as detailed above or to suit owner's individual requirements and drill the three holes $\frac{9}{32}$ in. diameter. Secure bracket to battery box with three set screws, nuts and spring washers to suit, fitting the three rubber buffers over the heads of the set screws. Attach foot dipper switch to bracket with two 2BA Bolts, nuts and spring washers.

Connect conversion cables to contacts on dipper switch and to connectors at back of hand operated dipper switch on the facia panel.

Replace battery, connect up positive and negative cables, and test dipper switch.

Material required:—

1 off bracket	Part No. 500799
1 off Dipper Switch	„ „ 138599
3 off Buffers	„ „ 129434
1 off Conversion Cables	„ „ 500825

SPEEDOMETERS. (TD Model.)

Later production Models are fitted with the magnetic speedometer and revolution counter.

Commencing at Chassis number TD.17548 a further modified speedometer was introduced, incorporating a headlamp main beam warning light.

COMBINED OIL PRESSURE AND WATER TEMPERATURE GAUGE. (TD Model.)

Commencing at Chassis number TD.13914, the above instrument previously an optional extra, was fitted as standard equipment.

Earlier production models, fitted with the oil gauge to part number A.903 can be converted to the later type of instrument by proceeding as follows:—

Disconnect oil pipe from back of oil pressure gauge and remove gauge. Fit combined instrument and run water temperature gauge pipe and element through hole in dash, through which the harness cables pass and run pipe along left hand side of Chassis. Remove plug from near side of radiator header tank, enter element and secure with union nut. Tape water pipe to near-side radiator support stay. Connect up oil pressure gauge pipe.

The most efficient running temperature for the engine is 70°-80°C.

from the 'M'

MG

to the 'TF'

CHAPTER VIII

BRAKING SYSTEM

BRAKING SYSTEM

DESCRIPTION. (M, D, J, F, L, P, K, KN and N models.)

The four wheel brake system of these models is simple and efficient, and periodical adjustment for wear an easy matter. The brakes are applied through the agency of steel cables which pass through metal sheathed rubber covered outer cables from either side of the centre of the chassis to the brake drums, the final application being shown in Fig.1.

Fig. 1. Method of actuating brakes.

A cross shaft is placed in the centre of the chassis, anchored at either end and supported in the centre by a tubular cross member of the frame. The footbrake is coupled to the brake actuating cross shaft by a rod, and either extremity of the brake cross shaft is provided with a pulley having holes drilled through the top and bottom to receive the end adaptors of

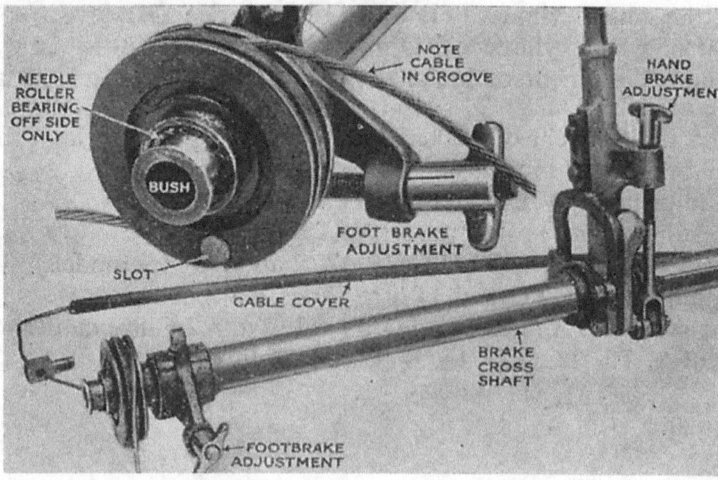

Fig. 2. *Hand and footbrake adjustment and needle bearing in end of cross shaft (M, D, F, J, L, P and K models).*

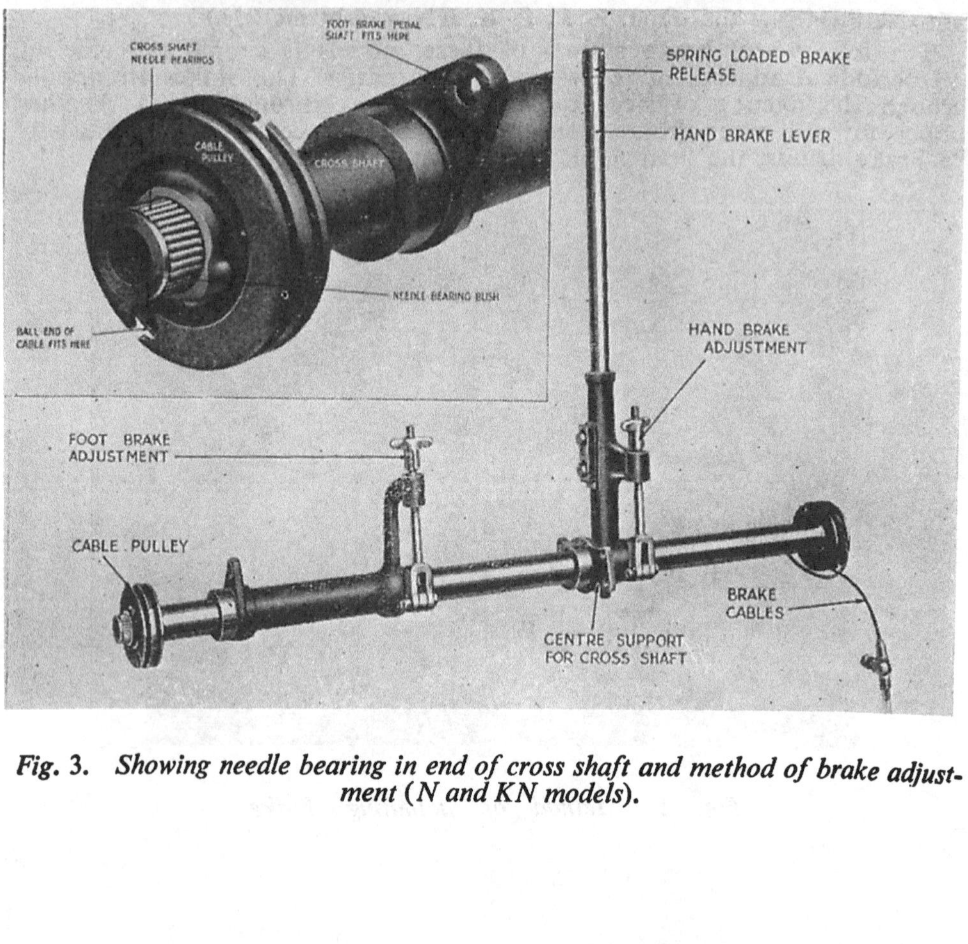

Fig. 3. *Showing needle bearing in end of cross shaft and method of brake adjustment (N and KN models).*

the brake cables. As either brake is applied, the brake cross shaft is rotated, pulling the rear brake cables forward and the front brake cables backwards. The whole of the brake cross shaft is lubricated from the dashwall nipples and in order to give perfect freedom of movement, the cross shaft is mounted on needle bearings.

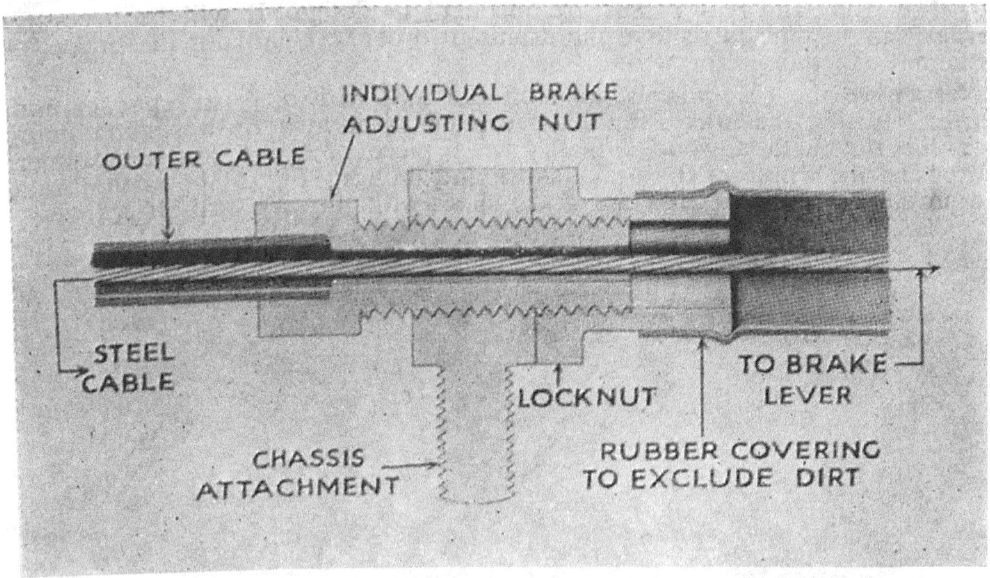

Fig. 4. The brake cable stop and individual adjustment fitted on the brake drum back plates.

BRAKE ADJUSTMENT. (M model.)

The main adjuster is incorporated in the rod from the pedal to the cross shaft, being accessible through the floorboards on the driver's side; consisting of a toggle and two lock nuts. These lock nuts are slackened back, one having a left-hand thread, and the toggle turned by a spanner. As one end is screwed right-hand and the other left, turning the toggle one way or the other will decrease or increase the effective length of the brake rod, and thus tighten or slacken the whole system.

Individual adjustment for each wheel is provided by Standard Bowden adjusters on the cables, and proves a simple matter and is worth attention if braking, steering and road holding are to be maintained in first class order.

The most simple method is as follows:—

With all four wheels jacked up, tighten up each yoke until the brake lining can just be heard scraping against the drum when the wheel is spun. Then *slacken* the yoke *one complete turn.* This will give an approximate setting, but final and accurate results can only be obtained by actual trial on the road.

BRAKE ADJUSTMENTS. (D, F, L, J, P, K, N and KN models.)

The main hand and footbrake adjustments take the form of thumb nuts, that for the footbrake being situated on the off side of the car, forward of the driver's seat, accessible through the trap door in the floorboards, and for the hand brake, in the extension at the base of the handbrake lever. They should be turned in a clockwise direction to tighten the brakes. Individual adjustment is provided for at each wheel and can be carried out by means of adjusting screws fitted to the cable stops on the axle back plates. At this point on the cables a rubber dust excluder is fitted in the form of a rubber tube, and this tube is mounted on what is actually the lock nut for this adjustment. By undoing the lock nut the cable stop can be adjusted on its thread by the amount required and the lock nut retightened.

BRAKE DRUMS. (M, D, F, L, J, P, K, KN and N models.)

The front and rear drums are identical in design. It will be necessary from time to time to remove the drums in order to clean out the brakes or examine the linings for wear.

Remove the rear wheels, take off the nuts with a $\frac{5}{16}$ in. spanner and, after releasing the brake, the drum can be withdrawn by a slight tapping on the ribs with a wooden mallet or a piece of wood and a hammer; exposing the whole of the brake shoes and linings. Fig. 5 shows the brake drum and components and the brake shoes with the two pull-off springs.

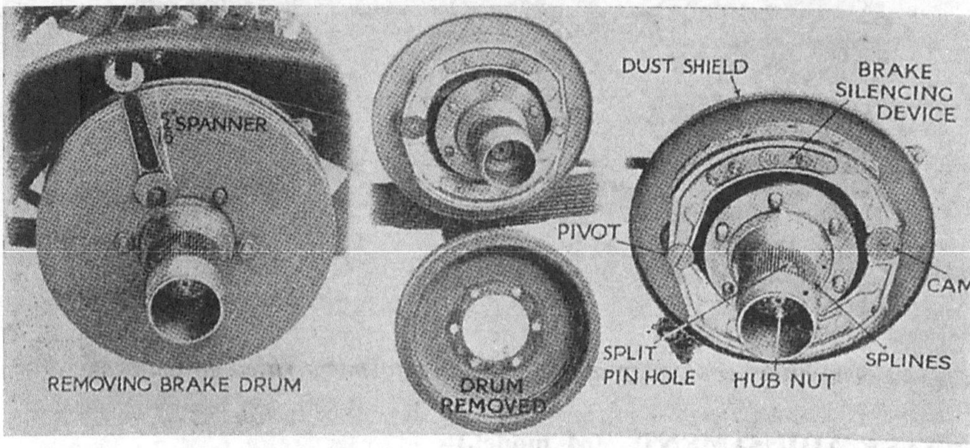

Fig. 5. Three views of brake being dismantled, first showing drums being removed, the centre view of the drum removed, and right, the brake shoes and silencing device (D, F, L, J, P K and KN models).

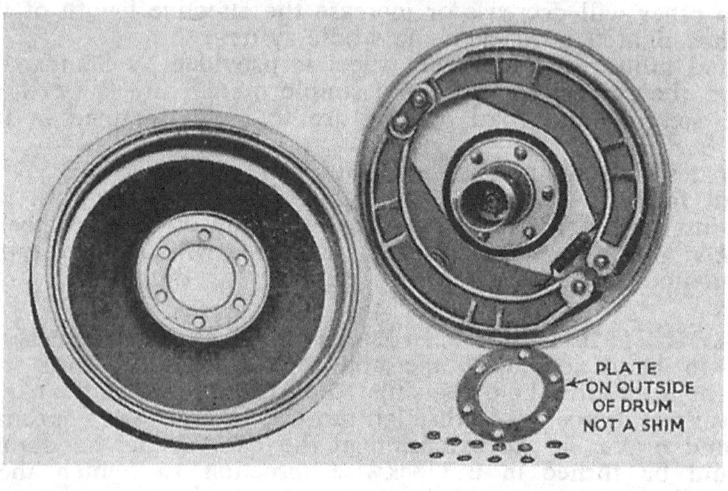

Fig. 6. View of front hub after brake drum has been removed showing type of brake removed (K models).

The countersunk screw in the fluted portion of the front hub is intended to permit the split pin of the hub nut being easily removed and replaced through the hole in the hub, when the complete hub is to be removed.

BRAKE ANTI-SQUEAK DEVICE. (D, F, L, J, P, K, KN and N models.)

The brake shoes are fitted with an anti-squeak device, consisting of lead alloy blocks fitted into the channel section of the brake shoes and held in position by means of clevis pins and small coil springs. When vibrations take place in the brake drum, they are communicated to the brake shoes and the shoes vibrate; the lead weights, however, are in effect " left behind " by the vibrating shoes and energy is wasted in friction between the weights and the shoes. The conditions under which the weights are mounted can in practice be made such that they can always waste more energy than the squeaking forces can generate and the squeak is prevented.

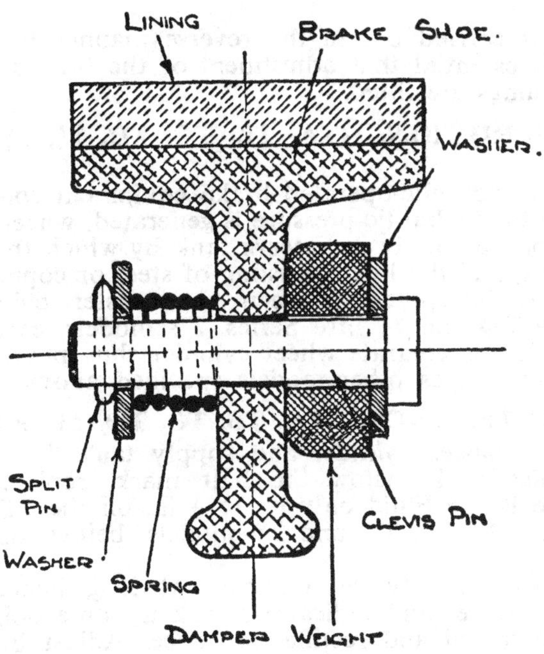

Fig. 7. Sectional view, showing brake shoe with the anti-squeak device attached.

REMOVAL AND REPLACEMENT OF FRONT AND REAR BRAKE CABLES. (M, D, J, F, L, P, N, KN and K models.)

Remove the carpets and floorboards, and disconnect the lubrication pipes from the cables and brake cross shaft.

Disconnect the cables from their anchorages on the brake back plates. Remove the two large bolts securing the brake cross shaft either side of the chassis, twist cross shaft slightly and brake cross shaft can be withdrawn clear of its location in chassis frame.

Remove brake cable nipple from the slot in the pulley on the cross shaft and pull cables clear of chassis.

Replacement is carried out in the reverse manner to that detailed for removal, care being taken to ensure that the cables are in their appropriate grooves in the pulley. Adjust brakes after new cables have been fitted.

REMOVAL AND REPLACEMENT OF BRAKE SHOES. (M, D, J, F, L, P, N, K and KN models.)

Jack up the car until the wheels are clear of the ground and remove the wheel and brake drums.

Release the brake-shoe-pull-off springs, the flat washer attached to the brake cam and the pivot pin, when the brake shoes can be lifted clear of the brake plate. (M, D, F, J, L, KN and N models.)

Release the brake-shoe-pull-off springs; the flat washer retaining the brake shoes in position on the brake cam and remove the two set bolt pivot pins; when the brake shoes can be lifted clear of the brake plate. (P model.)

Release the brake-shoe-pull-off springs and withdraw the split pins securing the shoes to the brake cam.

Remove the two nuts securing the shoes to the pivot pins, remove the retaining plate and brake shoes can then be lifted clear of the brake plate. (K model.)

Replacement is carried out in the reverse manner to that detailed for removal and it is essential that adjustment of the brakes be carried out if new shoes and linings are fitted.

GENERAL DESCRIPTION. (TA, TB, TC, TD, 1¼, 1½, 2 and 2.6 litre models.)

The Lockheed hydraulic-operated brake equipment consists of a master cylinder in which the hydraulic pressure is generated, wheel cylinders operating the brake shoes, a supply or reserve tank by which the system is maintained full of fluid; and the line consisting of steel or copper tubing, flexible hoses and unions interposed between the master cylinder and wheel cylinders. On the TD and 1¼ litre Series YB models, each brake shoe on the front wheels has a separate wheel cylinder thus providing two leading shoes. Rear wheels are as other models specified above.

ROUTINE MAINTENANCE. (TA, TB, TC, TD, 1¼ and 2.6 models.)

The combined master cylinder and supply tank fluid level should be examined periodically. If below the half mark, replenish with genuine Lockheed Orange Brake Fluid only up to ½ in. on the TD and 1¼ models, and 1 in. on the TA, TB, TC and 2.6 models, below the filler plug. (Do not fill completely.)

Topping up should only be necessary at long intervals. A rapid or considerable fall in the fluid indicates a leak at some point in the system which should be traced and rectified at once. Adjust brake shoes when the pedal travel becomes excessive, i.e., when the pedal travels to 1 in. of the floorboards before solid resistance is felt. The details under "Adjustment of the brake pedal" and "Footbrake adjustment."

ROUTINE MAINTENANCE. (1½ and 2 litre models.)

The supply tank containing the reserve supply of brake fluid is attached to the engine side of the bulkhead on the right hand side where it is accessible, and is connected to the master cylinder by a copper pipe. Examine the fluid level every month and keep topped up to within 1 in. of the cap with genuine Lockheed Orange Brake Fluid. Examine vent hole in filler cap periodically and keep free from obstruction. Other detail regarding topping up and brake shoe adjustment as above.

THE MASTER CYLINDER. (TA, TB & TC models.)

The Combination Barrel Type compensating Master Cylinder provides automatic compensation for expansion or contraction of the fluid due to temperature changes. Within the master cylinder is the piston (K), cupped washer (I), normally held in the "off" position by coiled springs (H). Immediately to the front of the cupped washer, with the piston in the "off" position, is a by-pass port hole connecting the cylinder interior with the supply tank. With any rise in temperature causing the fluid to expand in the system, fluid is allowed to pass through the port into the supply tank. With any drop in temperature causing the fluid to contract, the fluid flows back through the port. A constant volume of fluid is therefore maintained in the system.

By means of the push rod (O) attached directly to the brake pedal, pressure is applied to piston (K). To prevent ingress of dirt, a rubber boot is attached to the open end of the master cylinder. A combination inlet and

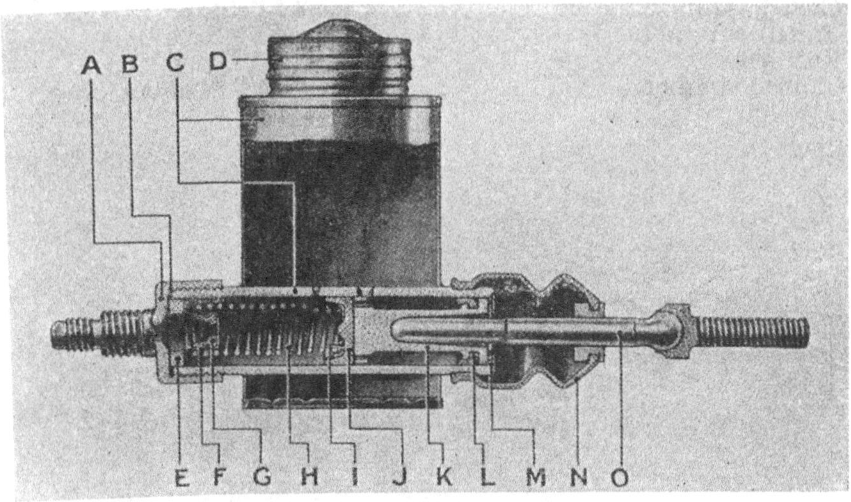

Fig. 8. *The master cylinder (TA, TB and TC models).*

outlet check valve (F-G) held in position by the return spring (H) is incorporated in the head of the master cylinder. This valve cuts off the return to the master cylinder of fluid pumped into the pipe line during the "bleeding" operation, ensuring a fresh charge of fluid being delivered at the next stroke of the pedal.

Application of the brake pedal pushes the master cylinder piston forward forcing fluid through holes in the metal valve body, deflecting the walls of the rubber cup and passing into the system. On releasing the brake pedal, the master cylinder return spring forces the piston back to the "off" position against the stop (M). As a result of the action of the brake shoe return springs, the pistons in the wheel cylinders are forcing back fluid, lifting the whole valve assembly off its seat, until the pressure balances the effect of the master cylinder return spring, and the inlet valve closes.

Leading from the interior of the supply tank to the annular space formed by the reduced skirt of the piston (K) is a large port. Through this port the annular space is kept at all times full of fluid from the supply tank, leakage at the rear of the piston being prevented by the secondary cup (L).

Should the return of the fluid from the pipe line be insufficient to equal the displacement caused by the return of the master piston, a vacuum is created in the master cylinder sufficient to cause the master piston cup (I) to turn in at the lip and allow the fluid to by-pass from the annular space through the small holes in the piston head, into the master cylinder.

Any excess fluid thus introduced into the system will pass freely into the supply tank through the by-pass port when the master cylinder piston returns to its "off" position.

THE MASTER CYLINDER. (1½ & 2 litre models.)

The single outlet panel type compensating master cylinder provides automatic compensation for expansion or contraction of the fluid due to temperature changes.

Its action is identical to that of the master cylinder as fitted to the TA, TB & TC models, although reference to Fig. 9 will show that the items indicated by letters vary. It is thought that it is worth while to index the parts as referred to in the illustration.

"A" Inlet and outlet check valve.
"B" Master cylinder.
"C" Coiled spring.
"D" Small port hole.
"E" Retainer.
"F" Cupped washer.
"G" Piston.
"H" Secondary cup.
"J" Stop.
"K" Retaining spring clip.
"L" Push rod.
"M" Rubber boot.

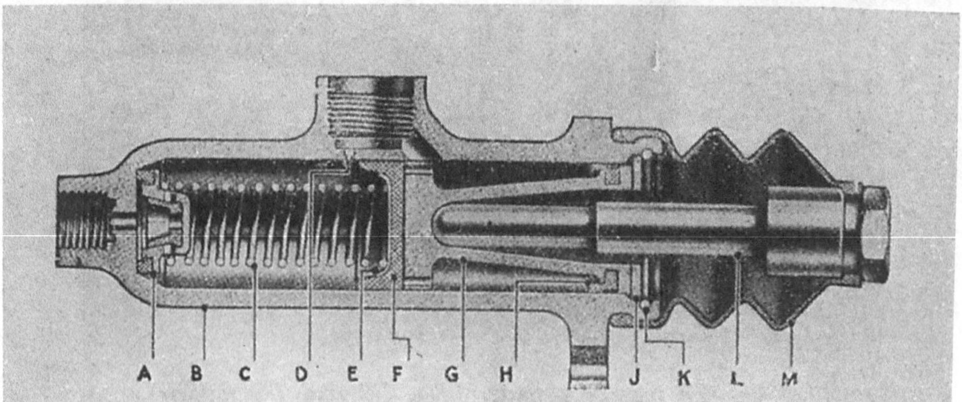

Fig. 9. The master cylinder.

THE MASTER CYLINDER. (2.6 litre model.)

The integral tandem master cylinder consists of a supply tank and two cylinders in line, one operating the front brakes and the other the rear brakes. There is no direct fluid communication between the two cylinders, each of which is fed by a separate compartment in the supply tank. The cylinders are known as primary and secondary. In the master cylinder "T" is the primary piston "E" and the secondary piston "J," backed by rubber cups "F," these being held in position by coil springs "H" and "L" which also serve to keep the pistons against their respective stops "C" and "K" when the system is at rest.

In front of the rubber cups are by-pass ports "P" and "Q" permitting full compensation for any expansion or contraction of fluid due to temperature changes, also serving to release additional fluid into the cylinders at each release of the brake pedal after brake application.

Secondary cups "D" are fitted to both pistons to prevent fluid leakage, whilst an additional cup of the type "F" is fitted to the secondary piston "J" to prevent fluid communication between the cylinders under pressure.

The open end of the cylinder is sealed by rubber boot "B" against the ingress of mud and water.

On depressing the foot pedal which is connected to the push rod "A," the primary piston "E" forces fluid through the check valve "X" (shown in scrap view of Fig. 10) to the front wheel cylinders until the clearance between the brake shoes and drums is taken up. Further movement of the primary piston generates pressure causing the secondary piston to move forward and force fluid through the check valve " Y " to the rear wheel cylinders. When sufficient fluid has been displaced to take up all the clearances further effect on the brake pedal generates equal pressure in both cylinders thus giving full compensated brakes. The check valves " X " & " Y " prevent the fluid pumped out of the cylinders during the " bleeding " operation from returning, thus ensuring a fresh charge is delivered at each stroke of the pedal.

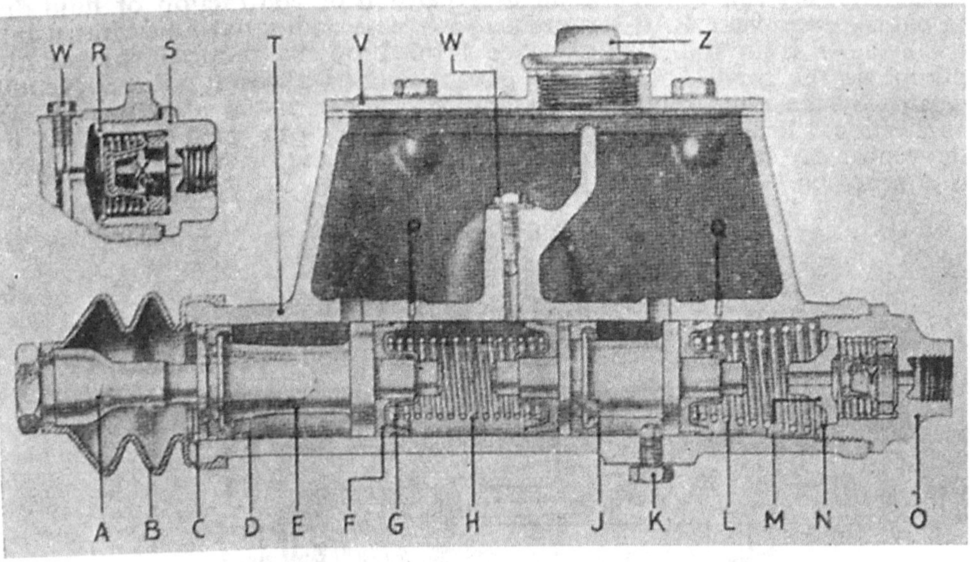

Fig. 10. *The integral tandem master cylinder.*

Application of the brakes pushes the master cylinder pistons forward, forcing fluid through the holes in the metal valve body, deflecting the walls of the rubber caps and so passing into the system. On releasing the pedal the master cylinder return springs force the pistons back to the " off " position against stops " C " and " K." At the same time the pistons in the wheel cylinders, as a result of the action of the brake shoe return springs, are forcing back fluid, so lifting the whole valve assemblies off their seats, until the fluid pressure balances with the effect of the master cylinder return springs, and the inlet valves close. Leading from the interior of the reserve oil supply above the cylinders to the annular space formed by the reduced skirt of the pistons " E " and " J " are large ports. Through these ports the annular space is kept full of fluid from the supply tank, leakage at the rear of the piston being prevented by the secondary cup " D." If the return of fluid from the pipe line is insufficient to equal the displacement caused by the return of the master piston, a vacuum is created in the master cylinder sufficient to cause the master piston cups " F " to turn in at the lip, allowing the fluid to by-pass from the annular space, through the small holes in the piston head into the master cylinder.

Excess fluid thus introduced into the system will pass freely into the supply tank through the ports " P " & " Q " when the master cylinder piston returns to its " off " position.

THE MASTER CYLINDER. (1¼ litre Series Y.)

The master cylinder is mounted on the right hand side of the chassis next to the brake pedal, and is integral with the supply tank. A combination inlet and outlet valve "K" is incorporated in the head of the master cylinder. Its function is to cut off the return to the master cylinder of the fluid pumped into the pipe line during the "bleeding" operation, ensuring a fresh supply of fluid being delivered into the pipe at each stroke of the pedal and a complete purge of air from the system.

In front of the main cup "G" when the system is at rest is the by-pass port "M." This ensures that the system is maintained full of fluid at all times, allowing full compensation in expansion or contraction of fluid due to temperature changes. It serves also to release additional fluid drawn into the cylinder from the annular space formed by the reduced skirt in the piston via the small holes in the piston head as a result of the vacuum created by the rapid return of the master cylinder piston after each brake application. Brakes will drag and bind if the by-pass port is covered by the rubber cap due to an incorrect adjustment of the push rod "A," which is attached to the brake pedal.

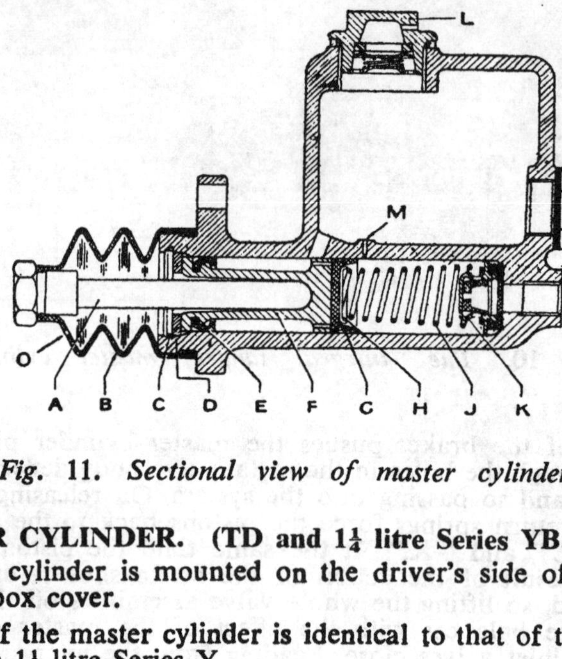

Fig. 11. *Sectional view of master cylinder.*

THE MASTER CYLINDER. (TD and 1¼ litre Series YB.)

The master cylinder is mounted on the driver's side of the car underneath the gearbox cover.

Operation of the master cylinder is identical to that of the master cylinder fitted to the 1¼ litre Series Y.

Fig. 12. *Master cylinder in section showing correct fluid level and disposition of components.*

Fig. 13. *Method of refilling master cylinder. (TD Midget.)*

ADJUSTMENT OF BRAKE PEDAL. (TA, TB, TC, TD, 1¼, 1½, 2 & 2.6 litre models.)

It is essential that the push rod should have a slight clearance where it seats in the piston, when in the " off " position. Should the push rod be adjusted tightly against the piston the port hole will be covered by the cup washer, thus preventing the compensating action of the master cylinder, causing binding brakes.

The correct amount of free movement is set during erection of the vehicle, and should never need attention.

If adjustment has been disturbed, adjust the effective length of the rod connecting the cylinder to the pedal until the pedal pad can be depressed approximately ½ in. before the piston begins to move; the clearance can be felt if the pedal is depressed by hand.

NOTE:—Before attempting any alteration it is important to ensure that neither the floorboards nor the floor carpet obstructs the pedal and that the piston has not stuck in the cylinder bore. In either case a false impression will be given, even though adjustment is correct.

REMOVAL OF MASTER CYLINDER. (TA, TB, TC, 1½, 2 & 2.6 litre models.)

Drain the supply tank by disconnecting the pipe connection at the master cylinder head, and depress the brake pedal slowly by hand, allowing the expelled fluid to flow into a clean container, until the tank is empty. Detach push rod from the foot pedal, remove nut securing the cylinder to the chassis and withdraw cylinder. On the 2.6 litre model it will be necessary to disconnect the two pipes, one at the master cylinder head and the other at the side, the master cylinder being secured to the chassis by three bolts as on the 1½ and 2 litre models.

REMOVAL OF MASTER CYLINDER. (1¼ litre Series Y.)

Remove driver's seat and front floorboard and remove the side cover from the pedal chassis box. Uncouple the pipe from the master cylinder, remove the cotter pin and pin securing the push rod to the brake pedal.

Remove securing bolts, nuts and spring washers and detach master cylinder from its mounting. Drain the supply tank by removing the filler plug.

REMOVAL OF MASTER CYLINDER. (TD and 1¼ litre Series YB.)

Remove the floorboard and disconnect the union nut connecting the pipe line to the master cylinder.

Remove the nuts and spring washers from the two bolts securing the master cylinder to the frame, and withdraw the bolts. Disconnect the brake pedal return spring from the frame, lift out brake pedal complete with push rod, rubber boot and pedal return spring when master cylinder complete with main supply pipe is easily removed. Drain the supply tank by removing the filler plug.

DISMANTLING THE MASTER CYLINDER. (TA, TB, TC, 1½, and 2 litre models.)

Remove the rubber boot and push rod, the piston retaining spring clip and washer and withdraw the piston from the panel. The cup, spring and valve are then easily removed. To remove secondary cup from piston, carefully stretch over end of flange, using the fingers only.

DISMANTLING THE MASTER CYLINDER. (2.6 litre model.) Fig. 10.

Remove the rubber boot "B," the circlip stop "C" and withdraw piston "E," cup "F" and spring "H." Unscrew cylinder head "O," withdraw spring "L," remove stop pin "K," when piston "J" can be withdrawn through either end of the cylinder. To remove the secondary cups "D" from the pistons carefully stretch over the flange, using the fingers only.

DISMANTLING THE MASTER CYLINDER. (1¼ litre Series Y.)

Remove the push rod "A" complete with rubber boot. Push the piston "F" down the cylinder bore, remove the circlips "C" when the remaining internal parts "D" to "K" can be withdrawn.

Using the fingers only, carefully stretch the cup "E" over the end flanges of the piston and remove secondary cup.

DISMANTLING THE MASTER CYLINDER. (TD and 1¼ litre Series YB.)

Remove the main feed pipe, union and copper washers, push the piston down the cylinder bore and remove the retaining circlip. Remove the remaining internal parts, i.e., the piston, piston master cup, return spring, valve cup assembly and valve seating washer.

To remove the secondary cup from the piston, carefully stretch it over the end flange, using the fingers only.

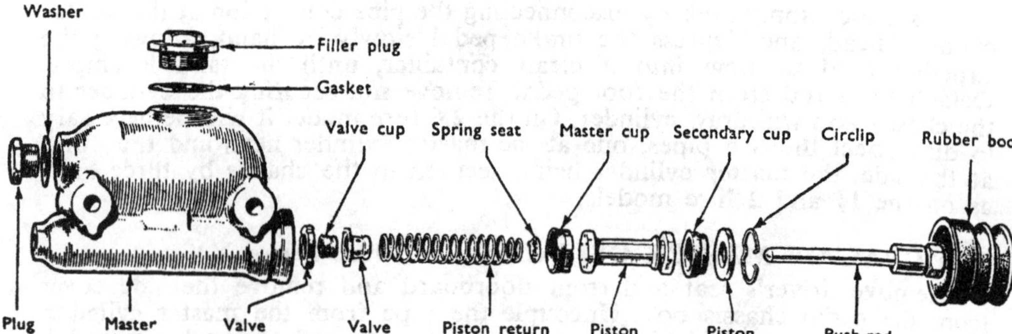

Fig. 14. The component parts of the master cylinder. (Later models have a thin protecting copper washer between the end of the piston and the master cup.)

ASSEMBLING THE MASTER CYLINDER. (TA, TB, TC, 1½ and 2 litre models.)

Clean all parts thoroughly using Lockheed Orange Brake Fluid, dip all parts in brake fluid and assemble wet.

Stretch the secondary cup over the end flange of the piston with the lip of the cup pointing towards the opposite end of the piston. When the cup is in position, work round gently with the fingers to ensure it is correctly seated.

Insert the valve body and cup assembly in one end of the spring and drop into the cylinder, holding the cylinder vertically to ensure the valve seats correctly. Insert the master cylinder cup, pressing firmly on to the end of the spring. Replace the piston and secondary cup assembly, piston stop, washer and circlip.

ASSEMBLING MASTER CYLINDER. (2.6 litre model.)

Clean all parts thoroughly in Lockheed Orange Brake Fluid and assemble wet.

Stretch the secondary cups over the end flange of the pistons with the lip of the cups pointing towards the opposite ends of the pistons. When the cups are in position, work round gently with the fingers to ensure they seat correctly.

Insert piston "J" complete with rubber cups so that space formed by the reduced piston skirt comes opposite stop pin hole "K," using new copper gasket under head. Insert spring "L" and screw home cylinder head "O," again using new copper gasket. Insert the spring "H," the piston "E" complete with cups and circlip "C" and fit rubber boot "B."

ASSEMBLING MASTER CYLINDER. (TD and 1¼ Litre Series YB)

Clean all parts thoroughly using genuine Lockheed Orange Brake fluid. Dip all internal parts in brake fluid and assemble wet. Stretch the secondary cup over the end flange of the piston with lip of the cup pointing towards the opposite end of the piston. When the cup is in the groove, work round gently with the fingers to make sure it is correctly seated.

Fit the check valve "K" on one end of the return spring "J" and retainer "H" into the opposite end, and insert, valve first, into the cylinder (1¼ litre Series Y).

Fit the valve washer, valve cup and body on to the return spring and insert the spring, valve first, into the cylinder. See that the spring retainer is in position. (TD and 1¼ litre Series YB).

Insert the main cup "G" or master cup, lip first, taking care not to damage or turn back the lip and press down on the spring retainer. Insert the piston taking care not to damage or turn back the lip of the secondary cup. Push the piston down the bore slightly, insert the piston stop and fix circlip in the groove of the cylinder.

Test the master cylinder by filling the tank and pushing the piston down the bore and allowing it to return; after one or two applications fluid should flow from the cylinder.

Insert the push rod and refit the boot; replacing if damaged or distorted.

It is essential that the vent hole "O" is at the bottom when assembled. (1¼ litre Series Y).

REPLACEMENT OF MASTER CYLINDER. (TA, TB, TC, TD, 1¼, 1½, 2 and 2.6 litre models.)

Replacement of the master cylinder is the reverse of that for removal, and the following points should be carefully checked. Check the adjustment of the push rod as described under heading "Adjustment of the Brake Pedal."

Connect the pipe lines and bleed the system as described under the heading "To Bleed the System."

Check the system for leaks with the brakes fully applied.

FRONT AND REAR WHEEL CYLINDER. (TA, TB, TC, 1¼ litre Series Y, 2 and 2.6 litre models.)

The front and rear wheel cylinders are mounted rigidly to the brake shoe back plate inside the brake drum, between the brake shoe tips. The opposed pistons act directly on the tips of the brake shoes; each piston being packed by rubber cups, within which are crown spring expanders. Coil springs hold the crown spring expanders in position and also keep the rubber cups in contact with the piston head. Rubber boots seal the ends of the cylinder against the ingress of dust, mud and water which may find a way into the brake drum. Located at the topmost position is the bleeder screw, through which air is expelled during the bleeding operation.

REMOVAL OF THE FRONT WHEEL CYLINDER. (TA, TB, TC, 1½, 2 and 2.6 litre models.)

It is advisable to unscrew the flexible hose at either end. Jack up car and remove wheel and brake drum. Disconnect the copper tube from the hose union "A" (Fig. 15) at the frame, remove the nut and lock washer "B," when the hose union may be removed from the bracket. Unhook the brake shoe return spring. Remove the two screws securing the cylinder to the back plate and withdraw with hose in place.

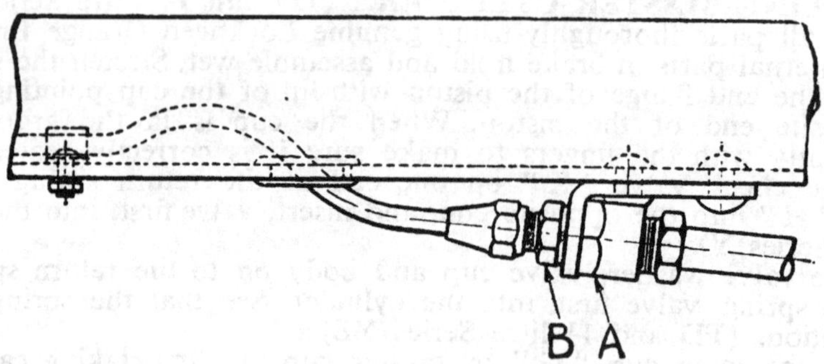

Fig. 15. The patent flexible hose connection.

REMOVAL OF THE REAR WHEEL CYLINDER. (TA, TB, TC, 1½, 2 and 2.6 litre models.)

Follow the instructions for the front wheel cylinder, except that the copper tubing must be disconnected at the cylinder inlet.

REMOVAL OF WHEEL CYLINDERS, FRONT AND REAR. (1¼ litre Series Y.)

Jack up the vehicle, remove the wheel and brake drum. Remove the flexible hoses by withdrawing the banjo bolt from the inlet fitting, unhook the brake shoe pull-off springs, and turn the brake shoe adjusting nuts until they are in the fully extended position. Remove the two set screws securing the cylinders to the brake plate and withdraw cylinder.

DISMANTLING WHEEL CYLINDERS. (TA, TB, TC, 1¼, 1½, 2 and 2.6 litre models.)

Remove the rubber boots, and withdraw internal parts, pistons, rubber cups and springs, complete with crown spring expanders.

ASSEMBLY OF WHEEL CYLINDERS. (TA, TB, TC, 1½, 2 and 2.6 litre models.)

Dip all parts in genuine Lockheed brake fluid and assemble wet. Insert one piston, flat face first, and insert one rubber cup, flat face first, from opposite end of the cylinder. Insert spring, complete with crown spring. Insert second cup, lip first, taking care not to damage or turn back the lip of the cup. Insert second piston, flat face first, and press down on to the rubber cup. Fit the rubber boots into the groove on the ends of the piston and then over the end of the cylinder.

REPLACEMENT OF WHEEL CYLINDER. (TA, TB, TC, 1¼, 1½, 2 and 2.6 litre models.)

Replacement is carried out in the reverse manner to that detailed for removal. Bleed the system as detailed on page 408 and adjust the brake shoe as detailed on page 407.

REMOVAL OF WHEEL CYLINDER. (TD and 1¼ litre Series YB.)

Front.—Raise the front of the car, remove hub cap, road wheel, brake drum and hub assembly. Draw brake shoes apart until assembly can be lifted from wheel cylinders and brake plate. Release flexible hoses. Remove flexible hose union bolt from the wheel cylinder. The copper washers on either side of the banjo union are of different sizes, the smaller washer next to the cylinder.

Remove the two ¼ in. nuts and spring washers securing the wheel cylinder to the brake plate and remove cylinder.

The other cylinder is removed after extracting the $\frac{5}{16}$ in. banjo union bolt and two bolts securing the cylinder to the brake plate.

Rear.—Raise rear of the car, remove hub cap, wheel and brake drum and hub assembly. Draw brake shoes apart until the assembly can be lifted from the brake plate.

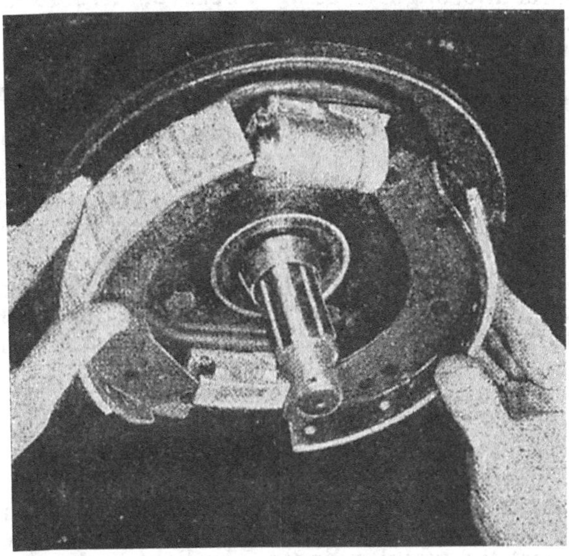

Fig 16. Removing rear brake shoes from brake plate after releasing return spring.

Release the metal feed pipe from the wheel cylinder by undoing the $\frac{3}{16}$ in. union nut. Remove the $\frac{5}{16}$ in. adaptor securing the bleed screw banjo union to the cylinder, noting that the large copper washer is fitted away from the cylinder. Remove the clevis pin from the hand brake cable yoke and disconnect the cable from the wheel cylinder lever. Remove the rubber boot and withdraw the lower half of the piston from the wheel cylinder. Extract the wheel cylinder from the brake plate.

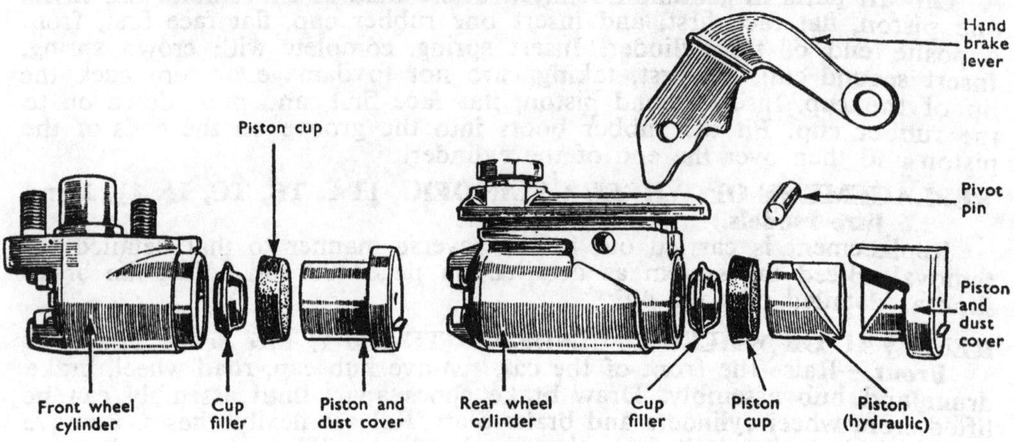

Fig. 17. *Front and rear wheel cylinder components.*

DISMANTLING THE WHEEL CYLINDER. (TD and 1¼ litre Series YB.)

Rear Cylinders.—Tap out the hand brake lever pivot pin and withdraw lever. Withdraw the upper half of the piston, the rubber cup and bakelite filler.

Front Cylinders.—Withdraw the piston, the rubber cup and cup filler.

ASSEMBLING THE WHEEL CYLINDERS. (TD and 1¼ litre Series YB.)

Clean all parts thoroughly, using Lockheed hydraulic brake fluid for rubber components.

All metal parts must be free of paraffin, petrol or trichlorethylene. Dip all parts in brake fluid and assemble wet. Insert the cup filler, shallow side first (this is important), and the rubber cup concave side first.

Replace the piston and in the case of the rear cylinders, insert the hand brake lever and its pivot pin.

REPLACEMENT OF WHEEL CYLINDER. (TD and 1¼ litre Series YB.)

The procedure for replacing is a reversal of the sequence given for removal, but the following points should be noted.

Front.—The front brake wheel cylinders are interchangeable but the link pipe banjo union must be fitted to them so that the flexible hose is connected to the forward cylinder and the bleed screw to the rear cylinder The brake shoes are interchangeable but the recessed ends must engage the " Micram " shoe adjusters and the brake shoe pull-off springs must be fitted between the shoes and brake plates.

Rear.—The wheel cylinder must be fitted on the forward side of the axle casing with the bleed screw vertical. The brake shoes are interchangeable but the recessed end of the lower or leading shoes must engage the " Micram " shoe adjuster. The other shoe should also be fitted with its recessed end against the wheel cylinder. The light brake shoe pull-off spring must be fitted away from the wheel cylinder, and both springs are fitted between the shoes and the brake plate.

REMOVAL AND REPLACEMENT OF BRAKE SHOES. (TA, TB, TC, 1¼, 1½, 2 and 2.6 litre models.)

Remove the wheel and brake drum. Unhook the brake shoe pull-off spring from the brake shoes, remove the split pins and washers from the brake shoe steady pins and withdraw the circlip from its groove in the shoe anchor pin.

In the case of the rear shoes the hand brake cable will need to be unhooked from the lever behind the brake shoe (TA, TB, TC, VA, SA and WA models).

In the case of the rear shoes remove the nut attaching the rear shoe to the fulcrum pin of the operating lever of the hand brake mechanism. (1¼ litre model.)

The shoes are now free and can be withdrawn from the anchor pin. Replacement is carried out in the reverse manner to that detailed for removal. On the 1½ litre, series VA, when reassembling care must be taken to ensure that the adjuster cam is in the position shown in Fig. 18.

REMOVAL AND REPLACEMENT OF BRAKE SHOES. (TD and 1¼ litre Series YB.)

Remove hub disc, wheel and brake drum and hub complete.

Draw the brake shoes apart until they can be removed from the brake plate.

IMPORTANT: When replacing the shoes, the pull-off springs must be between the shoes and the brake plate.

The brake shoes are interchangeable but the recessed end must engage the " Micram " shoe adjuster on the wheel cylinder. In the case of the rear brakes, the light pull off spring is fitted away from the wheel cylinder. The recessed end of the upper or trailing shoe is fitted against the end of the wheel cylinder.

RELINING THE BRAKES. (All models.)

Where brake linings become worn, new shoes complete with linings fitted by the makers should be employed. It is important that the material used for the lining is that as specified by the M.G. Car Co. Ltd., as any variation from this may give inferior braking performance. When new shoes and linings are employed considerable adjustment has to be made to hand

Fig. 18 *Position of double cam when reassembling brake shoes.* (*Upto Chassis no. VA*2519.)

and foot brake mechanism, entailing the return of the hand and foot brake adjusters to their normal positions before attempting to refit the brake drums over new linings.

FOOTBRAKE ADJUSTMENT. (VA model up to chassis No. VA 2519.)

When lining wear reaches a point where the foot pedal goes almost to the floorboard, it is necessary to adjust the brake shoes into closer relation to the drums.

Rotate the hand wheel, Fig. 19, in the direction of the arrow until the brake shoes come into contact with the drum, until it cannot be turned further by hand.

Never use tools for turning the hand wheel or the mechanism may be strained.

Fig. 19. Bleeding part of the brakings stem and showing foot brake adjustments.

TO ADJUST HAND BRAKE. (1½ litre series to chassis No. VA 2519.)

Jack up both rear wheels until they are clear of the ground and adjust foot brake as detailed. Screw up adjusting nut on front end of each cable, taking care not to over-adjust, beyond the shoe position obtained by the foot brake adjustment: Pull round each wheel by hand, hand brake partially applied, and readjust at cable nuts until an equal feel is obtained.

FOOTBRAKE ADJUSTMENT. (TA, TB, TC, 1¼, 2, 2½ and 1½, commencing at chassis VA 2520.)

When lining wear has reached a point where the pedal travels almost to the floorboards before the brakes come into action it becomes necessary to adjust the brake shoes into closer relation to the drums. Jack up the car until the wheels are clear of the ground. When adjusting rear brakes the hand brake should be in the " off " position. Adjust each shoe separately by rotating the adjustment cam " C," Fig. 20, against the stop-pin " B " on the shoe: Rotate adjustment nut " C " until the brake shoes come into contact with the drum then back off the adjustment slightly until the wheel rotates freely without any appreciable drag. One complete turn of the adjustment nut " C " is sufficient to take up all lining wear. The nut will require only a partial turn before the shoes are brought into contact with the drum.

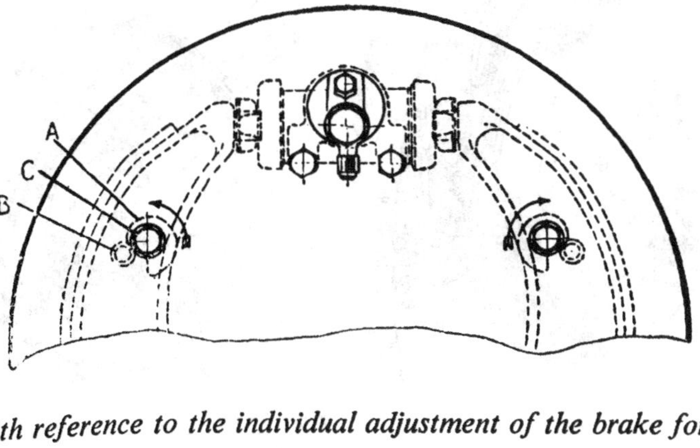

Fig. 20. *With reference to the individual adjustment of the brake for each wheel.*

BRAKE SHOE ADJUSTMENT. (TD and 1¼ litre Series YB.)

When lining wear reaches a point where the pedal travels to within 1 in. of the floorboards before the brakes operate, adjustment of the brake shoes is necessary.

Front Brakes.—Jack up the wheel which is to be adjusted: Remove hub cap and road wheel, rotating brake drum until both adjustment screws are visible through the holes provided in the face of the brake drums.

Fig. **21.** ***Showing location of holes in brake drum for front brake adjustment.***

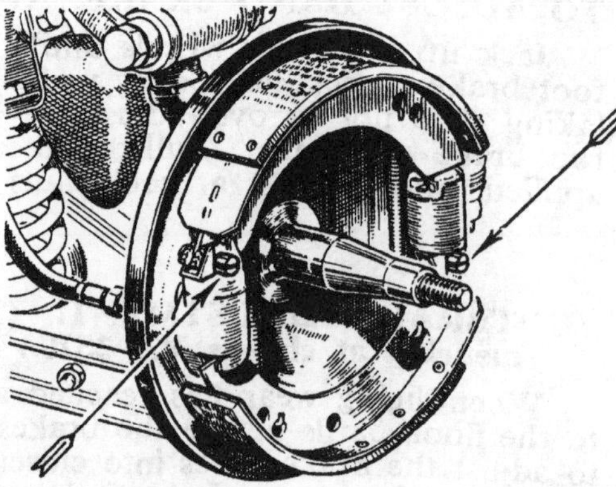

Fig. 22. *The front drum removed to show the two " Micram " adjusters.*

With a screwdriver, rotate both screws in a clockwise direction as far as they will go until the drum is locked solid, then turn anti-clockwise *ONE* notch only. The brake drum should then be free to rotate without the shoes rubbing and the adjustment on the wheel is complete. Repeat operation on the other front wheel.

Fig. 23. *Rear brake shoe adjustment.*

Fig. 24. *Rear hub and drum removed showing single " Micram " adjuster.*

Rear Brakes.—The procedure is similar to that detailed for the front brakes except that there is only one adjuster which controls both shoes and handbrake operation.

BLEEDING THE SYSTEM. (TA, TB, TC, TD, 1¼, 1½, 2 and 2.6 litre models.)

Bleeding the system is not a routine maintenance operation and only becomes necessary when some portion of the hydraulic equipment has been disconnected or the fluid drained off.

Fill the master cylinder or supply tank with genuine Lockheed Brake Fluid and keep at least half full during the bleeding operation, otherwise air will be drawn into the system, necessitating re-bleeding.

For the TA, TB, 1½, 2 and 2.6 litre models, remove the set screw from the end of the bleeder screw and screw in the bleeder drain tube, a screwed brass nipple fitted with a rubber tube, and for the TC, TD, and 1¼ litre models, attach the bleeder tube to the wheel cylinder bleeder screw, allowing the free end of the tube to hang into a clean glass jar, partially submerged in a small quantity of brake fluid.

Open the bleeder screw one complete turn, depress the pedal quickly, allowing pedal to return without assistance and repeating the pumping operation with a slight pause before each depression of the pedal.

Watch the flow of fluid into the glass jar and when bubbles cease to appear, hold the pedal firmly against the floorboards while the bleeder screw is securely tightened.

Repeat the operation on all wheels.

HAND BRAKE. (TA, TB, TC, TD, 1¼, 1½, 2 and 2.6 litre models.)

The hand brake is of the quick release type operating on the rear wheels only, through the medium of cables carried in protective casings, the ratchet knob being depressed after the handbrake has been applied. A slight pull on the lever will allow the handbrake to move to the " off " position without depressing the knob.

On all models except the TD and 1¼ litre Series YB ample and simple adjustment is provided by the thumb nut at the base of the handbrake lever on the TA, TB, TC, 1½, 2 and 2.6 litre models and by the two brass nuts at the brake cross shaft on the 1¼ litre Series Y. Exercise great care when adjusting to ensure that the adjustment is not taken up too tightly or the rear brake shoes may be held in contact with the drums when the hand brake is in the ' off " position. On the TD and 1¼ litre Series YB no adjustment is necessary, as adjustment of the hydraulic brakes automatically adjusts the hand brake.

CHAPTER IX

STEERING GEAR

STEERING GEAR

STEERING ADJUSTMENT. (M and D Midgets.)

If the steering column shows signs of end play, it may be overcome by slacking back the locknut situated immediately below the steering wheel and giving the adjustment nut a fraction of a turn. Make sure that the locknut is securely locked after adjustment.

The steering is provided with a complete wormwheel which may be set in a number of alternative positions. Should there be excessive movement of the steering wheel, examine all connections between the steering gearbox and front axle for looseness. If no looseness is apparent, there is probably wear in the steering wormwheel which should be turned to a new position. To do this, remove the stop plate spanning the steering arm and then remove the steering arm. Rotate the steering wheel, turning the steering wormwheel through 90° to a new position, and replace the steering arm, refitting the stop plate.

THE MARLES WELLER STEERING GEAR. Fitted to the J1, J2, PA (up to Chassis No. PA1286), L type Magna, K type Magnette.

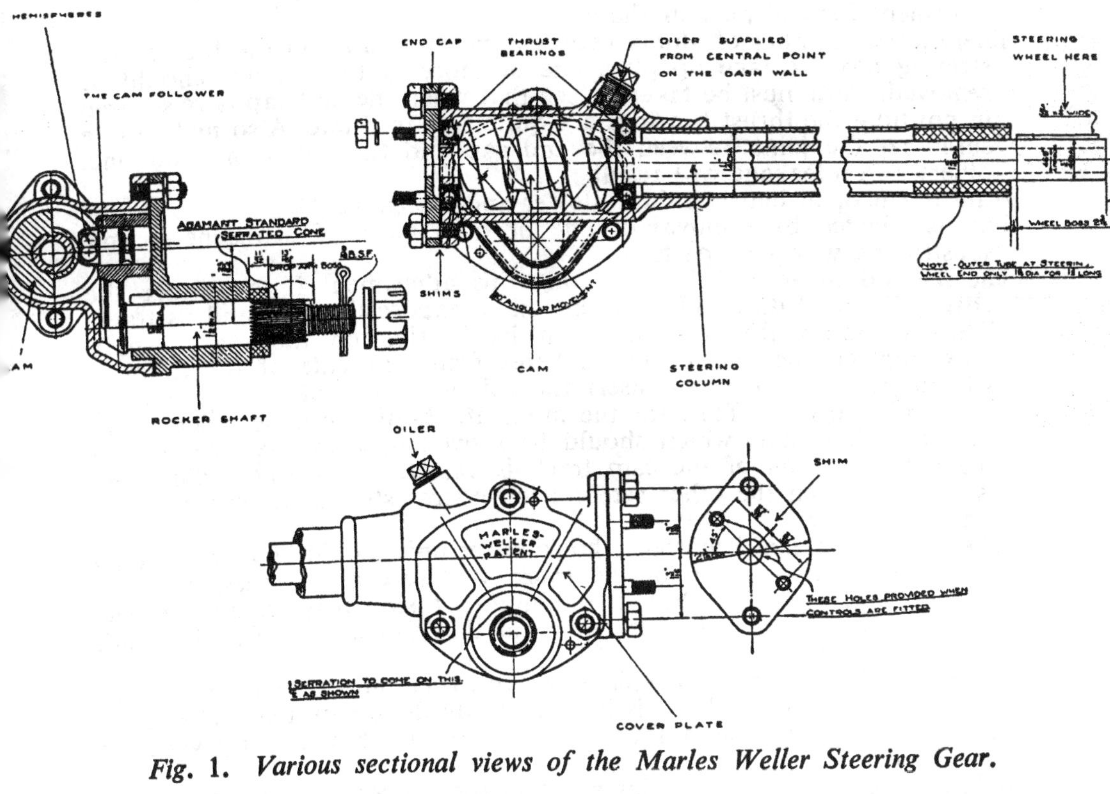

Fig. 1. *Various sectional views of the Marles Weller Steering Gear.*

STEERING.

On the Marles Weller Steering Gear, a hardened steel cam, in which a special groove is cut, is mounted on the shaft carrying the steering wheel. Into this groove is inserted a follower, on each side of which and deeply embedded in it are two steel hemispheres which make contact with the sides of the cam track. The follower is free to rotate in a bearing in the rocker shaft. This rocker shaft is attached to the drop arm which operates the drag link.

The cam is carried on two ball bearings and at the top end of the shaft there is a third ball bearing to give the necessary rigidity to the steering wheel. This bearing also, of course, prevents any bending of the shaft.

This bearing excludes all dust and dirt and is designed to damp out vibration of the steering column.

The rocker shaft which carries the drag link is carried in large phosphor bronze bearings.

The operation of the gear is very simple, for when the steering wheel and cam are rotated, the hemispheres and follower engaging the cam are moved backwards and forwards in the groove. At the same time, each hemisphere aligns itself in its seating to the side of the track engaging it, and the follower, complete with its four hemispheres, also adjusts itself with the bearing in the rocker shaft.

It will be seen from this that a considerable area is always evenly presented to the cam track by the hemispheres, and this area, together with the self-alignment of the contacting faces, is the fundamental difference between the Marles Weller and other types of steering gear.

Adjustment to the Steering Gearbox.—Two adjustments are provided and they control all that can be required, and they are as follows:—

1. Adjustment for end play of the cam.

   There are a number of shims between the end cap and the face of the steering box. To take up play, one or more of these shims should be removed. Care must be taken to see that when the end cap is re-secured in position, the thrust bearings are quite free to rotate. Also make quite sure that the joint between the end cap and the box is a good one, otherwise oil leakage will result.

2. Adjustment of fit between the follower and the came.

   This is effected by removal of one or both pairs of hemispheres and substituting with new parts.

   The method to adopt when effecting this adjustment is to remove the cover plate and drain off sufficient oil to expose the cam and its track. Then withdraw the follower from its bearing on the rocker shaft. Next replace one pair of hemispheres from one side of the follower with a new set and then insert the follower unit into the cam track in its mid position. This, for the moment, should be done independent of the rocker shaft which should be moved over to one side so that the centre portion of the cam track is exposed. If further adjustment is required, then the other set of hemispheres should be replaced in a similar manner.

   When the adjustment is correct, the follower unit should be a free sliding fit in the centre portion of the track to ensure a free passage of oil between the surfaces. Only a minimum of side play should be permitted, and it is wise to bear in mind that too close a fit will result in stiff steering.

   The next thing to do is to replace the follower unit through the rocker shaft and into the track. This is most easily done with the rocker shaft in the extreme locked position, because the track is slightly wider at these points.

   A little care is necessary at this point in the proceedings, and the hemispheres may be tilted slightly to give a lead-in effect.

   When inserting the new set of hemispheres, make sure that they are quite clean. A smear of grease in the seatings will check any tendency for them to fall out.

   The cover plate and its shims should next be replaced, taking care over the joint, and then the gearbox can be filled with new oil.

NOTE: 1. The shims between the cover plate and the box are placed there during initial assembly and should not be removed. A certain amount of end play in the rocker shaft is desirable, and no attempt should be made to remove this play by taking out the shims.

2. If the thrust bearings are removed at any time, make quite sure that they are replaced correctly. The correct way is when the word "thrust" which is stamped on one side of the inner race of the bearing is up against the end of the cam.

Fitting the Drop Arm to the Rocker Shaft.—Sometimes it may be necessary to remove the drop arm and it is essential that it is replaced in the same position from which it was removed. If this is not done, it is a certainty that lock will be limited in one direction and damage to the internal mechanism will result.

It is assumed that the steering assembly is attached securely to the chassis but that the drop arm is not fitted. It must be pointed out that all connections should be tight, including that on the dashboard. With the steering wheel in position the lower end of the drop arm carrying the ball pin should be fixed to the drag link, but the top end should not yet be reconnected to the rocker shaft of the steering box.

Now, jack up the front wheels and place them in the straight ahead position. Next, gentle rotation of the steering wheel will show that its movement is limited by internal stops located inside the steering box at each end of the travel of the internal mechanism. The number of turns of the steering wheel required to bring the gear from one end of its travel to the other should be counted. Divide this number by two and rotate the steering wheel to the central position. The drop arm may now be fitted to the rocker shaft, but before finally tightening up, the following check should be carried out: with the front wheels still jacked up, turn the steering wheel right over to one lock and, in this position, drop the drag link off the drop arm to see if the steering wheel can be moved any further in the same direction. If it can, then all is well, and the same procedure should be followed on the other lock.

Should further movement of the steering wheel be unobtainable, it means that the front axle stops are not functioning and some adjustment must be made as follows: if movement of the steering wheel is available on one lock and not on the other, the drop arm should be placed on the next serration and this may rectify matters. If no serration will give free movement of the steering wheel at both locks, then the front axle stops must be altered.

The amount of free movement available after dropping the drag link from the drop arm should be the same at both locks and this is the condition at which to aim.

It will be appreciated that the movement of the rocker shaft and drop arm is restricted by the internal stops in the steering box and it is therefore necessary to fit the drop arm and connect up to the front wheels on the one hand, and to the steering box mechanism on the other, so that the necessary movement of the front wheels from lock to lock is obtained at the same time as the follower is travelling from end to end of the cam track, but not far enough in either direction to hit the internal stops.

THE BISHOP CAM STEERING GEAR. Fitted to the PA (commencing at Chassis No. PA1287), PB, TA, TB and TC Midgets, 1½ Litre, 2 Litre, 2.6 Litre, KN Magnette and NA Magnette.

The operation of the Bishop steering gear is quite straightforward and a cam in which a special groove is cut is mounted on the steering wheel shaft. A follower makes contact with the cam track. The cam is mounted between special ball bearings and the whole mechanism is contained in an oil-tight casing.

Adjustments. 1. **The cam.**—Shims are inserted under each end cover of the steering box and provided wear has not taken place in the bearings, then no adjustment is necessary. If the gear is dismantled at any time, then make a careful note of the number of shims at each end. On reassembly, the shaft should spin quite freely with no end play.

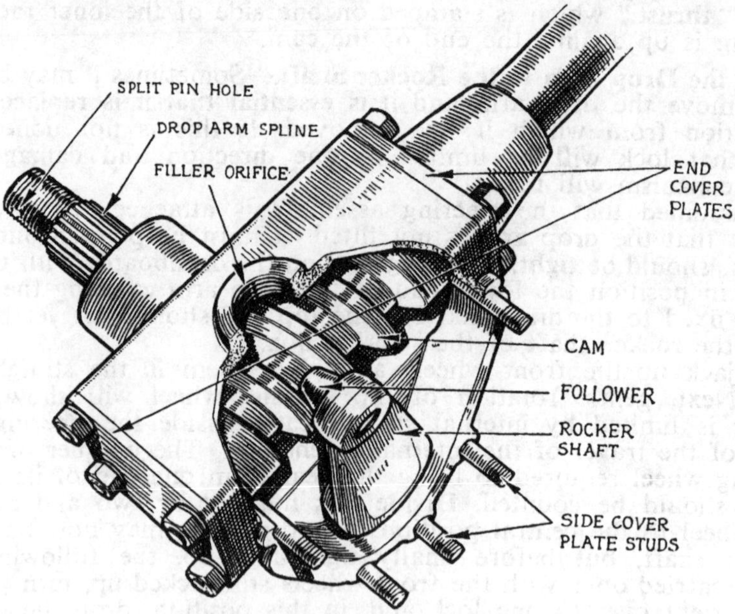

Fig. 2. The Bishop cam steering.

2. **The Rocker Shaft.**—The only adjustment likely to be needed is the removal of shims from underneath the side cover plate.

Movement of the steering wheel is transmitted to the road wheels through the cam and follower and in time a small amount of wear may occur. This will show as " lost motion " between the steering wheel and the road wheels. The remedy is to remove a shim or shims as mentioned above. The shims vary in thickness and a very fine adjustment can be obtained, it being essential that no stiffness is apparent when the wheels are in the straight ahead position. All adjustments should be carried out when the drag link is disconnected.

Gear Assembly.—The cam track is slightly relieved at either end so that there is no play at the centre position but a slight amount towards each lock. The amount of the clearance as checked by the motion of the drop arm should be equal on both locks and if it is greater at one end than the other, adjustment must be effected by removing one or more shims from under the end cover where most motion exists, and placing them at the other end until the clearance is equal.

Care should be taken to see that the flat face of the rocker shaft makes a good bearing against the hardened cover plate and that the cork gland is in good condition.

If a new gland is necessary, it can be cut radially and sprung into position without removing the retaining ring.

All adjustments should be made to the gear before the unit is filled with oil.

Removal of the Drop Arm.—The drop arm is splined to the rocker shaft and held in position by a nut and split pin. A special tool is generally necessary to pull the drop arm off its shaft, but if this special tool is not available, then the drive shaft may be driven through the drop arm after the side cover has been removed. This is much better than endeavouring to knock the drop arm off the shaft.

Fitting the Drop Arm.—The drop arm and the end of the rocker shaft are normally marked so that they may be reassembled in the correct position, but if, by some mischance, the marks have been obliterated, the procedure is as follows:—

With the steering wheel in position the lower end of the drop arm carrying the ball pin should be fixed to the drag link, but the top end should not yet be reconnected to the rocker shaft and steering box. Now, jack up the front wheels and set them in the straight ahead position. Next, gentle rotation of the steering wheel will show that its movement is limited by internal stops located inside the steering box at each end of the travel of the internal mechanism. The number of turns of the steering wheel required to bring the gear from one end of its travel to the other should be counted. Divide this number by two and rotate the steering wheel to the central position. The drop arm may now be fitted to the rocker shaft, but before finally tightening up, the following check should be carried out: with the front wheels still jacked up, turn the steering wheel right over to one lock and in this position, drop the drag link off the drop arm to see if the steering wheel can be moved any further in the same direction. If it can, then all is well and the same procedure should be followed on the other lock.

Should further movement of the steering wheel be unobtainable, it means that the front axle stops are not functioning and some adjustment must be made as follows: If movement of the steering wheel is available on one lock and not on the other, the drop arm should be placed on the next serration and this may rectify matters. If no serration will give free movement of the steering wheel at both locks, then the front axle stops must be altered. The amount of free movement available after dropping the drag link from the drop arm should be the same at both locks and this is the condition at which to aim.

REMOVAL OF STEERING GEAR. (2 Litre, 2.6 Litre and 1½ Litre.)

Disconnect the battery and undo the cables from the junction box which is located on the steering gearbox. Now remove the screws from the steering hub and the controls complete with the stator tube can be withdrawn. Removal of the steering wheel comes next and up to chassis No. SA902, it is secured by a nut and key. On the later telescopic types, the wheel can be detached from its splines after the circlip has been removed.

Remove the baffle board which extends from the lower edge of the facia to the foot ramp, the foot ramp plate, the column support and the drop arm. Now remove the bolts holding the steering gearbox to the chassis and withdraw the steering column complete.

REMOVAL OF THE STEERING GEAR. (TA, TB, and TC Midgets.)

Remove the drop arm and withdraw the steering wheel by slackening off the pinch bolt which secures it to the column. Remove the baffle board after undoing the screws that hold it in position. Remove the bolts securing the metal foot ramp. Now disconnect the steering column support bracket under the scuttle and remove the bolts which hold the steering column bracket to the frame, and withdraw the column complete with bracket from the car.

STEERING WHEEL ADJUSTMENT. (1½ Litre Series VA and 2.6 Litre Series WA.)

Two adjustments are provided to enable the steering wheel to be placed in the most convenient position.

1. **Height.**—Release the clamp holding the steering box to the frame and undo the bolt which locks the steering column to the body. It will then be possible to move the steering wheel into the desired position when the two attachments can be re-tightened.
2. **Length.**—Refer to Fig. 3.

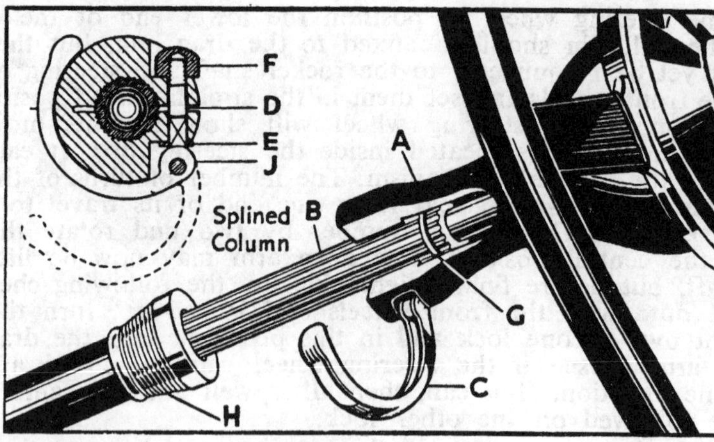

Fig. 3. *Diagrammatic view of adjustable steering wheel.* (1½, 2, *and* 2.6 *litre model*)

The adjustment of the steering column for length is quite simple and the wheel slides up and down an extension of the column. Both the steering wheel hub and the column are machined with longitudinal splines (see Fig. 3). This means that the hub " A " cannot rotate on the column " B," but up and down movement is possible. Adjustment is controlled by the lever " C " which, when it is raised, releases the wheel so that it can move on the column. The wheel can then be locked in the desired position by the cam action of the lever, which causes the split hub " D " to close and grip the splines on the column. An adjustment nut " F " is provided and this varies the amount of pressure exerted by the clamping lever. The pressure is correct if the lever can be fully closed by hand. If, after closing the clamp the wheel can be moved up and down, then the grip is insufficient and the lever pressure should be increased by screwing up the nut " F " a small amount until the wheel can be locked firmly in position. Occasionally the wheel should be raised to its highest position and the dust cover " H " moved aside. A few drops of engine oil can then be applied to the splines and also on the washer " E " where the cup " J " bears against it. If the clamp breaks at any time, then the assembly can be replaced by a ¼ in. nut and bolt until such time as new parts are available.

TO REMOVE THE STEERING WHEEL. (1½ and 2.6 Litre.)

Before the steering wheel can be taken off the central control column must bc removed, and this is done by slightly unscrewing the two grub screws which will be seen opposite each other on the outside of hub " A." It is not necessary to undo these screws more than ⅛ in. Next, disconnect the wiring at the bottom of the steering box and pull the control assembly up through the column " B." Make quite sure that the cables lie neatly alongside each other, otherwise you will have difficulty in getting them to pass up the central column.

The other screw heads on the outside of the hub " A " are for the trafficator switch and need not be disturbed. With the control column out of the way, the ring " G " can be taken off and the wheel pulled up the column " B." To replace, assemble the dust cover " H " as it came off the column and refit the wheel and circlip. Next carefully thread the wiring cable up through the column. It may be necessary to insert a length of copper wire upwards through the steering column inner mast to help pull the cable through. Care must be taken when introducing the tubing attached to the control unit and this must be turned until the telescopic joints half way down are properly engaged. No force must be used.

Screw up the two grub screws after you have made sure that the retaining plate is properly down on its seating in hub " A." This is important. If there is any doubt about the correct position of the plate, then take out one screw, when the edge of the plate can be seen through the tapped hole. The upper edge should be about half way across the hole or perhaps a little lower.

Note that the lower half of the control tube is fixed at the bottom of the steering box and need not be disturbed. This facilitates the removal of the unit when it is fitted to a saloon model.

Oil the joint between the two halves of this inner column when reassembling and sweet operation and minimum wear will result.

STEERING MODIFICATIONS. (1½ Litre Series VA commencing chassis No. VA 1257.)

There was a slight modification from this chassis number onwards and instead of the stator tube being held by a bolt and a clamp, a union nut and olive were used instead.

STEERING GEAR. (1¼ Litre Series Y and YB, Midget Series TD.)

The steering gear on these models is of the direct acting rack and pinion type providing light and accurate control under all conditions. It consists of a rack bar and helical-toothed pinion, and no adjustment for bearing wear in the bar is provided except by the fitting of new parts.

When new, the backlash between the pinion and the rack is extremely small, i.e., from 0—·003 ins.

The steering mast is attached to the steering gearbox by a rubber-bushed coupling in order to insulate the steering wheel from road shock. On the 1¼ litre car, a slip-ring assembly is mounted on the steering column in order to carry current to the centre mast for the horn and trafficator switches.

MAINTENANCE.

All the working parts are immersed in oil and an oil nipple is provided in the centre of the box. (See Fig. 4.)

Felt bushes are fitted to the steering column at the top end and during assembly they are impregnated with oil and graphite. If a squeak develops at the top end of the column, then a little oil on the felt bush will rectify matters.

REMOVAL AND REPLACEMENT OF THE STEERING WHEEL. (1¼ Litre Series Y and YB.)

Lift out the trafficator control and horn push complete by removing the set screw from the side of the steering wheel hub. Disconnect the four terminals underneath the control. Remove the steering wheel after undoing its retaining nut. Note: When reassembling, ensure that the wiring is pushed neatly down the column and that the terminals are tightened with wire running to the centre.

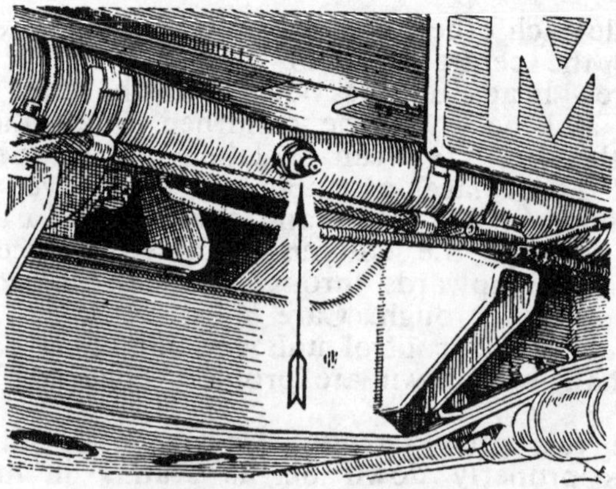

Fig. 4. Showing oil nipple for steering gearbox.

REMOVAL AND REPLACEMENT OF THE STEERING WHEEL. (TD Midget.)

Remove the clamping nut from the telescopic adjustment clamp and extract the clamp and bolt. This will allow the wheel to be withdrawn to its fullest extent and enable the plated helical sleeve to be contracted so that the key at the upper end of the column, which engages with the keyway in the splined adjustable shaft, may be withdrawn.

This will release the steering wheel and telescopic assembly, which can be removed for further dismantling if necessary.

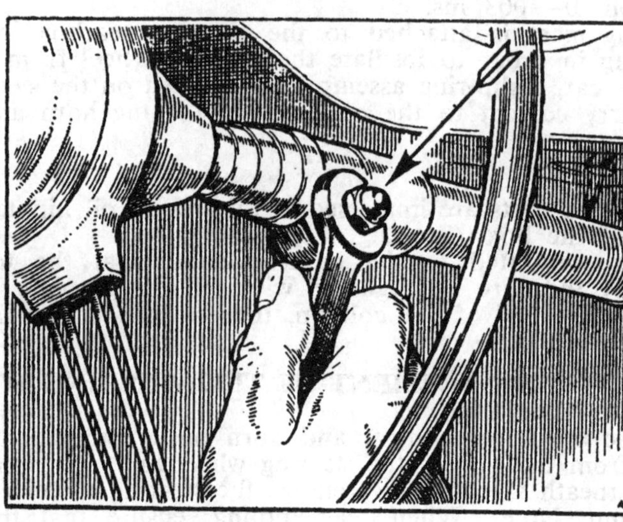

Fig. 5. Clamping nut for adjustable steering column (Midget series TD).

To dismantle the extension, support the hub of the wheel, and drive the splined shaft out, taking care not to lose the flat key in the process.

REPLACEMENT OF TOP AND BOTTOM FELT BUSHES. (TD Midget.)

Remove the steering wheel and extension as described on page 412. Pick out the old felt bush and feed in the new bush, which should be coated with graphite grease on the face which contacts the inner column. To change the bottom bush, the cover plate should be removed by undoing three small screws and the procedure mentioned above should be followed.

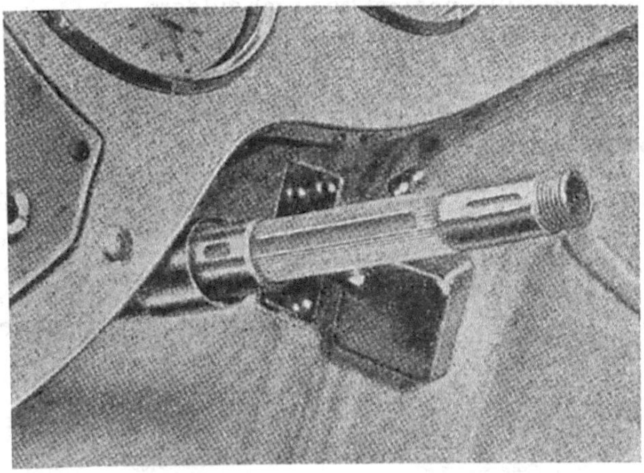

Fig. 6. The splined upper end of steering column indicating keyways.

REMOVAL AND REPLACEMENT OF THE STEERING COLUMN. (1¼ Litre Series Y and YB.)

Remove the steering wheel central control as explained on page 415. Remove the adjuster clamp bolt and slide back the clamp. Take out the key, and the steering wheel together with the adjustment assembly can be withdrawn. Disconnect the four wiring connectors to the steering column slip-ring unit. Remove the three flexible coupling bolts. Take out the three bolts holding the draught excluder plate at the column where it protrudes through the foot ramp. Slide back the draught excluder. Now remove the column clamp bolt, open the clip, and remove the distance piece. Slide the clip and the draught excluder up and over the top of the steering column.

The steering column assembly can now be lifted up at the coupling end and withdrawn towards the front through the space between the radiator and the wing.

When reassembling, do not forget to replace the earth spring between the coupling or, on later cars, the earth wire between the coupling bolt and the column clip.

The rubber-bushed couplings allow for only a small amount of universal action and it is important that when retightening the coupling bolts, the column be held in line with the gearbox pinion shaft. The bolts should be tightened evenly and this can be seen by the rubbers being displaced evenly on each side of the coupling and equally at each bolt. Also, see that the column is in line when tightening the clamp.

When bolting up the column at the clamp it will be noticed that the outer tube can be moved a small amount up and down. The collar must be fixed at the correct position to ensure that the slip-rings are functioning correctly.

REMOVAL AND REPLACEMENT OF THE STEERING COLUMN. (TD Midget.)

Remove the steering wheel as detailed on page 418. Remove the bolt and nut from the support clip under the dash. Now remove the bolt and nut holding the steering column to the body steady bracket (this is on the engine side of the bulkhead). Disconnect the universal joint and take care not to lose the rubber inserts. This will free the inner and outer columns which may be extracted through the space between the radiator and the wing.

When reassembling, make quite sure that the screws on the universal joint are tightened up fully against their shoulders.

REMOVAL AND REPLACEMENT OF THE STEERING GEARBOX. (1¼ Litre Series Y and YB.)

Disconnect the engine control link and remove the control link bracket from the frame (two bolts at the shock absorber and one at the front cross member). Remove the universal coupling bolts and unlock the steering tie-rod nuts, and unscrew the tie-rods out of the steering ball joints by means of the flats on the rods. Disconnect the wiring clips on the box and remove the electric horn. Raise the front of the car, block up under chassis and remove road wheels.

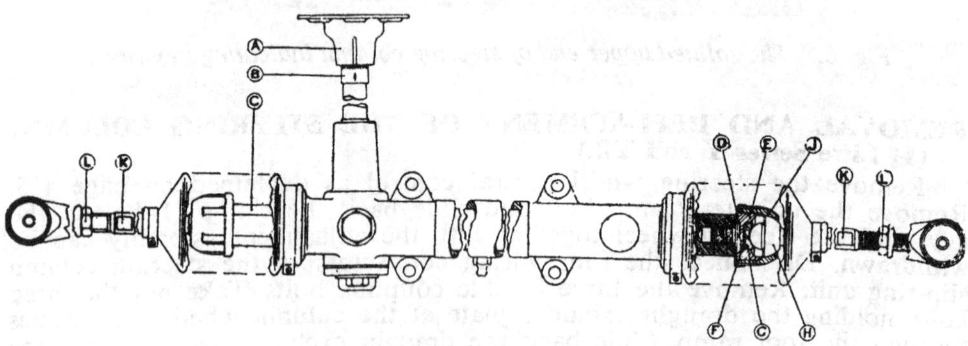

Fig. 7. Steering gearbox assembly. (1¼ *litre*).

Remove the four bolts holding the steering gearbox to the front cross member. Rotate the pinion shaft so that the rack is pushed over as far as it will go to the left.

Hold the pinion shaft coupling flange, lift upwards, and work the gearbox over to the left. The pinion shaft will then go between the front of the engine and the radiator. The right-hand steering tie-rod will now be clear of the hole in the right-hand chassis extension.

Now lower the gearbox on the right-hand side and withdraw the tie-rod from the hole in the left-hand side of the chassis.

When the unit is replaced on the chassis re-track the steering and adjust the engine control link.

REMOVAL AND REPLACEMENT OF THE STEERING GEARBOX. (TD Midget.)

Raise the car at the front end and block up under the chassis. Remove the wheels and disconnect the two track rods at their outer ends. Detach all electrical cables secured to the unit by clips.

Remove the outer ball joint on the same side as the steering column, taking care not to lose its position for reassembly. Undo the engine steady rod and remove the support bracket from the chassis.

Remove the three bolts and nuts at the steering column universal joint and remove the four bolts and nuts holding the steering gearbox to the frame. Now slide the complete unit to one side, when it will be possible to pull the track rod, from which the ball end has been removed, through the large hole in the chassis front extension. The unit may now be withdrawn to the front.

On reassembly, track up the front wheels and adjust the control link.

Fig. 8. *Method of withdrawing steering gearbox. (TD Midget.)*

DISMANTLING AND REASSEMBLING THE STEERING GEARBOX. (1¼ Litre Series Y and YB and TD Midget.)

Slacken off the jubilee clips and remove the two rubber dust excluders. Now unscrew the rack damper pad cap, when the damper spring and pressure pad can be removed. This will reveal a number of shims underneath the cap. (Fig. 9.)

Remove the pinion shaft tail bearing cap bolts and the tail bearing. Take off the coupling nut and slide off the coupling. Take off the circlip against which the coupling locates.

Now withdraw the pinion shaft by holding the gear with the pinion upwards and leaving the thrust washer behind. This thrust washer is trapped behind the rack teeth.

Now hold the rack bar in a vice, knock back the lock washers and undo the ball joint caps. A special spanner, No. T114, is necessary for this operation. The ball seat and the shims should then drop out.

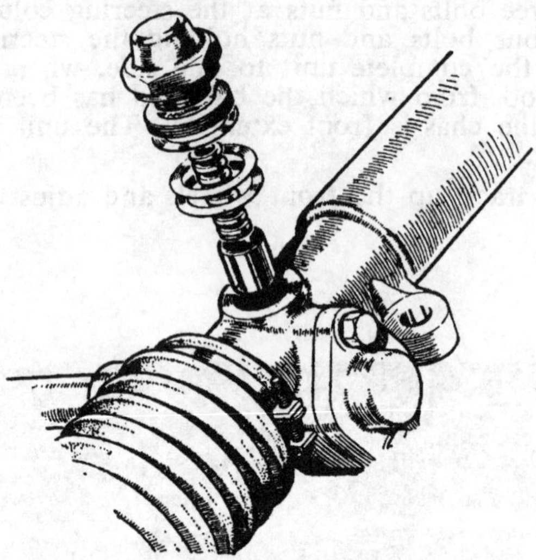

Fig. 9. Steering gearbox damper assembly. (TD Midget and 1¼ litre).

Next screw out the ball seat housing with a special peg spanner, No. T113.

NOTE: Should the ball joint caps come away complete with the ball seat housing it will be necessary to dismantle with a special tool, No. T122.

Now remove the rack damper and shims and withdraw the rack bar from its housing.

When replacing the pinion shaft, see that the thrust washers have their chamfered sides towards the pinion. End float should be .002 in.-.005 in. and is adjusted by means of shims. The oilway in the cap should be at the top. The method of adjusting the rack damper is described on page 423.

With the rack in the central position, engage the pinion with the arrow " B " uppermost (see Fig. 7).

Refit the coupling " A " (Fig. 7) with the coupling bolt in line with the arrow " B " on the shaft. This ensures that the steering wheel spokes are in the correct position.

Oil all parts before assembly and refill the steering gear box with ¾ pint of lubricant as specified on the lubrication chart.

Adjustment of Inner Ball Joint (Fig. 7).—Fit the lock plate " D " and screw home the ball seat housing " E " into the rack bar. Insert the shims " F " and the ball seat " G."

The ball cap " J " should be screwed home against its shroud after first inserting the ball ended tie-rod " H."

The ball should have no play but must be a free rolling fit. Adjustment can be effected by varying the shims " F," which are supplied in thicknesses of .003 and .005 in.

Adjustment of the Rack Damper.—This adjustment is provided to ensure the required amount of damping in the steering tie-rod and to maintain the minimum of backlash in the gear teeth. Adjustment is as follows: Check the damper spring which should have a free length of approximately 1.024 in. The outside diameter is $\frac{7}{16}$ in. and there are 8.6 coils of .08 in. sq. wire. This spring should give a load of 80 lb. when compressed to $\frac{3}{4}$ in.

When the steering gearbox is completely assembled, fit the cap, spring and plunger, but omit the shims. Now screw down the cap until the plunger bottoms. While screwing down the plunger rotate the pinion shaft. The plunger has bottomed when it is felt to just lock the rack bar in the housing. Take a measurement with feeler gauges of the gap which is left and add .051 in. to this measurement.

Now select shims to this total amount and insert them underneath the cap. This will give the standard condition of the pre-load. If, after a road test, the steering is felt to be incorrect, it is permissible to decrease the added measurement of .051 in. to .030 in. or increase it to .070 in.

Steering Arm Ball Joints.—If it is found necessary to renew the ball joints, the complete assembly must be changed, as no adjustment is provided. The dust excluder, however, may be renewed separately.

CHAPTER X

INBUILT JACKING SYSTEM

INBUILT JACKING SYSTEM

(1½ Litre, 2 Litre and 2.6 Litre.)

Operated on the hydraulic principle, the Jackall set consists of four single jacks mounted permanently on the axles. These are connected by means of piping and flexible joints to the self contained pump and distributor unit fitted beneath the floor boards in front of the passenger's seat on the nearside, and to which access is obtained after turning back the corner of the carpet and lifting off the cover.

This pump and distributor is supplied with fluid by a supply tank mounted on the engine shield, which should be examined every two to three months to maintain the fluid to the correct level.

SERVICE INSTRUCTIONS FOR THE "RED" JACKALL SYSTEM. (1¼ Litre.)

The "Red" Jackall hydraulic system requies no attention other than a periodical examination of the fluid supply tank which should be inspected monthly, a little fluid being added if necessary. A special fluid—"Red" Jackall—should be used. This is important as the use of any but the genuine Jackall fluid may result in damage to the system.

The distributor box and fluid supply tank are mounted beneath the left-hand side of the bonnet just in front of the forward bulkhead.

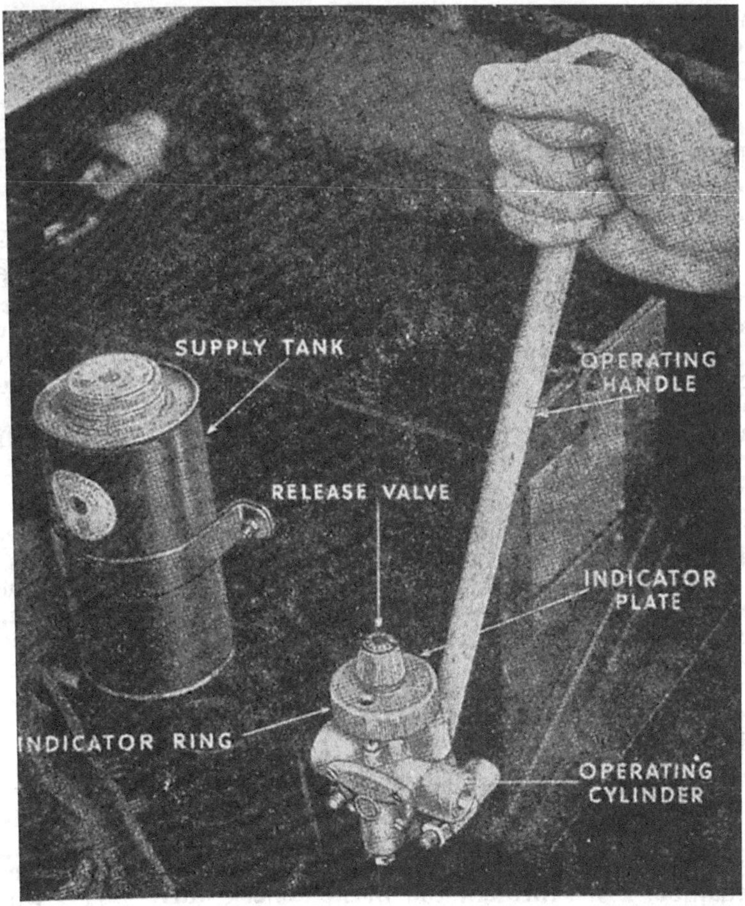

Fig. 1. *Showing position of distributor box and fluid supply tank* (1¼ *litre*).

OPERATING THE JACKS. (1½ Litre, 2 Litre, 2.6 Litre and 1¼ Litre.)

To bring the jacks into operation, the indicator pointer should first be turned towards the required position as marked on the indicator dial, i.e., towards " F " for " front " on the 1½ Litre, 2 Litre and 2.6 Litre models, and " Front " on the 1¼ Litre model if the front jacks only are required; towards " R " for " rear " on the 1½ Litre, 2 Litre and 2.6 Litre models and " Rear " on the 1¼ Litre model if the rear jacks only are required; and to " All " if all the jacks are required.

Before operating the jacks, make sure that the release valve is firmly, but not forcibly, screwed down.

Place the handle in position on the stub lever and operate the pump with a full stroke in each direction until the wheels are clear of the ground. No damage can be caused by continual pumping, as a relief valve is fitted to cope with any excessive pressure.

To release the jacks, remove the handle, unscrew the relief valve three turns opening the valve slowly at first to lower the car gently to the ground, and then, when the wheels have made contact with the ground, turn the indicator pointer to the " All " position, allowing the relief valve to remain well open.

The indicator pointer should always be kept at the " All " position, allowing the relief valve to remain open when not in use, to ensure that the jacks return to the inoperative position.

THE PUMP AND DISTRIBUTOR UNIT. (1½ Litre, 2 Litre and 2.6 Litre.)

The pump and distributor unit (see Fig. 2) consists of a neat, light, but exceptionally sturdy die-casting of a special alloy developed for the purpose. The double-ended plunger " A " is ground to a close fit in the highly finished cylinders " A1," no packing being necessary. The ends of the cylinders are sealed with the plugs " B " and the joint between the plug and body is made with copper washer " B1." The plunger is oscillated by the rocking arm " C," actuated by means of a detachable handle fitted to the rocking arm lever " D," which is keyed to the shaft by a special chrome nickel steel cotter pin " D1." (The ordinary cycle cotter is useless for this purpose.)

Leakage through the bearing is prevented by means of the moulded composition gland " D2." The orthodox type of spring-loaded inlet valve is dispensed with, as this is a feature of small high-pressure pumps which is prone to give trouble; instead, the cylinders are ported " E " and the plunger itself used to seal the port on the return stroke. By this system, so long as fluid is available, the pump cannot fail to function. The delivery valves, instead of being spring-loaded, are governed by the square-section gravity weights " F," which also serve the purpose of limiting the travel of the ball valve " F1 " to $\frac{1}{16}$ in., thus securing a high degree of efficiency and preventing damage to the valve seats by " pounding."

After passing the delivery valves the fluid enters the transverse passage " G," in communication with which is the pressure relief or safety valve " K," by means of which any excess pressure is avoided and surplus fluid allowed to return to the reservoir. The fluid under pressure, enters the annular chamber " J " through the master check valve " J1," which effectually prevents any return of the fluid back to the pump, thus relieving the ordinary delivery valves and safety valve of any duty other than their normal function. The safety or pressure relief valve " K " consists of a $\frac{7}{32}$ in. steel ball loaded with a laminated spring " H." No adjustment is necessary, as when the securing screw " K1 " in the centre is screwed firmly home, the spring resistance obtained gives the correct pressure.

A filter is incorporated in back cover-plate, through which the fluid passes after each operation.

THE DISTRIBUTOR. (1½, 2 and 2.6 Litre.)

The distributor consists of an annular chamber " J " cast integral with the pump body. In the " floor " of this chamber three ports are positioned, the outer ports " L " each supplying fluid to the jacks to which they are connected, the centre port " M " forming a passage for the return of the fluid to the reservoir after use.

The valve " member " consists of three valves " N1 " manufactured from special alloy, securely vulcanised in a moulded composition cup or bucket " N " which has on its reverse side three sockets " N2 " to receive the valve operating pins " N3." The outer pins are actuated by a distributor cam " O," by means of which the front and rear jacks can be selected or, if the pointer is placed in the " All " position, all jacks can be operated simultaneously.

THE JACKS. (1½, 2 and 2.6 Litre.)

The jacks (see Fig. 3) consist of two tubular steel members, the upper end "A3" of the outer tube "A2" being domed to form a cap or end, and the upper end "A4" of the inner tube "A1" being bulged to form a long guide or bearing which ensures stability when the jack is fully extended. The bottom cap "A5" contains the special moulded composition gland "A6" which is designed to prevent the escape of the fluid either under high pressure or whilst the jacks are in the normal out-of-use position.

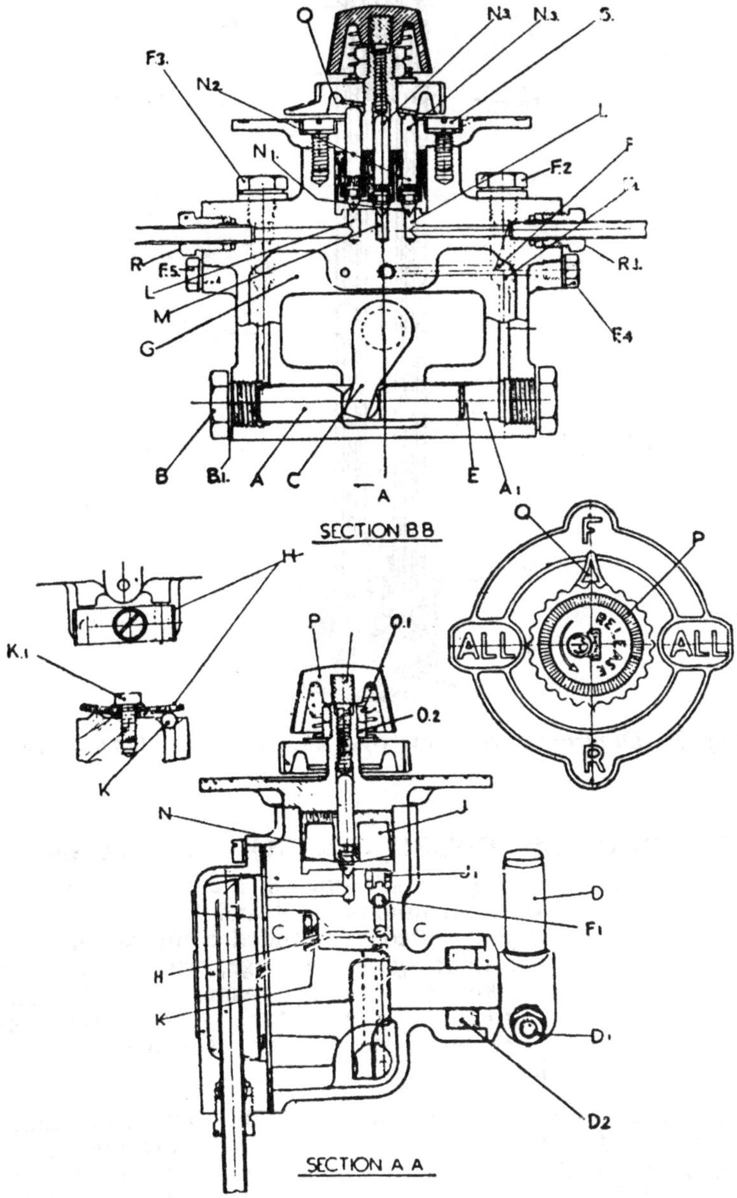

Fig. 2. Diagrammatic sketch of hydraulic jack pump and distributor unit. (1½, 2 *and* 2.6 *litre.*)

The return of the fluid is secured by the use of a compound spring of the highest quality " A7." There are two springs wound concentrically, with sturdy end hooks. These springs give an initial pull of over 30 lbs. when the jacks are in the inoperative or closed position, and from this it can be seen that there is no danger of the ram being extended whilst the vehicle is in motion.

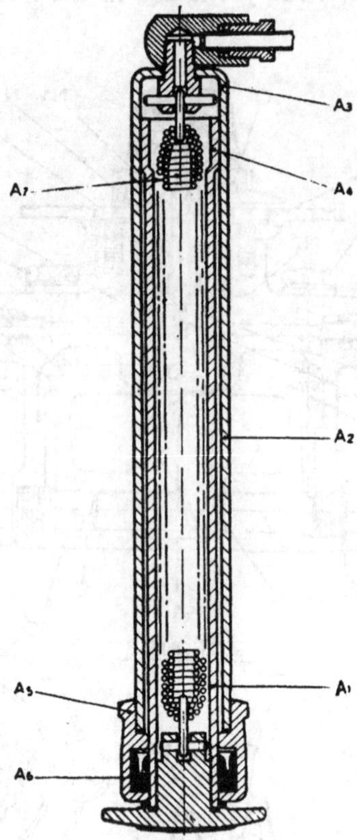

Fig. 3 Diagrammatic view of jack. (1½, 2 *and* 2.6 *Litre.*)

FLEXIBLE HYDRAULIC COUPLINGS. (1½, 2 and 2.6 Litre.)

The flexible couplings (Fig. 4) are one of the features of the Jackall system. They consist of die cast unions " 1 " and " 2 " in which the pipe end bushes " 3 " are flexibly mounted, a moulded composition " U " bucket " 4 " effectually sealing the joint against leakage, whilst permitting ample movement. The thrust is taken by a loose washer " 5." The unions are coupled together with a silico-manganese tube " 6 " suitably coiled to ensure flexibility and to deal with the movements caused by the varying degrees of spring deflection.

Two types of union are used, single and double outlet, the double outlet union being attached on the axle to distribute the fluid to the jacks mounted thereon, whilst the single outlet union is anchored firmly to the chassis and conveys the fluid from the pump and distributor unit to the flexible steel pipe. The outer end of each joint is sealed by a moulded rubber cap " 7 " which excludes grit, moisture and foreign matter.

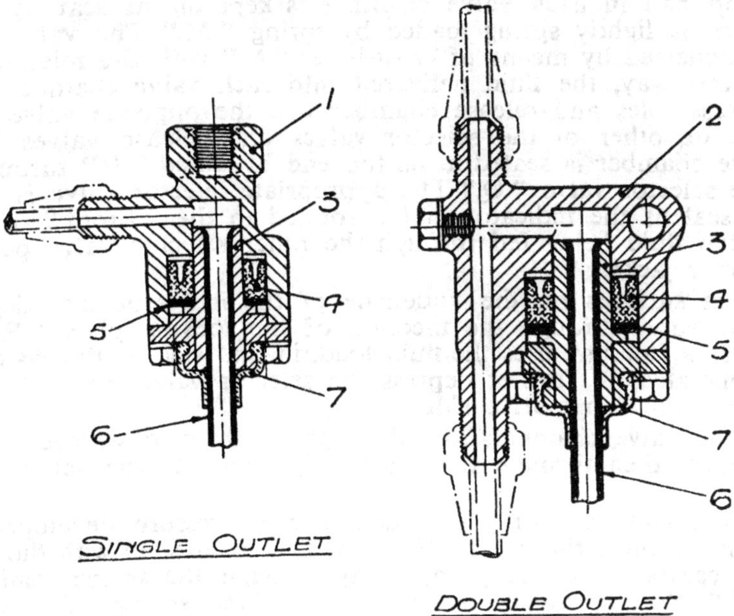

Fig. 4. *Flexible high-pressure joints.* (1½, 2 *and* 2.6 *Litre.*)

GENERAL INSTRUCTIONS. (1½ Litre, 2 Litre and 2.6 Litres.)

When it is necessary to drain fluid from the system for the purpose of repairs, disconnect either delivery pipe joint at distributor and pump the fluid into a clean receptacle. The fluid should be carefully filtered when refilling the system.

Where attention to a jack or a flexible hydraulic connection is necessary, the indicator pointer should be turned to the side opposite that which it is desired to disconnect. This closes the valve on the defective side and repairs can generally be effected without loss of fluid.

The valve bucket may be changed or removed for examination without loss of fluid. If it is desired to examine the bucket, use a cork or other suitable plug 1¼ in. diameter to seal the opening temporarily.

As a special rubber composition is used extensively throughout the system, use of mineral oil or the addition of even a small amount of mineral oil, would render the system wholly inoperative in a short time. Should the fluid become contaminated by mineral oil, the various units of the system should be removed from the chassis and thoroughly washed out with methylated spirit before replacing.

OPERATION OF DISTRIBUTOR BOX. (1¼ Litre.) (See Fig. 5.)

In operation, the fluid passes from the tank into the pump body through the filter " A " which is mounted in the cover.

By reciprocating the level " E " its full stroke in each direction, fluid is drawn through the ports " D " and forced along passages " F " into the valve chambers " G " and " H " via duplicated non-return ball valves " I " and " K." The valves have been duplicated to avoid any possibility of dirt entering and rendering the pump inoperative.

The top ball in each valve chamber is kept on its seat by a washer "L," which is lightly spring loaded by spring "M." The valve chambers are interconnected by means of cross-holes "N" with the release chamber "O." In this way, the fluid delivered into each valve chamber can pass via the cross-holes and release chamber into the opposite valve chamber, when one or other of the selector valves and release valves is closed. Each valve chamber is sealed at its top end by gland "P" through which passes the selector valve "Q." The appropriate selector valve is depressed on to its seat as the indicator dial is rotated to the "Front" or "Rear" position, the fluid being fed through the other selector valve port to the front or rear jacks.

The cam surfaces on the underside of the indicator dial depress the appropriate valve through the medium of the locking plate "R" which, by acting as a fulcrum, uses the fluid load, in that side of the system which is being operated, to further depress the selector valve on to its seat and thus cut off the inoperative side.

From the valve chambers, the fluid passes through whichever selector valve is open, then through the filter "U," and into the delivery line to the jacks.

When the jacks are fully extended, excess pressure developed in the pump passes through the holes "V," which are connected with the passages "F," and escapes into the pump body through the spring loaded relief valve "W." No adjustment is required to these springs, for when the return flow valve "X" is screwed firmly home, the spring resistance obtained gives the correct pressure.

When the release valve is opened to lower the car, fluid from the jacks again passes through the filter "U" (thus trapping any dirt from the jacks) into the valve chamber and via the release chamber through the return flow valve "X," into the pump body and thence back to the tank. The return flow valve "X" is fitted with a rubber sleeve around its reduced diameter which covers the outlet hole, and this prevents the head of the fluid in the tank standing on the jacks.

SERVICING THE DISTRIBUTOR BOX. (1¼ litre.)

When it is necessary to drain the fluid from the system for the purpose of repairs, close the release valve, set the indicator dial to the "All" position, disconnect one of the delivery pipes either at the bottom of the distributor or at one of the jacks, and pump the fluid into a suitable receptacle.

The fluid should be carefully filtered when refilling the system.

After the system has been refilled, the pump should be cleared of all air by opening the bleed screw "B" two turns. When the fluid flows freely from the vent hole "C," the screw should be tightened down firmly.

Whilst filters have been incorporated on both the inlets and outlets of the pump, every care should be taken to prevent dirt entering the system, either through the tank or through dirty connections.

During the repairs, the ends of the pipes should be plugged to keep out dirt. All parts of the distributor box should be thoroughly cleaned using "Red" Jackall fluid only. All traces of petrol or paraffin used for cleaning the metal parts must be removed and all parts smeared with "Red" Jackall fluid on assembly.

The correct level for the fluid with all the jacks retracted, is approximately ¾ in. from the top of the supply tank to the dotted line on the outside of the tank.

The fluid level should never be permitted to fall appreciably below this level and never rise above it, or there will be insufficient room for expansion.

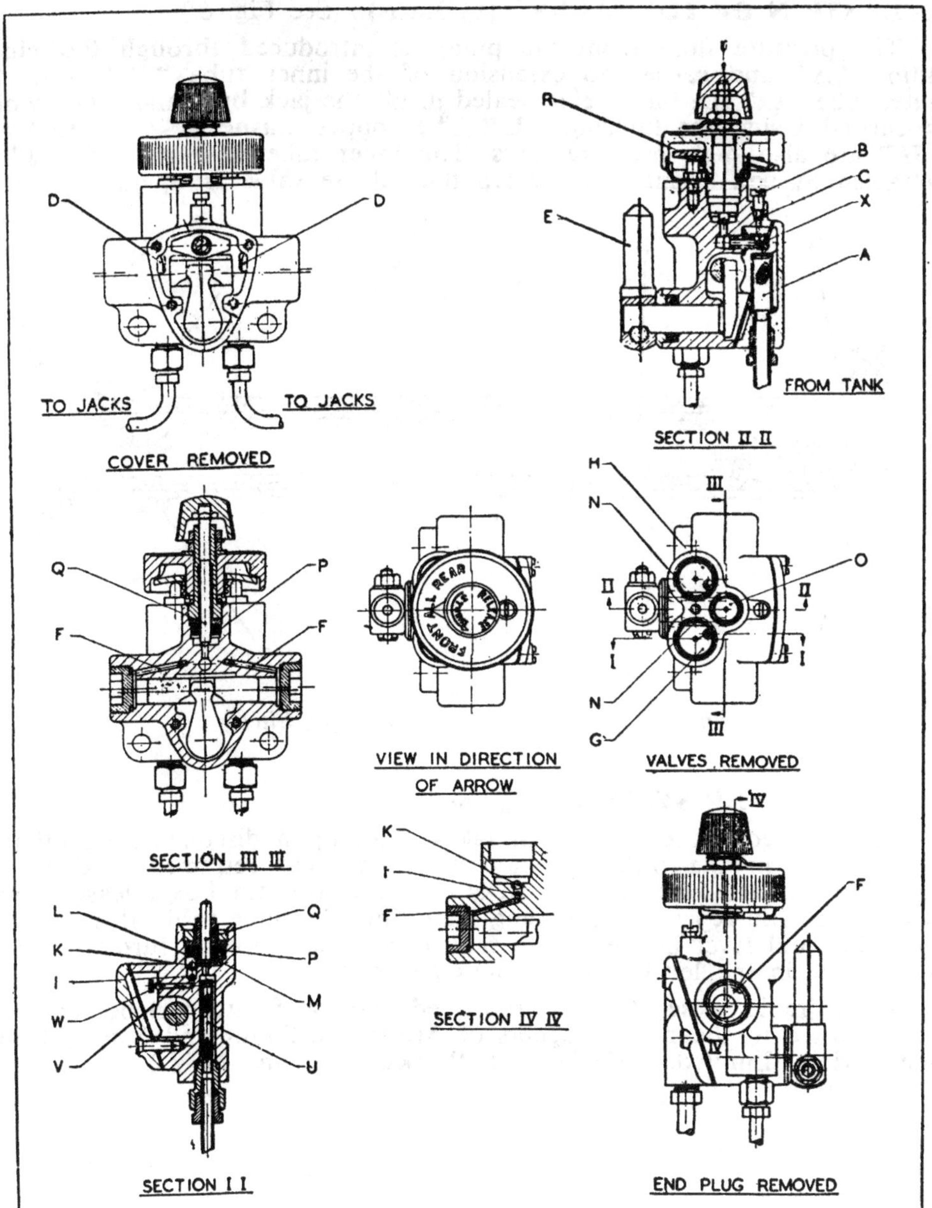

Fig. 5. Jackall distributor box. (1¼ *Litre.*)

REMOVING THE JACKS.

In the event of damage necessitating the removal of one or more of the jacks, the indicator pointer should be turned to the "Rear" position while the front jacks are dealt with and the "Front" while the rear jacks are removed.

To remove a jack cylinder, turn the indicator pointer for the opposite jacks as detailed above, disconnect the fluid delivery pipe union, remove the attachment bolts and withdraw the complete cylinder.

OPERATION OF THE JACKS. (1¼ Litre.) (See Fig. 6.)

The pressure fluid from the pump is introduced through the elbow fitting "A" and causes an extension of the inner tube "B" from the outer tube "C," the fluid being sealed inside the jack by gland "D" which is carried inside the housing "E." The copper washers "F," "G" and "H" are also high pressure seals. The inner tube with its foot "J" is retracted by coil spring "K" when the release valve is open.

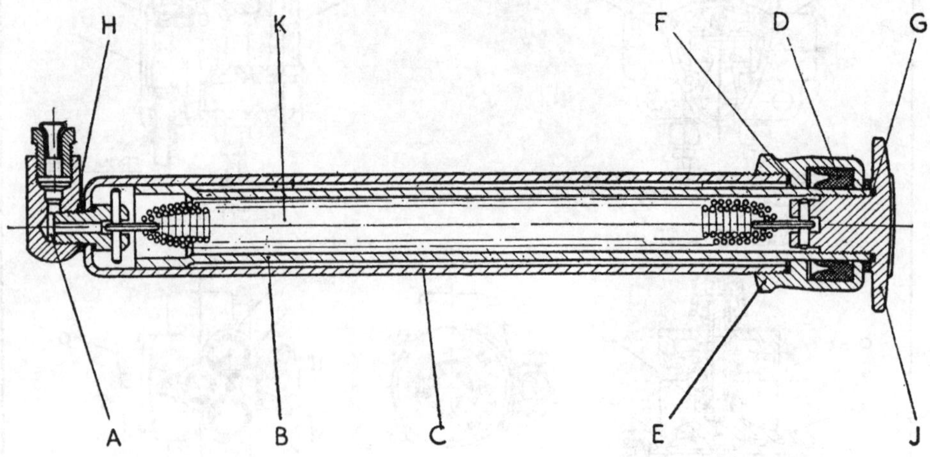

Fig. 6. Arrangement of jack. (1¼ *Litre.*)

REFILLING THE SYSTEM. (1¼ litre.)

To make good the loss which takes place upon disconnection, fill the supply tank in the ordinary way with the jacks retracted, and pump fluid into the jacks. Release the jacks by the distributor box release, screw down again and repeat the pumping operation. Continue to do this, making good the fluid level in the supply tank each time, until no further change takes place in the level with all jacks retracted.

Should faults arise, it is recommended that the faulty part be removed and returned to the nearest agents of Messrs. S. Smith and Sons (Motor Accessories), Limited, Cricklewood Works, London, N.W.2.

CHAPTER XI

SPECIAL TUNING FOR TRIALS WORK

Series M, D, J1, J2, PA, PB, TA, TB, TC and TD Midgets.
Series F and L Magnas.
Series K and N Magnettes.
Specification for Midget Series TD Mark II.

FOREWORD

The M.G., as delivered from the Factory in its standard form, is tuned to give maximum performance with " pump " petrol, consistent with complete reliability and reasonable freedom from pinking. There is, however, a more or less continuous demand from enthusiasts all over the world, for information on methods of improving the performance for competitive purposes, and it is to meet this demand that this section has been prepared.

It must clearly be understood that whereas it is a simple matter to increase the power output of the engine, this cannot be achieved without the use of fuel having better anti-detonation qualities than ordinary " pump " petrol. In addition, this increase in power must inevitably carry with it a tendency to reduce the reliability.

It is for this reason that the terms of the guarantee on a new M.G. expressly exclude any super-tuning of the kind described, but this does not mean that tuning in this way will necessarily make the car hopelessly unreliable. In fact, it may be assumed that it will be at least as reliable as other cars of similar performance.

The information is set out so as to give details of progressively increasing power. With the above ideas firmly in mind, the owner should select the simplest tuning method which will give him the performance he requires, remembering all the time that here, as elsewhere, POWER COSTS MONEY.

ENGINE. (P and N types, excluding PB)

When specially tuning the engine, it must be borne in mind that not only is it necessary to ensure that various small adjustments are checked and carried out more frequently, but additional care in handling must be exercised if the engine is to retain its reliability with the increased power produced. For this reason it is necessary, before carrying out any modifications, to consider the nature of the work the car will be called upon to do. For example, if the car is required for ordinary town work in addition to being used in trials, then it is not advisable to go to extremes.

If the car is required for town use, it is possible to raise the compression ratio from the standard figure of 6.2 to 1, to 6.7 to 1 and still retain its usual docility, but on the other hand, if it is desired to produce a car for trials work only then more power can be obtained by raising the compression ratio to 7.25 to 1.

COMPRESSION RATIO 6.7 to 1.

On T.D.C. the piston protrudes above the gasket face of the cylinder block from .0006 in. (approx.) to .0294 in. below the gasket face, which gives a mean dimension of .015 in. below the gasket face. This equals approximately 1 c.c.

The standard depth of the cylinder head is $3\frac{41}{64}$ in.

The capacity of the combustion space is 35.1 c.c.

$$\text{The compression ratio} = \frac{212 + 35.1\text{ c.c.} + 4.4\text{ c.c. (gasket)} + 1}{40.5}$$

This equals 6.2 to 1 approximately.

To raise the compression ratio to 6.7 to 1, the gasket face of the cylinder head must be machined to give an overall depth of $3\frac{19}{32}$ in. instead of the standard dimension of $3\frac{41}{64}$ in.

The standard copper and asbestos gasket is retained, and this is advisable and to be preferred to the employment of a thin steel gasket or no gasket at all. With both the latter, the making of a sound joint is very difficult unless the operation is carried out by a highly skilled man.

It will be appreciated that as the compression ratio is raised, the dimension between the dynamo and the cylinder head is reduced, necessitating the use of a shallower dynamo drive fork. The illustration shows the necessary modification.

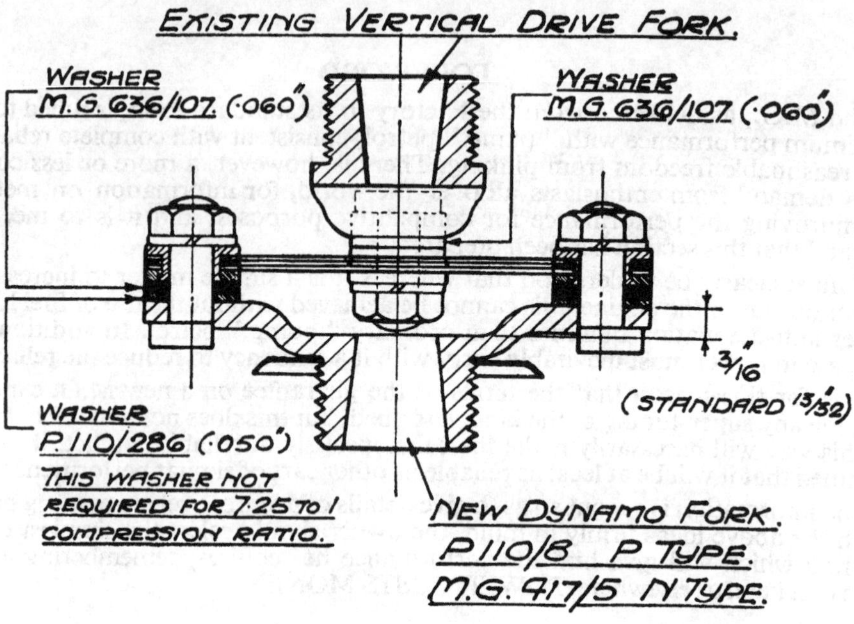

Fig. 1. Vertical drive coupling arrangement, 6.7 *to* 1 *compression ratio.*

COMPRESSION RATIO 7.25 to 1.

In this case the gasket face of the cylinder head should be machined to an overall depth of $3\frac{35}{64}$ in. The same remarks apply in connection with the cylinder head gasket and dynamo drive fork. The two packing washers P110/286 are not required with this compression. See Fig. 1.

SPARKING PLUGS.

With the above mentioned compression ratios, KLG Sparking Plugs Type LB1 should prove satisfactory. It is very important that the plug gaps be between .012 in. and .015 in. If the gaps are greater, then satisfactory results will not be obtained.

TAPPETS.

With the increase in compression ratio, the inlet tappet should be set to .008 in. and the exhaust to .010 in. ; (standard clearances are inlet .006 in. and exhaust .008 in.).

VALVE TIMING.

This is extremely important to obtain maximum power and details of the correct valve timing are given on Data Sheets.

IGNITION TIMING.

Full information on this important subject will be found on page 44.

CARBURETTER ADJUSTMENT.

Normally the standard needles will be found perfectly satisfactory, but if a richer mixture is required, use an M5 needle for the P type and H4 on the N type.

DISTRIBUTOR.

For full information on this item see page 44.

SPECIALLY BALANCING AND POLISHING CONNECTING RODS.

It is suggested that this important item of tuning should be placed in the hands of experts who have the necessary equipment. The interior of the cylinder heads and ports may be polished and exhaust and inlet ports checked to ensure that they line up with their respective manifolds.

GEAR RATIOS.

The standard gearbox ratios are :	First	4.18 to 1
	Second	2.32 to 1
	Third	1.36 to 1
	Top	Direct

Alternative gears can be used, which give the following ratios

First	3.58 to 1
Second	2.14 to 1
Third	1.36 to 1
Top	Direct

With the standard rear axle ratio of 5.375 to 1 on the P type and 5.125 to 1 on the N type, the standard gearbox will give the following overall ratios :

P type	First	22.48 to 1
	Second	12.46 to 1
	Third	7.31 to 1
	Top	5.373 to 1
N type	First	21.5 to 1
	Second	11.9 to 1
	Third	6.98 to 1
	Top	5.125 to 1

With the standard gearbox and standard tyres, the following m.p.h. per 1,000 r.p.m will be obtained :

P type	First	3.52
	Second	6.34
	Third	10.83
	Top	14.73
N type	First	3.79
	Second	6.83
	Third	11.66
	Top	15.86

With the alternative gears, the ratios are as follows :

P type	First	19.25 to 1
	Second	11.5 to 1
	Third	7.31 to 1
	Top	5.375 to 1
N type	First	18.35 to 1
	Second	10.97 to 1
	Third	6.98 to 1
	Top	5.125 to 1

With the alternative gears and standard tyres the following m.p.h. per 1,000 r.p.m. will be obtained:

P type	First	4.12
	Second	6.89
	Third	10.83
	Top	14.73
N type	First	4.43
	Second	7.41
	Third	11.66
	Top	15.86

For trials in which very steep gradients are encountered, a rear axle giving a ratio of 5.875 to 1 on the P type and 5.375 to 1 on the N type should be used.

With the standard gearbox the following ratios will be obtained :

P type	First	24.54 to 1
	Second	13.63 to 1
	Third	7.99 to 1
	Top	5.875 to 1
N type	First	22.48 to 1
	Second	12.46 to 1
	Third	7.31 to 1
	Top	5.375 to 1

With the standard gearbox and standard tyres, the following m.p.h. per 1,000 r.p.m. will be obtained :

P type	First	3.22
	Second	5.81
	Third	9.92
	Top	13.46
N type	First	3.59
	Second	6.46
	Third	11.00
	Top	15.00

With the alternative gears, the ratios are as follows :

P type	First	21.2 to 1
	Second	12.57 to 1
	Third	7.99 to 1
	Top	5.875 to 1
N type	First	19.24 to 1
	Second	11.5 to 1
	Third	7.31 to 1
	Top	5.375 to 1

With the alternative gears and standard tyres, the following m.p.h. per 1,000 r.p.m. will be obtained :

P type	First	3.77
	Second	6.3
	Third	9.92
	Top	13.46
N type	First	4.2
	Second	7.0
	Third	11.0
	Top	15.0

Differential.

If a locked differential is permitted by Trials Regulations then the star and planet wheels may be replaced in the differential unit by a differential lock. This device connects the axle shafts together and prevents loss of drive due to wheel-spin. However, very few, if any, organisers allow this method of drive these days and it is only mentioned as a matter of interest.

Steering Gear.

The ratio of the steering gear fitted to the standard models was 9.0 to 1 and this is considered quite high enough for general purposes. Under special circumstances, it may be considered advisable to use a higher ratio of 7.0 to 1. This will enable maximum lock to be obtained with less movement of the steering wheel.

Shock Absorbers.

No hard and fast rule can be laid down for setting of the shock absorbers, as this is a question of individual requirements. While it may be considered desirable to stiffen up the shock absorbers for certain events, it will be wise to bear in mind that this extra loading will impose a much greater strain on the shock absorbers and their mounting brackets, consequently, the settings should be returned to normal just as soon as possible after the event for which they have been tightened.

ENGINE. (M and D Midgets.)

The standard depth of the cylinder head is $3\frac{1}{2}$ in., giving an individual capacity of 43.5 c.c. This is equivalent to a compression ratio of 5.5 to 1. If $\frac{3}{32}$ in. is removed from the gasket face of the cylinder head, the individual capacity will become 36.9 c.c. and the compression ratio will be raised to 6.2 to 1. The standard copper and asbestos cylinder head gasket should be retained.

ENGINE. (PB Midget.)

The cylinder head fitted is identical to the Series PA having an overall depth of $3\frac{41}{64}$ in., but with a standard compression ratio of 6.7 to 1.

By removing $\frac{3}{64}$ in. from the gasket face of the cylinder head to give an overall depth of $3\frac{19}{32}$ in. and retaining the use of the standard copper and asbestos gasket, the compression ratio is raised to 7.25 to 1. This should be regarded as the maximum desirable.

ENGINE. (J1 and J2 Midgets.)

The standard depth of the cylinder head is $3\frac{7}{16}$ in. giving a compression ratio of 6.2 to 1. By removing $\frac{3}{64}$ in. from the gasket face of the cylinder head to give an overall depth of $3\frac{25}{64}$ in. and retaining the use of the copper and asbestos gasket, the compression ratio is raised to 6.8 to 1 and it is not advisable to increase beyond this point.

ENGINE. (TA Midget.)

The standard depth of the cylinder head is 3.386 in., giving a compression ratio of 6.4-6.6 to 1. By removing $\frac{3}{32}$ in. from the gasket face of the cylinder head, thereby reducing the overall depth to 3.292 in., and retaining the use of the copper and asbestos gasket, the compression ratio can be raised to 7.3 to 1. In this case, it is necessary to grind back the edges of the plug cavities to give a minimum metal thickness of $\frac{1}{16}$ in. This will prevent burning. It will also be necessary to insert packing washers under the cylinder head stud nuts to prevent the nuts bottoming on the stud threads. Special packing washers should be placed under the valve rocker shaft brackets. The thickness of the washers should be equal to two-thirds of the amount removed from the cylinder head, i.e., $\frac{1}{16}$ in.

ENGINE. (K type Magnette and L type Magna.)

The standard depth of the cylinder head is $3\frac{19}{32}$ in. The capacity of each combustion space is 29.1 c.c. The compression ratio is approximately 6.4 to 1, with a nominal ratio of 6.2 to 1. By removing $\frac{1}{16}$ in. from the gasket face of the cylinder head to give an overall depth of $3\frac{17}{32}$ in. and a combustion space of 24.7 c.c., the compression ratio is raised to 7.25 to 1.

By removing $\frac{3}{32}$ in. from the gasket face of the cylinder head to give an overall depth of $3\frac{1}{2}$ in. and a combustion space of 22.5 c.c., the compression ratio is raised to 7.75 to 1.

By removing $\frac{1}{8}$ in. from the gasket face of the cylinder head, the compression ratio could be increased to 8.65 to 1, but this is not considered advisable because the metal at the base of the sparking plug holes would be extremely thin.

NOTE :—In these four cases, the standard copper and asbestos cylinder head gasket should be used.

ENGINE. (F type Magna.)

The standard depth of the cylinder head is $3\frac{29}{64}$ in. and the capacity of the combustion space is 40 c.c. This gives a compression ratio of approximately 5.7 to 1.

By removing $\frac{3}{64}$ in. from the gasket face of the cylinder head to give an overall depth of $3\frac{13}{32}$ in., the compression ratio is raised to 6.2 to 1.

By removing $\frac{1}{8}$ in. from the gasket face of the cylinder head, to give an overall depth of $3\frac{21}{64}$ in., the compression ratio is raised to its maximum of 6.5 to 1.

NOTE :—In all three cases, the standard copper and asbestos cylinder head gasket should be used.

STANDARD ENGINE DATA—SERIES TC & TD.

Bore : 66.5 mm.
Stroke : 90 mm. 1,250 c.c. 4-cylinder.
Firing Order : 1, 3, 4, 2.
Sparking Plugs : Champion L.10.S.
Compression Ratio : 7.25 to 1.
Valve Timing : Inlet opens 11° B.T.D.C. Closes 57° A.B.D.C.
Exhaust opens 52° B.B.D.C. Closes 24° A.T.D.C.
Valve Lift : 8 mm. inlet and exhaust.
Tappet Clearance : .019 in. hot.
Contact Breaker Points : .010 in./.012 in.
Ignition Timing : T.D.C.
Octane Rating : Minimum requirements for knockfree operation =74 octane.
For maximum power .. .. =82 octane.
Carburetters : Twin " S.U." $1\frac{1}{4}$ in. bore.
Carburetter Jet : .090 in.
Carburetter Needles : Standard—" E.S." Richer—" E.M." Weaker—" A.P."
B.M.E.P. : 125 at 2,600 r.p.m.

B.H.P.	*R.P.M.*
11.00	1,000
23.50	2,000
36.00	3,000
47.00	4,000
54.00	5,000
54.00	5,200
52.50	5,500
47.00	6,000

Safe Maximum r.p.m. : 5,700.
Valve Crash r.p.m. : 6,000.
C.C. of Combustion Space : 45.5 c.c.
Cylinder Head Depth : Top to bottom face=76.75 mm.
Thickness of Cylinder Head Gasket : .045 in.
C.C. of Cylinder Head Gasket : Approximately 4.5 c.c. compressed.
Clutch Pressure Springs : 130/140 lb. Red.

STANDARD CAR DATA—SERIES TC MIDGET

Gear Ratios :

	Overall ratios.	*M.P.H. per 1,000 r.p.m.*
Top	5.125 to 1	15.84
3rd	6.93 to 1	11.74
2nd	10.00 to 1	8.13
1st	17.32 to 1	4.68

Rear Axle Ratio : 5.125 to 1.
Dimensions :

Overall length	11 ft. $7\frac{1}{2}$ in.
Overall width	4 ft. 8 in.
Overall height (hood up)	4 ft. 5 in.

Wheelbase	7 ft. 10 in.
Track	3 ft. 9 in.
Weight	15½ cwt.
Tyre size	4.50—19
Tyre pressures (front tyres)	24 lb.
Tyre pressures (rear tyres)	26 lb.
Front wheel toe-in	$\frac{3}{16}$ in.
Capacities :	
Capacity of water system	1¾ gallons.
Capacity of gearbox	1½ pints.
Capacity of rear axle	2 pints.
Capacity of petrol tank	13½ gallons.
Capacity of engine sump	9 pints.

STANDARD CAR DATA—SERIES TD MIDGET.

Gear Ratios :

	Overall ratios.	*M.P.H. per 1,000 r.p.m.*
Top	5.125 to 1	14.42
3rd	7.098 to 1	10.412
2nd	10.609 to 1	6.966
1st and Rev.	17.938 to 1	4.12

Rear Axle Ratio : 5.125 to 1.

Dimensions :	
Overall length	145 in.
Overall width	58⅝ in.
Overall height (hood up)	53 in.
Wheelbase	94 in.
Track—Front	47⅜ in.
Rear	50 in.
Weight	17¼ cwt.
Tyre size	5.50×15
Tyre pressures (front tyres)	18 lb.
Tyre pressures (rear tyres)	18 lb.
Front wheel toe-in	Nil.
Capacities :	
Capacity of water system	1¾ gallons.
Capacity of gearbox	1¼ pints } Hypoid "90"
Capacity of rear axle	2¼ pints } Hypoid "90"
Capacity of petrol tank	12½ gallons.
Capacity of engine sump	9 pints.

STAGE 1
HIGHER COMPRESSION RATIO TUNING (8.6 to 1)

The engine is raised to 8.6 to 1 compression ratio by removing $\frac{3}{32}$ in. from the cylinder head face.

The standard head is 76.75 mm. thick ; the finished thickness after machining should be 74.37 mm.

Remove any frazes left, and polish, but do not grind out the combustion chambers as these are already quite clean and are machined nearly all over.

Make sure the gasket edges do not overlap the combustion spaces.

The ports may be ground and polished, but should not be ground out so heavily that the shape or valve choke diameter is impaired.

The inlet port outer separating stud boss may be ground away slightly—about $\frac{1}{16}$ in. off each side (still maintaining its streamline shape)—so that oblong ports are obtained, 1 $\frac{3}{16}$ in. high, $\frac{21}{32}$ in. wide (minimum).

Do not remove this boss completely or it will affect mixture distribution.

Match up, by grinding, all the exhaust and inlet manifold ports with the cylinder head ports.

Grind out and polish the inlet manifold, also matching the carburetter bore.

(*a*) Use standard cylinder head gasket.

(*b*) Use $\frac{3}{32}$ in. thick × $\frac{3}{4}$ in. O.D. washers under cylinder head nuts (to correct for reduced head thickness).

c) Use 4 rocker-shaft bracket packing pieces. (Fitted under base of bracket to correct the rocker adjustment.) M.G. Part No. M.G. 862/459—$\frac{1}{16}$ in. thick mild steel with 3 holes to match base of bracket.

(*d*) Use fuel 50% petrol, 70 or 80 octane, and 50% benzol, or for sustained power use 25% petrol and 75% benzol.

(*e*) Plugs : Use Champion L.A.11 or Lodge R.49.

(*f*) Use carburetter needles E.S.—Jet .090 in.

(*g*) Tappet setting : .022 in.

(*h*) Ignition setting : T.D.C.

(*j*) Use standard 1$\frac{1}{4}$ in. carburetters.

The engine should then give the following brake horse-power :

B.H.P.	*R.P.M.*
11.3	1,000
23.5	2,000
38.0	3,000
49.0	4,000
58.0	5,000
60.0	5,500
60.0	6,000

STAGE 2

HIGH COMPRESSION RATIO TUNING (9.3 to 1), AND USING LARGER VALVES.

The standard engine is raised to 9.3 to 1 compression ratio by removing $\frac{1}{8}$ in. from the cylinder head face.

The finished thickness after machining should be 73.575 mm. This is the absolute maximum to remove.

Polish head, ports and manifolds as in Stage 1.

It will be noticed that a sharp edge is left on the combustion space profile at the end of the sparking plug hole.

File this sharp edge back vertically until it is a minimum of $\frac{1}{32}$ in. thick at the centre.

File this only locally at the plug hole (approximately $\frac{1}{2}$ in. wide scoop) and blend into the combustion chamber shape with a radius each side. Do not file back too far. Check that the combustion space edge still extends safely over the gasket edge.

Use larger inlet valves (Part No. M.G.862/460), having 36 mm. O.D. heads.

In fitting these valves it is necessary to cut away part of the combustion space wall to clear the valve head. To do this use a 38 mm. diameter (maximum) side and face cutter piloted off the valve guide.

The cutter corner should have a 1 mm. radius.

This may alternatively be done by careful grinding of the vertical wall until the valve head has a 1 mm. working clearance.

Then cut or bore out the valve choke in a similar manner to 33 mm. diameter.

Feather off by grinding any local ridge left in the valve port. Re-cut the seat to 30° × 34.9 mm. top diameter.

Fit larger exhaust valves (Part No. M.G.862/461), having 34 mm. O.D. heads.

Follow procedure as for inlet valves, but cut away combustion wall to clear head with 36 mm. diameter cutter, bore valve choke to 29 mm. diameter and re-cut seat 30° × 32.8 mm. top diameter.

Fit stronger valve springs (150 lb. open tension). Outer spring (Part No. M.G. 862/462) and inner spring (Part No. M.G.862/463).

These are interchangeable with the standard springs, but one point should be noticed : they are staggered pitch springs and the closed coil ends should be fitted next to the cylinder head. Valve crash occurs with these springs around 6,500 r.p.m.

(*a*) Use standard cylinder head gasket.

(*b*) Use ⅛ in. thick × ¾ in. O.D. washers under cylinder head nuts (to correct reduced thickness).

(*c*) Use 4 rocker-shaft bracket packing pieces (Part No. M.G.862/459) to correct rocker adjustment. These are $\frac{1}{16}$ in. mild steel with 3 holes to match base of bracket.

(*d*) Plugs : Use Champion L.A.11 or Lodge R.49.

(*e*) Tappet setting .022 in.

(*f*) Ignition setting : T.D.C.

(*g*) Use standard 1¼ in. carburetters.

Using fuel 75% benzol and 25% petrol, with carburetter needles E.S., Jet .090 in., the engine should then give the following brake horse-power :

B.H.P.	R.P.M.
61.00	5,000
65.00	5,500
63.00	6,000

Or using fuel 50% methanol, 20% petrol, 30% benzol, with carburetter needles R.O., Jet .090 in., the engine should give the following brake horse-power :

B.H.P.	R.P.M.
62.5	5,000
66.5	5,500
64.0	6,000

Alternative richer needle for above—R.L.S. or weaker No. 5.

An addition of 1% castor oil can be added to the methanol fuel.

When running on the above fuel, it is advisable to fit twin S.U. pumps, as explained at Item (*j*) in Stage 3.

NOTE :—The fan blades may be removed if the car is going to be used generally above 40 m.p.h., but for trials work, slow hill-climbing, and traffic work, they should be retained.

The fan takes approximately 1 b.h.p. to drive.

Tappet settings in all stages may be .019 in. if quietness is desired, with consequent loss of approximately 1 b.h.p.

In addition to the above, the 1½ in. diameter competition carburetters may be fitted, as described under Stage 3.

Using fuel 75% benzol and 25% petrol, with carburetter needles E.L., Jet .090 in., the engine should give the following brake horse-power :

B.H.P.	R.P.M.
63.00	5,000
68.00	5,500
66.00	6,000

Alternative richer needle for above—AA, or weaker—EO.

Using fuel 50% methanol, 20% petrol, 30% benzol, with carburetter needles R.L.S., Jet .090 in., the engine should give the following brake horse-power :

B.H.P.	R.P.M.
66.00	5,000
70.00	5,500
68.00	6,000

Alternative richer needle for above—C.S.2, or weaker—R.O.

If richer needles are required, change to the .100 in. range.

STAGE 3

HIGH COMPRESSION RATIO TUNING (12 to 1)

The standard engine, using the standard cylinder head with a depth of 76.75 mm., is raised to 12 to 1 compression ratio by the use of special pistons (Part No. M.G.862/458). (When ordering, please state exact bore sizes.)

These pistons can only be fitted one way round—that is, with the flame groove on the sparking plug side.

In using this high compression ratio it is necessary to run on a high content methanol base fuel and to carry out alterations as below :

(*a*) Use fuel as follows :
80% dry blending methanol. Specific gravity : .796 at 60° F.
10% benzol (90). Specific gravity : .8758 at 60° F.
10% petrol (Pool 70 octane or 80 octane).
1% castor oil may be added.

(*b*) Use standard 1¼ in. carburetters.

(*c*) Use carburetter jets size .100 in.

(*d*) Use carburetter needles G.K. or richer—R.C. and weaker—R.V.

(*e*) Sparking plugs : Use Champion L.A.14 or Lodge R.49.

(*f*) Use carburetter float-chamber needle and seat assemblies S.U. type T.3.

(*g*) Tappet setting : .022 in.

(*h*) Ignition setting 4° A.T.D.C. (flywheel).

(*j*) Use two S.U. petrol pumps for increased fuel delivery. Do not couple the pumps together, but use duplicate fuel lines. Run an additional pipe from the tank to the extra pump. Run an additional flexible pipe from this pump to the carburetters. On the carburetters use a double-feed banjo union on each float-chamber. One pump line to feed the rear float-chamber, and one pump line to feed the front one.
Now run a flexible fuel pipe between the two float-chamber banjo unions to balance the feed. The above arrangement will ensure getting full discharge from the pumps.

The engine should then give the following brake horse-power :

B.H.P.	R.P.M.
69.0	5,000
73.0	5,500
74.0	5,800
73.0	6,000

These figures can now be increased by the fitting of the larger inlet and exhaust valves and 150 lb. valve springs as in Stage 2.

The maximum power should then be 76 b.h.p. at 5,800 r.p.m.

Larger Carburetters

To increase still further, fit the 1½ in. diameter competition carburetters, grind out the inlet manifold at the outer bores to 1½ in. diameter to match these carburetters.

It is not possible to grind right through 1½ in. diameter, but it is necessary to taper off in about a ¾ in. length to 1⅜ in. diameter. Maintain a minimum of 1⅜ in. diameter or more right through, or otherwise a minimum area of 1.5 sq. in.

These carburetters will be found to be fitted with light aluminium pistons, with additional return springs fitted above in the dashpots.

For absolute maximum power these springs may be removed, but for good pick-up and general carburation smoothness they should be left in position.

Fit carburetters with .125 in. jets and V.E. needles, or richer—V.G., or weaker—V.A.

On the same fuel and with other conditions as stated for Stage 3 (including large valves), the engine should give the maximum power of 80 b.h.p. at 6,000 r.p.m.

Run on a fuel of 100% methanol, using .125 in. jets and V.J. needles, or richer—V.L., or weaker—V.I. The engine should then give a maximum power of 83 b.h.p. at 6,000 r.p.m.

NOTE :—When using methanol in the fuel it is necessary frequently to clean out the whole of the carburetter and fuel system.

STAGE 4

INCREASING PERFORMANCE BY FITTING THE "SHORROCK" SUPERCHARGER KIT

This kit is manufactured by "Shorrock" Superchargers, Ltd., Fletchams Tead Lane, Coventry.

It is an eccentric vane-type supercharger with balanced pressure lubrication system. It is mounted on the inlet manifold and driven by twin belts from the front end of crankshaft. It gives a large increase in power at the lower and medium engine speeds.

Details of Supercharger :

Drive ratio : 1.16 to 1 step-up on engine speed.

Swept volume : .72 litre per revolution.

Boost pressures :

R.P.M.	Lb. sq. in. boost (approx.)
1,000	1.5
2,000	2.5
3,000	3.8
4,000	5.5
5,000	6.0

Oil metering pin diameter : .304 in. diameter, fitted in a reamed housing bore of .3125 in. diameter.

Supercharger oil feed tank : Use oil S.A.E.30.

Carburetter : S.U. 1⅜ in.

Standard needle : R.L.S.

Jet size : .090 in.

With :

1. The supercharger kit fitted to a standard engine (see Standard Data).
2. Pool petrol (70 octane).
3. Plugs : Champion L.11.S or Lodge H.N.P.
4. Tappet setting .022 in.

the engine should give the following brake horse-power :

B.H.P.	R.P.M.
18.5	1,500
28.0	2,000
45.0	3,000
58.0	4,000
69.0	5,000
69.0	5,500

or with 80 octane fuel, the following brake horse-power :

B.H.P.	R.P.M.
19.0	1,500
28.5	2,000
45.5	3,000
61.0	4,000
70.0	5,000
70.0	5,500
70.0	6,000

If required for special purposes it may be run on a fuel of 50% methanol, 20% petrol, 30% benzol.

To do this, fit to the 1⅜ in. carburetter a .125 in. jet and use carburetter needle V.E. or richer—V.G., weaker—V.A.

Fit the carburetter float-chamber with a T.3 needle and seat, and it is advisable to fit twin-coupled S.U. petrol pumps (see Stage 3).

Sparking plugs : Use Champion L.11.S or Lodge H.N.P., or if harder plugs are required, Champion L.A.11 or Lodge R.49.

The engine should then give the following brake horse-power :

B.H.P.	R.P.M.
20.5	1,500
31.5	2,000
52.0	3,000
68.5	4,000
75.0	5,000
75.5	5,500
75.0	6,000

NOTE :—If carburetter vibration is experienced, which disturbs the mixture at high speeds, fit a ⅛ in. thick " Neoprene " gasket to the carburetter flange in place of the ordinary gasket.

Fit slotted nuts and double-coil spring washers to the carburetter fixing studs. Tighten these nuts only enough to grip carburetter firmly, drill the studs through the nut slots, and lock with wire from one stud to the other around the carburetter body.

STAGE 5
FITTING THE " SHORROCK " SUPERCHARGER IN CONJUNCTION WITH 9.3 to 1 COMPRESSION RATIO

Fit the " Shorrock " Supercharger Kit and raise the compression ratio to 9.3 to 1, as explained in Stage 2. Fit the larger exhaust and inlet valves and 150 lb. valve springs.

Fit a .125 in. jet to the carburetter and a T.3 needle and seat to the float-chamber.

Use carburetter needle V.G. (with ⅛ in. shank), or richer—V.1 (with ⅛ in. shank), or weaker—V.E. (with ⅛ in. shank).

Use twin-coupled S.U. petrol pumps.

Sparking plugs : Use Champion L.A.11 or L.A.14 or Lodge R.49 or R.51.

Use a fuel 50% methanol, 20% petrol, 30% benzol.

1% castor oil may be added.

The engine should then give the following brake horse-power :—

B.H.P.	R.P.M.
55.5	3,000
73.5	4,000
85.5	5,000
88.00	5,500
88.00	6,000

To obtain a further increase in power, fit a larger carburetter (S.U. Specification No. 538, Type H.6, 1¾ in. diameter) to the supercharger.

It will be necessary to make a new elbow (steel or aluminium) between the carburetter and supercharger with an inside diameter of 1¾ in. Make sure the inlet port to the supercharger matches up and is of the same diameter.

Use a .1875 in. diameter jet in the carburetter and needle R.M.7, or richer, R.M.8 ; or weaker, R.M.6.

With a fuel of 80% methanol, 10% petrol and 10% benzol, the engine should give the following brake horse-power :—

B.H.P.	R.P.M
74.5	4,000
82.00	4,500
89.00	5,000
94.25	5,500
97.5	6,000

STAGE 1A

HIGHER COMPRESSION RATIO TUNING (8.6 to 1), BUT PERMITTING USE OF LOW OCTANE FUEL

Raise the compression ratio to 8.6 to 1 and polish the head, ports and manifold as for Stage 1.

Fit larger inlet and exhaust valves as explained in Stage 2, using inlet valves (Part No. M.G.862/460), but sodium-cooled exhaust valves (Part No. M.G.862/466).

These exhaust valves have larger stems and require special bronze guides (Part No. MG.862/467), and special valve cotters (Part No. MG.862/468). Press the valve guides in so that they stand 24 mm. above the top of the cylinder head face. The bronze guides are longer than the standard ones and will protrude farther into the exhaust port. Due to the increase in valve weight fit the stronger 150 lb. valve springs (Part Nos. MG.862/462 and 463) as described in Stage 2.

The 150 lb. springs will prevent valve bounce up to 6,000 r.p.m.

To help compensate for the additional valve weight the valve stem shrouds may be cut off close to the top, leaving just the top collar and an ⅛ in. of the tubular portion to locate the inner valve spring. The valve stem shrouds are cyanide hardened and will have to be cut off with a grinding wheel.

Remove the thrust springs which are between the rockers on the rocker-shaft and replace them with steel distance tubes, leaving .003 in. to .005 in. end-float. This reduces some of the rocker friction.

If trouble is experienced with the cylinder head gasket a competition gasket (Part No. MG.862/472) can be supplied, or, if the car is used for racing purposes only, a gasket made from 20 S.W.G. silver finish auto body steel may be fitted. To accommodate this gasket it will be necessary to lap the cylinder head and block faces together with valve grinding compound. This makes a most reliable seal.

Fit the larger 1½ in. carburetters as described in Stage 3.

Remove the thermostat bellows and valve completely and plug up or blank off the small by-pass pipe which goes from the lower radiator outlet pipe up to the side of the thermostat body.

(*a*) Fuel of 70 octane may be used, some pinking will be experienced but will not be detrimental.

(*b*) Tappet setting .025 in. for maximum power .022 in. or .019 in. may be used but up to 2 b.h.p. may be lost.

(*c*) Ignition setting T.D.C.

(*d*) Use carburetter jets .090 in. and needles L.S.I. or weaker E.L., put in with the shoulder $\frac{1}{32}$ in. below the face of the dashpot piston, and mixture adjusting nuts screwed seven flats down, approximately. Remove the supplementary springs from the carburetter dashpots and remove the hydraulic damper pistons.

(*e*) Plugs : Use Champion L.11.S or for continuous high speed work use Champion L.A.11 or Lodge R.49.

(*f*) Leave off the air cleaner and connecting branch pipe for high speed work, but if it is felt necessary to fit it for road work bore out the port holes to match the outer carburetter flanges.

(*g*) Fit a Lucas B.R.12 coil.

(*h*) Plug setting .018 in. to .022 in.

(*j*) As a safety factor for high speed work and to ensure sufficient fuel supply under all conditions of vibration, etc. it is advisable to use two S.U. petrol pumps, as explained at Item (*j*) in Stage 3.

The engine on 70 octane fuel should give the following b.h.p. :—

R.P.M.	B.H.P.	R.P.M.	B.H.P.	R.P.M.	B.H.P.
2,500	29.50	4,000	50.00	5,500	64.00
3,000	40.50	4,500	56.00	5,700	64.5
3,500	43.50	5,000	61.00	6,000	63.00

If the ordinary large exhaust valves, Part No. MG.862/461 (as used in Stage 2) are used in place of the sodium-cooled exhaust valves the engine should be run on 80 octane petrol.

SPECIAL MATERIAL AVAILABLE FOR SERIES TB or TC MIDGET CAR

16 in. × 4.00 in. wheels, suitable for tyre sizes 5.00 in.—16 in., or 5.50 in.—16 in. (M.G. Part No. A.1338.)

Rear axle crown wheel and pinion 8/39=4.875 to 1. (M.G. Part No. MG689/10.)

The above axle ratio gives :—

16.67 m.p.h. per 1,000 r.p.m. with 4.50 in.—19 in. tyres.
100.02 m.p.h. at 6,000 r.p.m.
17.15 m.p.h. per 1,000 r.p.m. with 4.75 in.—19 in. tyres.
102.90 m.p.h. at 6.000 r.p.m.
17.6 m.p.h. per 1,000 r.p.m. with 5.00 in.—19 in. tyres.
105.6 m.p.h. at 6,000 r.p.m.

Steering wheel (Brooklands type spring arm). (M.G. Part No. A.1333.)

SPECIAL MATERIAL AVAILABLE FOR X.P.A.G. ENGINE — TB, TC, AND TD MIDGETS

	Part No.
Rocker-shaft bracket packing pieces ($\frac{1}{16}$ in. thick)	MG862/459
36 mm. inlet valves	MG862/460
34 mm. exhaust valves	MG862/461
High compression pistons—12 to 1 ratio (complete with special rings and gudgeon pin). (State exact bore size when ordering)	MG862/458
Competition cylinder head gasket	MG862/472
150 *lb. Valve Springs*	
Outer valve spring	MG862/462
Inner valve spring	MG862/463

The following are obtainable from S.U. Carburetter Co., Ltd., Wood Lane, Erdington, Birmingham, 24.:

Competition carburetters, 1 pair, 1½ in. diameter S.U. specification 532.
.100 in. bore jets. S.U. No. 1394—112/1.
.125 in. bore jets. S.U. No. 4185.
.1875 bore jets (fixed type). S.U.
Jet needles. As S.U. List.
Float-chamber seat and needle assemblies. S.U. No. T.3. (These are denoted by three grooves machined around the body.)
Booklet listing full range of S.U. needles, price 2s. 6d.
1¾ in. carburetter (for supercharger). S.U. specification No. 538.

Plugs — Champion

L.11.S.	Super Sports.
L.A.11	1st Step Racing.
L.A.14	2nd Step Racing.
L.A.15	3rd Step Racing.

Plugs — Lodge

H.N.P.	Super Sports.
R.49	1st Step Racing.
R.51	2nd Step Racing.
R.53	3rd Step Racing.

Lucas High Performance Coil, Type BR.12.
(The Model Q.12 coil fitted as standard is satisfactory up to 6,000 r.p.m., but the BR.12 is good up to 8,000 r.p.m.)

Lucas 4 V.R.A. Vertical Magneto (Lucas Part No. E.N.M.2002).
(This has a suitable advance curve for the X.P.A.G. engine.)
To fit this magneto it is necessary to indent valve chest plate to clear the body and move position of breather pipe elbow.

	Part No.
Competition Clutch Assembly	MG862/92
*Clutch Springs (Pressure Plate). Competition type, 150 lb. Light Blue	162515
*Clutch Driven Plate (Competition type)	MG862/93

The above clutch gives a quicker take up and increased torque or the standard clutch may be modified by fitting the items starred.

SPECIAL MATERIAL AVAILABLE FOR SERIES TD MIDGET CAR

Combined water temperature and oil pressure gauge (Part No. 500154) (replaces existing oil pressure gauge).

15 in. × 4.50 in. wheels suitable for tyre sizes 6.00 in. × 15 in. (M.G. Part No. 131904).

Rear axle crown wheel and pinion 8/39 = 4.875 to 1 (M.G. Part No. 101910 for the pair).

Gearbox speedo pinion for above (M.G. Part No. 24441).

Gearbox speedo gear for above (M.G. Part No. 24440).

Gear ratios with 8/39 *axle* :—

	Overall ratios	M.P.H. per 1,000 r.p.m. (with 15 in. × 5.00 in. tyres)
Top	4.875 to 1	15.195
3rd	6.752 to 1	10.97
2nd	10.09 to 1	7.34
1st and Rev. ..	17.06 to 1	4.34

Rear axle crown wheel and pinion 9/41 = 4.555 to 1 (M.G. Part No. 101593 for the pair).

Gearbox speedo pinion for above (M.G. Part No. 17212).

Gearbox speedo gear for above (M.G. Part No. 24442.)

Gear ratios with 9/41 *axle* :—

	Overall ratios	M.P.H. per 1,000 r.p.m. (with 15 in. × 5.00 in. tyres)
Top	4.555 to 1	16.259
3rd	6.309 to 1	11.81
2nd	9.429 to 1	7.85
1st and Rev. ..	15.942 to 1	4.64

Oil pump spring : (Supplementary, fits inside existing spring and raises pressure to 60/80 lbs/ins.^{2}	MG.706/226
Flexible oil pipe (Pump to filter, or filter to cylinder block) ..	MG.862/469

SPECIFICATION FOR MIDGET SERIES TD MARK II.

The Mark II engine is fitted with a cylinder head with an 8 to 1 compression ratio obtained by machining $\frac{1}{16}$ in. from the gasket face of the cylinder head. The larger valves and heavier valve springs as used in Stage 2, page 440, are introduced in place of the standard valves and springs. The head, ports and manifolds are polished as in Stage 1, page 439.

The thrust springs between the rockers on the rocker shafts are replaced by steel distance pieces, leaving .003 in. to .005 in. end-float.

1½ in. diameter competition carburetters and twin fuel pumps with twin fuel lines from the petrol tank, details as per Stage 3.

On the later T. Mark II Models, a larger oil bath air cleaner 9 in. in diameter is fitted thus providing more efficient breathing of the engine. Where this cleaner is fitted the carburetter needle L.S.1 should be used. On the earlier models where the 7 in. air cleaner is fitted, the E.M. needle should be used, when the use of the air cleaner is retained, but should the car be used without the air cleaner and pipe, the use of the L.S.1 needle is advocated.

The Champion N.A.8 plug is fitted as standard on the Export Models only. If the car is modified for use on the Home Market, it is recommended that the standard Champion L.10.S. plug be used, with the suitable alternatives, K.L.G. F70 or the Lodge H14 or HNP.

Andrex friction type, fully enclosed and adjustable dampers are fitted on both axles in addition to the Girling or Armstrong dampers, allowing the car to be cornered with imperceptible roll up to speeds at which it will begin to slide. The damper settings are: Front 24 lbs. Rear 22 lbs. Details of the Andrex Dampers are given at the end of Chapter 1, Chassis, page 36.

To ensure consistent ignition at high engine speeds, the standard coil is replaced by the Lucas B.R.12 High Performance Coil.

The alternative axle ratio 4.875 to 1 (8/39) is used in the rear axle in conjunction with the alternative gearbox speedometer pinion and speedometer gear.

Approximate brake horse power developed by the Mark II engine is given below:—

B.H.P.	R.P.M.	B.H.P.	R.P.M.
11.5	1,000	48.0	4,000
24.0	2,000	56.5	5,000
37.5	3,000	57.5	5,500

CHAPTER XII

SUPERCHARGERS

THE MARSHALL-NORDEC SUPERCHARGER.

The majority of readers will be familiar with the term "super-charging" which has in the past been referred more frequently as applying to racing vehicles than to normal touring cars. Racing cars, in order to obtain greatly increased power output have been supercharged with both single and two stage supercharger installations which frequently deliver to the engine a boost pressure of 20/30 lbs. P.S.I. It is not the intention however to deal with what may more correctly be termed as "high-pressure" supercharging, but to confine ourselves to "low-pressure" supercharging which is designed to produce a boost pressure more in keeping with the requirements of the normal touring and sports car. The following information therefore refers in the main to orthodox "low-pressure" supercharger installation.

Before proceeding with a description of what a supercharged engine *can* do it must be stressed that boosting a bad engine will not produce a good one. The range of MG engines manufactured in the past have all proved worthy of "blowing" and the present XPAG type fitted to the TB, TC, TD and Y models is no exception.

The Marshall-Nordec or "Roots" type of compressor is referred to as a constant displacement blower. No compression takes place within the air chamber itself. The mixture is merely trapped by the rotors, carried around from the inlet port and delivered at the outlet port (engine side) at substantially atmospheric pressure although this will vary according to throttle opening. This mixture then meets previously discharged gases and consequently a pressure is built up in the manifold, this process is of course, continuous. The size of compressor for any particular capacity engine is chosen for its displacement, in order that, when driven at a constant fixed ratio to engine speed, a suitable boost pressure is produced throughout the range of engine speeds. With the Roots compressor delivery is not determined by speed, the volume of free air delivered is a function of rotor speed, the volume of compressed delivered air is a function of speed and pressure ratio.

The Roots compressor is basically a gear type pump. The variation from the usual conception of a gear pump lies in the fact that only two "teeth" are provided on each "gear," although there are one or two types available which have three. The twin toothed or "lobed rotors" as they are usually referred to, revolve, in unison, with a constant and fixed clearance between themselves and the casing which houses them. This clearance and timing is maintained by means of a set of gears which operate with the minimum amount of backlash thus ensuring that the rotors will not come in contact with each other. This is particularly important in the case of the majority of Roots compressors in production at the present time where light alloy rotors are employed and contact at speed would result in scoring and partial seizing. Consider for a moment the requirements of an engine that is to produce a greater output than its otherwise identical counterpart. Firstly, the engine must be mechanically capable of dealing with the various increased stresses imposed upon it. Within the specified speed range these will not include by far the most serious inertia stresses caused by the speed of rotation. As any increase in output can fairly accurately be deduced from the amount of mixture burned it will follow that for a given increase it is necessary only to take into consideration the increased cylinder pressure. The increased pressure translated into terms of stresses on the principal mechanical moving parts, gudgeon pin and bush, connecting rod, crank and journal bearings will still only be a small percentage of the main inertia loadings for which the engine was designed, where boost pressure of up to 10 lbs. P.S.I. are employed. The term "low-pressure" supercharging is normally confined to boosts of up to 5 lbs. P.S.I. although it is becoming increasingly more common for this value to be exceeded in cars of the sporting type.

Apart from the obvious and primary reason for adopting a supercharger there are other not so obvious advantages. The increased flexibility of the engine when blown is not altogether a function of manifold pressure, variation of mixture distribution under certain throttle openings and engine speeds is completely eliminated, ensuring that each cylinder is "pulling its weight" in company with its neighbours. It has been demonstrated that a supercharged engine will operate for far longer periods without decarbonising than its normally aspirated counterpart. This is due in part to even distribution in the manifold but mainly to a consistently homogeneous and turbulent mixture in the combustion chamber which results in a constant and complete burning of the charge. Oil, which has normally a tendency to be drawn past the piston rings, is not exposed to the depression which exists in the cylinder during the induction stroke. A feature that is not altogether widely appreciated is that the supercharger, during the induction stroke, is tending to drive the engine pneumatically, and apart from returning to the engine cycle a part of the energy that has been expended in compressing the charge, counteracts to a degree the reversal of stresses which would otherwise normally occur in the connecting rod and bearings.

Another advantage in certain instances is that the supercharger will compensate for the loss of power which an engine will suffer when operating at high altitudes. Due to the gradually decreasing atmospheric pressure at increasing height above sea level the engine becomes starved and will lose power to the extent of 3% for each 1,000 ft. altitude. 3% does not appear to be very great amount, but at 10,000 ft., however, a loss of power to the tune of very nearly one-third of the total possible H.P. is a matter of some concern. Suitably geared, a normal sea level performance can be enjoyed at any altitude at which the car is operated when thus equipped.

The Marshall-Nordec installation marketed by the manufacturers, The North Downs Engineering Company, of Caterham, Surrey, include for the M.G. owners types to suit the TA, TC, TD and Y cars. The installation, which is supplied with fitting instructions and installation drawings, is complete to the last nut and bolt and can be fitted by the average intelligent mechanically minded owner, equipped with a modest tool kit.

The Marshall " Roots " type compressor is manufactured by Sir George Godfrey and Partners, Limited. For the 1,250 c.c. engine a compressor having a nominal displacement of 75 cub. ft. of free air per minute at a speed of 3,000 R.P.M. is used and this is designated as a type J75. There are four main castings which are in light alloy: a rotor case, finned for the dissipation of heat and added strength: two end plates similarly finned and machined to fine limits to house, at one end a pair of ball bearings for taking the journal loading and to locate the rotor shafts endwise: at the other, roller bearings purely for journal loads support the shafts adjacent to the timing gears. The gears, which are precision components, are hardened and have ground profile teeth and ensure exceptionally long life and maintenance of the correct operating clearances between the rotors. The twin lobed rotors, also made in light alloy, are of involute form and are cast directly on to the high tensile steel shafts and dynamically balanced after machining. The driving end of the compressor is enclosed in a nosepiece casting which is extended forward to a point at which it is convenient to pick up a support plate which is bolted to the front engine cross member. Incidentally, the drilling of two holes through this plate, on assembly, into the cross member is one of only two fitting operations which calls for any form of machining and this particular job can be carried out with the aid of an ordinary hand-brace quite easily. Enclosed in the nosepiece is a shaft, mounted and running on bearings similar to those used for the rotor shafts. A simple coupling device consisting of a hard fibre disc is provided with six holes equally spaced around two dissimilar but concentric pitch circles. Three dowels or driving pins are provided to a coupling which is secured to the end of one rotor shaft and engages with the disc. The shaft which protrudes from the nosepiece to pick up the drive is similarly provided, the pins engaging alternate holes in the disc. The feature of this type of arrangement lies in the fact that the compressor bearings are not subject to the driving loads and end thrust which would otherwise be transmitted from the driving shaft to the rotor shaft. Normally only one single driving belt is used for boost pressures up to 6 lbs. P.S.I. Light alloy is again used for both driving and driven pulleys. The driving pulley is designed to locate inside the existing M.G. crankshaft pulley and is secured to it by four recessed Allen cap head screws. The fitting of the pulley is the second of the two machining operations which are called for. A light skimming cut is required to clean up and centralise the case recess in the original crankshaft pulley. A protruding spigot provided on the driving pulley engages in the location thus provided, and using the pulley as a drilling jig, the four holes are transferred through and afterwards threaded to receive the screws. The belt, which is of ⅝ in. top width and readily replaceable, is tensioned by means of a jockey pulley also running on ball bearings and arranged to swing in a short arc sufficient to take up all practicable stretch or wear which might be expected before renewal becomes necessary. A large Simmonds nut locks the position of the jockey pulley, and this type of self locking nut is used almost exclusively throughout.

A specially designed manifold replaces the original induction system and is secured to the engine by means of the original clamps. The compressor is bolted to this manifold and occupies a position inclined at 20° towards the engine making a compact unit which harmonises with the rest of the power plant. Several types of carburettor mounting elbows are available to suit the various models which the XPAG is fitted. The carburettor on the TC for example is neatly tucked away to the rear of the compressor where the original choke and throttle controls are picked up without modification. Similarly in the case of the TD and Y models where a different elbow is designed to be used, the original air cleaner system has been ingeniously designed to be modified to suit the new layout. On TC models a new carburettor air chamber and piston of the oil damped type is necessary and this is supplied together with the correct needle for the satisfactory carburation at the boost for which the installation is designed. The carburettor, occupying a horizontal position when fitted to the supercharger requires a straight mounting float chamber to replace the original angular type, and here again this is provided by the manufacturers.

The lubrication system for the compressor is simplicity itself, a tee-piece union is inserted in the oil pressure gauge line at a point at which it leaves the cylinder block. Oil is then fed to an adjustable restrictor mounted on top of the extended nosepiece. The flow of oil into the unit is visible through a glass tube and adjusted by a knurled thumbscrew readily accessible when the bonnet is raised. As explained, the rotors operate with a clearance between themselves and it is not therefore necessary or desirable for oil to enter the actual air chamber. Oil is prevented from entering the air chamber by means of four oil seals, one at each end of the two rotor shafts. These are a standard size proprietary seal of the leather variety having a very long service life. Nevertheless, when replacement becomes necessary it is no great job for these to be replaced with new ones. The seals themselves are not allowed to bear on the shaft, instead, replaceable bushes are fitted, on the polished surfaces of which, the leather sealing elements of the oil seals make contact. These also are replaced when new oil seals are fitted, thus ensuring that the original sealing efficiency is restored. After the oil has circulated through the gears and bearings (integral oilways lead a supply to the bearings at the rear of the compressor) any excess above a certain predetermined sump level is allowed to drain away by means of a copper pipe into the engine via the tappet cover which is provided with a suitable union.

The following test data and performance figures relating to the TC model M.G. Midget are typical of results which have been obtained from these cars and represent average results from standard road tests. They fairly represent and clearly show the increase in output that can be obtained and the corresponding improvement in road performance that may be expected when the engine is fitted with this particular type of low-pressure supercharger installation.

Standard Form.—Maker's rating 54.4 B.H.P. at 5.200 R.P.M.

Supercharged.—1¼ in. carburettor and supercharger driven at 1.29 engine speed to produce 6 lbs. P.S.I. Maximum boost (70-72 octane fuel).

(*a*) See Power Curve below.

(*b*) See Acceleration Times also below.

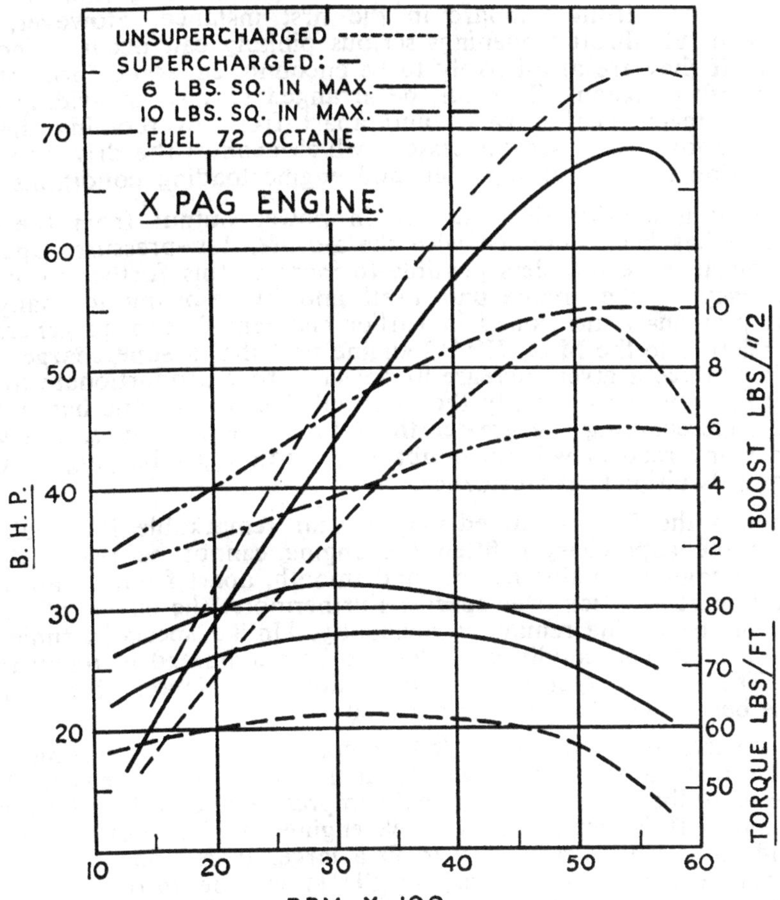

Fig. 1. *Showing Power Curve for Supercharged XPAG Engine*

ACCELERATION TIMES THROUGH GEARS.

Unsupercharged		*Supercharged*	
0-30 m.p.h.	6.5 secs.	0-30 m.p.h.	5.0 secs.
0-50 „	15.3 „	0-50 „	10.8 „
0-60 „	23.8 „	0-60 „	15.2 „
0-70 „	44.5 „	0-70 „	25.4 „

These figures represent the increase which is obtained by fitting a Marshall-Nordec low-pressure Supercharger Installation boosting at 6 lbs. P.S.I. on a standard M.G. XPAG engine unit. The power absorbed by the supercharger to produce this boost pressure is in the region of 5 BHP, but it is interesting to note that a proportionate of this energy is returned to the engine in the form of pressure in the cylinder during the normal induction cycle and is expended to useful purpose.

In the above examples it will be noted that fuels the equivalent to British Pool have been used which have octane ratings as shown. The ignition advance requirements of an engine fitted with a supercharger differs from standard, after the first initial advance it is necessary, as boost increases, for the ignition to be slightly retarded. Provided engine revs are high enough this setting does not become critical if the ignition has been set slightly retarded from standard in the first instance. However, at slow R.P.M. with full throttle openings serious pinking can occur. Under these conditions if they are at all likely to be encountered, some form of manual control is advantageous. This can be arranged to give overriding effect on the normal mechanical advance and apart from overcoming the trouble mentioned above, is a useful accessory which enables the driver to instantly adjust ignition to suit varying fuels and engine loading conditions.

Although a considerable increase in power output from the standard engine unit has been obtained with the aid of a low-pressure supercharger installation, it is nevertheless possible to increase this further by improving the efficiency of the engine unit itself and by resorting to many of the methods previously mentioned in earlier chapters. It can be generally said that in relation to the M.G. XPAG engine unit that a supercharger installation will produce a given increase in power output proportionate to original engine efficiency, consequently the more efficient the engine has been made prior to supercharging the greater the ultimate performance, and with this in mind many readers will no doubt look upon supercharging as a further stage towards ultimate performance.

Although the figures quoted may appear remarkable it is pointed out that with the supercharger fitted the engine cannot be regarded as race tuned. The engine remains flexible and smooth, apart from a barely noticeable hunting at low tick-over speeds. Furthermore, the effects of the supercharger on the engine remain permanently. Unlike specialist tuning which may result in critical conditions having to be maintained to retain increased output. The supercharger does in fact allow considerably longer intervals to elapse before decarbonising is required.

For owners primarily interested in competition and sporting events it is possible to obtain further power from the engine by increasing the boost pressure to 10 lbs. P.S.I. The identical compressor is used but the drive ratio from crankshaft is stepped up to 1.48 engine speed and is driven by twin belts and pulleys. It is necessary to fit a larger inlet elbow and to provide an alternative larger carburettor, an SUH4 can be recommended. In all other respects the installation is identical with the standard low-pressure installation previously referred to, but it will be seen from the power curve, Fig. 1, that a considerable power increase is obtained in this way, although in fairness to the engine it is desirable to use, wherever possible, a reasonable fuel of 80 octane or higher in order to prevent excessive pinking, and the installation whilst functioning in precisely the same manner as the standard low-pressure installation is primarily designed for competition use as distinct from normal touring.

Still further advantages are possible if an alternative back axle ratio is available, and in the case of the TC and TD these are available. Ratios of 4.875 may be fitted to both TC and TD models and for the latter an additional ratio of 4.5 can be supplied. With the increased torque available acceleration is not impaired by fitting one of these alternatives. Considerable advantages are gained in petrol consumption and normal maximum speed is possible at below the recommended maximum R.P.M.

THE SHORROCK SUPERCHARGER.

INTRODUCTION.

The design of the installation set is such that the work may confidently be carried out by any competent garage mechanic or owner driver.

The supercharger is mounted on the manifold side of the engine and is driven by twin vee-belts from the front end of the crankshaft.

FITTING THE SUPERCHARGER. (TB., TC. and TD. Midget.)

Remove the bonnet complete and drain the cooling system by opening the drain tap in the bottom of the radiator and in the right hand side of the cylinder block.

Remove the carburetter controls, air cleaner, carburetter and inlet manifolds. The air cleaner is not replaced as its capacity is not sufficient for the supercharged engine.

Slacken the dynamo mounting and adjusting bolts and remove the fan belt.

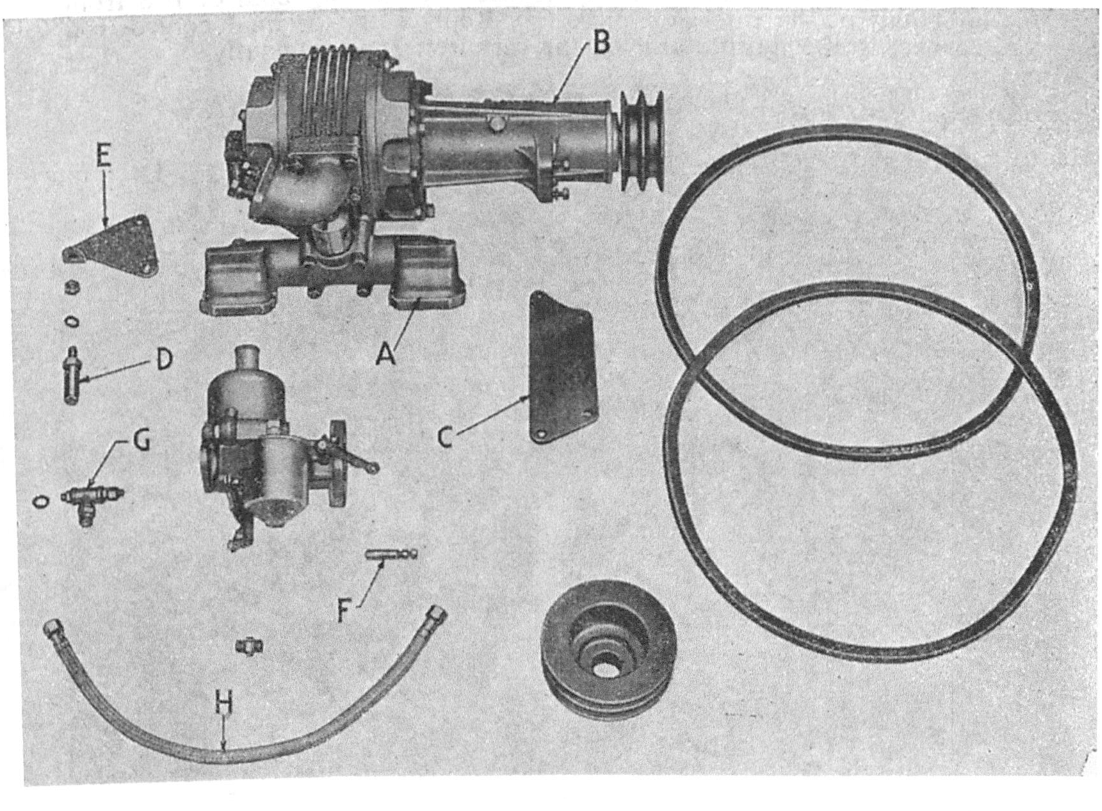

Fig. 2. *Shorrock Supercharger Kit*

Remove the starting handle dog nut and remove pulley from front end of the crankshaft. On the TB and TC Midgets, it will be necessary to remove the front engine mounting bolts and slightly raise the front of the engine before the crankshaft pulley can be removed.

Remove the bolts securing the right hand side radiator stay tube to the radiator and engine bulkhead and remove stay tube. This is not replaced.

Fit the new three groove pulley, placing the existing belt and the two new belts over the pulley before fitting. The pulley itself is made from cast iron and extreme care should be exercised when fitting as undue force may cause considerable damage. Replace the starting handle dog nut.

Fit the supercharger complete with the inlet manifold (A) and driving housing (B) both inlet and exhaust manifold being clamped by the existing clamps. The front plate (C) is fitted between the lug on the underside of the drive housing and the two bolts holding the square flange of the water pipe to the cylinder head.

Remove the cylinder head nut behind the rear inlet port and replace with the barrel nut (D) the back stay plate (E) being bolted to this and two of the supercharger end plate studs.

Fit the dynamo drive belt and the two supercharger drive belts over the supercharger pulley. It may be found necessary to cut away a small section of the radiator header tank support web to accomplish this.

Fit the carburetter to the flange on the supercharger and reconnect the existing jet control wire in its original position.

Connect the petrol pipe by linking the two existing pipes with the double ended union supplied.

To connect the throttle control, remove the existing control rod from the lever and fit the extension piece (F). Replace the original control rod and connect to the throttle arm on the carburetter, as originally.

Fig. 3.. *Shorrock Supercharger Installation Fitted to TD Series Midget*

LUBRICATION.

The supercharger lubrication is obtained from the engine oil supply by replacing the rocker oil feed banjo on the cylinder head with the union and tee piece (G).

The banjo on the oil feed pipe is removed and replaced with the union nut and nipple supplied attached to the tee piece.

The oil feed pipe to the overhead valve gear is connected to one branch of the tee piece, the other branch being connected to the supercharger by the flexible oil feed pipe (H).

Remove the oil filter plug from the top of drive housing and partially fill the drive housing with a good grade of 30 viscosity engine oil. The level of the oil is governed by the plug on the underside of the housing.

For Servicing and fitting of the spare pin (if necessary) see details under: SERVICING, LUBRICATING, below.

Adjust the dynamo and securely tighten in position.

Close the drain taps and refill system with water.

Refit bonnet and test car on road in order to obtain the correct carburettor and ignition settings.

SERVICING INSTRUCTIONS.

The supercharger is a precision mechanism, and although fully tested before despatch, requires running in and it should be driven gently for the first 250 miles. During the running in period the maximum road speed in top and third gears should not exceed 45 MPH and 30 MPH respectively.

Dismantling and major overhauls should only be undertaken by authorised Distributors, or Dealers but for preference the supercharger should be removed as a complete unit and returned direct to Shorrock Superchargers, Ltd., Fletchamstead Lane, Coventry, ENGLAND.

Lubrication.

It is essential that the lubrication oil be kept clean and free from sludge as dirty, unsuitable or graphited oil will clog the lubricating passage and eventually lead to serious damage.

When despatched the supercharger is fitted with a .309 in. diameter pin (E) in its lubricator. This should prove satisfactory for initial and permanent running. There should be a slight amount of smoke from the exhaust when starting up from cold, but this should clear after a few minutes running.

Should, however, signs of pronounced over-oiling arise (for instance clouds of exhaust smoke), then the lubrication pin should be replaced by the larger pin (C) having a diameter of .310 in. and supplied in the plastic capsule bolted to the supercharger.

If the larger diameter pin is used be sure to tighten down firmly the supercharger nuts after detaching the capsule, and, after fitting the new pin, check that the supercharger is getting a supply of oil.

Every 1,000 miles remove the lubricator pin and wipe with a soft rag. On no account should abrasives be used.

To remove and replace the lubricator pin proceed as follows:—

Superchargers with external lubricator (of brass). Unscrew the plug at the end of lubricator further from the shaft, when the plug, spring and pin will come out together.

To re-assemble, assemble pin, spring and plug together, insert, and screw down plug firmly.

For superchargers with the lubricator inside the drive shaft, unscrew the pipe union at the end of the supercharger, when the spring will push out the pin.

To reassemble insert spring first, then pin and then screw down pipe union firmly.

It must be remembered that the supercharger drive housing is not connected to the engine oil supply. The level should be checked every 1,000 miles.

NOTE:—The early type superchargers had a separate oil tank which was fitted to the engine bulkhead. The standard oil metering pin was .304 in. diameter with an alternative pin provided, .306 in. in diameter. This type of supercharger can, however, be very easily modified to the later type of supercharger now available.

THE WADE SUPERCHARGER. (TB, TC, TD Midget.)

DESCRIPTION.

The Rootes type of supercharger manufactured by the Wade Engineering Co., Ltd., Gatwick Airport, Horley, Surrey, England, consists basically of a casing, ported either side, containing two contra-rotating rotors driven from the engine. The mixture from the carburetter is carried round by the lobes of the two rotors, and expelled through the outlet port to the engine. With the delivery from the supercharger exceeding the swept volume of the engine pressure is built up, ensuring efficient filling of the cylinders and equal distribution of the mixture. With this construction, there are no sliding vanes or contacting parts, and whilst the rotors do run in very close proximity to each other, there is no actual contact.

The combination of the Helical port and four lobe rotors, with close operating clearances ensure a pulsation free delivery with the absolute minimum of operating noise. The frictionless oil seals have unlimited life, and the combination of steel to bronze gears provide long life and silent running.

The supercharger is mounted on a special cast inlet manifold, incorporating a blow-off safety valve. Independent mounting of the supercharger pulley on an outrigger bracket, with mechanical couplings between the shaft and pulley, relieve the rotor bearings of any load due to belt pull. Oil for lubrication of the rotor bearings and the gears is tapped from the main engine oil supply. The supercharger in standard form delivers at a maximum pressure of 4.75 lbs. per sq. in. above atmospheric.

FITTING THE SUPERCHARGER.

The supercharger is fitted on the induction side of the engine being driven by twin V-belts from the front of the crankshaft, and running at 1.1 engine speed. Remove carburetter controls, disconnect the exhaust pipe from the manifold, remove carburetters and induction and exhaust manifolds.

Drain the cooling system, remove all hoses and remove the radiator complete.

NOTE: On the TD Model, remove also the cylinder head water outlet pipe and thermostat, and fit the replacement casting supplied with the supercharger kit. The length of convolute rubber hose is for connecting the alternative elbow to the radiator header tank, as the thermostat is *not* replaced.

Slacken off the engine mountings and jack up the front of the engine.
Remove the pulley from the crankshaft and fit replacement pulley supplied in the kit. Lower the front of the engine and secure engine mountings.

NOTE: On the TD Midget reposition the horn alongside the petrol pump and lengthen the wire accordingly.

Fit the main supercharger assembly, fitting the supercharger assembly and exhaust manifold in position together. Check the gaskets between the manifolds and engine, replacing if there is any doubt as to their condition. Leakage between manifold and engine will result in loss of pressure and the smell of petrol in the car.

Fit the clamps and holding nuts and tighten up. Fit exhaust pipe to exhaust manifold. Fit the outrigger bearing assembly. The supporting plate is held by the two bolts securing the water outlet pipe on the left-hand side, looking from the front of the car. Remove these two bolts and offer the

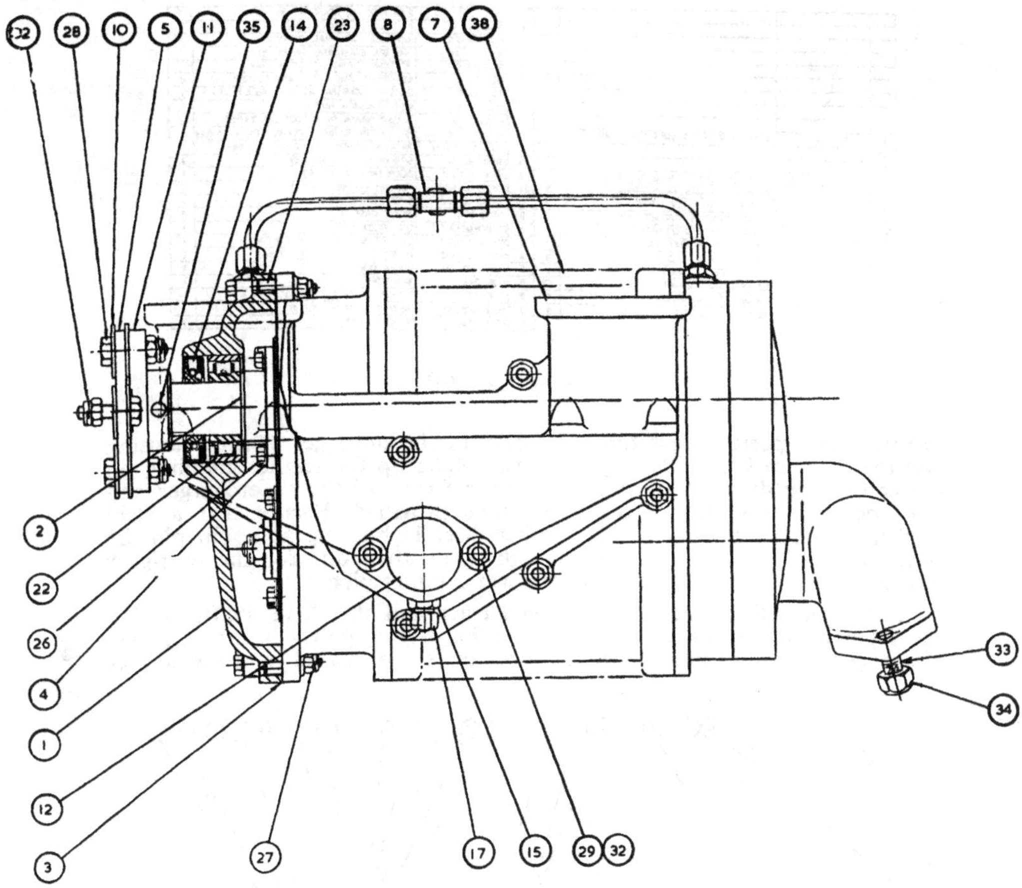

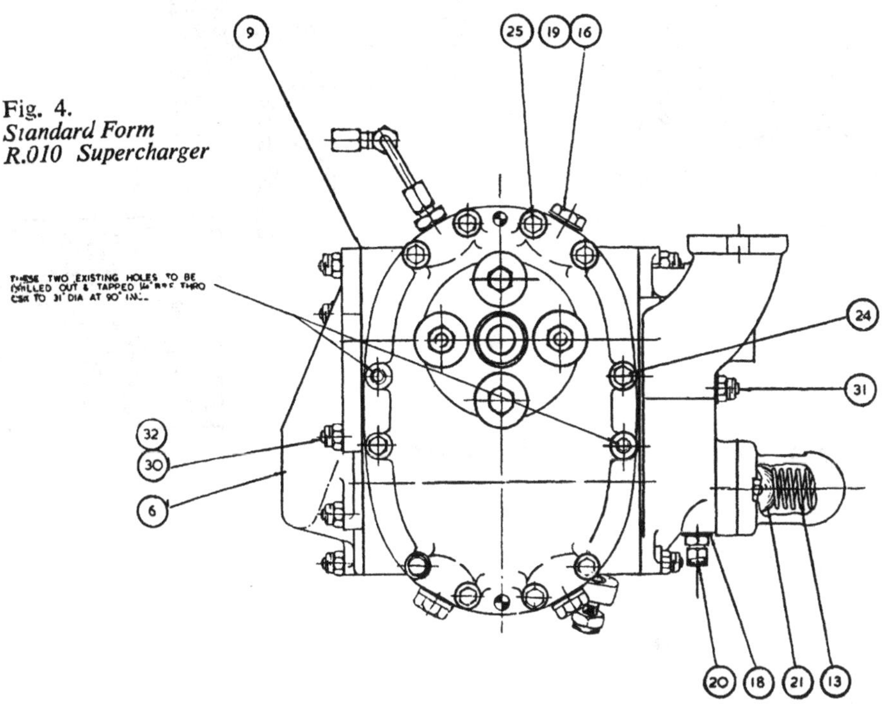

Fig. 4.
Standard Form
R.010 Supercharger

				24		2BA HEX HD BOLT X 1 1/8" LONG	2
				23	WGS 406-G	1/4" DIA X 5/8" LONG DOWEL PIN	2
				22	WGS 209-25	ROLLER JOURNAL BEARING	1
				21	WGS 115-28	7/8" DIA STANDARD STEEL BALL	1
				20	WGS 115-9	9/32" DIA. STANDARD STEEL BALL	1
				19	WGS 113-D	SEALING WASHER TO SUIT 1/4" BSP	4
				18	WGS 113-C	SEALING WASHER TO SUIT 7/16" X 19 T.P.I.	1
				17	WGS 111-B	1/8" BSP UNION NUT	1
				16	WGS 108-D	1/4" BSP BRASS PLUG (COLLAR TYPE)	2
38	2ROIO	STANDARD FORM SUPERCHARGER	1	15	WGS 103-BC	7/16" X 19 T.P.I TO 1/8" BSP DOUBLE ENDED UNION	1
37				14	2001 TYPE 'N'	3/8" WIDE CHROME LEATHER OIL SEAL	1
36				13	2ROIO-60-315	SPRING (RELIEF VALVE)	1
35		5/32" X 1 1/2" LONG - 1 IN 48 TAPER PIN	1	12	2ROIO-60-314	BODY (RELIEF VALVE)	1
34		3/8" BSF PHILIDAS INDUSTRIAL NUT	2	11	2ROIO-60-308	FLANGE (DRIVE SHAFT)	1
33		3/8" BSF STUD X 1 1/2" LONG	2	10	2ROIO-59-288	SPECIAL WASHER	8
32		1/4" BSF PHILIDAS TURRET NUT	18	9	ROIO-256	PORT GASKET	2
31		1/4" BSF STUD X 2 1/2" LONG.	1	8	2ROIO-60-249	OIL FEED PIPE	1
30		1/4" BSF STUD X 1 1/4" LONG.	11	7	2ROIO-60-233	OUTLET ELBOW (SCHR TO ENG.)	1
29		1/4" BSF STUD X 1" LONG	2	6	2ROIO-60-232	INLET ELBOW (CARB TO SCHR)	1
28		1/4" BSF HEX. HD. BOLT X 1" LONG	4	5	2ROIO-60-229	DRIVE DISC.	1
27		2BA PHILIDAS TURRET NUT	10	4	ROIO-226	TABWASHER	2
26		2BA HEX HD. SETSCREW X 0·6" LONG.	6	3	ROIO-223	GASKET (FRONT COVER TO HSG.)	1
25		2BA HEX. HD. BOLT X 1 1/2" LONG.	8	2	2ROIO-60-222	DRIVE EXTENSION PIECE	1
				1	2ROIO-200	FRONT COVER	1

(Key to Fig. 4 opposite)

assembly into position. Fit the ¼ in. stays to the outrigger bearing housing and the front casing of the supercharger. Make up the leather coupling on the end of the drive shaft with the driving flange of the supercharger and bolt together using the ¼ in. B.S.F. bolts provided. Ensure that a steel washer is fitted either side of the leather disc. Tighten up and split pin the securing bolts. Replace but do not tighten up the bolts for the outrigger bearing housing mounting plate to the water outlet pipe.

Check the alignment of the supercharger and the drive shaft, turning the supercharger by hand to ensure that the driving pulley is running true. If not, gently tap the pulley with a rubber or leather mallet until the correct

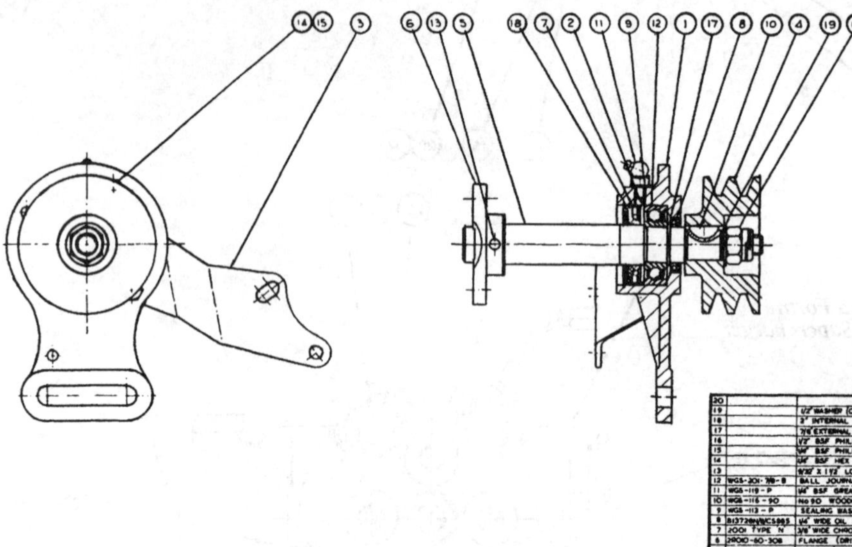

REF	PART No	PART NAME	QTY
20			
19		1/2" WASHER (OLD UNIT ST	1
18		2" INTERNAL CIRCLIP	1
17		7/8" EXTERNAL CIRCLIP	1
16		1/2" BSF PHILIDAS TURRET NUT	1
15		1/4" BSF PHILIDAS TURRET NUT	2
14		1/4" BSF HEX HD BOLT X 3/4" LONG	2
13		5/32" X 1 1/2" LONG - 1 IN 48 TAPER PIN	1
12	WGS-201-7/8-B	BALL JOURNAL BEARING	1
11	WGS-119-P	1/4" BSF GREASE NIPPLE	1
10	WGS-116-90	No 90 WOODRUFF KEY	1
9	WGS-113-P	SEALING WASHER TO SUIT 1/4" BSF	1
8	B13720N/U/CS005	1/4" WIDE OIL SEAL	1
7	2001 TYPE 'N'	3/8" WIDE CHROME LEATHER OIL SEAL	1
6	2ROIO-60-308	FLANGE (DRIVE SHAFT)	1
5	2ROIO-60-304	SHAFT (SCHGR DRIVE)	1
4	2ROIO-60-301	SUPERCHARGER PULLEY	1
3	2ROIO-60-235	FRONT SUPPORT BRACKET	1
2	2ROIO-60-234	DISTANCE PIECE	1
1	2ROIO-60-231	DRIVE HOUSING	1

Fig. 5. *Supercharger Drive Sub-Assembly*

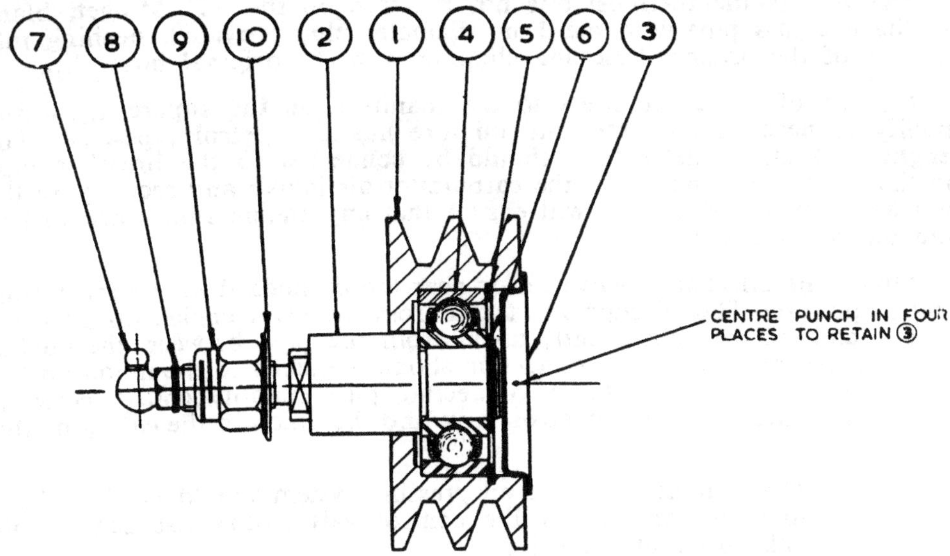

10		1/2" WASHER (OLD WHIT STD)	1
9		1/2" BSF PHILIDAS TURRET NUT	1
8	WGS 113 - P	SEALING WASHER TO SUIT 1/4" BSF	1
7	WGS 120 - P	1/4" BS.F. GREASE NIPPLE	1
6		3/4" EXTERNAL CIRCLIP	1
5		2" INTERNAL CIRCLIP	1
4	WGS 202 - 3/4 - B	BALL JOURNAL BEARING	1
3	2R010 - 60 - 306	COVER (JOCKEY PULLEY)	1
2	2R010 - 60 - 305	SHAFT (JOCKEY PULLEY)	1
1	2R010 - 60 - 302	JOCKEY PULLEY	1

Fig. 6.
Jockey Pulley Sub-Assembly

alignment is attained. The shaft must be correctly trued before the outrigger housing is finally tightened up.

The securing bolts for the mounting plate should be tightened up and the stays securing the housing and blower casing adjusted. Adjustment should be so carried out to make sure the outrigger bearing housing stays square otherwise the jockey pulley will run out of truth, leading to rapid belt wear.

Remove the four bolts securing the fan to the waterpump pulley, and withdraw pulley after removing the split pin and securing nut. Fit the replacement pulley provided. As the split pin in the fan spindle cannot be replaced, fit the $\frac{5}{16}$ in. shakeproof spring washer provided underneath the original castellated nut. Securely tighten the nut and refit the fan.

Fit the driving belts. The original belt is fitted as before, with the two new belts around the extra grooves in the crankshaft pulley, the extra grooves in the waterpump pulley and finally over the supercharger pulley.

Fit the jockey pulley, engaging with the slot in the outrigger bearing housing and adjust pulley to keep the belts tensioned. Refit radiator and connect up water hoses.

NOTE: As the thermostat is not replaced on the TD Midget, blank off the bye-pass pipe with the 1 in. diameter plug provided, pushing into the end of the original hose and clipping with the original hose clip.

Fit one of the carburetters to the manifold on the supercharger and modify as necessary, throttle and mixture linkages, petrol pipes, etc. The length of ⅜ in. diameter hose should be connected to the breather pipe on the valve cover and led to the carburetter air intake and secured by the bolt and clip provided. This will ensure that any engine fumes are sucked into the carburetter air intake.

Mount the oil control valve on or near the dashboard in the most convenient position. The oil supply is taken from the main engine oil pressure system, and can be conveniently taken from the pipe between the engine and oil pressure gauge. This connection should be taken to the union on the end of the control valve. Fit a connecting pipe suitably coiled, between the union on the side of the control valve and the union in the casing of the supercharger.

NOTE: The connection to the oil pressure system should *not* be joined up to the union on the side of the control valve, otherwise serious oil leakage will take place at the valve.

Remove the two brass plugs from the top of the supercharger and distribute a quarter of a pint of engine oil between the front and rear housings, replace and securely tighten both plugs. Start the engine and set to run at a fast tick-over. Slacken the connection of the oil supply pipe to the top of the supercharger, depress the control valve until the oil flows from the end of the pipe and reconnect pipe to supercharger.

Three spare needles are provided with the kit, DR being the richest, DQ the medium and DP the weakest. Test the car with the richer needle fitted, and should this prove to be excessively rich after the normal adjustment to the jet nut has been carried out, replace with a weaker needle.

Erratic running (lack of slow-running and pronounced flat spot on acceleration) denotes excessively weak mixture, probably caused by air leak. Check all manifold joints, oiling unions and plugs to see if they are tight. Only by correct carburation can full benefit be obtained from the supercharger.

MAINTENANCE.

Lubrication.—As there are no contacting parts no lubrication is required in the supercharger case. The end housings, however, contain the bearings and in the front, the gears. The amount of oil required by these components is negligible and they are lubricated by splash feed from the oil contained in these end housings. The oil level is maintained by "topping-up" from the tapping in the main engine supply. "Topping-up" is carried out by depressing the push button oiling valve for 15-20 seconds every 150 miles with the engine idling. Under no circumstances should the "topping-up" operation be neglected otherwise noisy running will be experienced.

Greasing.—The drive shaft outrigger bearing should be greased every 500 miles.

Care of V-Belts.—These should be checked periodically, about every 500 miles to ensure correct tension, otherwise belt slip will take place. They should be kept appreciably tighter than the normal dynamo belt. Adjustment is effected by moving the jockey pulley towards or away from the belt.

THE ARNOTT SUPERCHARGER.

The Arnott low pressure supercharger has as its primary object the supplying of petrol/air mixture to the engine under pressure so that 100% filling is effected, thereby overcoming the inherent volumetric efficiency associated with the unblown engine.

The capacity of the supercharger together with the speed at which it is driven are chosen so that approximately 33⅓% greater volume is available to the engine. This percentage is closely reflected in "before and after" performance figures, or differently expressed, the power/weight ratio is improved by 33⅓%. In addition to the improvement in acceleration the benefits of a well atomised and mixed gas is very apparent during part throttle conditions which is reflected in fuel economy. The engine also runs much more smoothly giving the feel of a six cylinder engine to a four cylinder one. The reduced atmospheric pressure at high altitudes causes a drop in the efficiency of any normally carburated engine. This loss can very easily be overcome by the use of a low pressure supercharger.

The use of V-belt drive and the freedom from any gearing within the supercharger, combined with well designed ports ensures quietness of operation.

LUBRICATION.

In the Arnott supercharger all vane and shaft journals run on ball races, and all races and components are lubricated via the hollow vane shaft to which is fed oil from a separate quart size oil tank. Oil passes to the supercharger through a calibrated orifice situated in the tank, being air bled from within the tank to ensure the free delivery of oil through the pipe lines and also to prevent any syphoning action. This calibrated orifice requires no attention as it is amply protected by gauze of a much finer mesh than would permit a blockage. A pressure balance pipe conveys the manifold pressure to the interior of the oil tank, the oil supply to the supercharger being based upon the fact that there is always less pressure within the rotor than in the manifold. This is so whether the throttle is closed or open, and therefore the inside of the tank is variably pressurised at all times in accordance with the conditions prevailing. The oil feed pipes are internally armoured P.V.C. tubing, thus providing flexibility between supercharger and tank. These should never be replaced by larger bore or rubber lined pipes, or the even delivery of the oil will be impaired.

THE ARNOTT LOW PRESSURE INSTALLATION FOR TB, TC AND TD SERIES MIDGET.

The Arnott supercharger, which is manufactured by Carburettors, Ltd., Grange Road, London, N.W.10, England, is a vane type pump, having an inbuilt feature which enables pressure to rise within the supercharger before transmitting to the manifold, this being attained by the late opening of the outlet port. Not only are pumping losses avoided, but this has a strong influence in preventing surge. A further reduction in power required to drive the vane type supercharger is due to progressive reduction of vane area exposed to pressure as compression rises within the supercharger.

The Model 1600 supercharger is used in the low pressure installation shown in Fig. 7. Kits are supplied with all parts required for the complete conversion and fitting instructions are enclosed to enable any competent amateur mechanic to do the work himself.

A special manifold replaces the twin carburetters and the supercharger mounts directly on the centre. The extended drive shaft is directly coupled to the supercharger, requiring no universal joints and, subsequently, no servicing. A grease pump connection is provided for the roller race drive pulley bearing.

Fig. 7. *Arnott Supercharger Installation*

In addition to the supercharger pulley two other pulleys are provided. The triple-V-cast iron crankshaft pulley replaces the standard one providing the twin drive to the supercharger. The light alloy fan pulley replaces the original giving clearance to the two supercharger belts. When fitting the fan pulley care must be taken to see that the ends of the 5 mm. bolts securing the fan do not protrude through the flange.

The front of the supercharger extended drive, which is encased in a light alloy housing, is supported from the front cylinder head stud and by a vertical stay from the front engine plate. The underside of the header tank support will need to be reduced slightly to enable easy belt changes, and the radiator support tube is refitted lower down the side of the radiator using the plate and spacers, etc., supplied.

MAINTENANCE.

The servicing of the supercharger installation is limited to three points:—

1. The oil supply requires replenishing every 800/900 miles.
2. The replacement of drive belts after long periods.
3. Application of a grease gun to the front bearing greaser when normal car servicing is being carried out.

It will be appreciated that the passage of oil through the supercharger, using the Arnott system described above, ensures a constant supply of upper cylinder lubricant giving a long life to the cylinder bores.

CARBURATION.

The front one of the original S.U. carburetters can be used with the standard Arnott installation. The jet control lever is moved to the outside and a special control, on the TD Midget set, is fitted to enable the opening of the throttle when starting. The standard throttle control link can be used with the extension clamp link provided. A double ended union connects the petrol pipes. The float chamber of this carburetter faces the *outside.*

Performance at the " top-end " can be increased by the fitting of the 1½ in. diameter S.U. carburetter, for which a special elbow can be supplied. As further alternatives, special Arnott carburetters are available, these being fitted directly to the inlet port of the supercharger, eliminating the use of an elbow and ensuring freedom from fuel vaporisation, difficult cold starting and giving improved performance throughout the range. Two sizes are available, the 1¼ in. diameter normally supplied and the 1⅝ in. diameter used for competitions and improved " top-end."

AIR CLEANER.

Unless the atmosphere is very dry and dusty, it is preferable to remove the air cleaner, as the capacity is insufficient for the efficient breathing of the engine at high revolutions. When the climate demands, however, it should be fitted and it is possible to enlarge the effective area of the filter.

IGNITION.

With a supercharged engine there is a distinct advantage in locking the automatic advance and retard mechanism fitted to the distributor. Removal of the distributor cover and the contact breaker plate, will expose the centrifugally operated weights, which are controlled by two springs hooked into a hole in each weight. There are usually additional holes into which the springs should be hooked. Where there are slots instead of holes, a small tag should be soldered across the weights to stop them moving. After replacing the contact breaker plate, etc., the ignition should be set to give maximum advance.

COMPETITION USE.

Although the Model 1600 can be speeded up to give greater boost, the efficiency at higher revolutions shows a definite tendency to fall away. Model 2800 is therefore specified for competition purposes and boost pressures of between 10 lbs. and 15 lbs. per sq. in. are easily available by simply changing the pulley sizes. The Arnott 1⅝ in. or the 1½ in. or 1¾ in. S.U. carburetters are used, and special jets and needles can be obtained from the respective manufacturers to enable Methanol mixtures to be used.

from the 'M'
MG
to the 'TF'

SERVICE DATA — MODEL "M" MIDGET — 1929 to 1932

Chassis Weight	8 cwts. 0 qrs. 23 lbs.
Weight Complete Car Fabric 2 str.	10 cwts. 0 qrs. 14 lbs.
Fabric Coupe	11 cwts. 2 qrs. 14 lbs.
Ground Clearance	6".
Wheelbase	6' 6".
Track	3' 6".
Water Capacity	16 pints.
Turning Circle	34'.
Steering Knuckle Angle	9°.
Castor Angle	3°.
King Pin Angle	6½°.
Toe-In	⅛".
Brake Drums, Front	8" dia
Brake Linings	Halo.
Standard Gear Ratio	9/44 4.891 : 1.
Oil Capacity, Rear Axle	1½ pints.
Diff. Gear	2 Star.
Brake Drums, Rear	8" dia.
Brake Linings, Rear	Halo.
Brakes, early models	Rod and Cable.
later models	Cable.
Steering	Adamant.
Steering Wheel	Bluemel.
Silencer	Burgess.
Petrol Tank	6 glns., incldg. 2 in res.
Wheels	3½" × 27".
Tyres	4.00" × 27" Dunlop.
Shock Aborbers, Front and Rear	Hartfords.
Tyre Pressures	27 lbs./sq. in. front and rear.
Battery	(One) Lucas 6 v. S.T.W.13E.
Amp. Hr. Cap. 10 Hr. Rate	75.
Speedometer	Smith's 3" dia.
Speedo. Reduction Box	Nil.
Petrol Gauge	Nil.
Contact Breaker Points	.015" — .020".
Engine Type	4 Cylinder, O.H.V. Camshaft.
R.A.C. Rating	8.05 H.P.
Bore, Stroke and Capacity	57 mm. × 83 mm. 847 cc.
Compression Ratio	5.4 : 1.
Cubic Capacity of Cylinder Hd.	43.5 cc.
Cubic Capacity of Gasket	4.5 cc.
Brake Horse Power	20 at 4,000 r.p.m.
M.P.H. per 1,000 R.P.M.	16.4 in top gear.
Firing Order	1 3 4 2

Ignition	Coil and Distributor. 6 volt.
Ignition Timing	Set at T.D.C. fully retarded, Points just breaking.
Sparking Plugs	Champion No. 7. K.L.G. M. 50. Lodge C. 3
Sparking Plug Gap	.020".
Tappet Clearance	.004".
Valve Timing Up to Engine No. 2023	Inlet Opens 9° A.T.D.C. Closes 35° A.B.D.C. Exh. Opens 43° B.B.D.C. Closes 7° A.T.D.C.
From Engine No. 2024 on	Inlet Opens at T.D.C. Closes at 50° A.B.D.C. Exh. Opens 43° B.B.D.C. Closes 7° A.T.D.C.
Carburettor	Single 1⅛" S.U. Horizontal.
Carburettor Needles	Standard M.5. G.2. Richer M.6. Weaker No. 2.
Piston Clearance	.0025" — .0035".
Connecting Rod	Steel Stamping. White Metal lined.
Length of Big End Bearings	1.12".
Dia. of Big End Bearings	1.50".
Main and Big End Clearances	.0015" — .0025".
Cam Lift	.20".
Valve Lift (with Rocker No. 173451)	.282".
Length from Valve Guide seat to Valve Port, top	2.5625".
Length from Boss in Port to top of Valve Port	1.40625".
Sump Capacity	8 pints.
Recomm. Engine Oil	N.O.L."30".
Oil Pressure (approx)	50-70 lbs./sq. in.
Maximum Revs.	5,000 r.p.m.
Clutch	Single Plate, Dry.
Clutch Springs	6 per set.
Gearbox	Manual, 3 Forward and 1 Reverse.
Gearbox Ratios	Top. 2nd. 1st Rev. 1 : 1 1.83 : 1 3.5 : 1 2.83 : 1
Overall Ratios	4.89:1. 8.95:1. 17.115:1. 13.84:1.
Oil Capacity	1½ pints.
Recommended Oil	N.O.L. "EP" Transmission Oil 140.

Model M Midget

SERVICE DATA — MODEL "F" MAGNA — 1931 to 1932

Chassis Weight	10 cwts. 2 qrs. 21 lbs.
Weight, Complete Car, Saloon	16 cwts. 0 qrs. 21 lbs.
4 Str.	15 cwts. 1 qr. 0 lbs.
Ground Clearance	6".
Wheelbase	7' 10".
Track	3' 6".
Turning Circle	Early cars 36', later 34'.
Steering Knuckle Angle	9°.
Castor Angle	6°.
King Pin Angle	6½°.
Toe-In	3/16".
Brake Drums, Front	Early cars 8", later 12".
Brake Linings	Ferodo M.R.
Water Capacity	19 pints.
Standard Gear Ratio	9/43. 4.78 : 1.
Oil Capacity	1½ pints.
Diff. Gear	2 Star.
Brake Drums, Rear	Early cars 8", later 12".
Brake Linings, Rear	Ferodo M.R.
Brakes	Cable.

Steering	Special.
Steering Wheel	Bluemel.
Silencer	Burgess.
Petrol Tank 2 Str.	12 gallons (including 3 reserve).
Saloon & 4 Str.	6 gallons (including 2 reserve).
Wheels	2.50" × 19".
Tyres	4.00" × 19".
Shock Absorbers, Front and Rear	Hartfords.
Battery	Lucas 12 volt.
Amp. Hr. Cap. 10 Hr. Rate	50.
Speedometer	Smith. MN and PC. 3"
Speedo Reduction Gear	Nil.
Petrol Gauge	Nil.
Contact Breaker Points	.015" — .020".
Engine Type	6 cylinder. O.H.V. Camshaft.
R.A.C. Rating	12.08 H.P.
Bore, Stroke and Capacity	57 mm. × 83 mm. 1,271 cc.

Model F Magna

(Continued on following page

SERVICE DATA **MODEL "F" MAGNA** **1931 to 1932** ***Continued.***

Compression Ratio	5.7 to 1.
Cub. Cap. Cylinder Head	40.22 + Gasket 4.5 cc.
Brake Horse Power	37.2 at 4,100 R.P.M.
M.P.H. per 1,000 R.P.M.	16.7 in top gear.
Firing Order ..	1 4 2 6 3 5.
Ignition	Coil and Distributor.
Ignition Timing ..	T.D.C. Fully Retarded, Points just breaking.
Sparking Plugs ..	Champion.
Sparking Plug Gap	.020".
Tappet Clearances ..	.004" inlet and exhaust.
Valve Timing Inlet.	Opens 9° A.T.D.C.
	Closes 35° A.B.D.C.
Exh.	Opens 43° B.B.D.C.
	Closes 7° A.T.D.C.
Carburettors ..	Twin 1" S.U. Horizontal.
Carburettor Needles	Standard M.5.
	Richer No. 1.
	Weaker M.6.
Piston	Aerolite.
Piston Clearance ..	.0015" — .0025".
Connecting Rod ..	Steel Stamp. White Metal Lined.
Dis. and Lgth. of Big-End Bearings	1.50" dia. × 1.125".
Main and Big End Brg. Clearance ..	.0015" — .0025".
Sump Capacity ..	12 pints.
Recommended Engine Oil	N.O.L. "30" Engine Oil.
Oil Pressure	50 - 70 lbs./sq. in. approx.
Maximum Engine Revs.	4,500 r.p.m.
Clutch	Single Plate, Dry.
Clutch Springs ..	6 per set.
Gearbox	E.N.V. Manual. 4 Forward speeds and 1 Reverse.
Gearbox Ratios Top	1 : 1.
3rd	1.36 : 1.
2nd	2.00 : 1.
1st	4.02 : 1.
Rev.	4.02 : 1.
Gearbox Overall Ratios Top	4.78 : 1.
3rd	6.4008 : 1.
2nd	9.56 : 1.
1st	19.2 : 1.
Rev.	19.2 : 1.
Gearbox Capacity ..	2 pints.
Recommended Gear-Box Oil	N.O.L. "E.P." Transmission Oil 140.

Model F Magna

SERVICE DATA **MODEL "D" MIDGET** **1931 to 1932**

Chassis Weight ..	9 cwts. 1 qrs. 21 lbs.
Weight of Complete Car 4 Str. Tourer	13 cwts. 1 qr. 0 lbs.
2 door Saloon	14 cwts. 2 qrs. 14 lbs.
Ground Clearance	6".
Wheel Base ..	First 100: 7' 0". Remainder 7' 2".
Track	3' 6".
Turning Circle ..	34' 0".
Water Capacity ..	16 pints.
Steering Knuckle Angle	9°.
Castor Angle ..	6°.
King Pin Angle ..	6½°.
Toe-In ..	3/16".
Brake Drums, Front	8" diameter.
Brake Linings Front	Halo.
Standard Gear Ratio	5.375 to 1 (8/43).
Alternative Gear Ratio	Nil.
Oil Capacity Rear Axle	1½ pints.
Diff. Gear	2 Star.
Brake Drums, Rear	8" diameter.
Brake Linings, Rear	Halo.
Steering	Adamant.
Steering Wheel ..	Bluemel.
Silencer	Burgess.
Petrol Tank Capacity	6 gals. incldg. 2 in reserves.
Wheels	2.50" × 19" W.B.
Tyres	4.00" × 19".
Tyre Pressures ..	27 lbs./ins^2 Front and Rear.
Shock Absorbers ..	Hartfords Front and Rear.
Battery	Lucas 12 volt.
Amp. Hr. Cap. 10 hr. Rate	50.
Speedometer ..	Smiths, P.N.
Speedo. Reduction Box	Nil.
Petrol Gauge ..	Nil.
Contact Breaker Points	.015"/.018".
Engine Type ..	4 cylinder. O.H.V. Camshaft.
R.A.C. Rating ..	8.05 H.P.
Bore	57 mm.
Stroke	83 mm.
Capacity of Combustion Space ..	Head 43.5 mm. Gasket 4.5 mm.
Compression Ratio	5.4 to 1.
Brake Horse Power	27 b.h.p. at 4,500 r.p.m.
M.P.H. per 1,000 r.p.m.	14.7 (Top Gear).
Firing Order ..	1 3 4 2.
Ignition	Coil and Distributor.
Ignition Timing ..	T.D.C. Fully Retarded, points just breaking.
Sparking Plugs ..	Champion No. 7.
	Lodge C.3.
	K.L.G. M.50
Sparking Plug Gap	.020".
Tappet Clearance ..	.004".
Valve Timing ..	Inlet Open T.D.C.
	Closes 50° A.B.D.C.
	Exhaust Opens 43° B.B.D.C.
	Closes 7° A.T.D.C.
Carburettor ..	Single 1" S.U. Horizontal.
Carburettor Needles	Standard M.5 or G.2.
	Richer No. 2.
	Weaker M.6.
Piston Clearance ..	.0025"/.0035".
Connecting Rod ..	Steel Stamping, white metal lined.
Length of Big-End Bearings	1.12".
Dia. of Big-End Bearings	1.50".
Main and Big-End Bearing Clearances	.0015"/.0025".
Cam Lift	.20".
Sump Capacity ..	8 pints.
Recommended Engine Oil	N.O.L. "Thirty".
Oil Pressure (approx.)	50 — 70 lbs./in.^{2}
Max. Engine Revs.	5,000 r.p.m.
Clutch	Single Plate, Dry.
Clutch Springs ..	6 per set.
Gearbox	Manual, 3 Forward, 1 Reverse.
Gearbox Ratios: Top	Direct.
2nd	1.83 to 1.
1st	3.50 to 1.
Rev.	2.83 to 1.
Gearbox Overall Ratios: Top	4.89 to 1.
2nd	8.95 to 1.
1st	17.115 to 1.
Rev.	13.84 to 1.
Gearbox Capacity ..	1½ pints.
Recommended Oil	N.O.L. "E.P." Transmission Oil 140.

Model D Midget

SERVICE DATA **MODEL "J.1." AND "J.2."** **1932 to 1934**

Wheelbase	7′ 2″.
Track	3′ 6″.
Steering Lock	1¾ turns of Strg. Wheel.
Steering Knuckle Angle	9°.
Castor Angle	6°.
King Pin Angle	6½°.
Toe-In	3/16″.
Brake Drums, Front	8″ diameter.
Brake Linings, Front	Ferodo M.R.
Standard Gear Ratio	8/43.
Alternative	8/47.
Oil Capacity, Rear Axle	1½ pints.
Diff. Gear	2 Star.
Brake Drums, Rear	8″.
Brake Linings, Rear	Ferodo M.R.
Propeller Shaft	Hardy Spicer, 1½″ dia., 39½″ long, 100½ joints.
Grade of Oil, Rear Axle	Duckham's Gear Oil "N."
Steering	Marles Weller, 10⅞ : 1.
Steering Wheel	Bluemel, 17″ dia. 3-spoke.
Silencer	Burgess, 3¼″ dia. × 30″ long.
Petrol Tank: 2 Str.	12 glns. (including 3 reserve).
Saloon and 4 Str.	6 glns. (including 2 reserve).
Wheels	19″ × 2.500″ W.B.
Tyres	19″ × 4.00″ Dunlop Standard.
Tyre Pressures	32 lbs./sq. in Front and Rear.
Battery	(One) Lucas 12 volt.
Amp. Hr. Cap. 10 Hr. Rate	50.
Speedometer	Combined with Rev. Counter.
Speedo. Reduc. Box Ratio	Nil.
Petrol Gauge	Nil.
Facia Board	Aluminium.
Contact Breaker Points	.015″ — .020″.
Distributor Type	Lucas DJ/4A-7-DA-O. Points set at T.D.C. Auto Advance commences at 1,000 r.p.m. and is fully advanced at 4,500 r.p.m. moving through 20° Crankshaft Angle; therefore ign. is fully advanced at 26° — 20° crankshaft angle, i.e. 46°.
Position of Toolbox	Under Bonnet.
Engine Type	4 cyl. O.H.V. Camshaft.
R.A.C. Rating	8.05 H.P.
Bore Stroke and Capacity	57 mm. × 83 mm. 847 cc.
Compression Ratio	6.2 : 1.
Cap. Combustion Space	Cyl. Head 36.1 cc. + Gasket 4.5 c.c.
Brake Horse Power	36 B.H.P. at 5,500 R.P.M.
M.P.H. per 1,000 R.P.M.	14.7 M.P.H. in top gear.
Firing Order	1 3 4 2.
Ignition & Timing	12 v. Coil and Distributor. Auto Adv. from 20° B.T.D.C. to 40° B.T.D.C. (Early engines—hand control).
Sparking Plug	14 mm. Lodge HD.14 or HNP.
Sparking Plug Gap	.018″.
Tappet Clearances	.006″.
Valve Timing Inlet	Opens 15° B.T.D.C. Closes 55° A.B.D.C.
Exh.	Opens 50° B.B.D.C. Closes 20° A.T.D.C.
Valve Timing measured round flywheel Inlet	Opens 1 5/16″ B.T.D.C. Closes 4 25/32″ A.B.D.C.
Exh.	Opens 4 11/32″ B.B.D.C. Closes 1¾″ A.T.D.C.
Carburettor	Twin 1″ S.U. Semi-D.D.
Carburettor Needle	Standard D.8. Weaker M.6. Richer No. 1.
Piston	Aerolite. 2 Comp. Rings, One Scraper. (Early engines, plain pistons).
Piston Clearance	.0025″ — .0035″ (Aerolite). .004″ (plain type).
Connecting Rod	Steel Stamp, White Metal Lined.
Dia. and Lgth. Big Ends	1½″ dia. 1.120″ Long.
Sump Capacity	1 gallon.
Grade of Engine Oil	Summer — Duckham's Adcol N.P.5. Winter—Duckham's Adcol NP.5.
Oil Pressure	40 — 60 lbs./sq. in.
Piston Ring Gap	.005″ — .007″.
Side Clearance of Rings in grooves	.001″ — .002″.
Cam Lift	.22″.
Valve Lift	.27″.
Depth of Head	3 7/16″ Standard.
Torque	34.363 lbs./ft. at 5,500 R.P.M.
Maximum Revs.	5,500 R.P.M.
Compression Hght. Piston	1.406″.
Rear Main Brg. Oil Groove	5/16″ wide × 7/64″.
Gudgeon Pin Slot	.125″ — .135″ deep. Earlier type has Flat 15/64″ deep from centre to bottom.
Valve Springs, Outer	Free Length 2.015″ compressed to 1.36″ under load of 25 lbs.
Valve Springs, Inner	Free Length 2.088″ compressed to 1.36″ under load of 15 lbs.
Clutch	Single Plate, Dry.
Clutch Springs	6 per set.
Gearbox	Wolseley, Manual, 4 Forward Speeds and Reverse.
Reduction for Speedo	8/14.
Gearbox Ratios Top	1 : 1.
3rd	1.36 : 1.
2nd	2.14 : 1.
1st	3.58 : 1.
Rev.	3.58 : 1.
Oil Capacity, Gearbox	1½ pints.
Grade of Oil, Gearbox	Duckham's Gear Oil "N".

	2 Str.	4 Str.	Saloon.
Overall Length	10′ 4″.	10′ 10″.	10′ 4½″.
Overall Height	4′ 4½″.	4′ 6½″.	4′ 7″.
Overall Width	4′ 3½″.	4′ 3½″.	4′ 3½″.
Chassis Weight	10 cwts.	10 cwts.	10 cwts.
Weight of Complete Car 1933	12¾ cwts.	13¼ cwts.	14½ cwts.
1934	13 cwts.	13¼ cwts.	14½ cwts.

Ground Clearance	6″.
Water Capacity	2 gallons.
M.P.H. per 1,000 R.P.M. with various tyre sizes	4.50″ × 19″ tyres — 15.53 m.p.h. 5.00″ × 19″ tyres — 15.72 m.p.h. (With 4.50″ × 19″ tyres, Speedo will read 10.05% slow).
Rear Engine Bearer Bolts	3/8″ B.S.F. × 3⅛ long. Cup washers fitted under bolt head, cup downwards.
Rear Hub Bearings	Ransome & Marles LJ.40.
Front Hub (Inner)	Ransome & Marles LJ.25.
Front Hub (Outer)	Ransome & Marles MJ.20.

Models J1 and J2

SERVICE DATA **MODEL "J.3."** **1932 to 1933**

Wheelbase	7′ 2″.
Track	3′ 6″.
Steering Knuckle Angle	9°.
Castor Angle ..	6°.
King Pin Angle ..	6½°.
Toe-In	3/16″.
Brake Drums, Front	8″ diameter.
Brake Linings, Front	Ferodo M.R.
Standard Gear Ratio	9/43.
Oil Capacity, Rear Axle	1½ pints.
Diff. Gear	2 Star.
Brake Drums, Rear	8″.
Brake Linings, Rear	Ferodo M.R.
Propeller Shaft ..	Hardy Spicer, 1½″ dia., 39½″ long, 100½ joints.
Grade of Oil, Rear Axle	Duckham's Adcol Gear Oil "N".
Steering	Marles Weller, Ratio 10⅞ : 1.
Steering Wheel ..	Bluemel, 17″ dia., 3-Spoke.
Silencer	Burgess, 3¼″ dia. × 44″ long.
Petrol Tank ..	12 glns. (incldg. 3 reserve).
Wheels	19″ × 2.50″ W.B.
Tyres	19″ × 4.00″.
Battery	(One) Lucas 12 volt.
Amp. Hr. Cap. 10 Hr. Rate	50.
Speedometer ..	Combined with Rev. Counter.
Speedo. Reduc. Box Ratio	Nil.
Petrol Gauge ..	Nil.
Facia Board ..	Aluminium.
Position of Toolbox	Nil.
Engine Type ..	4 cyl. O.H.V. Camshaft.
R.A.C. Rating ..	8.05 H.P.
Bore, Stroke and Capacity	57 mm. × 73 mm., 746 cc.
Compression Ratio	5.2 : 1.
Depth of Cyl. Head	3 17/32″.
Cap. Combustion Space	Cyl. Head 427 cc. + Gasket 2.7 cc.
Brake Horse Power	72.3 at 6,000 R.P.M.
M.P.H. per 1,000 R.P.M.	16.4 in top gear.
Firing Order ..	1 3 4 2.

Ignition and Timing	12 v. Coil and Dist. Auto Adv. from 20° B.T.D.C. to 40° B.T.D.C. on crankshaft.
Sparking Plug ..	14 mm. K.L.G. 718.
Sparking Plug Gap	.012″ — .015″.
Tappet Clearances ..	Inlet .008″. Exh. .010″.
Valve Timing Inlet	Opens 15° B.T.D.C.
	Closes 55° A.B.D.C.
Exh.	Opens 50° B.B.D.C.
	Closes 20° A.T.D.C.
Valve Timing measured round Flywheel Inlet	Opens 1 5/16″ B.T.D.C.
	Closes 4 25/32″ A.B.D.C.
Exh.	Opens 4 11/32″ B.B.D.C.
	Closes 1¾″ A.T.D.C.
Carburettor ..	Single S.U. 1¼″ dia. Horizontal.
Carburettor Needle	RLB.
Piston Special	Aerolite, 2 Comp. Rings, 1 Scraper.
Piston Clearance ..	.0025″ — .0035″.
Connecting Rod ..	Steel Stamp, White Metal Lined.
Dia. and Length of Big Ends ..	1⅞″ dia., 1.120″ long.
Sump Capacity ..	1 gallon.
Grade of Engine Oil	Aeroshell.
Supercharger ..	Powerplus No. 6A, running at engine speeds, 6-8 lbs. boost.
Supercharger Oil Consumption ..	3 cc. per minute at 1,000 R.P.M.
Clutch	Single Plate, Dry.
Gearbox	Manual, 4 Forward Speeds and Reverse.
Reduction for Speedo	8/14.
Gearbox Ratios Top	1 : 1.
3rd	1.36 : 1.
2nd	2.14 : 1.
1st	3.58 : 1.
Rev.	3.58 : 1.
Grade of Gearbox Oil	Duckham's Adcol Gear Oil "N"
Overall Length ..	10′ 4″.
Overall Height ..	4′ 4½″.
Overall Width ..	4′ 3½″.
Weight of Complete Car	13¼ cwts.
Ground Clearance ..	6″.
Water Capacity ..	2 gallons.

Model J3

SERVICE DATA **MODEL "J.4."** **1933**

Wheelbase	7′ 2″.
Track	3′ 6″.
Turning Circle ..	34′.
Steering Knuckle Angle	9°.
Castor Angle ..	6°.
King Pin Angle ..	6½°.
Toe-In	3/16″.
Brake Drums, Front	12″ dia.
Brake Linings, Front	Ferodo M.R.
Standard Gear Ratio	8/43.
Oil Capacity, Rear Axle	1½ pints.
Diff. Gear	2 Star.
Brake Drums, Rear	12″ dia.
Brake Linings, Rear	Ferodo M.R.
Propeller Shaft ..	Hardy Spicer, 2″ dia., length 37 1/16″.
Steering	Bishop Cam, 7 : 1 ratio.
Steering Wheel ..	Bluemel, 17″ dia., 4-Spoke.
Silencer	Brooklands Outside Exhaust.
Petrol Tank ..	12 glns. (incldg. 3 reserve).
Wheels	2.50″ × 19″.

Tyres	4.50″ × 19″.
Battery	(One) Lucas 12 volt.
Amp. Hr. Cap. 10 Hr. Rate ..	50.
Speedometer ..	Nil.
Petrol Gauge ..	Hobson Telegage.
Facia Board ..	Aluminium.
Position of Toolbox	Nil.
Engine Type ..	4 cyl. O.H.V. Camshaft.
R.A.C. Rating ..	8.05 H.P.
Bore, Stroke and Capacity	57 mm. × 73 mm. 746 cc
Compression Ratio	5.5 : 1.
Depth of Cylinder Head	3 15/32″.
Capacity of Combustion Space ..	Cylinder Head 38.3 cc. + Gasket 2.7 cc.
Ignition Timing ..	Set at 20° before T.D.C.
M.P.H. per 1,000 R.P.M.	15.2 m.p.h. in top gear.
Firing Order ..	1 3 4 2.

Model J4

Continued on following page)

SERVICE DATA — MODEL "J.4." — 1933 *Continued.*

Ignition	12 v. Coil and Distributor (Auto Adv.).
Sparking Plug ..	14 mm. K.L.G. 718.
Sparking Plug Gap	.012″ — .015″.
Tappet Clearances ..	.008″ Inlet, .014″ Exhaust.
Valve Timing Inlet	Opens 15° B.T.D.C.
	Closes 55° A.B.D.C.
Exh.	Opens 50° B.B.D.C.
	Closes 20° A.T.D.C.
Valve Timing measured round Flywheel Inlet	Opens 5/16″ B.T.D.C.
	Closes 4 25/32″ A.B.D.C.
Exh.	Opens 4 11/32″ B.B.D.C.
	Closes 1¾″ A.T.D.C.
Carburettor ..	Single S.U. 1⅝″ dia. Horizontal
Carburettor Needle	75.
Piston	Martlett.
Piston Clearance ..	.004″ (bottom of skirt).
Connecting Rod ..	Steel Stamp, White Metal Lined.
Dia. and Length of Big Ends	1⅝″ dia. × 1.120″ long.
Sump Capacity ..	1 gallon.
Fuel	80/20 Benzol, Ethyl.
Supercharger ..	Powerplus No. 7.
Clutch	Double Plate, Dry.
Clutch Springs ..	6 per set, compress to 1¼″ under 83 lbs.
Gearbox	E.N.V. Manual, 4 Forward Speeds and 1 Reverse.
Gearbox Ratios Top	1 : 1.
3rd	1.36 : 1.
2nd	1.86 : 1.
1st	2.69 : 1.
Rev.	2.69 : 1.
Overall Length ..	10′ 4″.
Overall Height ..	4′ 4½″.
Overall Width ..	4′ 3½″.
Ground Clearance ..	6″.
Water Capacity ..	2 gallons.

Model J4

SERVICE DATA — MODEL "K.3." — 1933 to 1934

Wheelbase	7′ 10″ (Actual 7′ 10 3/16″).
Track	4′.
Turning Circle ..	36′.
Steering Knuckle Angle	10½°.
Castor Angle ..	6°.
King Pin Angle ..	7½°.
Toe-In	3/16″.
Brake Drums, Front	13″ dia., Elektron.
Brake Linings, Front	Ferodo M.R.
Standard Gear Ratio	9/44 straight cut gears.
Alternatives ..	9/39, 9/52, straight cut gears.
Oil Capacity, Rear Axle	2 pints.
Diff. Gear	4 Star, E.N.V.
Brake Drums, Rear	13″ dia., Elektron
Brake Linings, Rear	Ferodo M.R.
Steering	Cam Gears, 7 : 1 ratio.
Steering Wheel ..	Bluemel 17″ dia., 3-Spoke.
Silencer	Brooklands, outside exhaust system.
Petrol Tank 1933	23½ glns. (incldg. 3 glns. res.).
1934	27½ glns. (incldg. 3 glns. res.).
Wheels	2.75″ W.B. × 19″.
Tyres	4.75″ × 19″ Dunlop Fort.
Shock Absorbers, Front	Hartford Duplex.
Shock Absorbers, Rear	Hartford Duplex.
Battery 1933	Two Lucas STW11E.
1934	Two Lucas STR11E.
Amp. Hr. Cap. 10 Hr. Rate	63.
Speedometer ..	Nil.
Rev. Counter ..	6″ dia., 8,000 r.p.m.
Petrol Gauge 1933	Hobson Telegauge.
1934	Nil.
Facia Board 1933	¼″ Alum, secured to body.
1934	3/16″ Dural secured to chassis.
Position of Toolbox	Nil.
Engine Type ..	6 cyl. Overhead Camshaft S/c.
R.A.C. Rating ..	12.08 H.P.
Bore, Stroke and Capacity	57 mm. × 71 mm. 1,086 cc.
Compression Ratio	6.2 : 1.
Capty. Combustion Space 1933	Cyl. Head 32 cc. + Gasket 3 cc.
1934	Cyl. Head 35 cc. + Gasket Nil.
M.P.H. per 1,000 R.P.M.	17.85 m.p.h. in top gear, 9/44 axle ratio.
Firing Order ..	1 4 2 6 3 5.
Ignition	B.T.H. Magneto.
Ignition Timing ..	T.D.C. Fully Retarded.
Sparking Plug ..	Racing—K.L.G. 718A. Warming up—Lodge H.D.14.
Tappet Clearances ..	Inlet .008″, Exh. .014″.
Valve Timing Inlet	Opens 15° B.T.D.C.
	Closes 55° A.B.D.C.
Exh.	Opens 50° B.B.D.C.
	Closes 20° A.T.D.C.
Valve timing measured round Flywheel Inlet	Opens 1 5/16″ B.T.D.C.
	Closes 4 25/32″ A.B.D.C.
Exh.	Opens 4 11/32″ B.B.D.C.
	Clcses 1¾″ A.T.D.C.
Carburettor ..	S.U. 1⅞″ dia.
Carburettor Needles	R.M.1. R.M.2.
Piston 1933	Martlett.
1934	Aerolite.
Piston Clearance ..	.004″ at bottom of skirt.
Connecting Rod ..	B.S. Special S.11. Racing Meta lined.
Dia. and Length of Big Ends	1⅝″ dia. × 1.125″ long.
Sump Capacity ..	2 gallons. Reserve tank on dash panel has capacity of 1 gallon.
Fuel	Pratt's M.G.1., or Shell N.3.
Supercharger 1933	Powerplus No. 9. Boost 14—16 lbs./sq. in. max.
1934	Marshall No. 85. Boost 14 lbs./sq. in. max.
Clutch	Incorporated in Gearbox.
Gearbox	E.N.V. Wilson Preselector.
Gearbox Ratios Top	1 : 1.
3rd	1.36 : 1.
2nd	2.00 : 1.
1st	3.40 : 1.
Rev.	5.07 : 1.
Gearbox Capacity ..	3 pints.

	1933	1934
Overall Length ..	12′ 0″	12′ 1″.
Overall Height ..	4′ 2½″.	3′ 10″.
Overall Width ..	4′ 11½″.	4′ 11½″.
Chassis Weight ..	17½ cwts.	17½ cwts.
Weight of Complete Car	18¼ cwts.	17¾ cwts.
Ground Clearance ..	6″.	6″
Water Capacity ..	3 gallons 3 pints.	

Model K3

SERVICE DATA **MODEL "K.N."** **1934 to 1935**

Wheelbase	9′ 0″.
Track	4′ 0″.
Turning Circle	38′ 0″.
Steering Lock	Approx. 1½ turns of steering wheel (full travel).
Steering Knuckle Angle	10½°.
Castor Angle	8° (incldg. 2° taper packing).
King Pin Angle	7½°.
Toe-In	$\frac{3}{16}$″.
Brake Drums, Front	Holden & Hunt, 13″ dia.
Brake Linings, Front	Ferodo M.R.
Standard Gear Ratio	9/52.
Alternatives	Nil.
Oil Capacity, Rear Axle	2 pints.
Diff. Gear	4 Star.
Brake Drums, Rear	Holden & Hunt, 13″ dia.
Brake Linings, Rear	Ferodo M.R.
Propeller Shaft	Hardy Spicer, 49$\frac{15}{16}$″ × 2½″ dia. 100½ joints.
Grade of Oil, Rear Axle	Duckham's N.O.L. "E.P." Transmission Oil 140.
Steering	Cam Gears, 9 : 1 ratio.
Steering Wheel	17″ dia. 3-Spoke Bluemel (Brown).
Silencer	Burgess, 44″ long.
Petrol Tank	11 glns. (incldg. 2 reserve).
Petrol Tap	Operated from facia board.
Wheels	2.50″ W.B. × 19″ Rudge-Whitworth.
Tyres	4.75″ × 19″.
Shock Absorbers, Front	Hartford Duplex.
Rear	Luvex.
Tyre Pressures: Front	30 lbs./sq. in.
Rear	30 lbs./sq. in.
Battery	(Two) Lucas 6 volt.
Amp. Hr. Cap. 10 Hr. Rate	63.
Speedometer	Mileometer, Jaeger, 1,750 Cable Revs.
Speedo, Reduction Box	1 : 1 (Angle Box).
Petrol Gauge	Hobson Telegage.
Facia Board	Walnut Faced, 9-ply, ½″ thick.
Position of Toolbox	Under bonnet, built in.
Contact Breaker Points	.015″ — .020″.
Engine Type	6 cyl. O.H. Camshaft.
R.A.C. Rating	12.08 H.P.
Bore, Stroke and Capacity	57 mm. × 84 mm. 1,286 cc.
Compression Ratio	6.2 : 1 (Depth of Head 3$\frac{41}{64}$″).
Capacity Combustion Space	Cyl. Head 35.1 cc. + Gasket 4.4 cc. Piston 1 cc.
Brake Horse Power	56 B.H.P. at 5,500 R.P.M.
M.P.H. per 1,000 R.P.M.	14.18 M.P.H. in top gear.
Firing Order	1 4 2 6 3 5.
Ignition	Coil and Dist. Auto 24° on Crank. Clockwise rotation.

Ignition Timing	Set at 20° B.T.D.C. Auto. comm. 2,000 ceases 4,000 r.p.m.
Sparking Plug	14 mm. K.L.G. F.70.
Tappet Clearances	Inlet .006″. Exh. .008″.
Valve Timing Inlet	Opens 15° B.T.D.C. Closes 55° A.B.D.C.
Exh.	Opens 50° B.B.D.C. Closes 20° A.T.D.C.
Valve Timing measured round Flywheel Inlet	Opens 1$\frac{5}{16}$″ B.T.D.C. Closes 4$\frac{25}{64}$″ A.B.D.C.
Exh.	Opens 4$\frac{11}{32}$″ B.T.D.C. Closes 1¾″ A.T.D.C.
Carburettor	Twin S.U. 1⅛″ Semi-D.D.
Carburettor Needles	Standard No. 3. Richer H.4. Weaker V.2.
Piston and Rings	Aerolite. 2 Comp. 1 Scraper.
Piston Clearance	.0025″ — .0035″.
Connecting Rod	Steel Stamp, white metal lined.
Dia. and Lgth. Big Ends	1⅝″ dia. × 1.120 long.
Sump Capacity	7 quarts.
Wght. Power Unit (no oil)	4 cwts.
Oil Pump By-pass Pressure	60 lbs./sq. in. — 80 lbs./sq. in.
Piston Ring Gap	.006″ — .010″.
Number of Main Bearings	4.
Grade of Engine Oil. Summer	Duckham's N.O.L. "30".
Winter	Duckham's N.O.L. "30".
Valve Lift	.307″.
Clutch	Roper & Wreaks. 8″ dia. Single Plate.
Clutch Springs	12 per set. Compress to ⅞″ long under load of 44/46 lbs.
Gearbox	Wolseley Manual. 4 Speeds and Reverse.
Reduction for Speedo	5/12.
Gearbox Ratios Top	1 : 1.
3rd	1.36 : 1.
2nd	2.32 : 1.
1st	4.18 : 1.
Rev.	4.18 : 1.
Grade of Oil, Gearbox	Duckham's N.O.L. "E.P." Transmission Oil 140.
Gearbox Capacity	2 pints.
Overall Length	13′ 0″.
Overall Height	5′. 0″.
Overall Width	4′ 11″.
Chassis Weight	13¾ cwts.
Weight Complete Car	21 cwts.
Ground Clearance	6″.
Water Capacity	3 gallons.
Cooling System	Pump.
Petrol Consumption	Approx. 23 m.p.g.
Plug Gap	.018″/.020″.

Model KN

SERVICE DATA MODELS "L.1." AND "L.2." MAGNA 1933 to 1934

Wheelbase	7′ 10 3/16″.
Track	3′ 6″.
Turning Circle ..	36′.
Steering Lock ..	1¾ turns of strg. wheel.
Steering Knuckle Angle	9°.
Castor Angle ..	7°.
King Pin Angle ..	6½°.
Toe-In	3/16″.
Brake Drum, Front	12″ dia.
Brake Linings, Front	Ferodo M.R.
Standard Gear Ratio	8/43.
Alternatives ..	8/47.
Oil Capacity, Rear Axle	1½ pints.
Diff. Gear	2 Star.
Brake Drums, Rear	12″ dia.
Brake Linings, Rear	Ferodo M.R.
Propeller Shaft ..	Hardy Spicer, 2″ dia. × 38⅜″. 100½ joints.
Grade of Oil, Rear Axle	Duckham's Gear Oil "N".
Steering	Marles Weller, 10⅞ : 1.
Steering Wheel ..	Bluemel 17″ dia. 3-Spoke.
Silencer	Burgess 3¼″ dia. × 30″ long.
Petrol Tank 2 Str.	12 glns. (incldg. 3 reserve).
4 Str.	10 glns. (incldg. 2 reserve).
Saloon	9 glns. (incldg. 2 reserve).
Coupe	13 glns. (incldg. 3 reserve).
Wheels	2.50″ W.B. × 19″.
Tyres	4.50″ × 19″ Dunlop.
Battery	(One) Lucas 12 volt.
Amp. Hr. Cap. 10 Rate	50.
Speedometer ..	Combined with Rev. Counter.
Speedo. Reduc. Box Ratio	100/143.
Petrol Gauge, 2 and 4 Str.	Archimedian.
Sal. and Coupe ..	Electric.
Facia Board 2 and 4 Str.	Aluminium.
Sal. and Coupe	
Position of Toolbox	2 Str. Nil. Under bonnet on remainder of models.
Engine Type ..	6 cyl. O.H.V. Camshaft.
R.A.C. Rating ..	12.08 H.P.
Bore, Stroke and Capacity	57 mm. × 71 mm. 1,087 cc.
Compression Ratio	6.4 : 1.
Capcty. Combustion Space	Cyl. Head 29 cc. + Gasket 4.5 cc.
Brake Horse Power	41 B.H.P. at 5,500 R.P.M.
M.P.H. per 1,000 R.P.M.	15.2 in top gear.
Firing Order ..	1 4 2 6 3 5.
Ignition	12v. Coil and Dist. Auto Advance.
Ignition Timing ..	Set at T.D.C.
Sparking Plug ..	14 mm. Lodge HD.14.

Tappet Clearances: Inlet	.006″.
Exh.	.008″.
Valve Timing: Inlet	Opens 15° B.T.D.C.
	Closes 55° A.B.D.C.
Exh.	Opens 50° B.B.D.C.
	Closes 20° A.T.D.C.
Valve Timing measured round Flywheel .. Inlet	Opens 1 5/16″ B.T.D.C.
	Closes 4 25/32″ A.B.D.C.
Exh.	Opens 4 11/32″ B.B.D.C.
	Closes 1¾″ A.T.D.C.
Carburettor ..	Twin S.U. Semi-D.D. 1⅛″ dia.
Carburettor Needles	Standard L.
	Richer R.3.
	Weaker 07.
Piston and Rings ..	Aerolite. 2 Comp. 1 Scraper.
Piston Clearance ..	.0025″ — .0035″.
Connecting Rod ..	Steel Stamp, white metal lined.
Dia. and Length, Big Ends.. ..	1⅝″ dia. × 1.120″ long.
Sump Capacity ..	1½ gallons.
Grade of Engine Oil, Summer	Duckham's Adcol N.P.3.
Winter	Duckham's Adcol N.P.2.
Sparking Plug Gap	.020″.
Contact Breaker Points	.015″/.018″.
Cam Lift	.220″.
Oil Pressure ..	50/70 lbs./sq. in. approx.
Valve Lift	.309″.
Clutch	Double Plate, Dry.
Clutch Springs ..	
Gearbox	Wolseley. Manual. 4 Forward speeds and Reverse.
Reduction for Speedo	8/14.
Gearbox Ratios Top	1 : 1.
3rd	1.36 : 1.
2nd	2.14 : 1.
1st	3.58 : 1.
Rev.	3.58 : 1.
Oil Capacity, Gearbox	1½ pints.
Grade of Oil, Gearbox	Duckham's Gear Oil N.

	2 Str.	4 Str.	Sal.	Coupe
Overall Length ..	11′ 4½″.	12′ 0″.	11′ 10″.	11′ 7″
Overall Height ..	4′ 4″.	4′ 4″.	4′ 8″.	4′ 8″.
Overall Width ..	4′ 4½″.	4′ 4½″.	4′ 4½″.	4′ 4½″
Chassis Weight in cwts.	12½.	12½.	12½.	12½.
Wght. Complete Car in cwts.	15¾.	16½.	17¼.	17.
Ground Clearance..	6″.	6″.	6″.	6″.
Water Capacity, all models	2¾ gallons.			

Models L1 and L2 Magna

SERVICE DATA **MODEL "PA" MIDGET** **1934 to 1935**

Wheelbase	7′ 3½″ (7′ 3 5/16″ Actual).
Track	3′ 6″.
Turning Circle	34′ 0″.
Steering Lock	1¾ Varles, 1½ Cam, turns of Steering Wheel (Ful Travel).
Steering Knuckle Angle	9°.
Castor Angle	8½° (includes 2½° taper packing).
King Pin Angle	6½°.
Toe-In	3/16″.
Brake Drums, Front	12″ dia. (.35 max., .30 min. Carbon Steel).
Brake Linings, Front	Ferodo M.R.
Standard Gear Ratio	8/43.
Alternatives	8/47 (for competition), 8/41.
Oil Capacity, Rear Axle	1½ pints.
Diff. Gear	4 Star.
Brake Drums, Rear	12″ dia. (.35 max., .30 min. Carbon Steel).
Brake Linings, Rear	Ferodo M.R.
Propeller Shaft	1¾″ dia. (1,002 Joints). Later 2″ dia. (1,100 Joints).
Grade of Oil, Rear Axle	Duckhams' Gear Oil "N".
Steering	Marles Weller (10⅞ : 1). Later Cam (9 : 1).
Steering Wheel	17″ dia., 3-Spoke Bluemel.
Silencer	Burgess 30″, later Burgess MG Design.
Petrol Tank	2 str.—12 gals. (3 res.). 4 str.—10 gals. (2 res.).
Petrol Tap	Operated from Facia Board.
Wheels	2.50″ × 19″ W.B. Rudge-Whitworth.
Tyres	4.00″ × 19″ Dunlop Std.
Shock Absorbers:	
Front	Single Hartfords.
Rear	Luvex.
Tyre Pressures:	
Front	2 str.—28 lbs./sq. in. 4 str.—28 lbs./sq. in.
Rear	2 str.—28 lbs./sq. in. 4 str.—28 lbs./sq. in.
Battery	(One) Lucas 12 volt.
Amp. Hr. Cap. 10 Hr. Rate	50
Speedometer	Mileometer. Jaeger. 1,609 cable revs.
Speedo. Reduc. Box Ratio	1 : 1 (Angle Box).
Petrol Gauge	Nil.
Facia Board	Wood. Veneered Sequoia ½″ thick.
Contact Breaker Points	.015″/.018″.
Distributor	Lucas. DK4A — ODA — 7/1.
Position of Toolbox	2 str. Nil. 4 str. Under Bonnet.
Engine Type	4 cyl. Overhead Camshaft.
R.A.C. Rating	8.05 H.P.
Bore, Stroke and Capacity	57 mm. × 83 mm. 847 cc.
Compression Ratio	6.2 : 1. (Depth of Head 3 41/64″).
Capacity of Combustion Space	Cyl. Head 35.1 cc., Gasket 4.4 cc. (Piston 1 cc.).
Brake Horse Power	36 b.h.p. at 5,500 r.p.m.
M.P.H. per 1,000 R.P.M.	14.73 m.p.h.
Firing Order	1 3 4 2.
Ignition	Coil Dist. Auto 20° Adv. on Crank. Clockwise Rotation.
Ignition Timing	Set at 20° B.T.D.C. Auto comm. 2,000 ceases 4,000 r.p.m.
Sparking Plug	(14 mm.) K.L.G. Type L.K.1.
Sparking Plug Gap	.018″.

Tappet Clearances	Inlet .006″., Exhaust .008″.
Valve Timing Inlet	Opens 15° B.T.D.C. Closes 55° A.B.D.C.
Exh.	Opens 50° B.B.D.C. Closes 20° A.T.D.C.
Valve Timing Measured round Flywheel .. Inlet	Opens 1 5/16″ B.T.D.C. Closes 4 25/32″ A.B.D.C.
Exh.	Opens 4 11/32″ B.B.D.C. Closes 1¾″ A.T.D.C.
Carburettors	Twin S.U. 1″ Semi-D.D.
Carburettor Needles	Standard M.6. Weaker S. Richer M.5.
Piston and Rings	Aerolite, 2 Comp. rings, 1 Scraper Ring.
Piston Clearance	.0025″ — .0035″.
Connecting Rod	Steel Stamp, white metal lined.
Dia. and Length of Big Ends	1⅝″ dia. 1.120″ long.
Sump Capacity	1½ gallons.
Weight of Unit Complete	3 cwts. 0 qrs. 26 lbs. (no oil).
Oil Pump By-Pass Pressure	60 - 80 lbs./sq. in.
Piston Ring Gap	.006″ — .010″.
Number of Main Bearings	Three.
Grade of Engine Oil:	
Summer	Duckham's Adcol NP5.
Winter	Duckham's Adcol NP3.
Valve Seat Angle	30°.
Valve Lift	.309″.
Mean Peak Piston Speed	2,995 ft./min.
B.H.P. per sq. in. of Piston Area	2.276 B.H.P.
Cam Lift	.220″.
Valve Seat	Length from valve guide seat to top of valve port 2.672″ ± .010″ inlet and exhaust.
Clutch	Roper & Wreaks Single Plate.
Clutch Diameter	8″.
Clutch Springs	12 per set. (Compress to ⅞″ long under load of 44-46 lbs.).
Gearbox	Wolseley. Manual. Four Speeds and Reverse.
Reduction for Speedo	4/10.
Gearbox Ratios Top	1 : 1.
3rd	1.36 : 1.
2nd	2.32 : 1.
1st	4.18 : 1.
Rev.	4.18 : 1.
Alternatives Top	1 : 1.
3rd	1.36 : 1.
2nd	2.14 : 1.
1st	3.58 : 1.
Rev.	3.58 : 1.
Gearbox Capacity	2 pints.
Gearbox Oil	Duckham's Gear Oil "N".
Overall Length	2 str. 10′ 11″. 4 str. 11′ 6″.
Overall Height	2 str. 4′ 3½″. 4 str. 4′ 3½″.
Overall Width	2 str. 4′ 4½″. 4 str. 4′ 4½″.
Chassis Weight	10¾ cwts.
Wght. of Complete Car	2 str. 14¾ cwts. 4 str. 15¼ cwts.
Ground Clearance	6″.
Water Capacity	19 pints.
Petrol Consumption	35 m.p.g. approx.
Cooling System	Thermo-Syphon.

Model PA Midget

SERVICE DATA **MODEL "PB" MIDGET** **1935 to 1936**

Wheelbase	7′ 3½″ (7′ 3$\frac{5}{16}$″ actual).
Track	3′ 6″.
Turning Circle ..	34′ 0″.
Steering Lock ..	1¼ turns of Steering Wheel (ful travel).
Steering Knuckle Angle	9°.
Castor Angle ..	8½° (includes 2½° taper packing).
King Pin Angle ..	6½°.
Toe-In	$\frac{3}{16}$″.
Brake Drums, Front	12″ dia. (.30-.35 Carbon Steel).
Brake Linings, Front	Ferodo M.R.
Standard Gear Ratio	8/43.
Alternatives ..	8/47 and 8/41.
Oil Capacity, Rear Axle	1½ pints.
Diff. Gear	4 Star.
Brake Drums, Rear	12″ dia. (.30-.35 Carbon Steel).
Brake Linings, Rear	Ferodo M.R.
Propeller Shaft ..	Hardy Spicer 2″ dia. 1,100 joints.
Grade of Oil, Rear Axle	Duckham's Gear Oil "N".
Steering	Cam. 8 to 1 Ratio.
Steering Wheel ..	17″ dia. 3-Spoke Bluemel.
Silencer	Burgess. M.G. Design.
Petrol Tank ..	2 str.—12 galls. (3 res.). 4 str.—10 galls. (2 res.).
Petrol Tap	Operated from Facia Board.
Wheels	2.50″ × 19″ W.B.
Tyres	4.00″ × 19″ Dunlop Std. "90" pattern.
Shock Absorbers:	
Front	Single Hartfords.
Rear	Luvex.
Tyre Pressures:	
Front	2 str.—28 lbs./sq. in. 4 str.—28 lbs./sq. in.
Rear	2 str.—28 lbs./sq. in. 4 str.—28 lbs./sq. in.
Battery	(One) Lucas 12 v. Type R.L.T.$\frac{41}{64}$. 11.A.
Amp. Hr. Cap. 10 Hr. Rate.. ..	50.
Speedometer ..	Jaeger (1,609 cable revs. per mile).
Speedo. Reduc. Box Ratio	1 : 1 (Angle Box).
Petrol Gauge ..	Nil.
Facia Board ..	Wood. Veneered Burr Walnut ½″ thick.
Distributor ..	Lucas DK4A—ODA—7/1.
Contact Breaker Points	.015″ — .018″.
Position of Tool Box	2 str.—Nil. 4 str.—Under Bonnet.
Position of Tool Box	
Engine Type ..	4 Cylinder O.H.V. Camshaft.
R.A.C. Rating ..	8.9 H.P.
Bore, Stroke and Capacity	60 mm. × 83 mm. 938.7 cc.
Compression Ratio	6.8 : 1 (Depth of Cyl. Head 3$\frac{41}{64}$″).
Capacity of Combustion Space ..	Cyl. Head 35.1 cc. Gasket 4.4 cc. Piston 1 cc.
Brake Horse Power	43 b.h.p. at 5,500 r.p.m.
M.P.H. per 1,000 R.P.M.	14.73 (in top gear).
Firing Order ..	1 3 4 2.
Ignition	Coil. Dist. auto 20° adv. on Crank. Clockwise Rotation.
gnition Timing ..	Set at 20° B.T.D.C. Auto. comm. 2,000, ceases 4,000 r.p.m.

Sparking Plug ..	(14 mm.) Champion L.10.
Sparking Plug Gap	.015″ — .018″.
Tappet Clearances ..	Inlet .006″, Exhaust .008″.
Valve Timing—Inlet	Opens 15° B.T.D.C. Closes 55° A.B.D.C.
Exh.	Opens 50° B.B.D.C. Closes 20° A.T.D.C.
Valve Timing Measured round Flywheel .. Inlet	Opens 1$\frac{5}{16}$″ B.T.D.C. Closes 4$\frac{25}{32}$″ A.B.D.C.
Exh.	Opens 4$\frac{11}{32}$″ B.B.D.C. Closes 1¾″ A.T.D.C.
Carburettors ..	Twin 1″ S.U. Semi-D.D.
Carburettor Needles	Standard M.6. Weaker S. Richer M.5.
Piston and Rings ..	Aerolite. 2 Comp. Rings, 1 Scraper Ring.
Piston Clearance ..	.0025″ — .0035″ (top skirt).
Connecting Rod ..	Steel Stamp, White metal Lined.
Dia. and Length of Big End	1⅛″ dia. 1.120″ long.
Sump Capacity ..	1½ gallons.
Piston Ring Gap ..	.006″ — .010″.
Valve Spring, Outer	Initial Comp. 25 lbs. at 1.36″.
Valve Spring, Inner	Initial Comp. 15 lbs. at 1.36″.
Oil Pump By-Pass Pressure	60 — 80 lbs./sq. in.
No. of Main Bearings	Three.
Grade of Engine Oil,	
Summer	Duckham's Adcol NP.3.
Winter	Duckham's Adcol NP.3.
Valve Seat Angle ..	30°.
Valve Lift	.309″.
Cam Lift	.220″.
Clutch	Roper & Wreaks Single Dry Plate
Clutch Diameter ..	8″.
Clutch Springs ..	12 per set. (Compress to ⅞″ long under a load of 44/46 lbs.).
Gearbox	Wolseley. Manual. Four Speeds and Reverse.
Reduction for Speedo	4/10.
Gearbox Ratios Top	1 : 1.
3rd	1.36 : 1.
2nd	2.14 : 1.
1st	3.58 : 1.
Rev.	3.58 : 1.
Alternatives, Top	1 : 1.
3rd	1.36 : 1.
2nd	2.32 : 1.
1st	4.18 : 1.
Rev.	4.18 : 1.
Gearbox Oil ..	Duckham's Gearbox Oil "N".
Gearbox Capacity..	2 pints.
Overall Length ..	2 str.—10′ 11″. 4 str.—11′ 6″.
Overall Height ..	2 str.—4′ 3½″. 4 str.—4′ 3½″.
Overall Width ..	2 str.—4′ 4½″. 4 str.—4′ 4½″.
Chassis Weight ..	10¾ cwts.
Weight of Complete Car	2 str.—14¾ cwts. 4 str.—15½ cwts.
Ground Clearances	6″.
Water Capacity ..	19 pints.
Petrol Consumption	32 m.p.g. approx.
Cooling System ..	Thermo-Syphon.

Model PB Midget

SERVICE DATA **MODEL "TB" MIDGET** **1939**

Wheelbase	7′ 10″.
Track	3′ 9″.
Turning Circle	37′ 0″.
Steering Lock	Full lock to lock approx. 1½ turns of steering wheel.
Steering Knuckle Angle	10½°
Castor Angle	6° (3° on base, 3° on chassis).
King Pin Angle	7⅛°
Toe-In	½″.
Brake Drums, Front	9″ dia.
Brake Linings, Front	Ferodo M.R.
Brake Operation	Lockheed Hydraulic.
Standard Gear Ratio	8/41.
Alternatives	Lower 8/43. Higher 8/39.
Oil Capacity, Rear Axle	2 pints.
Diff. Gear	4 Star.
Brake Drums, Rear	9″ dia.
Brake Linings, Rear	Ferodo M.R.
Handbrake Operation	By cable, rear wheels only.
Propeller Shaft	Hardy-Spicer 2½″ dia., 43 13/16″ long. 1,100 Flanges.
Grade of Oil, Rear Axle	Duckham's "XS" Press Oil.
Steering	Bishop Cam. Ratio 11 : 1.
Steering Wheel	Bluemel 17″, 3-Spoke Adjustable.
Grade of Oil, Steering	Duckham's "XS" Press Oil.
Steering Drop Arm	5½″ centres.
Silencer	Burgess Type.
Petrol Tank Capacity	13½ gallons (incldg. 3 reserve).
Wheels	19″ × 2.50″ W.B.
Tyres	19″ × 4.50″ Dunlop Standard "90".
Tyre Pressures:	
Front	24 lbs.
Rear	26 lbs.
Shock Absorbers:	
Front	Luvex Constant Pressure Type.
Rear	Luvex Constant Pressure Type.
Battery	(Two) Lucas 6 volt.
Amp. Hr. Cap. 10 Hr. Rate	50.
Speedometer	Jaeger 5″ dia. Green dial.
Speedo. Reduc. Box	Ratio 4/9.
Petrol Gauge	Nil.
Facia Board	Walnut.
Rev. Counter (with Clock)	Jaeger 5″ dia. Green dial.
Rev. Counter Reduc Box	On dynamo.
Position of Toolbox	Under bonnet.
Engine Type	4 cyl. O.H.V. Pushrod operated.
R.A.C. Rating	10.97 H.P.
Bore, Stroke and Capacity	66.5 mm. × 90 mm. 1,250 cc.
Compression Ratio	7.25 — 7.50 to 1.
Cacacity, Combustion Space	
Brake Horse Power	54.4 at 5,200 R.P.M.
M.P.H. per 1,000 R.P.M.	15.84 in top gear.
Firing Order	1 3 4 2.
Ignition	12 volt Coil and Distributor.
Sparking Plug	14 mm. Champion L.10.S.
Sparking Plug Gap	.018″.
Tappet Clearances	.019″ Hot.
Valve Timing, Inlet	Opens 11° B.T.D.C.
	Closes 57° A.B.D.C.
Exh.	Opens 52° B.B.D.C.
	Closes 24° A.T.D.C.

Valve Timing measured round Flywheel Inlet	Opens 23 mm. B.T.D.C.
	Closes 119.5 mm. A.B.D.C.
Exh.	Opens 108.9 mm. B.B.D.C.
	Closes 50.3 mm. A.T.D.C.
Carburettor	Twin S.U. 1¼″ Semi-D.D.
Carburettor Needle	E.5.
Piston	Aerolite 3 Ring.
Piston Clearances	.002″ — .003″.
Connecting Rod	178 mm. centres, loose white metal bearings.
Dia. and Length of Big Ends	45 mm. dia. × 28 mm. long.
Sump Capacity	9 pints.
Grade of Engine Oil,	
Summer	Duckham's NP.3.
Winter	Duckham's NP.3.
Piston Ring Gap	.006″ — .010″.
Main Bearings	Three. 52 mm. dia. × 38 mm. long.
Valve Throat diameter	Inlet 30 mm.
	Exhaust 26 mm.
Valve Spring Pressure	Shut 93 lbs. Open 125 lbs.
Valve Lift, Inlet and Exh.	8 mm.
Starter Motor	Ratio 9.3 : 1.
Dynamo	1.16 × Engine Speed.
Oil Pump	1 × Engine Speed.
Oil Pump By-Pass Pressure	70 lbs./sq. in.
Water Capcty. Block and Head	6.02 pints.
Fan diameter and speed	12¾″ dia. runs at engine speed.
Total wght. Power Unit	382 lbs.
Distributor Points	.010″ — .012″ gap.
Gudgeon Pin	18 mm. diameter.
Valve Seat Angle	30°.
Supercharger	Nil.
Wt. in lb. per B.H.P.	7.01 lbs.
Wt. in lb. per Litre	305.5 lbs.
B.H.P. per Litre	43.5.
Maximum Torque	765 lbs. in at 2,600 R.P.M.
Clutch	Borg & Beck, Single Plate, Dry.
Gearbox	Manual. 4 Speeds and Reverse. Synchro-Mesh on Top, 3rd and 2nd.
Reduction for Speedo	4/9.
Gearbox Ratios Top	1 : 1.
3rd	1.35 : 1.
2nd	1.95 : 1.
1st	3.38 : 1.
Rev.	3.38 : 1.
Gearbox Capacity	1¼ pints.
Grade of Oil, Gearbox	Duckham's Synchro Oil "S".
	2 str. / Coupe
Overall Length	11′ 7½″. / 11′ 7½″.
Overall Height	4′ 5″. / 4′ 7″.
Overall Width	4′ 8″. / 4′ 8″.
Weight of Complete Car	15 cwts. 2 qrs.
Ground Clearance	6″.
Water Capacity	1 gallon 6 pints.
Cooling System	Pump and Fan.
Weight of Body Complete	2 str. 3 cwts. 0 qrs. 14 lbs. (including seats, floorboards, carpets, windscreen, side screens, hood, facia, but no instruments).
Capacity of Lockheed System	1 pint.

Model TB Midget

SERVICE DATA **MODEL "VA" 1½-LITRE** **1937 to 1939**

Wheelbase	9' 0".
Track	4' 2".
Turning Circle ..	38' 0".
Steering Lock ..	2½ turns.
Steering Knuckle Angle	10°.
Castor Angle ..	4½° (includes 1½° Taper Packing).
King Pin Angle ..	8°.
Toe-In	¼".
Brake Drums, Front	10" dia.
Brake Linings, Front	Ferodo M.R.
Brake Operation ..	Lockheed Hydraulic.
Standard Gear Ratio	5.22 (9/47).
Rear Axle Oil Cap.	2 pints.
Diff. Gear	4 Star.
Brake Drums, Rear	10" dia.
Brake Linings, Rear	Ferodo M.R.
Brake Operation ..	Lockheed Hydraulic.
Handbrake.. ..	Cable, Rear Only.
Propeller Shaft ..	Hardy Spicer. 3" dia. Length 52 15/16".
Rear Axle Oil ..	Duckham's "XS" Press. Oil.
Steering	Bishop Cam. 14 : 1 Ratio.
Steering Wheel ..	17" dia. Bluemel. Telescopic.
Silencer	(Two) Burgess Type.
Petrol Tank ..	12 gallons (no reserve).
Wheels	Dunlop 3.00" W.B. × 19".
Tyres	Dunlop 5.00" × 19" Standard "90".
Battery	One 12 volt Lucas.
Amp. Hr. Cap. 10 Hr. Rate.. ..	60.
Speedometer ..	Jaeger 5" dia. Translucent.
Speedo. Reduc Box	5/11 Ratio.
Petrol Gauge ..	Jaeger Electric.
Facia Board ..	Walnut.
Position of Toolbox	Tourer—Under Bonnet. Saloon—in rubber tray in Boot Lid.
Engine Type ..	4 cyl. O.H.V. Push-Rod Operated.
R.A.C. Rating ..	12 h.p.
Bore, Stroke and Capacity.. ..	69.5 mm. × 102 mm. 1,549 cc.
Compression Ratio	6.5 : 1.
Cap. Comb. Space	
Brake Horse Power	55 at 4,400 r.p.m.
M.P.H. per 1,000 R.P.M.	16.29 in top gear.
Firing Order ..	1 3 4 2.
Ignition	12 v. Coil and Distributor.
Ignition Timing ..	T.D.C. to 28° advance on crank.
Sparking Plug ..	Champion L.10. 14 mm. × ½" reach.
Tappet Clearances..	Early cars .015" hot. Later cars .019" hot.
Valve Timing, Inlet	Opens 11° B.T.D.C. Closes 59° A.B.D.C.
Exh.	Opens 56° B.B.D.C. Closes 24° A.T.D.C.
Later Cars Inlet	Opens 11° B.T.D.C. Closes 57° A.B.D.C.
Exh.	Opens 52° B.B.D.C. Closes 24° A.T.D.C.
Valve Timing measured round Fly wheel .. Inlet	Opens 30.7 mm. B.T.D.C. Closes 164.7 mm. A.B.D.C.
Exh.	Opens 156.4 mm. B.B.D.C. Closes 67 mm. A.T.D.C.
Piston Clearance ..	.002" — .003".
Piston	Aerolite, 3 rings, Allum. Alloy.
Carburettor ..	Twin 1⅝" S.U. Semi-D.D.
Carburettor Needle	Standard CO. Richer BK. Weaker CP.
Connecting Rod ..	204 mm. centres, white metal lined.
Dia. and Length of Big Ends.. ..	48 mm. dia., 30 mm. long.

Sump Capacity ..	5½ quarts.
Grade of Engine Oil, Summer	Duckham's NP.5.
Winter	Duckham's NP.3.
Piston Ring Gap ..	.004" — .006".
Main Bearings ..	3. 55 mm. dia., 41 mm. long.
Valve Throat Diameter .. Inlet	32 mm.
Exh.	28 mm.
Valve Spring Pressure	Shut: 101 lbs. Open: 130 lbs.
Valve Lift	Inlet and Exhaust 8.4 mm.
Starter Motor ..	Ratio 9.46 to 1.
Dynamo	Ratio 1.34 × engine speed.
Oil pump.	½ engine speed.
By-Pass Pressure ..	60 lbs./sq. in.
Fan Diameter and Speed	12" dia. 1.1 × engine speed.
Total wght. of Power Unit	4 cwts. 1 qr. 8 lbs.
Distributor Points..	.010" — .012".
Gudgeon Pin ..	18 mm. diameter.
Valve Seat Angle ..	45°.
Main Bearings ..	55 mm. dia. × 41 mm. long.
Con. Rod Big Ends	48 mm. dia. × 30 mm. long.
Rear Hub Bearings	Ransome & Marles, LJ.40.
Max. Torque ..	892 lbs. in. at 2,600 r.p.m.
B.H.P. per Litre ..	35.6.
Clutch	Single Plate, Running in Oil.
Later	Single Plate 8" dia.
Clutch Springs: Oil type	Pressure (Pedal Free) 682.5 lbs. 15 per set.
Dry type.. ..	Pressure (Pedal Free) 840-876 lbs. 6 per set.
Gearbox	Manual. 4 speeds and rev. Synchro 1st., 2nd, 3rd, and top.
Gearbox Ratios Top	1 : 1.
3rd	1.35 : 1.
2nd	1.95 : 1.
1st	3.38 : 1.
Rev.	3.38 : 1.
Gearbox Oil Capacity	1¾ pints.
Grade of Oil, Gearbox	Duckham's Synchro Oil "S".
Overall Length of Car	Saloon. 14' 1". Tourer 14' 3". Coupe 14' 6".
Overall Height ..	Saloon 5' 1¾". Tourer 5' 0". Coupe 4' 11".
Overall Width ..	Saloon 5' 2". Tourer 5' 2". Coupe 5' 2".
Chassis Weight ..	18½ cwts.
Weight Complete Car	Saloon 24¼" cwts. Tourer 22 cwts. Coupe 25 cwts.
Ground Clearance..	7".
Water Capacity ..	2 gallons 3 pints.
Petrol Consumption	25 m.p.g. approx.
Cooling System ..	Pump and Fan.
Front Axle Weight	Saloon 12½ cwts. Tourer 11¾ cwts. Coupe 12¼ cwts.
Rear Axle Weight..	Saloon 11¾ cwts. Tourer 10½ cwts. Coupe 12¾ cwts.
Capacity of Lockheed System ..	1¼ pints.
Capacity of Jackall System	2¼ pints.
TPDG Engine (Police)	Bore 73 mm. Stroke 102 mm Cubic Capacity 1,707 cc.
Carburettor Needle	No. 69.
Sparking Plugs ..	Champion LA.11. with gaps at Standard.
Tyre Pressures: Saloon	30 lbs./sq. in.² front and rear.
Tourer	28 lbs./sq. in.² front and rear.

Model VA 1½-Litre

SERVICE DATA **MODEL "SA" 2-LITRE** **1936 to 1939**

Wheelbase	10′ 3″.
Track	4′ 5⅜″.
Turning Circle	40′.
Steering Lock	Approx. 2½ turns (Full Travel).
Steering Knuckle Angle	10°.
Castor Angle	4° (Inclination of Spring).
King Pin Angle	8°.
Toe-In	3/16″.
Brake Drums, Front	12″ diameter.
Brake Linings, Front	Ferodo M.R.
Brake Operation	Lockheed Hydraulic.
Standard Gear Ratio	12/57.
Alternative Ratios	11/49. 11/58.
Rear Axle Oil Capacity	2 pints.
Diff. Gear	4 Star.
Brake Drums, Rear	12″ diameter.
Brake Linings, Rear	Ferodo M.R.
Brake Operation	Hydraulic (Lockheed).
Handbrake (Rear Only)	by Cable.
Propeller Shaft	3½″ dia. Mean Length between Flanges—61″. Later 55″. 1,251 Joints.
Grade of Oil, Rear Axle	Duckham's "XS" Press. Oil.
Steering, Type and Ratio	Bishop Cam. 14 : 1 (Later 16 : 1).
Steering Wheel	17″ dia. 3-Spoke Bluemel. Later—Telescopic.
Steering Box Oil	Duckham's "XS" Press. Oil.
Drop Arm	7″ Centres. Later ; ½″ centres.
Silencer	(Two) Burgess 12″ long, 5″ dia.
Petrol Tank	10 gallons capacity. No res
Wheels	3.25″ × 18.00″ W.S.
Tyres	5.50″ × 18.00″ Dunlop Fort "90".
Tyre Pressures:	
Front	30 lbs./sq. in.
Rear	30 lbs./sq. in.
Shock Absorbers:	
Front	Luvex, Constant Pressure Type.
Rear	Luvex, Constant Pressure Type. Later, front and rear, Luvex Piston Type.
Battery	(Two) 6 volt Lucas.
Amp. Hr. Cap. 10 Hr. Rate	60.
Speedometer	Jaeger 5″ dia. (Translucent Dial).
Speedo. Reduc. Box Ratio	Nil.
Petrol Gauge	Jaeger Electric. with Oil Pressure and Oil Capacity Gauges.
Facia Board	Walnut.
Rev. Counter (with Clock)	Jaeger 5″ dia. (Translucent Dial).
Rev. Counter Reduc. Box	100/266 Ratio.
Position of Toolbox	Under Bonnet.
Engine Type	6 cyl. O.H.V. Push-Rod Operated.
R.A.C. Rating	17.71 h.p., later 17.97 h.p.
B M.E.P.	116 lbs./sq. in at 2,000 r.p.m.
Max. Torque	112.5 lbs./ft. at 2,400 r.p.m.
Peak Piston Speed	3,012 ft./min.
Compression Pressure	120 lbs./sq. in.
Piston Area sq. in./ton	15.9.
Litres per Ton Mile	2,700.
Bore, Stroke and Capacity	69 mm. × 102 mm. 2,288 cc. Later 69.5 mm. × 102 mm. 2,322 cc.
Compression Ratio	6.38 to 6.55 to 1.
Capacity of Comb. Space	Cyl. Head 64 cc. to 61.8 cc. Gasket 6.9 cc.
Brake Horse Power	78.5 at 4,200 r.p.m.
M.P.H. per 1,000 R.P.M.	18 m.p.h. in top gear.
Firing Order	1 5 3 6 2 4.
Ignition	12 v. Coil and Distributor (Auto.).

Ignition Timing (B.T.D.C.)	7° (19.5 mm. on Flywheel Rim) to 39° Advance on Crank.
Sparking Plug	14 mm. Champion L.10. ½″ Reach.
Tappet Clearances	.015″ hot.
Valve Timing, Inlet	Opens 11° B.T.D.C. Closes 59° A.B.D.C.
Exh.	Opens 56° B.B.D.C. Closes 24° A.T.D.C.
Valve Timing Measured round Flywheel .. Inlet	Opens 30.7 mm. B.T.D.C. Closes 164.7 mm. A.B.D.C.
Exh.	Opens 156.4 mm. B.B.D.C. Closes 67 mm. A.T.D.C.
Carburettor	Twin S.U. 1½″ D.D. (with Air Cleaner).
Carburettor Needles	Standard CH. Richer CL. Weaker CK.
Piston	Alum. Alloy 4 Ring, later 3 Ring (Aer.).
Piston Clearance	.11 mm. (4 Ring Piston). Later .002-″.003″ (3 Ring Piston).
Connecting Rod	204 mm. Centres, White Meta Lined.
Dia. and Length, Big Ends	48 mm. dia., 30 mm. long.
Sump Capacity	10 quarts.
Grade of Engine Oil, Summer	Duckham's NP.3.
Winter	Duckham's NP.3.
Piston Ring Gap	.004″ — .006″.
Main Bearings	4—55 mm. dia., 41 mm. ong.
Valve Throat dia.	Inlet—30.5 mm. Exhaust—26 mm.
Valve Spring Press.	Shut 60 lbs. Open 108.5 lbs. Later Shut 101 lbs. Open 130 lbs.
Valve Lift	10 mm. Inlet and Exhaust.
Starter Motor	Ratio 9.46 to 1.
Dynamo	Ratio 1.3 × engine speed.
Oil Pump	Half engine speed. By-Pass Pressure 80 lbs./sq. in.
Water Cap. of Block and Head	1 gallon 6 pints.
Fan. dia. and Speed	14¼″ dia. 11/12 engine speed.
Total weight. Power Unit	5 cwt. 2 qrs. 10 lbs.
Distributor Points	Set at .010″ — .012″ clearances.
Gudgeon Pin	18 mm. dia.
Valve Seat Angle	45°.
Main Bearings	55 mm. dia., 41 mm. long.
Connecting Rod	48 mm. dia., 30 mm. long.
Oil Pump Relief Valve Spring	Free Length—1.378″ to give 6½ lbs. when compressed to 1.063″.
Clutch	Single Plate running in oil. Spring Hub.
Clutch Springs	15 per set. Pressure, with pedal free—682.5 lbs.
Gearbox	Manual, 4 speeds and Rev. Later synchro-mesh on Top and 3rd.
Reduction for Speedo.	7/14.
Gearbox Ratios Top	1 : 1.
3rd	1.38 : 1.
2nd	2.13 : 1.
1st	3.76 : 1.
Rev.	4.76 : 1.
Gearbox Oil Capcty.	2¼ pints.
Grade of Oil, Gearbox.	Crash Box—Duckham's Gear Oil "N". Synchro—Duckham's Synchro Oil "S".
Overall Length of Car	Chassis 15′ 2″. Std. Saloon 16′ 1″ Tickford 16′ 5″.
Overall Height	Chassis 3′ 11½″. Std. Saloon 5′ 0″. Tickford 5′ 5½″.
Overall Width	Chassis 5′ 5″. Std. Saloon 5′ 6½″ Tickford 5′ 6½″.

Model SA 2-Litre

Continued on following page

SERVICE DATA **MODEL "SA" 2-LITRE** **1936 to 1939** *Continued.*

Chassis Weight ..	1 ton 1 cwts. 3 qrs. 0 lbs.	Front Axle Weight	Saloon 14 cwts. 0 qrs. 7 lbs.
Wght. Complete Car	Saloon 1 ton 9 cwts. 2 qrs. 21 lbs.	Rear Axle Weight..	Saloon 15 cwts. 2 qrs. 0 lbs.
Ground Clearance..	7".	Brake Lining Area.	99 sq. in./ton.
Water Capacity ..	2 gallons 7 pints.	Lockheed Capacity	1½ pints (complete system).
Petrol Consumption	20 m.p.g. approx.	Jackall Capacity ..	3 pints (complete system).
Cooling System ..	Pump and Fan.	Radiator Capacity..	7 pints.

Model SA Litre

SERVICE DATA **MODEL "WA" 2.6-LITRE** **1939**

Wheelbase	10' 3".
Track Front ..	4' 5⅛".
Rear ..	4'. 8¾".
Turning Circle ..	41'.
Steering Lock ..	2½ turns.
Steering Knuckle Angle	10°.
Castor Angle ..	6° (including 2° taper plate).
King Pin Angle ..	8°.
Toe-In	3/16"
Brake Drums, Front	14".
Brake Linings, Front	Ferodo M.R.
Brake Operation ..	Lockheed Hydraulic.
Standard Gear Ratio	9/43.
Alternatives ..	10/42. 9/41.
Oil Capacity, Rear Axle	2½ pints.
Diff. Gear	S.B. 2 Star.
Brake Drums, Rear	14" dia.
Brake Linings, Rear	Ferodo MR.
Brake Operation ..	Lockheed Hydraulic.
Handbrake.. ..	Cale, Rear Only.
Propeller Shaft ..	Hardy Spicer, Twin, Needle Bearing with Centre Bearing.
Grade of Oil, Rear Axle	Duckham's Adcol "XS" Press Oil.
Steering	Bishop Cam, Adjustable, 14 : 1.
Steering Wheel ..	Bluemel Adjustable, 17" dia., 4-Spoke.
Drop Arm	7 11/32" between centres.
Steering Box Oil ..	Duckham's Adcol "XS" Press Oil.
Silencer	(Two) Burgess, 12" long × 5" dia.
Petrol Tank ..	16.6 gallons, no reserve.
Wheels	3.25" W.B. × 18".
Tyres	5.50" × 18" Dunlop "Fort".
Battery	(Two) 6 volt Lucas.
Amp. Hr. Cap. 10 Hr. Rate.. ..	60.
Speedometer ..	Jaeger.
Speedo. Reduc. Box Ratio	Nil.
Facia Board ..	Veneered Walnut.
Rev. Counter ..	Jaeger.
Rev. Counter Reduc. Box	100/266 Ratio.
Position of Toolbox	Rubber Tool Tray in Boot Lid.
Engine Type ..	6 cyl. O.H.V. Push-Rod Operated.
R.A.C. Rating ..	19.82.
Bore, Stroke and Capacity.. ..	73 mm. × 102 mm. 2,561 cc.
Brake Horse Power	95.5 b.h.p. at 4,400 r.p.m.
M.P.H. per 1,000 R.P.M. Top	17.89 m.p.h.
3rd	13.62 m.p.h.
2nd	8.30 m.p.h.
1st	4.90 m.p.h.
Firing Order ..	1 5 3 6 2 4.
Ignition	12 v. Coil and Distributor, Automatic.
Ignition Timing ..	4° B.T.D.C.
Sparking Plug ..	Champion L.10.S. 14 mm.
Tappet Clearances..	Inlet and Exhaust .015" (hot).
Valve Timing Inlet	Opens 11° B.T.D.C. Closes 59° A.B.D.C.
Exh.	Opens 56° B.B.D.C. Closes 24° A.T.D.C.

Valve Timing Measured round Flywheel Inlet	Opens 30.7 mm. B.T.D.C. Closes 164.7 mm. A.B.D.C.
Exh.	Opens 156.4 mm. B.B.D.C. Closes 67 mm. A.T.D.C.
Compression Ratio	7.14 to 7.32 : 1.
Carburettor ..	Twin 1½" S.U. Semi-D.D.
Carburettor Needle	Standard EL. Richer EG. Weaker AM.
Piston	Aerolite. Alum. Alloy, 3 Ring.
Piston Clearance ..	.0027" — .0033".
Connecting Rod ..	204 mm. centres, detachable bearings.
Big Ends	48 mm. dia. × 30 mm. long.
Sump Capacity ..	10 quarts.
Grade of Engine Oil	Summer—Duckham's Aero NP.3. Winter—Duckham's Aero NP.3.
Main Bearings ..	4—55 mm. dia. × 41 mm. long.
Starter Motor ..	Ratio 10.4 × engine speed.
Dynamo	Ratio 1.34 × engine speed.
Oil Pump	Half engine speed, bye-pass 80 lbs./sq. in.
Fan Diameter and Speed	14½" × dia. 1.05 × engine speed.
Distributor Points..	.012" clearance.
Water cap. of Block and Head ..	1 gallon 6 pints.
Total wght. Power Unit	657 lbs.
Valve Lift	10 mm. Inlet and exhaust.
Valve Springs Press.	Shut 101.4 lbs., Open 139 lbs.
Valve Throat dia. ..	Inlet 30.5 mm. Exhaust 26 mm.
Gudgeon Pin ..	18 mm. dia.
Wt. in lb. per B.H.P.	6.87.
Wt. in lb. per Litre..	256.5.
B.H.P. per Litre ..	37.2.
Static Torque of crankshaft damper	70 lbs. in.
Max. Torque ..	1,605 lbs. in. at 2,800 r.p.m.
Clutch	Borg & Beck Single Dry Plate Spring Hub.
Clutch Springs ..	8 per set, 140-147 lbs.
Gearbox	Manual, 4 speeds and reverse.
Reduction for Speedo	7/14.
Gearbox Ratios Top	1 : 1.
3rd	1.418 : 1.
2nd	2.155 : 1.
1st	3.646 : 1.
Rev.	3.646 : 1.
Gearbox Oil Capacity	2½ pints.
Gearbox Oil ..	Duckham's Ascol "S" Synchro-Gear.
Overall Length of Car	Saloon 15' 11½". Coupe 15' 11½". Tourer 15' 10".
Overall Height ..	Saloon 5' 0". Coupe 5' 5½". Tourer 5' 2".
Overall Width ..	Saloon 5' 8". Coupe 5' 8". Tourer 5' 8".
Wt. of Complete Car	Saloon 32½ cwts. Tickford 32 cwts.
Ground Clearance..	7".
Water Capacity ..	3 gallons.
Jackall Capacity ..	2½ pints.
Petrol Gauge ..	Jaeger, Electric.

Model WA 2.6-Litre

SERVICE DATA — MODEL "T.C." MIDGET

COMM. CHASSIS TC.0251, 1945

POWER UNIT (Comm. Eng. No. XPAG. 883).

Type	4 cyl. OHV. Pushrod operated.
R.A.C. Rating	10.97 H.P.
Bore, Stroke and Capacity.	65.5 mm. × 90 mm., 1,250 cc.
Compression Ratio	7.25-7.5 to 1.
Piston Clearance	.002″-.003″ at thrust face.
Piston	"Aerolite" Alum. compensating.
Compression Rings	2. Width .0885 +.000″. − .0005″.
Scraper Rings	1, slotted. Width .1575″ + 000″. −.0005″.
Piston Ring Gap	.006″ to .010″.
Gudgeon Pin	18 mm. dia. +.015 mm. − .010 mm.
Gudgeon fit in Con. Rod.	Clamp Bolt.
Connecting Rod	Steel, 178 mm. long. Loose steel shell white metal bearings.

CRANKSHAFT

Main Journal, Front	Dia. 52 mm.— − .035 mm. Width 38 mm. − .015 mm. Rad. 2.5 + .15 mm. − .00 mm.
Intermain Journal	Identical to Main Journal, Front.
Main Journal, Rear	Dia. 52 mm.— − .035 mm. Width 40 mm. − .015 mm. Rad. 2.5 ± .15 mm.
Crankpin Journal	Dia. 45 mm.— −.035 mm. Wdh. 28 mm. +.015. −.015 mm. −.010. Rad. 2.5 ± .15 mm.

Crankshaft Regrind. (Max.)

Main Bearing, Front	Loose Steel Shell White Metal	52 .020″ dia. × 35.12 mm. 52 .005″ dia. × 34.88 mm.
Main Bearing, Inter		52 .020″ dia. × 37.955 mm. 52 .005″ dia. × 37.925 mm.
Main Bearing, Rear		52 .020″ dia. × 35.12 mm. 52 .005″ dia. × 34.88 mm.
Big End Bearing		1 .825 mm. thick × 22.46 mm. 1 .831 mm. thick × 22.21 mm.

Camshaft Drive	Chain.
Camshaft Bearing,— Front	White Metal.
Centre	Zinc Base Alloy.
Rear	Zinc Base Alloy.
Valve Throat dia.— Exh.	25 mm.
Inlet	30 mm.
Valve Seat Angle	30°.
Valve Spring Pressure.	Shut 93 lbs. Open 123 lbs. ± 6 lbs. (Double Spring).
Valve Lift	8 mm. Inlet and Exh.
Valve Timing Inlet	Opens 11° B.T.D.C. Closes 57° A.B.D.C.
Exh.	Opens 52° B.B.D.C. Closes 24° A.T.D.C.
Valve Timing measured round Flywheel Inlet	Opens 28.3 mm. B.T.D.C. Closes 119.5 mm. A.B.D.C.
Exh.	Opens 108.9 mm. A.T.D.C. Closes 503 mm. A.T.D.C.
Tappet Clearance	.019″ Hot, Inlet and Exh.
Oil Pump	Gear type, runs at ½ engine speed.
Oil Pressure	70 lbs./sq. in. Min. 40 lbs./sq. in.
Oil Filter	Full Flow type.
Carburettor	Twin S.U. 1¼″ S.U. Semi-D.D.
Carburettor Needle	Standard ES. Weaker AP. Richer EM.
Supercharger	Nil.
Fan	12¾″ dia. runs at engine speed.
Flywheel	298.9 mm. dia., 93 teeth.
Piston Area	21.57 sq. in.
Piston Area, sq. in./ton	27.8.

ELECTRICAL.

Distributor Rotation	Anti-clockwise.
Auto Advance	32° on Crankshaft.
Distributor Points Gap	.010″-.012″.
Starter Motor Ratio	9.3 to 1 (Lucas).
Dynamo	1.16 × engine speed. Comp. Volt. Control.
Sparking Plugs	14 mm. Champion L.10.S. K.L.G. F.80. Lodge H.N. or (better) H.N.P.
Sparking Plug Gaps	.020″-.022″.
Firing Order	1 3 4 2.
Ignition Timing	Points just breaking T.D.C.
Battery	Type SLXW.9A, Positive Earth, 51 amp. at 10 Hr. Rate.
Lighting	12 volt, earth return.
Ignition	12 volt, Coil and Distributor.
Sparking Plugs for 6 lbs. supercharge	Champion L.11.S. K.L.G. F.220. Lodge H.N.P.

CLUTCH AND GEARBOX.

Clutch Type	Dry. Borg & Beck.
Clutch Diameter	7¼″.
Gearbox Ratios Top	1 : 1.
3rd	1.36 : 1.
2nd	1.95 : 1.
1st	3.38 : 1.
Rev.	3.38 : 1.
Overall Ratio Top	5.125 : 1.
3rd	6.95 : 1.
2nd	10.0 : 1.
1st	17.32 : 1.
Rev.	17.32 : 1.
Synchro-mesh	Synchro-mesh on Top, 3rd and 2nd gears.

INSTRUMENTS.

Speedometer	5″, green dial, Jaeger.
Rev. Counter	5″, green dial, Jaeger. Clock.
Reduc. gear for Rev. Ctr.	On dyno. 100/225 Ratio.
Petrol Gauge	Nil. Warning light at 2½ glns. reserve.
Facia Board	Walnut. Later rexine covered.
Speedo. Reduc. Ratio	4/9.

FRONT AXLE AND STEERING.

Camber of Wheel	3°.
Castor Angle	8° (3° on Beam, 5° on Chassis) (5½° when 2½° taper plate is fitted).
King Pin Angle	7½°.
Toe-In	⅛″.
Steering	Bishop Cam. 11 : 1.
Turning Circle	37′.
Knuckle Angle	10½°.
Suspension	Semi-elliptical (orthodox Axles).
Steering Wheel	Bluemel 17″ 3-Spoke adjustable.
Wheels	19″ × 2.50″ W.B.
Tyre Size	19″ × 4.50″ (no alternative over-size).
Tyre Pressure	24 lbs./sq. in.
Left Hand Drive	Available.
Lock to Lock	1½ turns of Steering Wheel.

REAR AXLE.

Axle type	¾ floating.
Drive type	4 star diff. gear, Spiral Bevel.
Ratio	8/41.
Alternative ratios	7/38. 8/39.
Adjustment, Crown Wheel	By adjusting nuts.
Adjustment, Pinion	By shims.
End Play	Nil.
Lash	.007″ − .010″.
Wheels	19″ × 2.5″ W.B.
Tyre Size	19″ × 4.5″ (no alternative over-size).
Tyre Pressure	26 lbs./sq. in.
Ratio (Low)	7/38 (5.428 : 1) Wols. Pt. No. 100616 (Pair) (Series 2 Morris 10 cwt. Van). Wheel 100617. Pin. 100618.

SPRINGS, FRONT.

Type	Semi-elliptic.
Length, front	26¾″.
Width	1½″.

SERVICE DATA.

No. of leaves ..	6.
Thickness of leaves	$\frac{3}{16}$".
Camber	1.85 (flat at 500-520 lbs.).

SPRINGS, REAR.

Type	Semi-elliptic.
Length, front ..	36$\frac{1}{2}$".
Width	1$\frac{1}{4}$".
No. of leaves ..	9 (7 + 2 rebound).
Thickness of leaves	$\frac{7}{32}$" except Main $\frac{1}{4}$", 1st, Rebound $\frac{3}{16}$".
Camber	1.687 (flat at 385 lbs.).

BRAKES.

Types	Lockheed Hydraulic. Handbrake. by cable to rear only.
Linings	Ferodo M.R.
Brake Drums, front	9" dia. × 1$\frac{1}{2}$".
rear	9" dia. × 1$\frac{1}{2}$".
Brake linings, front and rear	26 sq. in. per pair, 104 sq. in., total 4.
Friction area, front	Drums 42 sq. in. each.
rear	drums 42 sq. in. each.
Total friction area..	168 sq. in.
Brake Lining area,sq. in./ton	134.

DIMENSIONS.

Wheelbase	7' 10".
Track	3'. 9" front and rear.
Overall length ..	11' 7$\frac{1}{2}$".
Overall Width ..	4' 8".
Overall Height ..	4' 5".
Ground Clearance..	6".

PERFORMANCE.

Brake Horse Power	54.4 at 5,200 R.P.M.
Max. Torque ..	765" lbs. at 2,600 R.P.M.
H.P. per sq. in. piston area ..	2.5 B.H.P.
Ft. per Min. of piston at maximum developed H.P.	3,068.
Litres per Ton Mile	Top Gear, laden 2,415. Unladen 3,053.
M.P.H. per 1,000 R.P.M.	15.64 in top gear.
Road Speed at 2,500 ft./min. of piston speed	57 M.P.H.
Engine speed per 10 M.P.H. ..	632 R.P.M. in top gear.
Petrol Consumption	30 to 35 miles per gallon.
Braking at 20 M.P.H.	14'.
30 "	26'.
40 "	44'.
Acceleration ..	10-30 M.P.H. in top gear—9.6 secs. 10-30 M.P.H. in 2nd gear—4 secs. 20-40 M.P.H. in top gear—8.5 secs.
Max. Fuel economy	2,500-3,000 R.P.M.
Piston Speed per 10 M.P.H.	374 ft. per min.
Brake Mean Effective Pressure ..	125 at 2,600 R.P.M.
Acceleration through gears	0-60 M.P.H.—18.5 secs.

WEIGHTS.

Complete Car, Dry	15 cwts. 2 qrs.
Laden	19 cwts. 2 qrs. 12 lbs.
Front, weight ..	7 cwts. 3 qrs. dry.
Rear, weight ..	7 cwts. 3 qrs. dry.
Chassis, as off line	10 cwts. 3 qrs.
as for export	13 cwts.
Body	3 cwts. 1 qr. (including floor boards, carpets, windscreen, hood).
Power Unit total ..	3 cwts. 1 qr. 6 lbs.
Weight for car per sq. in. of brake drum friction area	10.33 lbs. dry.
Power Unit less Gearbox	2 cwts. 3 qrs. 2 lbs.

MODEL " T.C." MIDGET. *Continued*

CAPACITIES.

Sump to level (oil)..	1 gallon 1 pint.
Gearbox (oil) ..	1$\frac{1}{4}$ pints.
Rear Axle (oil) ..	2 pints.
Block and Head (water)	6.02 pints.
Cooling system ..	1 gallon 6 pints.
Brake system fluid..	1 pint.
Fuel Tank	13$\frac{1}{2}$ gallons including 2$\frac{1}{2}$ reserve.

GRADES OF CONSUMABLES.

Engine oil	N.O.L. Thirty.
Engine oil, alternatives	Mobiloil " A." Double Shell. Castrol " XL."
Gearbox oil ..	N.O.L. " EP " Transmission oil 140.
Gearbox, alternatives	Sheel " EP " Spirax. Mobiloil " EP." Castrol Hi-Press.
Rear Axle oil ..	N.O.L. " EP " Transmission Oil 140.
Rear Axle, alternatives	Mobiloil " EP " Gear Oil. Shell " EP " Spirax. Wakefield Hi-Press.
Chassis Oil ..	Duckham's Soft Laminoid.
Steering Gear Oil ..	As for Rear Axle.
Hydraulic Brake Fluid	Lockheed " BR."
Hydraulic Damper Fluid	Luvax Girling Official Piston Type Hydraulic Damper Thin Fluid.

MISCELLANEOUS.

Silencer	Burgess Type.
Cooling System ..	Water Radiator, Pump and Fan.
Prop. Shaft.. ..	Hardy Spicer 2$\frac{1}{2}$" × 43$\frac{13}{16}$" long, 1,100 Flanges.
Hydraulic Damper, Front	" Luvax-Girling " Piston type pressure recuperation.
Hydraulic Damper, Rear	As for front dampers.
Position of battery..	Under bonnet.
Position of toolbox	Under Bonnet.
Bulb replacements..	Lucas. Head—54 & 176. Side 207. Rear—207. Stop—207. Panel Illum.—986 or 987. Map Reading—207. Trafficators—Nil. Foglamp, No. 2 Foglamp SFT.462—No. 162. Ignition Warning—970.
Centre of Gravity (Laden)	49$\frac{1}{2}$" behind front axle, 19" above ground line.

Torque Loading for :—

* Connecting Rod Bolts	230 lbs./ins.
* Main Bearing ..	750 lbs./ins.
Cylinder Head Nuts	600 lbs./ins.

* To the next pin hole.

OIL PUMP, ADDITIONAL DATA.

Max. gallons per min.	6 Imperial gallons at 4,000 R.P.M. (engine speed).

Maximum, minimum and average pressures depend upon pump and engine clearances.

Oil Relief Valve ..	Comes into operation at 60 lbs./sq. in.
Min. allowable pressure drop ..	15 lbs./sq. in.
Max. allowable temp.	Oil temp. of 188° F
Normal oil temp. ..	167° F.

SERVICE DATA. MODEL "N.A." MAGNETTE.

Chassis No. NA. 0251, 1934 to Chassis No. N.A. 0995, 1936.

Wheelbase	8′ 0″.
Track	3′ 9″.
Turning Circle ..	30′ 0″.
Steering Lock ..	Approx. 1½ turns of steering wheel (full travel).
Steering Knuckle Angle	10½°.
Castor Angle ..	8° (includes 2° taper packing).
King Pin Angle ..	7½°.
Toe-In	$\frac{1}{16}$°.
Brake Drums, Front	12″ dia. (.35 max. .30 min. Carbon Steel).
Brake Linings, Front	Ferodo M.R.
Standard Gear Ratio	8/41.
Alternatives ..	8/43, 8/47.
Oil Capacity, Rear Axle	1½ pints.
Diff. Gear	4 Star.
Brake Drums, Rear	12″ dia. (.35 max., .30 min. Carbon Steel).
Brake Linings, Rear	Ferodo M.R.
Propeller Shaft ..	2″ dia., 100½ Joints, later 1,100 Jts.
Grade of Oil, Rear Axle	Duckham's Gear Oil " N."
Steering	Cam Gears, 9 : 1 Ratio.
Steering Wheel ..	17″ dia., 3-Spoke Bluemel.
Silencer	Burgess 44″ long.
Petrol Tank Capacity	10 gallons (including 2 reserve).
Petrol Tap	Operated from facia board.
Wheels	2.50″ × 18″ W.B. Rudge-Whitworth.
Tyres	18″ × 4.75″ Dunlop Standard.
Shock Absorbers—	
Front	Hartford Duplex.
Rear	Luvax.
Tyre Pressures—	
Front	2 str. 26 lbs./sq. in., 4 str. 30 lbs./sq. in.
Rear	2 str. 26 lbs./sq. in., 4 str. 30 lbs./sq. in.
Battery	(Two) Lucas 6 volt.
Amp. Hr. Cap. 10 Rate	63.
Speedometer ..	Mileometer. Jaeger, 1,609 Cable Revs.
Speedo. Reduc. Box Ratio	1 : 1 (Angle Box).
Petrol Gauge ..	Electric (Smith's).
Facia Board ..	Wood, Veneered Sequoia, ½″ thick. Later, Burr Walnut, ½″ thick.
Position of Toolbox	Under Bonnet, built in.
Contact Breaker Points	.015″-.018″.
Engine Type ..	6 cyl. O.H.V. Camshaft.
R.A.C. Rating ..	12.08 H.P.
Bore, Stroke and Capacity ..	57 mm. × 84 mm. 1,286 cc.
Compression Ratio	6.2 : 1 (Depth of Head 3$\frac{41}{64}$″).
Capacity Combustion Space ..	Cyl. Head. 35.1cc. + Gasket 4.4 cc. (Piston 1 cc.).
Brake Horse Power	56 B.H.P. at 5,500 R.P.M.
M.P.H. per 1,000 R.P.M.	15.65 in top gear.
Firing Order ..	1 4 2 6 3 5.
Ignition	Coil, Dist. Auto. 25° on Crank, Clockwise Rotation.
Ignition Timing ..	Set at 20° B.T.D.C. Auto comm. 2,000, ceases 4,000 R.P.M.
Sparking Plug ..	14 mm. K.L.G. Type L.K.I.
Sparking Plug Gap	.018″.
Tappet Clearances..	Inlet .006″. Exh. .008″.
Valve Timing Inlet	Opens 15° B.T.D.C. Closes 55° A.B.D.C.
Exh.	Opens 50° B.B.D.C. Closes 20° A.T.D.C.
Valve Timing measured round Flywheel Inlet	Opens 1$\frac{5}{16}$″ B.T.D.C. Closes 4$\frac{25}{64}$″ A.B.D.C.
Exh.	Opens 4$\frac{11}{32}$″ B.B.D.C. Closes 1¾″ A.T.D.C.
Carburettor ..	Twin S.U. 1⅛″ dia. Semi-D.D.
Carburettor Needles	Standard No. 3. Richer H.4. Weaker V.2.
Piston and Rings ..	Aerolite. 2 Comp. 1 Scraper.
Piston Clearance ..	.0025″-.0035″.
Connecting Rod ..	Steel Stamp, white metal lined.
Dia. and Length, Big Ends	1⅝″ dia. 1.120″ Long.
Sump Capacity ..	7 quarts.
Weight Power Unit	Approx. 4 cwts. (no oil).
Oil Pump By-Pass Press.	60 lbs./sq. in.-80 lbs./sq. in.
Piston Ring Gap ..	.006″-.010″.
No. Main Bearings	4.
Grade of Engine Oil	Duckham's Adcol NP.3 for Summer and Winter.
Valve Springs (Inner)	Free Length 2.088″ to compress to 1.36″ under load of 15 lbs.
Valve Springs (Outer)	Free Length 2.015″, to compress to 1.36″ under load of 25 lbs.
Valve Lift	.307″.
Clutch	Roper & Wreaks Single Plate, 8″ dia.
Clutch Springs ..	12 per set. Compress to ⅞″ long under load 44-46 lbs.
Gearbox	Wolseley, Manual, 4 speeds and reverse.
Reductn. for Speedo	5/12.
Gearbox Ratios Top	1 : 1.
3rd	1.36 : 1.
2nd	2.32 : 1.
1st	4.18 : 1.
Rev.	4.18 : 1.
Alter. Ratios Top	1 : 1.
3rd	1.36 : 1.
2nd	2.14 : 1.
1st	3.58 : 1.
Rev.	3.58 : 1.
Grade of Oil, Gearbox	Duckham's Gear Oil " N "
Gearbox Capacity..	2 pints.
Overall Length ..	12′ 4″.
Overall Width ..	4′ 6″.
Overall Height ..	4′ 11″.
Chassis Weight ..	12¾ cwts.
Wght. Complete Car	2 str. 18¼ cwts. 4 str. 18½ cwts.
Ground Clearance..	6″.
Water Capacity ..	3 gallons.
Petrol Consumption	Approx. 25 m.p.g.
Cooling System ..	Pump.

PERFORMANCE DATA.

Acceleration Times through two upper ratios

	Top	3rd.
From 0-30 M.P.H.	13 secs.	9 secs.
20-40 M.P.H.	14⅜ secs.	9.8 secs.
30-50 M.P.H.	16⅜ secs.	11 secs.

Acceleration Times through gears :

From 0-50 M.P.H. 16.4 secs.
0-60 M.P.H. 22.8 secs.

Timed Speed over half mile, 80.72 M.P.H

SERVICE DATA.

MODEL "TA" MIDGET.

Chassis No. TA. 0251, 1936 to Chassis No. TA. 3253, 1939.

Wheelbase	7′ 10″.
Track	3′ 9″.
Turning Circle ..	37′ 0″.
Steering Lock ..	Approx. $1\frac{1}{8}$ turns of steering wheel (full travel).
Steering Knuckle Angle	$10\frac{1}{2}$°.
Castor Angle ..	6° (3° on beam + 3° on chassis).
King Pin Angle ..	$7\frac{1}{2}$°.
Toe-In	$\frac{1}{2}$″.
Brake Drums, Front	9″ dia.
Brake Linings Front	Ferodo MR.
Brake Operation ..	Lockheed Hydraulic.
Standard Gear Ratio	8/39.
Oil Capacity, Rear Axle	2 pints.
Diff. Gear	4 Star.
Brake Drums, Rear	9″ dia.
Brake Linings, Rear	Ferodo MR.
Brake Operation, Rear	Lockheed Hydraulic.
Handbrake (Rear only)	Cable.
Propeller Shaft ..	Hardy-Spicer $2\frac{1}{2}$″ dia., $43\frac{13}{16}$″ long, 1,100 Flanges.
Grade of Oil, Rear Axle	Duckham's "XS" Press Oil.
Steering ..	Bishop Cam, 8 : 1, later 11 : 1.
Steering Wheel ..	Bluemel 17″ 3-Spoke, later cars adjustable.
Grade of Oil, Steering Box	Duckham's "XS" Press Oil.
Steering Drop Arm	$5\frac{7}{16}$″ centres.
Silencer	Burgess Type.
Petrol Tank— Early Cars ..	15 galls. (including 3 reserve).
Later Cars ..	$13\frac{1}{2}$ galls.
Wheels	19″ × 2.50 W.B.
Tyres	19″ × 4.50 Dunlop Standard "90."
Tyre Pressure, Front	24 lbs.
Rear	26 lbs.
Shock Absorbers ..	Luvax, Constant Pressure Type front and rear.
Battery	(Two) 6 volt Lucas.
Amp. Hr. Cap. 10 Hr. Rate ..	50.
Speedometer ..	Jaeger, 5″ dia. Green dial.
Speedo. Reduc. Box, Ratio	7/15.
Facia Board ..	Walnut.
Rev. Counter (with Clock)	Jaeger, 5″ dia. Green dial.
Rev. Counter Reduc. Box	(on dynamo) 100/225 Ratio.
Position of Toolbox	Under bonnet.
Contact Breaker Points	.010″-.012″.
Engine Type ..	4 cyl. O.H.V. Pushrod Operated.
R.A.C. Rating ..	10 H.P.
Bore, Stroke and Capacity ..	63.5 mm. × 102 mm. 1,292 cc.
Compression Ratio	6.4-6.6 to 1.
Capcty. Combustion Space	Cyl. Head 51 cc. + Gasket 6.34cc. + Piston below block 1.45 cc.
Brake Horse Power	50 at 4,500 R.P.M.
M.P.H. per 1,000 R.P.M.	16.66 M.P.H. in top gear.
Firing Order ..	1 3 4 2.
Ignition	12v. Coil and Distributor.
Ignition Timing ..	T.D.C. to 32° advance on Crank.
Sparking Plug ..	14 mm. Champion L.10 ($\frac{1}{2}$″ reach).
Sparking Plug Gap	.018″.
Tappet Clearances— Inlet	.010″ Hot.
Exh.	.015″ Hot.
Valve Timing Inlet	Opens 11° B.T.D.C. Closes 59° A.B.D.C.
Exh.	Opens 56° B.B.D.C. Closes 24° A.T.D.C.
Valve Timing measured round Flywheel Inlet	Opens 27.8 mm. B.T.D.C. Closes 149.3 mm. A.B.D.C.
Exh.	Opens 141.7 mm. B.B.D.C. Closes 60.7 mm. A.T D.C.

Carburettor ..	Twin S.U. 1″ dia. Semi-D.D.
Carburettor Needle	Standard AC. Richer M.1. Weaker 5.
Piston Clearance ..	.11 mm. (4 Ring Piston).
Piston Early	Aluminium Alloy 4 Ring.
Later	Aerolite 3 Ring.
Piston Clearance (3 Ring)	.002″-.003″ (Aerolite).
Connecting Rod ..	190 mm. centres, white metal lined.
Dia. and Length, Big Ends	45 mm. dia., 28 mm. long.
Sump Capacity ..	$5\frac{1}{4}$ quarts.
Grade of Engine Oil	Duckham's NP.3, summer and winter.
Piston Ring Gap ..	.004″-.006″.
Main Bearings ..	3. 52 mm. dia. × 38 mm. long.
Valve Throat Dia.— Inlet	30.5 mm.
Exh.	26 mm.
Valve Spring Pressure	Shut 60 lbs. Open 108.5 lbs.
Later	Shut 101 lbs. Open 170 lbs.
Valve Lift	10 mm. Inlet and Exhaust.
Starter Motor ..	$4\frac{3}{16}$″ dia., Ratio 9.5 : 1.
Dynamo	Ratio 1.16 × engine speed.
Oil Pump	Runs at half engine speed.
Oil Pump By-Pass Pressure	60 lbs./sq. in.
Water Capty. Block and Head ..	7 pints.
Fan dia. and Speed	12″ dia. runs at engine speed.
Total Weight Power Unit	3 cwts. 3 qrs. 0 lbs.
Distributor Points..	.010″-.012″.
Gudgeon Pin ..	16 mm. dia.
Valve Seat Angle ..	45°.
Main Bearings ..	52 mm. dia. × 33 mm. long.
Piston Area ..	7.855 sq. in.
Road Speed at 2,500 feet per minute piston speed ..	66.5 M.P.H. in top gear.
M.P.H. per 1,000 R.P.M. with 5.00″ × 19″ tyres ..	15.90 M.P.H.
Supercharger ..	Nil.
Clutch	Single Plate, running in oil, cork insert, spring hub.
Clutch Springs ..	Pressure, pedal free 652.2 lb. 12 springs per set.
Gearbox	Manual, 4 speeds and reverse. Synchro-Mesh Top & 3rd Gears.
Reductn. for Speedo	7/15.
Gearbox Ratios ..	To Engine No. MPJG.683 :— Top 1 : 1. 3rd 1.42 : 1. 2nd 2.2 : 1. 1st 3.715 : 1. Rev. 4.78 : 1.
Gearbox Ratios ..	From Engine No. MPJG.684 :— Top 1 : 1. 3rd 1.32 : 1. 2nd 2.04 : 1. 1st 3.454 : 1. Rev. 4.44 : 1.
Grade of Oil, Gearbox	Duckham's Synchro Oil "S."
Gearbox Capacity..	2 pints.

	2 str.	Coupe.
Overall Length ..	11′ $7\frac{1}{2}$″	11′ $7\frac{1}{2}$″.
Overall Height ..	4′ 5″	4′ 7″.
Overall Width ..	4′ 8″	4′ 8″.
Weight of Complete Car	15 cwt. 3 qrs.	17 cwt. 2 qrs.

Chassis Weight (in condition as supplied for overseas)	13 cwts.
Ground Clearance..	6″.
Water Capacity ..	1 gallon 7 pints.
Petrol Consumption	30 m.p.g. approx.
Cooling System ..	Pump and Fan.
Weight of Body Complete ..	3 cwts. 0 qrs. 14 lbs. (inc. seats, floorboards, carpets, windscreen, side screens, hood, facia board but no instruments.
Capacity of Lockheed System ..	1 pint.

SERVICE DATA. MODEL "T.D." MIDGET.

Comm. Chassis TD.0251, 1949.	
POWER UNIT. Comm. Eng. No. XPAG/TD/501.	
Type	4 cyl. OHV. Pushrod operated.
Bore, Stroke and Capacity	66.5 mm. × 90 mm. 1,250 cc.
Compression Ratio	7.25 : 1.
Depth of Head	76.75 mm.
Cap. Combustion Space	45.5 cc. ± 1 cc.
Piston Clearance ..	.002"-.003" at thrust faces.
Piston	" Aerolite " Alum. compensating.
Compression Rings	2. Width .0885" +.000". −.0005".
Scraper Ring ..	1, slotted. Width .1575"+.000". −.0005".
Piston Ring Gap	.006"—.010".
Gudgeon Pin Dia..	18 mm. − .015 mm. − .010 mm.
Fit in Piston ..	Double Thumb Press Fit.
Fit in Con. Rod ..	Clamp Bolt.
Con. Rod	Steel, 178 mm. long.

CRANKSHAFT.

Main Journal, Front	Dia. 52 mm. − .035 mm. − .015 mm. Width 38 mm.
Intermain Journal ..	Dia. 52 mm. − .035 mm. − .015 mm. Width 38 mm.
Main Journal, Rear	Dia. 52 mm. − .035 mm. − .015 mm. Width 40 mm.
Crankpin Journal ..	Dia. 45 mm. − .035 mm. − .015 mm.
Crankshaft Regrind.	Width 28 mm.

		Max.
Main Bearing, Front	Loose	52.020 mm. dia. × 35.12 mm.
		52.005 mm. dia. × 34.88 mm.
Main Bearing, Inter	Steel Shell	52.020 mm. dia. × 37.955 mm.
		52.005 mm. dia. × 37.925 mm.
Main Bearing, Rear	White Metal	52.020 mm. dia. × 35.12 mm.
		52.005 mm. dia. × 34.88 mm.
Big End Bearing ..		22.46 mm. rad. × 1.825 mm. thick.
		22.21 mm. rad. × 1.831 mm.

Camshaft Drive ..	Chain.
Camshaft Bearing— Front	White Metal.
Rear and Centre	Zinc base Alloy.
Valve Throat dia.— Exh.	26 mm.
Inlet	30 mm.
Valve Seat Angle ..	30°.
Valve Spring Pressure	Shut 93 lbs. Open 123 ± 6 lbs. (Double Spring).
Valve Lift	8 mm. Inlet and Exhaust.
Valve Timing Inlet	Opens 11° B.T.D.C.
	Closes 57° A.B.D.C.
Exh.	Opens 52° B.B.D.C.
	Closes 24° A.T.D.C.
Valve Timing measured round Flywheel .. Inlet	Opens 28.3 mm. B.T.D.C.
	Closes 119.5 mm. A.B.D.C.
Exh.	Opens 108.9 mm. B.B.D.C.
	Closes 50.3 mm. A.T.D.C.
Tappet Clearance ..	.019" Hot.
Oil Pump	Gear type, runs at $\frac{1}{2}$ engine speed.
Oil Pressure ..	70 lbs./sq. in. Min. 40 lbs./sq. in.
Oil Filter	Full Flow type.
Carburettor ..	Twin S.U. $1\frac{1}{4}$" dia. Semi-D.D.
Carburettor Needle	Standard ES.
	Richer EM.
	Weaker AP.
Fan	$12\frac{3}{4}$" dia. runs at engine speed.
Flywheel	298.9 mm. dia. 120 teeth.
Piston area ..	21.57 sq. in.

ELECTRICAL.

Distributor rotation	Anti-clockwise.
Auto advance ..	32° on crankshaft.
Distributor Points Gap ..	.010"-.012".
Starter Motor Ratio	13.33 : 1 (Lucas), 9 tooth.
Dynamo	1.16 engine speed Comp. Volt. Control.
Plug make	14 mm. Champion L.10.S.
Alternatives	K.L.G. F.80.
	Lodge H.N. or H.N.P.
Sparking Plug Gap	.020"-.022".
Firing Order ..	1 3 4 2.
Ignition Timing ..	Points just breaking T.D.C.
Battery Type ..	6TW9A Positive Earth. ± .51 amp. at 10 hr. rate.
Lighting	12 volt, earth return.
Ignition	12 volt Coil and Distributor.

CLUTCH AND GEARBOX.

Clutch Type ..	Dry, $7\frac{1}{4}$" Borg & Beck.
Facing	$7\frac{1}{4}$" dia. RYZ.
Gearbox Ratios, Top	1 : 1.
3rd	1.385 : 1.
2nd	2.07 : 1.
1st	3.5 : 1.
Rev.	3.5 : 1.
Overall Ratios Top	5.125 : 1.
3rd	7.098 : 1.
2nd	10.609 : 1.
1st	17.938 : 1.
Rev.	17.938 : 1.
Layshaft End Float	.48 mm. to .77 mm.

INSTRUMENTS.

Speedometer ..	5" dia. Jaeger. 1,600 T.P.M.
Reduc. Gear for Speedo	6/15.
Rev. Counter ..	5" dia. Jaeger.
Reduc. Gear for Rev. Counter ..	On dynamo 100/225 ratio.
Petrol Gauge ..	Warning Light on panel at $2\frac{1}{2}$-3 gallons reserve.
Facia Board ..	Rexine covered.

FRONT AXLE AND STEERING.

Camber of Wheel ..	1° Positive to $-\frac{1}{2}$° on full bump.
Castor Angle ..	2° ± .5° (with side member parallel to road).
King Pin Angle ..	9° to $10\frac{1}{2}$° on full bump.
Toe-In	Nil.
Steering	Rack and Pinion.
Turning Circle ..	31' 3".
Steering Knuckle Angle	10°.
Suspension ..	Independent.
Steering Wheel ..	$16\frac{1}{2}$" dia.
Wheels	Vent disc. 4.00" × 15.00". 4J Rim.
Tyre size	5.50" × 15".
Tyre Pressure ..	18 lbs./sq. in.
Left Hand Drive ..	Available.
Lock to Lock ..	$2\frac{3}{4}$ turns of strg. wheel.

REAR AXLE.

Axle type	Semi-floating.
Drive type	Hypoid 1" offset.
Axle ratio	8/41 (5.125 : 1).
Alternatives	8/39 (4.875 : 1). Pt. No. (101910) pair.
	9/41 (4.555 : 1). Pt. No. (101593) pair.
Adjustment ..	See Workshop Manual.
Wheels	Vent Disc. 4.00" × 15.00". 4J Rim.
Tyre size	5.50" × 15".
Tyre pressure ..	18 lbs./sq. in.

SPRINGS, FRONT.

Type	Coil.
Length, Front ..	(Free) 9.59" ± $\frac{1}{16}$".
Mean dia.	3.238".
No. effective Coils..	7.5.
Dia. of wire (ground)	.498".
Max. Deflection ..	4.24".

SPRINGS, REAR.

Type	Semi-elliptic.
Width	$1\frac{1}{2}$".
No. of leaves ..	7.
Thickness of leaves	$\frac{7}{32}$".
Camber	(Free) 4.10".

BRAKES.

Types	Lockheed Hydraulic and Cable Handbrake (rear only).
Brake Drums, Front	9" dia., $1\frac{1}{2}$" wide. 2LS.
Brake Drums, Rear	9" dia., $1\frac{1}{2}$" wide.
Brake Linings, Front	Ferodo M.R.19.
Brake Linings, Rear	Ferodo M.R.19.
Friction Area, Front	49.74 sq. in.
Friction Area, Rear	49.74 sq. in.
Friction Area, Total	99.48 sq. in.

SERVICE DATA.

DIMENSIONS.

Wheelbase	7′ 10″.
Track, Front ..	3′ 11⅜″.
Rear ..	4′ 2″.
Overall Length ..	12′ 1″.
Overall Width ..	4′ 10⅝″.
Overall Height, hood up	4′ 5″.
Height, screen up hood down ..	4′ 2″.
Height, screen down	3′ 9″.
Ground Clearance	6″.

PERFORMANCE.

B.H.P.	54.4 at 5,200 R.P.M.
Max. Torque ..	765 in./lb. at 2,600 R.P.M.
H.P. per sq. in. piston area ..	2.5.
Piston speed at max. H.P.	3,068 ft. per min.
Litres per ton mile..	2,858.
M.P.H. per 1,000 R.P.M. in Top	14.42.
3rd	10.412.
2nd	6.966.
1st	4.12.
M.P.H. at 2,500 f.p.m. of piston speed	61.18.
Engine speed per 10 M.P.H. ..	693 R.P.M. (top gear).
Average petrol consumption ..	30-35 miles per gallon.
Consumption at 30 M.P.H.	39 miles per gallon.
Max. Fuel Economy	42 M.P.G. at 20 M.P.H. (1,385 R.P.M.).
Acceleration (top gear)	10-50 M.P.H. in 22.15 secs.
Piston speed per 10 M.P.H.	409 ft. per min.

WEIGHTS.

Complete Car— Fully laden ..	18 cwts. 0 qrs. 18 lbs.
Front	8 cwts. 3 qrs. 0 lbs.
Rear	9 cwts. 1 qr. 18 lbs.
Rear (no fuel) ..	8 cwts. 2 qrs. 0 lbs.
Chassis, as off line with oil and water	10 cwts. 3 qrs. 14 lbs.
Body with seats, floor, etc. (no scuttle or bumpers)	3 cwts. 0 qrs. 14 lbs.
Power Unit total ..	3 cwts. 1 qr. 16 lbs.
Laden Weight with two 12-stone passengers, 12½ galls. fuel and 100 lbs. luggage	22 cwts. 0 qrs. 0 lbs.
Front	9 cwts. 3 qrs. 0 lbs.
Rear	12 cwts. 1 qr. 0 lbs.

CAPACITIES.

Sump to level (oil)..	9 pints.
Gearbox (oil) ..	1½ pints.
Rear Axle (oil) ..	2¼ pints.
Block and Head (water)	6.02 pints.
Cooling System ..	12 pints.
Brake System Fluid	1 pint.
Fuel Tank	12½ gallons.

GRADES OF CONSUMABLES.

Engine Oil	N.O.L. Thirty.
Alternatives	Mobiloil " A."
	Double Shell.
	Castrol " XL."
Gearbox Oil ..	Hypoid " 90."
Alternatives	Castrol Hypoy.
	Esso Expee Compound 90.
	Mobilube GX.90.
	Energol EP. S.A.E.90.
	Shell Spirax 90 E.P.
	Hypoid Filtrate 90.
	Sternol Liquid Ambroleum Hypoid 90.
Rear Axle Oil ..	As for Gearbox above.
Alternatives	As for Gearbox above.
Steering Gear Oil ..	As for Gearbox above.
Alternatives	As for Gearbox above.

MODEL " T.D." MIDGET. *Continued.*

Chassis Oil ..	Duckham's Laminoid Soft, or Duckham's Adcol H.P.G. Grease.
Alternatives	Shell Retinax " C."
	Mobilgrease No. 4.
	" Castrolease " Medium.
Hydraulic Brake Fluid	Lockheed Fluid No. 33.
Hydraulic Damper Fluid	Luvax-Girling Official Piston-type Damper Fluid.

MISCELLANEOUS.

Silencer	Burgess Type.
Cooling System ..	Radiator, Pump and Fan.
Propeller Shaft ..	Hardy Spicer.
Hydraulic Dampers	Luvax-Girling, Piston Type.
Position of Battery	Under Bonnet.
Position of Toolbox	Under Bonnet.

Bulb Replacements :	Volts.	Watts.	Cap.	Lucas Number
Head, Home O/S ..	12	36	Pre-Focus	162
Head, Home N/S (Dip Left) ..	12	36/36	Pre-Focus	300
Head, RHD Export (Dip Left) ..	12	36/36	Pre-Focus	300
Head, LHD Export (Dip Right) ..	12	36/36	Pre-Focus	301
Head, for U.S.A. (Dip Right) ..	12	36/36	Pre-Focus	301
Side Lamps	12	6	S.C.C.	207
Stop Tail Lamp (1st Type)	12	6/24	S.B.C.	189
Stop Lamp (Index Bulb, 2nd Type)..	12	6/24	S.B.C.	353
Number Plate Illumination	12	6	M.C.C.	989
Panel Light ..	12	2.2	M.E.S.	987
Ignition Warning Light	2.5	0.5	M.E.S.	970
Fuel Warning Light	2.5	0.5	M.E.S.	970

Torque Loading for—

*Connecting Rod Bolts	230 lbs./ins.
*Main Bearing ..	750 lbs./ins.
Cylinder Head Nuts	600 lbs./ins.

* To the next pin hole.

OVERALL GEAR RATIOS (USING ALTERNATIVE AXLE RATIOS).

8/39 axle (4.875 : 1) :

Top	4.875 : 1.
3rd	6.752 : 1.
2nd	10.09 : 1.
1st	17.06 : 1.
Rev.	17.06 : 1.

9/41 axle (4.555 : 1) :

Top	4.555 : 1.
3rd	6.309 : 1.
2nd	9.429 : 1.
1st	15.942 : 1.
Rev.	15.942 : 1.

M.P.H. PER 1,000 R.P.M. (USING ALTERNATIVE AXLE RATIOS).

8/39 axle (4.875 : 1) :

Top	15.195 M.P.H.
3rd	10.97 M.P.H.
2nd	7.34 M.P.H.
1st	4.34 M.P.H.

9/41 axle (4.555 : 1) :

Top	16.259 M.P.H.
3rd	11.81 M.P.H.
2nd	7.85 M.P.H.
1st	4.64 M.P.H.

KILOMETERS PER 1,000 R.P.M. (USING ALTERNATIVE AXLE RATIOS).

8/39 axle (4.875 : 1) :

Top	24.453 K.P.H.
3rd	17.654 K.P.H.
2nd	11.812 K.P.H.
1st	6.98 K.P.H.

9/41 axle (4.555 : 1) :

Top	26.165 K.P.H.
3rd	18.89 K.P.H.
2nd	12.639 K.P.H.
1st	7.469 K.P.H.

SERVICE DATA.
MODEL "N.E." MAGNETTE.

Wheelbase	8′ 0″.
Track	3′ 0″.
Turning Circle	35′.
Steering Lock	Approx. 1 turn of steering wheel.
Steering Knuckle Angle	10½°.
Castor Angle	8° (includes 2° taper packing).
King Pin Angle	7½°.
Toe-In	3/16″.
Brake Drums, Front	12″ dia. (.35 max., .30 min. Carbon Steel).
Brake Linings, Front	Ferodo M.R. on steel brake shoes.
Standard Gear Ratio	8/39 (straight cut gears).
Alternatives	8/36 (straight cut), 8/33 (straight cut).
Oil Capacity, Rear Axle	1½ pints.
Diff. Gear	4 Star.
Brake Drums, Rear	12″ dia. (.35 max., .30 min. Carbon Steel).
Brake Linings, Rear	Ferodo M.R. on steel brake shoes.
Propeller Shaft	2″ dia. 100½ Joints. Balanced to 7,000 R.P.M.
Grade of Oil, Rear Axle	Castrol "R."
Steering	Cam Gears, 7 : 1 Ratio.
Steering Wheel	17″ dia., 3-Spoke Bluemel.
Silencer	Twin Brooklands or Ulster. Outside Exhaust System.
Petrol Tank Capacity	18 gallons (including 2 reserve).
Wheels	2.50″ W.B. × 18″ with racing rims, Rudge-Whitworth.
Tyres	4.75″ × 18″ Dunlop Fort.
Shock Absorbers—Front	Hartford Duplex and Luvax.
Rear	Luvax.
Battery	(Two) Lucas 6 volt.
Amp. Hr. Cap. 10 Hr. Rate	63.
Speedometer	Nil.
Speedo. Reduc. Box Ratio	Nil.
Petrol Gauge	Nil.
Facia Board	Alum. 1/8″ thick, mounted on chassis.
Revolution Counter	Jaeger 5″ dia., 8,000 R.P.M., Black Dial.
Position of Toolbox	Nil.
Engine Type	6 cyl. O.H.V. Camshaft.
R.A.C. Rating	12.08 H.P.
Bore, Stroke and Capacity	57 mm. × 84 mm., 1,286 cc.
Compression Ratio	9.8 : 1 (Depth of Head 3 41/64″).
Capacity Combustion Space	Cyl. Head 34 cc. Piston 10 cc.
Brake Horse Power	68 B.H.P. at 6,500 R.P.M.
M.P.H. per 1,000 R.P.M.	16 M.P.H. in top gear (8/39 axle).
Firing Order	1 4 2 6 3 5.
Ignition	Distributor. Lucas special Bosch type.
Sparking Plug	K.L.G. 690 Long Reach.
Tappet Clearances—Inlet	.010″.
Exh.	.016″.
Valve Timing Inlet	Opens 25° B.T.D.C. Closes 60° A.B.D.C.
Exh.	Opens 60° B.B.D.C. Closes 25° A.T.D.C.
Valve Timing measured round Flywheel .. Inlet	Opens 2 3/16″ B.T.D.C. Closes 5 3/16″ A.B.D.C.
Exh.	Opens 5 3/16″ B.B.D.C. Closes 2 3/16″ A.T.D.C.
Carburettor	Twin S.U. 1 3/8″ dia. Horizontal (Hiduminium).
Carburettor Needle	C.I.
Piston and Rings	Aerolite Mk.V. (raised top). 2 Compression, 1 Scraper.
Piston Clearance	.0035″-.0045″.
Connecting Rod	Steel Stamp, lined with racing metal.
Dia. and Length of Big Ends	1 5/8″ dia., 1.120″ long.
Sump Capacity	7 quarts.
Fuel	50/50 Benzol Ethyl, or 30/70 Benzol Ethyl.
Number of Main Bearings	4.
Oil Pump By-Pass Pressure	60-80 lbs./sq. in.
Grade of Engine Oil	Castrol "R."

(NOTE.—On this type of engine the capacity of the combustion space varies slightly. The above compression ratio of 9.8 : 1 is taken as an average.)

Clutch	Double Plate.
Clutch Springs	6 per set. Compress to 1⅛″ under load of 90 lbs.
Gearbox	Wolseley. Manual. 4 Speeds and Reverse.
Gearbox Ratios Top	1 : 1.
3rd	1.36 : 1.
2nd	2.14 : 1.
1st	3.58 : 1.
Rev.	3.58 : 1.
Gearbox Oil Capacity	2 pints.
Grade of Oil, Gearbox	Castrol "R."
Overall Length	12′ 7″.
Overall Height	3′ 6″ (screen flat).
Overall Width	4′ 6″.
Weight of Complete Car	16 cwts. 1 qr. 7 lbs.
Ground Clearance	6″.
Water Capacity	3 gallons.
Cooling System	Pump.

SERVICE DATA.
MODEL "QA" MIDGET.

Chassis No. QA.0251 to Chassis No. QA 0258.

Wheelbase	7′ 10 3/16″.
Track	3′ 9″.
Turning Circle	35′ 0″.
Steering Lock	Approx. 1 turn of strg. wheel (full travel).
Steering Knuckle Angle	10½°.
Castor Angle	8° (includes 2° taper packing).
King Pin Angle	7½°.
Toe-In	3/16″.
Brake Drums, Front	12″ dia. Holden & Hunt.
Brake Linings, Front	Ferodo M.R.
Standard Gear Ratio	8/36 straight cut gears.
Alternatives	8/39, 8/33, straight cut gears.
Oil Capacity, Rear Axle	1½ pints.
Diff. Gear	4 star.
Brake Dums, Rear	12″ dia. Holden & Hunt.
Brake Linings, Rear	Ferodo M.R.
Propeller Shaft	3″ dia.
Steering	Cam Gears, 7 : 1 ratio.
Steering Wheel	Bluemel, 17″ dia. 3-Spoke.
Silencer	Brooklands. Outside Exhaust System.
Petrol Tank Capacity	19 gallons (including 2 reserve).
Wheels	2.50″ × 18″ Racing Rims.
Tyres	4.75″ × 18″ Dunlop Fort.
Shock Absorbers—Front	Hartford Duplex.
Rear	Luvax.
Battery	(Two) Lucas 6RW.9.
Amp. Hr. Cap. 10 Hr. Rate	51.
Speedometer	Nil.
Rev. Counter	5″ dia., 8,000 R.P.M.
Speedo. Reduc. Box	Nil.
Petrol Gauge	Nil.
Facia Board	Dural supported by bracket mounted on frame.
Position of Toolbox	Nil.
Engine Type	4 cyl. O.H. Camshaft, Supercharged.
R.A.C. Rating	8.05 H.P.

SERVICE DATA.
MODEL "QA" MIDGET. *Continued.*

Bore, Stroke and Capacity ..	57 mm. × 73 mm., 746 cc.
Compression Ratio	6.2 : 1.
Capacity Combustion Space ..	Cyl. Head 35 cc. + 1 cc. Piston.
M.P.H. per 1,000 R.P.M.	18.5 M.P.H. (8/36 axle ratio).
Firing Order ..	1 3 4 2.
Ignition	Distributor, Delco-Remy.
Sparking Plug ..	K.L.G.718 for road work. K.L.G.689 for racing.
Tappet Clearances..	Inlet .008". Exh. .014".
Valve Timing Inlet	Opens 15° B.T.D.C. Closes 55° A.B.D.C.
Exh.	Opens 50° B.B.D.C. Closes 20° A.T.D.C.
Valve Timing measured round Flywheel .. Inlet	Opens 1 5/16" B.T.D.C. Closes 4 25/32" A.B.D.C.
Exh.	Opens 4 11/32" B.B.D.C. Closes 1 3/4" A.T.D.C.
Carburettor ..	S.U. 1 7/8" dia.
Carburettor Needle	R.M.2.
Piston and Rings ..	Aerolite Special (No. 14 Cam form). 2 Comp. 1 Scraper.
Piston Clearance ..	.004"-.005".
Connecting Rod ..	B.S. Special, s.11 material. Racing metal lined.
Dia. and Length Big Ends	1 5/8" dia. 1.120" long.
Sump Capacity ..	1 1/4 gallons.
No. of Main Bearings	3.
Oil Pump By-Pass Pressure	60-80 lbs./sq. in.
Supercharger ..	McEvoy. Zoller Q4.
Clutch	M.G. Design. Double Plate Inoperative.
Clutch Springs ..	6 per set. Compress to 1 1/8" under load of 90 lbs.
Gearbox	E.N.V. Wilson Preselector Type QA., or Type QH.
Gearbox Ratios— Type QA. Top	1 : 1.
3rd	1.36 : 1.
2nd	2.00 : 1.
1st	3.40 : 1.
Rev.	5.07 : 1.
Type QH. Top	1 : 1.
3rd	1.31 : 1.
2nd	1.84 : 1.
1st	3.097 : 1.
Rev.	4.15 : 1.
Overall Length ..	11' 11".
Overall Height ..	3' 7" (screen down).
Overall Width ..	4' 5".
Weight of Complete Car	13 cwts. 21 lbs. in racing trim.
Ground Clearance..	6".
Water Capacity Cooling System ..	2 gallons 3 pints.

SERVICE DATA.
MODEL "R.A." MIDGET.

Chassis No. RA.0251, 1935 to Chassis No. RA.0260, 1935.

Wheelbase	7' 6 1/2".
Track .. Front	3' 10 3/8".
Rear	3' 9 1/4".
Turning Circle ..	28' 0".
Steering Lock ..	1 1/3 turns of steering wheel (full travel).
Steering Knuckle Angle	10 1/2°.
Castor Angle ..	5°.
King Pin Angle ..	7 1/2°.
Toe-In	1/16".
Brake Drums, Front	12" dia. Holden & Hunt.
Brake Linings, Front	Ferodo M.R.
Standard Gear Ratio	8/33.
Alternatives ..	8/36. 8/39.
Oil Capacity, Rear Axle	1 quart.
Diff. Gear	4 Star.
Brake Drums, Rear	12" dia. Holden & Hunt.
Brake Linings, Rear	Ferodo M.R.
Propeller Shaft ..	2 1/4" dia. 1.200 K.R. Joints. Balanced to 8,000 R.P.M.
Steering	Cam Gears, M.G. Design. Ratio 9 : 1.
Steering Wheel ..	Bluemel 17" dia. 3-Spoke, Special Semi-rigid.
Silencer	Twin Brooklands. Outside Exhaust system.
Petrol Tank Capacity	21 gallons (3/4 gallon res.).
Wheels	18" × 3.00 Racing Rims.
Tyres	18" × 4.75" Dunlop Fort.
Shock Absorbers— Front	Luvax.
Rear	Luvax.
Battery	(Two) 6 volt.
Amp. Hr. Cap. 10 Hr. Rate ..	12.
Speedometer ..	Nil.
Rev. Counter ..	8,000 R.P.M.
Petrol Gauge ..	Nil.
Facia Board ..	M.S. Sheet, carried on tubular frame mounted on chassis.
Position of Toolbox	Nil.
Engine Type ..	4 cyl. Overhead Camshaft, S/c.
R.A.C. Rating ..	8.05 H.P.
Bore, Stroke and Capacity ..	57 mm. × 73 mm., 746 cc.
Compression Ratio	6.2 : 1.
Capacity Combustion Space ..	Cyl. Head 35 cc. + Piston 1 cc.
M.P.H. per 1,000 R.P.M.	20.2 in top gear (8/33 axle ratio).
Firing Order ..	1 3 4 2.
Ignition	Magneto (Lucas).
Sparking Plug ..	K.L.G. 689.
Tappet Clearances..	Inlet .008". Exh. .014".
Valve Timing Inlet	Opens 15° B.T.D.C. Closes 55° A.B.D.C.
Exh.	Opens 50° B.B.D.C. Closes 20° A.T.D.C.
Valve Timing measured round Flywheel Inlet	Opens 1 5/16" B.T.D.C. Closes 4 25/32" A.B.D.C.
Exh.	Opens 4 11/32" B.B.D.C. Closes 1 3/4" A.T.D.C.
Carburettor ..	S.U. 1 7/8" dia.
Carburettor Needle	R.M.2.
Piston and Rings ..	Aerolite Special (No. 14 Cam form). 2 Comp. 1 Scraper.
Piston Clearance ..	.004"-.005".
Connecting Rod ..	B.S. Special S.11 material, Racing metal lined.
Dia. and Length, Big Ends ..	1 5/8" dia. × 1.120" long.
Sump Capacity ..	1 1/4 gallons.
No. of Main Bearings ..	3.
Oil Pump By-Pass Pressure	120 lbs./sq. in.
Supercharger ..	McEvoy. Zoller Type 4RA.
Clutch	M.G. design. Double Plate, Inoperative.
Clutch Springs ..	6 per set. Compress to 1 1/8" under load of 120 lbs.
Gearbox	E.N.V. Wilson Preselector Type RH.
Gearbox Ratios Top	1 : 1.
3rd	1.31 : 1.
2nd	1.84 : 1.
1st	3.097 : 1.
Rev.	4.15 : 1.
Overall Length ..	11' 9".
Overall Height ..	3' 9".
Overall Width ..	4' 8 1/4".
Weight of Complete Car	12 cwts. 1 qr. 16 lbs.
Ground Clearance..	5".
Water Capacity ..	2 1/2 gallons.
Cooling System ..	Pump

SERVICE DATA.

Comm. Chassis. No. Y.0251 ,1947

POWER UNIT (Comm. Eng. No. XPAG/SC/10001).

Engine Type ..	4 cyl. O.H.V. Pushrod operated.
Bore, Stroke and Capacity ..	66.5 mm. × 90 mm. 1,250 cc.
R.A.C. Rating ..	10.97 H.P.
First Oversize Bore	66.75 mm.
Later Oversize Bores	67-67.25-67.5-67.75 mm.
Max. Oversize Bore	68.00 mm.
Compression Ratio	7.2-7.4 : 1.
Depth of Cylinder Head	76.75 mm.
Capacity Combustion Space ..	45.5 cc.
Piston	Aerolite 3-Ring.
Piston Weight ..	9 oz.
Compression Rings	2. Width .0885″ + .000″. − .0005″.
	Radial Thickness { .109″. .101″.
Scraper Rings ..	1. 8 Slots. Width .1575″-.0005″.
	Radial Thickness { .105″. .097″.
Piston Ring Gap ..	.006″ to .010″.
Gudgeon Pin ..	18 mm. dia., Clamp type.
Gudgeon Pin Weight	2½ oz.
Gudgeon Pin Fit in Piston	Bore in Piston. 18 mm. − .0001″. − .0003″.
Gudgeon Pin Fit in Con. Rod ..	Clamp bolt.
Bore of Con. Rod ..	18 mm. + .03 mm. − .01 mm.
Connecting Rod ..	Steel, 178 mm. length. Loose Steel Shell White Metal Bearings.

CRANKSHAFT.

Main Journal, Front	Dia. 52 mm. − .035. − .015.
	Width 38 mm.
	Radius 2.5 ± .15.
Intermain Journal ..	Dia. 52 mm. − .035. − .015.
	Radius 2.5 ± .15.
Main Journal, Rear	Dia. 52 mm. − .035. − .015.
	Radius 2.5 ± .15.
Crankpin Journal ..	Dia. 45 mm. − .035. − .015.
	Width 28 mm.
	Radius 2.5 ± .15.

Crankshaft Regrind.—

Main Bearing, Front	Loose Steel Shell White Metal Lined	{ 52.020 mm. 52.005 mm.
Main Bearing, Inter		{ 52.020 mm. 52.005 mm.
Main Bearing, Rear		{ 52.020 mm. 52.005 mm.

Big End Bearing ..	1.825 mm. thick × 22.46 mm.
	1.831 mm. thick × 22.21 mm.
Camshaft Drive ..	Chain.
Camshaft Bearings	Front—White Metal.
	Rear—Zinc base Alloy.
Valve Throat Dia.— Exh.	26 mm.
Inl.	30 mm.
Valve Seat Angle ..	30°.
Valve Spring Pressure	Shut Inner 31 lbs.
	Outer 62 lbs.
	Open Inner 43 lbs.
	Outer 82 lbs.
Valve Lift	6.5 mm.
Valve Timing Inlet	Opens 11° B.T.D.C.
	Closes 57° A.B.D.C.

MODEL " Y " SALOON.

Exh.	Opens 52° B.B.D.C.
	Closes 24° A.T.D.C.
Valve Timing measured round Flywheel .. Inlet	Opens 28.7 mm. round Flywheel Periphery.
Tappet Clearances— Cold	.037″.
Hot	.019″.
Oil Pump	Gear type, runs at ½ engine speed.
Oil Pressure ..	70 lbs./sq. in. to 40 lbs./sq. in. min. 60 lbs./sq. in. By-Pass.
Oil Filter	Full Flow Type, 82 dia. × 160 dia. W. Breedon.
Carburettor ..	S.U. 1¼″ Semi-D.D.
Carburettor Needle	Standard FI.
	Weaker EF.
	Richer DK.
Fan	12¾″ dia., runs at 9/10ths engine speed.
Flywheel	298.9 mm., 93 teeth (Later : 120 teeth).
Crypton Compression Pressure Test	90 lbs.-100 lbs. (throttle open and at starter speed).
Valve Lift	.256″.
Length Engine Dipstick	8½″ from underneath cap to " Full " mark.
Dipstick Hole ..	6.5 mm. dia., counterbored for tube 2.5 mm. deep—.375-.001-.002 dia.

ELECTRICAL.

Distributor Rotation	Anti-clockwise.
Distributor Points Gap	.010″-.012″.
Starter Motor Ratio	9.3 : 1 (Later : 13.33).
Starter Motor Type	Lucas, 4 1/16″ dia.
Dynamo	Runs at 1.16 engine speed.
	4 9/16″, C.V.C.
Sparking Plug ..	14 mm. Champion L.10S.
Alternatives	K.L.G. F.80.
	Lodge H.N. (or better H.N.P.).
Sparking Plug Gap	.020″-.022″.
Ignition Timing ..	Points just breaking T.D.C.
Battery Type ..	Lucas STXW9A.
	Positive Earth.
	51 amp. at 10 Hr. Rate.
Lighting system ..	12 volt, earth return.
Sparking Plug for 6 lb. supercharge	Champion L.11S.
	K.L.G. F.220.
	Lodge H.N.P.
Distributor Cam Angle	(Contact Breaker Closed Period) 45° ± 4°.

CLUTCH AND GEARBOX.

Clutch Type ..	Dry, 7¼″ Borg & Beck.
Facing	Fabric.
Gearbox Ratios Top	1 : 1.
3rd	1.385 : 1.
2nd	2.07 : 1.
1st	3.5 : 1.
Rev.	3.5 : 1.
Overall Ratios Top	5.143 : 1.
3rd	7.121 : 1.
2nd	10.646 : 1.
1st	18.00 : 1.
Rev.	18.00 : 1.

INSTRUMENTS.

Speedometer ..	British Jaeger
Reduc. Gear for Speedo.	6/15 ratio.
Rev. Counter ..	None.

SERVICE DATA.
MODEL " Y " SALOON. *Continued*

Petrol Gauge	British Jaeger.
Facia Board	Walnut.

FRONT AXLE AND STEERING.

Camber of Wheel	Nil to $-1\frac{1}{2}°$ on Bump.
Castor Angle	$1° \pm \frac{1}{2}°$.
King Pin Angle	$10°$ to $11\frac{1}{2}°$ on Bump.
Toe-In	Nil.
Steering	Rack and Pinion, $2\frac{5}{8}$ turns, lock to lock.
Turning Circle	35′ on either lock.
Knuckle Angle	10°.
Suspension	Independent, Coil Springs.
Steering Wheel	$16\frac{1}{2}$″ dia. Wire Spokes, 3″ adjustment.
Wheels	Vent Disc, 3.00 × 16″.
Tyre Size	5.25 × 16″ E.L.P.
Left Hand Drive	Not available.

REAR AXLE.

Axle type	$\frac{3}{4}$ Floating.
Drive type	Spiral Bevel.
Axle Ratio	7/36.
Lash	.006″ to .008″.
Wheels	Vent Disc, 3.00 × 16″.
Tyre Size	5.26 × 16″ E.L.P.

TYRE PRESSURES.

Front Tyres	23 lbs./sq. in.
Rear Tyres	25 lbs./sq. in.

SPRINGS, FRONT.

Type	Coil.
Free Length	9.82″ ± $\frac{1}{16}$″.
Effective Coils	7.34.
Mean Diameter	3.238″.

SPRINGS, REAR.

Type	Semi-elliptic.
Length, Front	$42\frac{1}{2}$″ ± $\frac{1}{8}$″.
Width	$1\frac{1}{2}$″.
No. of Leaves	7.
Thickness of Leaves	$\frac{1}{4}$″.
Camber	4″ free.

BRAKES.

Type	Lockheed Hydraulic.
Handbrake	By cable to rear only.
Type Linings	Ferodo M.R.19.
Brake Drums, Front	9″ dia. Pressing.
Rear	9″ dia. Pressing.
Brake Lining Area— Front	26 sq. in. per pair.
Rear	26 sq. in. per pair.
Total	104 sq. in.
Friction Area, Front	11.31 sq. in. per shoe.
Rear	11.31 sq. in. per shoe.
Total	90.48 sq. in.
Brake Lining Area	107 sq. in. per ton.

DIMENSIONS.

Wheelbase	8′ 3″.
Track — Front	3′ $11\frac{3}{8}$″.
Rear	4′ 2″.
Overall Length	13′ $5\frac{3}{16}$″.
Overall Width	4′ 11″.
Overall Height	4′ 9″.
Ground Clearance	6″ (lowest point Rear Check Strap bolts).

PERFORMANCE.

B.H.P.	46 B.H.P. at 4,800 R.P.M.
Maximum Torque	702 lb./in. at 2,400 R.P.M.
H.P. per sq. in. Piston Area	2.13 B.H.P.
Piston Speed at max. B.H.P.	2,835 ft. per min.
Litres per ton mile	2,627 in top gear.
M.P.H. per 1,000 R.P.M.	14.6 in top gear.
Road Speed at 2,500 F.P.M. Piston Speed	61.78 in top gear.
Engine Speed per 10 M.P.H.	685 R.P.M. in top gear.
Petrol Consumption, at	30 M.P.H., 36.5 M.P.G. 40 M.P.H., 32.7 M.P.G. 50 M.P.H., 29.0 M.P.G.
Braking on dry concrete at	30 M.P.H., 31 feet. 50 M.P.H., 92 feet. 70 M.P.H., 227 feet.
Acceleration	0-30 M.P.H., 7.5 secs. 0-50 M.P.H., 18.8 secs. 0-60 M.P.H., 29.5 secs.
B.H.P. per Litre	S.C. 35.62., D.C. 42.65.
Maximum Speed	71.4 M.P.H. average.
Handbrake Braking on dry tarmac	30 M.P.H., 45 feet.
Piston Area	21.57 sq. in.
B.M.E.P.	116 at 2,800 R.P.M.
Piston Area, sq. in. per ton	22.3.

WEIGHTS.

Complete Car (with water, oils but no petrol)	19 cwt. 2 qrs. 0 lbs.
Front	9 cwt. 2 qrs. 0 lbs.
Rear	10 cwt. 0 qrs. 0 lbs.
Chassis as off line, with bumpers but no wings or lamps	10 cwt. 3 qrs. 22 lbs.
Power Unit total	3 cwt. 2 qrs. 5 lbs.
Chassis with bonnet, front wings, oil and water but less bumpers	13 cwt. 0 qrs. 0 lbs.
Loose Kit	$1\frac{3}{4}$ cwt. 0 qrs. 0 lbs.
Total	$14\frac{3}{4}$ cwts.

CAPACITIES.

Sump to level (oil)	9 pints.
Gearbox (oil)	$1\frac{1}{4}$ pints.
Rear Axle (oil)	$1\frac{1}{2}$ pints.
Block and Head (water)	6.02 pints.
Cooling System	$13\frac{1}{2}$ pints.
Fuel Tank	8 gallons.

MISCELLANEOUS.

Cooling System	Water Radiator, Pump and Fan.
Prop. Shaft	Hardy Spicer $2\frac{1}{2}$″ tube, $46\frac{3}{8}$″ face to face, Joint KR1111.
Hydraulic Dampers	Luvax-Girling, Piston Type.
Position of Battery	Under Bonnet.
Position of Toolbox	Spare Wheel Locker.
Jacks, Type	Red Jackall.

Weight of Car, fully equipped and laden, with four passengers, 1-11 stone ; 2-12 stone ; 1-14 stone :—

Total	1 ton, 5 cwts. 3 qrs. 0 lbs.
Front	11 cwts. 0 qrs. 14 lbs.
Rear	14 cwts. 2 qrs. 7 lbs.

SERVICE DATA.
MODEL " Y.T." TOURER.

Commencing 1948.
POWER UNIT. Type XPAG/TR or TL.

Engine type	4 cyl. O.H.V. Pushrod Operated.

SERVICE DATA.
MODEL "Y.T." TOURER. *Continued.*

Treasury Rating	10.97 H.P.
Bore, Stroke and Capacity	66.5 mm. × 90 mm., 1,250 cc. 2.62″ × 3.54″, 76.3 cu. in.
Bore/Stroke Ratio	1.353 : 1.
First Oversize Bore	66.75 mm.
Later Oversize Bores	67 mm. 67.5 mm. 67.75 mm.
Max. Oversize Bore	68 mm.
Compression Ratio	7.2-7.4 : 1.
Depth of Cylinder Head	76.75 mm.
Capacity of Comb. Chamber	45.5 cc.
Cylinder Head Gasket	Copper and Asbestos .045″ thick.

PISTONS.

Type	"Aerolite" Alum. Alloy, tin coated, controlled expansion.
Compression Ring Grooves	Width .0886/.0887″. Dia. 2.381/2.385″.
Scraper Ring Groove	Width .1576/.1577″. Dia. 2.385/2.389″.
Side Clear. of rings in groove	.001″-.002″.
Compression Rings	2. Width .0885″, .109/.101 thick.
Scraper Ring (8 Slot)	1. Width .1576/.1577″. Thickness .105/.097″.
Piston Ring Gap	.006″-.010″.
Gudgeon Pin Bore	18 mm. −.0001″/.0003″.
Piston Area	21.57 sq. in.
Piston area, sq. in. per ton	23.
Gudgeon Pin	Clamp type. Dia. 18 mm. Length 58.5 mm.

CONNECTING RODS.

Type and Length	Steel, 178 mm. long, with loose steel shell white metal bearings.
Bore at Big End	46.658/46.671 mm.
Bore at Small End	18 mm. + .03 mm. − .01 mm.
Width at Big End	27.685/27.890 mm.
Width at Small End	19 mm.

CRANKSHAFT.

Main Journal, Front	Dia. 52 mm. − .035 mm. − .015 mm. Width 38 mm. Rad. 2.5 mm. ± .015 mm.
Main Journal, Inter	Dia. 52 mm. − .035 mm. − .015 mm. Width 38 mm. Rad. 2.5 mm. ± .015 mm.
Main Journal, Rear	Dia. 52 mm. − .035 mm. − .015 mm. Width 40 mm. Rad. 2.5 mm. ± .15 mm.
Crankpin Journal	Dia. 45 mm. − .035 mm. − .015 mm. Width 28 mm. Rad. 2.5 mm. ± .15 mm.
Main Bearing, Front	Dia. 52.020/52.005 mm. Width 35.12/34.88 mm.
Main Bearing, Inter	Dia. 52.020/52.005 mm. Width 37.955/37.925 mm. Radius 2.75 mm. U/cut. .25 mm.
Main Bearing, Rear	Dia. 52.020/52.005 mm. Width 35.12/34.88 mm.
Bore size of crankcase for main bearings	56.34 mm. ± .01 mm.
Main Bearing clearance on crankshaft dia.	.020/.055 mm.
Main Bearing (inter) end float on crankshaft	.035/.095 mm.
Big End Bearing	Dia. 45 mm. Width 22.46/22.21. Thick 1.825/1.831 mm.
Big End clearance on crankshaft dia.	.011/.056 mm.
Crankshaft First Regrind	51.7 mm.
Later Regrinds	51.5 mm. 51.25 mm. 51 mm. Max. 50.75 mm.
Camshaft Drive	By chain. Roller ⅜ pitch. 60 pitches, endless.
Camshaft Bearing, Front	White Metal. O/Dia. 43.61 mm. I/Dia. 41 mm. ± .05/.08 mm. ream in pos. in block.
Camshaft Bearing, Centre and Rear	O/Dia. 43.5 mm. + .00. − .02. I/Dia. 23 mm. + .015 mm. − .005 mm.
Bore of Cylinder Block for Bearing	43.5 mm. + .02 mm./−.01 mm.
Camshaft End Float	.005″/.013″ (.125 mm./.325 mm.).
Diameter of Shaft for Bearing	Front 41 mm. + .01 mm. − .02 mm. Centre and Rear 23 mm.
Cam Lift	5.785 mm., equals 8 mm. at valve.

VALVES AND GUIDES.

Valve Port Throat dia.	Inlet 30 mm. Exhaust 26 mm.
Valve Head dia.	Inlet 33 mm. Exhaust 31 mm.
Valve Stem dia.	8 mm. + .01 mm./ .01 mm.
Valve Guides	Bore 8.06/8.08 mm. O/Dia. 14 mm. + .02 mm. + .04 mm.
Valve Guide Length	Inlet 59 mm. Exhaust 54 mm.
Valve Seat Angle	30°.
Valve Spring Pressure	Shut 93 lbs. Open 123 lbs. (Double Spring).
Valve Lift	8 mm.
Valve Timing, Inlet	Opens 11° B.T.D.C. Closes 57° A.B.D.C.
Exh.	Opens 52° B.B.D.C. Closes 24° A.T.D.C.
Valve Timing measured round Flywheel .. Inlet	Opens 28.7 mm. round flywheel periphery.
Valve Opens Inlet	Piston traverse 1.19 mm. B.T.D.C.
Valve Clearance	.019″, engine hot. .037″, engine cold.
Ignition Timing	T.D.C.
Firing Order	1 3 4 2.
Sparking Plugs	Champion L.10.S.
Alternatives	K.L.G. F.80. Lodge HN. Lodge HNP.
Sparking Plug Gap	.020″-.022″.
Oil Pump Speed	Gear type, runs at half engine speed.
Oil Pressure	70 lbs./sq. in.-40 lbs./sq. in.
Oil Pump By-Pass	60 lbs./sq. in.
Oil Filter	Full flow type.
Flywheel	Dia. 298.9 mm. 93 teeth (later 120).

PROPELLER SHAFT.

Type	Hardy Spicer needle bearing.
Length	Face to face, 46¾″.
Tubular Shaft	2½″ dia.
Joint Size	KR.1111. GB.22/1118. GB.65

CLUTCH AND GEARBOX.

Type	Single plate, dry, Borg and Beck.
Diameter	7¼″.
Facings	Fabric.

SERVICE DATA. MODEL "Y.T." TOURER. *Continued.*

Clutch Springs ..	132 to 143 ins./lbs. Colours : Red. 6 off. 1.93" free length.
Clutch Plate Damper Springs	3-Blue on "drive." 3-Black on "overrun."
Clutch Linings ..	Ext. dia. 7⅛". Int. dia. 4⅞". Thickness ⅛".
Gearbox	4 speed, synchromesh on three higher ratios.
Gearbox Ratios Top	1 : 1.
3rd	1.385 : 1.
2nd	2.07 : 1.
1st	3.5 : 1.
Rev.	3.5 : 1.
Overall Ratios Top	5.143 : 1.
3rd	7.121 : 1.
2nd	10.646 : 1.
1st	18.00 : 1.
Rev.	18.00 : 1.
Gearbox Oil Capacity	1¼ pints.

INSTRUMENTS.

Speedometer ..	British Jaeger.
Speedo. Reduc. Gear Ratio	6/15.
Revolution Counter	British Jaeger.
Electric Clock ..	British Jaeger.
Revolution Counter Reduc. Gear ..	On dynamo, 100/225 ratio.
Petrol Gauge ..	British Jaeger.
Oil Gauge	British Jaeger.
Facia Board ..	Rexine covered.

FRONT AXLE AND STEERING.

Camber of Wheel ..	Nil to − 1½° on bump.
Castor Angles ..	1° ± ½°.
King Pin Angles ..	10° to 11½° on bump.
Toe-In	Nil.
Knuckle Angle ..	10°.
Steering	Direct acting Rack and Pinion. 2⅝ turns lock to lock.
Steering Wheel ..	Bluemel 16½", wire spokes, 3 in. adjustment.
Turning Circle ..	35".
Suspension ..	Independent.
Left Hand Drive ..	Available.

REAR AXLE.

Type	¾ floating.
Drive	4 Star Diff. Spiral Bevel.
Ratio	7/36 (5.143 : 1).
Adjustment ..	Only Crown Wheel by adjusting nuts.
End play	Nil.
Crown Wheel and Pinion backlash..	.006"-.008".
Oil Capacity of Rear Axle	1½ pints.

SPRINGS (REAR).

Type	½ elliptic interleaved with rubber.
Length	42" + .000" − 125" between spring eyes, when flat.
Width	1½".
Number of Leaves..	8.
Thickness of Leaves	¼".
Camber	2.94" free.
Working Load ..	600 lbs.
Max. Deflection ..	4.44".
Mean Rate ..	248 lbs./in.
Static Deflection ..	2.44".

SPRINGS (FRONT).

Type	Coil.
Length	9.82" ± 1/16" free.
Width	3.238" mean diameter.
Number of Leaves..	7.34 effective coils.
Thickness	.538" wire dia. (ground).
Working Load ..	629 lbs. at 7.47" ± 1/32". 1390 lbs. at 6.63" ± 1/32".
Max. Deflection ..	5.005" solid height.
Mean Rate ..	435 lbs./ins.

WHEELS AND TYRES.

Type	Dunlop CDM.317. Ventilated Disc.
Wheel Size	3.00 × 16".
Tyre Size	5.25 × 16" ELP.
Tyre Pressures Front	23 lbs./sq. in.
Rear	25 lbs./sq. in.

BRAKES.

Type	Lockheed Hydraulic, handbrake by cable to rear only.
Type Lining ..	Ferodo MR.19.
Brake Drums, Front	Pressing, 9" dia.
Brake Drums, Rear	Pressing, 9" dia.
Length of Linings ..	8½" ; width 1½" ; thickness 3/16".
Brake Lining Area— Front	26 sq. in.
Rear	26 sq. in.
Friction Area Front	26 sq. in.
Rear	26 sq. in.
Brake Lining Area, per ton	112 sq. in. per ton.

FUEL SYSTEM.

Petrol Tank Capacity	8 gallons.
Fuel Delivery ..	S.U. Electric Pump, Type L.
Carburettors ..	Twin S.U. 1¼" dia. Semi-D.D.
Carburettor Needles	Standard FI. Richer DK. Weaker EF.
Air Cleaner ..	A.C. Sphinx.
Home	Dry 1527131.
Export	Oil Bath E/AC.2563.

COOLING SYSTEM.

System	Water, radiator, fan and pump with thermostat.
Capacity of System	13½ pints.
Capacity of Radiator	7½ pints.
Circulation ..	By pump fitted at front of cylinder block.
Drain Taps.. ..	1—bottom of radiator. 1—front end of cylinder block below exhaust manifold.
Fan	12¾", runs at .9 engine speed.

HYDRAULIC DAMPERS.

Type	Girling Piston type front and rear

HYDRAULIC DAMPERS.

Front	Luvax Girling PR5X/2 No. S87/30 Luvax Girling PPR5/7 L.H. S87/31y. Luvax Girling PPR5/8 R.H. S87/31x.

ELECTRICAL.

Lighting	12 volt, earth return.
Distributor Rotation	Anti-clockwise.
Manual advance in degrees	Auto ; advance—32° on crankshaft.
Advance Starts (crank R.P.M.) ..	400-800.
Maximum Advance (crank R.P.M.) ..	4,400.

SERVICE DATA.
MODEL "Y.T." TOURER. *Continued.*

Firing Point ..	Points just breaking T.D.C.
Contact Breaker Gap	.012".
Distributor Cam Angle (contact breaker closed period)	45° ± 4°.
Starter	Lucas, Model M418C. No. 255378.
Starter Motor Ratio	9.3 : 1 (Later 13.33).
Starter Motor Speed	400 R.P.M. at 8 volts. 1,000 R.P.M. at 9 volts. 1,500 R.P.M. at 10 volts. 3,000 R.P.M. at 12 volts.
Starter Torque ..	3.2 lbs. at 3,000 R.P.M.
Dynamo	Lucas, Model C45Y-V3. No. 228139 C.V.C.
Dynamo Speed ..	1.16 engine speed.
Battery	Type STXW9A, Lucas. Positive. Earth. Rate 51 amp. at 10 hrs. 58 amp. at 20 hrs.

DIMENSIONS.

Overall Length ..	164".
Overall Height ..	58½".
Overall Width ..	59".
Wheelbase	99".
Track, Front	47⅜".
Rear	50".
Ground Clearance..	6" (lowest point—rear check strap bolt).

PERFORMANCE.

Brake Horse Power	54.4 at 5,200 R.P.M.
Brake Mean Effective Press.	125 at 2,600 R.P.M.
Maximum Torque..	765 lbs./ins. at 2,600 R.P.M.
H.P. per sq. in. piston area	2.5 B.H.P.
F.P.M. of Piston at max. H.P.	3,068.
Litres per ton mile..	2,732 in top gear, dry weight.
B.H.P. per Litre ..	43.62.
M.P.H. per 1,000 R.P.M.	14.6.
Speed at 2,500 f.p.m. of piston speed ..	61.78.
Engine speed per 10 M.P.H.	685 R.P.M. in top gear.
Petrol Consumption	30 M.P.H.—35.99 M.P.G. 40 M.P.H.—31.55 M.P.G. 50 M.P.H.—27.00 M.P.G.
Braking, Tar/Flint surface	30 M.P.H.—36.5'. 50 M.P.H.—127.3'. 70 M.P.H.—286.2'.
Handbrake Braking	30 M.P.H.—91' on tar and flint surface.
Acceleration ..	0-30 M.P.H. (2nd)— 7.2 secs. 0-50 M.P.H. (3rd)—19.4 secs. 0-60 M.P.H. (Top)—31.1 secs. 0-70 M.P.H. (Top)—52.5 secs.
Maximum Speed (average) ..	73.2 M.P.H.

WEIGHTS.

Complete Car, water and oil	18 cwts. 3 qrs. 20 lbs. (no petrol).
Front	9 cwts. 2 qrs. 14 lbs.
Rear	9 cwts. 1 qrs. 4 lbs.
Chassis, as off line..	11 cwts. bumpers fitted, no wings and lamps.
Power Unit Total ..	3 cwts. 2 qrs. 10 lbs.
Weight for Car per sq. in. of Drum Friction Area ..	12.5 lbs. dry.

GRADES OF CONSUMABLES.

Engine Oil . ..	N.O.L. "Thirty" Engine Oil.
Gearbox	N.O.L. "E.P." Transmission Oil 140.
Rear Axle	N.O.L. "E.P." Transmission Oil 140.
Chassis Oil ..	Duckham's Laminoid Soft, or H.P.G. Grease.
Steering Gear Oil ..	N.O.L. "E.P." Transmissio Oil 140.
Hydraulic Brake Fluid	Lockheed "Orange."
Hydraulic Dampers	Luvax-Girling Official Piston-type Hydraulic Damper Fluid, Thin.
Hydraulic Jacks ..	"Red Jackall" Fluid.

MISCELLANEOUS AND ACCESSORIES.

Engine Sump Capacity	9 pints.
Silencer	Burgess Type, Straight Through.
Hydraulic Jacks ..	Red Jackall.
Position of Battery	Under bonnet, on dash.
Position of Tool Box	Spare wheel compartment, below boot.
Fuses	35 amperes.

BULB REPLACEMENTS.
Lucas—all cars except for U.S.A.

Headlamps ..	No. 171 l.h. No. 54 r.h.
Sidelamps	No. 207.
Rear	No. 207.
Stoplamp	No. 207.
Foglamp	No. 87.
Reverse Lamp ..	No. 1.
Panel Illumination..	No. 207.
Trafficators.. ..	No. 256.
Ignition Warning Lamp	No. 970.

FRONT SPRING MODELS. M. TYPE 2 SEATER & COUPE. PART No: P. 103.	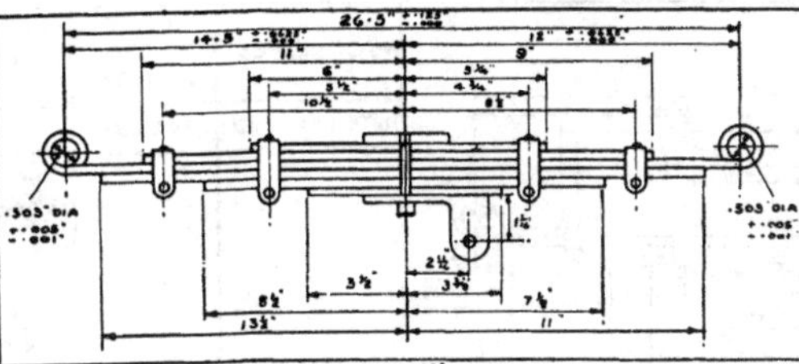	No OF LEAVES	1 x 5/32" 4 x 3/16" 1 x 7/32"	SPRING TO BE FLAT AS DRAWN WITH 280 LBS LOAD
		UNLOADED CURVATURE.	2·1/2"	
		RATE PER INCH	112 LBS	
REAR SPRING MODELS M. TYPE 2 SEATER & COUPE. (CEASED ON COUPE AFTER M 2090) PART No: P. 106.	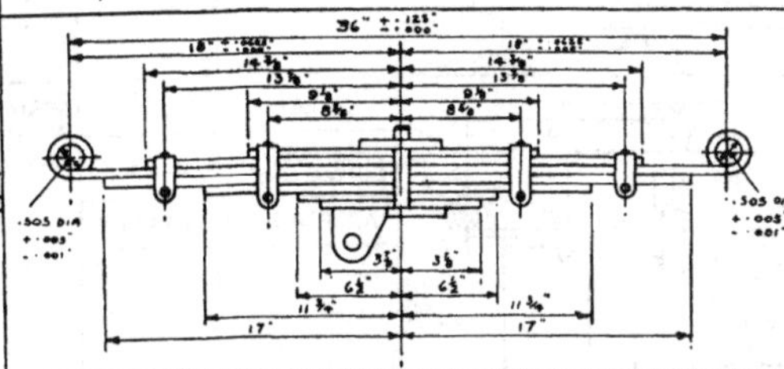	No OF LEAVES	1 x 3/16" 5 x 7/32" 1 x 1/4"	SPRING TO BE FLAT AS DRAWN WITH 500 LBS LOAD
		UNLOADED CURVATURE.	3.1/8"	
		RATE PER INCH	160 LBS	
REAR SPRING MODEL. M. TYPE COUPE. (COMM^D M. 2090) P. 175.	IDENTICAL WITH SPRING No P106 BUT WITH 1/2" MORE FREE CAMBER.			
FRONT SPRING MODELS. F1. F2. F3. J3. L1. L2. (COMMENCED ON F1. AT F. 0522.) PART No: P 313.	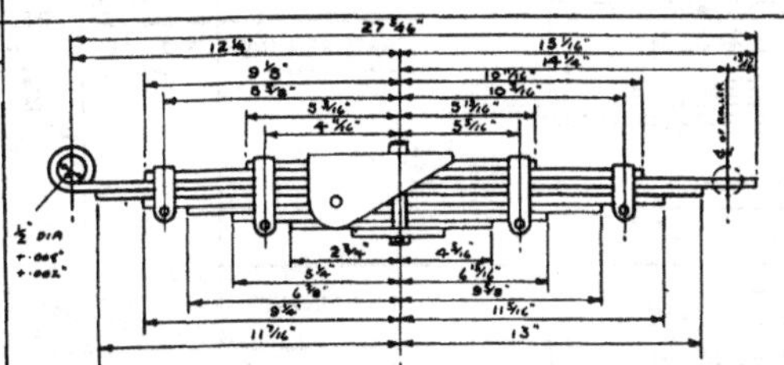	No OF LEAVES	1 x 5/32" 5 x 3/16" 2 x 7/32"	SPRING TO BE FLAT AS DRAWN WITH 545 LBS LOAD.
		UNLOADED CURVATURE.	1.1/4"	 1¼" FREE CAMBER MEASURED FROM ℄ OF ROLLER TO MAIN LEAF AT EYE.
		RATE PER INCH	436 LBS	
REAR SPRING MODELS D. J1. J2. J3. F1. F2. F3. L2. PART No: P. 237.	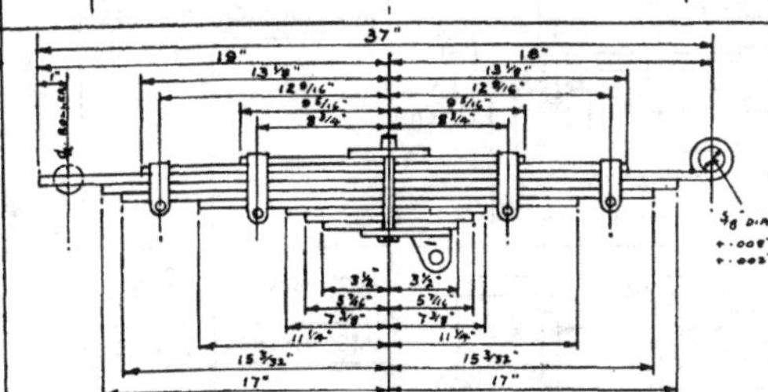	No OF LEAVES.	1 x 3/16" 7 x 7/32" 1 x 1/4"	SPRING TO BE FLAT AS DRAWN WITH 474 LBS LOAD.
		UNLOADED CURVATURE.	2.1/16"	
		RATE PER INCH	228 LBS	
FRONT SPRING MODELS D. J1. J2. F1. (CEASED ON F1. AT F. 0521.) PART No: P. 236.	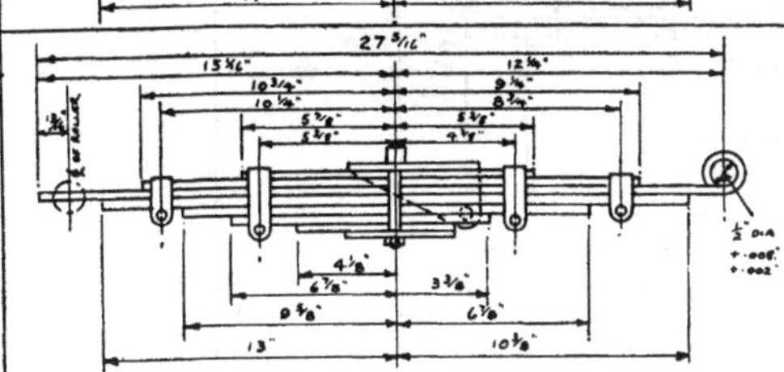	No OF LEAVES	1 x 5/32" 5 x 3/16" 1 x 7/32"	SPRING TO BE FLAT AS DRAWN WITH 410 LBS LOAD.
		UNLOADED CURVATURE.	1·41"	
		RATE PER INCH	290 LBS.	

Spring Data—Sheet 1

FRONT SPRING MODELS C. J4. PA. PB. TA. PART No: P. 285.	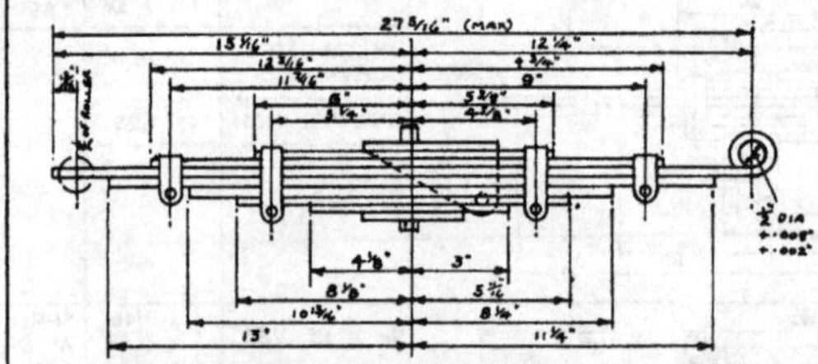	No OF LEAVES RATE PER INCH	1 x 5/32" 5 x 3/16" 1 x 7/32" 344 LBS	SPRING TO BE FLAT AS DRAWN WITH 515 LBS LOAD. 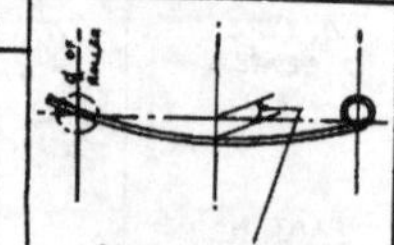 1·5" FREE CAMBER. FROM ℄ OF ROLLER TO MAIN LEAF AT EYE.
REAR SPRING MODELS C. J4. PART No: P. 286.	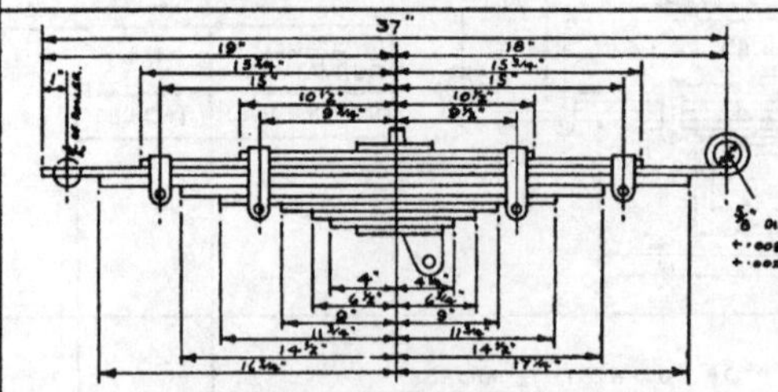	No OF LEAVES UNLOADED CURVATURE. RATE PER INCH	8 x 7/32" 1 x 1/4" 2.1/4" 294 LBS.	SPRING TO BE FLAT AS DRAWN WITH 650 LBS LOAD.
FRONT SPRING MODELS K1. K2. KN. NA. ND. NE. (CEASED ON NA AT NA. 0585.) PART No MG 392.	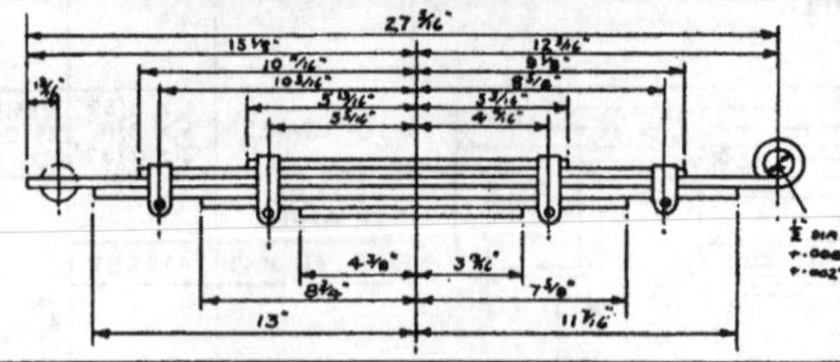	No OF LEAVES. RATE PER INCH	1 x 5/32" 3 x 3/16" 2 x 7/32" 337 LBS.	SPRING TO BE FLAT AS DRAWN WITH 665 LBS LOAD. 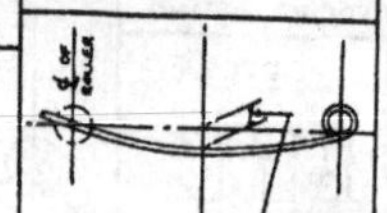 1 31/32" FREE CAMBER. MEASURED FROM ℄ OF ROLLER TO MAIN LEAF AT EYE.
REAR SPRING MODELS K1. KN. PART No: MG 393.	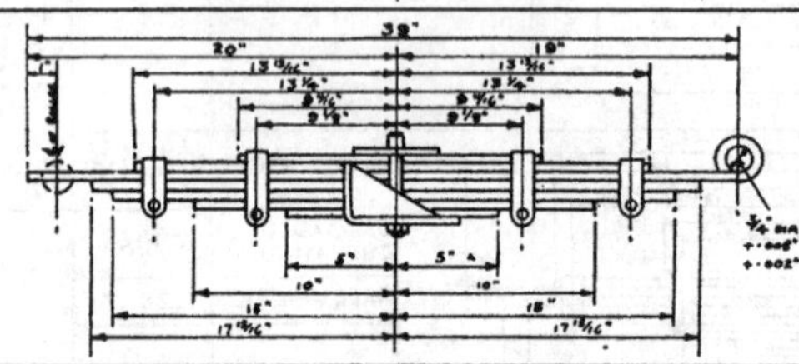	No OF LEAVES RATE PER INCH	1 x 3/16" 4 x 7/32" 2 x 1/4" 227 LBS.	SPRING TO BE FLAT AS DRAWN WITH 825 LBS LOAD.  3 5/8" FREE CAMBER. MEASURED FROM ℄ OF ROLLER TO MAIN LEAF AT EYE.
REAR SPRING MODELS NA. ND. NE. (CEASED ON NA AT NA. 0585.) PART No: MG. 504.	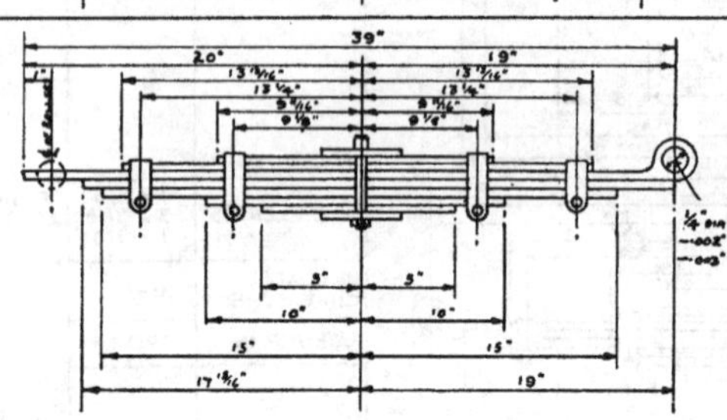	No OF LEAVES. RATE PER INCH	1 x 3/16" 4 x 7/32" 2 x 1/4" 227 LBS.	SPRING TO BE FLAT AS DRAWN WITH 596 LBS LOAD. 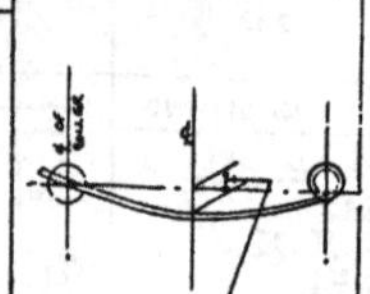 2 5/8" FREE CAMBER. MEASURED FROM ℄ OF ROLLER TO MAIN LEAF AT EYE.

Spring Data—Sheet 2

REAR SPRING				
MODELS PA. PB PART No: MG. 505	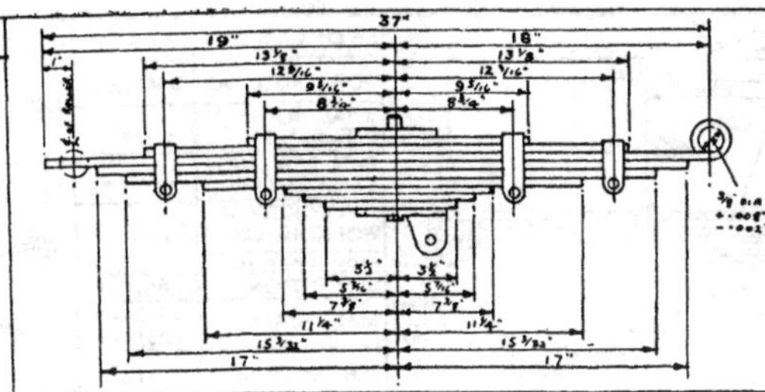	No OF LEAVES	1 x 1/4" 1 x 3/16" 7 x 7/32"	SPRING TO BE FLAT AS DRAWN WITH 385 LBS LOAD.
		RATE PER INCH	228 LBS	
		UNLOADED CURVATURE.	1-11/16"	

FRONT SPRING				
MODELS NA. COMMᴸ NA 0586 PART No. MG. 598.	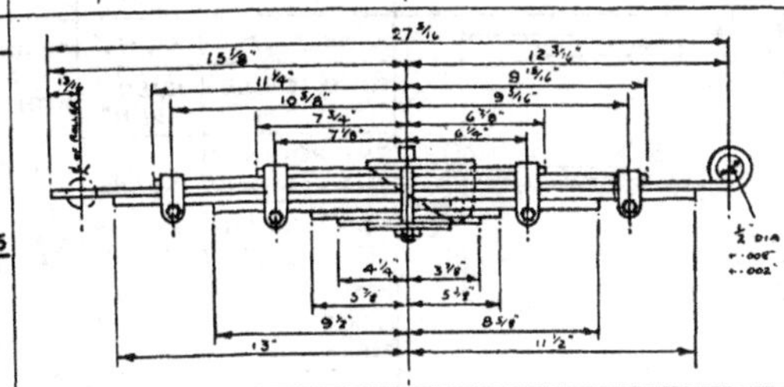	No OF LEAVES.	6 x 3/16" 1 x 7/32"	SPRING TO BE FLAT AS DRAWN WITH 583 LBS LOAD.  1 21/32" FREE CAMBER, MEASURED FROM ℄ OF ROLLER TO MAIN LEAF AT EYE.
		RATE PER INCH	350 LBS.	

REAR SPRING				
MODELS NA. COMMᴸ NA-0586 PART No. MG. 604.	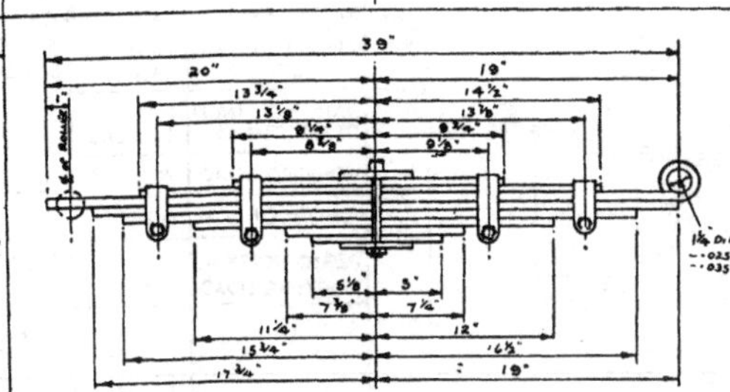	No OF LEAVES	6 x 5/32" 1 x 7/32" 1 x 1/4"	SPRING TO BE FLAT AS DRAWN WITH 592 LBS LOAD.  4" FREE CAMBER. MEASURED FROM ℄ OF ROLLER TO MAIN LEAF AT EYE.
		RATE PER INCH	148 LBS	

FRONT SPRING				
MODELS SA. PART No. MG. 841.	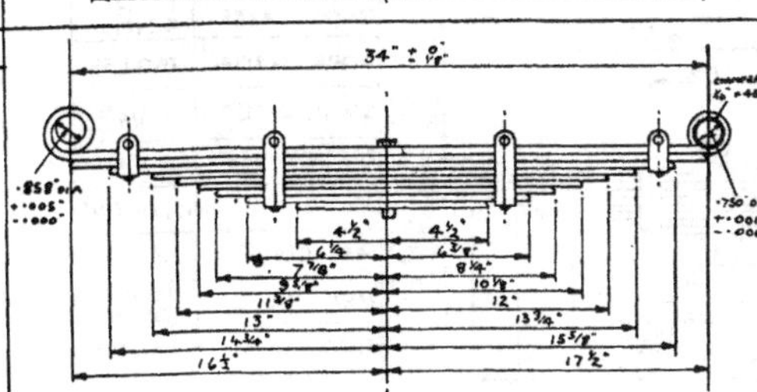	No OF LEAVES	1 x 15/64" 3 x 7/32" 5 x 3/16"	
		WORKING LOAD	700 LBS	
		STATIC DEFLECᴺ	2·17"	
		MEAN RATE	322 LBS/IN.	
		FREE CAMBER	2·37"	
		MAX DEFLECᴺ	5"	
		MAX STRESS	121,500 LBS/IN².	
		FREQUENCY	127·5 C.P.M.	

REAR SPRING				
MODELS SA. PART No. MG. 677.	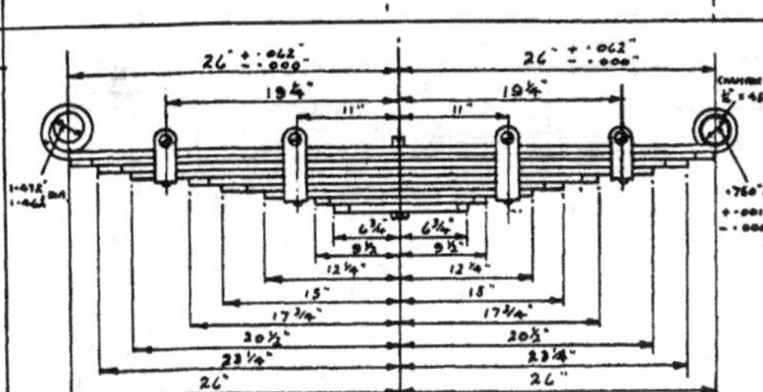	No OF LEAVES	1 x 1/4" 8 x 15/64"	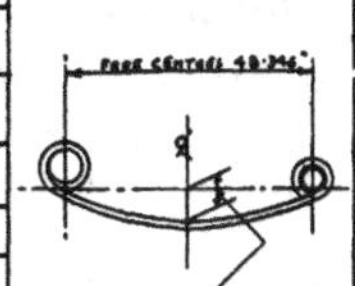 FREE CAMBER 7·375"
		DEFLᴺ PER 100 LBS. IN SITU.	·740"	
		DEFLᴺ PER 100 LBS POINT LOAD	·7605"	
		STRESS PER 100 LBS LOAD	8884	
		WORKING LOAD	750 LBS	
		DEFLᴺ AT W/L	5·550"	
		MAX PERMISSIBLE STATIC LOAD	900 LBS	
		DEFLᴺ AT M/P/L	6·660"	

Spring Data—Sheet 3

FRONT SPRING

MODELS. VA.

PART No. MG 733/1 R.H. MG 733/2 L.H.

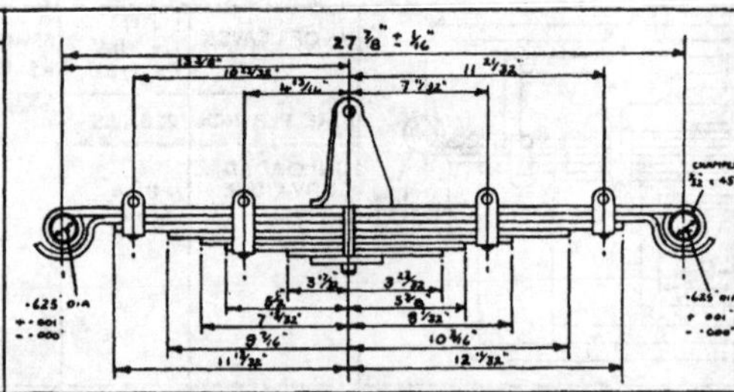

No OF LEAVES	7 x 3/16"
WORKING LOAD	560 LBS
TOTAL POSS: DEFLECTION	3·875"
RATE PER INCH DEFLECTION	280 LBS
PERIODICITY AT WORKING LOAD	132
STRESS PER INCH DEFLECTION	27,600 LBS/IN²
MAX STRESS	107,000 LBS/IN²
LENGTH OF SEAT	3·1/8"

CAMBER TO BE 19/32 ± 1/16" WITH 560 LBS LOAD.

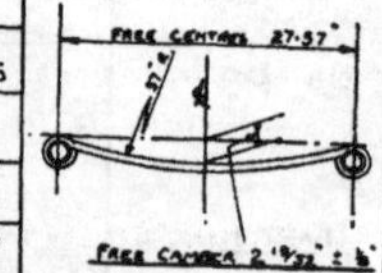

REAR SPRING

MODELS. VA

PART No MG. 734/1 R.H. MG. 734/2 L.H.

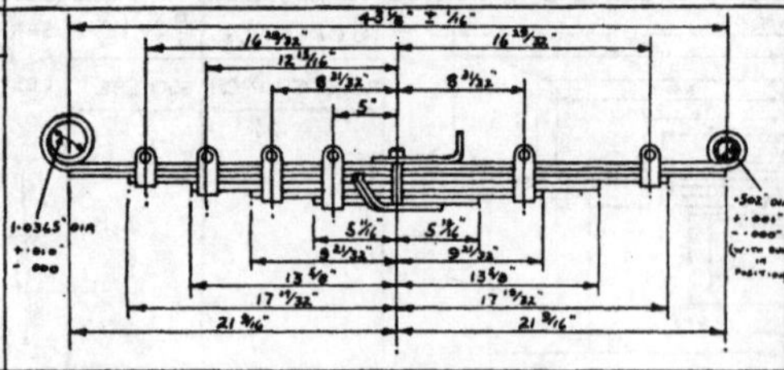

No OF LEAVES	6 x 1/4"
STRESS PER INCH DEFLECTION.	15,150 LBS/IN²
RATE PER INCH DEFLECᴺ (IN SITU)	154 LBS
WORKING LOAD.	660 LBS
TOTAL POSS: DEFLECTION	7 11/16"
LENGTH OF SEAT	3·13/32"
PERIODICITY AT W L.	90.

SPRING TO BE FLAT ± 1/16" AS DRAWN WITH 600 LBS LOAD

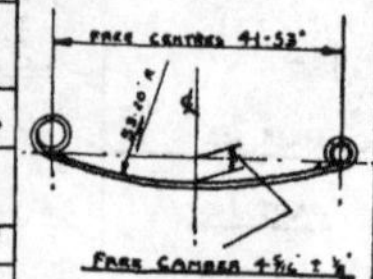

FRONT SPRING

MODELS. WA

PART No. MG 808 R.H. MG 829 L.H

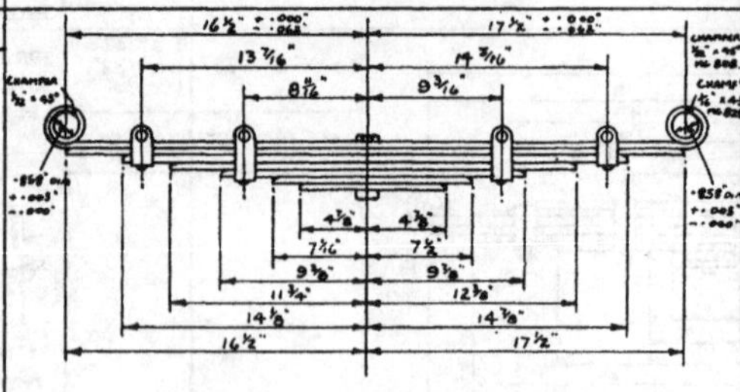

No OF LEAVES	7 x 4G(·232)
STRESS PER INCH DEFLECTION.	23,600 LBS/IN²
RATE PER INCH DEFLECTION.	364 LBS.
WORKING LOAD	700 LBS
LENGTH OF SEAT	3·750"
PERIODICITY AT WORKING LOAD.	135.

CAMBER TO BE 17/64" ± 1/16" WITH 700 LBS LOAD.

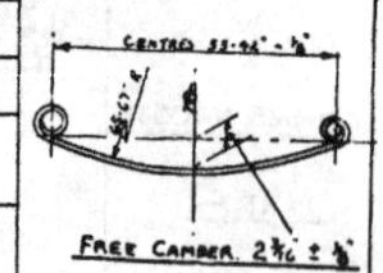

REAR SPRING

MODELS. WA.

PART No. MG. 816/1 R.H. MG. 816/2 L.H.

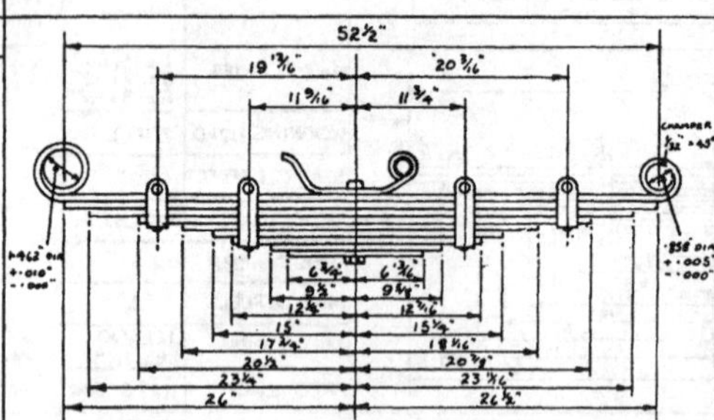

No OF LEAVES	1 x 1/4" 8 x 15/64"
WORKING LOAD	760 LBS
STATIC DEFLᴺ	5.3/4"
FREE CAMBER.	5.3/4"
RATE	131 LBS/IN²
MAX DEFLᴺ	10.3/4"
MAX STRESS	116000 LBS/IN²
FREQUENCY	78 CPM
LENGTH OF SEAT	3·75"

FRONT SPRING

MODELS. TB.

PART No. MG. 785/1 R.H. MG. 785/2 L.H.

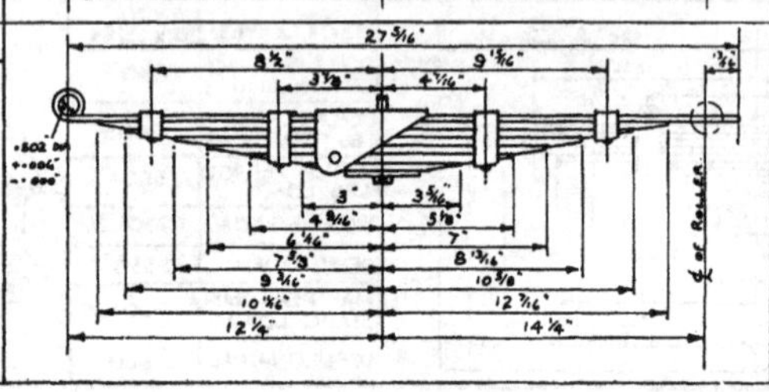

No. OF LEAVES.	7 x 7/32"
WORKING LOAD	515 LBS.
STATIC DEFLᴺ	1·24"
FREE CAMBER	1·24"
MEAN RATE	416 LBS/INS.
MAX DEFLᴺ	2"
MAX STRESS	74000 LBS/IN²
FREQUENCY	169 C.P.M

Spring Data—Sheet 4

REAR SPRING **MODELS** TA TB. PART NO. MG. 694.	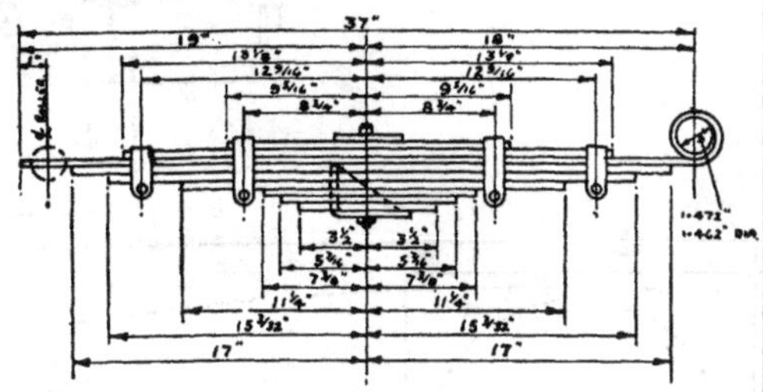	Nº OF LEAVES	1 x 3/16" 1 x 1/4" 7 x 7/32"	SPRING TO BE FLAT AS DRAWN AT 385 LBS
		UNLOADED CURVATURE	1 11/16"	
		RATE PER INCH	228 LBS	
FRONT SPRING **MODELS.** TC. PART NO. MG. 875.	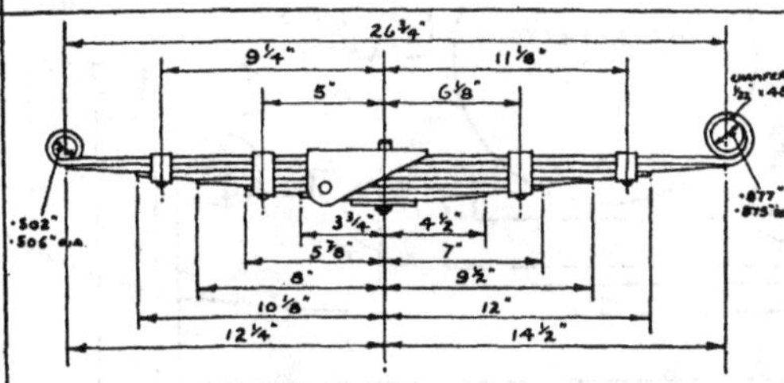	Nº OF LEAVES.	6 x 3/16"	SPRING TO BE FLAT AS DRAWN AT 500 LBS 520 LBS.
		STATIC LOAD.	434 LBS.	
		STATIC DEFLECN.	1·6"	
		FREE CAMBER	1·85"	
		MEAN RATE.	270 LBS/INS	
		MAX DEFLECN.	2·94"	
		MAX STRESS	100,000	
		FREQUENCY	148	
REAR SPRING **MODELS** TC. PART NO. MG. 876.	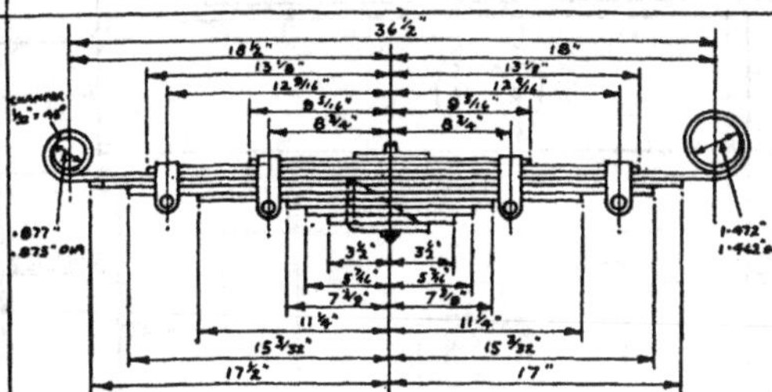	Nº OF LEAVES.	1 - 3/16" 1 - 1/4" 7 - 7/32"	SPRING TO BE FLAT AS DRAWN AT 385 LBS.
		STATIC LOAD	476 LBS	
		STATIC DEFLECN.	2·08"	
		FREE CAMBER	1·687"	
		MEAN RATE	228 LBS/INS	
		MAX DEFLECN.	4·187"	
		MAX STRESS	97,000	
		FREQUENCY	130 CPM.	
REAR SPRING **MODELS** Y. PART NO. MG. 894.	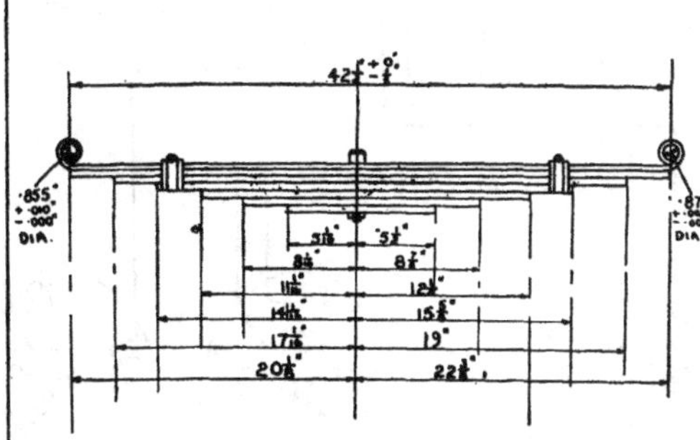	Nº OF LEAVES.	7 x 1/4.	
		STATIC DEFLECN.	3"	
		FREE CAMBER.	4"	
		MEAN RATE	216·6 LBS/INS	
		MAX DEFLECN.	6 1/4"	
		MAX STRESS	111500 LBS/SQ IN	
		FREQUENCY	108 CPM.	
		WORKING LOAD.	660 LBS	
		SPRING SEAT LGTH	4"	

Spring Data—Sheet 5

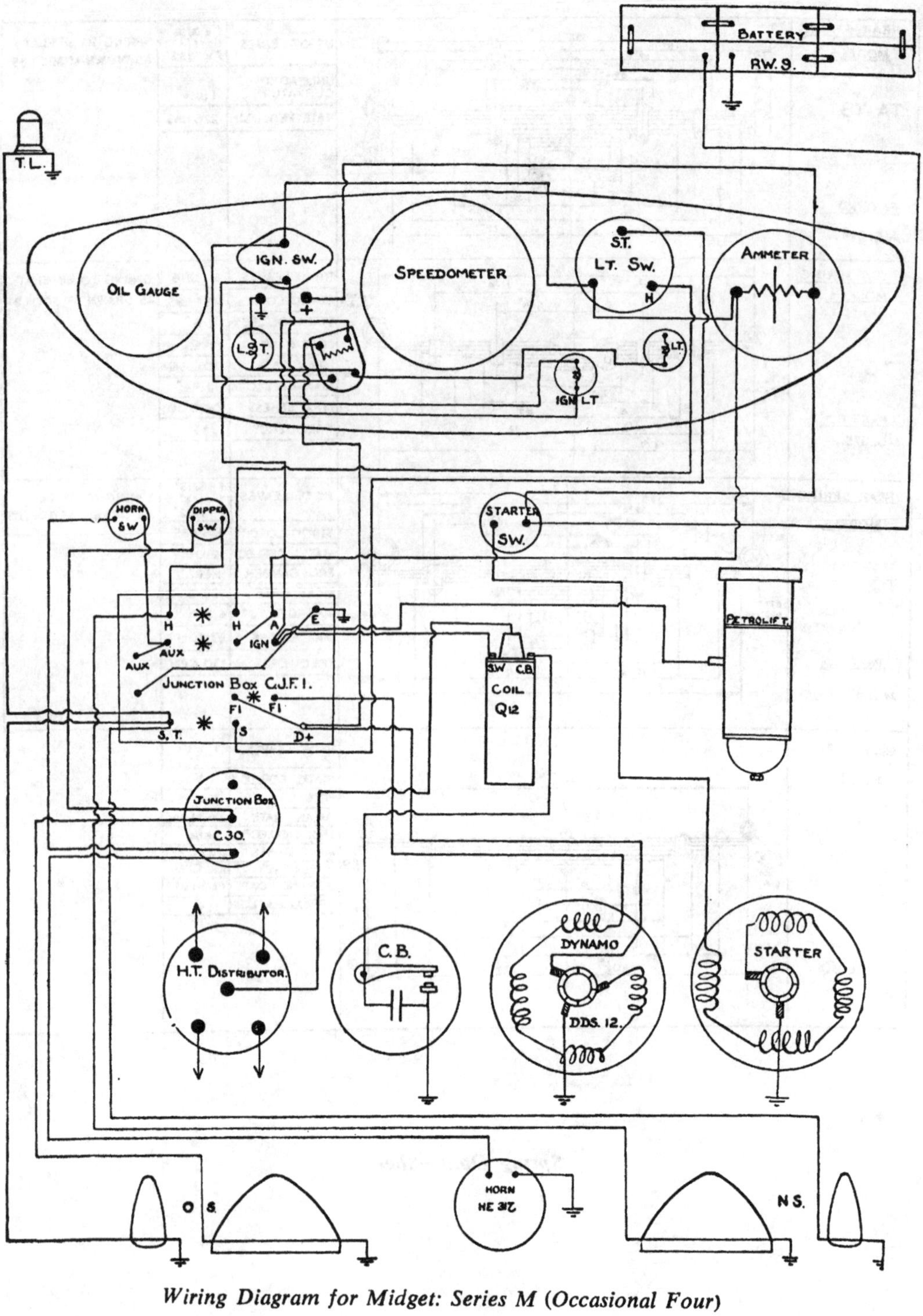

Wiring Diagram for Midget: Series M (Occasional Four)

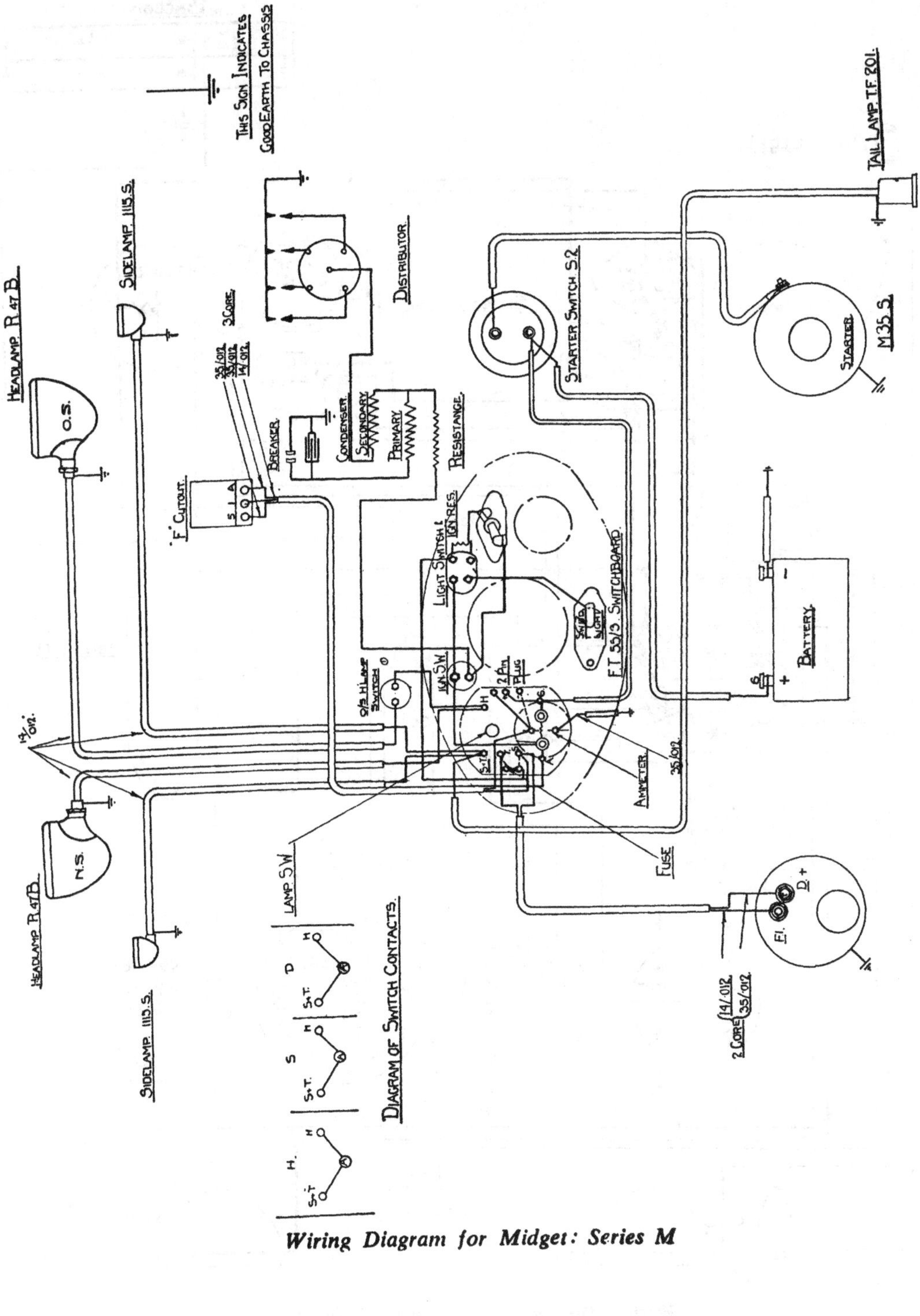

Wiring Diagram for Midget: Series M

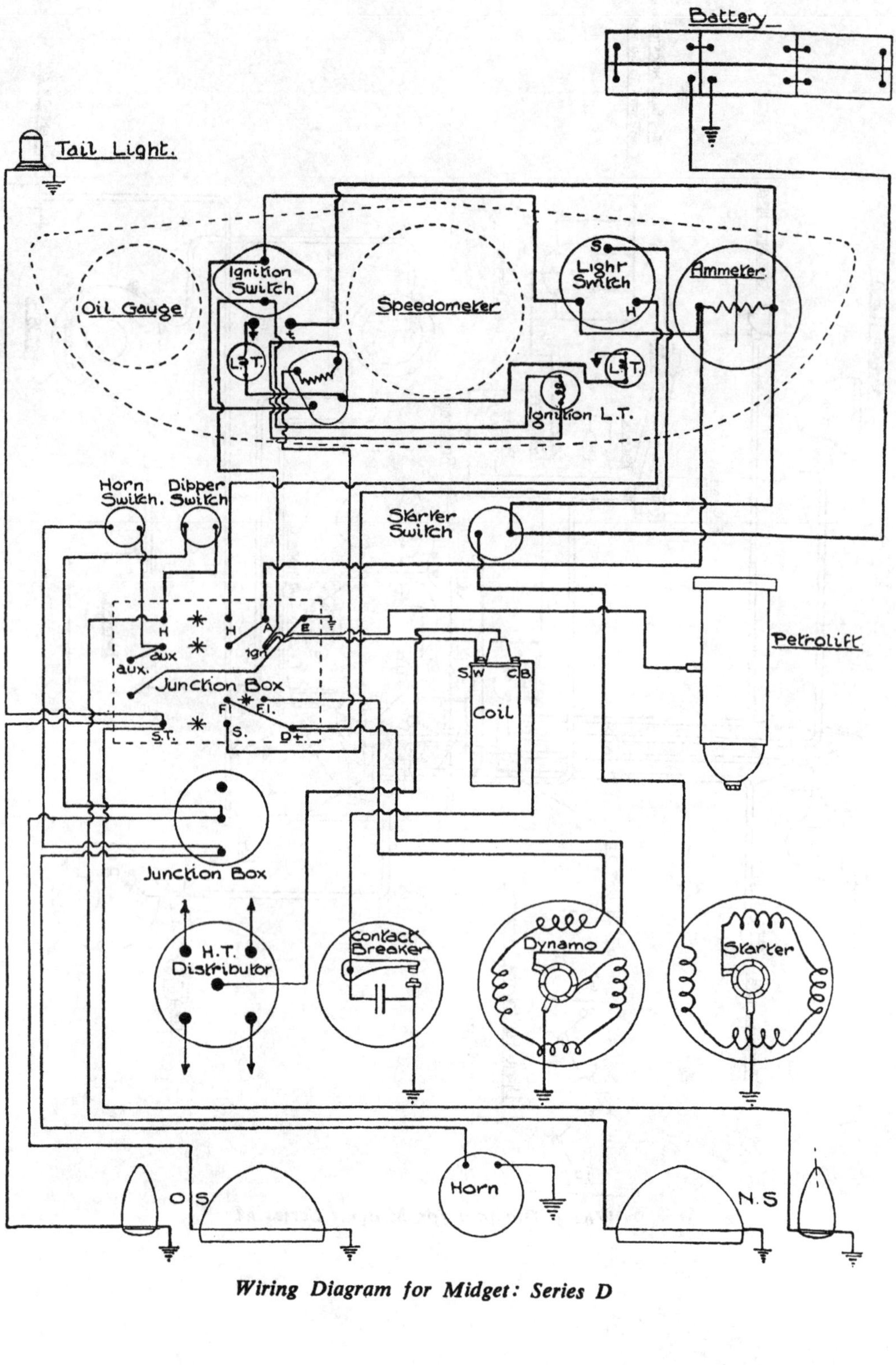

Wiring Diagram for Midget: Series D

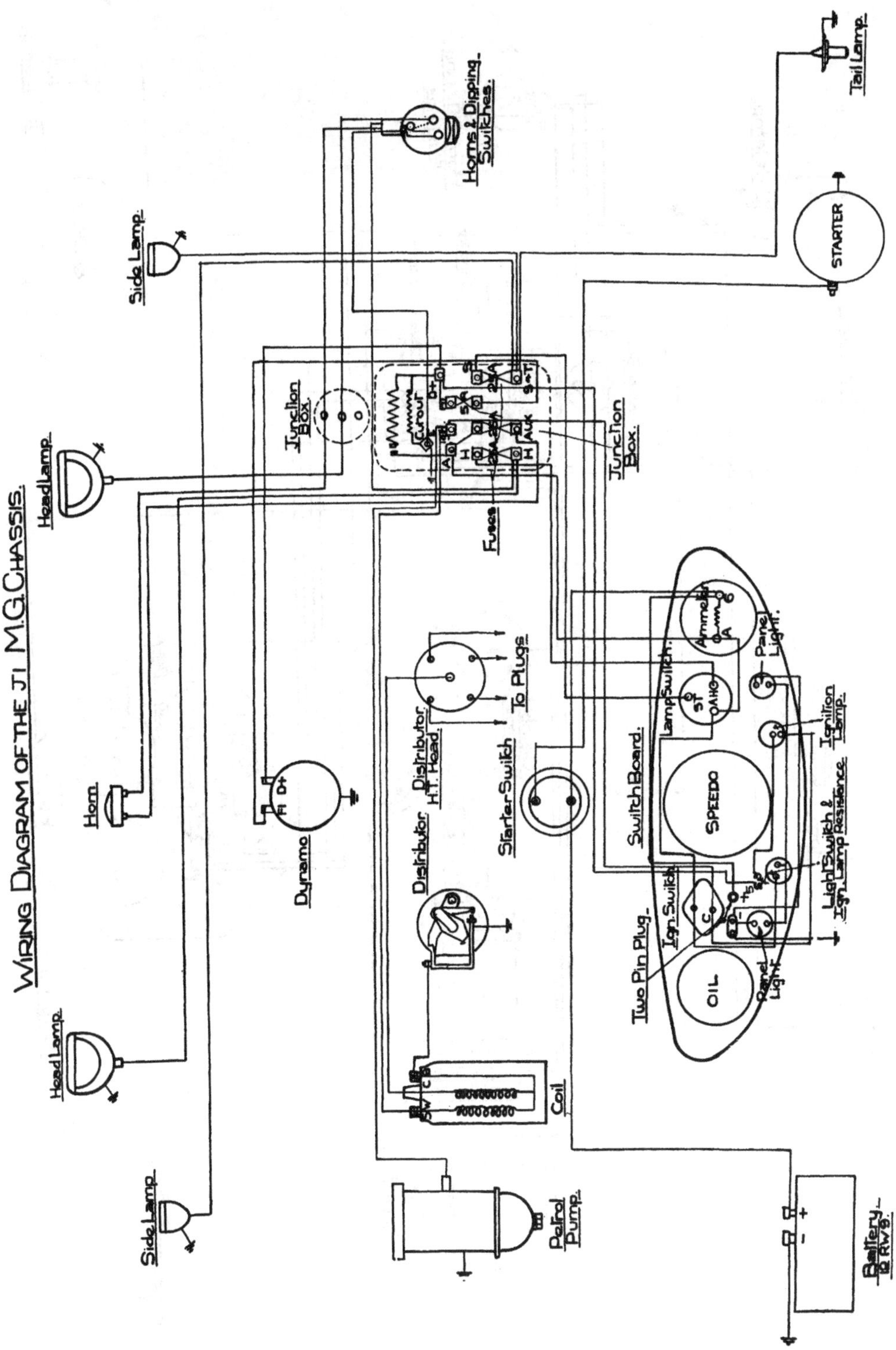

Wiring Diagram for Midget: Series J1

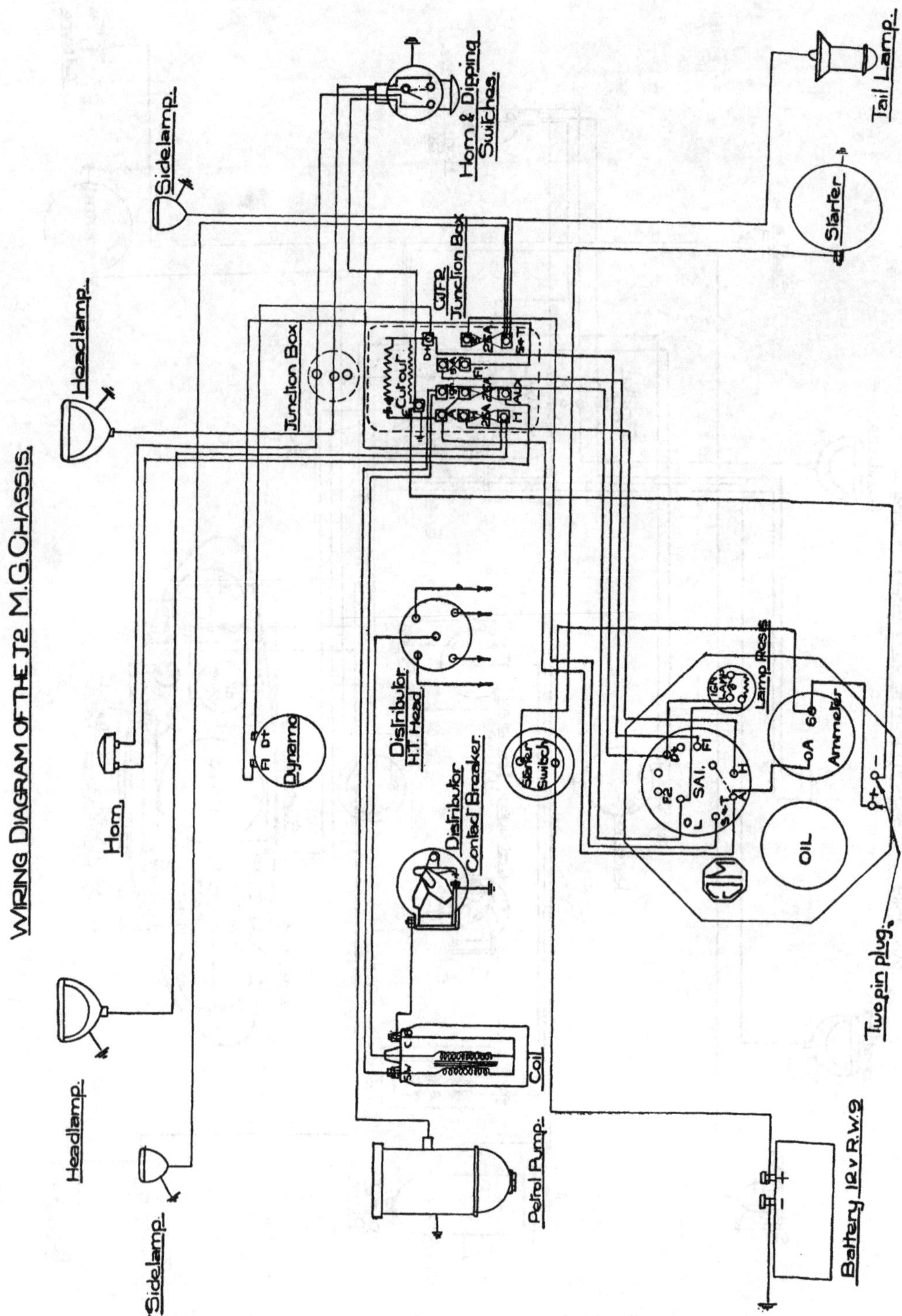

Wiring Diagram for Midget: Series 12

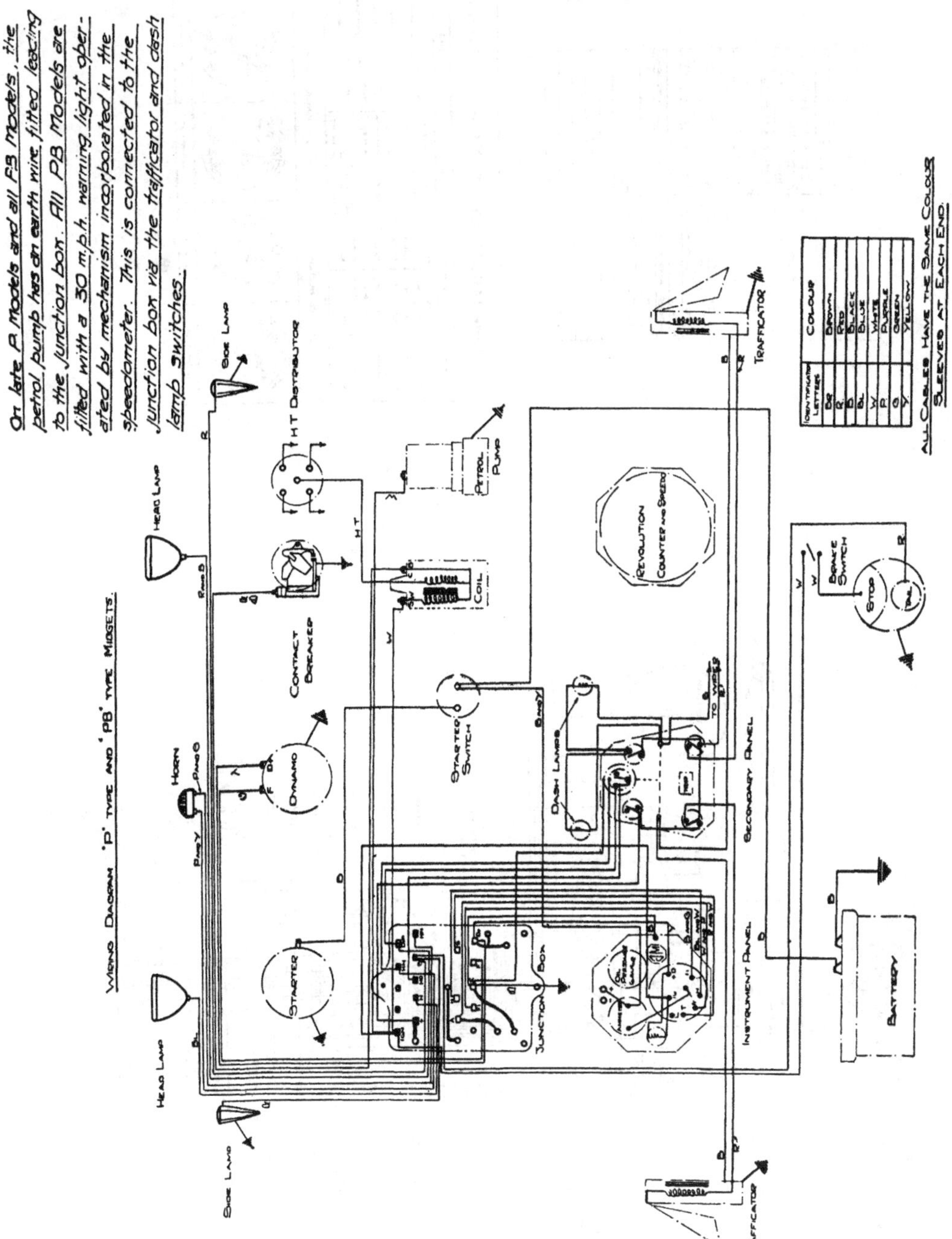

Wiring Diagrams for P type and PB type Midgets

KEY TO CABLE COLOURS

1	RED	18	BLUE AND PURPLE
2	RED AND YELLOW	19	BLUE AND BLACK
3	RED AND BLUE	20	WHITE
4	RED AND WHITE	21	WHITE AND GREEN
5	RED AND GREEN	22	WHITE AND BROWN
6	RED AND BROWN	23	WHITE AND PURPLE
7	RED AND BLACK	24	WHITE AND BLACK
8	YELLOW	25	GREEN
9	YELLOW AND BLUE	26	GREEN AND BROWN
10	YELLOW AND GREEN	27	GREEN AND PURPLE
11	YELLOW AND BROWN	28	GREEN AND BLACK
12	YELLOW AND PURPLE	29	BROWN
13	YELLOW AND BLACK	30	BROWN AND PURPLE
14	BLUE	31	PURPLE
16	BLUE AND WHITE	32	PURPLE AND BLACK
17	BLUE AND GREEN	33	BLACK
18	BLUE AND BROWN		

M.G 10 H.P. MIDGET SERIES T CARS (1937-8)

SPECIFICATION OF EQUIPMENT	MODEL		SERVICE No.
DYNAMO	C45HV	CA21	228194
STARTER	M418G	A04	239667
SWITCH	PLC3	L14	344130
CUT-OUT & FUSE UNIT	CJR3	L39	356181
BATTERY	2 UNITS 6STW 11E		
DISTRIBUTOR	DK4A	DA27	405996
COIL	Q12	L	401650

BULBS	
HEADLAMP	LUCAS No. 54
SIDE STOP & TAIL LAMPS.–	LUCAS No. 207
IGNITION WARNING LIGHT	LUCAS No. C252A
PANEL LAMPS.–	LUCAS No. 1224M
FOGLAMP.–	LUCAS No. 2
30 M.P.H. WARNING LAMP MAP READING LAMP & DASH LAMPS.–	LUCAS No. 207

FUSES
DYNAMO FIELD No. FA45 4.5 AMPERES
HEADLAMPS – No. FA25 25 AMPERES
SIDELAMPS.– No. FA25 25 AMPERES
ACCESSORIES – No. FA25 25 AMPERES
DIPPING REFLECTOR:– No. FA6 6 AMPERES

INTERNAL CONNECTIONS OF LIGHTING, CHARGING & IGNITION SWITCH.

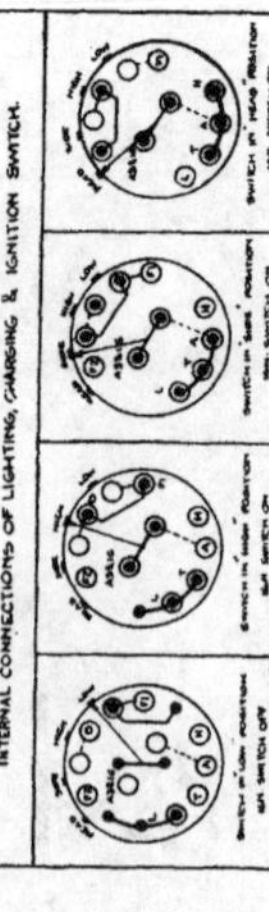

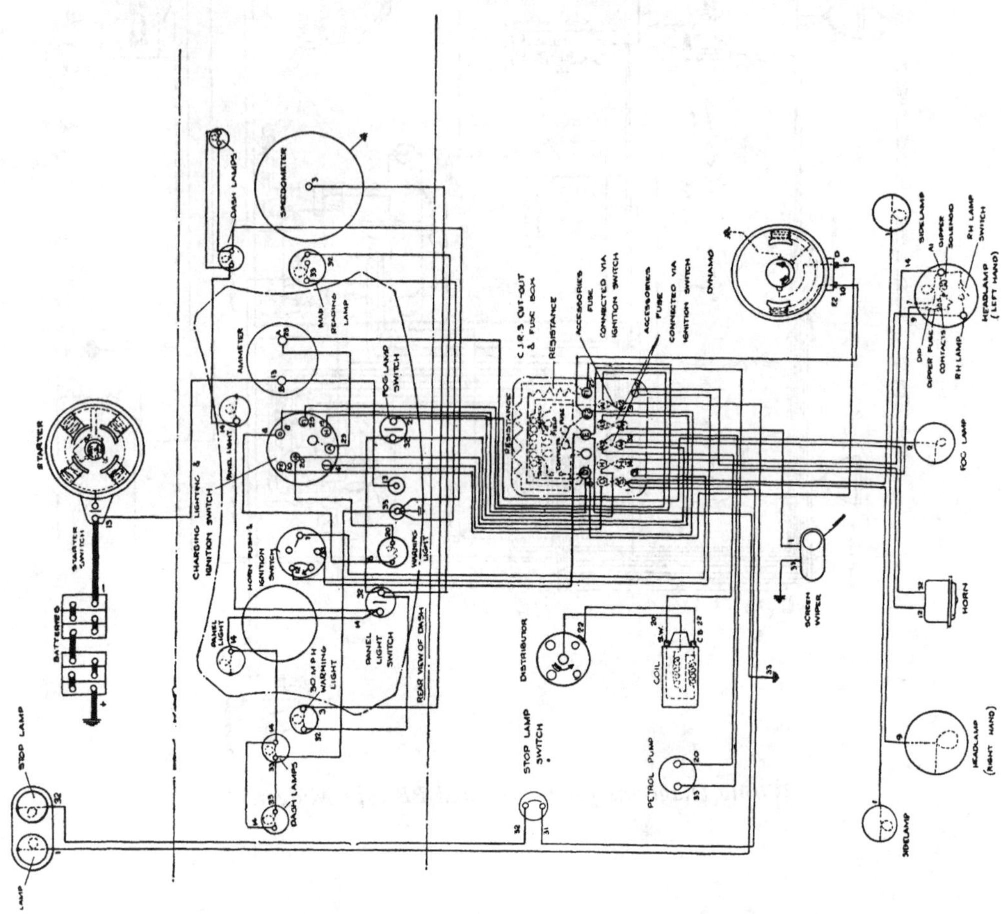

Wiring Diagram for Midget: Series TA

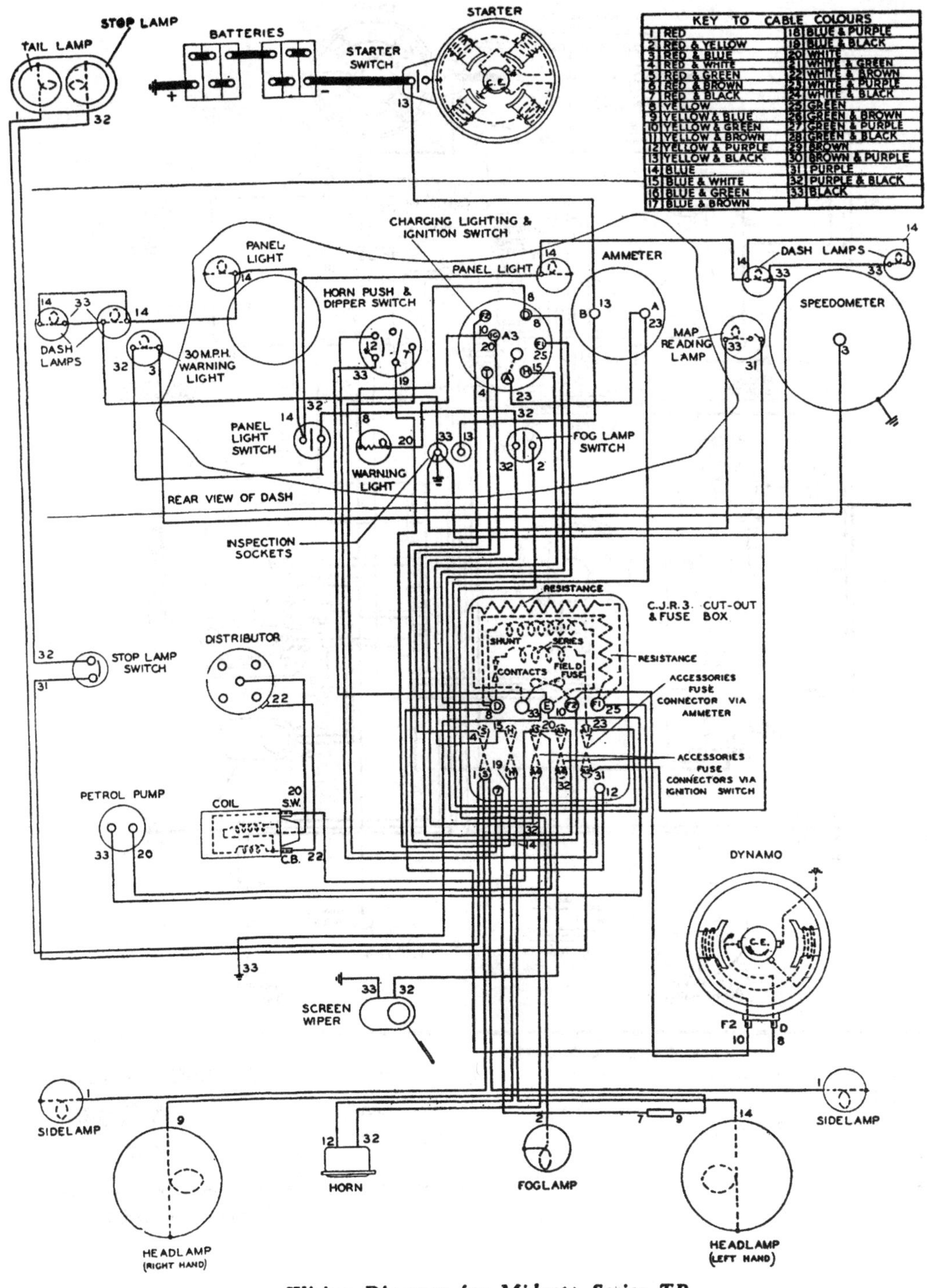

KEY TO CABLE COLOURS			
1	RED	18	BLUE & PURPLE
2	RED & YELLOW	19	BLUE & BLACK
3	RED & BLUE	20	WHITE
4	RED & WHITE	21	WHITE & GREEN
5	RED & GREEN	22	WHITE & BROWN
6	RED & BROWN	23	WHITE & PURPLE
7	RED & BLACK	24	WHITE & BLACK
8	YELLOW	25	GREEN
9	YELLOW & BLUE	26	GREEN & BROWN
10	YELLOW & GREEN	27	GREEN & PURPLE
11	YELLOW & BROWN	28	GREEN & BLACK
12	YELLOW & PURPLE	29	BROWN
13	YELLOW & BLACK	30	BROWN & PURPLE
14	BLUE	31	PURPLE
15	BLUE & WHITE	32	PURPLE & BLACK
16	BLUE & GREEN	33	BLACK
17	BLUE & BROWN		

Wiring Diagram for Midget: Series TB

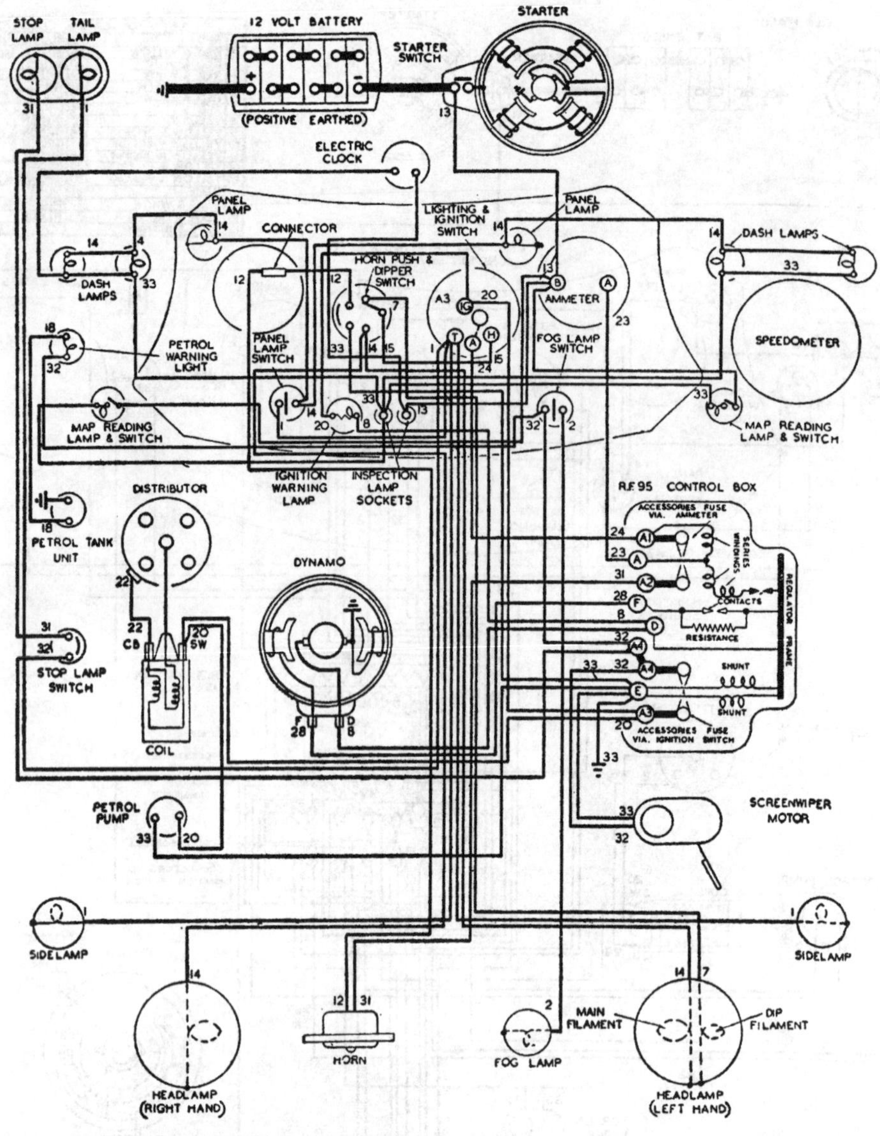

KEY TO CABLE COLOURS

1 RED	10 YELLOW & GREEN	18 BLUE & PURPLE	27 GREEN & PURPLE
2 RED & YELLOW	11 YELLOW & BROWN	19 BLUE & BLACK	28 GREEN & BLACK
3 RED & BLUE	12 YELLOW & PURPLE	20 WHITE	29 BROWN
4 RED & WHITE	13 YELLOW & BLACK	21 WHITE & GREEN	30 BROWN & PURPLE
5 RED & GREEN	14 BLUE	22 WHITE & BROWN	31 PURPLE
6 RED & BROWN	15 BLUE & WHITE	23 WHITE & PURPLE	32 PURPLE & BLACK
7 RED & BLACK	16 BLUE & GREEN	24 WHITE & BLACK	33 BLACK
8 YELLOW	17 BLUE & BROWN	25 GREEN	
9 YELLOW & BLUE		26 GREEN & BROWN	

WIRING DIAGRAM
No. W70482
12 VOLT.

Wiring Diagram for Midget: Series TC. (Chassis No. TC 3414 onwards)

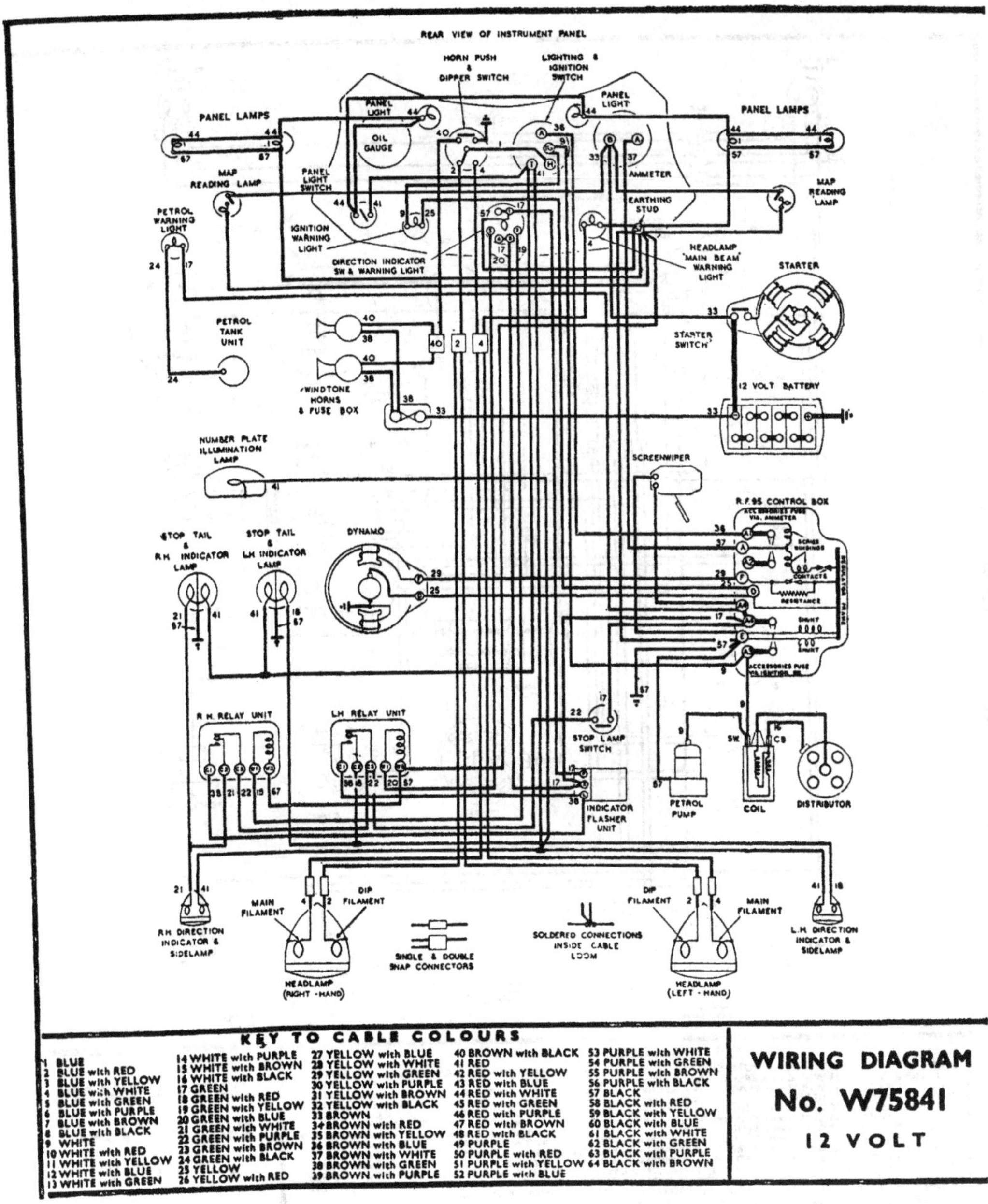

KEY TO CABLE COLOURS

1 BLUE	14 WHITE with PURPLE	27 YELLOW with BLUE	40 BROWN with BLACK	53 PURPLE with WHITE
2 BLUE with RED	15 WHITE with BROWN	28 YELLOW with WHITE	41 RED	54 PURPLE with GREEN
3 BLUE with YELLOW	16 WHITE with BLACK	29 YELLOW with GREEN	42 RED with YELLOW	55 PURPLE with BROWN
4 BLUE with WHITE	17 GREEN	30 YELLOW with PURPLE	43 RED with BLUE	56 PURPLE with BLACK
5 BLUE with GREEN	18 GREEN with RED	31 YELLOW with BROWN	44 RED with WHITE	57 BLACK
6 BLUE with PURPLE	19 GREEN with YELLOW	32 YELLOW with BLACK	45 RED with GREEN	58 BLACK with RED
7 BLUE with BROWN	20 GREEN with BLUE	33 BROWN	46 RED with PURPLE	59 BLACK with YELLOW
8 BLUE with BLACK	21 GREEN with WHITE	34 BROWN with RED	47 RED with BROWN	60 BLACK with BLUE
9 WHITE	22 GREEN with PURPLE	35 BROWN with YELLOW	48 RED with BLACK	61 BLACK with WHITE
10 WHITE with RED	23 GREEN with BROWN	36 BROWN with BLUE	49 PURPLE	62 BLACK with GREEN
11 WHITE with YELLOW	24 GREEN with BLACK	37 BROWN with WHITE	50 PURPLE with RED	63 BLACK with PURPLE
12 WHITE with BLUE	25 YELLOW	38 BROWN with GREEN	51 PURPLE with YELLOW	64 BLACK with BROWN
13 WHITE with GREEN	26 YELLOW with RED	39 BROWN with PURPLE	52 PURPLE with BLUE	

WIRING DIAGRAM

No. W75841

12 VOLT

Wiring Diagram for Midget: Series TC. Export to U.S.A.

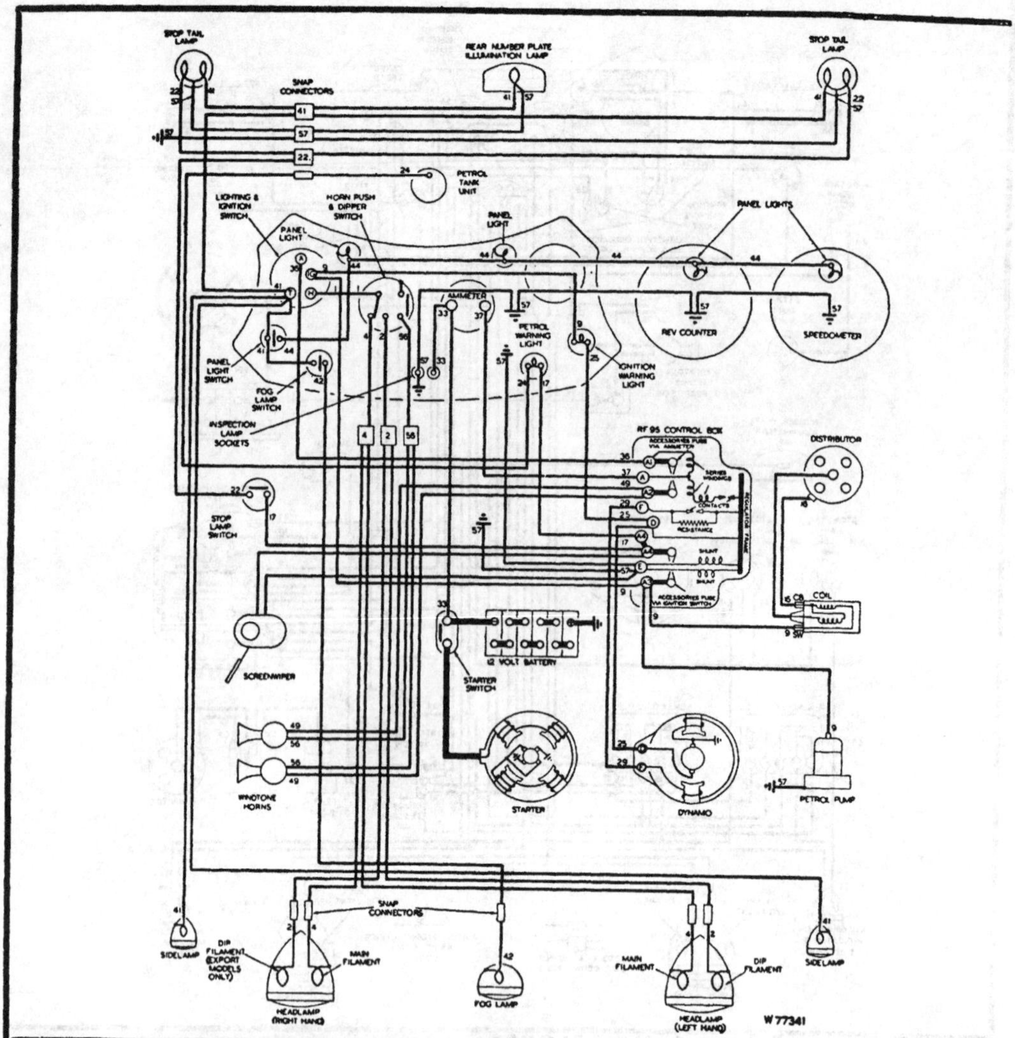

KEY TO CABLE COLOURS

1 BLUE	14 WHITE with PURPLE	27 YELLOW with BLUE	40 BROWN with BLACK	53 PURPLE with WHITE
2 BLUE with RED	15 WHITE with BROWN	28 YELLOW with WHITE	41 RED	54 PURPLE with GREEN
3 BLUE with YELLOW	16 WHITE with BLACK	29 YELLOW with GREEN	42 RED with YELLOW	55 PURPLE with BROWN
4 BLUE with WHITE	17 GREEN	30 YELLOW with PURPLE	43 RED with BLUE	56 PURPLE with BLACK
5 BLUE with GREEN	18 GREEN with RED	31 YELLOW with BROWN	44 RED with WHITE	57 BLACK
6 BLUE with PURPLE	19 GREEN with YELLOW	32 YELLOW with BLACK	45 RED with GREEN	58 BLACK with RED
7 BLUE with BROWN	20 GREEN with BLUE	33 BROWN	46 RED with PURPLE	59 BLACK with YELLOW
8 BLUE with BLACK	21 GREEN with WHITE	34 BROWN with RED	47 RED with BROWN	60 BLACK with BLUE
9 WHITE	22 GREEN with PURPLE	35 BROWN with YELLOW	48 RED with BLACK	61 BLACK with WHITE
10 WHITE with RED	23 GREEN with BROWN	36 BROWN with BLUE	49 PURPLE	62 BLACK with GREEN
11 WHITE with YELLOW	24 GREEN with BLACK	37 BROWN with WHITE	50 PURPLE with RED	63 BLACK with PURPLE
12 WHITE with BLUE	25 YELLOW	38 BROWN with GREEN	51 PURPLE with YELLOW	64 BLACK with BROWN
13 WHITE with GREEN	26 YELLOW with RED	39 BROWN with PURPLE	52 PURPLE with BLUE	

WIRING DIAGRAM

No. W77341

12 VOLT

Wiring Diagram for Midget: Series TD

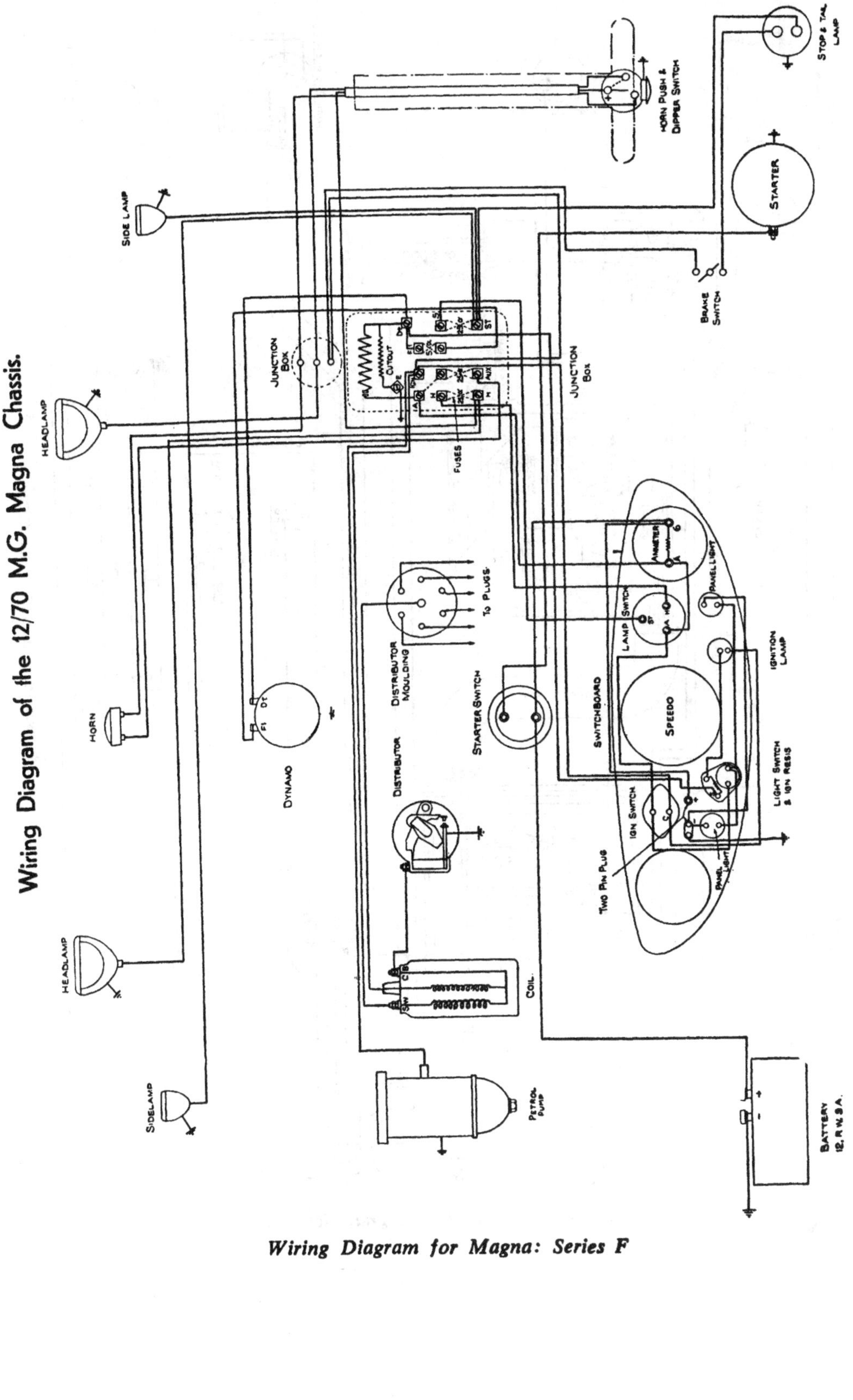

Wiring Diagram for Magna: Series F

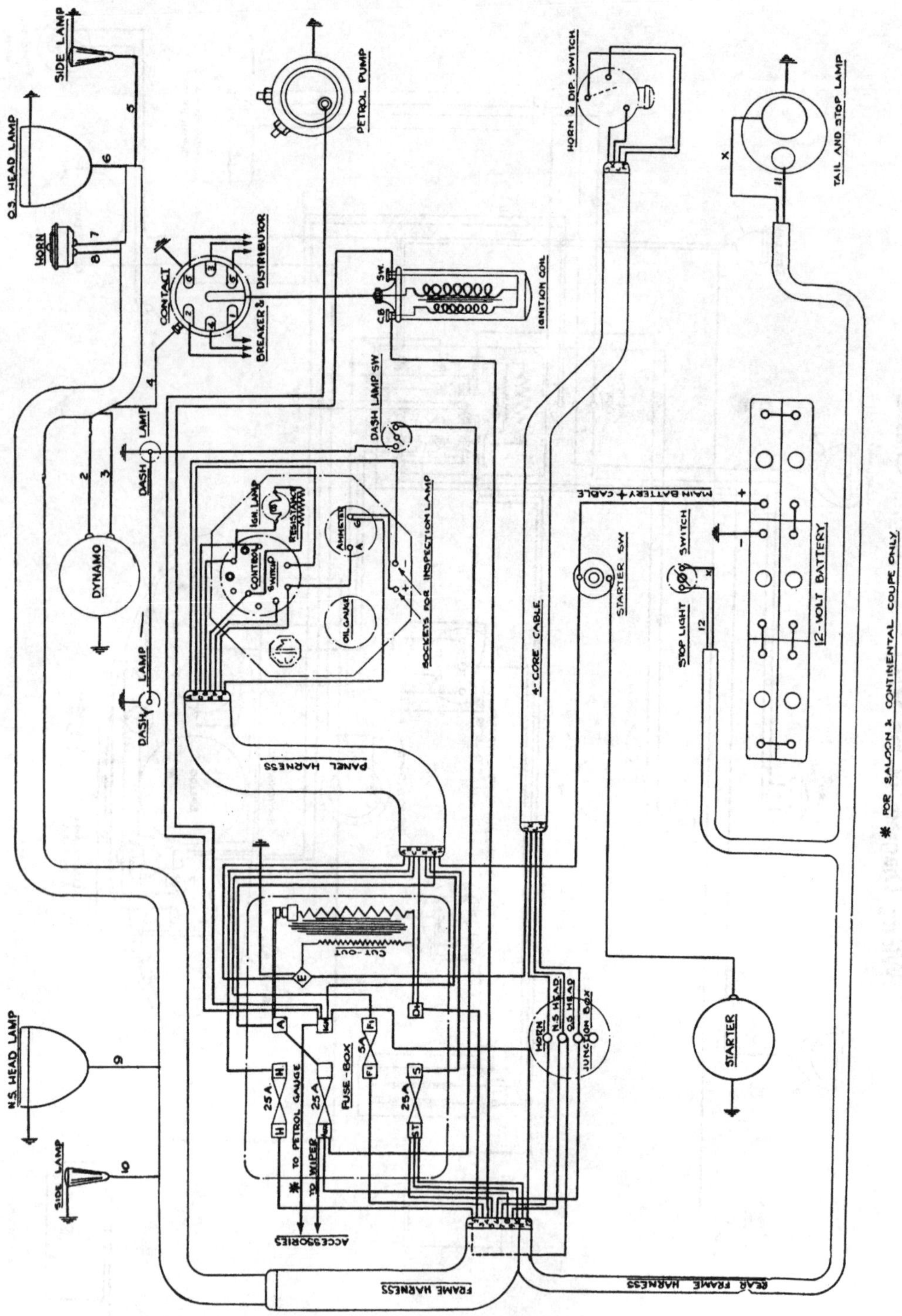

Wiring Diagram for Magna: Series L

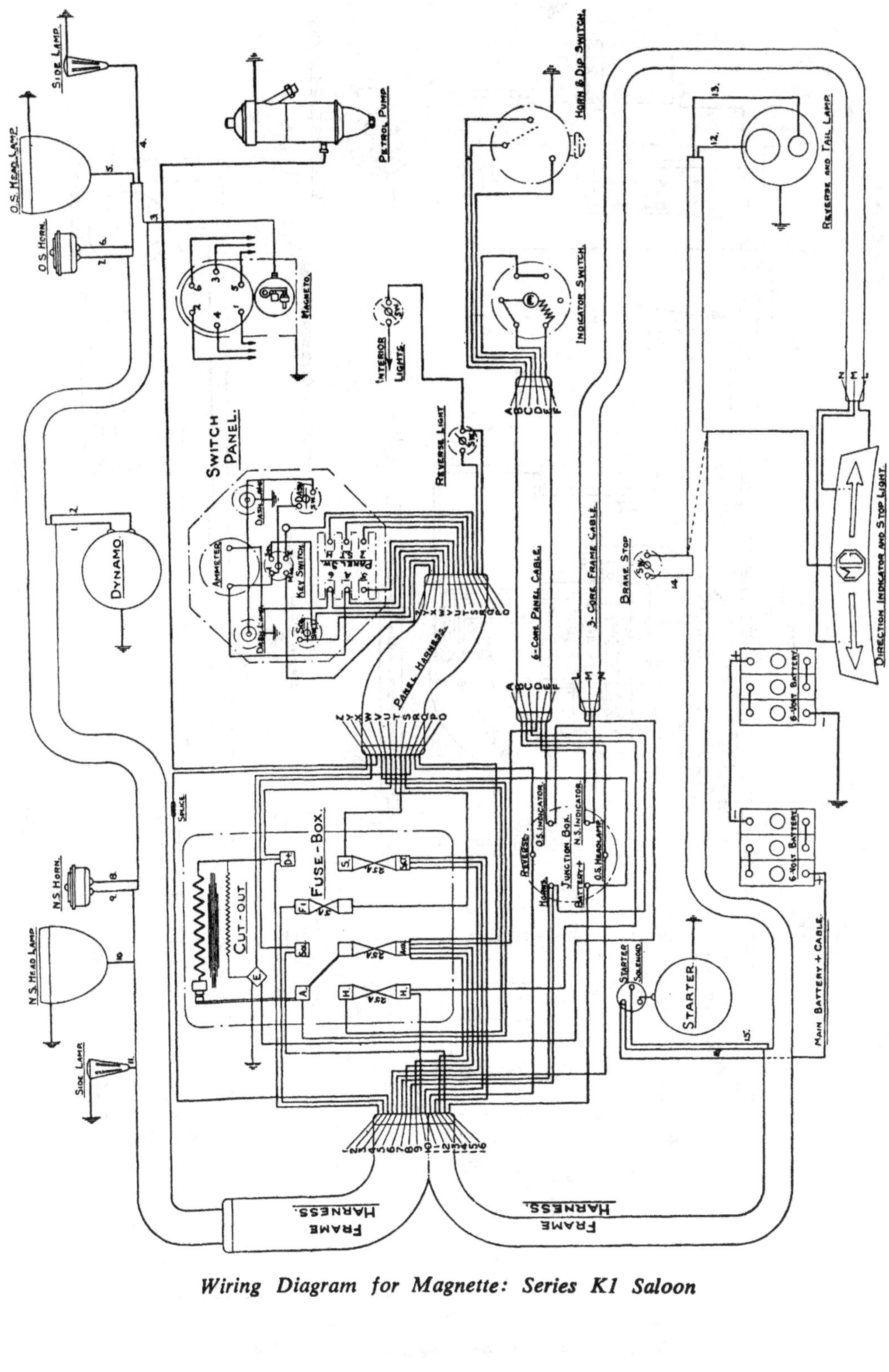

Wiring Diagram for Magnette: Series K1 Saloon

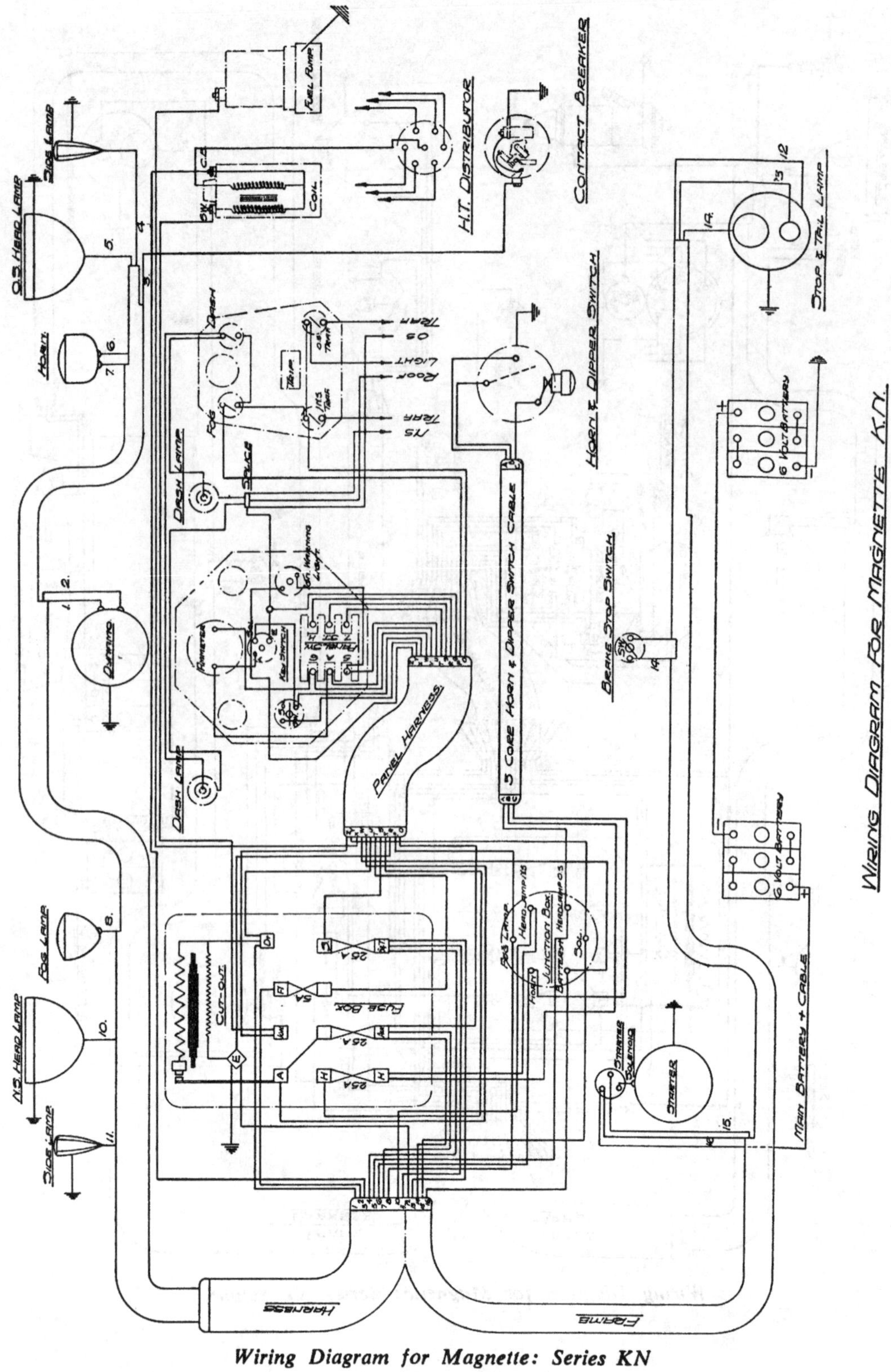

Wiring Diagram for Magnette: Series KN

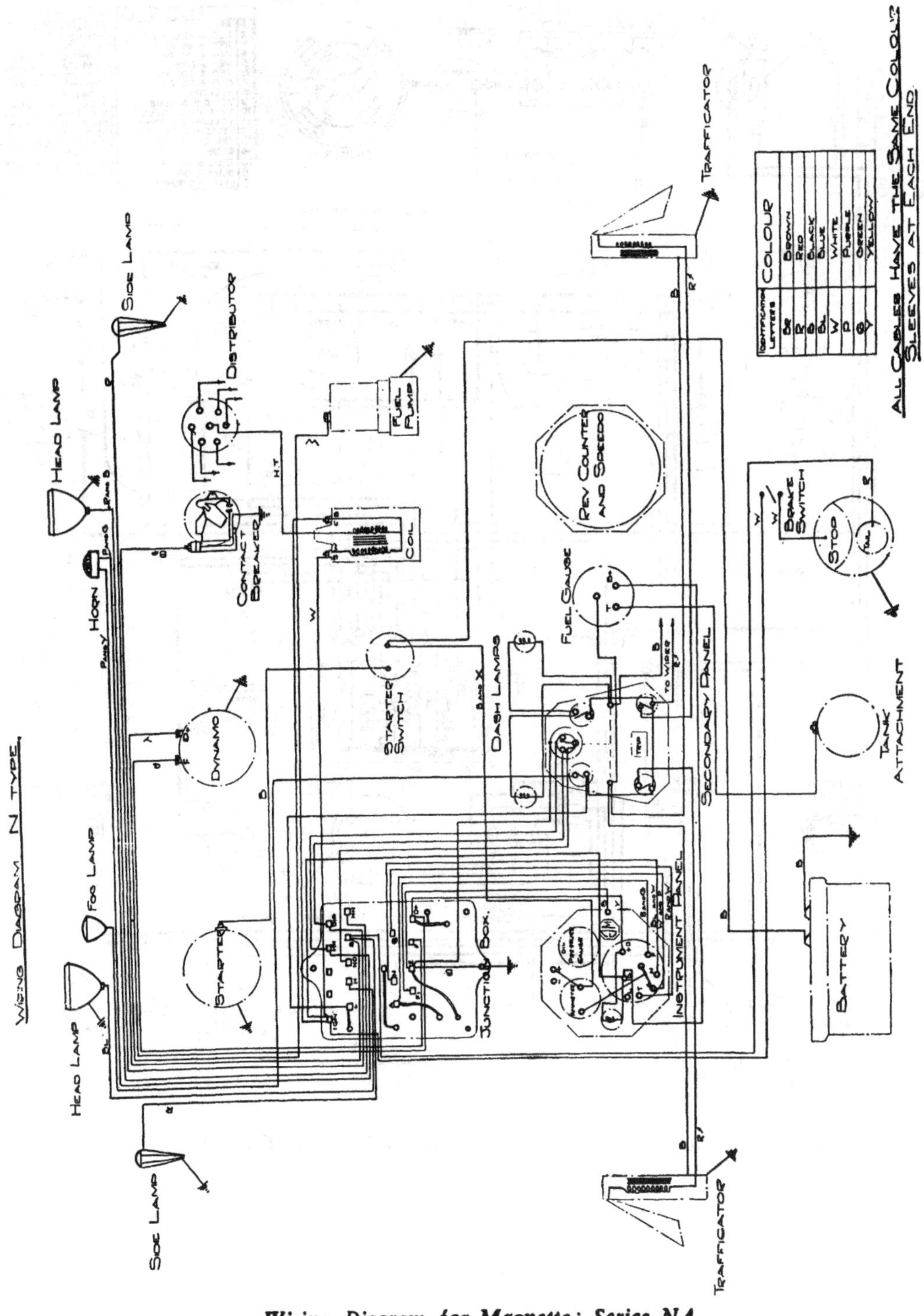

Wiring Diagram for Magnette: Series NA

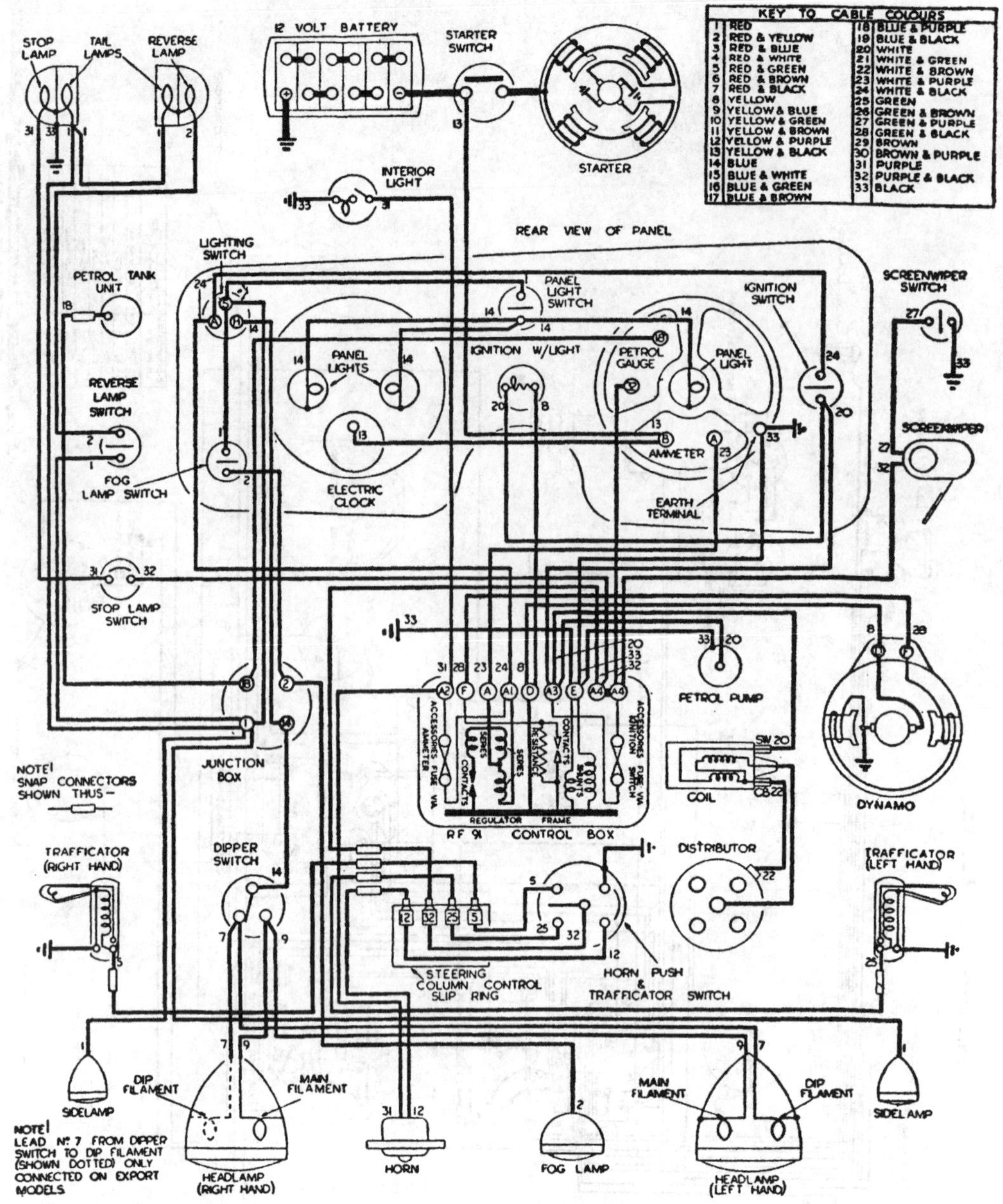

KEY TO CABLE COLOURS			
1	RED	18	BLUE & PURPLE
2	RED & YELLOW	19	BLUE & BLACK
3	RED & BLUE	20	WHITE
4	RED & WHITE	21	WHITE & GREEN
5	RED & GREEN	22	WHITE & BROWN
6	RED & BROWN	23	WHITE & PURPLE
7	RED & BLACK	24	WHITE & BLACK
8	YELLOW	25	GREEN
9	YELLOW & BLUE	26	GREEN & BROWN
10	YELLOW & GREEN	27	GREEN & PURPLE
11	YELLOW & BROWN	28	GREEN & BLACK
12	YELLOW & PURPLE	29	BROWN
13	YELLOW & BLACK	30	BROWN & PURPLE
14	BLUE	31	PURPLE
15	BLUE & WHITE	32	PURPLE & BLACK
16	BLUE & GREEN	33	BLACK
17	BLUE & BROWN		

Wiring Diagram for 1¼-Litre Saloon: Series Y (up to Chassis No. Y0583)

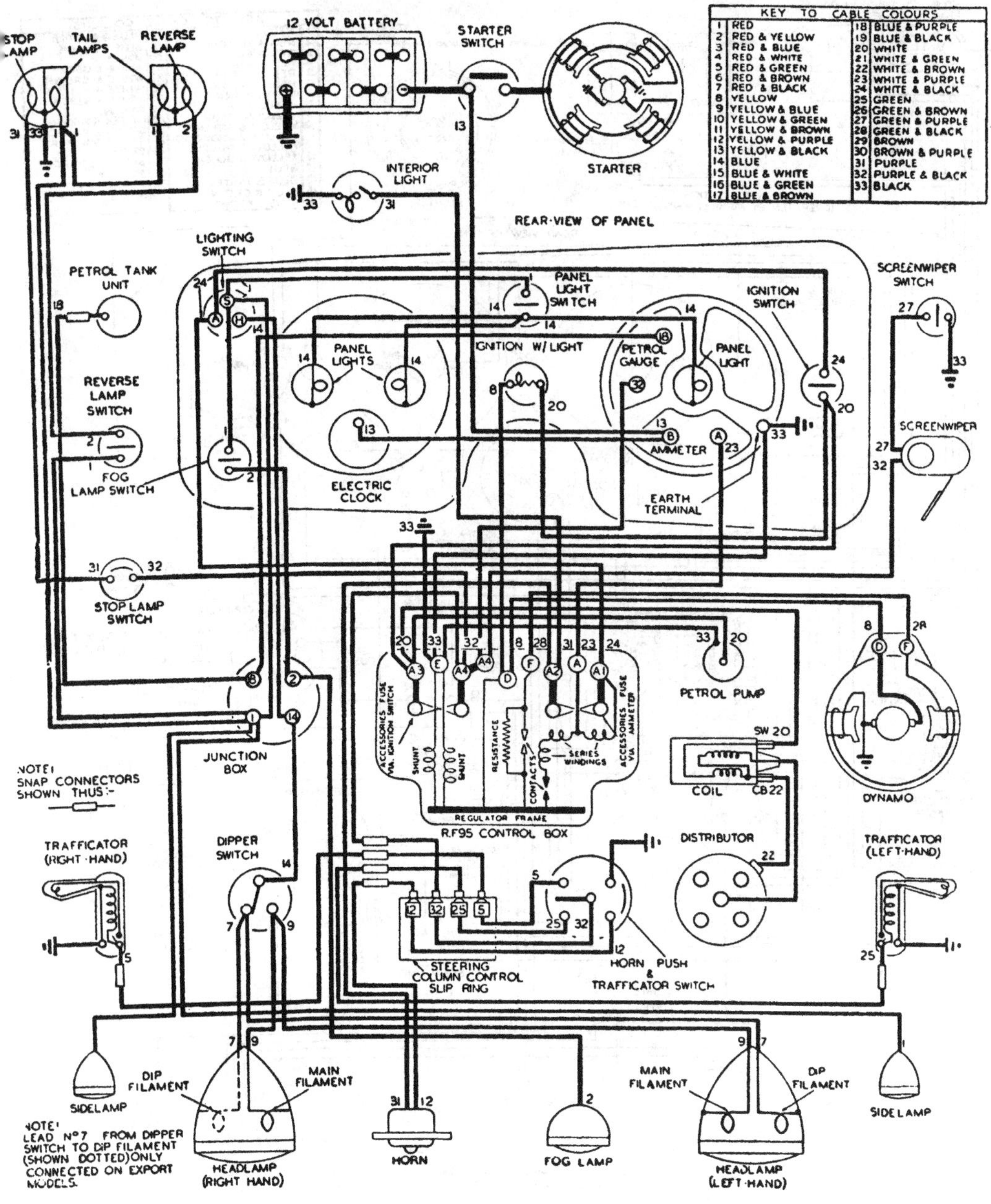

Wiring Diagram for 1¼-Litre Saloon: Series Y (Chassis No. Y0584 and onward)

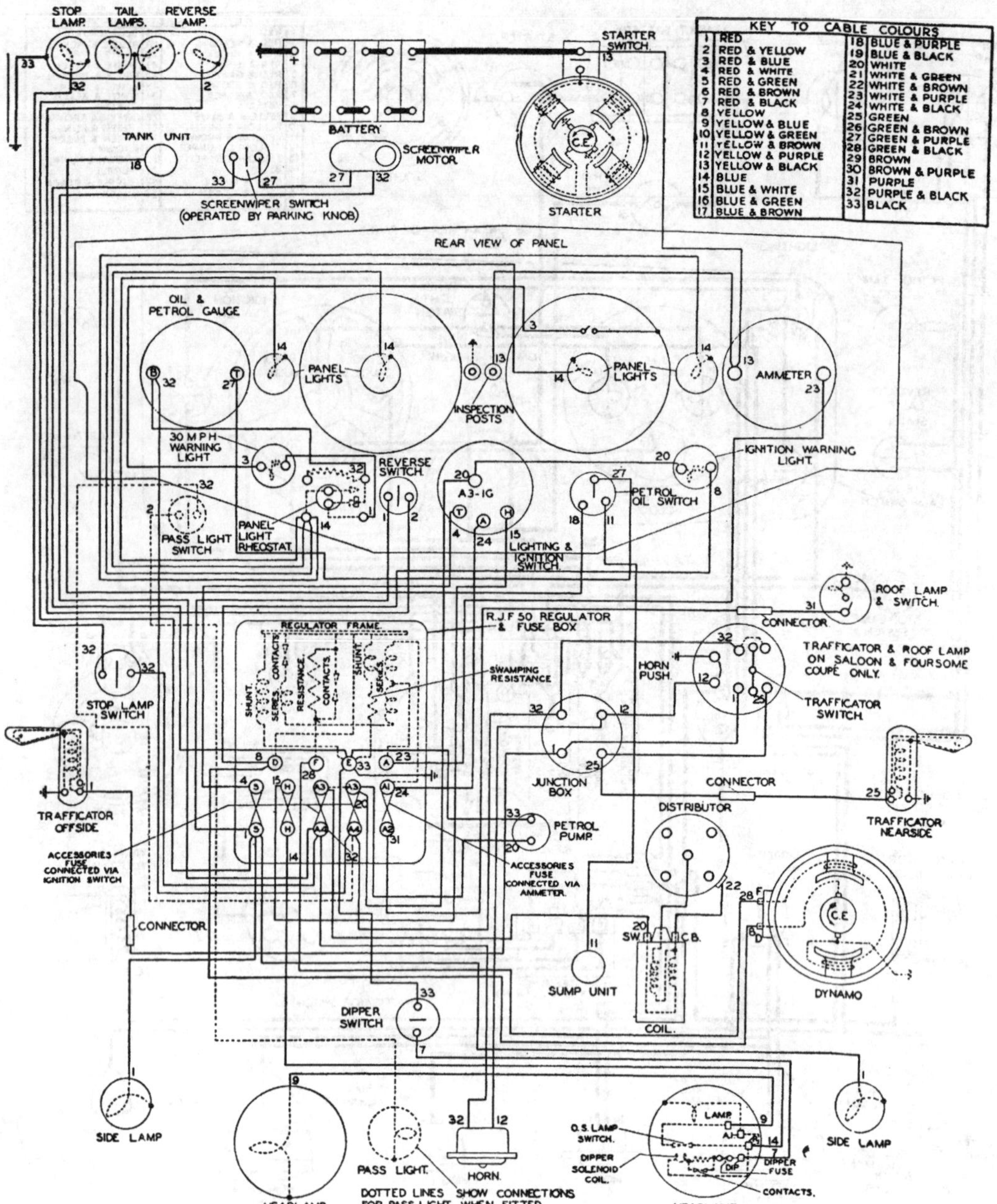

KEY TO CABLE COLOURS

1	RED	18	BLUE & PURPLE
2	RED & YELLOW	19	BLUE & BLACK
3	RED & BLUE	20	WHITE
4	RED & WHITE	21	WHITE & GREEN
5	RED & GREEN	22	WHITE & BROWN
6	RED & BROWN	23	WHITE & PURPLE
7	RED & BLACK	24	WHITE & BLACK
8	YELLOW	25	GREEN
9	YELLOW & BLUE	26	GREEN & BROWN
10	YELLOW & GREEN	27	GREEN & PURPLE
11	YELLOW & BROWN	28	GREEN & BLACK
12	YELLOW & PURPLE	29	BROWN
13	YELLOW & BLACK	30	BROWN & PURPLE
14	BLUE	31	PURPLE
15	BLUE & WHITE	32	PURPLE & BLACK
16	BLUE & GREEN	33	BLACK
17	BLUE & BROWN		

Wiring Diagram for 1½-Litre: Series VA

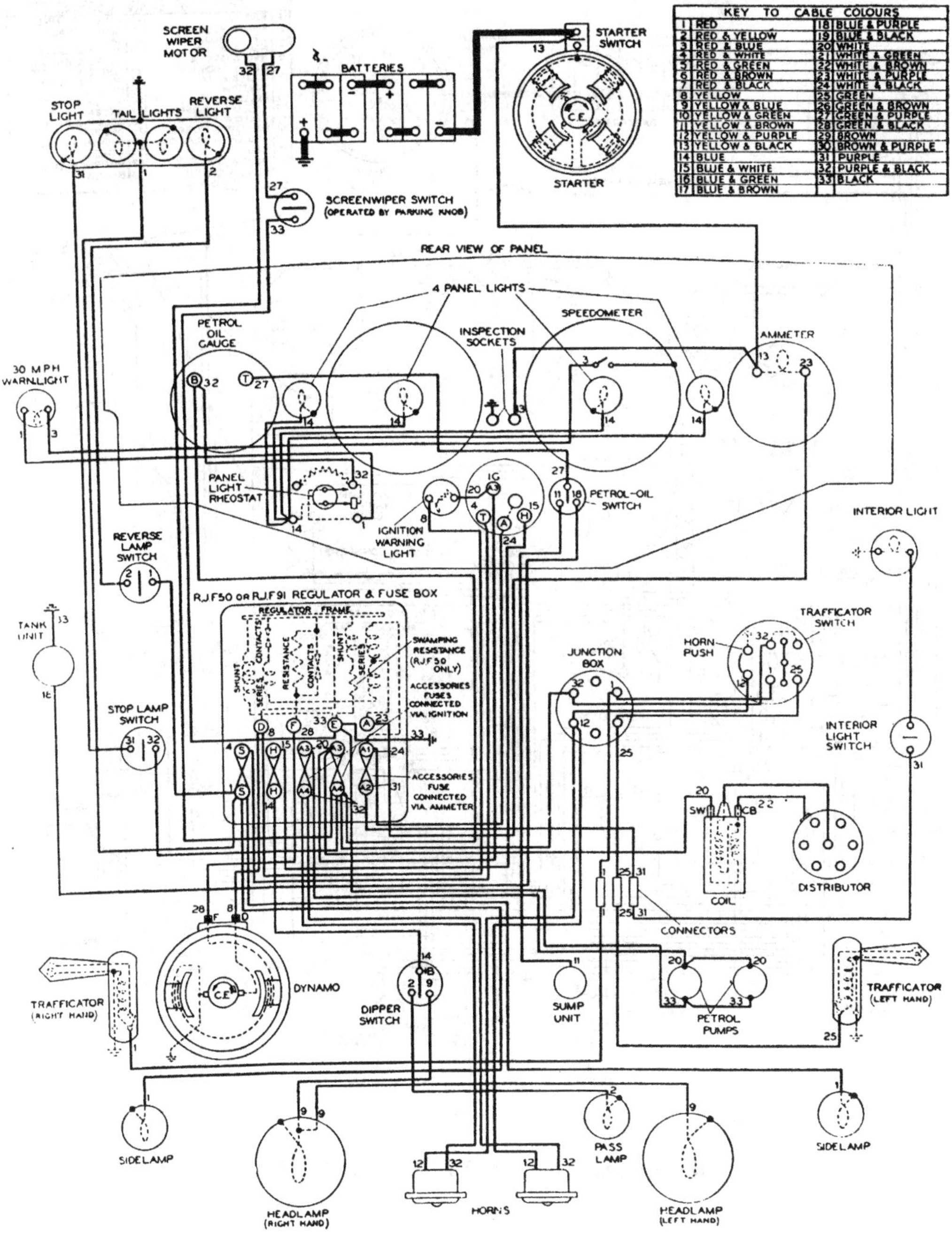

KEY TO CABLE COLOURS			
1	RED	18	BLUE & PURPLE
2	RED & YELLOW	19	BLUE & BLACK
3	RED & BLUE	20	WHITE
4	RED & WHITE	21	WHITE & GREEN
5	RED & GREEN	22	WHITE & BROWN
6	RED & BROWN	23	WHITE & PURPLE
7	RED & BLACK	24	WHITE & BLACK
8	YELLOW	25	GREEN
9	YELLOW & BLUE	26	GREEN & BROWN
10	YELLOW & GREEN	27	GREEN & PURPLE
11	YELLOW & BROWN	28	GREEN & BLACK
12	YELLOW & PURPLE	29	BROWN
13	YELLOW & BLACK	30	BROWN & PURPLE
14	BLUE	31	PURPLE
15	BLUE & WHITE	32	PURPLE & BLACK
16	BLUE & GREEN	33	BLACK
17	BLUE & BROWN		

Wiring Diagram for 2-Litre: Series SA

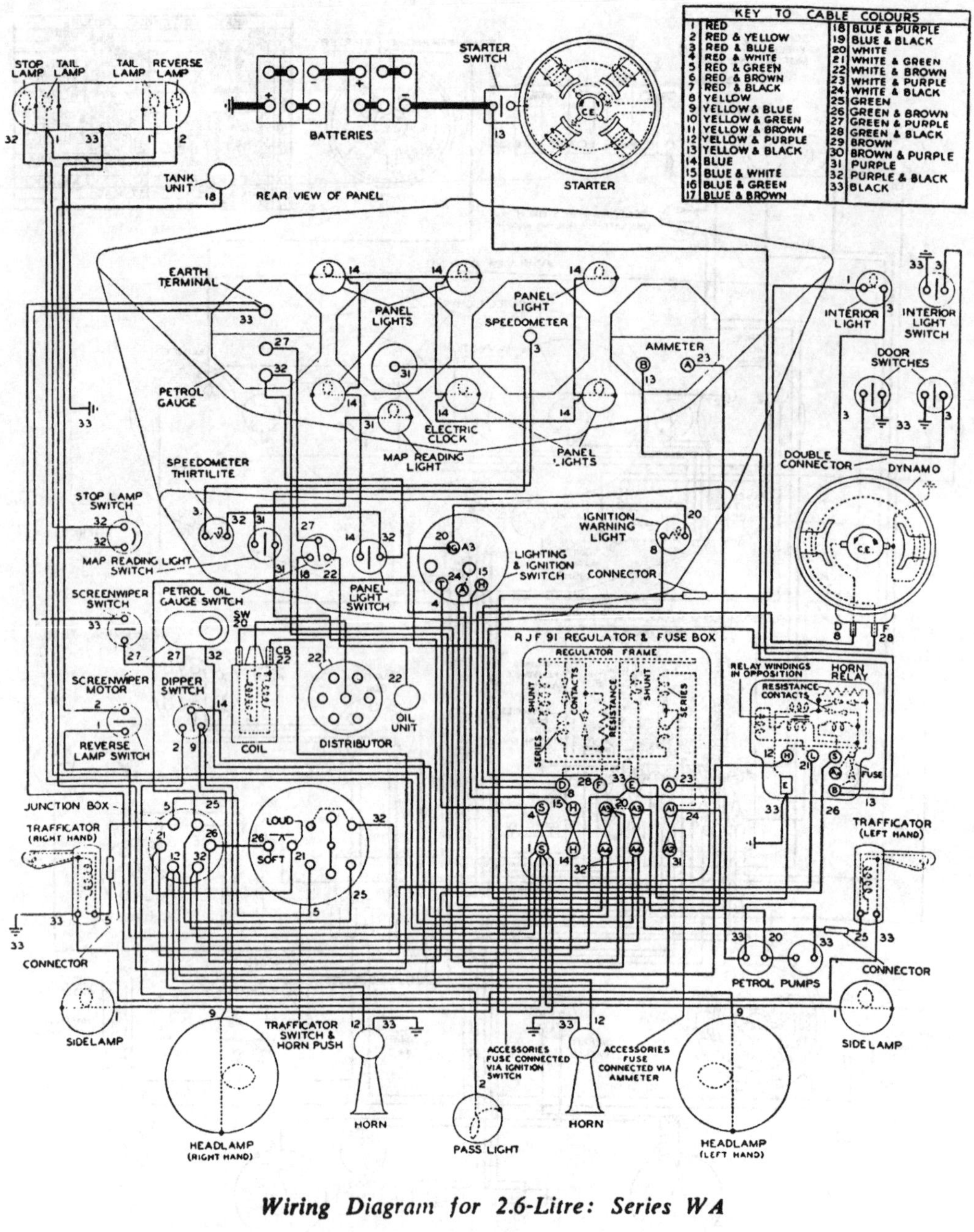

Wiring Diagram for 2.6-Litre: Series WA

SUPPLEMENTS

SUPPLEMENT No. 1

IDENTIFICATION OF "UNIFIED" SCREW THREADS: (TD Midget Rear Axle).

The general standardisation of "Unified" screw threads makes it necessary to identify all nuts, bolts, and set screws with these threads in order to ensure their correct use with correspondingly threaded components, and the fitting of correct replacements. Later Midget axles have A.N.F. or "Unified" screw threads.

Identification has been standardised and is affected in the following manner:

Nuts. By a circular groove turned on the end face of the nut or by connected circles stamped on one flat of the **Hexagon.**

Bolts and Set Screws. By a circular depression turned on the head or by connected circles stamped on one flat of the hexagon.

Wheel Stud Nuts, by a notch cut in all the corners of the hexagon.

These identification marks are clearly shown in the illustration below and it is obviously of the utmost importance that any nuts, bolts, or set screws so marked are used only in conjunction with their associated components having "Unified" threads and that any replacement parts with "Unified" threads are used, as these are *not* interchangeable with Whitworth B.S.F. or Metric threads.

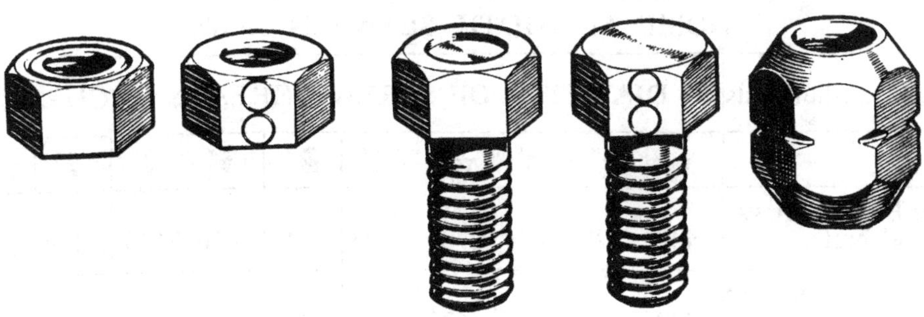

Fig. 1. The identification marks for components with "Unified" threads.

The "Unified" thread is however interchangeable with the American National fine (A.N.F.) thread for all practical purposes.

As an interim measure, prior to the standardisation of the "Unified" thread, certain of the Hypoid axles fitted to Nuffield vehicles have been manufactured with A.N.F. threaded components and such axles are identifiable from the fact that they are fitted with wheel stud nuts having the notch type identification on the corner of the hexagon. Care must be taken, in the case of these axles, to use the correct nuts, bolts, and set screws when re-assembling or when fitting new components; With the exception of the wheel nuts (which are notched) there is no identification mark on A.N.F. threaded bolts, nuts and set screws by which they can be readily identified. Components and assemblies which have "Unified" threads or A.N.F. threads will be identifiable by the new part Nos. which have been allocated to them.

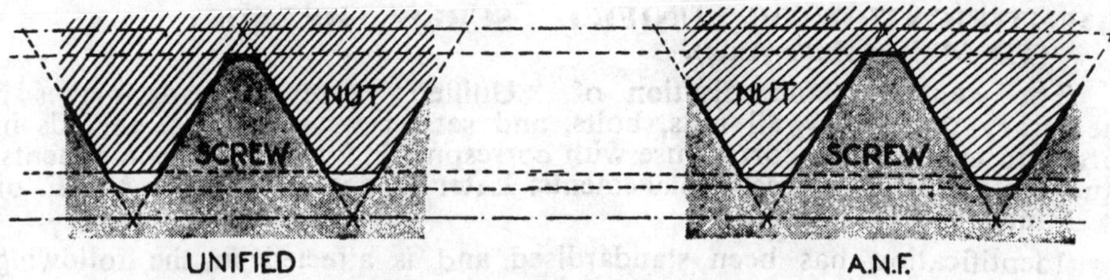

Fig. 2. *Illustration of the " Unified " thread and A.N.F. thread to the same scale indicates their close relationship.*

Certain Service Parts supplied with B.S.F. threads will bear no part numbers. Thus—Assemblies with the old part numbers or without part numbers have B.S.F. threads.

Assemblies with the new part Nos. have " Unified " threads or A.N.F. threads.

Spanners. It is to be noted that all A.N.F. and " Unified " threaded nuts and hexagon headed bolts are made to the standard American and hexagon sizes and that spanners of the appropriate size must be used when tightening or loosening them.

KEY TO SPANNER SIZES.
NOMINAL WIDTH BETWEEN JAWS.

Nuffield Standards	DIAMETER OF SCREW THREADS (INCHES)									
	1/4"	5/16"	3/8"	7/16"	1/2"	7/16"	5/8"	3/4"	7/8"	1"
For B.S.F. Screws and Nuts ..	.448	.529	.604	.705	.825	.925	1.016	1.207	1.309	1.489
For A.N.F. Screws and Nuts ..	.440	.504	.566	.629	.755	.880	.944	1.132	1.320	1.508
For " Unified " Screws ..	.440	.504	.566	**.630**	.755	**.817**	.943	1.132	1.321	1.509
For " Unified " Nuts (Normal)	.440	.504	.566	**.629**	.755	**.880**	.943	1.132	1.321	1.509
For " Unified " (Heavy) ..	—	—	—	—	—	—	**1,069**	**1.258**	**1.446**	—

NOTE :—In the case of some " Unified " threaded components the size of the hexagon for the nut is different from that of the bolt. Where this occurs the spanner size is shown in heavy type in the above table.

MODIFICATIONS. MIDGET SERIES " TD."
Cylinder Head Block and Gasket.

Commencing at engine N. XPAG/TD2/17969 a cylinder block to part No. SA.2404/11 with revised water circulating holes was fitted in conjunction with gasket to part No. X.24481, with pear shaped water holes and large rear water hole.

Commencing at Engine No. XPAG/TD2/22735 a cylinder head to part No. SA.2203/10 and cylinder head gasket part No. 168423 was introduced. This head and gasket have water holes to suit the above cylinder block, the gasket having round water holes and a large rear water hole with reinforcing bar across it. In addition the Champion L.10S Sparking Plugs have been replaced by Champion N.A.8. 14 m/m, ¾ in. reach plugs.

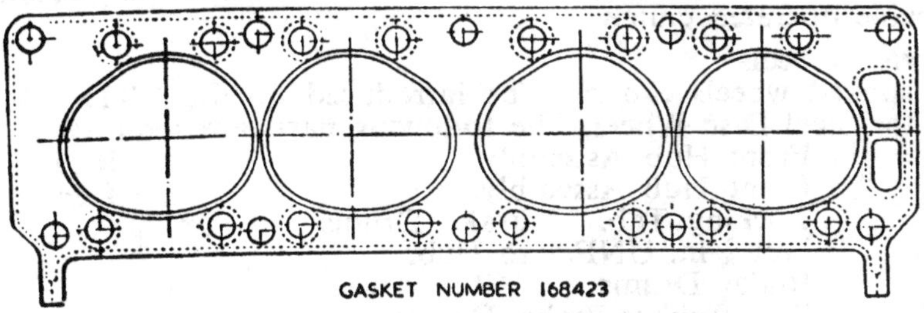

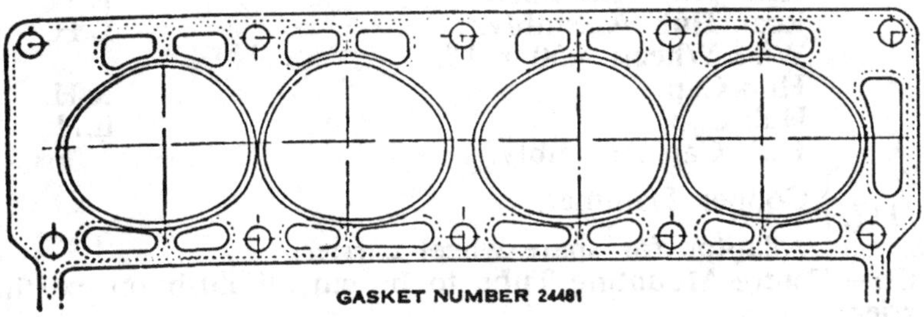

Fig. 3. *The modified gaskets. They can be readily identified by the difference in shape of the water passage aperture.*

Camshaft.

Commencing at Engine No. XPAG/TD2/24116 a new camshaft was introduced with a view to improving wear on the tappet faces.

With this Camshaft, part No. 168553 which has replaced the earlier type camshaft M.G. 862/171, this now no longer being obtainable, the tappet clearance has now been altered from .019 in. to .012 in.

The new camshaft also alters the valve timing, details as follows:

Old Timing.			**New Timing.**		
Inlet Valve opens	11°	B.T.D.C.	Inlet Valve opens	5°	B.T.D.C.
,, ,, closes	57°	A.B.D.C.	,, ,, closes	45°	A.B.D.C.
Exhaust Valve opens	52°	B.B.D.C.	Exhaust Valve opens	45°	B.B.D.C.
,, ,, closes	24°	A.T.D.C.	,, ,, closes	5°	A.T.D.C.

Remote Control Assembly.

Commencing at Engine No. XPAG/TD2/28167, the heads of the bolts for the support cap, part No. 168221, have been drilled in order that the bolts may be locked in position by means of lockwire. This is to prevent the bolts from slacking back.

Clutch Control.

Commencing at Chassis No. TD.22251, the Clutch operating cable has been replaced by a clutch control rod. This means that the clutch cable part No. S.96/6, yoke pin S.47/17, bracket for clutch, and brake pedal R.H.D. and L.H.D. part Nos. 139776, and 139797 respectively, and abutment bracket for clutch cable M.G.900/156 have been deleted and replaced by.

500786.	Clutch Rod Assembly.	R.H.D.	1.off
500787.	Clutch Rod Assembly.	L.H.D.	1 „
500596.	Stop for clutch pedal.		1 „
500597.	Yoke Pin.		1 „
500586.	Bracket Clutch and Brake Pedal	R.H.D.	1 „
500589.	Bracket Clutch and Brake Pedal	L.H.D.	1 „

Adjustments are carried out in the same manner as details given for the cable operated clutch.

Wire Spoked Wheels.

Wire spoked wheels can now be introduced on the "TD" Midget in place of the Steel Disc Wheel. The following parts are required.

102398.	Front Hub Assembly.	R.H.	1.
102399.	Front Hub Assembly.	L.H.	1.
102404.	Locking Tab. } Brake Drums		12.
UFN.206.	Nut ⅜ in. UNF. } to Hub.		12.
102394.	Brake Drums.		4.
ACG.5012.	Plug-Rubber-Brake Drums.		4.
102400.	Rear Hub Assembly.	R.H.	1.
102401.	Rear Hub Assembly.	L.H.	1.
500878.	Wire Wheels 5.50 x 15.		5.
1123.	Hub Cap.	R.H.	2.
1756.	Hub Cap.	L.H.	2.
ACG.5014.	Hub Cap Assembly.		1.
ACG.5037. or P.330/117.	Copper Hammer.		1.
ACG.5162.	Adaptor for spare wheel carrier.		1.

NOTE :—Centre Mounting Tube to be cut off flush on existing spare wheel carrier.

Tail Lamps. (Fig. 4.)

Commencing at Chassis No. TD.21303 circular stop and tail lamps were fitted in place of the moulded type previously used. They are mounted on chromium plated plinths, the plinths being attached to the rear wings by Phillips screws.

The plated rim and glass are held in grooves in the rubber surround. Move aside the rubber ring and remove the plated rim (1) and then in a similar manner remove the glass (2). The bulb which is then accessible is of the double filament type (6/18 Watts). The smaller wattage filament is connected to the tail lamp, and the larger wattage filament to the stop lamp circuit. (This filament is also used for the directional indicator lights where fitted).

When fitting replacement bulbs no difficulty should be experienced in fitting these correctly as the bulbs are designed so that the locating pegs are offset and therefore can only be fitted into the bulb holder in the correct position.

Flashing Equipment.

From Chassis No. TD.22371 all cars for the American Market were equipped with flashing side and tail lamps. The equipment included the round tail lamps introduced at Chassis No. TD.21303, modified side lamps, relay box, flashing light switch, bracket for-do, warning light for flasher, the flasher equipment being introduced on arrival in America.

Details of the wiring are shown on the wiring diagram on page 582 to Lucas Drawing No. W.23300.

Windscreen Wiper.

A new centrally mounted windscreen wiper was introduced at Chassis No. TD.22315, this modification entailing alterations to the windscreen frame and glass, the arm, and blade assembly and coupling bar.

Fig. 4. *Stop and Tail lamps fitted to later models.*

MODIFICATIONS. M.G. 1¼ LITRE SERIES "YB."

Camshaft.

The XPAG/SC/2 engine had a camshaft fitted to part No. 22329 which altered the valve timing to the following although the tappet clearance remained at .019 in. hot.

Inlet Valve opens	5° B.T.D.C.	Exhaust Valve opens	45° B.B.D.C.
Inlet Valves closes	45° A.B.D.C.	Exhaust Valve closes	5° A.T.D.C.

Commencing at engine No. XPAG/SC2/18097, a new camshaft part No. 168553 was fitted to improve wear on tappet faces, the valve clearance being altered from .019 in. to .012 in. The valve timing is not affected by this change.

Cylinder Head Assembly and Gasket.

A new cylinder head assembly part No. SA.2203/10 and cylinder head gasket 168423 were introduced at engine No. XPAG/SC2/17994. Reasons for the change are clearly indicated on page 523.

The Champion L.10 S. plugs have been replaced by the Champion N.8.B. 14 mm. ¾ in. reach plugs.

MODIFICATION TO MIDGET SERIES "TD" AND 1¼ LITRE SALOON "YB."

Valves.

In order to increase exhaust valve life, the material specification for the exhaust valve has been altered and they are now made from "YB" quality. This type of valve was introduced at engine No. XPAG/TD2/18291 on the "TD" Series Midget and at engine No. XPAG/SC2/17500 on the 1¼ Litre Saloon Series "YB."

Oil Pump.

Commencing at engine No. XPAG/TD2/20972 on the "TD" Midget and at engine No. XPAG/SC2/17680 on the 1¼ Litre Series "YB" a priming plug has been fitted in the oil pump cover to improve accessibility of pump priming for use after engine draining. In addition a modification has been made to the oil pump cover to take the plug.

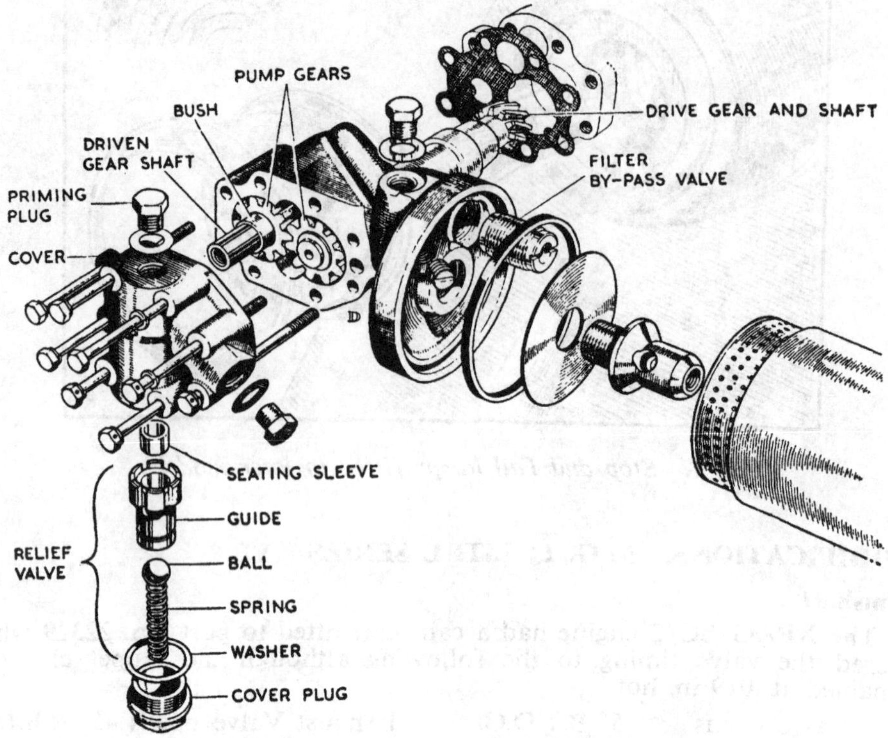

Fig. 5. The components of the modified pump and purolator oil filter.

Distributor Fixing.

A new standardised method of distributor fixing was incorporated on the "TD" Midget and 1¼ Litre Models having an improved fixing arrangement in place of the adjusting clip. The improved fixing details were introduced at engine No. XPAG/TD2/20942 on the "TD" Midget and engine No. XPAG/SC2/17670 on the 1¼ Litre Series "YB."

The modified fixing comprises a tapered cotter bolt passing through the distributor housing boss in the cylinder block and contacting the stem of the distributor, and it replaces the split adjusting clip hitherto employed.

Distributors using the cotter pin attachment are released by slackening the cotter bolt inwards to free its tapered surface from the stem of the distributor.

While the location of the distributor rotor will not be disturbed owing to the action off the offset driving tongue, release of the distributor body and stem will affect the ignition point, and it is therefore essential to mark the distributor body and the face of its housing before removal, to ensure correct ignition timing on replacement. The distributor housing face is marked with a scale to facilitate this.

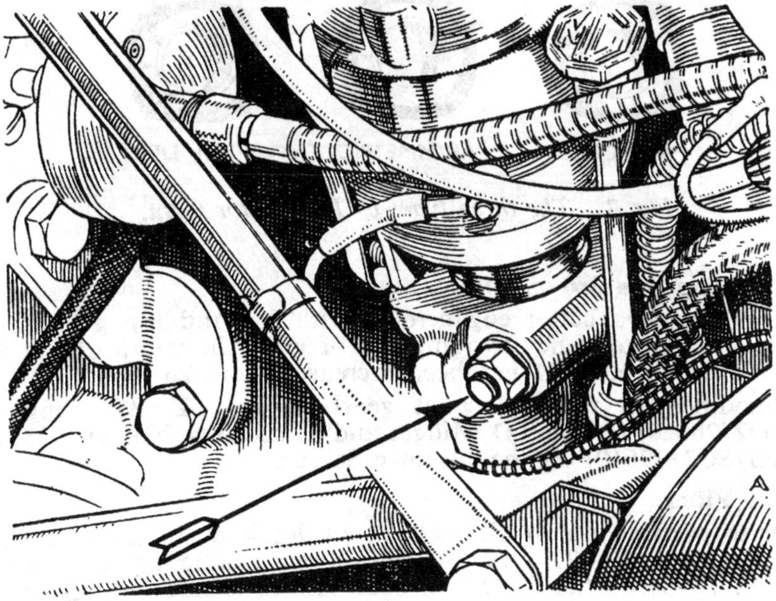

Fig. 6. Later type Distributor Fixing consisting of cotter bolt engaging distributor body and stem.

Distributors.

New distributors were introduced on the M.G. Midget "TD" at engine No. XPAG/TD2/24489 and on the 1¼ Litre Saloon Series "YB" at engine No. XPAG/SC2/18097.

These later type distributors are fitted with high lift cams. Owing to the shape of these cams the contact breaker gap must be set at .014 in. to .016 in. (.36 m/m. to .41 m/m.). Previous distributor cams were of two types, symmetric and asymmetric and both these types necessitated a contact breaker gap of .010 in. to .012 in. (.25 mm. to .030 mm.).

The wider gap of the high lift cam together with the steep angle of the cam face, gives more accurate ignition timing and controls pitting and piling action which limits useful contact life.

When setting contact gaps with the high lift cam more care is needed when checking that the fibre heel is in the highest point of the cam rise, because maximum separation of the contact points is only obtained over a small angular movement of the distributor shaft.

All three types of cams are illustrated in Fig. 7. Apart from the appearance of the cams, distributors fitted with the high lift cams can be identified by reference to the suffix letter which follows the Service No. Distributors bearing the suffix "E" or any letter subsequent to "E" after the service No. 40162 are fitted with high lift cams, and those with suffix letters previous to "E" are of the symmetric or asymmetric type.

The contact breaker setting for the symmetric and asymmetric type is .010 in. to .012 in. (.25 mm. to .30 mm.) gap, and for the high lift cam type is .014 in. to .016 in. (.36 mm. to .41 mm.).

It is interesting to note that this type of distributor has a built in suppressor, the loose suppressor to part No. 24464 on the "TD" and part No. 163439 on the 1¼ Litre Saloon Series "YB" now no longer being required.

Fig. 7. The three types of Distributor Cam.

Gearbox Striking Dog 3rd and 4th.

In order to give a better engagement with 3rd and top gear dog teeth and to alleviate the tendency to jump out of gear, the width of the striking dog part No. M.G. 900/161 has been increased from 25 mm. to 27 mm.

This modification took place at gearbox No. TZ. 355 (engine No. XPAG/TD2/20026) on the TD Midget and at gearbox No. SJ. 528 (engine No. XPAG/SC2/17610) on the 1¼ Litre Saloon

Oil Suction Filter.

In order to prevent loss of oil pressure on deceleration, a new oil suction filter to part No. 168727 was introduced on production at engine No. XPAG/TD2/26364 on the "TD" Midget, and at engine No. XPAG/SC2/18097 on the 1¼ Litre Saloon.

All engines fitted with the 10½ pint capacity sump, part No. 168137, are to have this oil suction filter fitted.

Retrospective action should be taken from engine No. XPAG/TD2/14948 on the "TD" and XPAG/SC2/17383 on the 1¼ Litre.

Crankshaft.

Commencing at engine No. XPAG/TD2/27551 on the TD Midget and engine No. XPAG/SC2/18272 on the 1¼ Litre Saloon a new crankshaft made from EN 100 Steel for additional strength has been fitted.

This crankshaft to part Nos. 168632 with bush and bearings, 168628 crankshaft with bush and 168537, crankshaft bare, can be identified by means of the part number stamped on one of the webs. From a Service point of view, the new crankshaft will replace the old when stocks of the latter are absorbed.

Cylinder Heads.

Commencing at engine No. XPAG/TD2/27867, on the "TD" Midget and XPAG/SC2/19037 on the 1¼ Litre Saloon, the height of the valve spring faces has been reduced by .5 mm. in order to ensure that the valve springs do not become coil bound with the latest camshaft introduced, at engine No. XPAG/TD2/24116 on the TD and XPAG/SC2/18097 on the 1¼ Litre.

The Mark II competition engine had this modification introduced at engine No. XPAG/TD3/27996.

Oil Pump Assembly.

Commencing at engine No. XPAG/TD2/26635 on the TD Midget and engine No. XPAG/SC2/18120 on the 1¼ Litre Model, a new type of oil pump body was fitted which has an increased oil level to enable the pump to be kept primed.

Parts deleted are:

S.A.2448/1.	Oil Pump Assembly	1.
SA.2442/1.	Oil Pump Body (Bushed)	1.
168118.	Oil Pump Body (Bare)	1.

The above has been replaced by:—

SA.2448/2.	Oil Pump Assembly	1.
S.A.2442/2.	Oil Pump Body (Bushed)	1.
168576.	Oil Pump Body (Bare)	1.

Oil Pump Priming.

Although a modification has been made to the oil pump bodies on the "TF" Midget, in order to replace an air lock and make the pump self priming commencing at engine No. XPAG/TF/31263 it can be incorporated on the TD Midget and 1¼ Litre Saloon, fitted with the combined oil pump and Purolator filter.

The modification consists of the addition of a 3.5 mm. diameter hole drilled through from the hole which takes the 12 mm. diameter plug adjacent to the filter, into the hole for the balance valve.

Existing oil pumps and oil pump bodies may be altered if so desired. When engines are started up for the first time, or after engine has been stationary for a considerable period the following routine is to be followed.

1. Remove the 12 mm. dia. x 1.5 mm. pitch plug from the top of the oil pump body adjacent to the filter.
2. Start up engine and run with no load at 1,500 to 2,000 r.p.m. until oil appears at the hole.
3. Replace plug.

Fig. 8. Midget series "TF" fitted with wire wheels.

Fig. 9. Midget series " TF " showing weatherproofing equipment.

M.G. MIDGET SERIES "TF".

INTRODUCTION.

MECHANICAL.

The M.G. Midget Series " TF " introduced during the 1953 International Motor Show is a more streamlined version of the competition Midget Series " TD " Mark II. Basically the chassis lay-out and engine design are identical.

The " TF ", however, does not incorporate the additional friction type damper on the front and rear axles and employs a single fuel pump instead of twin fuel pumps.

The engine is produced with a compression ratio of 8.0/8.2 to 1, obtained by removing $\frac{1}{16}$ in. from the gasket face of the cylinder head, and larger valves 36 mm. O/D heads on the inlet valves and 34 mm. O/D heads on the exhaust valves are used in conjunction with the 150 lb. open tension competition valve springs.

The depth of the radiator has been decreased some 3 in. in order to affect the reduction in frontal area and streamline appearance, and to incorporate the modified front wings with built in headlamps. With the reduction in frontal height, the 9 in. diameter oil bath air cleaner as employed on the TD Mark II has been replaced by individual pan-cake type air cleaners and silencers, attached directly to the mouth of the carburetters.

With the decrease in the depth of the radiator the width has been increased slightly, in order to retain, as near as possible, the same coolant capacity as the TD Midget. The cooling system is now pressurised and with the bonnet sloping to the front a modified type of water outlet pipe has been introduced, the pipe now extending vertically from the cylinder head and housing the thermostat. The water connecting pipe is bolted directly to the thermostat housing, the connection to the header tank being made by a length of concertina rubber hose.

The bulb for the radiator thermometer is now entered into the thermostat housing and not in the radiator header tank as has been the normal practice.

CARBURATION.

Carburation is by means of two S.U. semi-downdraught 1½ in. diameter carburetters to specification H.4. They vary from the competition carburetters employed on the TD Mark II insofar that they are somewhat shorter in the body, light blue springs are fitted but no hydraulic piston dampers. Dampers were introduced commencing at Chassis No. TF.3495.

Up to and including Chassis No. TF.1509 the standard " L " type petrol pump was fitted, being attached to the side of the tool box.

Commencing at Chassis No. TF.1510 a high pressure petrol pump is fitted, this being mounted on a bracket on the R.H. rear side of the chassis frame just forward of the upswept section of the chassis frame side member.

BODY.

Accessibility to the engine has been somewhat restricted by discarding the removable bonnet top and sides. The one-piece bonnet sides are now fitted by means of bolts and screws at both top and bottom corners. Screws, one at each corner at the bottom, secures the side in position at these points. Special brackets are attached to the top of the engine bulkhead and radiator casing, and a bolt at each top corner secured by a Simmonds locking nut secures the bonnet side at these points. The sides are also secured at the bottom to the chassis frame by four bolts and flat washers and a form of tie rod is used to connect the sides, fitted just behind the radiator, being secured by two bolts and nuts on each side.

These bolts and nuts also are used for securing the front wings to the bonnet sides. The bonnet top comprises two removable panels, hinged in the centre by means of a chromium bonnet hinge, the panels being secured to the bonnet sides by means of special Amal type quick release push button fasteners. The body has been lowered on the chassis in order to maintain a low waist line, which has brought about slight reduction in the petrol tank capacity, the capacity being approximately half a gallon less than the TD Midget. Modified front and rear valances are now fitted.

Weatherproofing equipment comprises a double texture mercerized cotton rep/Grandelle hood and sidescreens, the sidescreens and hood rear light being of Vibrac plastic.

INSTRUMENTS AND CONTROLS ETC.

Instruments comprise a large speedometer dial with clock, large revolution indicator with headlamp main beam warning light, and combined instrument comprising ammeter, oil pressure gauge and water temperature gauge, all instruments being surrounded by attractive octagonal chromium plated bezels.

Controls comprise starter pull, choke or mixture control, ignition switch, light switch, panel light switch, and auxiliary switch, ignition warning light, petrol reserve warning light, and flashing indicator warning light, all instruments, controls and warning lights being mounted on a central instrument panel, attached to a pressed steel facia board. Two glove trays are provided in the facia board.

The control knobs for the wiper blades are situated inside the glove boxes either side of the instrument panel.

The switch for the flashing indicator is fitted either end of the facia board, according to the driving position, with the horn push fitted on the scuttle just above this switch, adjacent to the driver's L.H. or R.H., again according to the driving position.

Twin map reading lamps are fitted on either side of the scuttle underneath the trim roll. The double dipping headlamps, fitted with pre-focus bulbs and block lenses, are controlled by a foot operated dipping switch

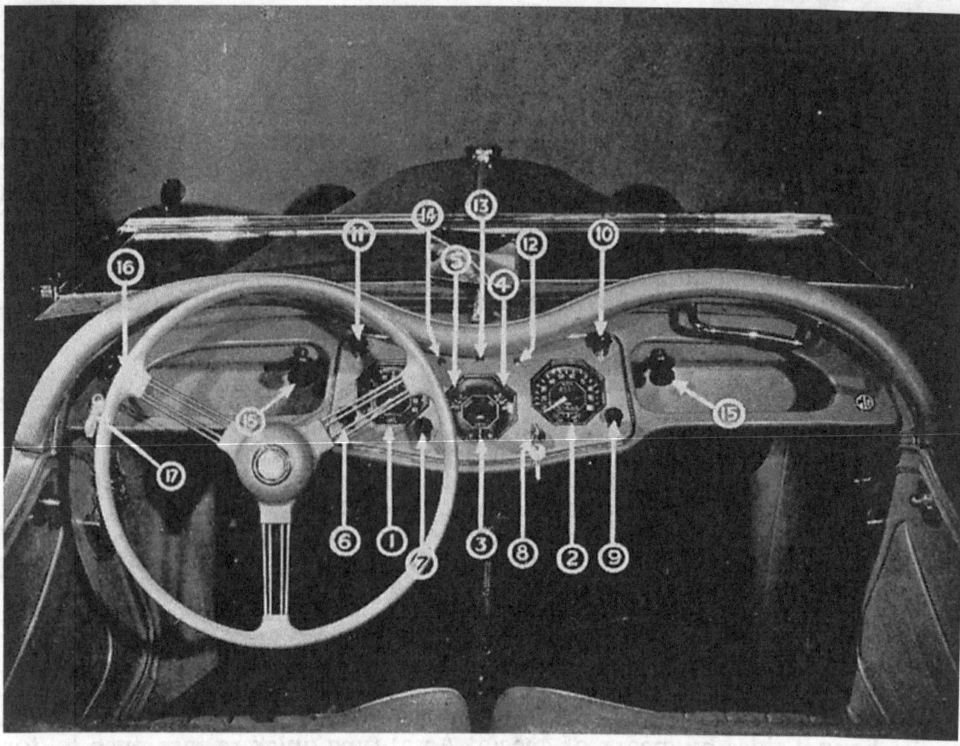

Fig. 10. *Showing position of Instruments and Controls, etc* (*L.H. Drive Model.*)

Key to Fig. 10.

1. 4½ in. Revolution counter with main beam warning light
2. 4½ in. Speedometer with clock, mileage and trip recorder.
3. Ammeter.
4. Oil Pressure Gauge.
5. Water Temperature Gauge.
6. Starter pull. Pull outwards and upwards to operate.
7. Light switch. Pull out to switch on sidelamps. Give quarter turn and pull outwards again to switch on headlamps.
8. Ignition switch.
9. Choke or Mixture control. Pull outwards and twist in clockwise direction to engage notches.
10. Auxiliary switch for foglamp when fitted.
11. Panel lamp switch. Pull outwards to switch on panel lamps, give quarter turn and pull outwards again to switch on map reading lamps.
12. Flashing director indicators warning light.
13. Ignition warning light.
14. Petrol reserve warning light. Comes into operation when petrol level reaches 2½ gallons.

15. Windscreen wiper control knobs.
16. Horn push.
17. Flashing director indicator switch.

On the R.H. drive Model the position of 1 and 2 and 6 and 9 are reversed.

The Trip recorder is set by means of a Bowden cable with a light spring loaded knob, the cable being secured in a small clip bolted to the base of the R.H. glove compartment on the L.H. drive Model, and to the L.H. Glove compartment on the R.H. drive Model.

To set Trip recorder to zero press spring loaded knob and turn anti-clockwise.

The clock hands are set by means of a knurled knob on the back of the instrument, access to the knob is by means of a small hole cut in the scuttle masking board at a point just behind the instrument.

Many of the descriptions, dismantling and reassembly details given for the Midget Series TD apply to the Series "TF". Where, however, the procedure differs, full details of the procedure to adopt will be given.

THE CHASSIS

REAR SUSPENSION. Details as TD and 1¼ Litre Models.

REMOVAL AND REPLACEMENT OF REAR SPRINGS. Details as TD and 1¼ Litre Models.

FRONT SUSPENSION. Details as TD Midget.

REMOVING FRONT SUSPENSION. Details as TD Midget.

REPLACING THE FRONT SUSPENSION. Details as TD Midget.

REMOVAL OF FRONT BRAKE DRUM AND HUB. Details as TD Midget.

REPLACEMENT OF FRONT HUB. Details as TD Midget.

REMOVAL AND REPLACEMENT OF FRONT COIL SPRINGS. Details as TD Midget.

SHOCK ABSORBERS.

The TF Model is fitted with Armstrong double acting, self-regulating shock absorbers, resistance being offered to the compression and recoil of the road springs, the shock absorbers being accurately set before the car leaves the works, to give the amount of damping most suitable for the car. No adjustments are therefore necessary or provided for.

MAINTENANCE OF SHOCK ABSORBERS.

When in position on the car, maintenance is confined to periodic examination of the fluid level and of the secure anchorage to the chassis, the fixing bolts being tightened as required.

TOPPING UP ARMSTRONG SHOCK ABSORBERS.

When necessary, the shock absorbers should be topped up with Armstrong Shock Absorber oil of the recommended grade.

Before removing the filler plug, positioned on top of the shock absorber, carefully wipe the exterior with a clean rag, to prevent dirt entering through the filler hole.

The operation of topping up must not be neglected, otherwise air may enter the pressure cylinder and seriously impair the action of the shock absorber.

REMOVAL AND REPLACEMENT OF FRONT SHOCK ABSORBERS.

Details as in Section "A" under TD Midget and 1¼ Litre Models.

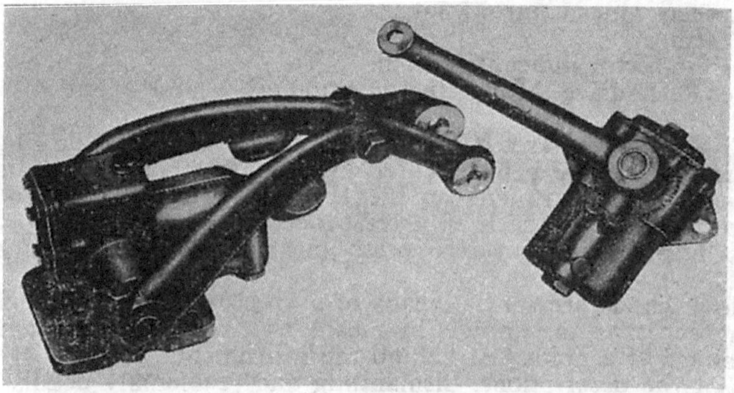

Fig. 11. *The Armstrong Shock Absorber.*

DETAIL CONSTRUCTION OF ARMSTRONG SHOCK ABSORBERS.

Fig. 12 shows the construction of both the standard double acting (D.A.S.) and independent (I.S.) types of shock absorbers.

The shock absorber body " A " is bolted to the car chassis and the lower end of the link " L " is connected to the axle. In the case of the I.S. type, the arms form the upper wishbone of the independent suspension, no link being used. When the axle moves relative to the chassis (this movement being allowed by the car spring) arm " H " is moved up or down, and since it is splined to spindle " S " the latter is rotated. The spindle in turn is splined to crank assembly "C " which is connected by means of the connecting rods " B " to pistons " P " (in which are situated lightly spring loaded recuperating valves " R ").

Thus pressure is built up in one cylinder or the other and since the cylinders are connected by ports in the body to valve chamber " V " this pressure is dependent on valve setting (see valve operation).

The shock absorber is filled up to the bottom of the filler plug boss which prevents overfilling and maintains the necessary air space essential to correct operation. It is sealed by filler plug " F ", thus completely submerging the working mechanism in oil. Oil is prevented from leaking along the spindle by means of oil seal " O ".

WHAT A SHOCK ABSORBER DOES.

If a car has ideal suspension the body would follow a straight horizontal path whilst the wheels would move up and down to follow irregularities in the road (bumps, holes, etc.). In a conventional car, leaf or coil springs are interposed between the axles and the chassis in an attempt to achieve this ideal. Unfortunately, however, a spring merely stores energy (although to some extent a leaf spring dissipates energy by friction between the leaves) and therefore to prevent this stored energy being transferred to the car body (in the form of an upward bump) damping devices must be added between the axles and the chassis to absorb the energy.

This is the purpose of shock absorbers, which in addition have the function of damping out general spring vibrations.

HOW THE ARMSTRONG SHOCK ABSORBER DOES IT.

The Armstrong shock absorber is of the hydraulic double acting type,

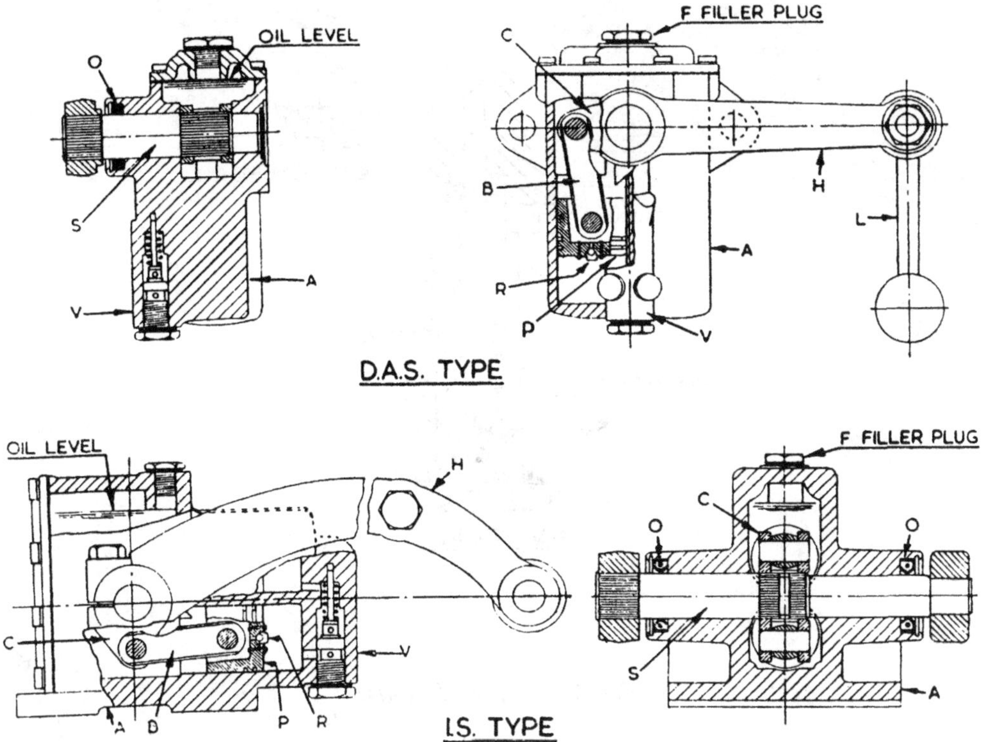

Fig. 12. *Detailed construction of the Armstrong Shock Absorbers.*

and works on the principle of pumping oil backwards and forwards between two cylinders through suitable valves set to give the required degree of restriction in each direction.

This design has advantages in that, firstly, all the working parts are submerged in oil and, secondly, it is only necessary to "recuperate" the almost negligible amount of oil which is forced past the pistons by pressure built up as the shock absorber operates.

VALVE OPERATION.

To accomplish general damping of the car spring, a small "bleed" is built into the valve. This operates both in compression (axle moving up) and on rebound (axle moving down). As bumps become more severe on compression pressure builds up in the left hand cylinder of the D.A.S. type, and in the lower cylinder in the case of the I.S. type (Fig. 13) and blows compression valve "VC" off its seat at a pre-determined pressure controlled by spring "SC". As the speed of rebound increases pressure is built up in the right hand cylinder, or upper cylinder in the case of the I.S. type, and blows rebound valve "VR" off its seat at a pre-determined pressure controlled by spring "SR".

It will be clear that by suitable selection of springs, any range of blow-off from zero to the maximum rating of the shock absorber can be obtained in either direction.

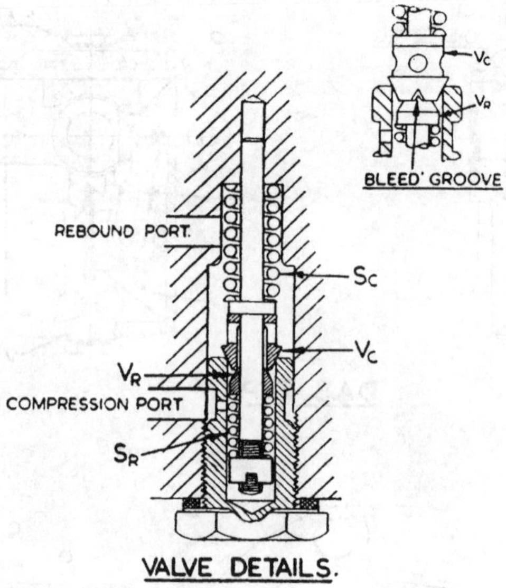

Fig. 13. *Valve Details.*

OIL.

These shock absorbers are only guaranteed when filled with Armstrong shock absorber oil No. 549.

SPARES.

Details can be obtained on direct application to Armstrong Patents Co., Ltd., Service Dept., Eastgate, Beverley, Yorkshire, England.

REPLACEMENT OF SHOCK ABSORBER VALVES COMPLETE.

Remove the shock absorber from the chassis, removal of front shock absorber as details given for TD Midget.

Carefully clean all dirt from the vicinity of the valve chamber. Unscrew the hexagon headed screw " A " (Fig. 14), until the complete valve assembly comes clear of the shock absorber body.

Ensure that the loose compression spring " D " is withdrawn with the rest of the assembly.

NOTE.

On no account must the shock absorber arm be moved once the valve assembly has been removed, otherwise oil will be expelled through the valve chamber. Ensure that the valve chamber is free from all dirt or grit, then screw in the new clean valve assembly, operate the shock absorber arm up and down until an even resistance is felt through the full stroke, then remove the filler plug and top up with Armstrong shock absorber oil of the recommended grade. Should a valve change be contemplated with the shock absorber in situ on the chassis, disconnect the link before commencing above procedure.

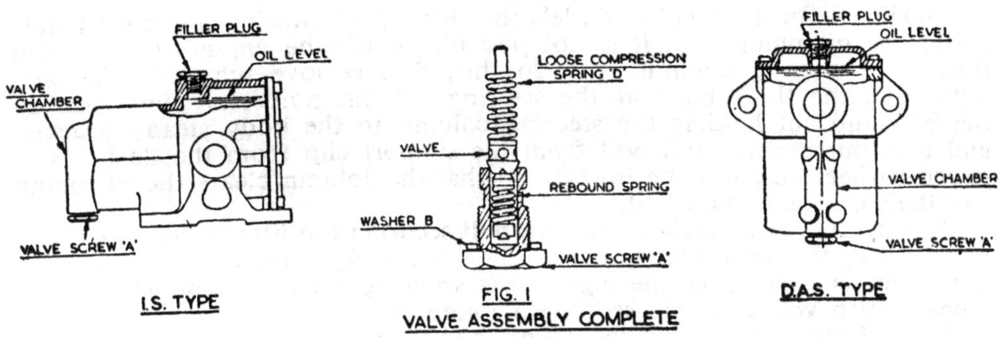

Fig. 14. *Valve Assembly complete.*

NOTE.

As all valve assemblies are pre-set before leaving the Factory, it is essential that no parts are interchanged and that the new washer "B" is used only with a new valve assembly, otherwise incorrect settings will result.

PROPELLER SHAFT.

Details as TD Midget.

DISMANTLING UNIVERSAL JOINTS.

As details on page 32.

THE ENGINE

DESCRIPTION.—Details at TD Midget.

ENGINE LUBRICATION.

In the main the function of the lubrication system is identical to that for the TD Midget, but it must be realised that with the combined oil pump and Purolator oil filter there is no by-pass valve in the crankcase immediately behind the oil pump.

A by-pass valve is now incorporated in the filter. If the filter element becomes clogged through neglect, the filter by-pass valve comes into operation allowing the oil to by-pass the filter, maintaining engine lubrication until a new filter element is fitted. Full details of the oil pump and filter are shown in Fig. 5.

REMOVAL AND REPLACEMENT OF SUMP.—Details at TD Midget.

REMOVAL OF THE OIL PUMP.

Drain the radiator, slack off the bottom water hose and remove the top water hose. Remove the front engine mounting bolts holding the engine bracket to the rubber block. Slightly jack up the engine at the front to allow the pump to clear the frame member.

NOTE:—On the L.H.D. Models the steering column passes immediately above the oil pump, and it is not possible to lift the engine without first lifting the steering column. To do this, first remove the split pins and nuts from the three bolts at the steering column universal joint, slacken the bolt and nut holding the steering column to the body steady bracket, and take out the nut and bolt from the support clip from the dash. The steering wheel may now be lowered so that the column clears the oil pump and the engine may be lifted.

Remove the large centre retaining bolt securing the filter to the oil pump body, break the joint and withdraw the filter body complete with element, top cover, etc. Remove the eight bolts securing the pump to the cylinder block, which will also release the oil pump cover.

Lift off the cover from the pump body. This will release the driven gear, which can easily be withdrawn.

Remove the pump by gently tapping the side of the body and withdrawing it downwards. If required, the driven gear shaft can be withdrawn from the cylinder block by screwing a suitable extractor into the end.

DISMANTLING, REASSEMBLING AND REPLACING THE OIL PUMP.

After withdrawing the pump from the cylinder block remove the circlip securing the driving gear to the oil pump shaft and vertical gear. By employing a suitable drift, the oil pump shaft and gear can be tapped partly through the driving gear. Extract the key and gear before completely removing the shaft, otherwise the key will damage the bush.

Clean all parts, examine and check for wear.

The gear depth is 1.378 in. − .0016 in. − .0024 in. (35 mm. − .04 mm. − .06 mm.). With a diameter of 1.2678 in. + .001 in. (32.1 mm. + .025 mm.).

The housing depth is 1.378 in. + .0012 in. (35 mm. + .03 mm.) with a bore of 1.2795 in. + .001 in. − .0006 (32.5 mm. + .025 mm. − .015 mm.).

This results in a gear end float of .0016 in. to .0035 in. (.04 mm. − .09 mm.) with the end cover fitted, and a radial clearance of .0056 in. to .0064 in. (.145 mm. to .162 mm.).

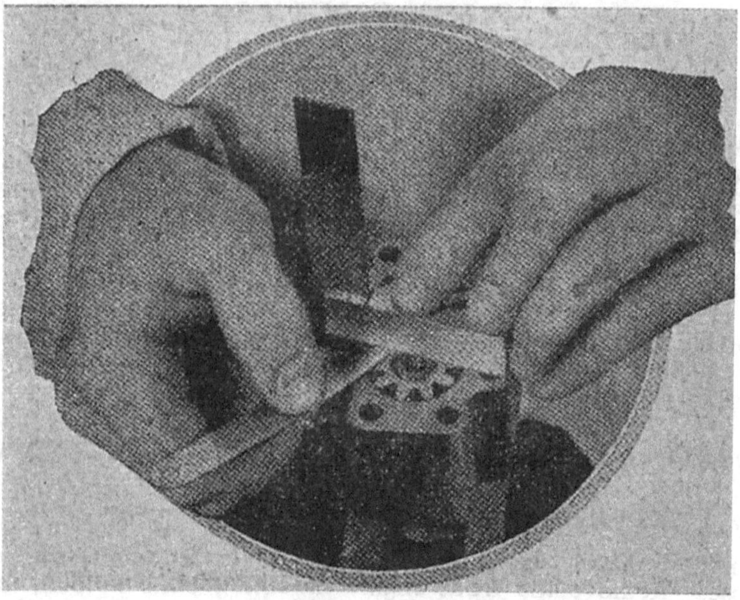

Fig. 15. Testing End clearance of oil pump gears.

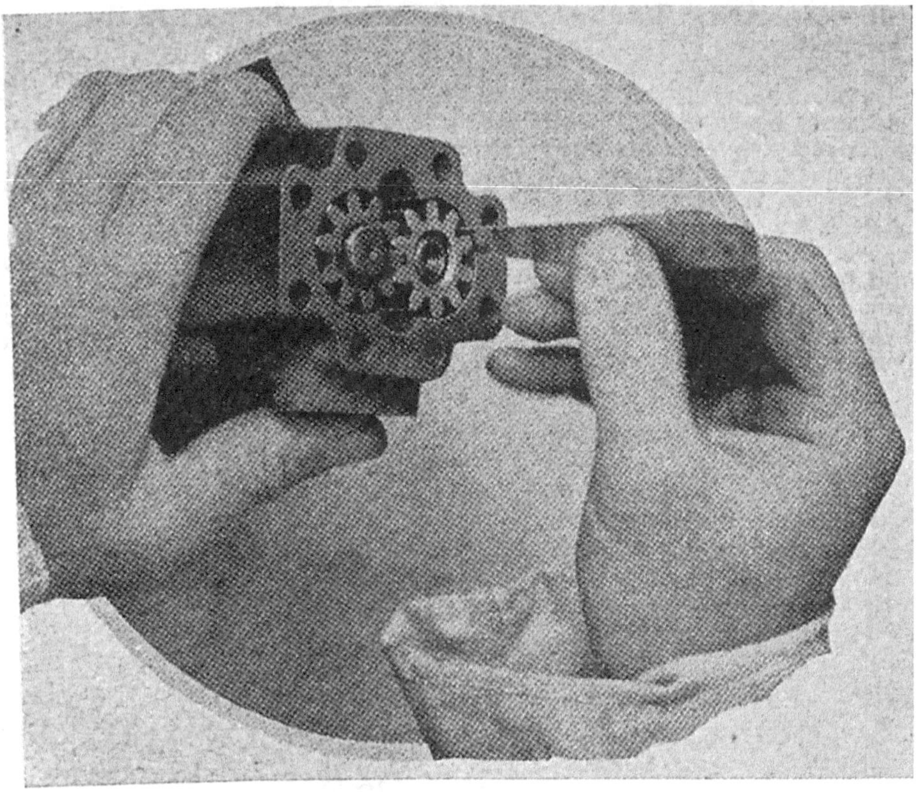

Fig. 16. Testing Clearance between oil pump gears and oil pump body.

The backlash between the teeth is approximately .020 in. to .025 in. (.51 mm. to .63 mm.). (See Figs. 15 and 16.)

Renew worn parts as necessary. The pump housing and driven gear are fitted with renewable bushes: the oil pump is reassembled and replaced on the engine in the reverse manner to that detailed for removal and dismantling.

Ensure that the abutting faces of the cylinder block and the pump body are clean before replacing, and that the paper gasket between these faces is in good condition. Renew the gasket if there is any doubt as to its condition. There is no gasket between the pump body and the pump cover.

REMOVAL AND REPLACEMENT OF THE OIL PRESSURE RELIEF VALVE. (See Fig. 5.).

The oil pump relief valve comprising the ball, spring and ball guide is incorporated in the oil pump cover. The spring should be 17 gauge or .056 in. (1.42 mm.), .500 in. (12.7 mm.) overall diameter, 1.476 in. (37.5 mm.) free length. Total number of coils 13½, giving a load of 7 lbs. (3.17 kilograms) when compressed to 1.063 in. (27 mm.).

No adjustment is provided for and should only be dismantled for cleaning and examination; access to the parts being obtained by unscrewing the retaining plug in the bottom side of the oil pump cover. Ensure when replacing the components of the relief valve that the seating, sleeve, ball guide, spring and fibre washer for the plug are correctly replaced, and the fibre washer is in good condition. Renew if there is any doubt as to its condition.

REMOVAL AND REPLACEMENT OF MAIN AND BIG END BEARINGS.

Big end bearings can be renewed without removing the engine from the frame, after removal of the sump. To renew the main bearings the engine must be removed. Both the main and big end bearings are loose, steel backed, white metal lined bearings.

It is imperative that no adjustment be made to the bearings. Bearings which are worn should be renewed, and should it be found that the crankshaft journals are in a worn condition, a service reground crankshaft complete with main and big end bearings as supplied from the Works should be fitted.

Small tags on each half bearing, locate the big end bearings in position and the bearings are fitted so that the tags come on the same side of the bearing housing as shown in Fig. 17.

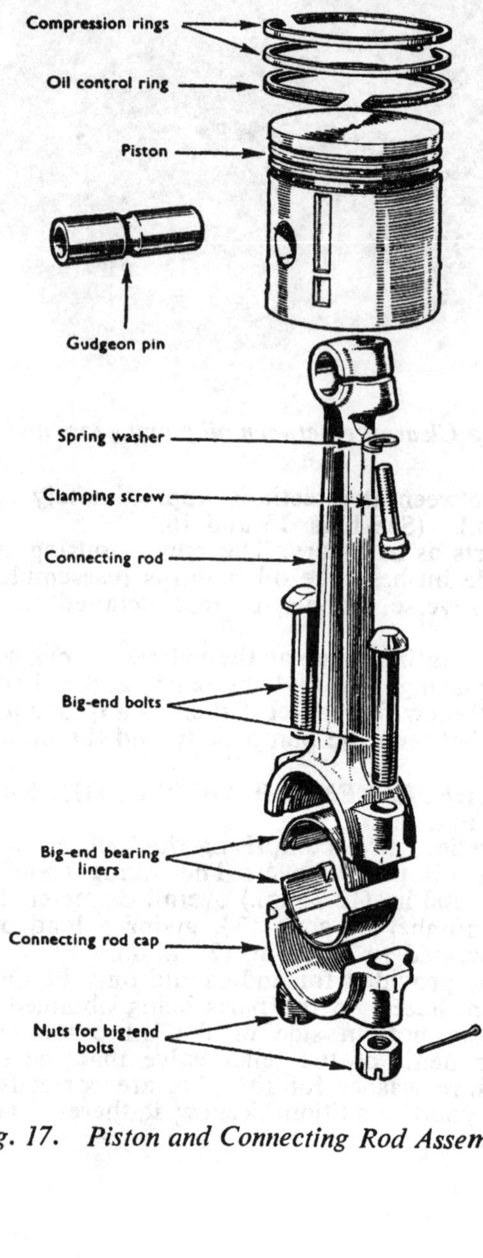

Fig. 17. Piston and Connecting Rod Assembly.

The main bearings are located in position by dowels in the bearing caps, and in the crankshaft housing.

To remove the big end bearings drain and remove the engine sump.

Extract the split pins from the big end bolts and undo the nuts. Remove the connecting rod caps and extract the bearings. Care should be exercised to see that the bearing journals and other parts are cleaned before installing new bearings. No scraping is required as the bearings are machined to give the correct diametrical clearance of .0005 in. (.011 mm.) to .002 in. (.056 mm.) and a side clearance of from .004 in. to .006 in. (.10 mm. to .15 mm.).

To renew the main bearings it is essential the engine be removed from the frame, the sump be removed, and connecting rods and pistons, etc., be removed.

Remove the split pins from the retaining studs, unscrew the slotted nuts and withdraw the main bearing caps. This will enable the crankshaft to be lifted from the engine.

Clean the oil ways drilled in the crankshaft and bearing journals, place the new bearings in position on the locating dowels, the shouldered bearing being fitted in the centre position and replace the crankshaft. Again no scraping is required as the bearings are machined to give the correct diametrical clearance of from .0008 to .003 in. (.02 mm. to .075 mm.) and a side clearance of from .0014 in. to .0037 in. (.035 mm. to .095 mm.) on the centre bearing. The end bearings have no end location.

In the case of a run bearing it is always advisable to clean out all oilways in the crankshaft and block. Then wash out the engine base with paraffin.

REMOVAL AND REPLACEMENT OF ROCKER SHAFT AND ROCKERS. Details as TD Midget.

REMOVAL AND REPLACEMENT OF CYLINDER HEAD.

Remove the bonnet top after extracting the two screws securing the rear hinge bracket.

Disconnect positive and negative leads to the battery.

Drain the water system by opening the drain tap in the rear, bottom, left hand corner of the radiator and the tap in the cylinder block, immediately below and in front of the exhaust manifold.

Detach the high tension leads to the sparking plugs, and remove the sparking plugs taking care not to break or damage the porcelain insulator.

Remove the water temperature gauge bulb from the left hand side of the thermostat housing and disconnect the oil feed pipe to the cylinder head.

Remove the hose from the cylinder head cover to the front air cleaner.

Remove the two bolts securing each air cleaner to the carburetters and withdraw air cleaners.

Disconnect the throttle and mixture controls.

Remove the carburetter overflow pipes from the carburetter float chambers.

Up to and including chassis No. TF.1509 disconnect the fuel pipe from the fuel pump.

Commencing at chassis No. TF.1510 disconnect the flexible petrol pipe for the carburetters at the connection for the main petrol pipe from the pump.

Remove the bolts securing the carburetters to the induction manifold and withdraw carburetters.

Uncouple the exhaust pipe from the exhaust manifold and remove the bolt securing the exhaust pipe support clip to the gearbox casing.

Remove the four nuts securing the induction and exhaust manifolds to the cylinder head and withdraw the clamps and manifolds.

Loosen the clips securing the concertina hose between the radiator header tank and metal water pipe on the thermostat housing and remove the hose.

Loosen the top clips securing the by-pass pipe hose to the right hand side of the thermostat housing.

Slacken the engine fume pipe and remove the tappet inspection cover.

Remove the valve cover and rocker gear, details as for **TD** Midgets. Lightly tap each push rod with a screwdriver to ensure they are free from the tappets and gently withdraw the push rods.

NOTE:—If this operation is neglected there is the possibility that the cam followers will be lifted with the push rod and drop into the engine.

Release the ten cylinder head nuts a partial turn at a time until they are free for complete removal by hand.

Replace the two cylinder head cover nuts on the cover studs and lift the head from the cylinder block.

NOTE:—To facilitate breaking the cylinder head joint, tap each side of the head with a hammer using a piece of wood interposed to take the blow. When lifting the head a direct pull should be given so that the head is pulled evenly up the studs.

REFITTING THE CYLINDER HEAD.

Ensure that the surfaces of both cylinder head and cylinder block are clean. Coat the gasket with grease and slip over the studs.

Lower head into position and fit the cylinder head stud nuts finger tight.

It is essential that they should be tightened down gradually half a turn at a time, in order that a good joint be achieved and to avoid distortion of the cylinder head.

NOTE:—Ensure that the gasket is fitted with the elongated hole for the waterways to the rear of the cylinder head.

Replace push rods and refit the rocker gear assembly. Check to ensure tappet clearances are adequate.

Place valve cover gasket in position and replace cylinder head cover, taking care to see that the cover fits squarely.

Refit the tappet inspection plate and fume pipe, renewing the gasket if necessary. Care should be taken to see that the oil drain holes in gasket are at the bottom.

Clean out exhaust manifold, refit induction and exhaust manifolds, renewing the gaskets if necessary.

Reconnect exhaust pipe to manifold and replace support clip on side of gearbox casing.

Replace carburetters, fit mixture and throttle controls and replace carburetter overflow pipes.

Reconnect petrol pipe.

Replace radiator top hose and tighten both hose clips and clip at top of by-pass hose and replace oil feed pipe to cylinder head.

Replace water temperature gauge bulb in thermostat housing.

Check, clean and replace sparking plugs and replace high tension leads to the plugs.

Switch on the ignition and check fuel connections; and then fill cooling system.

Start the engine and allow it to run briskly until the water temperature rises to between 70°C. and 80°C., or to 160°F. to 175°F.

Remove valve cover, check and adjust tappet clearances, and finally tighten cylinder head stud nuts.

Refit valve cover, replace air cleaners in position and fit hose between cover and front air cleaner.

REMOVAL AND REPLACEMENT OF CONNECTING RODS AND PISTONS (ENGINE IN SITU).—Details as TD Midget.

REMOVING PISTON FROM CONNECTING ROD.—Details as TD Midget.

FITTING THE GUDGEON PINS.

Gudgeon pins must be a thumb push fit for three quarters of their travel, to be finally tapped home with a rawhide mallet. For this operation the piston and gudgeon pin must be cold.

Gudgeon pin bores in the piston should not be reamed out because oversize pins are not available or permissible. Gudgeon pins must be fitted by selection.

REMOVAL AND REPLACEMENT OF PISTON RINGS.—Details as all Models.

REMOVAL AND REPLACEMENT OF CARBURETTERS.

Slacken clips securing hose from front air cleaner to cylinder head cover and remove hose.

Remove two bolts securing each air cleaner to each carburetter and remove air cleaners.

Disconnect carburetter over-flow pipes at float chamber end.

Detach petrol pipe from either fuel pump or main copper pipe from fuel pump.

Disconnect the mixture control wire from its attachment to the carburetter levers.

Detach the throttle control link rod at its attachment to the end of the accelerator lever.

Remove the four bolts holding the carburetters to the induction manifold and lift carburetter assembly clear of engine.

Replacement is carried out in the reverse manner to that detailed for removal.

REMOVAL AND REPLACEMENT OF EXHAUST AND INLET MANIFOLD.

Remove carburetters as detailed above.

Release the exhaust pipe from the manifold flange.

Undo the four nuts securing the manifolds to the cylinder head and remove the complete assembly.

REMOVAL OF TIMING CHAIN.—Details as TD Midget.

REPLACEMENT OF TIMING CHAIN.—Details as TD Midget.

THE TIMING CHAIN TENSIONER.

This device consists of an hydraulically damped, spring loaded plunger and combined slipper block encased in a housing which is bolted to the cylinder block. The slipper is held against the chain by tension of the spring and oil pressure.

The spring, which is 22 gauge (.17 mm.) wire, has a free length of 2.795 in. (71mm.), an outside diameter of .295 in. (7.5 mm.), and has 35 effective coils. It gives a load of 1.25 lb. (.57 KG) when compressed to 48 mm. (1.89 in.).

The plunger is fed with oil from the crankshaft front main bearing via an oil way drilled through the cylinder block, mating with an oil way in the tensioner housing or feed block. This oil way is then reduced in diameter to .04 in. (1 mm.), and the oil feeds through to the stem of the plunger, which is .43 in. (11 mm.) in diameter, and then through the bleed hole in the plunger, which is .10 in. (2.5 mm.) diameter. This causes an increased pressure and produces a cushioning effect between the chain and the slipper. Oil passing through the bleed hole passes on to the chain and slipper ensuring effective lubrication between the two.

REMOVAL OF CHAIN TENSIONER.

Remove radiator and timing chain case, as details given.

Break the lockwire at the two set screws securing the chain tensioner feed block to the cylinder block and unscrew them, taking care to hold the assembly against the chain to overcome the tension of the spring.

Before replacing, examine the bore of the feed block for wear (this should be .43 in. $\pm$ .004 in. (11 mm. $\pm$ .01 mm.) and make sure that the oil way is clear. Look out for any wear on the chain tensioner, the outside diameter of the stem should be .43 in. $-$ .0004 in. to $-$ .0012 in. (11 mm. $-$ .01 mm. $-$ .03 mm). Finally check the tensioner spring to the details already given. When refitting ensure rewiring the set screws securing the feed block.

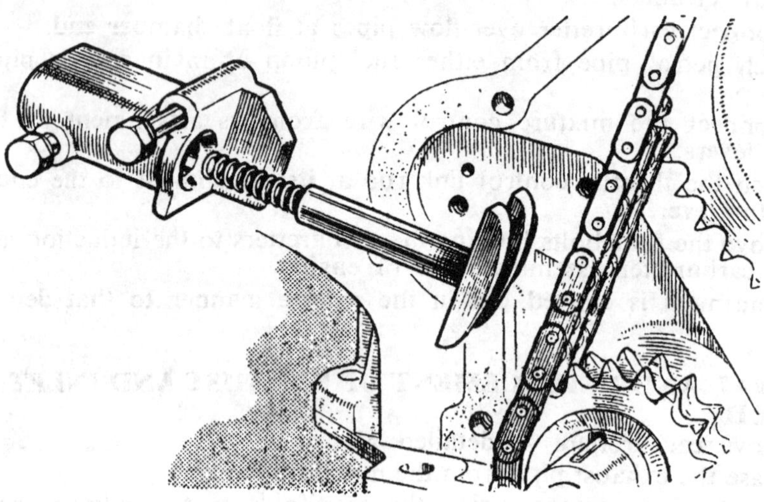

Fig. 19. *The Timing Chain Tensioner.*

REMOVAL AND REPLACEMENT OF TAPPETS.

Remove hose to front air cleaner.

Remove cylinder head cover. Unbolt the fume pipe from the tappet inspection cover and remove the cover.

Remove the rocker gear assembly and withdraw push rods after lightly tapping with a screw driver to disengage them from the tappets.

Lift the tappets from the cylinder block.

Replacement is carried out in the reverse manner to that detailed for removal, and both the tappets and push rods, should be fitted in the same position they occupied before removal to avoid the need for extensive adjustment.

REMOVAL AND REPLACEMENT OF CAMSHAFT.—Details as TD Midget.

TAPPET ADJUSTMENT.—Details as for TD Midget.
Bear in mind that clearance must be set at .012 in.

REMOVAL OF VALVES.—Details as (All Models).

REPLACING THE VALVES.—Details as (All Models).

REMOVING CARBON FROM HEAD.—Details as (All Models).

CLEANING THE PISTON HEAD.—Details as (All Models).

GRINDING IN THE VALVES.—Details as (All Models).

REMOVAL AND REPLACEMENT OF VALVE GUIDES.

Remove the cylinder head and valves as already detailed.

Rest the head with machined face downwards on a clean, flat surface and drive the valve guides downwards into the combustion chamber using a suitably sized shouldered drift.

This should take the form of a hardened steel punch ½ in. (12.7 mm.) in diameter and not less than 6 in. (15 CM) in length, with a locating spigot $\frac{5}{16}$ in. (8 mm.) in diameter, machined on one end for a length of 1 in. (2.57 cm.) to engage the bore of the guide.

When fitting the new valve guides press them in until .945 in. (24 mm.) is protruding above the machined surface of the cylinder head.

NOTE:—Although the inlet valve guides are $\frac{7}{32}$ in. (5.56 mm.) longer than the exhaust valve guides all valve guides project the same distance above the valve seating, namely .945 in. (24 mm.).

Recut the valve seat from the new guide, if necessary, to ensure that the valves seat correctly.

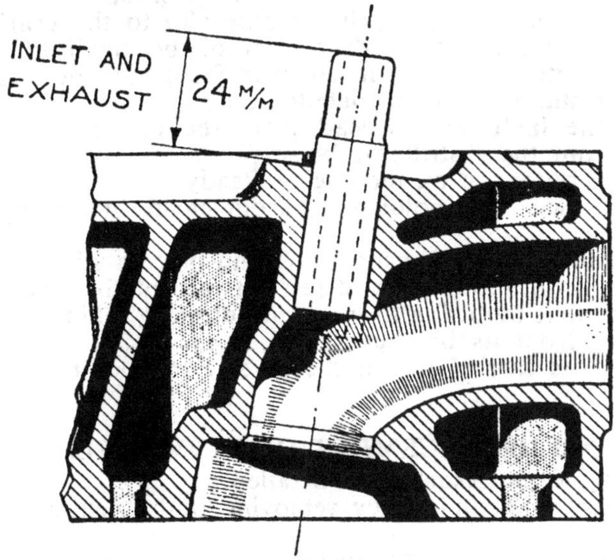

Fig. 20. Showing correct position of Valve Guide in the cylinder head.

REMOVAL AND REPLACEMENT OF ENGINE C/W GEARBOX.

Drain the cooling system as detailed previously.

Remove the bonnet tops and radiator as previously detailed.

Disconnect battery leads.

Detach flexible petrol pipe from petrol pump or main feed pipe from petrol pump.

Remove air cleaners.

Disconnect the mixture control inner and outer cables at the rear carburetter attachments.

Disconnect the throttle ball joint at the forward end of the accelerator pedal arm.

Detach the accelerator control spring and release the starter cable from the starter switch on the bulkhead.

Remove the carburetters complete.

Remove the brass nuts holding the exhaust pipe to the manifold.

Remove the four bolts, nuts and flat washers, securing the bonnet side tie rod in position.

These bolts are also utilised for securing the wings to the bonnet sides.

Remove the steering assembly detailed on page 572.

Undo the oil pressure gauge pipe at the cylinder block and detach the revolution counter cable from the reduction box on the rear of the dynamo.

Remove the water temperature gauge element from the left hand side of the thermostat housing.

Disconnect the dynamo leads from the terminals on the rear of the dynamo.

Remove the seats from their runners.

Lift the carpets and detach the toeboard on the passengers' side.

Remove both floorboards, the gear level knob and the gearbox cowl.

Take off the gearbox remote control assembly, and cover the aperture with a piece of cardboard to prevent dirt entering the inside.

Unscrew the screws round the pedal draught excluder and disconnect the speedometer drive at the gearbox.

Detach the forward end of the propellor shaft marking the flanges so that they can be reassembled in the same position.

Undo the clip holding the exhaust pipe clip to the gearbox. This will allow the front exhaust pipe to drop clear of the manifold.

Detach the clutch operating mechanism from the side of the sump.

Remove the starter motor as detailed on page 340.

Uncouple the high tension lead from the coil and remove the low tension wire from the distributor body.

Take off the outer nut on the engine steady.

Remove the two bolts holding the front engine mounting to its rubber block.

Disconnect the rear engine mounting and place a sling around the unit, just behind the front mounting and just forward of the flywheel housing.

Remove the unit by lifting forward and upward, taking care to disengage the steady link from its bracket.

Replacement is carried out in the reverse manner to that detailed for removal.

REMOVAL AND REPLACEMENT OF FLYWHEEL.

Remove engine and gearbox as detailed.

Remove gearbox complete, by removing bolts securing clutch housing to the crankcase.

Remove clutch assembly complete.

Remove the locking wire on the flywheel fixing bolts and withdraw fixing bolts.

Use a suitable extractor and pull off flywheel, taking care not to damage locating dowels.

Replacement is carried out in the reverse manner to that detailed for removal.

Check the flywheel for accuracy. It should be no more than .002 in. out of truth at any point when rotated with a dial gauge in contact with the clutch face.

REMOVAL AND REPLACEMENT OF CRANKSHAFT (ENGINE OUT OF CHASSIS).—Details as TD Midget.

TOP DEAD CENTRE MARK.—Details as TD Midget.

REFITTING ENGINE CONTROL LINK.—Details as TD Midget.

LOCATING ENGINE TROUBLES.—Details as TD Midget.

THE COOLING SYSTEM

The pressurised cooling system is of the thermo-syphon impeller assisted with a thermostatic control. The water is circulated from the base of the radiator by the pump, and after passing round the cylinders and cylinder head it reaches the header tank via the thermostat top water connecting pipe and top water hose.

From the header tank it passes downwards through the cooling ducts to the radiator base tank.

Air is drawn through the radiator by a fan which is mounted on the water pump spindle, and both are driven by a belt from the crankshaft which also drives the dynamo.

DRAINING AND FILLING THE SYSTEM.

Lift the L.H. side top bonnet panel.

Drain the water from the cooling system by means of the drain tap on the left hand side of the radiator bottom tank and the tap on the right hand side of the engine at the front of the cylinder block. The system cannot be fully drained from the radiator tap alone owing to the location of the water impeller in the circulating system.

Fig. 21. Showing position of radiator drain tap.

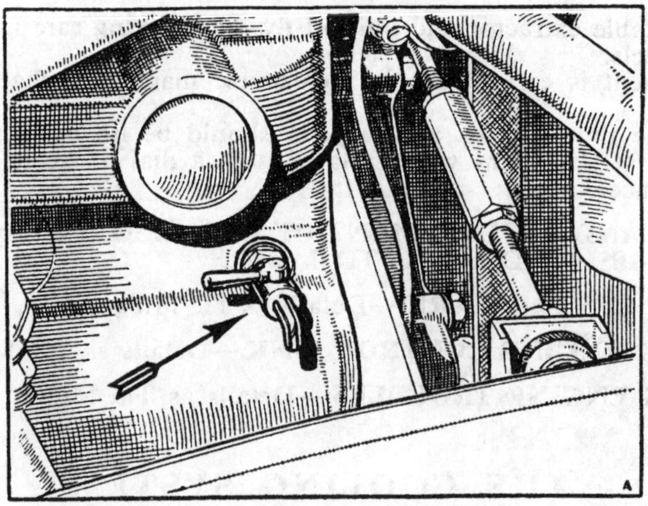

Fig. 22. Showing position of water drain tap on cylinder block.

If " Bluecol " or any other anti-freeze mixture is used the water should be drained into a suitable clean container and kept for future use. To fill the system, close both drain taps and fill through the filler device in the radiator header tank until the water is approximately ½ in. below the bottom of the filler neck.

NOTE:—The filler cap on the radiator case cannot be removed. When possible rain water should be used for filling.

If the filler cap is to be removed while the engine is hot, unscrew the filler cap slowly.

The filler cap is retained by a bayonet catch with a graduated cam which permits release of internal pressure prior to removal. A lobe on the end of the cam guards against accidental release of the cap before the internal pressure is relieved. **Protect your hand against escaping steam.**

Fig. 23. Showing position of cooling system filler cap.

COLD WEATHER PRECAUTION.

If the car is not stored in a warm building, steps must be taken to prevent the cooling water from freezing during frosty weather. Water upon freezing expands with the result that there is considerable risk of bursting either the radiator or the cylinder block by the pressure generated. As a precautionary measure when frost is anticipated the water should be drawn from the radiator and engine, as details given, or better still, an anti-freezing solution may be used in the radiator.

We recommend the use of Smith's " Bluecol," Filtrate " Nevafreeze " or " Shell " " Snowflake " anti-freeze in order to protect the cooling system during frosty weather and reduce corrosion to a minimum.

The correct quantities of anti-freeze for varying degrees of frost resistance in the M.G. Midget Series TF are:—

7°F (−14°C.).	0°F. (−18°C.).
15% Solution:	20% Solution:
Quantity 2 Pints (1.14 Litres).	Quantity 2¾ Pints (1.56 Litres).

If temperatures below 0°F. or −18°C. are likely to be encountered a solution of at least 25 per cent. anti-freeze must be used.

First decide what degree of frost protection is required before adding the anti-freeze to the radiator.

Before introducing anti-freeze mixture to the radiator it is advisable to clean out the cooling system thoroughly by swilling out the passages with a hose inserted in the filler cap aperture, keeping the two drain taps open. Only top up when the cooling system is at normal running temperature in order to avoid losing anti-freeze due to expansion.

Make sure that the cooling system is water tight and examine all joints, renewing any defective rubber hose.

REMOVAL AND REPLACEMENT OF RADIATOR.

Drain the radiator by opening the tap on the left hand side radiator bottom tank.

Remove the two screws securing the rear bonnet hinge bracket, withdraw the two bonnet tops rearwards from the front hinge and lift bonnet clear of car.

Detach the forward end of the radiator stays.

Remove the two set bolts securing the front bumper, and withdraw bumper complete with number plate and distance pieces.

Remove the two bolts, nuts and washers, one in each corner, securing the front valance to the lower corners of the front wings.

Remove the bolts, nuts and washers securing the front valance to the radiator and withdraw front valance.

Remove the two self tapping screws one in each front bottom corner of the bonnet sides, securing the bonnet sides to the radiator casing.

Remove, one each side, the bolts, nuts and washers securing the bonnet sides and front wings to the radiator casing.

Remove, one each side, the bolt and locking nut and flat washer securing the top front corners of the bonnet sides to the brackets on the radiator case.

Disconnect the by-pass hose at the thermostat housing at the front of the cylinder head.

Disconnect the hose on the elbow at the pump and the main hose at the metal water connecting pipe bolted to the top of the thermostat housing. Remove the fixing nuts and locknuts from the mounting brackets. Note the position of the rubber buffers, retaining ring and washers.

The shell and radiator block can then be lifted clear of the car. Replacement is carried out in the reverse manner to that detailed for removal, bearing in mind the drain tap on the radiator must be closed prior to filling with soft water.

FAN BELT ADJUSTMENT.

The adjustment of the dynamo, fan and water pump belt tension is affected by slackening the two bolts on which the dynamo pivots, releasing the set screw (at the top front end on the dynamo) securing the slotted link and pulling the dynamo bodily outwards until the belt tension is correct. Tighten up the set screws with the dynamo in this position.

Fig. 24. Pivot bolts and adjusting bolt for dynamo mounting and fan belt tension adjustment are clearly shown by arrows.

NOTE:—Only gentle hand pressure must be exerted on the dynamo or the belt tension will be excessive and cause rapid wear.

REMOVAL AND REPLACEMENT OF WATER PUMP.

This operation is greatly facilitated if the radiator complete is drawn from the car details as given on page 549.

Release the dynamo bolts and set screw, push dynamo towards engine and withdraw the driving belt.

Remove the bolts and spring washers securing the water pump body to the cylinder block and remove the nuts from the water pump body studs. Note that the bolts withdrawn are of different lengths.

Withdraw the pump from the studs and remove water pump unit complete from front of engine.

Replacement is carried out in the reverse manner to that detailed for removal, but ensure that the flange washer is in good order and that the pump bolts are replaced in their correct position.

DISMANTLING AND REASSEMBLING THE WATER PUMP.

The pump is fitted with a special integral seal which provides an efficient water seal and needs no lubrication or adjustment. Should it be necessary to dismantle the unit, proceed as follows:—

Remove radiator and water pump assembly as previously detailed.

Withdraw the three bolts securing the fan blades to the fan centre and remove fan blades. Tap out the taper pin attaching the impeller vane to the pump spindle taking care to see that it is tapped out in the right direction. Withdraw the impeller vane from the spindle which will give access to the gland seating washer and integral seal.

Remove the gland seating washer and seal, taking care not to damage the seal during removal.

The pump spindle is carried on two ball races which should give no trouble unless they have been neglected.

Remove the attachment nut and flat washer securing the fan pulley to the pump spindle.

Withdraw the fan pulley from the spindle and withdraw the pulley key.

Remove the felt sealing ring and felt retaining cover.

Remove the outer bearing circlip by engaging the lugs with a pair of long-nosed pliers.

Pour a little paraffin into the impeller body around the outer bearing and tap the inner end of the spindle on a piece of wood until the outer bearing can be withdrawn.

This will release the distance piece between the bearings, which can be withdrawn giving access to the inner race.

Remove the inner circlip by contracting the ring and inserting a screw-driver behind it to ease it out of its groove; after removal of the circlip the retaining bearing and the impeller spindle can be withdrawn.

If the front or rear felt oil sealing rings are badly worn or the bearings unduly slack, they should be renewed.

Carefully examine the seal for wear or damage and renew if necessary.

The face of the gland sealing washer should be checked for flatness and all edges freed from burrs.

Reassembly is carried out in the reverse manner to that detailed for dismantling noting that the sealing for the integral seal is against the boss of the impeller vane.

The space between the two races should be partially filled with grease and the felt washers liberally soaked in engine oil or grease before replacement.

The slotted nut retaining the pulley should not be over-tightened; as long as it is just firm it will be satisfactory.

1. Spindle.
2. Fan Pulley.
3. Felt washer (front).
4. Retaining cover (front).
5. Bearing circlip.
6. Bearings.
7. Distance tube.
8. Retaining cover.
9. Felt Washer (rear).
10. Retaining washer (rear).
11. Pump spindle circlip.
12. Pump body.
13. Seal.
14. Impeller Vane.
15. Taper Pin.
16. Gland seating washer.

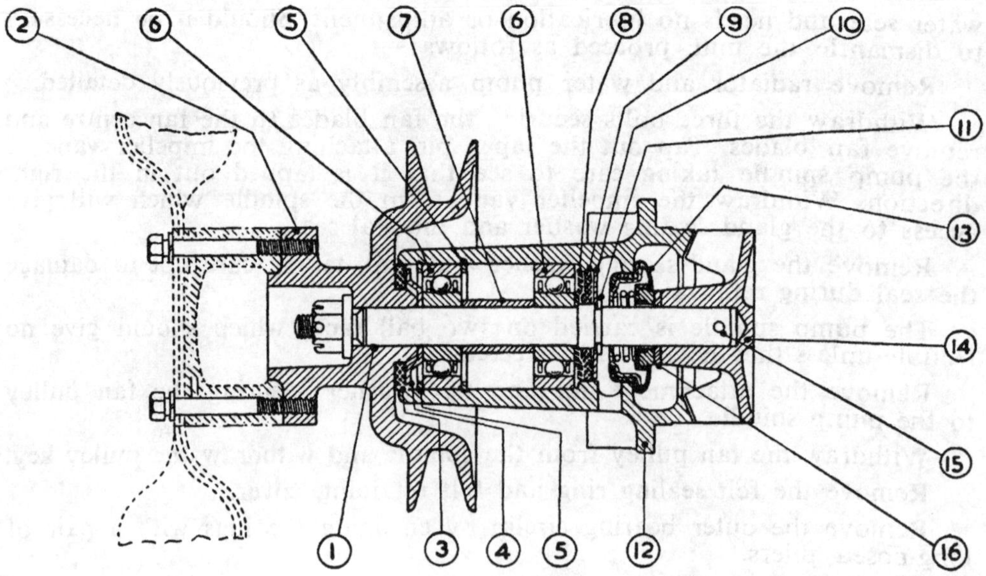

Fig. 25. The assembly of the water pump.

CLUTCH

The clutch fitted is a single plate Borg & Beck 8 in. diameter model, fitted with 6 light brown springs, 150/160 lb., the driven plate having wound yarn facings.

GENERAL DESCRIPTION OF THE CLUTCH. Details as TD Midget.

CLUTCH RUNNING ADJUSTMENT.

The clutch operation is by means of a clutch operating rod instead of cable, but details of adjustment as for " TD " Midget apply.

REMOVAL OF CLUTCH. Details as TD Midget.

DISMANTLING THE CLUTCH. Details as TD Midget.

ASSEMBLY OF CLUTCH. Details as TD Midget.

ADJUSTING THE RELEASE LEVERS. Details as TD Midget.

REFACING THE DRIVEN PLATE. Details as TD Midget.

REFITTING THE CLUTCH. Details as TD Midget.

GEARBOX

The gearbox has four forward and one reverse gear, synchromesh being incorporated on second, third and top gear.

Top gear is direct, third and second are in constant mesh, while first and reverse are obtained through sliding gears.

REMOVAL AND REPLACEMENT OF GEARBOX. Details as TD Midget.

DISMANTLING THE GEARBOX. Details as TD Midget.

REMOVING THIRD SPEED CONSTANT MESH GEAR FROM MAINSHAFT. Details as TD Midget.

REMOVING SECOND SPEED CONSTANT MESH GEAR FROM MAINSHAFT. Details as TD Midget.

DISMANTLING SYNCHROMESH MECHANISM. Details as TD Midget.

REASSEMBLY OF SYNCHROMESH MECHANISM. Details as TD Midget.

REASSEMBLY OF THE GEARBOX. Details as TD Midget.

REAR AXLE

GENERAL DESCRIPTION. Details as TD Midget.

LUBRICATION. Details as TD Midget.

REMOVAL AND REPLACEMENT OF BRAKE DRUM AND HUB. Details as TD Midget.

REMOVAL AND REPLACEMENT OF BRAKE PLATE ASSEMBLY. Details as TD Midget.

REMOVAL AND REPLACEMENT OF AXLE SHAFT. Details as TD Midget.

REMOVAL OF REAR AXLE. Details as TD Midget.

IMPORTANT POINTS CONCERNING AXLE ATTENTION. Details as TD Midget.

DISMANTLING THE AXLE AND REMOVING DIFFERENTIAL ASSEMBLY. Details as TD Midget.

DISMANTLING THE DIFFERENTIAL AND CROWN WHEEL ASSEMBLY. Details as TD Midget.

ASSEMBLING THE DIFFERENTIAL AND CROWN WHEEL. Details as TD Midget.

REMOVAL OF THE PINION. Details as TD Midget.

REPLACEMENT OF THE PINION. Details as TD Midget.

REASSEMBLING THE AXLE. Details as TD Midget.

REFITTING THE REAR AXLE. Details as TD Midget.

FUEL SYSTEM

CARBURETTERS.

In the main the construction is identical to the $1\frac{1}{2}$ in. competition carburetters employed on the Mark 11 TD and the standard $1\frac{1}{4}$ in. carburetters employed on the standard TD Midget. It must be borne in mind, however, that the hydraulic damper fitted to the hollow piston rod was not introduced until Chassis No. TF.3495 and a blue spring is also installed.

MAINTENANCE AND ADJUSTMENTS. Details as given in Fuel System, Chapter V1.

CARBURETTER TROUBLES. Details as given in Fuel System, Chapter V1.

SYNCHRONISATION OF THE CARBURETTERS. Details as given in Fuel System, Chapter V1.

FUEL PUMP.

Up to and including Chassis number TF.1509 the Petrol Pump fitted was an S.U. 12volt, type " L " Electric Pump, being mounted on dash wall or tool box.

The Petrol Pipe outside diameter is $\frac{5}{16}$ in.

The maximum output for this type of pump is 8 gallons (36 litres) per hour with a suction lift of approximately 48 in. (122 cm.) and an output lift of approximately 24 in. (61 cm.).

Commencing at Chassis number TF.1510 this Pump was replaced by the S.U. 12 volt " H.P." High Pressure Pump, which is mounted on the inside of the R.H. Chassis Frame Member, just on the upward sweep of the Chassis Section.

The petrol pipe outside diameter is $\frac{5}{16}$ in.

The maximum output for this Pump is 10 gallons (45 litres) per hour, with a suction lift of 33 in. (81 cm.) approximately and an output lift of approximately 48 in. (122 cm.).

REMOVAL AND REPLACEMENT OF PETROL PUMP. (Up to and including Chassis No. TF.1509).

Details as given in Fuel Section, Chapter V1.

REMOVAL AND REPLACEMENT OF PETROL PUMP. (Commencing at Chassis No. TF.1510.)

Jack up the rear end of the car, or place the car over a pit or on a hoist.

Disconnect the battery, drain the Petrol Tank.

Remove the Petrol Pipes from the inlet and outlet union on the Pump.

Remove the cable to the pump terminal and remove the earth wire.

Remove the two bolts securing the pump mounting plate to the bracket on the Chassis and withdraw the pump complete with mounting plate, taking care not to lose the rubber mounting between the bracket and mounting plate.

Replacement is carried out in the reverse manner to that detailed for removal.

CONSTRUCTION OF FUEL PUMP.

The construction of the " HP " Pressure Pump is similar to the " L " Type pump, details of which are given in the Fuel System, Section V1.

On the "HP" Pressure Pump the coil housing unit and diaphragm unit are slightly different and the length of the cast iron housing is increased about $\frac{5}{8}$ in.

CONTACT BREAKER MECHANISM. (Type "L" and "HP" High Pressure).
Details as given in Fuel System, Section V1.

FUEL PUMP ACTION. (Type "L" and "HP" High Pressure.)
Details as given in Fuel System, Section V1.

TRACING TROUBLES. (Type "L" and "HP" High Pressure.)
Details as given in Fuel System, Section V1.

A NOISY PUMP. (Type "L" and "HP" High Pressure.)
Details as given in Fuel System, Section V1.

FAILURE TO DELIVER FUEL. (Type "L" and "HP" High Pressure.)
Details as given in Fuel System, Section V1.

FUEL PUMP ADJUSTMENTS. (Type "L" and "HP" High Pressure.)
Details as given in Fuel System, Section V1.

ADJUSTING THE MAGNET ASSEMBLY. (Type "L" and "HP" High Pressure.)
Details as given in Fuel System, Section V1.

MAINTENANCE. (Type "L" and "HP" High Pressure.)
Details as given in Fuel System, Section V1.

AIR CLEANERS.

A.C. or Vokes circular type air cleaners are employed, the air cleaners being attached to the mouth of the Carburetters.

Every 3,000 miles (5,000 km.) wash the Filter element in petrol and allow to dry. Re-oil the element with S.A.E.20 engine oil and allow to drain before reassembling.

When servicing it is only necessary to withdraw the two hexagon headed screws and lift off the outer cover to release the corrugated element. Re-assemble the front element with the corrugations clear of the breather spigot in the main filter case.

Fig. 26. Sectional view of air cleaners.

IGNITION AND ELECTRICAL EQUIPMENT

GENERAL DESCRIPTION. Details as TD Midget.

DISTRIBUTOR TYPE.

The distributor is a Lucas Model D.2.A.4, No. 40367, with high lift cam. These identification marks are stamped on the side of the distributor. When ordering replacements, always quote these numbers.

IGNITION COIL TYPE. Details as TD Midget.

SPARKING PLUGS.

The standard sparking plugs for the M.G " TF " are Champion N.A.8, 14 mm., ¾ in. reach.

LOCATING THE CAUSE OF UNEVEN FIRING.

To Test with Sparking Plugs in Position. Details as TD Midget.

TESTING LOW-TENSION CIRCUIT.

Low-tension Circuit. Testing in Position.

(a) Spring back the securing clips on the distributor and remove the moulded cap and rotor. If the rotor is a tight fit, it can be levered off carefully with a screwdriver.

(b) Check that the contacts are clean and free from pits, burns, oil or grease. Turn the engine and check that the contacts are opening and closing correctly and that the clearance when the contacts are fully opened is between .014 in. and .016 in. (.36 mm. and .41 mm.). Correct the gap if necessary.

(c) Disconnect the cable at the contact breaker terminal (CB) of the coil and at the low-tension terminal of the distributor, and connect a test lamp between these terminals. If the lamp lights when the contacts close and goes out when the contacts open, the low-tension circuit is in order.

Low-tension Circuit. To locate fault.

(a) Having determined, by testing as previously described, that the fault lies in the low-tension circuit, switch on the ignition and turn the engine until the contact breaker points are fully opened.

(b) Refer to the wiring diagram and check the circuit with a Voltmeter (0-20 volts) as follows:—

NOTE:—If the circuit is in order the reading on the Voltmeter should be approximately 12 volts.

(c) **Battery to starter switch terminal.**

Connect a voltmeter between the starter switch terminal and a good earthing point. No reading indicates a damaged cable or loose connections.

(d) **Starter switch to ammeter (brown lead).**

Connect a voltmeter to another ammeter terminal, and earth. No reading indicates faulty cable or loose connection.

(e) **Ammeter.**

Connect a voltmeter to another ammeter terminal and earth. No reading indicates fault in ammeter which must be replaced.

(f) **Ammeter to control box terminal "A" (brown with white lead).**

Connect voltmeter to control terminal "A" and earth. No reading indicates faulty cable or loose connections.

(g) **Control box terminal " A1 ".**

Connect voltmeter to control box terminal " A1 " and earth. No reading indicates fault in series winding of control box.

(h) **Control box terminal " A1 " to lighting and ignition switch terminal "A" (brown with blue lead).**

Connect voltmeter to "A" terminal on switch and earth. No reading indicates faulty cable or loose connections.

(j) **Lighting and ignition switch terminal " 1G ".**

Connect voltmeter to " 1G " terminal on switch and earth. No reading indicates fault in switch which must be replaced.

(k) **" 1G " terminal to fuse unit terminal " A3 " (white lead).**

Connect voltmeter to " A3 " terminal on fuse unit and earth. No reading indicates faulty cable or loose connections.

(l) **Control box terminal " A3 " to ignition coil terminal " SW " (white lead).**

Connect a voltmeter to the ignition coil terminal " SW " and to earth. No reading indicates a damaged cable or loose connections.

(m) **Ignition coil.**

Disconnect the cable from the " CB " terminal of the ignition coil and " connect " a voltmeter to this terminal and to earth. No reading indicates a fault in the primary winding of the coil and a replacement coil must be fitted. If the correct reading is given, remake the connections to the coil terminal.

(n) **Ignition coil to distributor (white with black lead).**

Disconnect the cable from the low tension terminal on the distributor and connect the voltmeter to the end of this cable and to earth. No reading indicates a damaged cable or loose connections. If the correct reading is given, remake the connections to the distributor " LT." terminal.

HIGH TENSION CABLES.

Details as for TD Midget apply except for the specification of high tension cables. The 7 mm. rubber covered ignition cable is replaced by 7 mm. UM 837 Neoprene Ozone tested ignition cable.

ATTENTION TO SPARKING PLUGS: CHAMPION N.A.8. Details as TD Midget.

CONTACT BREAKER MECHANISM.

After the first 500 miles (800 km.) and subsequently every 3,000 miles (5,000 km.) check the contact breaker as follows:—

(a) Turn the engine until the contact breaker points are fully opened, and check the gap with a gauge having a thickness of from .014 in. to .016 in. If the gap is correct, the gauge should be a sliding fit. Do not alter the setting unless the gap varies considerably from the gauge thickness. To adjust the setting, keep the engine in the position which gives maximum opening of the contacts, and then slacken the two screws securing the fixed contact plate. Adjust the position of the plate until the gap is set to the thickness of the gauge and then tighten the two locking screws.

Remember that the cam only keeps the contact points open over a small angular movement of the distributor shaft so that special care must be taken when setting the gap to see that the fibre heel is on the highest point of one of the cam lobes.

(b) Details as TD Midget.

(c) Details as TD Midget.

DISTRIBUTOR LUBRICATION. Details as TD Midget.

REMOVAL OF THE DISTRIBUTOR.

Details as TD Midget, but it must be borne in mind that the distributor is fitted with a cotter bolt attachment. It will first be necessary to mark the base of the distributor body and the face of the housing in the cylinder block to ensure correct replacement, before unscrewing the cotter bolt lock-nut one or two turns and carefully tapping it inwards to release the distributor body stem.

Details of this distributor are given in the front section of the supplement, under modifications introduced on the Midget Series TD.

DISMANTLING THE DISTRIBUTOR.

Details as for TD Midget: Under paragraph (e) the sentence "Soften the solder securing the condenser in its clip with a hot iron and remove the condenser by applying pressure at one end" should be deleted.

THE CONDENSER. Details as TD Midget.

FITTING NEW DISTRIBUTOR BUSH.

In order to ensure easy running of the distributor shaft when the shank has been rebushed, the new bush must be fitted so that it is in correct alignment.

The fitting instructions as for TD Midget apply, bearing in mind that the two bearing bushes have, on the D.2.A.4 distributor, been replaced by a single long-reach porous bronze bush. This bush may be pressed into the shank using a shouldered, highly polished mandrel.

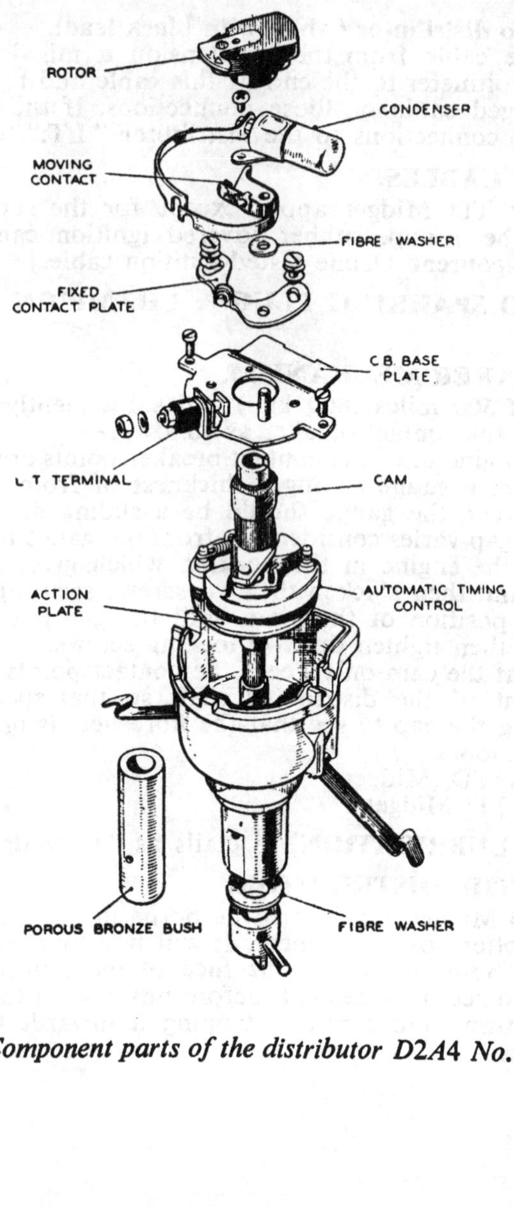

Fig. 27. Component parts of the distributor D2A4 No. 40367.

REASSEMBLING THE DISTRIBUTOR.

The details given for the TD Midget apply but under paragraph (a) the sentence " Two holes are provided in each toggle; the springs must be fitted to the inner hole in each case " should be deleted.

Under paragraph (e) the contact breaker points should be set to .014 in.-.016 in.

ELECTRICAL EQUIPMENT GENERAL DESCRIPTION. Details as TD Midget.

THE BATTERY. Model GTW.9.A. Capacity (10-hour rate) 51 ampere hours. Details as TD Midget.

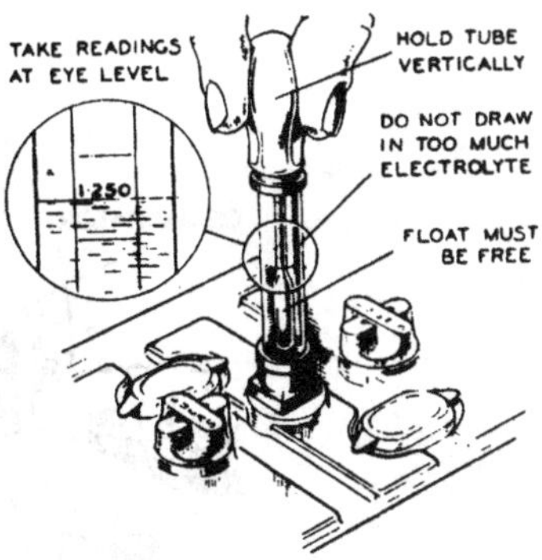

Fig. 28. *The hydrometer in use to determine the condition of the individual cells*

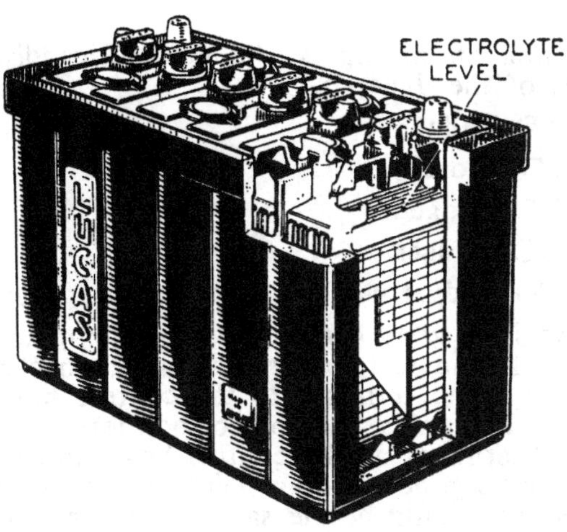

Fig. 29. *Sectioned view of battery showing plates and separators and correct electrolyte level.*

THE DYNAMO.

Type.—The dynamo is a Lucas model C.39.P.V.2. Service No. 22265.B. These identification marks are stamped on the yoke. When ordering replacements always quote these numbers.

To Test on Vehicle.

The cutting in speed is from 1,050 to 1,200 r.p.m. at 13 dynamo volts.

The output is 19 amps; at 1,850-2,100 r.p.m. at 13.5 dynamo volts, taken on a resistance load of 0.7 ohm without regulator.

(a) Details as TD Midget.

(b) Check that the dynamo and control box are connected correctly. The larger dynamo terminal should be connected to the control box terminal " D " and the smaller dynamo terminal connected to the control box terminal " F."

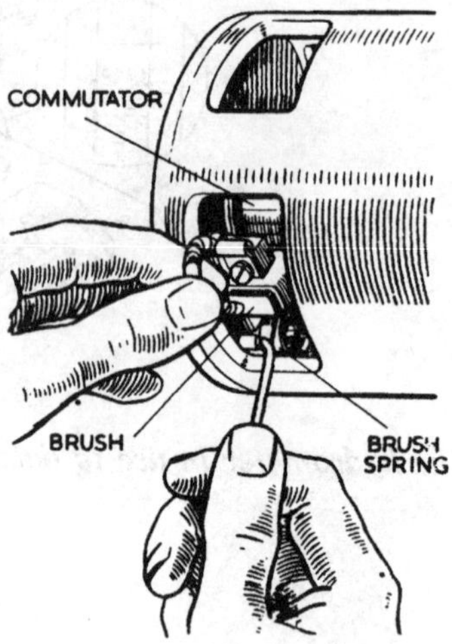

Fig. 30. To release brushes hold back the brush spring with a piece of bent wire as shown.

(c) After switching off all lights and accessories disconnect the cables from the terminals of the dynamo.

(d) Details as TD Midget.

(e) Details as TD Midget.

(f) Details as TD Midget.

(g) Details as TD Midget.

(h) Details as TD Midget with the final paragraph amended to read:

" If the dynamo is in good order, leave the temporary link in position between the terminals and restore the original connections, taking care to connect the larger dynamo terminal to the control box terminal ' D ' and the smaller dynamo terminal to the control box terminal " F." Remove the lead from the " D " terminal on the control box and connect the voltmeter between the cable and a good earthing point on the vehicle. Run the engine as before. The reading should be the same as that measured directly at the dynamo. No reading on the voltmeter indicates a break in the cable to the dynamo. Carry out the same procedure for the ' F ' terminal connecting the voltmeter between cable and earth. Finally remove link from dynamo."

If the reading is correct test the control box, details on page 564.

REMOVAL AND REPLACEMENT OF DYNAMO.

Disconnect the dynamo leads from the terminals on the dynamo.

Disconnect the revolution counter reduction box from the rear of the dynamo.

Remove the top bolts, nuts and washers, one each side, securing the bonnet side support rod brackets. These bolts go right through and secure the wing to the bonnet side.

Slacken the bottom bolts securing the bonnet side stiffening rod bracket, which will enable the stiffening rod to be swung clear of the fan belt.

Slacken the three dynamo attachment bolts and swing the dynamo towards the engine to give maximum belt slackness. Carefully free the belt from the dynamo pulley. Supporting the dynamo, completely remove its attachment bolts, enabling it to be lifted from the engine.

Replacement is carried out in the reverse manner to that detailed for removal.

DISMANTLING THE DYNAMO.

Remove the dynamo pulley.

Remove the cover band, hold back the brush springs and remove brushes from their holders.

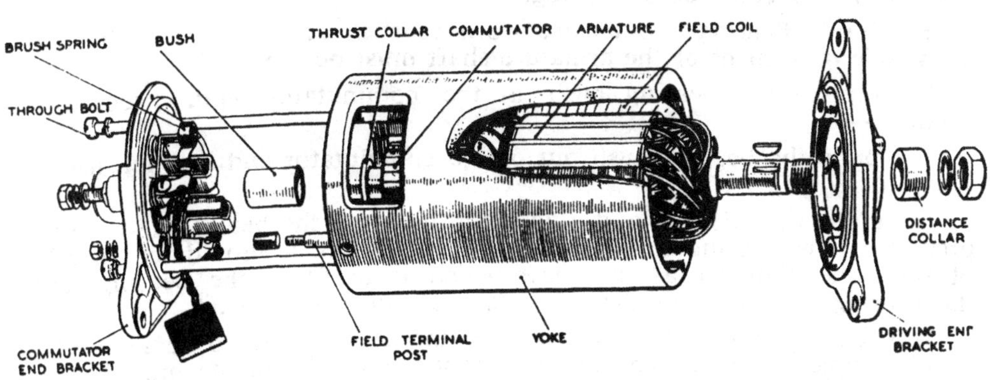

Fig. 31. *The dynamo components.*

Withdraw the two through bolts from the commutator end

Remove the bracket from the dynamo yoke.

The driving end bracket, together with the armature, can be now lifted out of the yoke.

A driving end bracket which on removal from the yoke, has withdrawn with it the armature and armature shaft ball bearing, need not be separated from the shaft unless the bearing is suspected, and requires examination; in which event the armature should be removed from the end bracket by means of a hand press.

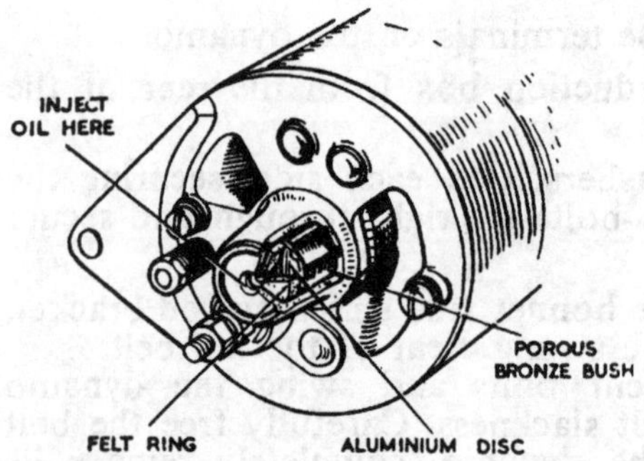

Fig. 32. Showing method of lubricating dynamo, to be carried out every 2,000 miles.

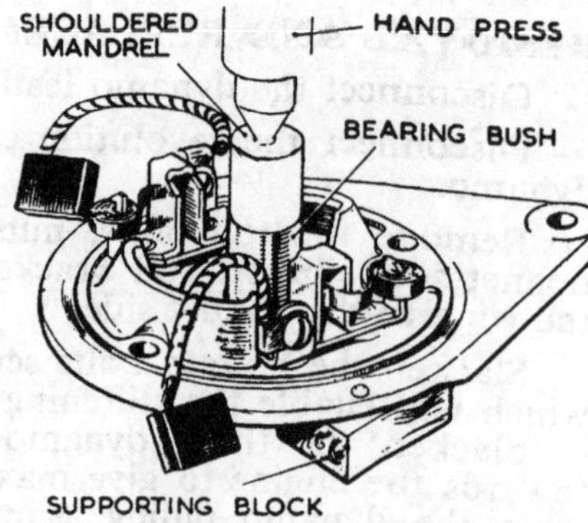

Fig. 33. Showing correct method of fitting the bronze bearing bush on the dynamo. Note the supporting blocks.

SERVICING THE DYNAMO.

Brushes. Details as TD Midget.

Commutator. Details as TD Midget.

Field Coils. Details as TD Midget.

Bearings. Bearings which are worn to such an extent that they will allow side movement of the armature shaft must be replaced.

To replace the bearing bush at the commutator end, proceed as follows:—

(a) Draw the bearing bush out of the commutator end bracket, using a $\frac{5}{8}$ in. tap

(b) Insert the felt ring and aluminium disc into the bearing housing and press the new bearing into the end bracket (using a shouldered mandrel of the same diameter as the shaft which is to fit in the bearing) until the bearing is flush with the inner face of the end bracket.

NOTE:—Before fitting the new bearing bush it should be allowed to stand completely immersed for twenty-four hours in engine oil. This will allow the pores of the bush to be filled with lubricant.

The ball bearing at the driving end is replaced as follows:—

(a) Knock out the rivets which secure the bearing retaining plate to the end bracket and remove the plate.

(b) Press the bearing out of the end bracket and remove the corrugated washer, felt washer and oil retaining washer.

(c) Before fitting the replacement bearing see that it is clean and pack it with a high melting point grease.

(d) Place the oil retaining washer, felt washer and corrugated washer in the bearing housing in the end bracket.

(e) Locate the bearing in the housing and press it home.

(f) Fit the bearing retaining plate. Insert the new rivets from the inside of the end bracket and open the rivet by means of a punch to secure the plate rigidly in position.

REASSEMBLY.
In the main the reassembly of the dynamo is carried out in the reverse manner to that as detailed for dismantling.

Before the refitting of the dynamo to the vehicle inject Oiline BBB, or any good quality medium viscosity (S.A.E.30) oil, into the hole marked OIL in the C.E. Bracket bearing housing. This lubrication should be carried out every 12,000 miles (See Fig. 32).

THE STARTER.
Type. The starter is a Lucas Model M.35G. Service No. 25022. These identification marks are stamped on the yoke. When ordering replacements always quote these numbers.

Its lock torque is approximately 9.3 lb.-ft. (1.3 m/KG), with 370-390 amps. and 7.7—7.3 volts.

It is interesting to note that the voltage drop increases with increase of current.

To Test on Vehicle. Details as TD Midget.

REMOVAL AND REPLACEMENT OF STARTER. Details as TD Midget.

STARTER SWITCH.
The starter switch is fitted on the engine side of the bulkhead and is operated by pulling the control knob in the instrument panel. The knob is connected to the switch by means of a cable. At the back of the starter switch are two terminals, from one of which three cables (one feed is also taken to fuse unit terminal A1 to supply the horns) are connected, the heavy one to the battery and the other to the ammeter. From the second terminal a heavy cable is taken to the terminal on the commutator end bracket of the starter.

THE STARTER COMMUTATOR AND BRUSH GEAR.
Inspect the commutator and brush gear in the manner described for the dynamo. The brush spring tension should be 15-20 oz. This operation should be carried out every 12,000 miles or 20,000 kilometers. In the unlikely event of brush replacement being necessary this must be carried out by a Lucas Service Agent, as the operation entails dismantling of the starter.

THE STARTER DRIVE. Details as TD Midget.

SERVICING THE STARTER.
Examination of Commutator and Brush Gear. Details as TD Midget.
Dismantling. Details as TD Midget.
Brushes. Details as TD Midget.

Bear in mind that the correct tension of the brush springs when tested with a spring scale should be 15-20 oz.

Drive.
(a) If the pinion is tight on the screwed sleeve, wash away any dirt with paraffin.

(b) If any parts are worn or damaged they must be replaced.

(c) Remove the split pin " A " (Fig. 34) from the shaft nut at the end of the starter drive. Hold the squared end of the starter shaft at the commutator end by means of a spanner and unscrew the shaft nut " B."

(d) Lift off the main spring " C " (Fig. 34) and remove the retaining ring " E."

(e) The corrugated washer " L," control nut " G," sleeve " H " and retaining spring " J " will now slide off.

(f) Withdraw the splined washers " K " from the armature shaft and remove the pinion and barrel assembly " F."

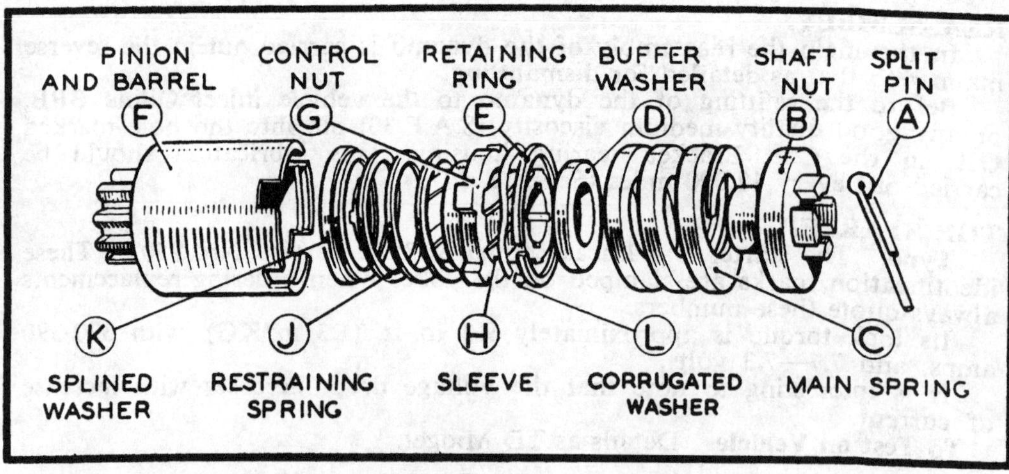

Fig. 34. Exploded view of starter drive.

Commutator. Details as TD Midget.
Field Coils. Details as TD Midget.
Armature. Details as TD Midget.
Bearings. Details as TD Midget.
Reassembly. Details as TD Midget.

NOTE:—Secure the body of the starter in a vice and test by connecting it with heavy gauge cables to a battery of the correct voltage. One cable must be connected to the starter terminal and the other held against the starter body or end bracket. Under these light load conditions the starter should run at 8,500-10,000 r.p.m. at 45 amperes (maximum).

THE CONTROL BOX.

Regulator Adjustment.

The regulator is carefully set before leaving the Works to suit the normal requirements of the standard equipment, and in general it should not be necessary to alter it. If however, the battery does not keep in a charged condition, or if the dynamo output does not fall when the battery is fully charged, it may be advisable to check the setting and if necessary re-adjust it.

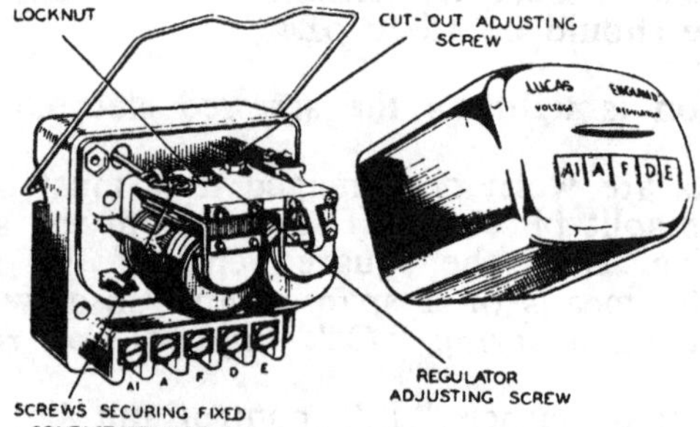

Fig. 35. The control box showing cut-out and regulator adjusting screws.

It is important, before altering the regulator setting, when the battery is in a low state of charge, to check that its condition is not due to a battery defect or to the dynamo belt slipping or to loose connections in the charging circuit.

HOW TO CHECK AND ADJUST ELECTRICAL SETTING.

The regulator setting can be checked without removing the cover on the control box. Withdraw the cables from the terminals marked " A " and " A1 " at the control box and join them together. Connect the negative lead of a moving coil voltmeter (0-20 volts full scale reading) to control box terminal " D " and connect the other lead from the meter to a convenient chassis earth.

Slowly increase the speed of the engine until the voltmeter needle " flicks " and then steadies; this should occur at a voltmeter reading between the limits given below for the appropriate temperature of the regulator. Open circuit setting of the regulator at 20°C. and 1,500 dynamo R.P.M. should be 15.6-16.2 volts. For ambient temperatures other than 20°C., the following allowances should be made:—

For every 10°C. (18°F.) above 20°C. subtract 0.3 volt.

For every 10°C. below 20°C add 0.3 volt.

If the voltage at which the reading becomes steady occurs outside the approximate limits, the regulator must be adjusted.

Shut off the engine, remove the control box cover, release the locknut " A " (Fig. 36) holding the adjusting screw " B."

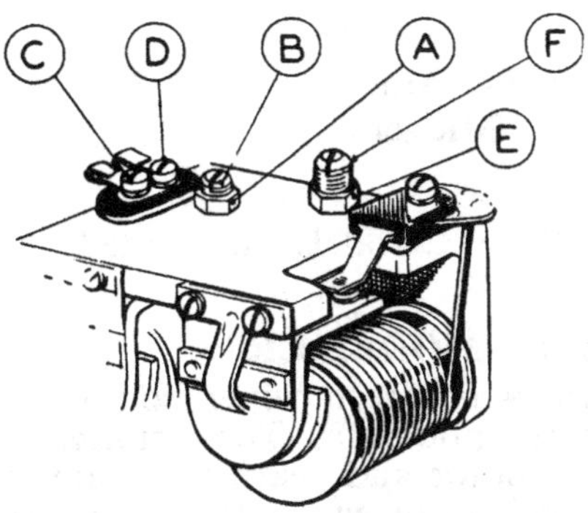

Fig. 36. The cut-out and regulator assembly.

The screw turns in a clockwise direction to raise the setting, or in an anti-clockwise direction to lower the setting.

Turn the adjusting screw a fraction of a turn in the required direction and tighten the locknut.

When the dynamo is run at a high speed on open circuit it builds up a high voltage. When adjusting the regulator do not run the engine up to more than half throttle or a false voltmeter reading will be obtained.

Mechanical Setting. Details as TD Midget.

(i) Slacken the two armature fixing screws (Fig. 37). Insert a .020 in. feeler gauge between the back of the armature and the regulator frame.

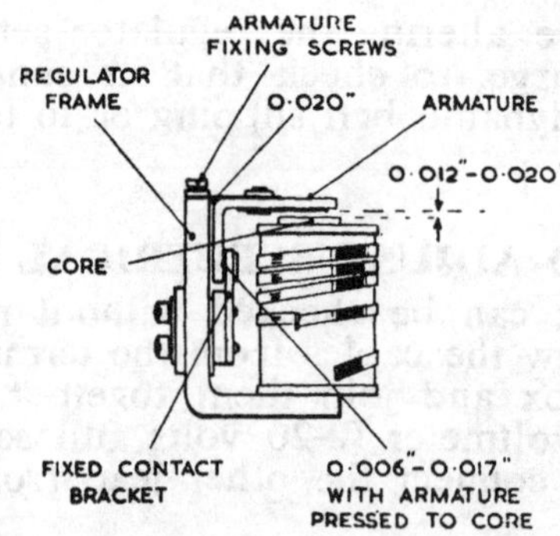

Fig. 37. The mechanical adjustment for the regulator.

It is permissible for this gap to taper upwards or downwards, between the limits of .018 in.-.022 in.

(ii) Details as TD Midget.

(iii) Check the gap between the under side of the arm and the top of the bobbin core; this must be .012 in.-.020 in. (.30-.50 mm.). If the gap is outside these limits correct by carefully bending the fixed contact bracket.

(iv) Details as TD Midget.

Cleaning Contacts.

Details as TD Midget, but amend final sentence to read:—

"Finally re-position the fixed contact bracket and tighten the securing screw."

CUT OUT.

Adjustment. Details as TD Midget.

Cleaning. Details as TD Midget.

RADIO SUPPRESSION.

The Series TF Midget Models are fitted with suppressors at the Factory.

WINDSCREEN WIPER. MODEL CRT.

The CRT.12 electric windscreen wiper combines a powerful and silent action with unobtrusive mounting. The arrangement as detailed below consists essentially of a motor and gearbox mounted under the bonnet and a flexible cable rack mechanism which transmits the drive from the motor to the wiper spindles, these being mounted at the bottom of the screen. Incorporated in the motor is a protective device in the form of a thermostat which, under conditions of excessive heating, cuts off the current supply to the motor until normal conditions are restored. If the blades are prevented from moving (as for example by ice or packed snow on the windscreen, in extreme winter conditions) the thermostat will operate as soon as the motor becomes hot; normal working will be resumed if the obstruction has been removed, when the motor has cooled.

The cooling down period will depend on general operating conditions, and may be as long as ten minutes. Excessive heating from any other cause will also temporarily stop the wiper, but again the wiper will automatically re-start as soon as the temperature falls.

The flexible cable rack consists of an inner core of steel wire, around which is wound a wire helix forming the rack. The rack, which is contained in an armoured outer casing, is given reciprocating motion by means of a crank in the wiper gearbox, and engages with gears on the wiper spindles to drive the wiper arms.

The wiper consists of motor and gearbox, and two wheelboxes, one for each of the two wiper arms. The wheelbox spindle housings are threaded so that the wheelboxes may be screwed directly to the scuttle or windscreen by means of a suitable nut and packing pieces.

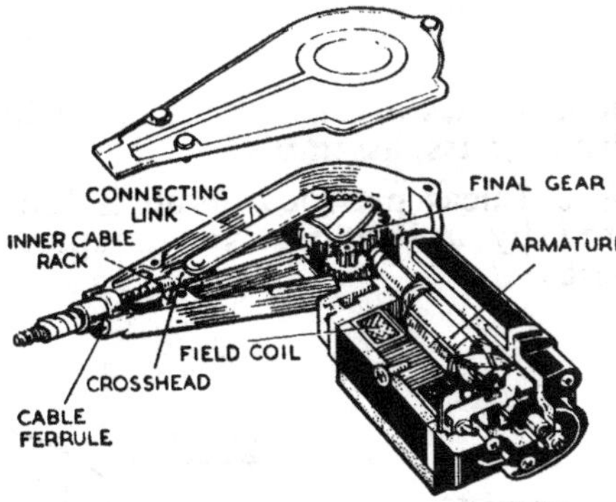

Fig. 38. The windscreen wiper CRT.12 with cover removed showing internal mechanism.

Routine Maintenance.

The moving parts are packed with grease during assembly and need no further lubrication in service; no adjustments are necessary. The only maintenance required consists of an occasional inspection of the rubber wiper blades and renewing when necessary.

If the wiper fails to operate or operates unsatisfactorily, proceed as follows:

Ascertain if the battery is supplying current to the wiper by switching on and noting if the ammeter responds.

Examine the fuse protecting the wiper. If it has blown examine the wiring of the wiper and associated circuits for evidence of short circuit or chafed leads.

THE ELECTRIC HORNS. MODEL WT. 614. Details as TD Midget.

THE FUSES.

The fuses are clipped into a Lucas Fuse Unit, Model S.F.6.

Fuse Connection A1 and A2.—This fuse protects the horn circuits.

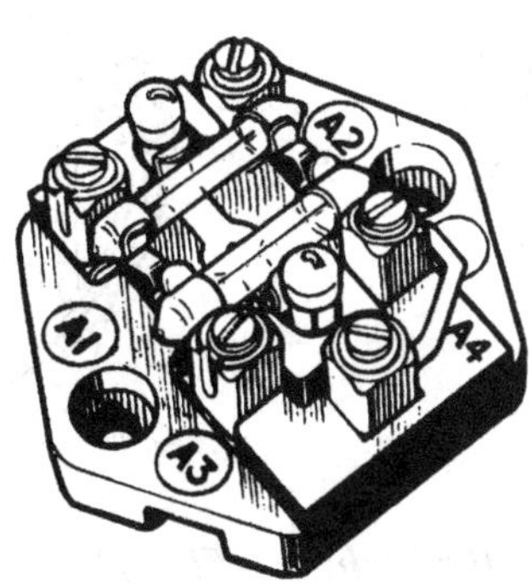

Fig. 39. The fuse unit. Two spare fuses are carried in recesses in the fuse box

Fuse Connections A3 and A4.—This fuse protects the accessories which are connected so that they operate only when the ignition is switched on.

Units Protected.—The units which are protected by each fuse can readily be identified by referring to the wiring diagram Lucas No. W.25080. Page 581.

Blown Fuses. Details as TD Midget.

THE HEADLAMPS.

Each headlamp incorporates a Lucas light unit which consists essentially of a reflector and front glass assembly.

The bulb which is a Lucas pre-focus type is located accurately in the reflector and is secured by a bayonet-fixed adaptor which also provides the contact to the bulbs.

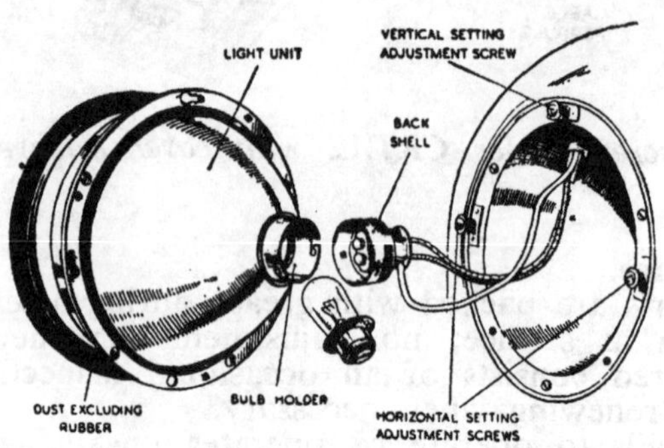

Fig. 40. The headlamp separated showing component parts and vertical and horizontal setting adjustment screws.

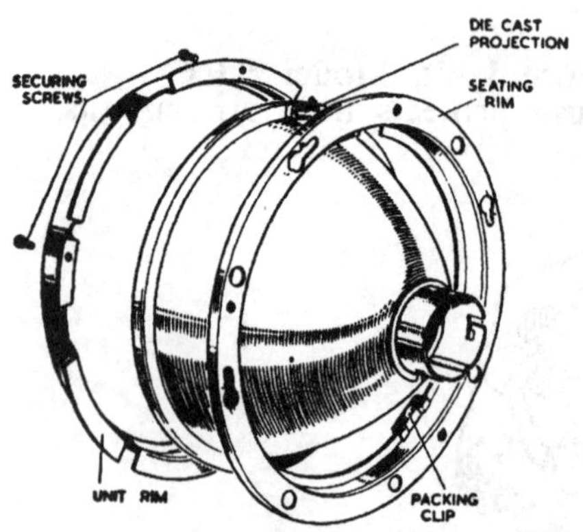

Fig. 41. Component parts of the Light Unit.

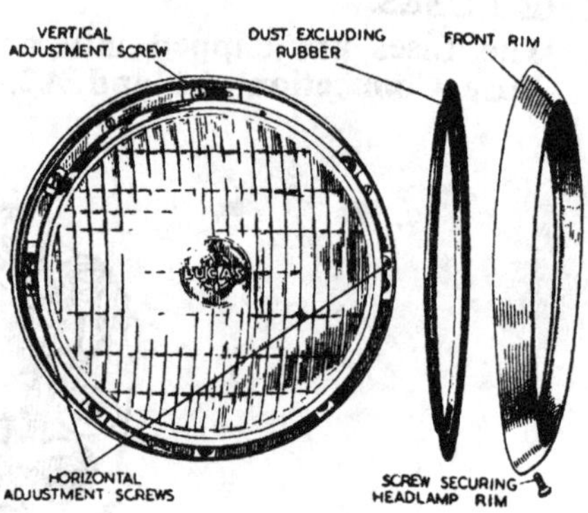

Fig. 42. Front view of Headlamp showing front rim and dust excluding rubber removed.

Bulb Replacement.

Slacken the screw fitted at the bottom of the front rim and lift off the rim and dust excluder rubber. To remove the light unit, press it inwards against the three adjusting screw springs and turn it in an anti-clockwise direction until the heads of the adjusting screws can be disengaged through the slotted holes of the light unit rim. Do not disturb the screws as this will affect the beam setting.

Access to the bulb is gained by turning the back shell in an anti-clockwise direction and withdrawing it. Identical double filament bulbs are fitted in each lamp. In the dipped position, the beams from each lamp are deflected downwards, and to the left, right or vertically according to local anti-dazzle legislation. Insert the replacement bulb in the bulb-holder, making sure that the slot in the periphery of the bulb flange engages the projection in the holder. Engage the projections on the back shell with the slots in the holder. Press it on and twist it clockwise until it engages with its catch.

Setting the Headlamp Beams.

To set the lamps remove the front rim and dust excluder rubber, when three spring loaded adjusting screws will be visible. Turning the top screw clockwise will serve to raise the beam and anti-clockwise, will serve to lower the beam. Adjustment in the horizontal plane is affected by turning the two screws at the sides of the light units.

The lamps should be set so that the main beams are straight ahead and parallel to each other and to the road **when the vehicle is normally loaded.** All Lucas Service Depôts and many Agents include among their special testing facilities a Lucas Beam setter. In order to obtain maximum efficiency from the headlamps the use of a Beam Setter is strongly recommended. Where, however, this is not possible the vehicle should be placed in front of a blank wall at a distance of at least 25 ft., taking care that the surface on which the vehicle stands does not slope relative to the wall. Whilst setting the beam of one headlamp, the other should be masked by covering with a coat or rug.

The design of the bulb and its holder is such that the bulb is correctly positioned in relation to the reflector and no focussing is required when a replacement bulb is fitted.

It is to be noted that the block pattern lenses are different on L.H.D. Models from those on R.H.D. models, and that it is essential to fit correct replacement glasses if the beam formation is to be retained.

The Anti Dazzle Device.

On Export and Home Models both headlamps are fitted with double filament bulbs. Operation of the dip-switch extinguishes the main driving beam in each headlamp and brings the dipping beam into action.

Replacing a Light Unit.

In the event of damage to the front lens or reflector a replacement light unit can be fitted as follows:—

Remove the defective light and bulb. Withdraw the three screws from the unit rim and remove the seating rim and unit rim from the light unit. Position the replacement light unit on the seating rim, taking care to see that the locating projection at the edge of the Light Unit fits into the slot in the seating rim. Ensuring that the unit rim is correctly positioned before fitting the three fixing screws.

Refit the bulb, adaptor, etc.

THE SIDE LAMPS. Details as TD Midget.

THE TAIL AND STOP LAMPS AND FLASHERS. Details of these lamps are given on page 524, these lamps being introduced on later TD Midget models.

THE NUMBER PLATE ILLUMINATION BULB. Details as TD Midget.

REMOVAL OF INSTRUMENT PANEL.

Remove the bolts, nuts, and washers and Phillips screws securing the facia panel undershield and remove undershield.

This will give access to the rear of the instrument panel which is held by six nuts.

Disconnect the speedometer and revolution counter cables from the instruments.

Disconnect the oil pipe at the oil pressure gauge.

Disconnect the water temperature gauge element from the thermostat housing.

Disconnect the starter and mixture controls.

Remove the panel attachment nuts and withdraw panel from facia board.

Disconnect all electrical cables and remove panel from facia board complete with all instruments and controls.

THE IGNITION, FUEL WARNING AND DIRECTIONAL INDICATOR WARNING LIGHTS.

Access to the bulbs for the ignition and fuel warning and directional indicator lights is obtained by removing the facia board undershield.

Location and Remedy of Faults. Details as TD Midget.

THE HEADLAMP BEAM WARNING LIGHT.

In order to give the driver visual warning to dip his headlamp beams on approaching other vehicles a warning light is fitted in the revolution counter which glows when the headlamp beams are in the raised position and is extinguished when the beams are dipped.

DIPPING SWITCH.

M.G. Midget Series 'TF' cars are provided with a foot operated headlamp beam dipping switch. The wiring diagram covering this equipment is W.25080.

THE CORRECT "ACID LEVEL" DEVICE.

Correct acid-level-devices are now fitted as standard to the majority of Lucas batteries, one of these devices being located in each cell filler hole.

The method of topping up a battery fitted with these is shown in Fig. 43. It will be seen that the well-known "air-trap" is the principle underlying their operation.

The device consists of a central tube with a perforated flange which rest on a ledge in the filling orifice. When topping up pour distilled water **into the flange, not down the central tube,** until no more drains through into the cell and the water begins to rise in the flange. This will happen when the electrolyte level reaches the bottom of the central tube, thereby preventing further escape of air displaced by the topping up water introduced By lifting the tube slightly the small amount of water held in the flange will drain into the cell and the electrolyte level will then be correct.

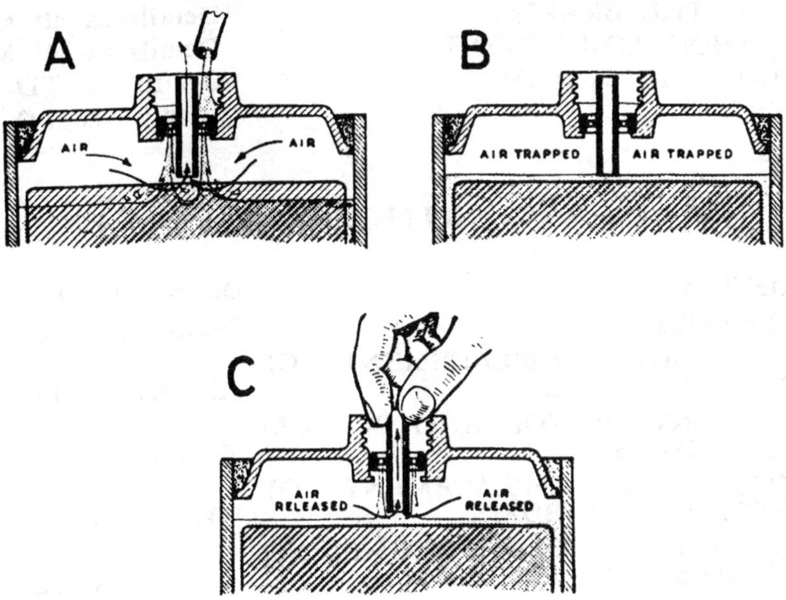

Fig. 43. *The operation of " The Correct acid Level " device.*

Fig. 43. *Shows* :—

(a) Electrolyte level below normal. Pour distilled water into the flange around the central tube.

(b) Cease pouring when the water ceases to flow through the flange and begins to rise above it. The electrolyte level is now controlled by the air lock.

(c) Release the water which has collected in the flange by lifting the tube slightly. The electrolyte level is then correct.

BRAKING SYSTEM

GENERAL DESCRIPTION.	Details as TD Midget.
ROUTINE MAINTENANCE.	Details as TD Midget.
THE MASTER CYLINDER.	Details as TD Midget.
ADJUSTMENT OF BRAKE PEDAL.	Details as TD Midget.
REMOVAL OF MASTER CYLINDER.	Details as TD Midget.
DISMANTLING THE MASTER CYLINDER.	Details as TD Midget.
ASSEMBLING THE MASTER CYLINDER.	Details as TD Midget.
REPLACEMENT OF MASTER CYLINDER.	Details as TD Midget.
REMOVAL OF WHEEL CYLINDER.	Details as TD Midget.
DISMANTLING THE WHEEL CYLINDERS.	Details as TD Midget.
ASSEMBLING THE WHEEL CYLINDERS.	Details as TD Midget.
REPLACEMENT OF WHEEL CYLINDERS.	Details as TD Midget.
REMOVAL AND REPLACEMENT OF BRAKE SHOES.	Details as TD Midget.

RELINING THE BRAKES. Details as all Models.
BRAKE SHOE ADJUSTMENT. Details as all Models.
BLEEDING THE SYSTEM. Details as TD Midget.
HANDBRAKE. Details as TD Midget.

STEERING GEAR

DESCRIPTION. Details as TD Midget.
MAINTENANCE. Details as TD Midget.
REMOVAL AND REPLACEMENT OF STEERING WHEEL. Details as TD Midget.
REPLACEMENT OF TOP AND BOTTOM FELT BUSHES. Details as TD Midget.
REMOVAL AND REPLACEMENT OF STEERING GEARBOX. Details as TD Midget.
DISMANTLING AND REASSEMBLING THE STEERING GEARBOX. Details as TD Midget.
REMOVAL AND REPLACEMENT OF THE STEERING COLUMN. Details as TD Midget.

Remove radiator as detailed on page 549.

Remove carburetters and air cleaner as detailed on page 543 (R.H.D. Model only).

Remove dynamo as detailed on page 561 (L.H.D. Model only).

Remove steering wheel and telescopic extension detailed on page 415.

Remove the bolt and nut from the support clip under the dash. Then remove the bolt and nut holding the steering column to the body steady bracket (this is on the engine side of the bulkhead).

Remove the split pins, take off the nuts and unscrew the three bolts at the universal joint. Do not lose the rubber inserts.

This will free the steering column assembly from the pinion flange which can be pulled out towards the front of the car.

Replacement is carried out in the reverse manner to that detailed for removal bearing in mind that the screws on the universal joint should be tightened fully up against their shoulders.

BODY

The open body is of normal coachbuilt construction with metal panels on a wood framework and is equipped with detachable side screens and a folding hood for complete weather protection when required.

The body is attached to the chassis frame by a series of bolts, and it is removed without difficulty by following the directions given.

THE SIDESCREENS.

The sidescreens are fitted to the door by socket fittings and slotted brackets which engage locking screws. Their fitting and removal from the doors is straightforward and presents no difficulties. When not in use they are stored in a special compartment in the back of the body behind the seats.

THE HOOD. Maintenance and Cleaning.

Much unnecessary damage is done to the hood by incorrect and careless folding when stowing.

The hood should never be folded when wet.

Always wait for the hood to dry before folding it.

Before folding the hood back release the press buttons at each side. Make sure that no hood material is trapped betweeen the hood sticks and that the rear panel of the hood is pulled well forward, as at (1).

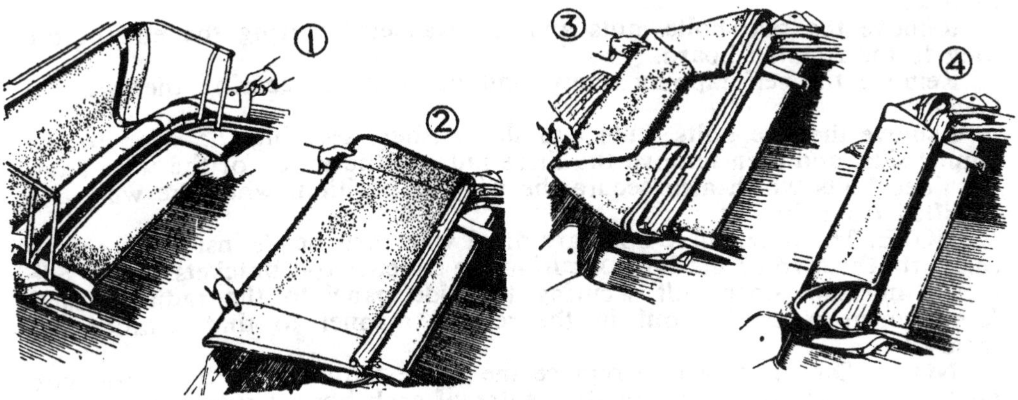

Fig. 44. *Correct method of folding hood.*

Then fold the hood sticks right down and gently pull the hood material out as at (2), now fold the two corners in at right angles, and fold the hood as illustrated in (3), the hood material is then folded over again as shown in (4) and it is then ready for the tonneau cover to be fitted.

The hood may be cleaned as follows:—

The hood is made from quality R.P.68 double texture Mercerised Cotton Repp/Grandrelle and small grease or oil marks can be removed by using a proprietary brand of cleaner named "Dabitoff." Although this is similar to a number of other cleaners, it is marketed in a suitable form for the purpose required. The glass container is fitted with a stopper covered with canvas and will not permit more than the requisite amount of liquid to be imparted to the cloth, just sufficient to remove the stain without injury to the rubber combining agent. If the liquid is used as per the directions on the bottle, there should be no detrimental results whatsoever.

The appearance of the hood after exposure or a very long journey can be improved by brushing with a soft brush immersed in luke warm water. ON NO ACCOUNT SHOULD SOAP OR ANY OTHER DETERGENT BE APPLIED.

CLEANING UPHOLSTERY.

The upholstery may be cleaned by wiping with a damp cloth and polishing it with a clean soft cloth when it is dry. In cases of badly soiled upholstery it may be cleaned by the additional use of a little pure soap, but caustic soaps must on no account be used.

LUBRICATION.

An oil can filled with a good quality three-in-one oil should be used sparingly on the door hinges, door stop pin, bonnet hinge, bonnet locks, and door locks every 1,000 miles.

Coat the door lock slam plates lightly with grease at the same time.

REMOVAL AND REPLACEMENT OF FRONT WINGS.

Disconnect the battery, jack up the front of the car and remove the road wheels.

Disconnect the wiring to the headlamps by withdrawing the cables from their snap connections inside the engine compartment, these cables running along the bonnet side.

Remove the six nuts and washers securing the chromium plated tread strips to the wing.

Remove the one bolt, nut and washer securing the wing to the front valance.

Remove the two bolts, nuts and flat washers securing the end of the wing to the running board.

Remove the self tapping screw, and flat washer securing the wing to the body.

Remove the five bolts, nuts and flat washers securing the wing to the bonnet side, and then remove the three nuts on the inside of the wing from the three bolts which also secure the wing in position, when the wing can be lifted from the car.

NOTE. The latter three bolts are fitted with their heads inside the engine compartment, two of the bolts securing the stiffener rod brackets to the side panel, and the other bolt securing the side panel to the radiator case. Replacement is carried out in the reverse manner to that detailed for removal.

NOTE. Do not forget to replace the spade terminal for the earth wire on the bolt, approximately in the centre of each bonnet side panel.

REMOVAL AND REPLACEMENT OF REAR WINGS.

Remove the Phillips screws and flat and spring washers securing the tail and stop lamps to the wing and disconnect the cables, noting the respective location of each to ensure correct operation of the stop lamp on reassembly.

Remove nut, bolt and two flat washers securing the wing to the chassis frame and the eight bolts and flat washers securing the wing to the wheel arch and body.

Remove the nut, bolt and flat washer securing the wing to the running board and lift off wing.

Replacement is carried out in the reverse manner to that detailed for removal.

REMOVAL AND REPLACEMENT OF RUNNING BOARDS.

Remove the seven nuts securing the chromium tread strips to the running boards leaving the six nuts securing the strips to the front wing in position.

Remove the two bolts, nuts and flat washers securing the running board to the front wing.

Remove the one bolt, nut and flat washer securing the running board to the rear wing.

Remove the two bolts and flat washers (one in the central position and one at the rear end) securing the running board to the body and lift the running board clear of the car.

Replacement is carried out in the reverse manner to that detailed for removal.

REMOVAL AND REPLACEMENT OF BONNET SIDES.

Remove front wings as detailed above.

Remove the four bolts and flat washers securing the bottom of the side panel to the chassis frame.

Remove the small bolt, nut and spring washer securing the clip for the light cables to the bonnet side.

Remove the two self tapping screws securing the bottom corners of the side panel to the radiator casing and to the body scuttle.

Remove the two Simmonds nuts, bolts and washers securing the top corners of the bonnet side, one to a bracket on the scuttle and one securing to a bracket on the radiator.

The bonnet side is now clear of all attachments and can be lifted clear of the car.

Replacement is carried out in the reverse manner to that detailed for removal.

NOTE. Ensure when fitting the front wings that the cable clips and spade terminal for the earth wire are fitted under their respective bolts going through the wings to the bonnet side.

REMOVAL AND REPLACEMENT OF BODY.

Extract the two screws securing the rear bonnet support to the scuttle and remove the bonnet tops clear of the car.

Disconnect the positive and negative leads from the battery.

Release the battery securing bolts and lift the battery from the battery carrier.

Remove the front wings as detailed on page 574.

Remove the running boards as detailed on page 574.

Remove the bonnet sides as detailed on page 574.

Remove the rear wings as detailed on page 574.

Remove the rear number plate lamp front and disconnect and withdraw the cable through the back of the lamp.

Remove the two bolts, spring and plain washer securing the rear bumper to the chassis frame and withdraw the bumper complete with number plate, rear lamp and distance tubes.

Remove the spare wheel and extract the two Phillips screws and washers at each end of the rear valance and the two bolts, nuts and washers in the centre and lift the rear valance clear of the car.

Drain the petrol tank, disconnect the delivery pipe and petrol tank attachment cable.

Remove the two bolts, nuts and plain washers securing the bottom ends of the petrol tank straps to the brackets at the base of the outer tubes of the spare wheel carrier.

Remove the two bolts, nuts and plain washers securing the spare wheel carrier to the chassis frame side members and the bolt, nut and washer securing the centre tube of the carrier to the rear cross member.

NOTE. This bolt has the spade terminals for the earth wires under the head.

Disengage the petrol tank straps from the brackets at the back of the body, remove straps and rubber packing strips and lift the petrol tank clear of the car.

Withdraw the sidescreens from their stowage, remove the bucket seats from their runners and lift seats from the car.

Remove the seat slide runners and remove carpets and underfelts.

Remove the steering wheel adjustment clamp bolt and slide back the clamp to expose the locking key.

Lift the key out of the keyway by inserting a strip of thin metal down the keyway, underneath it, and withdraw the steering wheel and splined shaft from the column. Release the steering column from the support bracket.

Remove the Phillips screws and nuts, bolts and washers securing the cockpit undershield in position and withdraw the shield.

Remove the gearbox cover, pedal draught excluder retaining plate and floorboards.

Extract the pins and bolts securing the front portion of the side trim panel and allow trim panel to spring back clear of the dash stiffener assembly.

Slacken the draught excluder clip at the foot of the steering column and extract the three bolts securing the retainer plate to the foot ramp.

Disconnect accelerator pedal and remove the bolts, nuts, spring and flat washers securing the foot ramp to the body and body stiffener bracket and remove foot ramp.

Slacken off and remove the hand brake cable adjusting nuts and springs.

Release the outer cables from the abutment brackets and thread them through the propeller shaft tunnel flange. Extract the bolts securing the propeller shaft tunnel to the body and chassis frame.

Extract the two bolts from the attachment plates securing each side of the scuttle to the dash stiffener assembly below the dash.

Disconnect at the starter switch end, the cable from starter switch to starter motor.

Remove speedometer cable at the gear box end and remove from clip on chassis frame side member.

Disconnect the rev. counter cable from the reduction box on the rear end of the dynamo.

Remove the earth wire to the distributor and remove clips securing the radiator thermometer tube to the radiator support tube.

Remove the radiator thermometer bulb from the L.H. side of the thermometer housing at the front of the cylinder head.

Remove cables on rear end of dynamo and disconnect main ignition lead at coil on bulkhead.

Disconnect the oil pipe to the oil pressure gauge at the two-way connector on the L.H. side of the engine.

Remove bolt securing panel harness cable clip to the bulkhead.

Disconnect, at snap connectors, panel harness cables from the main loom.

The body is secured to the chassis frame by eight bolts. The two rear are located at the rear end of the body and the two front are inserted through the dash panel flanges into the "A" brackets on the chassis frame. The remaining four are inserted through the body floor.

Place a suitable sling around the body and hoist it clear of the chassis.

The body will come away from the chassis complete with facia board. windscreen wipers, controls and panel harness.

Replacement is carried out in the reverse manner to that detailed for removal.

:RVICE DATA
ODEL "T.F." MIDGET

)WER UNIT.

pe XPAG/TF ..	4 cyl. O.H.V. Pushrod operated.
easury Rating ..	10.97 H.P.
re	66.5 mm. (2.62 ins.). Stroke 90 mm. (3.54 ins.).
ıpacity	1,250 cc. (76.3 cu. ins.).
rst O/size Bore ..	67 mm. (2.640").
cond O/size Bore	67.5 mm. (2.660").
aximum O/size Bore	Linered back to Standard.
mpression Ratio	8.1 to 1. Depth of Cylinder Head 75.16 mm. (2.99").
ıpacity of Combustion Space.	39.3 cc.
linder Head Gasket.	Copper and asbestos .045" thick.
ıpacity of Cylinder Head Gasket.	4.5 cc.
ın, 4 blade ..	Dia. 12¾". Speed .9 of engine speed.

STONS

pe	"Aerolite" aluminium alloy, tin coated, controlled expansion.
ompression Ring Grooves.	Width .0886"/.0887". Diameter 2.381"/2.385".
l Control Ring Grooves.	Width .1576"/.1577". Diameter 2.385"/2.389".
de Clearance of Rings in Groove.	.001"—.002".
mpression Rings	2 off. Width .0885". Thickness .109"—.101".
l Control Ring (8 Slot).	1 off. Width .1575". Thickness .105"—.097".
ston Ring Cap ..	.006"—.010".
dgeon Pin Bore ..	18 mm.—.0001"/.0003".
ston Area, sq. ins. per ton.	23.
ston Clearance in Bore.	.002"—.003" at thrust faces.
dgeon Pin.	Dia. 18 mm.—.010/.015 mm. Length 58.5 mm.
t in Piston ..	Double thump press fit.
t in Connecting Rod.	Clamp bolt.
ston Area ..	21.57"².

ONNECTING RODS.

pe, Steel ..	Length 178 mm. with loose steel shell white metal, lined bearings.
re at Big End ..	48.658 mm./48.671 mm.
re at Little End ..	18 mm.—.01/+.03 mm.
idth at Big End ..	27.885/27.890 mm.
idth at Little End	19 mm.

RANKSHAFT.

ain Journal (front)	Dia. 52 mm. − .015. .035. Wdh. 38 mm. Rad. 2.5 mm. ± .015.
termediate Journal	Dia. 52 mm. − .015. .035 Whd. 38 mm. Rad. 2.5 mm. ± .015.
ain Journal (Rear)	Dia. 52 mm. − .015. .035. Wdh. 40 mm. Rad. 2.5 mm. ± .015.
ankpin Journal ..	Dia. 45 mm. − .015. .035. Wdh. 28 mm. Rad. 2.5 mm. ± .015.
ain Bearing (front)	Dia. 52.020/52.005 mm. Width 35.12/34.88 mm.
ain Bearing (inter)	Dia. 52.020/52.005 mm. Width 37.955/37.925 mm. Rad. 2.75 mm. U/cut. .25 mm.
ain Bearing (rear)	Dia. 52.020/52.005 mm. Width 35.12/34.88 mm.
Bore Size of Crankcase for Main Bearing.	56.34 mm. ± .01 mm.
Main Bearing Clearance of Crankshaft Diameter.	.020/.055 mm. (.001" − .002").
Main Bearing (inter) End - Float on Crankshaft.	.035/.095 mm. (.0015" − .0037").
Big-end Bearing ..	Dia. 45 mm. Width 22.46/22.21 mm. Thickness 1.825/1.831 mm.
Big - end Clearance on Crankshaft Diameter.	.011/.056 mm. (.0005" − .0025").
Crankshaft 1st Regrind.	51.7 mm. Later 51.5 mm 51.25 mm., 51 mm. Max 50.75 mm.

CAMSHAFT.

Camshaft Drive ..	Chain: Type Roller ⅜ pitch, 60 pitches endless.
Camshaft Bearing (front).	White metal O/dia. 43.61 mm I/dia. 41 mm. ±.05/.08 mm Ream in position in block.
(centre and rear).	Zinc Alloy: O/dia. 43.5. mm +.03. −.02. I/dia. 23 mm. +.015. −.005.
Bore of Cylinder Block for Bearings	43.5 mm. +.02 mm. −.01 mm.
Camshaft End Float	.125/.325 mm. (.005"/.013").
Diameter of Shaft for Bearing	Front : 41 mm. +.01. −.02. Centre and rear : 23 mm.
Cam Lift	5.785 mm. Valve Lift 8.3 mm.

VALVES AND GUIDES.

Valve Port Throat Diameter.	Inlet 33 mm. ; Exhaust 29 mm.
Valve Head Dia. ..	Inlet 36 mm. ; Exhaust 34 mm.
Valve Stem Dia. ..	8 mm. +.01. −.01.
Valve Guides ..	Bore 8.06 mm./8.08 mm. 0/dia. 14 mm. +.02. +.04.
Valve Guide Length	Inlet 59 mm. ; Exhaust 54 mm.
Valve Seat Angle ..	30°.
Valve Spring Pressure.	Shut 114 lbs. Open 150 lbs. (Double spring).
Valve Lift	8.3 mm.
Valve Timing in Degrees	Inlet valve opens 5° B.T.D.C. Inlet valve closes 45° A.B.D.C. Exhaust valve opens 45° B.B.D.C Exhaust valve closes 5° A.T.D.C.
Valve Clearance—Engine Hot.	.012".
Ignition Timing ..	T.D.C. fully retarded, points just breaking.
Order of Firing ..	1, 3, 4, 2.
Sparking Plugs ..	Champion N.A.8. 14 mm. ¾" reach.
Sparking Plug Gap	.020" − .022".
Oil Pump Speed ..	1/2 engine speed. Gear type.
Oil Pressure ..	40 lbs.-50 lbs. per sq. in.
Oil Pump Bye-Pass	60 lbs. per sq. in
Oil Filter .. .	Full flow, percolator type with renewable element, attaches direct to pump, blow-off valve in filter.
Flywheel .. .	273 mm. dia. No. of teeth 120.

PROPELLER SHAFT.

Type	Hardy Spicer needle Bearings.

SERVICE DATA.

MODEL "T.F." MIDGET—*continued.*

CLUTCH AND GEARBOX.

Type	Borg and Beck. Single plate, dry.
Diameter	8".
Facings	Wound yarn.
Clutch Springs	Colour brown. 150/160 lbs. No. 7. S.W.G. Free height 2" approx. Assembled height 1.56".
Clutch Plate Damper Springs.	6— Black/lt. green.
Clutch Linings	External dia. 8". Internal dia. 5.7".

COMPETITION CLUTCH.

Type	Borg and Beck, single plate, dry.
Diameter	8".
Facings	Ferodo. RAD. 41. Cemented.
Clutch Springs	Colour, light grey. 195/205 lbs. Free height 2.27". Assembled height 1.56". Rate per inch deflection. 266 lbs.

GEARBOX.	4 speed. Synchromesh, Second, third and fourth.
Gearbox Ratio	Top Direct. Third. 1.385 to 1. Second 2.07 to 1. First 3.5 to 1.
Overall Ratio	Top. 4.875 to 1. Third 6.752 to 1. Second 10.09 to 1. First 17.06 to 1.
Layshaft End Float	.46 mm./.77 mm. (.018"/.030").
Overall Gear Ratio with Alternative Axle Ratio. 4.55 to 1.	Top. 4.555 to 1. Third. 6.309 to 1. Second 9.429 to 1. First 15.942 to 1.
Overall Gear Ratio with Alternative Axle Ratio 5.125 to 1.	Top. 5.125 to 1. Third. 7.098 to 1. Second 10.609 to 1. First. 17.938 to 1.
Oil Capacity of Gearbox.	1¼ pints.

INSTRUMENTS.

Speedometer	British Jaeger. 4½" dia. With clock. 1,525 T.P.M.
Reduction Gear for Speedometer.	6/15.
Revolution Counter	British Jaeger. 4½" dia. with main beam warning light.
Reduction Gear for Rev. Counter.	On rear end of dynamo. 100/225.
Petrol Gauge	Warning light on panel.
Oil Pressure Gauge	British Jaeger.
Water Temperature Gauge.	British Jaeger. Approx. temperature 70°/80°C.
Facia Board	Steel pressing. Painted. All instruments and controls on Central panel.

FRONT AXLE AND STEERING.

Camber of Wheel	1° positive to ½° negative on full bump.
Castor Angle	2° ± .5° (with side member parallel with road).
King Pin Angle	9° to 10½° on full bump.
Toe-In	Nil. Knuckle angle 10°.
Steering	Direct acting rack and pinion.
Steering Wheel	"Bluemell" 16½ dia. Wire spokes.
Turning Circle	31' – 3".
Lock to Lock	2¾ turns.
Steering Ratio	13.75 to 1.
Left-hand Drive Available.	Yes.
Suspension	Independent. Incorporating coil springs.

REAR AXLE.

Type	Semi-floating.
Drive	Hypoid. 1" offset.
Ratio	8/39=4.875 to 1.
Alternatives	9/41=4.55 to 1. 8/41. 5.125 t
Adjustment	See Text.
Crown Wheel and Pinion Back Lash.	.006"/.008".
Oil Capacity of Rear Axle.	2¼ pints.

SPRINGS (REAR).

Type	½ elliptic. Interleaved with rubb
Length	42". Stress rate 14,880 lbs./ins.
Working Camber	.50" positive.
Working Deflection	3.60". Working load 500 lbs.
Working Stress	42,670 lbs./ins.². Max. stre 90,700 lbs./ins.².
Laden Frequency	98.5 C.P.M. No. of leaves 7.
Size of Leaf	7/32" × 1½". Load rate 139/lbs./i
Free Camber	4.10". Flat load 570 lbs.
Max. Deflection	6.09 lbs.

SPRINGS (FRONT).

Type	Coil. Length 9.59" ± 1/16".
Mean Coil Dia.	3.238". Wire dia. .498".
No. of Effective Coils.	7.5. Rate 348 lbs./ins.
To Carry 1,095 lbs. at.	6.44" ± 1/32". Max. deflect 4.24".
Max. Stress (Corrected).	121,060 lbs./ins.².
Solid Height Based on Load at Wheel.	500 lbs.
Rate	73 lbs./ins. on wheel.

WHEELS AND TYRES.

Tyre	Vent disc 4.J rim, or wire.
Wheel Size	4.00" × 15".
Tyre Size	5.50" × 15". Dunlop Gold Se
Tyre Pressure	18 lbs./ins.² front and rear.

BRAKES.

Type	Lockheed hydraulic. Handbr cable to rear only.
Type Linings	Ferodo.
Brake Drums (Front)	Composite hub and drum 9" d 2 leading shoe. 1½" wide.
Brake Drums (Rear)	Composite hub and drum 9" d 1½" wide.
Brake Linings (Front).	Ferodo D.M.7.
Brake Linings (Rear)	Ferodo D.M.7.
Friction Area (Front).	49.74 ins.² } Total friction area 99.48 ins.².
Do. (Rear)	49.74 ins.²
Length of Linings	8.46" + .06" – .03" × 1½".

FUEL SYSTEM.

Petrol Tank Capacity.	12 gallons.
Fuel Delivery	To chassis No. TF. 1509. S electric pump type "L" mounted on tool box. From chassis No. TF. 1510. S electric pump type "H. high pressure pump mounted plate at rear o/s of chassis.
Carburetters	Twin S.U. 1½" dia. Semi-dov draught. Specification H.4.
Carburetter Needles	"Standard" GJ. "Richer" "Weaker" G.L.
Air Cleaners	A.C. or Vokes. D.47937-D.479

RVICE DATA.

ODEL "T.F." MIDGET—*continued.*

OLING SYSTEM.

stem	Water, Radiator, Fan and Pump with thermostat control. Pressurised system.
pacity of System	10¼ pints.
pacity of Block and Head.	6 pints.
rculation ..	By pump fitted at front of cylinder block.
ain Taps ..	1 bottom of radiator. 1 front end cylinder block below exhaust manifold.
n	4 blade. Dia. 12¾″. Speed .9 of engine speed.

YDRAULIC DAMPERS.

pe	Armstrong.

ECTRICAL.

stributor Type ..	Lucas. D.2A4. 40367.
stributor, Rotation.	Counter clockwise.
anual Advance in Degrees.	None.
utomatic Advance	25° engine at 3,100 R.P.M.
nition Timing ..	Points just breaking T.D.C.
stributor Points Gap.	.014″—.016″.

TARTER.

ype	Lucas. M.35G2. 25022A.
arter Motor Ratio	13.33 : 1. 9 tooth.
arter Torque ..	4.9 lbs./ft. C. 1,000 R.P.M.

YNAMO.

ype	Lucas. C.39PV2. 22265B.
ynamo Speed ..	1.16 engine speed.

ATTERY.

ype	Lucas. GTW 9A/2. Earth Positive.
ate	51 amp at 10 hr. Rate.
ighting	12-volt. Earth return.
nition	12-volt coil and distributor.
iring Order ..	1, 3, 4, 2.

IMENSIONS.

Vidth	59″. Overall length 147″.
verall Height ..	52½″. Wheelbase 94″.
rack	Front 47⅞″. Rear 50″ with disc wheels.
	Front 48 3/16″. Rear 50 13/16″ with wire wheels.
round Clearance..	6″.

ERFORMANCE.

rake Horse Power	57 @. 5,500 R.P.M.
aximum Torque..	65 ft./lbs. @. 3,000 R.P.M.
.P. per sq. in. Piston Area.	2.64.
eet per min. of Piston at max. H.P.	3,248.
itres per ton Mile..	2,235.
oad Speed at 2,500 ft. per Minute Piston Speed.	64.3 M.P.H.
Engine Speed per 10 M.P.H.	658 R.P.M. (top gear).
Petrol Consumption	30.35 M.P.G.
Piston Speed per 10 M.P.H.	388.6 ft./min.

M.P.H. PER 1,000 R.P.M.

	4.875 to 1 Axle Ratio.	5.125 to 1 Axle Ratio.	4.55 to 1 Axle Ratio.
Top	15.195.	14.42	16.259
Third	10.97	10.412	11.81
Second	7.34	6.966	7.85
First	4.34	4.12	4.64

KILOMETRES PER 1,000 R.P.M.

	4.875 to 1 Axle Ratio.	5.125 to 1 Axle Ratio.	4.55 to 1 Axle Ratio.
Top	24.453	23.1	26.165
Third	17.654	16.66	18.89
Second	11.812	11.15	12.639
First	6.98	6.59	7.469

WEIGHTS.

Complete Car ..	17 cwts. 1 qr. 0 lbs.
Front	8 cwts. 3 qrs. 14 lbs.
Rear	8 cwts. 1 qr. 14 lbs. (Fully equipped, with oil and water, but ½ gallon petrol only.)
Chassis, as off Line	10 cwts. 3 qrs. 14 lbs.
Body	3 cwts.
Power-Unit Total ..	3 cwts. 1 qr. 14 lbs.

CAPACITIES.

Sump to Level (Oil)	10½ pints. Block and Head (water). 6 pints.
Gear Box (Oil) ..	1¼ pints. Cooling system. 10¼ pints.
Rear Axle (Oil) ..	2¼ pints. Brake system Fluid. 1 pint.
Fuel Tank	12 gallons.

GRADES OF CONSUMABLES.

Engine Oil	N.O.L. "Thirty" engine oil.
Alternatives ..	"Shell" × – 100 S.A.E. 30. "Energol" S.A.E. 30. Medium "Filtrate." "Sternol" W.W. 30. "Castrol" X.L. "Essolube" 30. Mobiloil "A."
Gearbox Oil ..	Duckham's Hypoid. 90.
Alternatives ..	"Shell" Spirax 90 E.P. "Energol." "E.P." S.A.E. 90. Hypoid "Filtrate" 90. "Sternol" Ambroleum E.P. 90. "Castrol" Hypoy. "Esso" Expee Compound 90. Mobilube "G.X." 90.
Rear Axle Oil	As above.
Alternatives ..	As above.
Steering Gear Oil ..	As above.
Alternatives ..	As above.
Wheel Hubs and Fan Bearing.	Duckham's H.B.B. Grease.
Alternatives ..	"Shell' Retinax A. "Energrease" C3. "Ambroline" R.B. Grease. "Castrolease" Heavy. "Filtrate" R.B. Grease. Esso Grease. Mobil Hub Grease.
Steering Connections, King Pins, Propeller Shaft, Clevis Pins and Lever Fulcrums.	Duckham's H.P.G. Grease.

SERVICE DATA.

MODEL "T.F." MIDGET—*continued.*

Alternatives ..	"Shell" Retinax A. "Energrease" C. 1. H.P. Solidified "Filtrate." "Ambroline" M.M. Grease. "Castrolease" Medium. "Esso" Pressure Gun Grease. Mobilgrease No. 2 or 4.
Cables and Vital Control Points.	Duckham's "Keenol" K.G. 16 Grease.
Alternatives ..	"Shell" Retinax A. "Energrease" C. 1. "Filtrate" A.F. Grease. "Ambroline" A. F. Grease. "Castrolease" Brake Cable Grease. "Esso" Pressure Gun Grease. Mobilgrease No. 2 or 4.
Brake Fluid ..	Lockheed "Orange."
Shock Absorbers ..	Armstrong Damper fluid.

MISCELLANEOUS AND ACCESSORIES.

Silencer	Burgess Type straight through.
Position of Battery	Under bonnet.
Position of Toolbox	Under bonnet.

BULB REPLACEMENTS.

Head Lamp (Home)	L.H. Dip. 12-volt 42/36 wa Lucas. 354.
Head Lamp (Export)	R.H. Dip. 12-volt 36/36 wa Lucas. 301.
Head Lamp (Export)	Europe (except France). Vertic Dip. 12-volt 45/35 watt. Luca 360.
Side Lamps ..	12-volt. 18/6 watt. Lucas. 36
Tail Lamps ..	12-volt. 18/6 watt. Lucas. 36
No. Plate Lamp ..	12-volt. 4 watt. Lucas. 22
Panel Warning ..	12-volt. 2.2 watt. Lucas. 98
Panel Illumination..	12-volt. 2.2 watt. Lucas. 98
Map Reading ..	12-volt. 2.2 watt. Lucas. 98

SPECIAL NOTE CONCERNING THE M.G. MODEL "T.F. 1500"

The M.G. Model "T.F. 1500", which was the last of the "T-Type" M.G. cars, is in every way essentially identical to the M.G. "T.F." Midget described in preceding parts of this supplement. The significant difference is the larger piston displacement, obtained by an increase in the cylinder bore.

The bore increase, which was necessary in order to produce the Model "T.F. 1500", was accomplished by an ingenious method that obviated any need for further modification to the earlier engine. The cylinder bore centerlines remain unchanged, as do the locations of the cylinder head studs. However, the cylinder bore portions of the engine bl casting which, perforce, had to be increased in diame became of the "siamesed" type; that is, the water passa were eliminated between cylinders No. 1 and No. 2 a cylinders No. 3 and No. 4

The engine of the Model "T.F. 1500" is designated as power unit Type XPEG/TF, whereas the earlier "125 engine is designated as a power unit Type XPAG/TF. T following specifications apply particularly to the engine of M.G. Model "T.F. 1500" car:

POWER UNIT

Type XPEG/TF ..	4 cyl. O.H.V. Pushrod operated
Bore	72.0 mm. (2.835").
Stroke	90 mm (3.54")
Capacity	1466 cc. (89.4 cu. ins.).
First 0/size Bore ..	72.5 mm. (2.854").
Second 0/size Bore ..	73.0 mm. (2.874").
Maximum 0/size ..	Linered back to Standard

PERFORMANCE

Brake Horse Power ..	65 at 5,500 R.P.M.
Maximum Torque ..	76 ft./lbs. at 3,000 R.P.M.

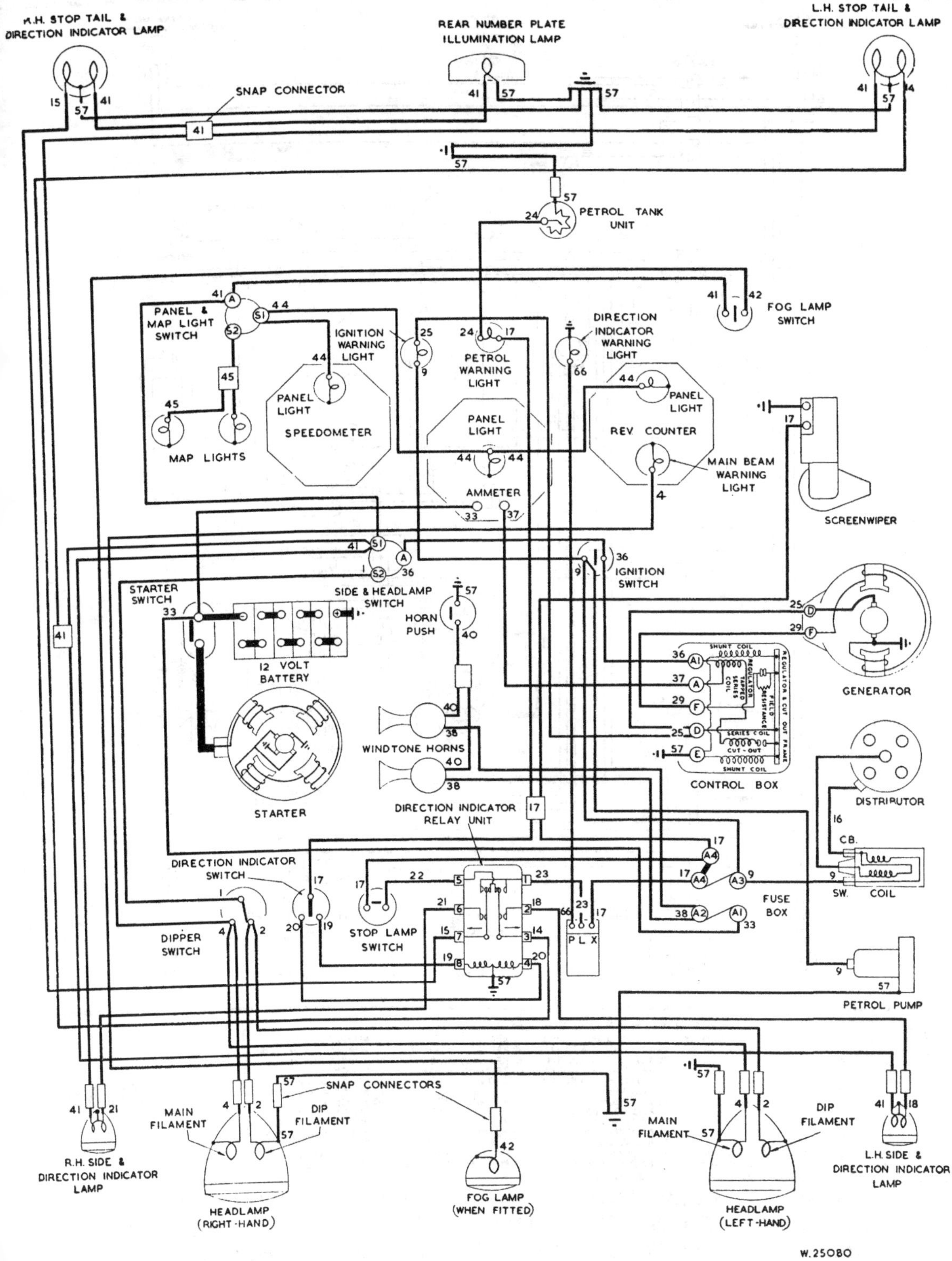

WIRING DIAGRAM FOR MIDGET: SERIES "TF."

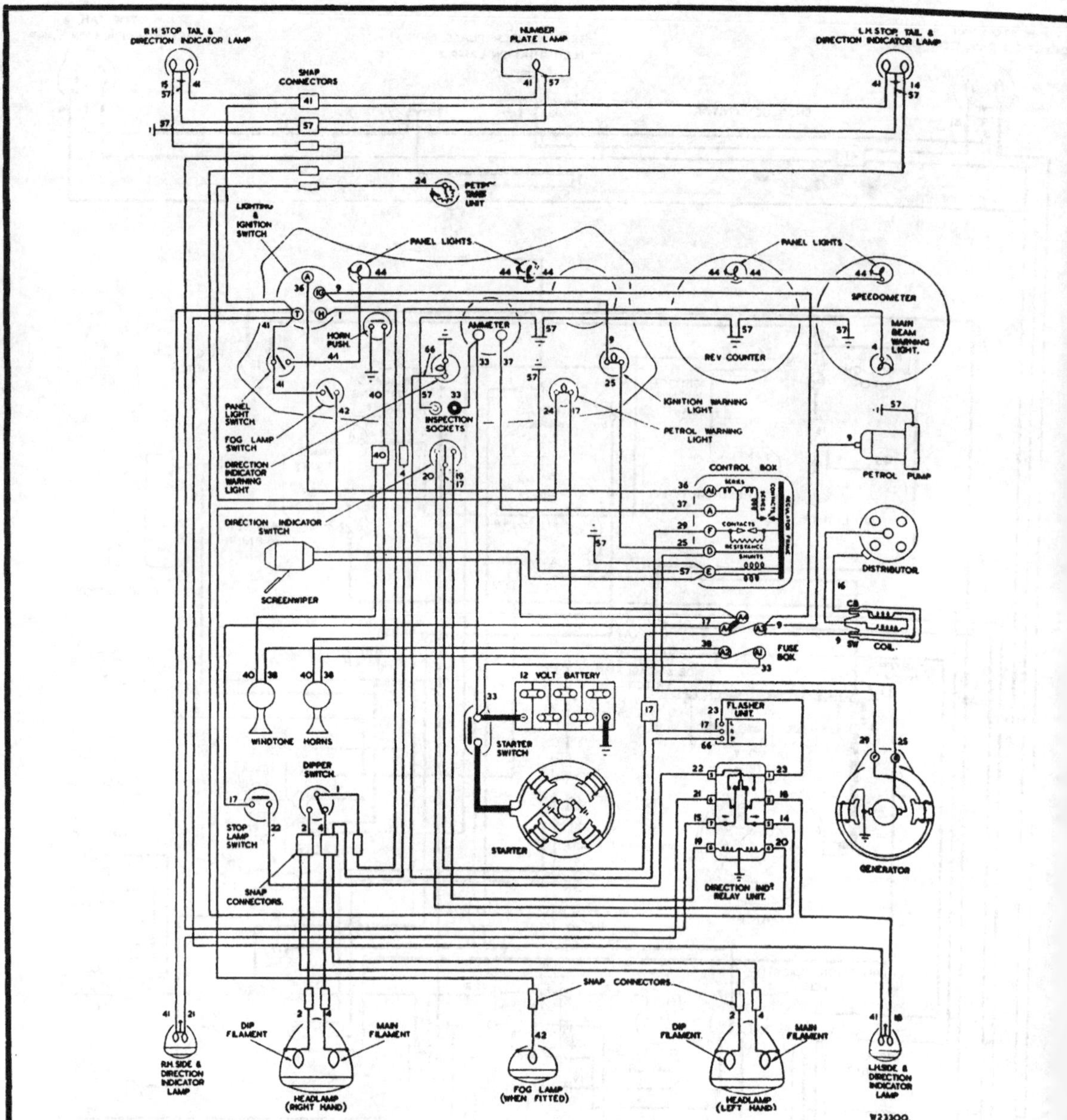

KEY TO CABLE COLOURS

1 BLUE
2 BLUE with RED
3 BLUE with YELLOW
4 BLUE with WHITE
5 BLUE with GREEN
6 BLUE with PURPLE
7 BLUE with BROWN
8 BLUE with BLACK
9 WHITE
10 WHITE with RED
11 WHITE with YELLOW
12 WHITE with BLUE
13 WHITE with GREEN
14 WHITE with PURPLE
15 WHITE with BROWN
16 WHITE with BLACK
17 GREEN
18 GREEN with RED
19 GREEN with YELLOW
20 GREEN with BLUE
21 GREEN with WHITE
22 GREEN with PURPLE
23 GREEN with BROWN
24 GREEN with BLACK
25 YELLOW
26 YELLOW with RED
27 YELLOW with BLUE
28 YELLOW with WHITE
29 YELLOW with GREEN
30 YELLOW with PURPLE
31 YELLOW with BROWN
32 YELLOW with BLACK
33 BROWN
34 BROWN with RED
35 BROWN with YELLOW
36 BROWN with BLUE
37 BROWN with WHITE
38 BROWN with GREEN
39 BROWN with PURPLE
40 BROWN with BLACK
41 RED
42 RED with YELLOW
43 RED with BLUE
44 RED with WHITE
45 RED with GREEN
46 RED with PURPLE
47 RED with BROWN
48 RED with BLACK
49 PURPLE
50 PURPLE with RED
51 PURPLE with YELLOW
52 PURPLE with BLUE
53 PURPLE with WHITE
54 PURPLE with GREEN
55 PURPLE with BROWN
56 PURPLE with BLACK
57 BLACK
58 BLACK with RED
59 BLACK with YELLOW
60 BLACK with BLUE
61 BLACK with WHITE
62 BLACK with GREEN
63 BLACK with PURPLE
64 BLACK with BROWN
65 DARK GREEN
66 LIGHT GREEN

W23300

WIRING DIAGRAM FOR TD MIDGET: WITH SEPARATE CONTROL AND FUSE BOXES, HEADLAMP BEAM WARNING LIGHT, FOOT OPERATED DIPPER SWITCH AND PROVISION FOR FLASHING TRAFFIC INDICATOR LIGHTS.

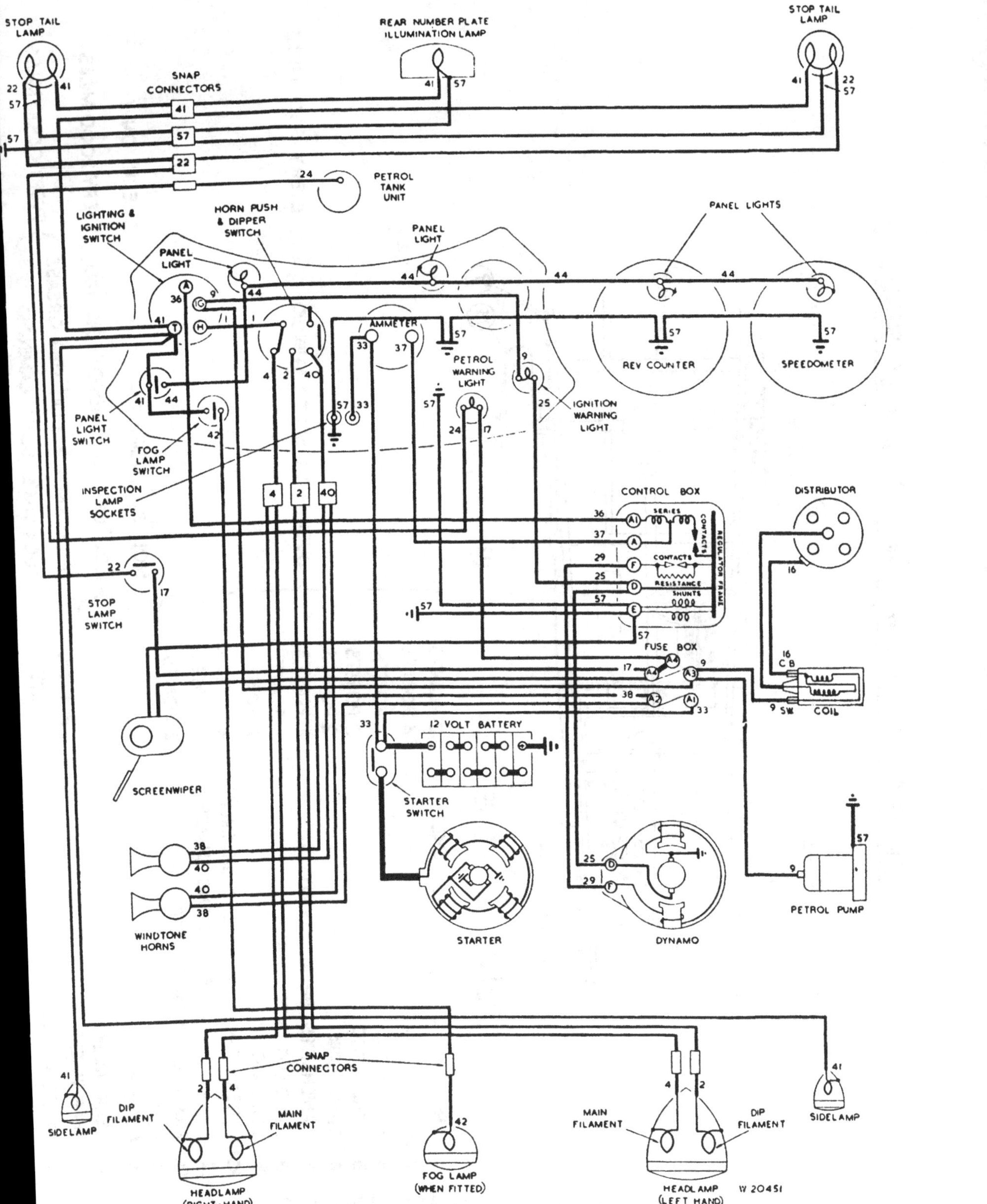

…ING DIAGRAM FOR TD MIDGET WITH SEPARATE CONTROL AND FUSE BOXES

LUBRICATION CHART FOR THE M.G. MIDGET SERIES "M."

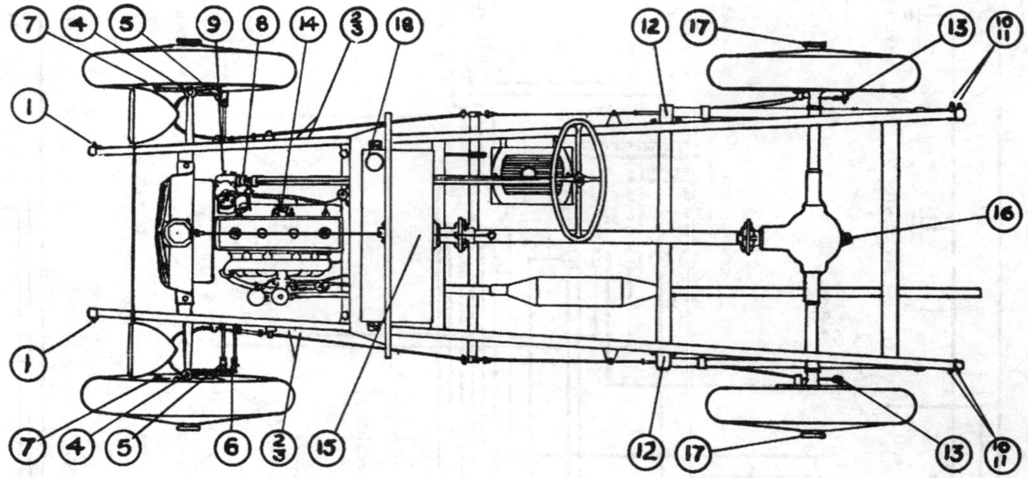

1	FRONT SPRING FRONT SHACKLES	2 POINTS.	DUCKHAM'S SOFT LAMINOID	EVERY 500 MILES
2&3	" " REAR "	4 POINTS.	DUCKHAM'S SOFT LAMINOID	EVERY 500 MILES
4	KING PINS	4 POINTS.	DUCKHAM'S SOFT LAMINOID	EVERY 500 MILES
5	TRACK ROD	2 POINTS.	DUCKHAM'S SOFT LAMINOID	EVERY 500 MILES
6	DRAG LINK	2 POINTS.	DUCKHAM'S SOFT LAMINOID	EVERY 500 MILES
7	CAMSHAFTS BRAKE FRONT	2 POINTS.	DUCKHAM'S SOFT LAMINOID	EVERY 500 MILES
8	STEERING BOX THRUST	1 POINT.	DUCKHAM'S SOFT LAMINOID	EVERY 500 MILES
9	STEERING BOX	1 POINT.	DUCKHAM'S SOFT LAMINOID	EVERY 500 MILES
10&11	REAR SPRING REAR END SHACKLES	4 POINTS.	DUCKHAM'S SOFT LAMINOID	EVERY 500 MILES
12	" " FRONT END SHACKLE	2 POINTS.	DUCKHAM'S SOFT LAMINOID	EVERY 500 MILES
13	BRAKE CAMSHAFT, REAR	2 POINTS.	DUCKHAM'S SOFT LAMINOID	EVERY 500 MILES
14	SUMP		NOL "30" ENGINE OIL	CHECK DAILY
15	GEAR BOX		NOL "EP" TRANSMISSION OIL 140	EVERY 2000 MILES
16	REAR AXLE		NOL "EP" TRANSMISSION OIL 140	EVERY 2000 MILES
17	REAR WHEEL HUB	2 POINTS.	DUCKHAM'S "ADCOL" HBB GREASE	EVERY 10,000 MILES
18	BRAKE PEDAL	1 POINT	DUCKHAM'S SOFT LAMINOID	EVERY 10,000 MILES

Lubrication Chart for the M.G. Midget: Series M

Oiling Chart of the 8/33 M.G. Midget Long Chassis.

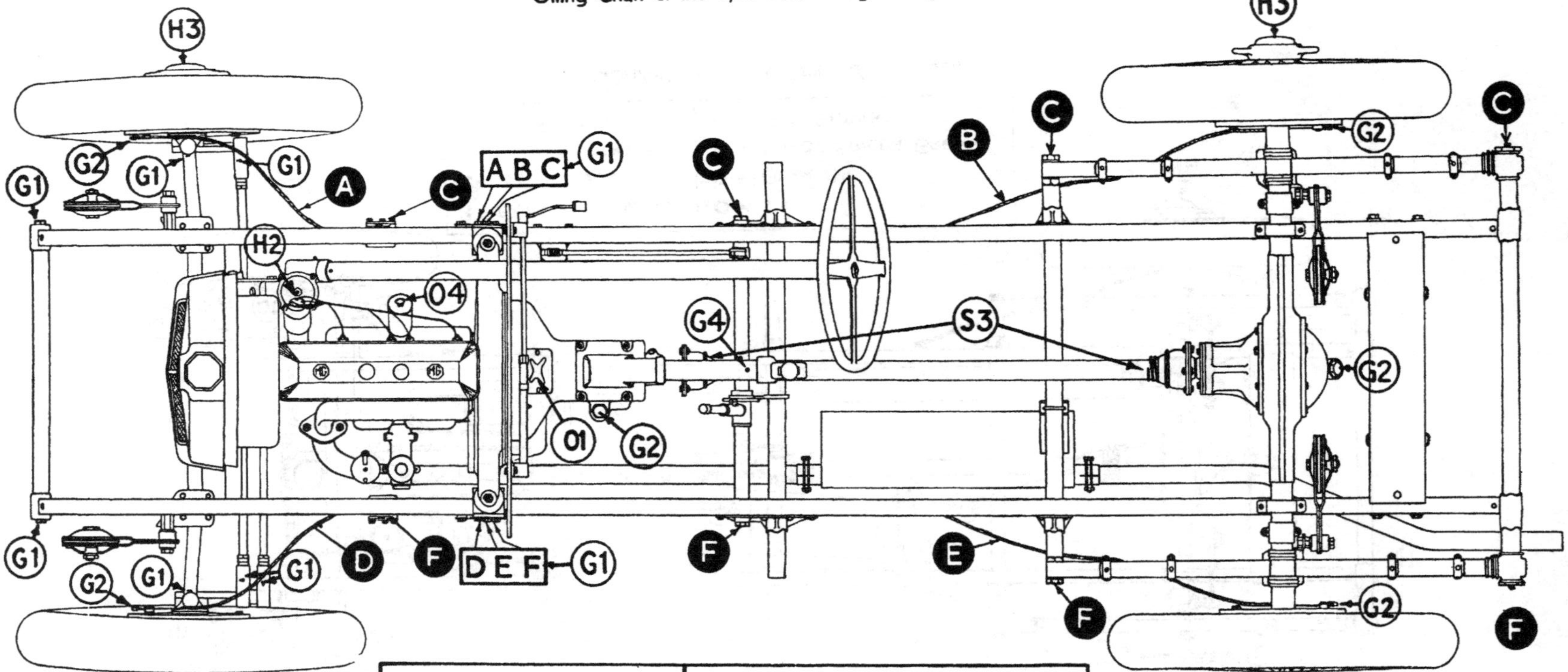

1 = Every 500 Miles	O = Oil
2 = " 2000 "	G = Gear Oil
3 = " 10000 "	S = Hardy Spicer Grease
4 = As Required	H = Hub Grease

Lubrication Chart for Midget Series D.

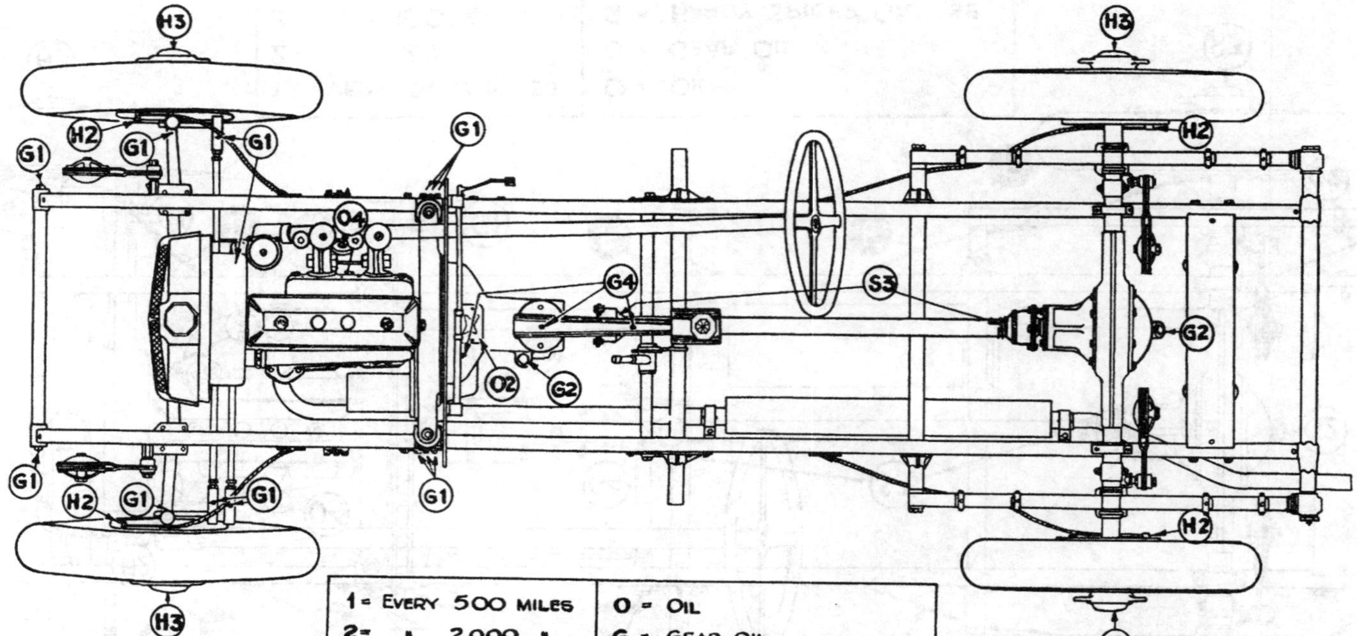

1 = Every 500 miles	O = Oil
2 = " 2,000 "	G = Gear Oil
3 = " 10,000 "	S = Hardy Spicer Grease
4 = As Required	H = Hub Grease

Lubrication Chart for J type M.G. Midget.

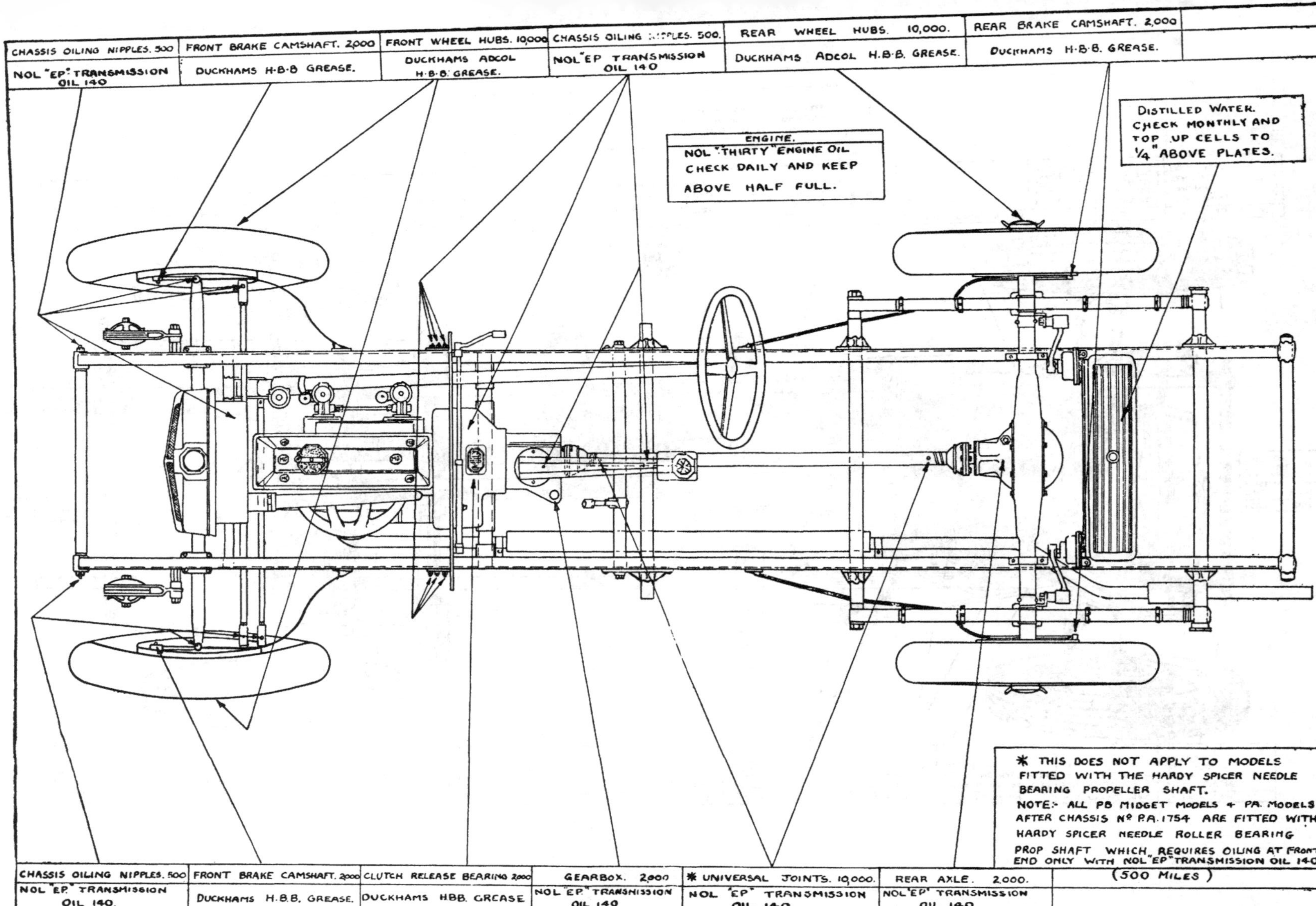

Lubrication Chart for Midget : Series P.

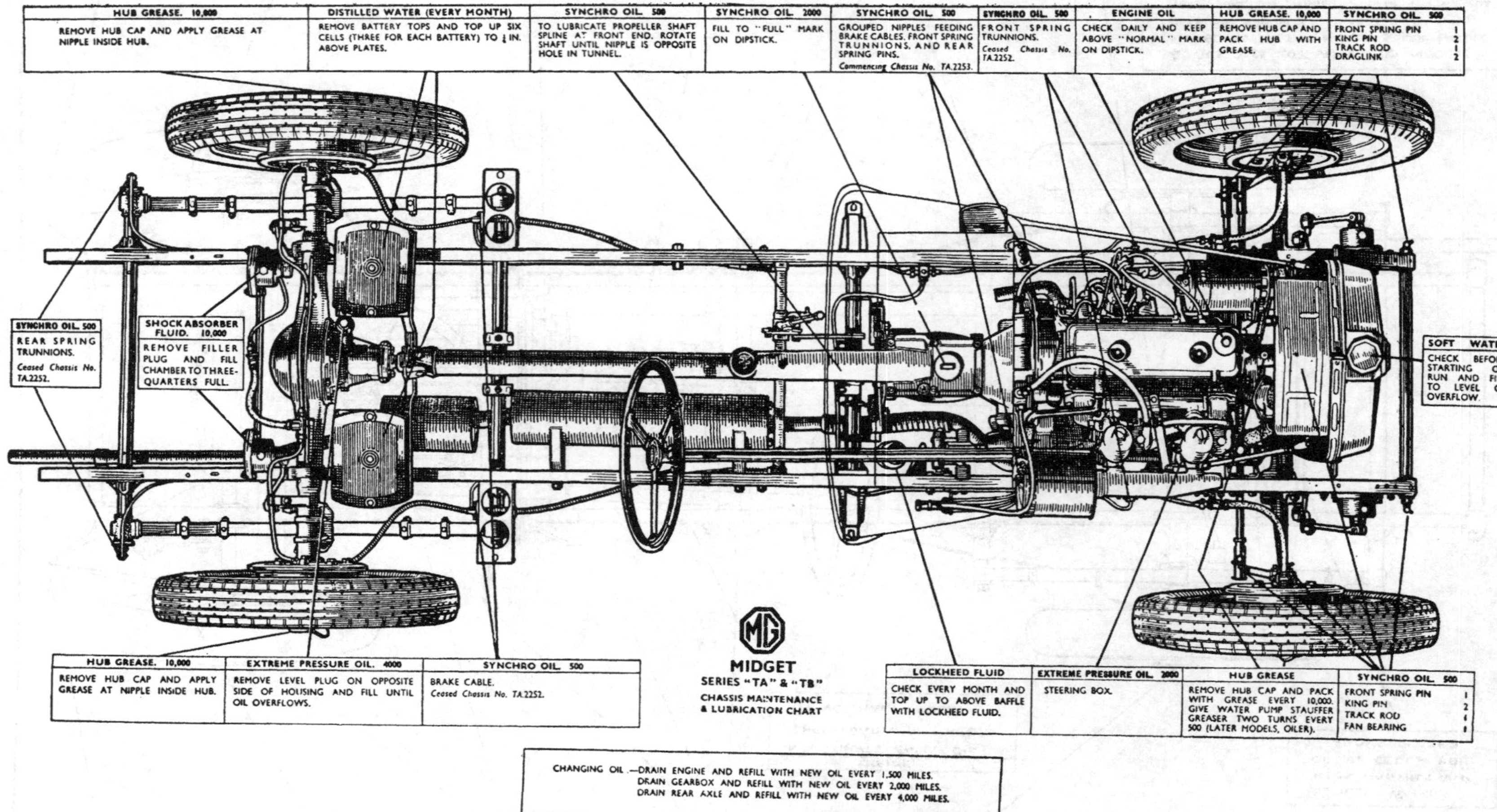
HUB GREASE. 10,000
REMOVE HUB CAP AND APPLY GREASE AT NIPPLE INSIDE HUB.
DISTILLED WATER (EVERY MONTH)
REMOVE BATTERY TOPS AND TOP UP SIX CELLS (THREE FOR EACH BATTERY) TO ¼ IN. ABOVE PLATES.
SYNCHRO OIL. 500
TO LUBRICATE PROPELLER SHAFT SPLINE AT FRONT END, ROTATE SHAFT UNTIL NIPPLE IS OPPOSITE HOLE IN TUNNEL.
SYNCHRO OIL. 2000
FILL TO "FULL" MARK ON DIPSTICK.
SYNCHRO OIL. 500
GROUPED NIPPLES FEEDING BRAKE CABLES, FRONT SPRING TRUNNIONS, AND REAR SPRING PINS.
Commencing Chassis No. TA.2253.
SYNCHRO OIL. 500
FRONT SPRING TRUNNIONS.
Ceased Chassis No. TA.2252.
ENGINE OIL
CHECK DAILY AND KEEP ABOVE "NORMAL" MARK ON DIPSTICK.
HUB GREASE. 10,000
REMOVE HUB CAP AND PACK HUB WITH GREASE.
SYNCHRO OIL. 500
FRONT SPRING PIN 1
KING PIN 2
TRACK ROD 1
DRAGLINK 2
SOFT WATER
CHECK BEFORE STARTING ON RUN AND FILL TO LEVEL OF OVERFLOW.
SYNCHRO OIL. 500
REAR SPRING TRUNNIONS.
Ceased Chassis No. TA.2252.
SHOCK ABSORBER FLUID. 10,000
REMOVE FILLER PLUG AND FILL CHAMBER TO THREE-QUARTERS FULL.
HUB GREASE. 10,000
REMOVE HUB CAP AND APPLY GREASE AT NIPPLE INSIDE HUB.
EXTREME PRESSURE OIL. 4000
REMOVE LEVEL PLUG ON OPPOSITE SIDE OF HOUSING AND FILL UNTIL OIL OVERFLOWS.
SYNCHRO OIL. 500
BRAKE CABLE.
Ceased Chassis No. TA.2252.
MG
MIDGET
SERIES "TA" & "TB"
CHASSIS MAINTENANCE & LUBRICATION CHART
LOCKHEED FLUID
CHECK EVERY MONTH AND TOP UP TO ABOVE BAFFLE WITH LOCKHEED FLUID.
EXTREME PRESSURE OIL. 2000
STEERING BOX.
HUB GREASE
REMOVE HUB CAP AND PACK WITH GREASE EVERY 10,000. GIVE WATER PUMP STAUFFER GREASER TWO TURNS EVERY 500 (LATER MODELS, OILER).
SYNCHRO OIL. 500
FRONT SPRING PIN 1
KING PIN 2
TRACK ROD 1
FAN BEARING 1
CHANGING OIL.—DRAIN ENGINE AND REFILL WITH NEW OIL EVERY 1,500 MILES.
DRAIN GEARBOX AND REFILL WITH NEW OIL EVERY 2,000 MILES.
DRAIN REAR AXLE AND REFILL WITH NEW OIL EVERY 4,000 MILES.

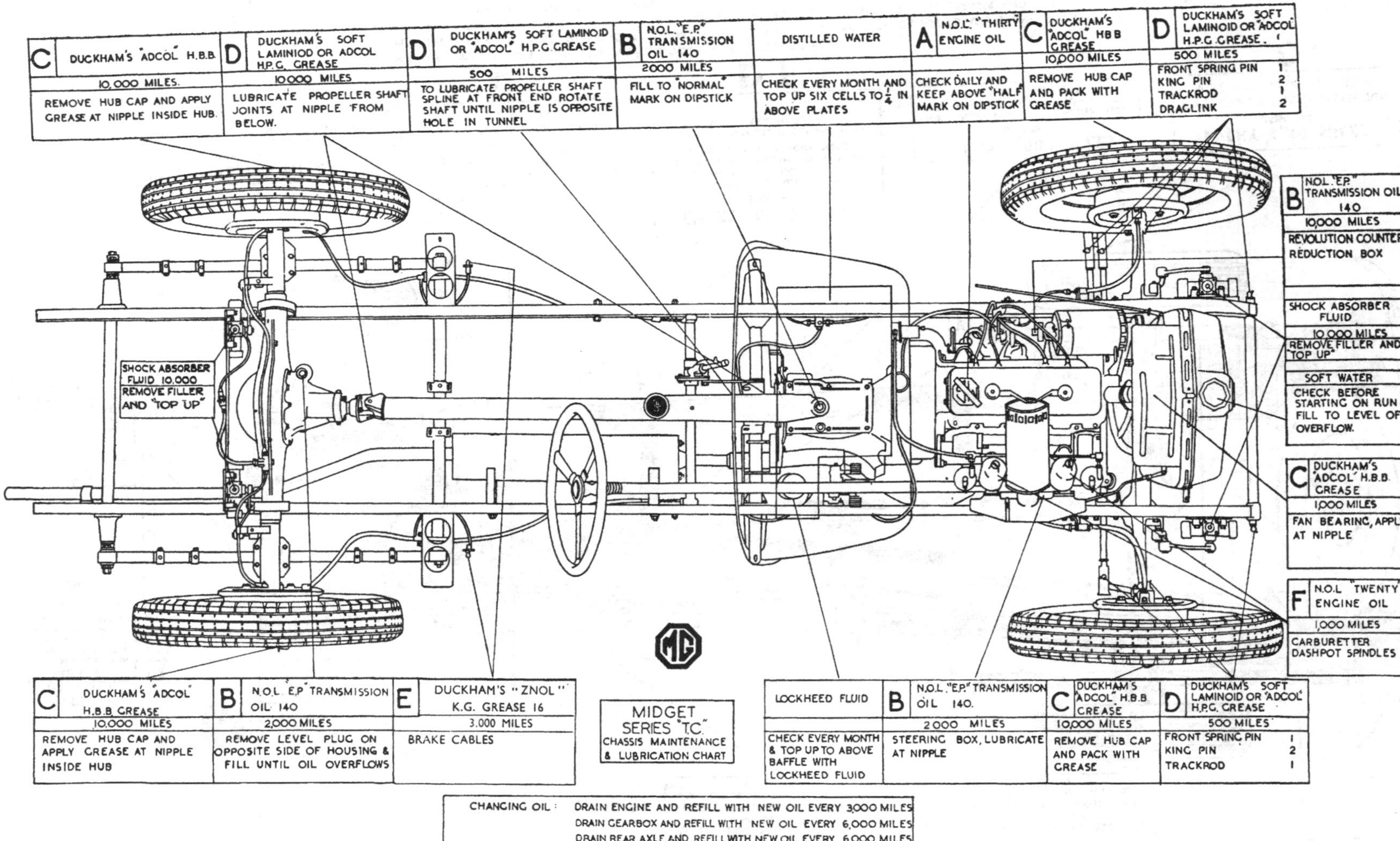
C DUCKHAM'S "ADCOL" H.B.B.
10,000 MILES.
REMOVE HUB CAP AND APPLY GREASE AT NIPPLE INSIDE HUB.
D DUCKHAM'S SOFT LAMINIOD OR ADCOL H.P.G. GREASE
10,000 MILES
LUBRICATE PROPELLER SHAFT JOINTS AT NIPPLE FROM BELOW.
D DUCKHAM'S SOFT LAMINOID OR "ADCOL" H.P.G. GREASE
500 MILES
TO LUBRICATE PROPELLER SHAFT SPLINE AT FRONT END ROTATE SHAFT UNTIL NIPPLE IS OPPOSITE HOLE IN TUNNEL
B N.O.L. "E.P." TRANSMISSION OIL 140
2000 MILES
FILL TO "NORMAL" MARK ON DIPSTICK
DISTILLED WATER
CHECK EVERY MONTH AND TOP UP SIX CELLS TO 1/4 IN ABOVE PLATES
A N.O.L. "THIRTY" ENGINE OIL
CHECK DAILY AND KEEP ABOVE "HALF" MARK ON DIPSTICK
C DUCKHAM'S "ADCOL" HBB GREASE
10,000 MILES
REMOVE HUB CAP AND PACK WITH GREASE
D DUCKHAM'S SOFT LAMINOID OR "ADCOL" H.P.G. GREASE.
500 MILES
FRONT SPRING PIN 1
KING PIN 2
TRACKROD 1
DRAGLINK 2
B N.O.L. "E.P." TRANSMISSION OIL 140
10,000 MILES
REVOLUTION COUNTER REDUCTION BOX
SHOCK ABSORBER FLUID
10,000 MILES
REMOVE FILLER AND "TOP UP"
SOFT WATER
CHECK BEFORE STARTING ON RUN & FILL TO LEVEL OF OVERFLOW.
C DUCKHAM'S "ADCOL" H.B.B. GREASE
1,000 MILES
FAN BEARING, APPLY AT NIPPLE
F N.O.L "TWENTY" ENGINE OIL
1,000 MILES
CARBURETTER DASHPOT SPINDLES
SHOCK ABSORBER FLUID 10,000 REMOVE FILLER AND "TOP UP"
C DUCKHAM'S "ADCOL" H.B.B. GREASE
10,000 MILES
REMOVE HUB CAP AND APPLY GREASE AT NIPPLE INSIDE HUB
B N.O.L. E.P. TRANSMISSION OIL 140
2,000 MILES
REMOVE LEVEL PLUG ON OPPOSITE SIDE OF HOUSING & FILL UNTIL OIL OVERFLOWS
E DUCKHAM'S "ZNOL" K.G. GREASE 16
3,000 MILES
BRAKE CABLES
MIDGET SERIES "T.C."
CHASSIS MAINTENANCE & LUBRICATION CHART
LOCKHEED FLUID
CHECK EVERY MONTH & TOP UP TO ABOVE BAFFLE WITH LOCKHEED FLUID
B N.O.L. "E.P." TRANSMISSION OIL 140.
2000 MILES
STEERING BOX, LUBRICATE AT NIPPLE
C DUCKHAM'S "ADCOL" H.B.B. GREASE
10,000 MILES
REMOVE HUB CAP AND PACK WITH GREASE
D DUCKHAM'S SOFT LAMINOID OR "ADCOL" H.P.G. GREASE
500 MILES
FRONT SPRING PIN 1
KING PIN 2
TRACKROD 1
CHANGING OIL: DRAIN ENGINE AND REFILL WITH NEW OIL EVERY 3,000 MILES
DRAIN GEARBOX AND REFILL WITH NEW OIL EVERY 6,000 MILES
DRAIN REAR AXLE AND REFILL WITH NEW OIL EVERY 6,000 MILES

THE M.G. MIDGET (Series "TD") LUBRICATION CHART

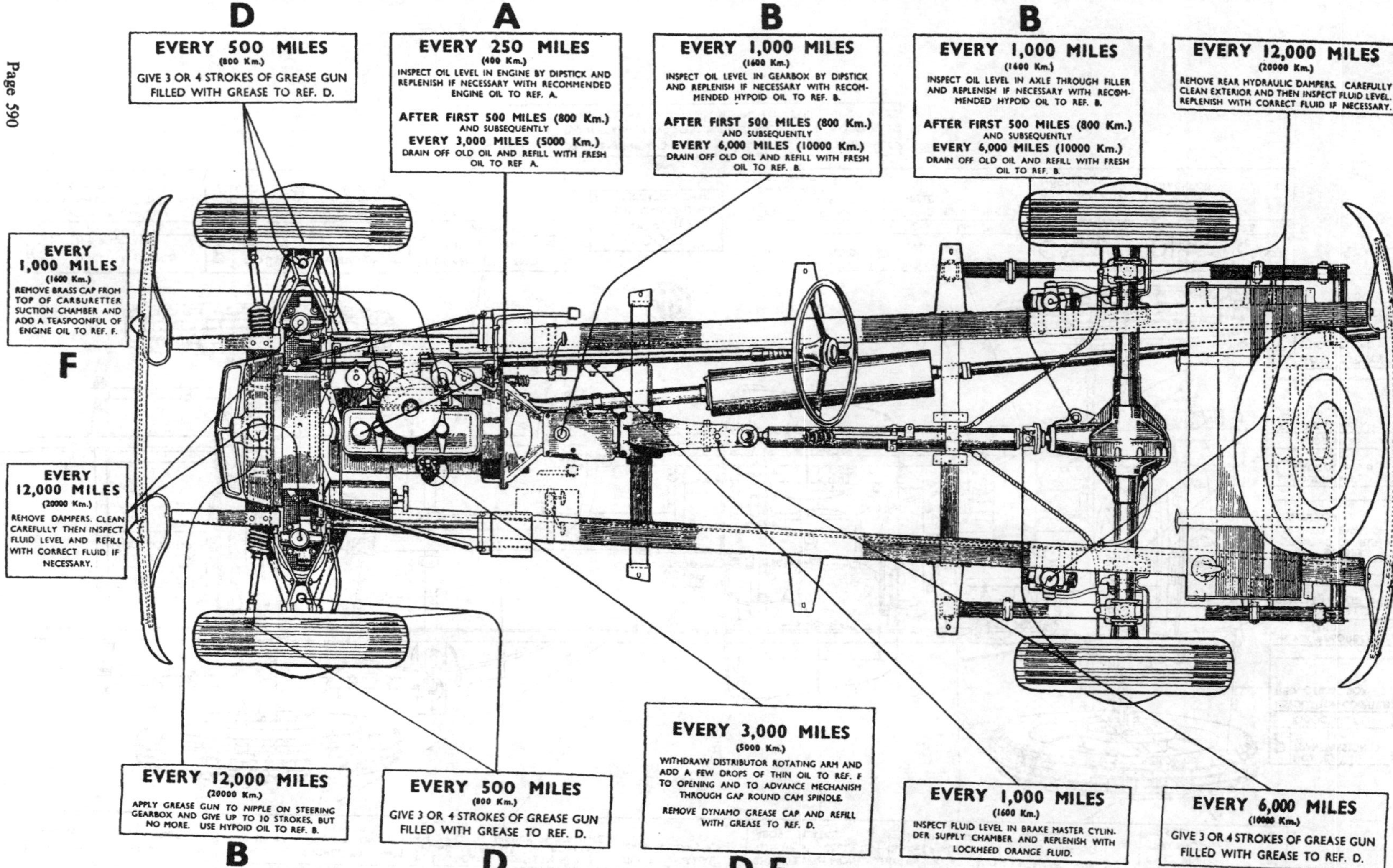

EVERY 1,000 MILES (1600 KM.). USE OILCAN ON ALL CONTROL JOINTS, DOOR LOCKS, HINGES TO REF. F.
EVERY 6,000 MILES (10000 KM.). REMOVE FRONT WHEEL HUB DISC FROM HUB. APPLY GREASE GUN FILLED WITH GREASE TO REF. C AND GIVE ONE STROKE.
EVERY 1,000 MILES (1600 KM.). APPLY GREASE GUN FILLED WITH GREASE TO REF. D TO PROPELLER SHAFT SLIDING JOINT GREASER AND GIVE TWO OR THREE STROKES.
EVERY 3,000 MILES (5000 KM.). CLEAN AND RE-OIL AIR INTAKE CLEANER.
EVERY 1,000 MILES (1600 KM.). APPLY GREASE GUN FILLED WITH GREASE TO REF. C TO FAN BEARING LUBRICATOR AND GIVE TWO STROKES.
EVERY 12,000 MILES (20000 KM.). APPLY GREASE GUN FILLED WITH GREASE TO REF. D TO ...

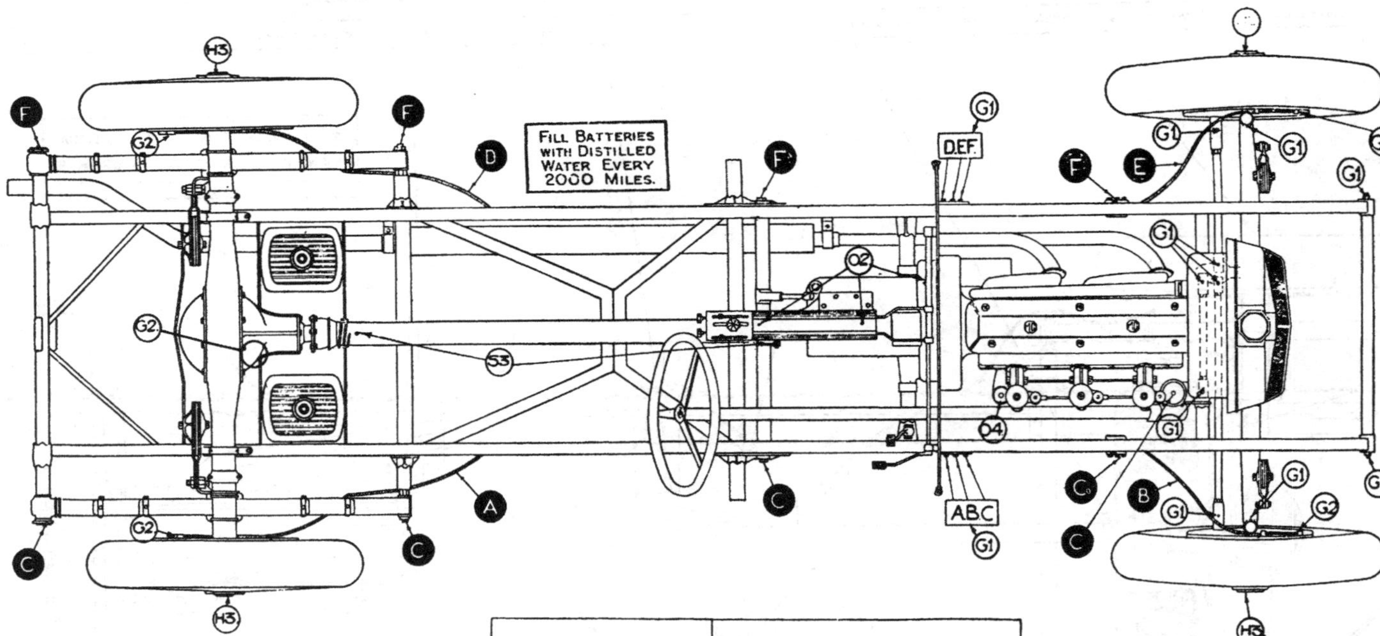

1	EVERY 500 MILES.	O. – OIL
2	„ 2000	G. – GEAR OIL
3	„ 10000 „	S. – HARDY SPICER GREASE.
4	AS REQUIRED.	H. – HUB GREASE.

ON CARS WITH THE MANUAL TYPE GEARBOX THE CLUTCH THRUST RACE SHOULD BE LUBRICATED WITH NON-SEPARATING GREASE EVERY 2000 MILES.

Lubrication Chart for K1, KA, KB and KD type Magnettes.

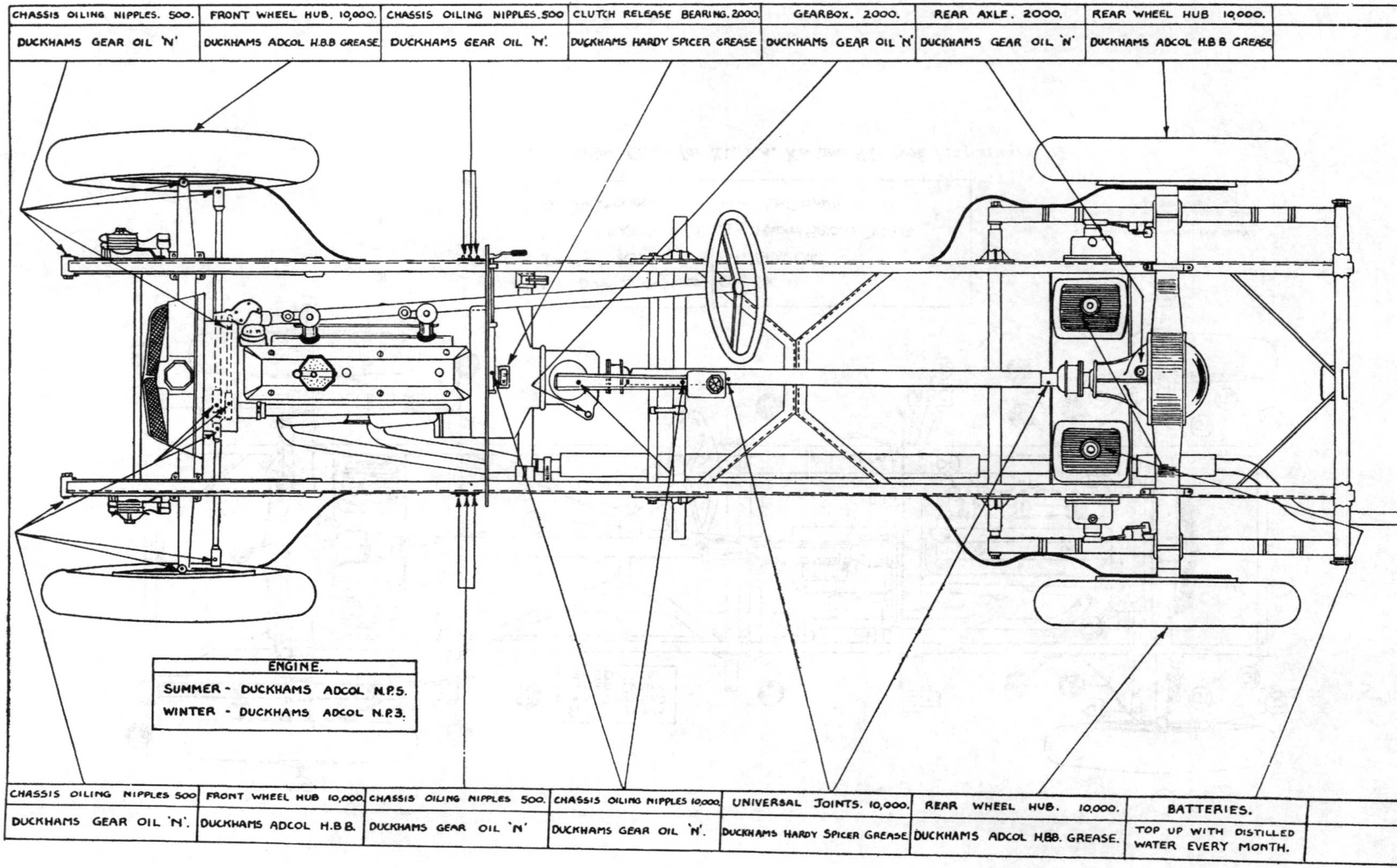

Lubrication Chart for KN Magnette

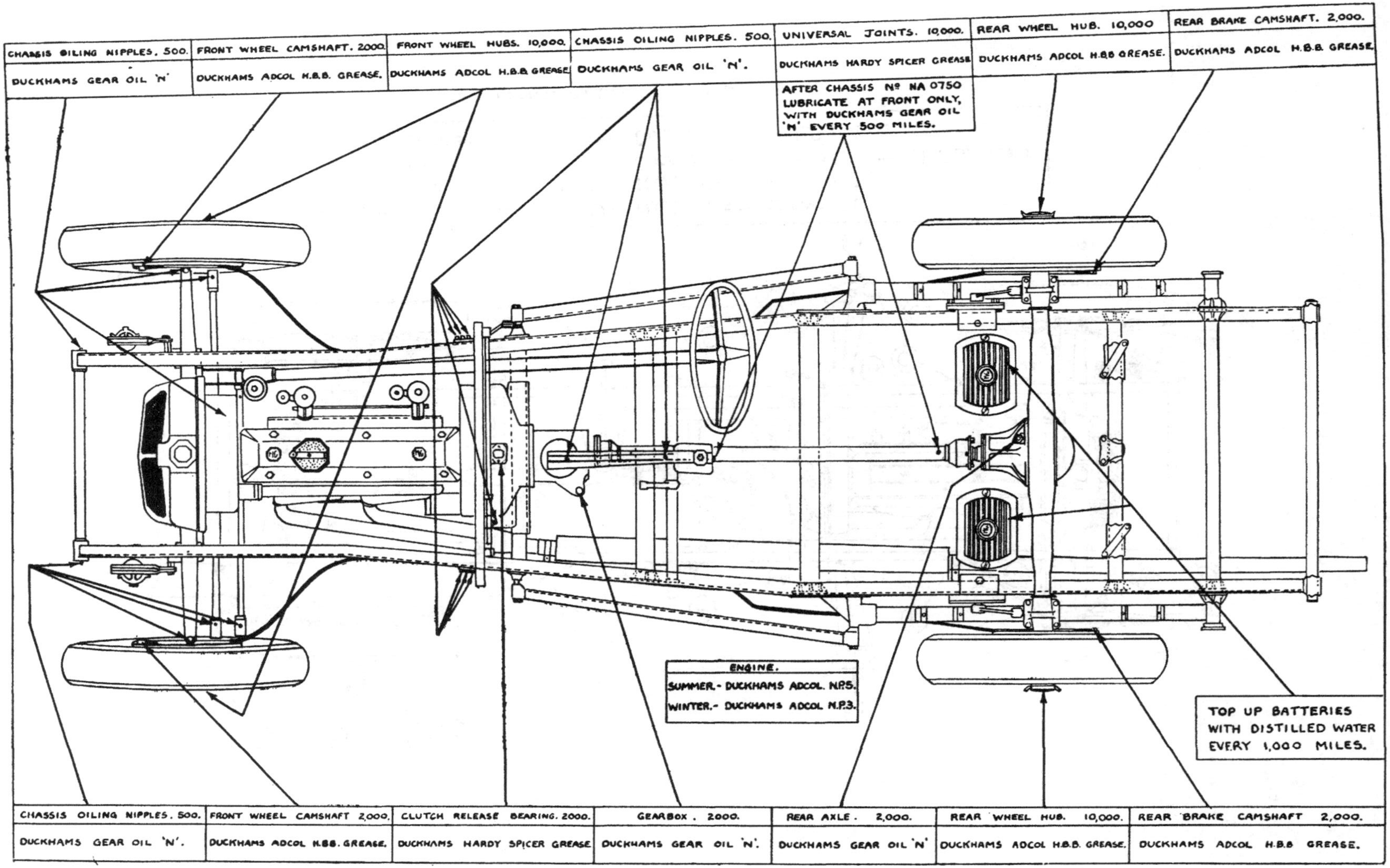

Lubrication Chart for NA Magnette.

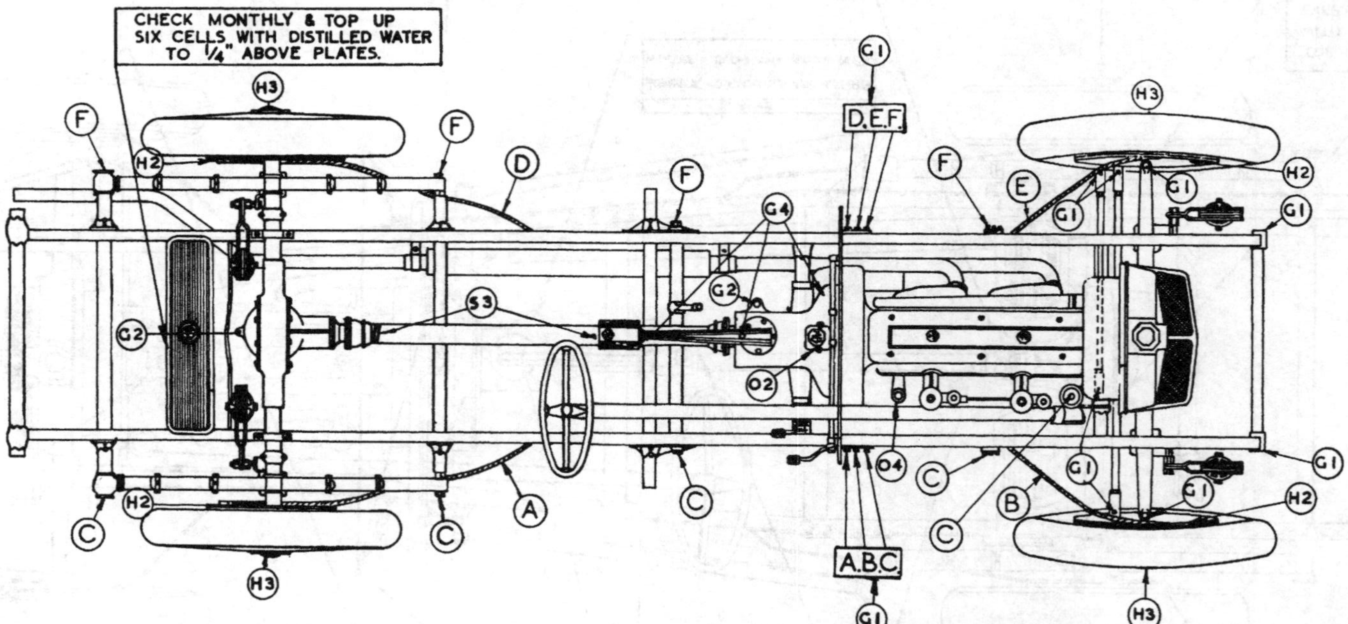

1 Every 500 Miles	O — N.O.L. Thirty Engine Oil
2 " 2000 "	G — N.O.L. 'EP' Transmission Oil 140
3 " 10000 "	S — Duckham's Laminoid Soft or 'Adcol' H.P.G. Grease
4 As Required.	H — Duckham's 'Adcol' H.B.B. Grease

Lubrication Chart for L type Magna.

Oiling Chart of the 12/70 M.G. Magna Chassis.

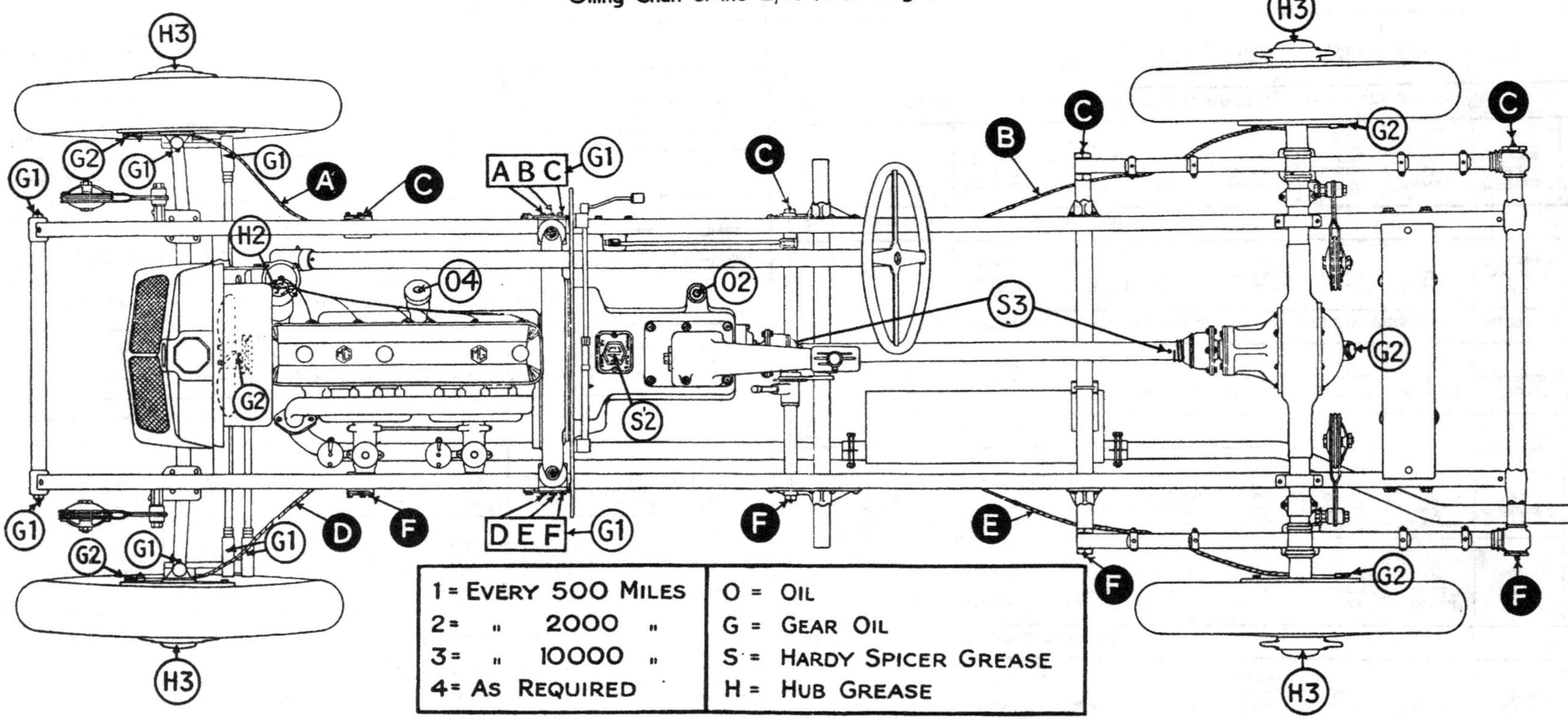

Lubrication Chart for F type Magna.

KEY TO RECOMMENDED LUBRICANTS

Climatic Conditions	Duckham's Alexander Duckham & Co. Ltd.	Castrol C. C. Wakefield & Co. Ltd.	Essolube Anglo-American Oil Co. Ltd.	Mobiloil Vacuum Oil Co. Ltd.	Shell Shell Mex & B.P. Ltd.	Energol Price's Lubricants Ltd.	Filtrate Edward Joy & Sons Ltd.	Sternol Sternol Ltd.
A ENGINE AND AIR CLEANER								
Tropical and Temperate down to 32° F. (0° C.)	Duckham's N.O.L. "Thirty"	"Castrol" X.L.	"Essolube" 30	Mobiloil "A"	Double "Shell"	"Energol" S.A.E. 30	Medium "Filtrate"	"Sternol" W.W. 30
Cold and extreme cold down to 0° F. (−18° C.)	Duckham's N.O.L. "Twenty"	"Castrolite"	"Essolube" 20	Mobiloil "Arctic"	Single "Shell"	"Energol" S.A.E. 20	Zero "Filtrate"	"Sternol" W.W. 20
Arctic below 0° F. (−18° C.)	Duckham's N.O.L. "Ten"	"Castrol" Z	"Essolube" 10	Mobiloil "Arctic" Special	Silver "Shell"	"Energol" S.A.E. 10	Sub-Zero "Filtrate"	"Sternol" W.W. 10
B GEARBOX, STEERING GEARBOX AND REAR AXLE (HYPOID GEARS)								
Tropical and Temperate down to 10° F. (−12° C.)	Duckham's Hypoid 90	"Castrol" Hypoy	"Esso" Expee Compound 90	Mobilube "G.X." 90	"Shell" Spirax 90 E.P	"Energol" "E.P." S.A.E. 90	Hypoid "Filtrate" 90	"Sternol" Liquid Ambroleum Hypoid 90
Extreme cold below 10° F. (−12° C.)	Duckham's Hypoid 80	"Castrol" Hypoy 80	"Esso" Expee Compound 80	Mobilube "G.X." 80	"Shell" Spirax 80 E.P.	"Energol" "E.P." S.A.E. 80	Hypoid "Filtrate" 80	"Sternol" Liquid Ambroleum Hypoid 80
C WHEEL HUBS AND FAN BEARINGS								
All conditions	Duckham's H.B.B. Grease	"Castrolease" Heavy	Essogrease	Mobil Hub Grease	"Shell" Retinax R.B.	"Belmoline" C	"Filtrate" R.B. Grease	"Ambroline" R.B. Grease
D STEERING CONNECTIONS, KING-PINS, PROPELLER SHAFT, CLEVIS PINS AND LEVER FULCRUMS								
All conditions	Duckham's H.P.G. Grease	"Castrolease" Medium	"Esso" Pressure Gun Grease	Mobilgrease No. 2 or 4	"Shell" Retinax C	"Belmoline" D	H.P. Solidified "Filtrate"	"Ambroline" M.M. Grease
E CABLES AND VITAL CONTROL JOINTS								
All conditions	Duckham's "Keenol" K.G. 16 Grease	"Castrolease" Brake Cable Grease	"Esso" Pressure Gun Grease	Mobilgrease No. 2 or 4	"Shell" Retinax C	"Belmoline" D	"Filtrate" A.F. Grease	"Ambroline" A.F. Grease
F UTILITY LUBRICANT, S.U. CARBURETTER DASHPOT, OILCAN POINTS, ETC.								
All conditions	Duckham's N.O.L. "Twenty"	"Castrolite"	"Essolube" 20	Mobiloil "Arctic"	Single "Shell"	"Energol" S.A.E. 20	Zero "Filtrate"	"Sternol" W.W. 20

EVERY 1,000 MILES (1600 Km.). Use oilcan on all control joints, door locks, hinges.

SPECIAL NOTE. Only Hypoid oils must be used in the rear axle.

Key to recommended Lubricants detailed on Lubrication Chart for TD Midget.

KEY TO RECOMMENDED LUBRICANTS

Climatic Conditions	Duckham's Alexander Duckham & Co. Ltd.	Castrol C. C. Wakefield & Co. Ltd.	Essolube Anglo-American Oil Co. Ltd.	Mobiloil Vacuum Oil Co. Ltd.	Shell Shell Mex & B.P. Ltd.	Energol Price's Lubricants Ltd.	Filtrate Edward Joy & Son Ltd.	Sternol Sternol Ltd.
A ENGINE AND AIR CLEANER								
Tropical and Temperate down to 32° F. (0° C.)	Duckham's N.O.L. "Thirty"	"Castrol" X.L.	"Essolube" 30	Mobiloil "A"	Double "Shell"	"Energol" S.A.E. 30	Medium "Filtrate"	"Sternol" W.W. 30
Cold and extreme cold down to 0° F. (−18° C.)	Duckham's N.O.L. "Twenty"	"Castrol-ite"	"Essolube" 20	Mobiloil "Arctic"	Single "Shell"	"Energol" S.A.E. 20	Zero "Filtrate"	"Sternol" W.W. 20
Arctic—below 0° F. (−18° C.)	Duckham's N.O.L. "Ten"	"Castrol" Z	"Essolube" 10	Mobiloil "Arctic" Special	Silver "Shell"	"Energol" S.A.E. 10	Sub-Zero "Filtrate"	"Sternol" W.W. 10
B REAR AXLE, GEARBOX, STEERING GEAR								
Tropical and Temperate down to 10° F. (−12° C.)	Duckham's N.O.L. "E.P." Transmission 140	"Castrol" Hi-Press	"Esso" Expee Compound 140	Mobilube "G.X." 140	"Shell" Spirax 140 E.P.	"Energol" S.A.E. 140	E.P. "Filtrate"	"Sternol" Liquid Ambroleum E.P. 140
Extreme cold below 10° F. (−12° C.)	Duckham's N.O.L. "E.P." Transmission 80	"Castrol" Hypoy 80	"Esso" Expee Compound 80	Mobilube "G.X." 80	"Shell" Spirax 80 E.P	"Energol" "E.P." S.A.E. 80	E.P. "Filtrate" 80	"Sternol" Liquid Ambroleum E.P. 80
C WHEEL HUBS AND FAN BEARINGS								
All conditions	Duckham's H.B.B. Grease	"Castrol-ease" Heavy	Essogrease	Mobil Hub Grease	"Shell" Retinax R.B.	"Belmoline" C	"Filtrate" R.B. Grease	"Ambroline" R.B. Grease
D STEERING CONNECTIONS, KING-PINS, PROPELLER SHAFT, CLEVIS PINS AND LEVER FULCRUMS								
All conditions	Duckham's H.P.G. Grease	"Castrol-ease" Medium	"Esso" Pressure Gun Grease	Mobilgrease No. 2 or 4	"Shell" Retinax C	"Belmoline" D	H.P. Solidified "Filtrate"	"Ambroline" M.M. Grease
E CABLES AND VITAL CONTROL JOINTS								
All conditions	Duckham's "Keenol" K.G. 16 Grease	"Castrol-ease" Brake Cable Grease	"Esso" Pressure Gun Grease	Mobilgrease No. 2 or 4	"Shell" Retinax C	"Belmoline" D	"Filtrate" A.F. Grease	"Ambroline" A.F. Grease
F UTILITY LUBRICANT, S.U. CARBURETTER DASHPOT, OILCAN POINTS, ETC.								
All conditions	Duckham's N.O.L. "Twenty"	"Castrol-ite"	"Essolube" 20	Mobiloil "Arctic"	Single "Shell"	"Energol" S.A.E. 20	Zero "Filtrate"	"Sternol" W.W. 20

EVERY 1,000 MILES (1600 Km.). Use oilcan on all control joints, door locks, hinges.

Key to recommended Lubricants detailed on Lubrication Chart for 1¼ litre Saloon.

F	C	A	DISTILLED WATER	B	D	E	HYDRAULIC DAMPER FLUID C	C
1,000 MILES.	6,000 MILES			2,000 MILES	1,000 MILES	6,000 MILES	12,000 MILES	6,000 MILES
CARBURETTER DASHPOT SPINDLES.	REMOVE HUB DISC & GREASE AT NIPPLE.	CHECK DAILY & KEEP ABOVE "HALF" MARK ON DIPSTICK.	CHECK EVERY MONTH AND TOP UP SIX CELLS LEVEL WITH THE TOPS OF THE SEPARATORS	FILL TO "NORMAL" MARK ON DIPSTICK.	TO LUBRICATE PROP SHAFT JOINTS, ROTATE UNTIL NIPPLE IS OPPOSITE HOLE IN TUNNEL.	BRAKE CABLES.	REMOVE FILLER & TOP UP	REMOVE HUB DISC & GREASE AT NIPPLE.

D — 1,000 MILES. KING PIN (TOP & BOTTOM) OUTER STEERING BALL SOCKET.

C — 1,000 MILES. FAN BEARING, APPLY AT NIPPLE.

SOFT WATER. CHECK BEFORE STARTING ON RUN FILL TO OVERFLOW LEVEL.

B — 2,000 MILES. STEERING BOX, LUBRICATE AT NIPPLE.

D — 1,000 MILES. KING PIN (TOP & BOTTOM) OUTER STEERING BALL SOCKET.

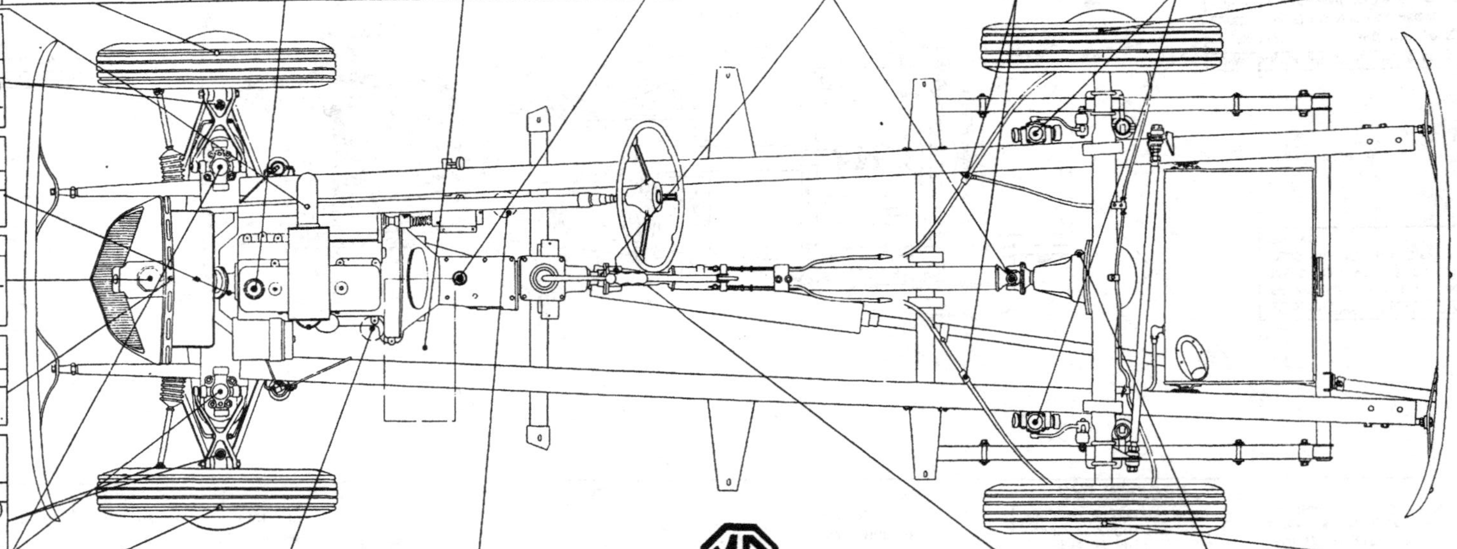

HYDRAULIC DAMPER FLUID.	C	"RED" JACKALL FLUID.	LOCKHEED FLUID.
12,000 MILES	6,000 MILES		
REMOVE FILLER AND TOP UP.	REMOVE HUB DISC & GREASE AT NIPPLE.	TOP UP SUPPLY TANK MONTHLY	CHECK MONTHLY AND TOP UP

MG 1¼ LITRE SERIES Y

CHASSIS MAINTENANCE & LUBRICATION CHART.

D	B	C
1,000 MILES.	2,000 MILES.	6,000 MILES
TO LUBRICATE PROP SHAFT SPLINE, ROTATE UNTIL NIPPLE IS OPPOSITE HOLE IN TUNNEL.	FILL TO LEVEL INDICATED ON DIPSTICK.	REMOVE HUB DISC & GREASE AT NIPPLE

CHANGING OILS :— DRAIN ENGINE AND REFILL WITH NEW OIL EVERY 3,000 MILES (5000 Km.)
DRAIN GEARBOX AND REFILL WITH NEW OIL EVERY 6,000 MILES (10000 Km.)
DRAIN REAR AXLE AND REFILL WITH NEW OIL EVERY 6,000 MILES (10000 Km.)
FIT NEW EXTERNAL OIL FILTER EVERY 12,000 MILES (20000 Km.)

NOTE :— FOR KEY TO LETTER REFERENCES, SEE LIST OF RECOMMENDED OILS IN LUBRICATION SECTION.

1½ LITRE (VA TYPE)
CHASSIS MAINTENANCE CHART

GEAR OIL 1000	GEAR OIL 2000	GEAR OIL 500	LOCKHEED FLUID	SYNCHRO-OIL 2000	GEAR OIL 500	GEAR OIL 1000
REMOVE WHEELS AND APPLY OILGUN, GIVE TWO OR THREE STROKES ONLY.	REMOVE PLUG FROM STEERING BOX AND FILL TO LEVEL OF PLUG.	GROUPED NIPPLES, FEEDING BRAKE CABLES, REAR SPRING SHACKLES AND FRONT SPRING PIN.	CHECK EVERY MONTH AND TOP UP TO LEVEL MARK ON CONTAINER. USE LOCKHEED ORANGE FLUID.	DRAIN AND REFILL WITH SYNCHRO-OIL TO 'FULL' MARK ON DIPSTICK.	TO LUBRICATE PROPELLER SHAFT SPLINES AT FRONT END ROTATE SHAFT UNTIL NIPPLE IS POINTING TO THE GROUND AND APPLY GUN FROM UNDERNEATH.	REMOVE WHEELS AND APPLY OILGUN, GIVE TWO OR THREE STROKES ONLY.

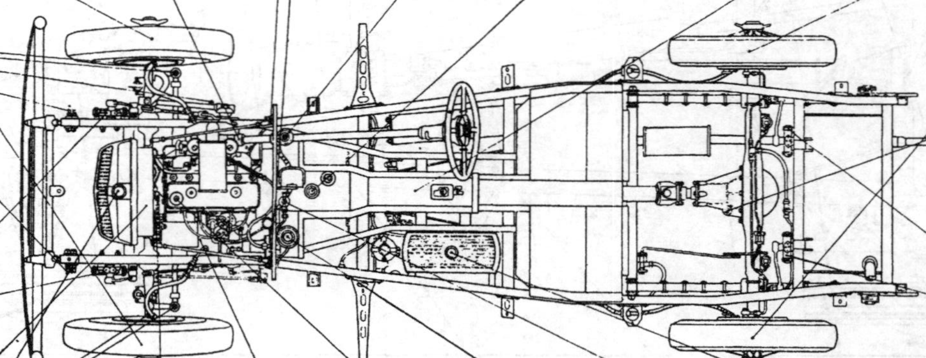

GEAR OIL 500
FRONT SHACKLES 2
TRACK ROD 1
KING PIN 2
DRAGLINK 1

LUVAX SHOCK ABSORBER FLUID 8000
CHECK EVERY MONTH AND TOP UP TO LEVEL OF PLUG, DO NOT FILL ABOVE THIS LEVEL (SEE MAINTENANCE OF SHOCK ABSORBERS IN INSTRUCTION MANUAL).

EXTREME PRESSURE OIL 4000
DRAIN AND REFILL EVERY 4000 MILES TO LEVEL OF FILLER PLUG. DO NOT MIX DIFFERENT BRANDS.

LUVAX SHOCK ABSORBER FLUID 8000
CHECK EVERY MONTH AND TOP UP TO LEVEL OF PLUG. DO NOT FILL ABOVE THIS LEVEL (SEE MAINTENANCE OF SHOCK ABSORBERS IN INSTRUCTION MANUAL)

GEAR OIL 500	SOFT WATER	ENGINE OIL 500	VASALINE	ENGINE OIL	JACKALL FLUID	SPECIAL OIL	DISTILLED WATER (EVERY MONTH)
FRONT SHACKLES 2 TRACK ROD 1 KING PIN 2 FAN BEARING 1	CHECK DAILY AND TOP UP. PREFERABLY WITH SOFT WATER.	FILL THE SPRING CAP OILER ON WATER PUMP.	REMOVE CAP OF LUBRICATOR ON DYNAMO AND REFILL YEARLY.	CHECK DAILY AND KEEP ABOVE ¾ MARK. RENEW EVERY 1500 MILES.	CHECK EVERY MONTH AND TOP UP TO LEVEL MARK ON CONTAINER. THIS MUST BE DONE WITH	CHECK EVERY MONTH AND TOP UP TO WITHIN ONE INCH OF TOP.	REMOVE VENT PLUG AND TOP UP CELLS TO A LEVEL ¼" ABOVE PLATES.

Ref.	Lubricant	Interval	Instructions
C	DUCKHAM'S "ADCOL" H.B.B. GREASE.	10,000 MILES	REMOVE HUB CAP AND APPLY GREASE AT NIPPLE INSIDE HUB.
	DISTILLED WATER		CHECK EVERY MONTH AND TOP UP SIX CELLS TO 1/4 IN ABOVE PLATES.
D	DUCKHAM'S LAMINOID SOFT OR "ADCOL" H.P.G. GREASE.	500 MILES.	TO LUBRICATE PROPELLER SHAFT SPLINE AT FRONT END, ROTATE SHAFT UNTIL NIPPLE IS OPPOSITE HOLE IN TUNNEL.
E	DUCKHAM'S "ZNOL" K.G. GREASE 16.	3,000 MILES.	REAR BRAKE CABLES.
D	DUCKHAMS LAMINOID SOFT OR "ADCOL" H.P.G. GREASE.	500 MILES.	FRONT SPRING PIN. REAR SPRING SHACKLE.
	JACKALL FLUID.		CHECK MONTHLY AND TOP TO LEVEL OF MARK ON CONTAINER, RELEASE VALVE OPEN.
A	N.O.L. "THIRTY" ENGINE OIL	SEE BELOW.	CHECK DAILY AND KEEP ABOVE 3/4 MARK ON DIPSTICK
C	DUCKHAM'S "ADCOL" H.B.B. GREASE.	1,000 MILES.	FAN BEARING AND WATER PUMP, APPLY AT GREASER.
C	DUCKHAM'S "ADCOL" H.B.B. GREASE.	10,000 MILES.	REMOVE HUB CAP AND PACK WITH GREASE.
D	DUCKHAM'S LAMINOID SOFT OR "ADCOL" H.P.G. GREASE.	500 MILES.	FRONT SHACKLES 2. TRACK ROD. 1. KING PIN. 2.

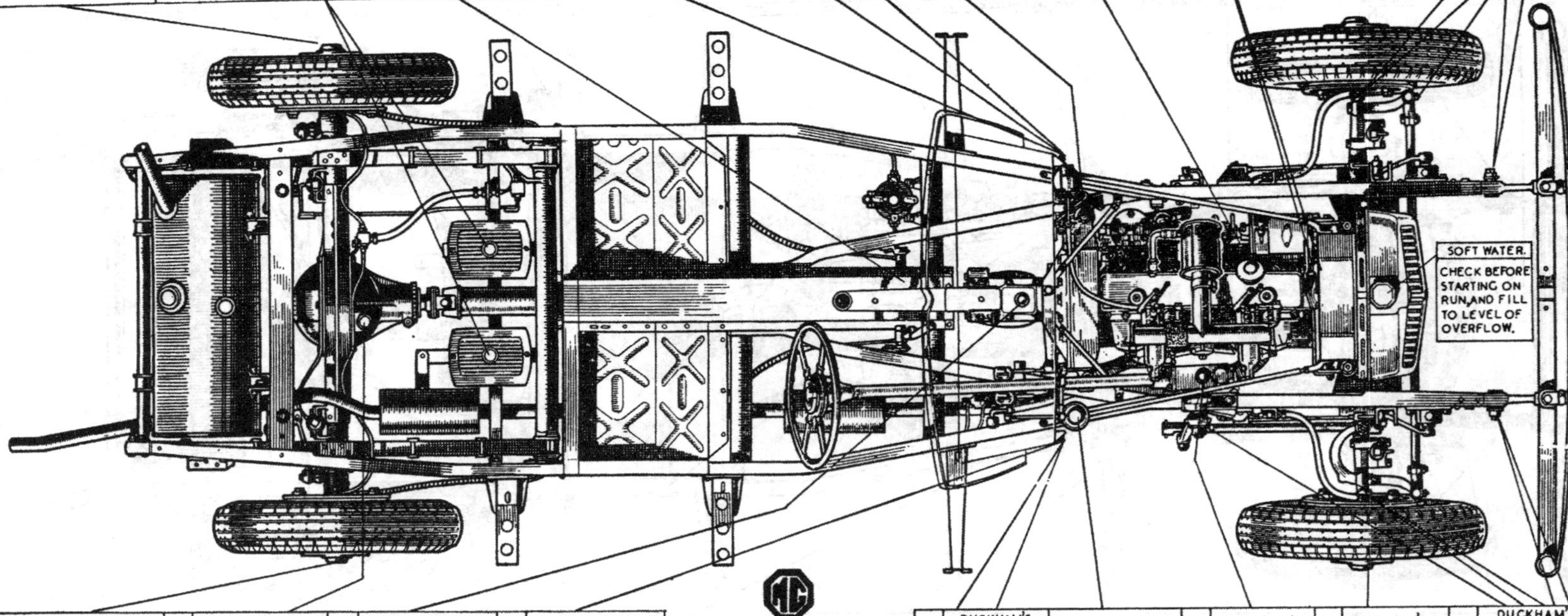

Ref.	Lubricant	Interval	Instructions
C	DUCKHAM'S "ADCOL" H.B.B. GREASE.	10,000 MILES	REMOVE HUB CAP AND APPLY GREASE AT NIPPLE INSIDE HUB.
B	N.O.L. "E.P." TRANSMISSION OIL 140.	2,000 MILES	FILL TO "FULL" MARK ON DIPSTICK. DO NOT OVER-FILL.
B	N.O.L. "E.P." TRANSMISSION OIL 140.	2,000 MILES	FILL TO "FULL" MARK ON DIPSTICK.
E	DUCKHAM'S "ZNOL" K.G. GREASE 16.	3,000 MILES.	REAR BRAKE CABLES.

2 LITRE SERIES "S"
CHASSIS MAINTENANCE AND LUBRICATION CHART.

Ref.	Lubricant	Interval	Instructions
D	DUCKHAM'S LAMINOID SOFT OR "ADCOL" H.P.G. GREASE.	500 MILES.	REAR SPRING SHACKLE. FRONT SPRING PIN
	LOCKHEED FLUID		CHECK MONTHLY AND TOP TO LEVEL OF MARK ON CONTAINER.
B	N.O.L. "E.P." TRANSMISSION OIL 140.	2,000 MILES.	REMOVE NUT AND TOP TO LEVEL OF FILLER.
C	DUCKHAM'S "ADCOL" H.B.B. GREASE.	10,000 MILES.	REMOVE HUB CAP AND PACK WITH GREASE.
D	DUCKHAM'S LAMINOID SOFT OR "ADCOL" H.P.G. GREASE.	500 MILES.	FRONT SHACKLES. 2. TRACK ROD. 1. DRAG LINK 1 KING PIN. 2.

CHANGING OIL:— DRAIN ENGINE AND REFILL WITH NEW OIL EVERY 3,000 MILES.
DRAIN GEARBOX AND REFILL WITH NEW OIL EVERY 6,000 MILES.
DRAIN REAR AXLE AND REFILL WITH NEW OIL EVERY 6,000 MILES.

NOTE:— FOR KEY TO LETTER REFERENCES SEE LIST OF RECOMMENDED OILS IN LUBRICATION SECTION.

M.G. 2.6 LITRE CHASSIS MAINTENANCE CHART

LUVAX PISTON TYPE THIN FLUID. 8000	EXTREME PRESSURE OIL. 2000	EXTREME PRESSURE OIL. 10,000	JACKALL FLUID	ENGINE OIL	VASELINE	EXTREME PRESSURE OIL. 2000	EXTREME PRESSURE OIL. 500	LUVAX PISTON TYPE THIN FLUID. 8000
CHECK AND TOP UP TO LEVEL OF PLUG (SEE MAINTENANCE OF SHOCK ABSORBERS IN INSTRUCTION MANUAL)	REMOVE HUB CAP AND APPLY OIL-GUN SPARINGLY AT NIPPLE INSIDE HUB	PROPELLER SHAFT JOURNALS. ROTATE SHAFT UNTIL NIPPLE POINTS TO GROUND. (CENTRE NIPPLE FITTED UP TO CHASSIS NO WA.0495 ONLY.)	CHECK EVERY MONTH AND TOP UP TO LEVEL MARK ON CONTAINER. THIS MUST BE DONE WITH RELEASE VALVE OPEN	CHECK DAILY AND KEEP ABOVE ½ MARK RENEW EVERY 1500 MILES.	REMOVE CAP OF LUBRICATOR ON DYNAMO AND REFILL YEARLY	REMOVE WHEEL AND APPLY OILGUN SPARINGLY AT NIPPLE ON HUB	FRONT SHACKLES 2 TRACK ROD 1 KING PIN 2 FAN BEARING 1	CHECK AND TOP UP TO LEVEL OF PLUG (SEE MAINTENANCE OF SHOCK ABSORBERS IN INSTRUCTION MANUAL.)

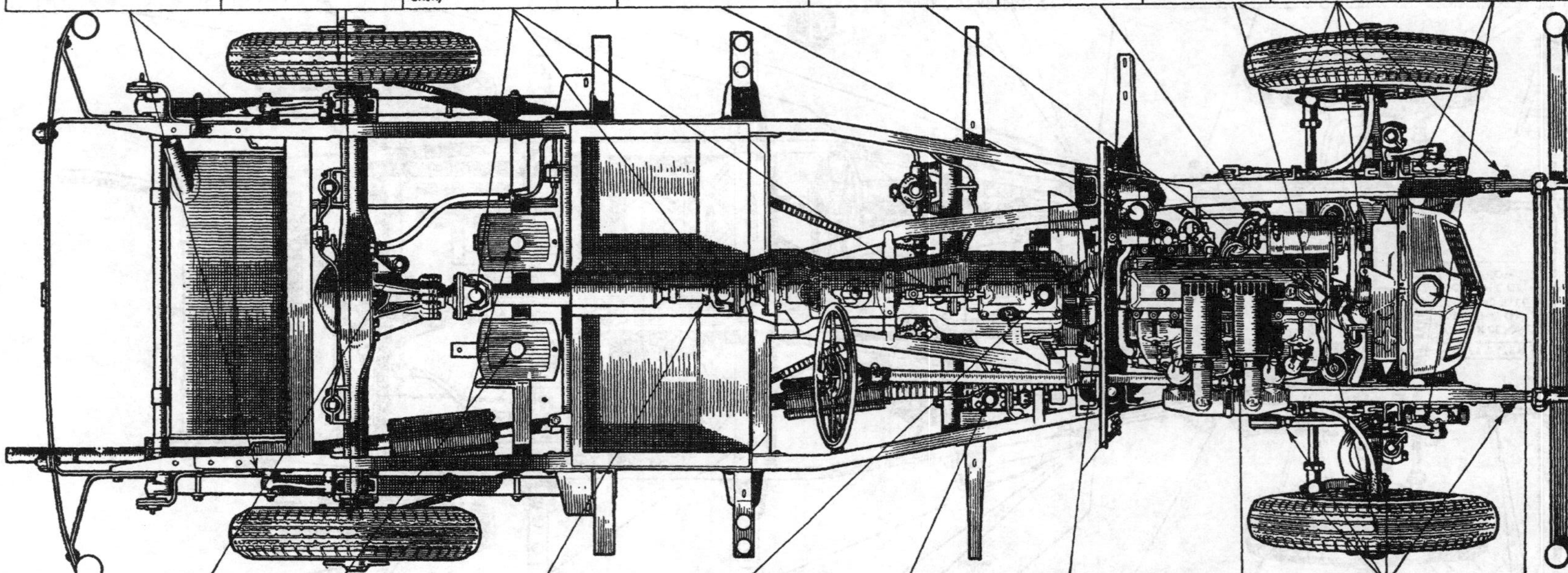

EXTREME PRESSURE OIL. 4000	DISTILLED WATER (MONTHLY)	EXTREME PRESSURE OIL. 500	SYNCHRO-GEAR OIL. 2000	LOCKHEED ORANGE FLUID	EXTREME PRESSURE OIL. 500	EXTREME PRESSURE OIL. 2000	EXTREME PRESSURE OIL. 500	SOFT WATER
DRAIN AND REFILL EVERY 4000 MILES TO LEVEL OF FILLER PLUG DO NOT MIX DIFFERENT BRANDS	REMOVE VENT PLUGS AND TOP UP CELLS TO ¼ IN ABOVE PLATES	PROPELLER SHAFT SPLINES. TO LUBRICATE SPLINES ROTATE SHAFT UNTIL NIPPLE POINTS TO GROUND. (NIPPLE BEHIND GEARBOX COMMENCING CHASSIS NO. WA.0496.)	DRAIN AND REFILL WITH SYNCHRO-GEAR OIL TO "FULL" MARK ON DIP-STICK.	CHECK EVERY MONTH AND TOP UP TO WITHIN ¼ IN OF TOP COVER.	GROUPED NIPPLES FEEDING BRAKE CABLES, REAR SPRING SHACKLES, FRONT SPRING PINS. ALSO PROPELLER SHAFT CENTRE BEARING UP TO CHASSIS NO. WA.0495.	REMOVE PLUG FROM STEERING BOX AND FILL TO LEVEL OF PLUG	FRONT SHACKLES 2 TRACK ROD 1 KING PIN 2 DRAGLINK 2	CHECK DAILY AND TOP UP TO LEVEL OF OVERFLOW

Selected Titles From Bentley Publishers

Engineering

Robert Bentley's Repair Manual for British Cars *John Organ* ISBN 978-0-8376-0041-3

Vintage Racing British Sports Cars *Terry Jackson* ISBN 978-0-8376-0153-3

Scientific Design of Exhaust & Intake Systems *Philip H. Smith and John C. Morrison* ISBN 978-0-8376-0309-4

The Design and Tuning of Competition Engines *Philip H. Smith* ISBN 978-0-8376-0140-3

New Directions in Suspension Design: Making the Car Faster *Colin Campbell* ISBN 978-0-8376-0150-2

Racing Car Design and Development *Len Terry with Alan Baker* ISBN 978-0-8376-0080-2

Austin-Healey

Austin-Healey 100/6 NS 3000 Workshop Manual: 1956-1968
British Leyland Motors
ISBN 978-0-8376-0133-5
Softcover, 416 pages
Includes: Workshop Manual

The Complete Official 1275 cc Sprite/Midget: 1967-1974
British Leyland Motors
ISBN 978-0-8376-1748-0
Hardcover, 400 pages

MG

The Complete Official MGB: 1962-1974 *British Leyland Motors*
ISBN 978-0-8376-1761-9
Hardcover, 480 pages
Includes: Driver's Handbook, Workshop Manual, Special Tuning Manual

The Complete Official MGB: 1975-1980 *British Leyland Motors*
ISBN 978-0-8376-1760-2
Hardcover, 304 pages
Includes: Driver's Handbook, Workshop Manual

The Complete Official MG Midget 1500: 1975-1979
British Leyland Motors
ISBN 978-0-8376-0131-1
Softcover, 384 pages
Includes: Driver's Handbook, Workshop Manual

The MG Workshop Manual: 1929-1955 *W.E. Blower*
ISBN 978-0-8376-1746-6
Hardcover, 608 pages

Triumph

The Complete Official Triumph GT6, GT6+ & GT6 MK III: 1967-1973 *British Leyland Motors*
ISBN 978-0-8376-1749-7
Hardcover, 480 pages
Includes: Driver's Handbook, Workshop Manual

The Complete Official Triumph TR6 & TR250: 1967-1976
British Leyland Motors
ISBN 978-0-8376-1762-6
Hardcover, 608 pages
Includes: Driver's Handbook, Workshop Manual

The Complete Official Triumph Spitfire MK III, MK IV & 1500: 1968-1974 *British Leyland Motors*
ISBN 978-0-8376-1763-9
Hardcover, 480 pages
Includes: Driver's Handbook, Workshop Manual

The Complete Official Triumph Spitfire 1500: 1975-1980
British Leyland Motors
ISBN 978-0-8376-1745-9
Hardcover, 536 pages
Includes: Driver's Handbook, Workshop Manual

The Complete Official Triumph TR7: 1975-1981
British Leyland Motors
ISBN 978-0-8376-0116-8
Softcover, 544 pages
Includes: Driving Manual, Workshop Manual

The Complete Official Triumph TR4 & TR4A: 1961-1968
British Leyland Motors
ISBN 978-0-8376-0121-2
Softcover, 404 pages
Includes: Driving Manual, Workshop Manual, Competition Preparation Manual by R.W. "Kas" Kastner

The Complete Official Triumph TR2 & TR3: 1953-1961
British Leyland Motors
ISBN 978-0-8376-1750-3
Hardcover, 454 pages
Includes: Driver's Instruction Book, Service Instruction Manual

Jaguar

The Complete Official Jaguar "E" Series 1 & Series 2: 1961-1971
British Leyland Motors
ISBN 978-0-8376-1764-0
Hardcover, 792 pages
Includes: Driver's Handbook, Workshop Manual, Special Tuning Manual